Integrated Interconnect Technologies for 3D Nanoelectronic Systems

For a list of recent titles in the *Artech House Integrated Microsystems Series*,
please turn to the back of this book

Integrated Interconnect Technologies for 3D Nanoelectronic Systems

Muhannad S. Bakir
James D. Meindl

Editors

ARTECH
HOUSE

BOSTON | LONDON
artechhouse.com

Library of Congress Cataloging-in-Publication Data
A catalog record for this book is available from the U.S. Library of Congress.

British Library Cataloguing in Publication Data
A catalogue record for this book is available from the British Library.

ISBN-13: 978-1-59693-246-3

Cover design by Igor Valdman

685 Canton Street
Norwood MA 02062

10 9 8 7 6 5 4 3 2 1

This book is dedicated to my mom and dad,
and brothers, Tariq and Basil
and for the never ending support and inspiration

M.S.B.

Contents

Foreword

It has long been predicted that key issues related to interconnects and packaging will increasingly limit overall chip- and system-level performance as device scaling continues. Indeed, as industry has struggled to keep pace with Moore's law and is employing system-on-a-chip solutions for high-performance applications, it now recognizes that thermal management has become a challenge of paramount importance in order to achieve optimal performance. The ability to evolve novel thermal management techniques for cost-effective heat removal from packaged chips is increasingly critical to reducing the cost-per-function targets of the International Technology Roadmap for Semiconductors (ITRS). In addition, issues associated with power delivery and chip reliability continue to grow with successive chip generations. Understanding and solving the on-chip and chip-to-chip wiring issues, along with the challenges associated with various packaging approaches, will be critical to harnessing the full potential of future gigascale systems.

It was recognized that these complex problems cannot be solved in isolation. In response, U.S.-based university research centers were initiated through the Semiconductor Research Corporation (SRC) to address the growing interconnect and packaging challenges. An example of one such research effort is the Interconnect Focus Center (IFC), funded through the SRC's Focus Center Research Program (FCRP). The FCRP was launched in 1998, supported by both the semiconductor industry and the U.S. Department of Defense. The result: a multidisciplinary, collaborative, university-directed research effort formed to further stimulate creativity and innovation and to enable delivery of critically important solutions to intractable industry problems. One of the first FCRP centers to be initiated, the Interconnect Focus Center was created, with Professor James D. Meindl as its director, to evaluate new frontier concepts and innovative approaches for ultra-high- performance nanoscale interconnect architectures, as well as for high-speed, low-power applications. The IFC research program embraces optical interconnects as well as novel electrical interconnects and supports a strong thrust for creation of thermal and power management solutions for future ICs. Professor Muhannad S. Bakir has been a key IFC contributor as well.

Through my experience in the industry as a leader of teams delivering successful interconnect solutions for successive generations of high-performance semiconductor technologies, and particularly through my present role, I can attest to the ever-increasing level of anxiety associated with the issues of system-on-a-chip (SoC), system-in-package (SIP), and three dimensional (3D) integration, along with the challenges of thermal dissipation, power delivery, I/O bandwidth, chip-to-package interaction, latency, and reliability. The convergence of back-end processing, packaging, and design has impact on global interconnects, multicore system applications, and thermal and power management solutions. The challenges are truly daunting.

For that reason, I am pleased that Professors Bakir and Meindl have compiled this timely and critically important book to address these challenges head on, with both a historical perspective and a sound assessment of current approaches to mitigating or resolving the issues, as well as with a realistic look to future solutions.

Betsy Weitzman
Executive Vice President
Semiconductor Research Corporation
Research Triangle Park, North Carolina
November 2008

Preface

The most important economic event of the past century has been the information revolution (IR). It has given us the personal computer, the multimedia cell phone, the Internet, and countless other electronic marvels that influence our lives continuously. The silicon integrated circuit (IC) has been the most powerful driver of the IR throughout its history. During the initial era of the IC beginning in the early 1960s, bipolar and then complementary metal oxide semiconductor (CMOS) transistors were the principal determinants of IC cost and performance. In the second era of the IC commencing in the early 1990s, interconnects became the dominant determinants of the cost, performance, and energy dissipation of gigascale ICs. Currently, the cost and performance of products dominated by gigascale ICs have become limited by "ancillary" technologies that surround virtually every IC in a product setting. The central purpose of this book is to elucidate the extension of core wafer level processing (as practiced in front end of the line (FEOL) and back end of the line (BEOL) IC manufacturing) to electrical, optical, and thermal input/output interconnects for 2D and 3D nanoelectronic integrated systems. The intent is simply to begin extending the potent advantages of wafer level processing to these ancillary technologies and thus enable the third era of the IC.

This book is the cumulative effort from international industry researchers at IBM, Intel, Samsung, Rambus, Cypress Semiconductor, Texas Instruments, The Dow Chemical Company, and NanoNexus and academic researchers at Georgia Tech, MIT, Stanford University, The University of Texas at Austin, and Keio University. To our knowledge, no other book covers silicon ancillary technologies in the scope, depth, and approach of this book. This book contains five major topics relating to chip I/O interconnects relevant today and in the future: (1) forming reliable mechanical interconnection between the chip and substrate (Chapters 2 and 3), (2) delivering power to the chip (Chapters 4 and 5), (3) providing high-bandwidth chip-to-chip (electrical and optical) communication (Chapters 6–8), (4) cooling chips (Chapters 9–12), and (5) creating three-dimensional (3D) integrated systems and the interface of a chip to a probe substrate (Chapters 13–16).

This book is the result of more than two-and-a-half years of planning, preparation, and editing. We are gratified to see the final result and trust that this book will serve as a valuable reference on the challenges and opportunities for silicon ancillary and chip I/O interconnect technologies to enable ultimate performance gains from silicon nanotechnology in the third era of the IC. We wish to thank and gratefully acknowledge the hard work of the authors of the book chapters for their time, effort, and valuable contributions. We also wish to express our most sincere gratitude and thanks to Betsy Weitzman for her very profound, insightful Foreword, and the Semiconductor Research Corporation (SRC), the Interconnect Focus Center Research Program (IFC), and the National Science Foundation (NSF) for their generous and critical support of our research on silicon ancillary technologies over the past 10 years.

Muhannad S. Bakir
James D. Meindl
Editors
Atlanta, Georgia
November 2008

CHAPTER 1

Revolutionary Silicon Ancillary Technologies for the Next Era of Gigascale Integration

Muhannad S. Bakir and James D. Meindl

1.1 Introduction

The performance gains from metal-oxide-semiconductor field-effect transistor (MOSFET) scaling are beginning to slow down, and the need for more revolutionary innovations in integrated circuit (IC) manufacturing will only increase in the future in order to sustain the best possible rate of progress. But that's not all. The semiconductor industry has entered an era in which chip input/output (I/O) interconnects have become the critical bottleneck in the realization of the ultimate performance gains of a microsystem and will be a key driver for innovation and progress in the future. The three main bottlenecks that chip I/O interconnections impose on high-performance chips are their inability to: (1) maintain low chip-junction temperature with increasing and non-uniform power dissipation across a chip that effectively limits frequency of chip operation, increases leakage power dissipation, and degrades reliability, (2) provide low-latency, low-energy/bit, and massive off-chip bandwidth resulting in reduced system throughput and power efficiency, and (3) deliver power with high efficiency and low-supply noise with increasing current consumption and decreasing timing margins resulting in reduced clock frequency, increased transistor performance variability, and false logic switching. The central thesis of this effort is that revolutionary innovations in silicon (Si) ancillary, or support, technologies are urgently needed to realize the ultimate capabilities of intrinsic silicon technology and will be key factors to the next era of gigascale integration (GSI) for the semiconductor industry.

This book addresses five major topics relating to chip I/O interconnects of high relevance to current and future GSI silicon technologies: (1) forming reliable mechanical interconnection between the chip and substrate (Chapters 2 and 3), (2) delivering power to the chip (Chapters 4 and 5), (3) providing high-bandwidth chip-to-chip (electrical and optical) communication (Chapters 6 to 8), (4) enabling heat removal from chips (Chapters 9 to 12), and (5) creating three-dimensional (3D) integrated systems and the interface of a chip to a probe substrate (Chapters 13 to 16). Each chapter of this book begins with an introduction to the chip I/O topic that it covers, followed by a discussion of the limits and opportunities of the most promising technologies that address the needs of that particular chip I/O topic. The objective of this first chapter is to describe and emphasize the importance of the topics covered in the book. We seek to put the theme of the book in perspective by portraying where the semiconductor industry stands today, as well as its most promising opportunities for further advances.

This chapter is organized as follows: Section 1.2 discusses the role of innovation in sustaining Moore's law over the past five decades. In Section 1.3, we discuss the three eras of silicon technology (transistor centric, on-chip interconnect centric, chip I/O centric). Challenges to current silicon ancillary technology and the most promising solutions to address these limits are discussed in Section 1.4. This section also provides an overview of the chapters covered in the book. Section 1.5 discusses recent advances in chip I/O technology by describing "trimodal" chip I/Os. Finally, Section 1.6 is the conclusion.

1.2 The Role of Innovation in Sustaining Moore's Law

The inventions of the transistor in 1947 and the integrated circuit in 1958 were the intellectual breakthroughs that revealed the path for the exponential growth of semiconductor technology. Today, semiconductor technology is approximately a \$270 billion market and has penetrated essentially every single segment of human life. The industry is a far cry from its early days when the military was the only consumer and driver of transistor technology and miniaturization [1, 2]. The foundation upon which the semiconductor industry is based is the silicon crystal, which forms the heart of the silicon microchip.

The silicon microchip is the single most powerful engine driving the information revolution for two compelling reasons: productivity and performance. For example, from 1960 through 2008, the productivity of silicon technology improved by a factor of more than a billion. This is evident from the fact that the number of transistors contained in a microchip increased from a handful to two billion [3], while the cost of a microchip remained virtually constant. Concurrently, the performance of a microchip improved by several orders of magnitude. These simultaneously sustained exponential rates of improvement in both productivity and performance are unprecedented in technological history.

The exponential productivity of silicon technology was first pointed out by Gordon Moore in 1965 when he projected that the number of transistors in a given area doubles every 12 months; the rate was adjusted to every 24 months in 1975 to account for growing chip complexity [4]. This simple observation, which is known as Moore's law, still holds and is projected to continue in the foreseeable future. Figures 1.1 and 1.2 illustrate the average transistor minimum gate length and cost versus year, respectively, over a period of 40 years [5]. Figure 1.1 clearly illustrates that the average minimum gate length has scaled by more than three orders of magnitude in the period from 1970 to 2000. Today, patterns smaller than 45 nm (0.045 μm) are fabricated in high-volume manufacturing. For reference, the silicon atom is 0.234 nm in diameter, and the distance between the two point contacts in the 1947 point contact transistor was approximately 50 μm. Coupled with this decrease in size is the fact that the average cost of a transistor has dropped by a factor of a million in the period between 1970 and 2000 [5]. According to Moore, "this unprece-

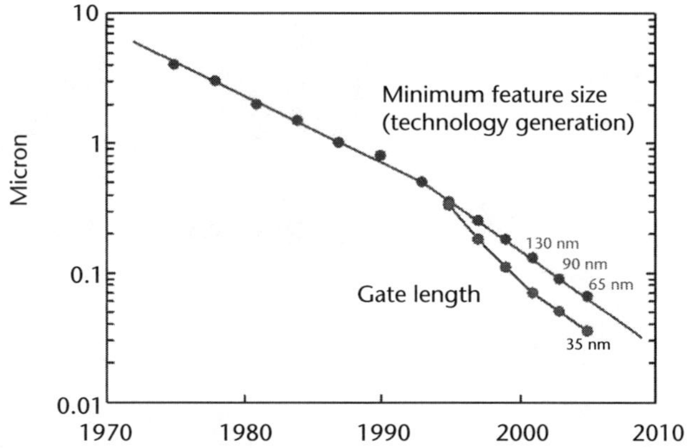

Figure 1.1 Average minimum gate length of a transistor versus year [6]. (*Source:* IEEE.)

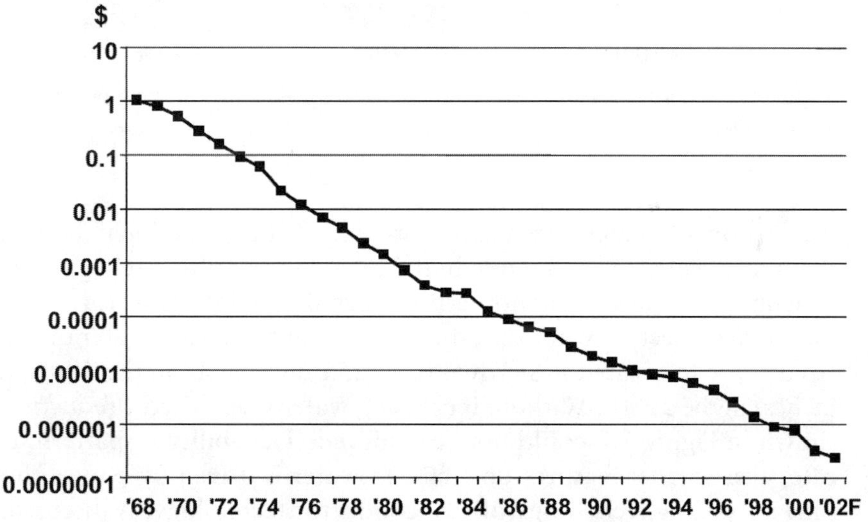

Figure 1.2 Average cost of a transistor versus year [5]. (*Source:* IEEE.)

dented decrease in the cost of a manufactured product, combined with the growth in consumption, constitute the engine that has driven the [semiconductor] industry" [5].

The principal enabler in advancing silicon technology throughout the past five decades is *never-ending innovation.* Innovation, in this case, is a combination of both evolutionary and revolutionary innovation (advances) in every aspect of the microchip: lithography technology used to define the features of the microchip, diameter of the wafer on which the microchip is fabricated, materials used to construct the microchip, and structures and devices used to form the microchip. Figure 1.3 illustrates a simplified schematic of the microchip fabrication process featuring the following major process steps: (1) transistor fabrication, or semiconductor front-end-of-line (FEOL) processing; (2) on-chip multilayer interconnect network fabrication, or back-end-of-line (BEOL) processing; (3) electrical chip I/O

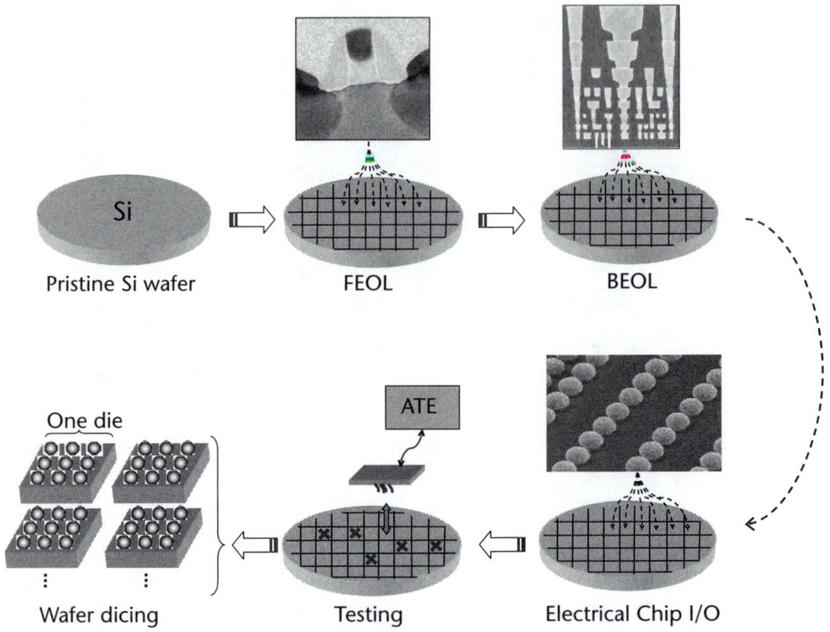

Figure 1.3 Schematic illustration of the processes used to transform a pristine silicon wafer into microchips.

fabrication; (4) wafer-level probe testing to identify functional dice; and (5) dicing of the silicon wafer to yield individual dice, which are ultimately packaged, tested, and assembled on the system motherboard, as shown in Figure 1.4.

With respect to wafer size, the diameter of the silicon wafer on which integrated circuits are fabricated has grown from 25 mm (1 inch) in the 1960s to 300 mm (12 inches) in the 2000s. Without increasing wafer size, the cost reduction per transistor shown in Figure 1.2 could not be achieved. The ability to pattern, therefore fabricate, ever-smaller features on a silicon wafer is driven by photolithography and is thus the pacing technology of semiconductor technology. A discussion of advances in photolithography is beyond the scope of this book, and the interested reader is referred to [7].

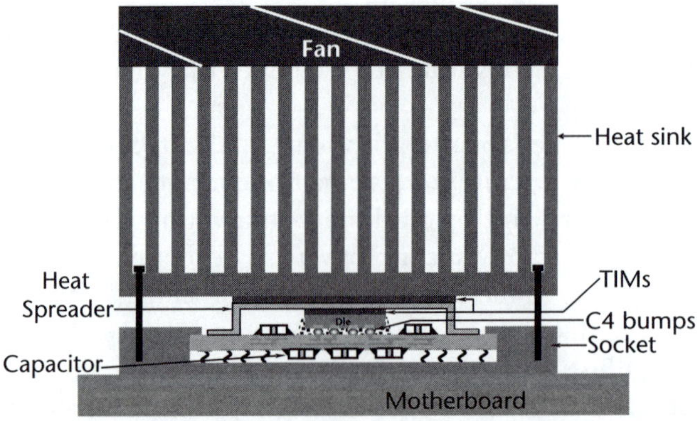

Figure 1.4 Schematic illustration of a silicon die after packaging and assembly on a motherboard. The figure highlights current silicon ancillary technologies.

The materials and structures used to fabricate a microchip have also undergone revolutionary changes in order to sustain Moore's law [8, 9]. This is true both for the transistors and the wires that interconnect them. With respect to the transistor, both its structure and the materials used for its fabrication have changed over the years. For example, transistor technology changed from bipolar junction transistor (BJT) in the 1960s, to n-channel metal-oxide-semiconductor (NMOS) field effect transistor in the 1970s, to complementary metal-oxide semiconductor (CMOS) devices in the 1980s (up to today) in order to reduce power dissipation. In 2007, the materials used to fabricate a MOSFET underwent a revolutionary change when high-k dielectrics and metal gates in MOSFETs replaced the silicon-dioxide dielectric and polysilicon gates. This provided more than 20 percent improvement in transistor switching speed and reduced transistor gate leakage by more than tenfold [3]. Moreover, a number of technologies that include strained-silicon growth, use of SiGe source drains [6], and dual-stress liner [10] have been pursued to provide improved electron-hole mobility in scaled transistors. Since the invention of the IC, one of the most significant revolutions for on-chip interconnects occurred in 1997 when the semiconductor industry began to replace aluminum wires (which had existed in the 1960s) with copper wires [11] to reduce latency, power dissipation, and electromigration in interconnects.

The critical point of the above discussion is that *never-ending innovation* has been and will always be the key to performance and productivity gains of the microchip with technology scaling. Without doubt, more revolutionary than evolutionary innovation will be needed in the future to sustain the best possible rate of progress of the microchip. Moreover, this revolutionary innovation mind-set must now be extended to the silicon ancillary technologies to enable the ultimate performance gains of the silicon microchip. The overarching strategy to accomplish this must be to *extend low-cost wafer-level batch processing*, the key to the success of Si technology, to the ancillary technologies that have now become the *millstone around the neck of Si technology itself*. In the next section, we discuss the three eras of silicon technology in more detail and focus on the chip I/O era.

1.3 Silicon Technology: The Three Eras

1.3.1 First Era: Transistor Centricity (1960s Through 1980s)

The improvement of microchip performance in the first three decades since the invention of the integrated circuit was driven by improvements in transistor performance. In the first three decades, this was achieved through the fabrication of smaller devices, as well as by migrating from BJT technology to CMOS technology. The fabrication of physically smaller transistors during each generation yielded significant enhancements in transistor frequency of operation, power dissipation, cost, and density of devices. While Moore's law provided the semiconductor industry with transistor integration density targets [6], R. Dennard et al. in 1974 helped define the path for actually achieving such scaling for MOSFETs by deriving guidelines for how to best scale the physical dimensions, silicon crystal doping, and bias voltages of a transistor [12]. Table 1.1 summarizes the scaling theory proposed by Dennard et al. and its impact on performance. Constant electric field (CE) scaling of

Table 1.1 R. Dennard's Constant Electric Field Scaling of MOSFETs

Device Parameter	*Scaling Factor*	*Device Parameter*	*Circuit Behavior*
Device dimension: t_{ox}, L, W	$1/k$	Delay time: CV/I	$1/k$
Doping concentration: N_a	k	Power dissipation: VI	$1/k^2$
Voltage, current: V, I	$1/k$	Power-delay product: CV^2	$1/k^3$
Capacitance: $\varepsilon A/t_{ox}$	$1/k$	Power density: VI/A	1

MOSFETs begins with the definition of a device scaling factor $k > 1$. All lateral and vertical device dimensions are scaled down by the same factor l/k. In addition, drain supply voltage is scaled as l/k in order to maintain a constant electric field intensity and consequently undiminished safety margins for device operation. The principal benefits of CE scaling are a device delay time that decreases as $1/k$, a power density that remains constant, a packing density that increases as k^2, and a power-delay product that decreases as $1/k^3$. MOSFET scaling has yielded tremendous improvements in microchip performance and productivity. However, continued performance gains from scaling past the 65 nm technology are slowing down. Table 1.2 illustrates past and projected performance gains of scaling. Some of the critical issues to address at the transistor level in order to maintain the historic rate of progress of the microchip are: (1) field-effect transistor (FET) gate tunneling currents that serve only to heat the microchip and drain energy and that are increasing rapidly due to the compelling need for scaling gate insulator thickness, (2) FET threshold voltage that rolls off exponentially below a critical value of channel length and consequently strongly increases FET subthreshold leakage (off) current without benefit, (3) FET subthreshold swing that rolls up exponentially below a critical channel length and consequently strongly reduces transistor drive current and therefore switching speed, and (4) critical dimension tolerance demands that are increasing with scaling and therefore endangering large manufacturing yields and low-cost chips.

1.3.2 Second Era: On-Chip Interconnect Centricity (1990s)

Scaling of transistors reduces their cost, intrinsic switching delay, and energy dissipation per binary transition. On the other hand, scaling of on-chip interconnects increases latency in absolute value and energy dissipation relative to that of the transistors [14]. Thus, although scaling down transistor dimensions yields improvements in both transistor cost and performance, the scaling down of interconnect cross-sectional dimensions unfortunately degrades performance. As a result, the performance of an IC becomes limited by the on-chip interconnects rather than by the transistors [11, 14–16].

Table 1.2 Performance Scaling: Past and Future Projections of Si MOSFET Scaling [13]

	2004	*2006*	*2008*	*2010*	*2012*	*2014*	*2016*	*2018*
Technology node (nm)	90	65	45	32	22	16	11	8
Delay scaling (CV/I)	0.7	~0.7	>0.7	Delay and energy scaling will slow down				
Energy scaling	>0.35	>0.5	>0.5					
Variability		Medium		High			Very High	

Until approximately the past decade, designers largely neglected the electrical performance of on-chip interconnections beyond a cursory accounting for their parasitic capacitance. They effectively addressed this problem simply by increasing transistor channel width to provide larger drive currents and thus enable transistor-level circuit performance. Unfortunately, this simple fix is no longer adequate for two salient reasons: both interconnect latency and energy dissipation now tend to dominate key metrics of transistor performance. For example, in the 100 nm technology, the latency of a 1-mm-long copper interconnect benchmark is approximately six times larger than that of a corresponding benchmark transistor [11, 14]. Moreover, the energy dissipation associated with a benchmark interconnect's binary transition is approximately five times larger than that of a corresponding transistor. This "tyranny of interconnects" escalates rapidly for future generations of silicon technology. Consequently, in the near and medium-term future, exponential rates of increase in transistors per chip will necessarily require advances in on-chip interconnect technology. These advances will be extremely diverse and include new interconnect materials and processes, optimal reverse scaling, repeater circuits, microarchitectures that shorten interconnects, and more powerful computer-aided design tools for chip layout and interconnect routing. Advances in chip I/O interconnects, addressed in the next section, have also become critical and are the focus of this book.

The urgency of interconnect-centric design can only increase as scaling continues. A simple reasoning process elucidates this assertion. For small wires, latency is the most challenging performance metric. Latency is given by the expression $\tau = RC$, where R and C are an interconnect's total resistance and capacitance. A wire's resistance is commonly expressed as $R = \rho(L/WH)$, where ρ is the metal's resistivity, and L, W, and H are the metal conductor's length, width, and height, respectively. Assuming that ρ remains constant, and W and H are scaled proportionately for a wire of constant length, R increases quadratically as $1/WH$. Neglecting fringing, an isolated wire's capacitance is approximately $C = \varepsilon(WL/T)$, where ε is the insulator's permittivity, and T is its thickness. Assuming that ε remains constant, and W and T are scaled proportionately for a wire of constant length, C remains constant. Consequently, $\tau = RC$ increases quadratically as $1/HT$. However, there is more to the problem. First, surface and grain boundary scattering impose rapid increases in effective resistivity, ρ, when wire cross-sectional dimensions and copper grain size become smaller than the bulk copper electron's mean free path length. Also, the thin, relatively high resistivity liners, which must surround a copper interconnect to prevent copper atoms' migration into the silicon, become comparable in thickness to the copper interconnect itself. This effectively reduces the copper's cross-sectional area and thus increases wire resistance and hence latency. Second, power dissipation causes temperature increases in the wires; therefore, resistivity increases. Finally, high-speed operation creates a greater current density near the wires' periphery than in their central region. This so-called "skin effect" further increases wire resistance and hence latency; combining this effect with scattering causes anomalous skin effect [17], significantly increasing wire resistance and latency.

For the moment, let us assume that a high-temperature superconductive material with resistivity $\rho \to 0$ is discovered. For $\rho \to 0$, we can no longer calculate an interconnect's latency using the approximation $\tau = RC$. When the RC product is extremely small, two mechanisms determine interconnect latency. For relatively short interconnects, latency is the time a driver transistor requires to charge its load capacitance according to the relationship $t_d = C_t V/I$, where C_t is the total transistor and interconnect capacitance, V is the interconnect voltage swing, and I is the transistor drive current. For longer interconnects with $\rho \rightarrow 0$, an electromagnetic wave's sheer time of flight fundamentally constrains interconnect latency. An approximate time-of-flight expression is $ToF = (\varepsilon_r)^{1/2} L/c_o$, where ε_r and L are the relative insulator permittivity and interconnect length, respectively, and c_o is light's velocity in free space (a fundamental limit). A brief calculation reveals interconnects' inescapable tyranny even for this extreme case of superconductive behavior or $\rho \rightarrow 0$. A simple model for the switching time of a 10 nm channel length transistor is $t_d = L_{ch}/v_{th}$, where L_{ch} is channel length and $v_{th} = 10^7$ cm/s is the channel's average carrier velocity, which we assume is an electron's thermal velocity at room temperature. Therefore, for a 10 nm generation transistor and an average channel carrier velocity equal to the thermal velocity, transistor switching time delay $t_d = 0.1 \times 10^{-12}$ sec = 0.1 ps.

For an ideal interconnect with $\rho = 0$ and $\varepsilon_r = 1$, the interconnect length traveled by an electromagnetic wave front in 0.1 ps is $L = ToF\ c_o/(\varepsilon_r)^{1/2} = 30\ \mu$m. Thus, an ideal superconductive interconnect with a vacuum insulator whose length exceeds 30 μm will have latency exceeding that of a 10 nm transistor! Moreover, the interconnect's switching energy transfer will be much larger than that of a minimum-size 10 nm transistor. This simple example reveals the limits on-chip interconnects impose on the future of nanoelectronics.

1.3.3 Third Era: Chip I/O Centricity (2000s)

As gigascale silicon technology progresses beyond the 45 nm generation, the performance of the microchip has failed by progressively greater margins to reach the "intrinsic limits" of each particular generation of technology. The root cause of this failure is the fact that the capabilities of monolithic nanosilicon technology per se have vastly surpassed those of the ancillary or supporting technologies that are essential to the full exploitation of a high-performance GSI chip. The most serious obstacle that blocks fulfillment of the ultimate performance of a GSI chip is power dissipation and inferior heat removal. The increase in clock frequency of a GSI chip has been virtually brought to a halt by the lack of an acceptable means for removing, for example, 200W from a 15 × 15 mm die. With a cooling limit of ~100 W/cm^2, it is expected that the maximum frequency at which a single core microprocessor can operate is approximately 4 GHz [18]. In addition, the inability to remove more than 100 W/cm^2 per stratum is the key limiter to 3D integration of a microprocessor stack. A huge deficit in chip I/O bandwidth due to insufficient I/O interconnect density and poor off-chip interconnect quality is the second most serious deficiency stalling high-performance gains. The excessive access time of a chip multiprocessor (CMP) for communication with its off-chip main memory is a direct consequence of the lack of, for example, a low-latency 100 THz aggregate bandwidth I/O signal network. Lastly, GSI chip performance has been severely constrained by inadequate I/O interconnect technology capable of supplying, for example, 200A to 300A at 0.7V to a CMP with ever-decreasing noise margins. With respect to reliability, the use of low-k interlayer dielectrics (ILDs) to reduce on-chip interconnect parasitic capacitance has exacerbated the difficulty of maintaining high thermomechanical reliability of dice assembled on organic substrates. Due to the fragile nature of low-k ILDs and their relatively poor adhesion to the surrounding materials, it is becoming progressively more critical to minimize thermomechanical and mechanical stresses imparted on the chip during thermal cycling and wafer-level probing, respectively.

In the following subsections, we expand the discussion for the above listed issues and present that discussion in the sequential order of these issues' coverage in the book.

1.3.3.1 Mechanical Interconnection Challenges

The motivation for the use of low-k ILD within the on-chip multilayer interconnect network is driven by the need to reduce the *RC* delay and power dissipation of on-chip interconnects [11]. However, the gains in the electrical performance of the on-chip interconnect network come at the expense of the complexity of the ILD's process integration within BEOL processing, mechanical strength, adhesion properties, and thermal conductivity. As a result, it is critical to minimize mechanical/thermomechanical stresses imparted on the silicon chip to prevent Cu/low-k interfacial crack formation, ILD delamination, and the mechanical failure of the IC. Therefore, a current area of research is how to package chips with low-k ILD without inducing any damage to the BEOL interconnect networks as a result of the packaging process. The coefficient of thermal expansion (CTE) mismatch between silicon dice (CTE ≈ 3 ppm/°C) and standard organic boards (CTE ≈ 17 ppm/°C, which is matched to copper) is a common cause of failure in electronic components that use an area-array distribution of solder bumps [19, 20]. The physical deformation due to the CTE mismatch can induce stresses in the Cu/low-k structure, resulting in the formation and propagation of interfacial cracks. Chapter 2 discusses the experimental and modeling studies to investigate chip-package interaction and its impact on low-k interconnect reliability.

A common solution to this CTE mismatch problem is the use of underfill [20], which is an epoxy-based material, to distribute the stresses imparted on the solder bumps during thermal cycling. However, underfill is time-consuming to process and difficult to dispense at fine pitches due to the low stand-off height, does not allow easy chip rework, and degrades electrical performance of high-frequency signal interconnects [21]. The need for underfill can be eliminated by augmenting the bumps with mechanically compliant chip I/O leads, which are designed to compensate for the CTE mismatch between the chip and the printed wiring board (PWB): the compliant leads are displacement-absorbing interconnect structures that undergo strain during thermal cycling. These compliant interconnects are fabricated between the die pads and their respective solder bumps. Thus, as the pad on the PWB experiences a relative displacement with respect to the die pad during thermal cycling, the compliant lead easily and elastically changes shape to compensate. As a result, this minimizes the force at the die pads and consequently reduces the stresses there and in the surrounding low-k dielectric during thermal cycling. There are several wafer-level-compliant I/O interconnection technologies that differ greatly in their electrical and mechanical performances, size, cost, fabrication, and I/O density. Unlike wire bonding, for example, the fabrication of most compliant interconnections does not require the application of a high force load on the chip because they are lithographically fabricated; they are a "third-era" interconnect. Chapter 3 discusses the potential application of mechanically compliant leads (to replace or augment solder bumps) to address the low-stress interconnection requirements on the die pads.

1.3.3.2 Power Delivery Challenges

Challenges in power delivery are very daunting and require innovative solutions and technologies now and more so in the future. Power dissipation, supply voltage, and current drain trends for high-performance microprocessors are illustrated in Figures 4.1 and 4.2 (see Chapter 4). As power dissipation increases and supply voltage decreases (historically), the supply current has increased, reaching a value greater than 100A for some microprocessors. Runtime power management techniques to reduce power dissipation of circuit blocks on a microprocessor are common today. Two common examples are power gating and clock gating. The former is used to disconnect idle circuit blocks from the power distribution network to reduce static power dissipation, while the latter is used to disable the clock signal to reduce dynamic power dissipation. The primary objective of the power delivery network is to distribute power efficiently to all transistors on a chip while maintaining an acceptable level of power-supply noise. The empirical acceptable power-supply noise is typically equal to 10% to 15% of the supply voltage (see Chapter 5).

As supply voltage continues to scale (although at a much slower pace in the future), logic circuits become increasingly sensitive to power-supply noise. Excessive supply noise can severely degrade the performance of the system by introducing gate delay variation, logic failures, false switching, and signal integrity challenges (on- and off-chip). As discussed in Chapter 4, every millivolt of voltage drop yields approximately a 3 MHz decrease in the maximum operating frequency of the Core 2 Duo microprocessor, for example. Power-supply noise comprises the *IR*-drop across the power distribution network and the simultaneous switching noise (SSN), which is equal to the product of the inductance (L) of the power delivery network and the rate of change of the current (di/dt), or Ldi/dt. Given the large current demands and switching speeds of modern microprocessors, combined with the fact that modern microprocessors may have large circuit blocks that switch from idle/sleep mode to active mode, the value of di/dt can be very large. For reference, the power delivery network of the Intel Xeon processor is designed to handle a maximum slew rate (di/dt) of 930 A/μs [22]. The need for high-efficiency (low-loss), compact, fast, and cost-effective voltage-regulator modules requires new architectures and solutions, making the design of the voltage regulator an area of very active research today. This topic is discussed in Chapter 4.

The increase in current consumption of modern microprocessors will impose ever-increasing demands on the power delivery network and need to reduce the resistive losses (and problems relating to electromigration) through the motherboard, socket, package, and chip I/Os, which can become quite excessive [23]. It is common to allocate more than two-thirds of the total number of die pads to power and ground interconnection [24]. In the Intel 1.5 GHz Itanium 2 microprocessor, which dissipates 130W (worst case) at 1.3V supply voltage, 95% of the total 7,877 die pads are allocated for power and ground interconnection [25]. However, with very high demand for signal bandwidth in multicore microprocessors, the number of signal I/Os will also be large. In order to maintain acceptable power-supply noise at the die and provide efficient power delivery, trade-offs and proper resource allocation and codesign must be made at the die, package, and motherboard levels. On-die resources include the on-die decoupling capacitors, the width and height (although the latter is typically determined by process technology) of the on-chip power distribution network, and the number of power and ground pads. At the package level, total decoupling capacitors and the use of sufficient power and ground planes with enough second-level power/ground pads is critical. Chip-package codesign of the power delivery network is discussed in Chapter 5. At the motherboard level, properly designed dc-dc converters and proper allocation of capacitors are critical to the overall performance of the power distribution network and are discussed in Chapter 4.

1.3.3.3 Signaling Challenges

A corollary of Moore's law is that the off-chip bandwidth doubles every 2 years. Today, aggregate off-chip bandwidth for a microprocessor is ~100 GB/s with an I/O power efficiency of ~10 mW/GB/s [26]. The transition to multicore microprocessors will introduce an unprecedented appetite for off-chip bandwidth that will easily be on the order of several terabits per second in the short term to fully utilize its computational capacity [26]. An off-chip communication link consists of three elements: (1) the transmitter block, (2) the channel over which the signal is transmitted, and (3) the receiver block. In order to meet future multi-hundred- terabit-per-second off-chip bandwidth while meeting the power, latency, circuit size, and cost constraints for off-chip communication, it is important to optimize all three elements simultaneously [27].

It is well known that the quality of the channel impacts the complexity, area, and power dissipation of the transmitter and receiver blocks. Today, inadequate numbers of electrical signal I/Os coupled with frequency-dependent losses (dispersion), impedance mismatches, and crosstalk encountered on a typical organic substrate impose severe constraints on the performance of the I/O link that becomes exacerbated as off-chip bandwidth per channel increases and signal noise budget decreases. It has recently been shown that the resistive losses due to copper wire roughness on the substrate can increase the total loss in the range from 5.5% to 49.5% [28]. Improvements in the physical channels (e.g., length reduction, improved impedance matching, lower dielectric losses, lower crosstalk) can greatly reduce the power dissipation, latency, circuit size, and cost of the overall electrical link. A discussion of challenges and opportunities in multigigabit per second signaling covered in Chapter 6.

Microphotonic interconnect technology has been proposed to address these limitations [29, 30]. However, the use of chip-to-chip optical interconnects will greatly extend the technical and economical challenges of chip I/O interconnects due to fabrication, packaging, alignment, and assembly requirements [31]. Chapter

7 presents an overview of optical chip-to-chip interconnection networks as well as motivation for optical signaling. The most challenging aspect of optical interconnection to a silicon chip is the integration of high-performance optical devices (optical source, possibly a modulator, and detectors) using CMOS compatible processing. This is discussed in Chapter 8.

1.3.3.4 Thermal Interconnect Challenges

In order to maintain constant junction temperature with increase in power dissipation, the size of the heat sink used to cool a microprocessor has been steadily increasing. A plot illustrating the power dissipation and heat sink size (volume) of various Intel microprocessors is shown in Figure 11.1 (see Chapter 11). It is clear that the size of the heat sink has been increasing with each new microprocessor, thus imposing limits on system size, chip packing efficiency, and interconnect length between chips. While the minimum feature size of a silicon transistor has been decreasing, the thermal I/O (heat sink) has scaled in the opposite direction in order to attain smaller junction-to-ambient thermal resistance (R_{ja}). It is projected that the junction-to-ambient thermal resistance at the end of the roadmap will be less than 0.2°C/W [24]. For reference, the Intel 386 SX microprocessor in 1986 operated with an R_{ja} value of 22.5°C/W under an air-flow rate of 2.03 m/s; no heat sink was needed for the Intel 386 microprocessor to operate at the needed chip junction temperature because of the small power dissipation [32].

Using the best available materials for the various thermal interconnects between the silicon die and the ambient [the heat spreader, the heat sink, and the thermal interface materials (TIMs) at the die/heat spreader and heat spreader/heat sink interfaces], the lowest attainable thermal resistance from an air-cooled heat sink is approximately 0.5°C/W. Although increasing the air-flow rate can help in reducing the thermal resistance of the heat sink to a certain extent, an important constraint on fans used to cool processors is set by the acoustic noise and serves as an important constraint for today's electronic devices. Not only does the TIM account for a large fraction of the overall thermal resistance, but it also presents many reliability problems [33]. If the temperature of a microprocessor is not maintained below a safe level (typically 85°C), a number of undesirable effects occur that include: (1) increase in transistor leakage current, leading to increases in static power dissipation; (2) increase in the electrical resistance of on-chip interconnects that increases RC delay and I^2R losses and decreases bandwidth; (3) decrease in electromigration mean time to failure; and (4) degraded device reliability and decreases in carrier mobility. Challenges in cooling are exacerbated by the fact that the power-dissipation density is nonuniform across a microprocessor. This can result in very large thermal (temperature) nonuniformity across the chip, leading to device performance variation across the chip. Regions that dissipate the highest power density are called hot spots. The power density of a hot spot can reach as high as 400 W/cm^2, although this is usually over an area equal to a few hundred micrometers squared. Such hot spots require very high-quality heat spreader solutions and cooling. This topic is covered in Chapter 9.

Revolutionary cooling technologies (thermal interconnects) will undoubtedly be needed in the future that can: (1) eliminate/improve the TIM, (2) reduce the thermal resistance of the heat sink, (3) maintain low junction temperature over high-power-density regions, (4) improve heat spreading, and (5) reduce the dimensions of the chip cooling hardware from inches to microns. Thermal interconnect opportunities to address the above challenges are discussed in Chapters 10 to 12. Chapter 10 discusses the benefits of liquid cooling, and Chapter 11 discusses on-die liquid-cooling implementation technologies. In Chapter 12, carbon nanotubes are explored as a potential thermal interconnect to replace/augment current TIM materials.

1.3.3.5 Three-dimensional System Integration

Today, it is widely accepted that three-dimensional (3D) system integration is a key enabling technology that has gained significant momentum in the semiconductor industry. Three-dimensional integration may be used either to partition a single chip into multiple strata to reduce on-chip global interconnect length [34] and/or to stack chips that are homogenous or heterogeneous. An example of 3D stacking of homogenous chips is memory chips, while an example of heterogeneous chip stacking is memory and multicore microprocessor chips.

In order to highlight the benefits of 3D technology, increasing the number of strata from one to four, for example, reduces the length of a distribution's longest wires by 50%, with concurrent improvements of up to 75% in latency and 50% in interconnect energy dissipation [35].

The origins of 3D integration can date back to 1960, when James Early of Bell Laboratories discussed 3D stacking of electronic components and predicted that heat removal would be the primary challenge to its implementation [36]. This has indeed proven to be the case for today's high-performance integrated circuits. Aside from the form-factor issue, an air-cooled heat sink (and heat spreader), at best, provides a junction-to-ambient thermal resistance of 0.5°C/W. When two 100 W/cm^2 microprocessors are stacked on top of each other, for example, the net power density becomes 200 W/cm^2, which is beyond the heat-removal limits of air-cooled heat sinks. This is the key reason why stacking of high-performance (high-power) chips has not been demonstrated so far, because, simply put, it is hard enough to cool a single chip. Thus, apparently cooling is the key limiter to the stacking of high-performance chips today. Power delivery to a 3D stack of high-power chips also presents many challenges and requires careful and appropriate resource allocation at the package level, die level, and interstratal interconnect level [37]. Finally, the prospects of photonic device integration (through monolithic or heterogeneous integration) with CMOS technology require the support of optical interconnect networks between 3D stacks and potentially within a stack. As a result, a number of challenges have yet to be addressed to enable the 3D integration of high-performance chips (see Figure 1.5).

Figure 1.6 is a representative schematic of the 3D integration technologies that have been proposed to date and illustrate three categories. The first category consists of 3D stacking technologies that do not utilize through-silicon vias (TSVs); these are shown in Figure 1.6(a–c). The second category consists of 3D integration technologies that require TSVs [Figure 1.6(d, e)], and the third category consists of monolithic 3D systems that make use of semiconductor processing to form active levels that are vertically stacked (with on-chip interconnects between). Of course, combinations of these technologies are possible.

The non-TSV 3D systems span a wide range of different integration methodologies. Figure 1.6(a) illustrates stacking of fully packaged dice. Although this may offer the advantages of being low cost, simplest to adopt, fastest to market, and providing modest form-factor reduction, the overhead in interconnect length and low-density interconnects between the two dice do not enable one to fully exploit the advantages of 3D integration. Figure 1.6(b) illustrates the stacking of dice, based on the use of wire bonds. Naturally, this 3D technology is suitable for low-power and low-frequency chips due to the adverse effect of wire-bond length, low density, and peripheral pad location for signaling and power delivery. Figure 1.6(c) illustrates the

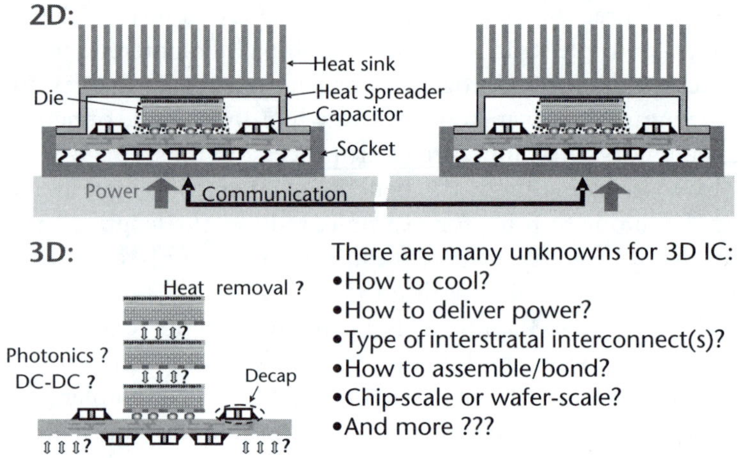

Figure 1.5 Schematic illustration of challenges associated with the stacking of GSI high-performance chips.

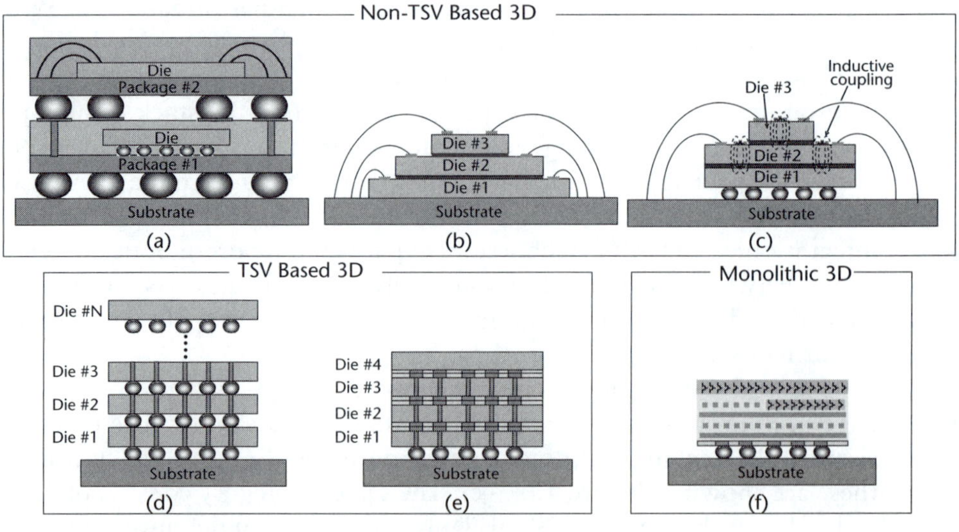

Figure 1.6 (a–f) Schematic illustration of various 3D integration technologies.

use of wireless signal interconnection between different levels using inductive coupling (capacitive coupling is also possible) [38], the details of which are discussed in Chapter 15. There are several derivatives of the topologies described above, such as the dice-embedded-in-polymer approach [39]. This approach, although different from others discussed, makes use of a redistribution layer and vias through the polymer film and is thus a hybrid die/package-level solution. It is important to note that all non-TSV approaches rely on stacking at the die/package level (die-on-wafer possible for inductive coupling and wire bond) and thus do not utilize wafer-scale bonding. This may serve to impose limits on economic gains from 3D integration due to the cost of the serial assembly process.

Figure 1.6(d, e) illustrates 3D integration based on TSVs. The former figure illustrates bonding of dice with C4 bumps and TSVs. The short interconnect lengths and high density of interconnects that this approach offers are important advantages. Compared to wire bonding, it is possible to have several orders of magnitude more interconnects. Although it is possible to bond at the wafer level, this approach is most suitable for die-level bonding (using a flip-chip bonder) and thus faces some of the same economic issues described above. Figure 1.6(e) illustrates 3D stacking based on thin-film bonding (metal-metal or dielectric-dielectric) [40–42]. Not only are solder bumps eliminated in this approach, but increased interconnect density and tighter alignment accuracy can also be achieved as compared to the previous approach due to the fact that these approaches are based on wafer-scale bonding. Thus, they utilize semiconductor-based alignment and manufacturing techniques. These technologies are discussed in Chapters 13 and 14.

Finally, Figure 1.6(f) illustrates a purely semiconductor manufacturing approach to 3D integration. The main enabler to this approach is the ability to deposit an amorphous semiconductor film (Si or Ge) on a wafer during the IC manufacturing process, then to recrystallize it to form a single-crystal film using a number of techniques [43, 44]. The silicon layer can also be grown from the underlying silicon ("seed") layer, as discussed in Chapter 13. Ultimately, this approach may offer the most integrated system while needing the fewest interconnects, but it may be restricted to forming "silicon islands" in the upper strata. Results from this 3D integration technology for memory devices are discussed in Chapter 13.

It is important to note that none of the above described 3D integration technologies addresses the need for cooling in a 3D stack of high-performance chips. This is a significant omission and imposes a constraint on the ability to fully utilize the benefits of 3D technology. As such, new 3D integration technologies are needed for such applications. These are discussed in Chapters 10 and 11.

Finally, wafer-level probing of dice represents a form of 3D technology (based on temporary vertical interconnection) and is discussed in Chapter 16. The elementary purpose of testing in microchip manufacturing is to ensure that only known-good-die (KGD) are shipped to a customer. Unfortunately, the process of screening bad dice from good ones is a time-consuming and increasingly difficult task. Shrinking device geometries, increasing frequencies of operation, and the sheer magnitude of transistors and I/Os on a chip are all factors contributing to the increasing complexity of IC testing. From an IC manufacturer's point of view, the basic function of a probe card is to interface with the I/Os of the die. It should not load the die or cause any signal degradation. In addition, it should be able to do this repeatedly (hundreds of thousands of touchdowns) without damaging the chip I/Os. Of course, this delicate combination of high-quality electrical design and mechanical robustness needs to be achieved at the minimum possible cost. The challenges in testing are broadly categorized under electrical, mechanical, and reliability requirements. With the prospects of microphotonic integration onto CMOS wafers, the complexity of testing increases, and extraordinary innovations in probe card technology will undoubtedly be needed to meet these diverse and complex requirements.

1.4 Need for Disruptive Silicon Ancillary Technologies: Third Era of Silicon Technology

Addressing the highly interdisciplinary and complex needs of chip I/Os will require revolutionary approaches and solutions that integrate these requirements in a bottom-up approach. In order to provide all critical interconnect functions for a gigascale chip, fully compatible, low-cost, and microscale electrical, optical, and fluidic ("trimodal") chip I/O interconnects have recently been proposed [45, 46]. A schematic illustration of a cross section of a gigascale chip with trimodal I/Os is shown in Figure 1.7. Scanning electron microscope (SEM) images of the various I/Os are also shown in the figure. A key feature of the I/Os under consideration is that they are manufactured using wafer-scale batch fabrication, which is the key to the success of silicon technology. Thus, these technologies provide microscale solutions fully compatible with CMOS process technology and batch fabrication. Although the electrical I/Os are implemented using conventional solder bumps in the figure, mechanically compliant leads, such as those described in [47], can be used instead to address the thermomechanical reliability requirements of chips with low-k interlayer dielectric [48].

The optical I/Os are implemented using surface-normal optical waveguides and take the form of polymer pins (or "pillars") [31, 49]. A polymer pin, like a fiber optic cable, consists of a waveguide core and a cladding. The polymer pin acts as the waveguide core, and unlike a fiber optic cable, the cladding is air. A key feature of the optical pins is that they are mechanically flexible and can thus bend to compensate for the CTE mismatch between the chip and substrate. It has been shown that the optical pins can provide less than 1 dB of optical loss for a displacement compensation of 15 μm [31]. Additional performance details are discussed in Chapter 7.

The fluidic I/Os are implemented using surface-normal *hollow-core* polymer pins, or micropipes [45, 50]. Unlike prior work on microfluidic cooling of ICs that require millimeter-sized and bulky fluidic inlets/outlets to the microchannel heat sink, the proposed micropipe I/Os are microscale, wafer-level batch fabricated, area-array distributed, flip-chip compatible, and mechanically compliant. Using the fluidic I/Os, a silicon chip has been flip-chip bonded on a substrate and used to cool 300 W/cm^2 with a chip junction temperature rise of ~40°C. An in-depth discussion of the fluidic I/Os is presented in Chapter 11.

The process used to fabricate the trimodal I/Os is shown in Figure 1.8. It is assumed that the optical devices (detectors or sources) are monolithically or heterogeneously integrated on the CMOS chip (Chapter 8 discusses such technologies).

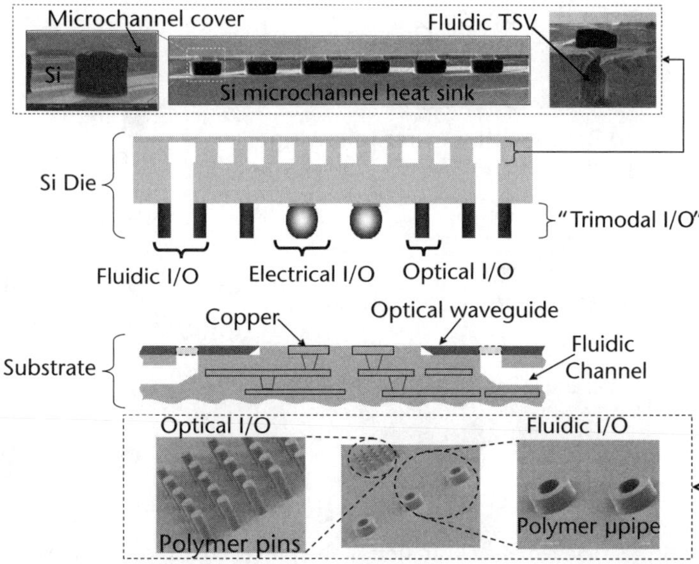

Figure 1.7 Schematic illustration of a chip with electrical, optical, and fluidic I/O interconnects. SEM images are also shown.

The fabrication process begins by etching through-wafer fluidic vias starting from the back side of the chip (side closest to the heat sink) and trenches into the silicon wafer [Figure 1.8(b)]. Following the silicon etch, the microchannel heat sink is enclosed using any of a number of techniques [Figure 1.8(c)] [51] (See Chapter 1). This completes the fabrication of the microchannel heat sink. Next, solder bumps are fabricated on the front side of the chip using standard processes [Figure 1.8(d)]. Next, a photosensitive polymer film, equal in thickness to the height of the final optical and fluidic I/Os, is spin coated on the front side of the wafer (and over the solder bumps), as shown in Figure 1.8(e). Finally, the polymer film is photodefined to yield the optical and fluidic I/Os *simultaneously*. Essentially, the trimodal I/Os are an extension of the wafer-level, batch-fabricated, on-chip multilayer interconnect network and represent a "third-era" chip I/O technology to address the tyranny of limits current silicon ancillary technologies impose.

It is clear that in order to assemble a chip with trimodal I/Os, it is critical to have a substrate with trimodal planar interconnects. While substrate-level optical waveguides have been widely studied [52–56], integrated electrical, optical, and fluidic interconnects have not been reported until recently [46]. Figure 1.9 illustrates optical and cross-sectional SEM micrographs of integrated electrical, optical, and fluidic interconnects at the substrate level. Using such substrate for the assembly of chips with trimodal I/Os has been demonstrated recently [45, 46], and details are discussed in Chapters 7 and 11.

1.5 Conclusion

Silicon technology has evolved from being transistor centric, to being on-chip interconnect centric, to being at present chip I/O centric. With the paradigm shift to this

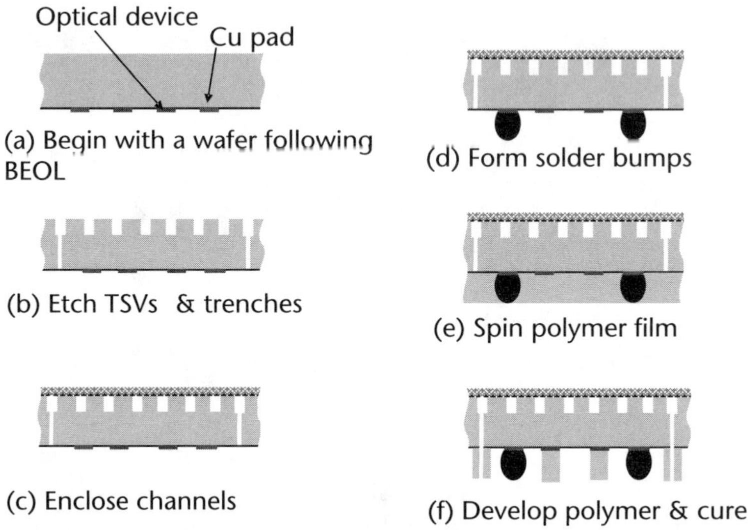

Figure 1.8 (a–f) Schematic illustration of the process used to fabricate electrical, optical, and fluidic chip I/O interconnects and the monolithically integrated silicon microchannel heat sink.

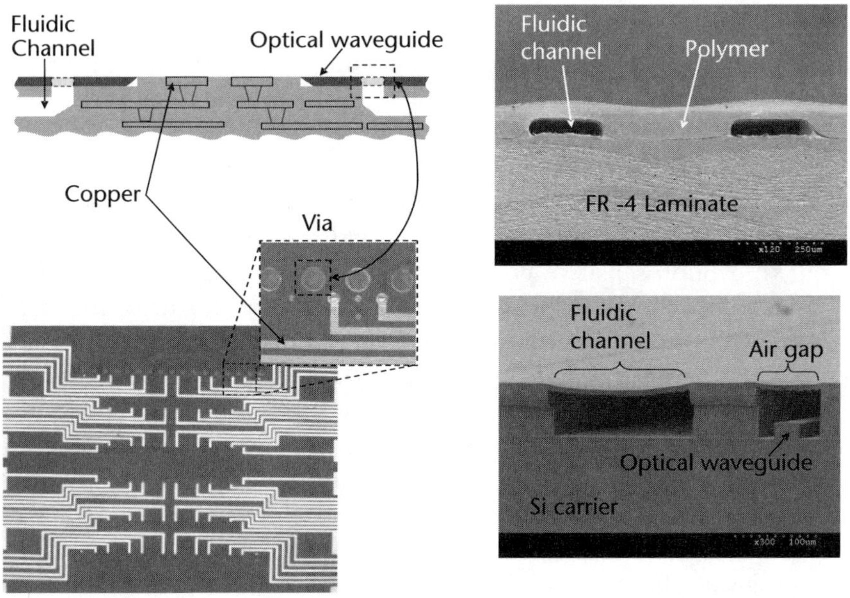

Figure 1.9 Micrographs of substrates with electrical, optical, and fluidic interconnects.

third era of gigascale silicon technology, this book provides readers with a timely, highly integrated, and valuable reference for all important topics relating to chip I/O interconnects: mechanical interconnection, power delivery, electrical and optical signaling, thermal management, three-dimensional system integration, and probe card technology. Integrated interconnect technologies to address this "tyranny of limits" on gigascale chips are urgently needed.

References

[1] Keonjian, E., *Microelectronics: Theory, Design, and Fabrication*: New York: McGraw-Hill, 1963.

[2] Eckert, M., and H. Schubert, *Crystals, Electrons, Transistors: From Scholar's Study to Industrial Research*, Melville, NY: American Institute of Physics, 1990.

[3] See Intel at www.intel.com.

[4] Moore, G. E., "Progress in Digital Integrated Electronics," *Proc. IEEE Int. Electron Devices Meeting*, 1975, pp. 11–13.

[5] Moore, G. E., "No Exponential Is Forever: But 'Forever' Can Be Delayed!" *Proc. IEEE Int. Solid-State Circuits Conf.*, 2003, pp. 20–23.

[6] Bohr, M., "A 30 Year Retrospective on Dennard's MOSFET Scaling Paper," *IEEE Solid-State Circuits Society Newsletter*, vol. 12, 2007, pp. 11–13.

[7] Campbell, S. A., *The Science and Engineering of Microelectronic Fabrication*. Oxford: Oxford University Press, 2001.

[8] Thompson, S. E., et al., "In Search of 'Forever,' Continued Transistor Scaling One New Material at a Time," *IEEE Trans. Semiconductor Manufacturing,* vol. 18, 2005, pp. 26–36.

[9] Nowak, E. J., "Maintaining the Benefits of Scaling When Scaling Bogs Down," *IBM J. Res. Dev.,* vol. 46, 2002, pp. 169–180.

[10] Narasimha, S., et al., "High Performance 45-nm SOI Technology with Enhanced Strain, Porous Low-k BEOL, and Immersion Lithography," *Proc. IEEE Electron Devices Meeting*, 2006, pp. 1–4.

[11] Davis, J. A., and J. D. Meindl, *Interconnect Technology and Design for Gigascale Integration*, Norwell, MA: Kluwer Academic Publishers, 2003.

[12] Dennard, R. H., et al., "Design of Ion-Implanted MOSFETs with Very Small Physical Dimensions," *IEEE J. Solid-State Circuits,* vol. 9, 1974, pp. 256–268.

[13] Mooney, R., "Multi-Gigabit I/O Design for Microprocessor Platforms," *IEEE Custom Integrated Circuits Conf., Educational Session Tutorial*, 2007.

[14] Meindl, J. D., "Beyond Moore's Law: The Interconnect Era," *IEEE Computing in Science and Engineering,* vol. 5, 2003, pp. 20–24.

[15] Meindl, J. D., et al., "Interconnect Opportunities for Gigascale Integration," *IBM J. Res. Dev.,* vol. 46, March–May 2002, pp. 245–263.

[16] Bohr, M. T., "Interconnect Scaling—the Real Limiter to High Performance ULSI," *Proc. IEEE Int. Electron Devices Meeting*, 1995, pp. 241–244.

[17] Sarvari, R., and J. D. Meindl, "On the Study of Anomalous Skin Effect for GSI Interconnections," *Proc. IEEE Int. Interconnect Technol. Conf.*, 2003, pp. 42–44.

[18] Shahidi, G. G., "Evolution of CMOS Technology at 32 nm and Beyond," *Proc. IEEE Custom Integrated Circuits Conf.*, 2007, pp. 413–416.

[19] Uchibori, C. J., et al., "Effects of Chip-Package Interaction on Mechanical Reliability of Cu Interconnects for 65nm Technology Node and Beyond," *Proc. Int. Interconnect Technol. Conf.*, 2006, pp. 196–198.

[20] Tummala, R. R., *Fundamentals of Microsystems Packaging*, New York: McGraw-Hill, 2001.

[21] Zhiping, F., et al., "RF and Mechanical Characterization of Flip-Chip Interconnects in CPW Circuits with Underfill," *IEEE Trans. Microwave Theory and Techniques,* vol. 46, 1998, pp. 2269–2275.

[22] See "Voltage Regulator Module (VRM) and Enterprise Voltage Regulator-Down (EVRD) 10.0 Design Guidelines," Intel, www.intel.com.

[23] Mallik, D., et al., "Advanced Package Technologies for High Performance Systems," *Intel Technol. J.*, vol. 9, 2005, pp. 259–271.

[24] International Technology Roadmap for Semiconductors (ITRS), 2007.

[25] Stinson, J., and S. Rusu, "A 1.5 GHz Third Generation Itanium 2 Processor," *Proc. IEEE Design Automation Conference*, 2003, pp. 706–709.

[26] Balamurugan, G., et al., "A Scalable 5–15 Gbps, 14–75 mW Low-Power I/O Transceiver in 65 nm CMOS," *IEEE J. Solid-State Circuits,* vol. 43, 2008, pp. 1010–1019.

[27] Casper, B., et al., "Future Microprocessor Interfaces: Analysis, Design and Optimization," *Proc. IEEE Custom Integrated Circuits Conference*, 2007, pp. 479–486.

[28] Deutsch, A., et al., "Prediction of Losses Caused by Roughness of Metallization in Printed-Circuit Boards," *IEEE Trans. Advanced Packaging,* vol. 30, 2007, pp. 279–287.

[29] Miller, D. A. B., "Rationale and Challenges for Optical Interconnects to Electronic Chips," *Proceedings of the IEEE,* vol. 88, 2000, pp. 728–749.

[30] Dawei, H., et al., "Optical Interconnects: Out of the Box Forever?" *IEEE J. Selected Topics in Quantum Electronics,* vol. 9, 2003, pp. 614–623.

[31] Bakir, M. S., et al., "Mechanically Flexible Chip-to-Substrate Optical Interconnections Using Optical Pillars," *IEEE Trans. on Advanced Packaging,* vol. 31, 2008, pp. 143–153.

[32] See Intel for various microprocessor datasheets at www.intel.com.

[33] Prasher, R., "Thermal Interface Materials: Historical Perspective, Status, and Future Directions," *Proceedings of the IEEE,* vol. 94, 2006, pp. 1571–1586.

[34] Joyner, J. W., P. Zarkesh-Ha, and J. D. Meindl, "Global Interconnect Design in a Three-Dimensional System-on-a-Chip," *IEEE Trans. Very Large Scale Integration (VLSI) Systems,* vol. 12, 2004, pp. 367–372.

[35] Meindl, J. D., "The Evolution of Monolithic and Polylithic Interconnect Technology," *IEEE Symp. on VLSI Circuits*, 2002, pp. 2–5.

[36] Early, J., "Speed, Power and Component Density in Multielement High-Speed Logic Systems," *Proc. IEEE Int. Solid-State Circuits Conf.*, 1960, pp. 78–79.

[37] Huang, G., et al., "Power Delivery for 3D Chip Stacks: Physical Modeling and Design Implication," *Proc. IEEE Conf. on Electrical Performance of Electronic Packaging*, 2007, pp. 205–208.

[38] Ishikuro, H., N. Miura, and T. Kuroda, "Wideband Inductive-Coupling Interface for High-Performance Portable System," *Proc. IEEE Custom Integrated Circuits Conf.*, 2007, pp. 13–20.

[39] Moor, P. D., et al., "Recent Advances in 3D Integration at IMEC," *Proc. Materials Research Society Symp.*, 2007.

[40] Lu, J. Q., et al., "A Wafer-Scale 3D IC Technology Platform Using Dielectric Bonding Glues and Copper Damascene Patterned Inter-Wafer Interconnects," *Proc. IEEE Int. Interconnect Technol. Conf.*, 2002, pp. 78–80.

[41] Tan, C. S., et al., "A Back-to-Face Silicon Layer Stacking for Three-Dimensional Integration," *Proc. IEEE Int. SOI Conference*, 2005, pp. 87–89.

[42] Burns, J. A., et al., "A Wafer-Scale 3-D Circuit Integration Technology," *IEEE Trans. Electron Devices,* vol. 53, 2006, pp. 2507–2516.

[43] Witte, D. J., et al., "Lamellar Crystallization of Silicon for 3-Dimensional Integration," *Microelectronic Engineering,* vol. 84, 2007, pp. 1186–118.

[44] Feng, J., et al., "Integration of Germanium-on-Insulator and Silicon MOSFETs on a Silicon Substrate," *IEEE Electron Device Lett.,* vol. 27, 2006, pp. 911–913.

[45] Bakir, M., B. Dang, and J. Meindl, "Revolutionary Nanosilicon Ancillary Technologies for Ultimate-Performance Gigascale Systems," *Proc. IEEE Custom Integrated Circuits Conf.*, 2007, pp. 421–428.

[46] Bakir, M., et al., "'Trimodal' Wafer-Level Package: Fully Compatible Electrical, Optical, and Fluidic Chip I/O Interconnects," *Proc. Electronic Components Technol. Conf.*, 2007, pp. 585–592.

[47] Bakir, M. S., et al., "Sea of Leads (SoL) Ultrahigh Density Wafer-Level Chip Input/Output Interconnections for Gigascale Integration (GSI)," *IEEE Trans. Electron Devices,* vol. 50, 2003, pp. 2039–2048.

[48] Bakir, M. S., et al., "Dual-Mode Electrical-Optical Flip-Chip I/O Interconnects and a Compatible Probe Substrate for Wafer-Level Testing," *Proc. Electronic Components and Technol. Conf.*, 2006, pp. 768–775.

[49] Bakir, M. S., et al., "Sea of Polymer Pillars: Compliant Wafer-Level Electrical-Optical Chip I/O Interconnections," *IEEE Photonics Technol. Lett.,* vol. 15, 2003, pp. 1567–1569.

[50] Dang, B., et al., "Integrated Thermal-Fluidic I/O Interconnects for an On-Chip Microchannel Heat Sink," *IEEE Electron Device Lett.*, vol. 27, 2006, pp. 117–119.

[51] Dang, B., et al., "Wafer-Level Microfluidic Cooling Interconnects for GSI," *Proc. IEEE Int. Interconnect Technol. Conf.*, 2005, pp. 180–182.

[52] Mule, A. V., et al., "Polylithic Integration of Electrical and Optical Interconnect Technologies for Gigascale Fiber-to-the-Chip Communication," *IEEE Trans. Advanced Packaging,* vol. 28, 2005, pp. 421–433.

[53] Ishii, Y., et al., "SMT-Compatible Large-Tolerance 'OptoBump' Interface for Interchip Optical Interconnections," *IEEE Trans. Advanced Packaging,* vol. 26, 2003, pp. 122–127.

[54] Mederer, F., et al., "3-Gb/s Data Transmission with GaAs VCSELs over PCB Integrated Polymer Waveguides," *IEEE Photonics Technol. Lett.*, vol. 13, 2001, pp. 1032–1034.

[55] Chen, R. T., et al., "Fully Embedded Board-Level Guided-Wave Optoelectronic Interconnects," *Proceedings of the IEEE,* vol. 88, 2000, pp. 780–793.

[56] Choi, C., et al., "Flexible Optical Waveguide Film Fabrications and Optoelectronic Devices Integration for Fully Embedded Board-Level Optical Interconnects," *IEEE/OSA J. Lightwave Technol.,* vol. 22, 2004, pp. 2168–2176.

CHAPTER 2

Chip-Package Interaction and Reliability Impact on Cu/Low-k Interconnects

Xuefeng Zhang, Se Hyuk Im, Rui Huang, and Paul S. Ho

2.1 Introduction

The exponential growth in device density has yielded high-performance microprocessors containing two billion transistors [1]. The path toward such integration continues to require the implementation of new materials, processes, and design for interconnect and packaging structures. Since 1997, copper (Cu), which has a lower resistivity than aluminum (Al), has been selected as an interconnect material to reduce the RC delay. At the 90 nm technology node, dielectric materials with k (dielectric constant) lower than silicon dioxide (SiO_2, k ~ 4) were implemented with Cu interconnects [2, 3]. As the technology advances, the interconnect structure continues to evolve with decreasing dimensions and an increasing number of layers and complexity. At this time, the effort of the semiconductor industry is focused on implementing ultralow-k (ULK) porous dielectric material (k < 2.5) in Cu interconnects to further reduce the RC delay (Figure 2.1) [4]. However, mechanical properties of the dielectric materials deteriorate with increase in the porosity, raising serious concerns about the integration and reliability of Cu/low-k interconnects.

For advanced integrated circuits (ICs), the packaging technology is mainly based on the area-array packages, or the flip-chip solder interconnects. This type of first-level structure interconnects the active device side of the silicon (Si) die face

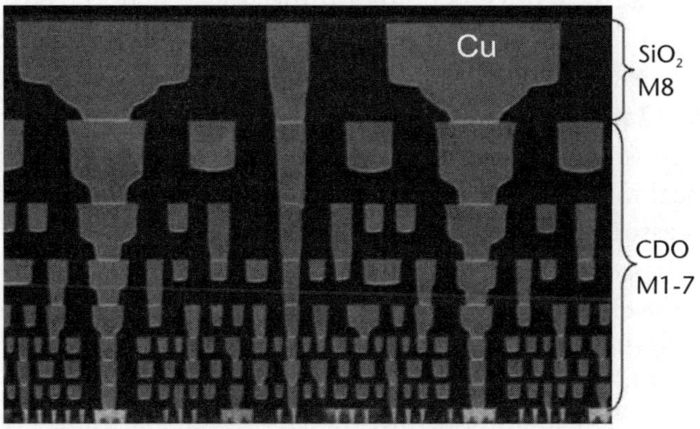

Figure 2.1 SEM image of Intel 45 nm Cu/low-k interconnect structure [4].

down via solder bumps on a multilayered wiring substrate. The area-array configuration has the ability to support the required input/output (I/O) pad counts and power distribution due to the improvement of the device density and performance. With the implementation of Cu/low-k interconnects, the flip-chip package has evolved, including the implementation of organic substrates with multilayered high-density wiring and solder bumps with pitch reducing from hundreds of microns to tens of microns. Furthermore, environmental safety mandates the change from Pb-based solders to Pb-free solders, which are more prone to thermal cyclic fatigue failures and electromigration reliability problems [5, 6].

Structural integrity is a major reliability concern for Cu/low-k chips during fabrication and when they are integrated into high-density flip-chip packages. The problem can be traced to the thermomechanical deformation and stresses generated by the mismatch in thermal expansion between the silicon die with Cu/low-k interconnects and the organic substrate in the package [7]. Although the origin of the stresses in the interconnect and packaging structures is similar, the characteristics and the reliability impact for the low-k interconnects are distinctly different. At the chip level, the interconnect structure during fabrication is subjected to a series of thermal processing steps at each metal level, including film deposition, patterning, and annealing. The nature of the problem depends to a large degree on the thermal and chemical treatments used in the fabrication steps. For instance, for deposition of metal and barrier layers, the temperature can reach 400°C and for chemical-mechanical polishing (CMP), the chip is under mechanical stresses and exposed to chemical slurries simultaneously. When subjected to such process-induced stresses, the low-k interconnects with weak mechanical properties are prone to structural failure. Such mechanical reliability problems at the chip level have been extensively investigated [8].

When incorporated into the organic flip-chip package, the fabrication of the silicon die containing the interconnect structure is already completed, so the interconnect structure as a whole is subjected to additional stresses induced by the packaging and/or assembly processes. Here the maximum temperature is reached during solder reflow for die attach. Depending on the solder materials, the reflow temperature is about 160°C or higher for eutectic Pb alloys and about 250°C for Sn-based Pb-free solders. During accelerated or cyclic thermal tests, the temperature varies from –55°C to 125°C or 150°C. Although the assembly or test temperatures of the package are considerably lower than the chip processing temperatures, the thermomechanical interaction between the chip and the package structures can exert additional stresses onto the Cu/low k interconnects. The thermal stress in the flip-chip package arises from the mismatch of the coefficients of thermal expansion (CTEs) between the chip and the substrate, which are 3 ppm/°C for Si and about 17 ppm/°C for an organic substrate. The thermally induced stresses on the solder bumps increase with the distance to the die center and reach a maximum at the outermost solder row. By using underfills, the stresses at the solder bumps can be effectively reduced to improve package reliability [9]. However, the underfill causes the package to warp, resulting in large peeling stresses at the die-underfill interfaces [10, 11]. The thermomechanical deformation of the package can be directly coupled into the Cu/low-k interconnect structure, inducing large local stresses to drive interfacial crack formation and propagation, as shown in Figure 2.2. This has generated exten-

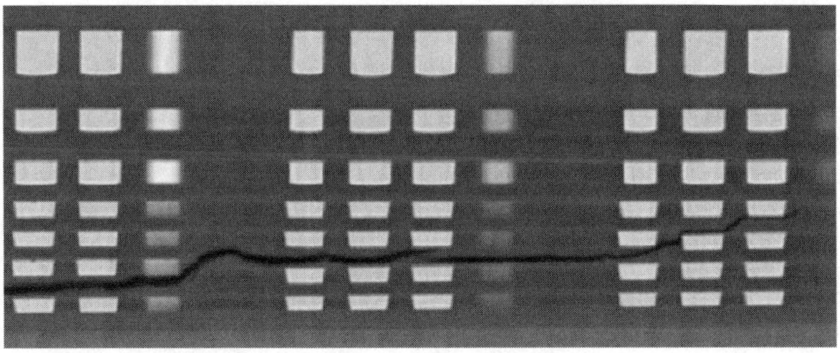

Figure 2.2 Crack propagation in a multilevel interconnect.

sive interest recently in investigating chip-package interaction (CPI) and its reliability impact on Cu/low-k structures [12–19].

In this chapter, we first review two experimental techniques important for the study of CPI and reliability, followed by a general discussion of fracture mechanics in Section 2.3. Then, a three-dimensional (3D), multilevel, submodeling method based on finite element analysis (FEA) is introduced in Section 2.4 to calculate the CPI-induced crack-driving force for interfacial delamination in the low-k interconnect structure. The chip-package interaction was found to be maximized at the die-attach step during packaging assembly and most detrimental to low-k chip reliability because of the high thermal load generated by the solder reflow process before underfilling. The discussion of the chip-package interaction in Sections 2.5 and 2.6 is first focused on the effects of dielectric and packaging materials, including different low-k dielectrics and Pb-based and Pb-free solders. The discussion is then extended in Section 2.6 to the study of the scaling effect, where the reduction of the interconnect dimension is accompanied by more metal levels and the implementation of ultralow-k porous materials. Finally, some recent results on CPI-induced crack propagation in the low-k interconnect and the use of crack-stop structures to improve chip reliability are discussed.

2.2 Experimental Techniques

2.2.1 Thermomechanical Deformation of Organic Flip-Chip Package

Thermal deformation of a flip-chip package can be determined using an optical technique of moiré interferometry. This is a whole-field optical interference technique with high resolution and high sensitivity for measuring the in-plane displacement and strain distributions [20]. This method has been successfully used to measure the thermal-mechanical deformation in electronic packages to investigate package reliability [7, 10, 21]. The sensitivity of standard moiré interferometry is not sufficient for measuring thermal deformation in high-density electronic packages, particularly for small features, such as solder bumps. For such measurements, a high-resolution moiré interferometry method based on the phase-shifting technique was developed, which measured the displacement field by extracting the phase angle as a function of position from four precisely phase-shifted moiré interference patterns [7, 11]. Once the phase angle is obtained, the continuous displacements in the horizontal (u) and vertical (v) directions can be determined. The strain components can then be evaluated accordingly:

$$\varepsilon_x = \frac{\partial u}{\partial x}, \quad \varepsilon_y = \frac{\partial v}{\partial y}, \quad \gamma_{xy} = \frac{\partial u}{\partial y} + \frac{\partial v}{\partial x} \tag{2.1}$$

The high-resolution moiré analysis was carried out for an experimental flip-chip package. The package was first sectioned and polished to reach the cross section of interest. A schematic of the experimental flip-chip package with the cross section that was analyzed is shown in Figure 2.3.

The moiré experiment was performed at room temperature (22°C), and the grating was attached to the cross section of the specimen at the temperature of 102°C, providing a reference (zero) deformation state and a thermal loading of −80°C, which can generate good deformation signals without introducing large noises from the epoxy [22]. An optical micrograph of the right half of the package cross section is shown in Figure 2.4.

The displacement field (u and v) phase-contour maps generated from the phase-shifting moiré interferometer with fringe spacing of 208 nm are shown in Figure 2.5. An outline of the interfaces obtained from the optical micrograph is superimposed onto the phase contour to highlight the local change of the displacement field in various packaging components. The global deformation of the u and v

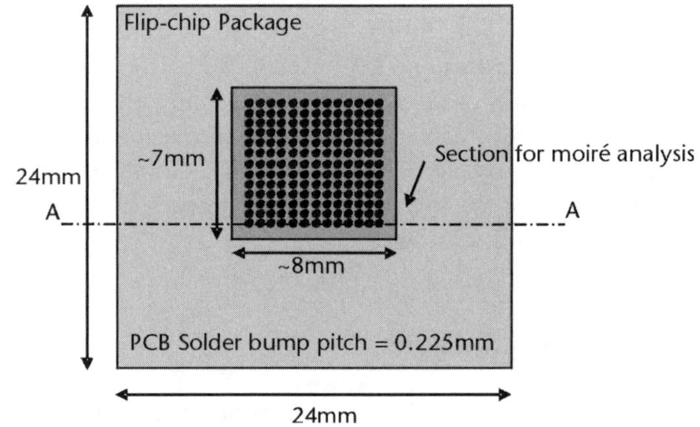

Figure 2.3 Schematic of a flip-chip package for moiré interferometry study, where the optical grating was attached to the cross section as indicated for moiré measurements.

Figure 2.4 Optical micrograph for the package cross-section used for moiré interferometry study.

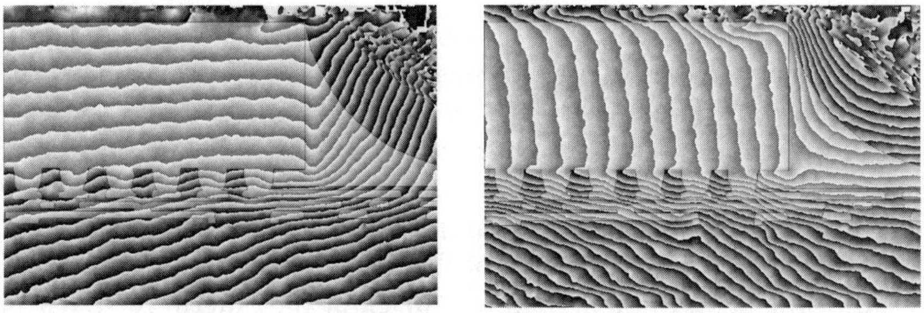

(a) (b)

Figure 2.5 Phase contour maps obtained by high-resolution moiré interferometry for the flip-chip package in Figure 2.3: (a) *u* field and (b) *v* field.

fields shows overall bending contours of the package due to warpage. This gives rise to the *u* field with relatively smooth horizontal (x) displacement distribution, while the *v* field displays high-density fringes in the solder bump/underfill layer, which is caused by the high coefficient of thermal expansion (CTE) in this layer. The die corner at the lower right has the highest shear strain, which can be seen from the large displacement gradient along the vertical (y) direction in the *u* field.

The phase contours in Figure 2.5 were used to map the displacement and strain distributions in the flip-chip package. The results are illustrated in Figure 2.6, where the displacement and strain distributions are determined along three lines: the sili-

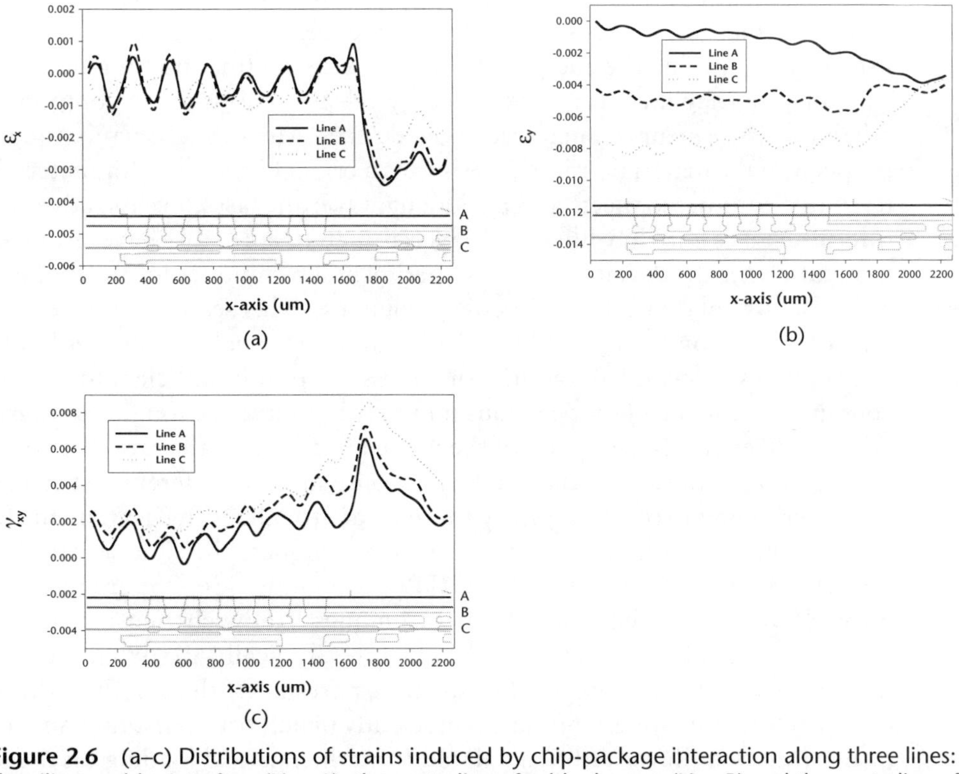

(a) (b) (c)

Figure 2.6 (a–c) Distributions of strains induced by chip-package interaction along three lines: the silicon-solder interface (Line A), the centerline of solder bumps (Line B) and the centerline of the high density interconnect wiring layer above the substrate (Line C).

con-solder interface (line A), the centerline of solder bumps (line B), and the centerline of the high-density interconnect wiring layer above the bismaleimide triazine (BT) substrate (line C). Overall, the normal strains ε_x and ε_y show the existence of a positive peeling stress in the bottom fillet area, while the shear strain γ_{xy} reaches a maximum in the fillet of the underfill near the lower die corner, corresponding to the most critical stress concentration in the package. The strain components generally increase toward the edge of the packaging as expected and can reach a value as high as 0.6% under a thermal load of –80°C for the outermost solder bump. Thus, the strain induced by the package deformation is about three to five times larger than the thermal strain caused by thermal mismatch between the die and Cu/low-k interconnect. It can be directly coupled into the low-k interconnect structure near the outermost solder bumps to drive crack formation. This underscores the importance of chip-package interaction in causing interfacial delamination in the interconnect structure, particularly with the incorporation of the low-k dielectric with weak thermomechanical properties.

2.2.2 Measurement of Interfacial Fracture Toughness

As a thermodynamic process, crack growth is driven by the release of stored strain energy in the material. The driving force for fracture is hence defined as the amount of strain energy released per unit area of crack growth, namely, the *energy release rate* (ERR). On the other hand, the resistance to crack growth is the energy required to break the bonds, create new surfaces, and generate dislocations or other defects near the crack tip. The total energy required to grow the crack by a unit area is defined as the *fracture toughness* of the material. A fracture criterion is thus established by comparing the energy release rate with the fracture toughness [8].

Fracture toughness (or critical energy release rate) is a key component for the reliability assessment of microelectronic devices. Measuring fracture toughness as a property of the material or interface is thus a critical procedure for materials characterization for interconnects and packaging. Over the last 20 years, advances in fracture mechanics for thin films and layered materials [8, 23] have provided a solid foundation for the development of experimental techniques for the measurement of both cohesive and interfacial fracture toughness. This section discusses experimental techniques commonly used to measure fracture toughness of low-k interfaces.

While a single-valued fracture toughness is typically sufficient for characterizing cohesive fracture in a homogeneous material, the interface toughness must be properly characterized as a function of the mode mix, namely, the ratio between shearing and opening stresses near the crack tip. Consequently, different test structures and load conditions are often necessary for interface toughness measurements [23, 24].

Among many different measurement techniques, the double cantilever beam (DCB) [25, 26] and four-point bend (FPB) techniques [26–28] are most popular in microelectronics applications. Both techniques sandwich one or more layers of thin-film material between two thick substrates (typically Si) so that the whole specimen is easy to load. Because the substrates are much thicker than the films, the energy release rate for an interfacial crack advancing between a film and a substrate or between two films can be calculated from the far-field loading on the substrates (i.e., the homogeneous solutions given by Hutchinson and Suo [23]), neglecting the thin films. For the DCB test (Figure 2.7), the energy release rate (J/m^2) under a symmetric loading (i.e., $F_1 = F_2 = P$) is given by

$$G = \frac{12\left(1-\upsilon^2\right)P^2 a^2}{EB^2 H^3} \tag{2.2}$$

where E and υ are the Young's modulus (N/m^2) and Poisson's ratio of the substrate, respectively; P is the applied force (N); a is the crack length (m); H is the substrate thickness (m); and B is the beam width (m). With a predetermined crack length, a critical load P_c to advance the crack can be determined from the load-displacement curve, and the interface toughness is then calculated by (2.2) as the critical energy release rate (i.e., $\Gamma = G(P_c)$). For the FPB test (Figure 2.8), the crack growth along the interface reaches a steady state with the energy release rate independent of the crack length:

$$G = \frac{21\left(1-\upsilon^2\right)P^2 L^2}{16EB^2 H^3} \tag{2.3}$$

where L is the distance (m) between inner and outer loading points. The load P at the steady state can be determined from the plateau in the load-displacement diagram.

The mode mix for the sandwich specimen depends on the local conditions, including the materials and thickness of the thin films. It is rather cumbersome to

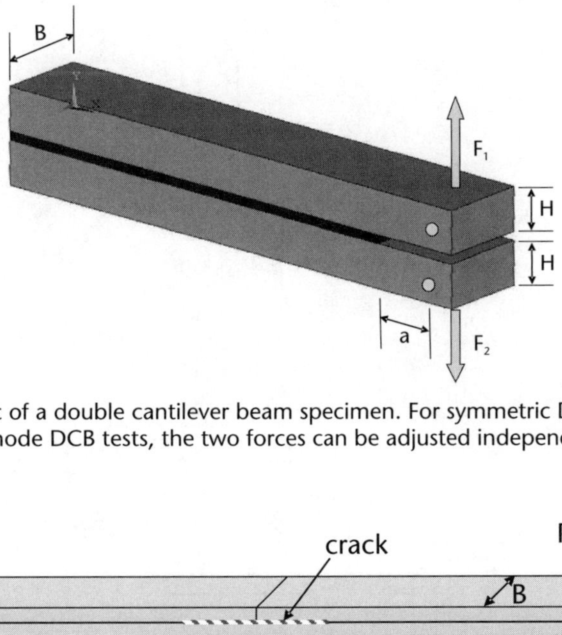

Figure 2.7 Schematic of a double cantilever beam specimen. For symmetric DCB tests, $F_1 = F_2 = P$; for mixed-mode DCB tests, the two forces can be adjusted independently (see Figure 2.9).

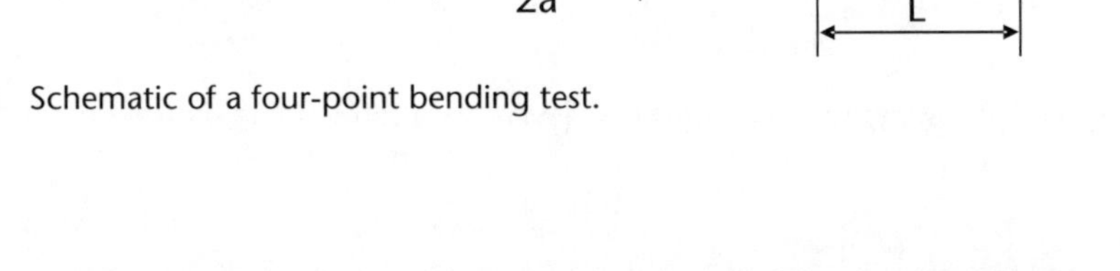

Figure 2.8 Schematic of a four-point bending test.

calculate the local mode mix when several films are sandwiched. A common practice has been to specify the mode mix for the sandwich specimens by the far-field phase angle, $\psi_\infty = \tan^{-1}(K_{II}^\infty / K_I^\infty)$, where K_I^∞ and K_{II}^∞ are, respectively, the opening and shearing modes stress-intensity factors at the crack tip [8]. For the symmetric DCB test, $\psi_\infty = 0$, hence a nominally mode I far field. For the FPB test, $\psi_\infty \approx 41°$. Other mode mixes can be obtained by using generalized laminated beam specimens loaded under cracked lap shear (mixed mode) or end-notched flexure (mode II) conditions [29] or by a modified DCB test configuration as described later.

An instrument to measure interfacial fracture energy under arbitrarily mixed-mode loading was developed using the approach originally conceived by Fernlund and Spelt [30]. This instrument utilizes a double cantilever beam (DCB) sample with a loading fixture as illustrated in Figure 2.9. By changing the positions of the different links in the link-arm structure, the forces, F_1 and F_2, applied respectively on the upper and lower beams, can be changed to adjust the mode mix. The instrument allows interfacial fracture measurements for phase angles ranging from 0° (pure tension, $F_1 = F_2$) to 90° (pure shear, $F_1 = -F_2$). Additionally, multiple tests can be run on the same sample. The challenge of this technique resides in the crack length measurement, which is required for deducing the fracture energy for the DCB configuration. The energy release rate can be calculated as

$$G = \frac{6(1-v^2)F_1^2 a^2}{EB^2H^3}\left[1+\left(\frac{F_2}{F_1}\right)^2 - \frac{1}{8}\left(1-\frac{F_2}{F_1}\right)^2\right] \tag{2.4}$$

The phase angle varies as a function of the ratio F_1/F_2:

$$\psi = \arctan\left(\frac{\sqrt{3}}{2}\frac{\left(\frac{F_1}{F_2}-1\right)}{\left(\frac{F_1}{F_2}+1\right)}\right) \tag{2.5}$$

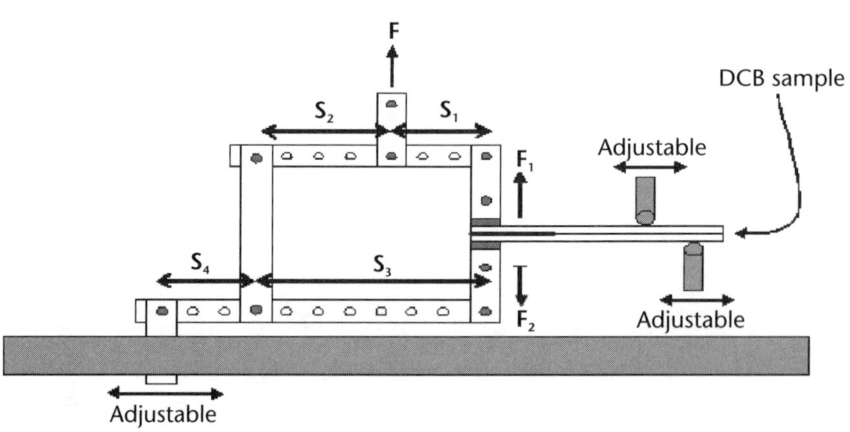

Figure 2.9 Mixed-mode double cantilever beam test loading fixture.

This mixed-mode DCB test can measure the interface toughness as a function of the phase angle (from 0° to 90°), as shown in Figure 2.10 for a porous low-k (k ~ 1.9) thin-film structure. The measured interface toughness in general exhibits a trend to increase as the phase angle increases. It is understood that the shearing mode promotes inelastic deformation in the constituent materials and near-tip interface contact/sliding, both contributing to the energy dissipation during the crack growth [31].

The measurements of interface fracture toughness provide a tool for materials selection and process control in the microelectronics industry. One typically measures the fracture toughness for specific interfaces under various process conditions, then selects the material and condition that gives an adequate toughness. In the development of Cu interconnects, new barrier layers were required to prevent copper diffusion into dielectrics and to provide adhesion of copper to the dielectrics. Using the FPB technique, Lane et al. [32] measured the interface toughness and subcritical cracking for a range of Tantalum (Ta) and Tantalum Nitride (TaN) barrier layers and showed that the presence of N significantly improves the adhesion and resistance to subcritical cracking. Moreover, a cap layer is typically used to suppress mass transport and thus improve the electromigration (EM) reliability of the Cu interconnects. A correlation between the EM lifetime and interface toughness was demonstrated so that the interface toughness measurements can be used as a screening process to select cap-layer materials and processes [33, 34]. Sufficient interface toughness is also a requirement for the integration of low-k dielectric materials in interconnect structures. Recently, the FPB technique has been adapted to quantitatively determine the effective toughness of different designs of crack-stop structures to prevent dicing flaws at the edge of chips from propagating into the active areas under the influence of thermal stresses during packaging [35].

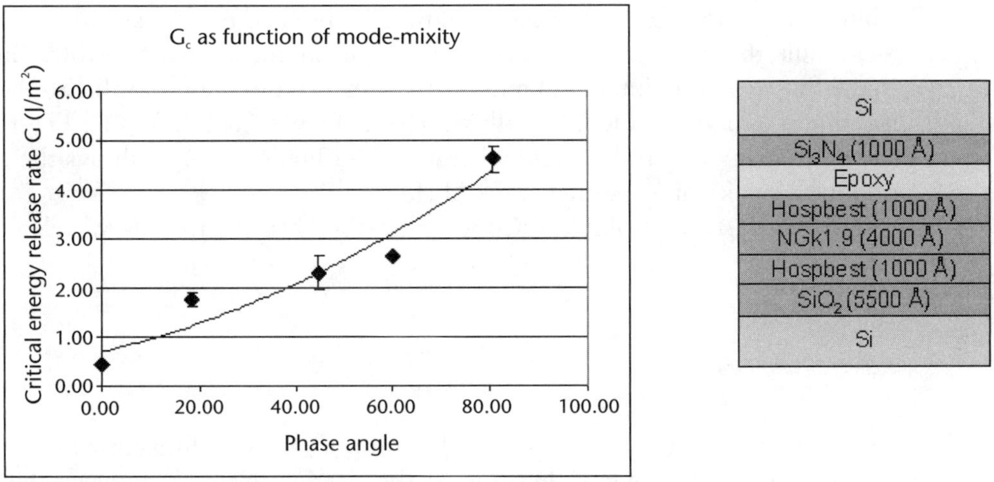

Figure 2.10 Interface toughness as a function of the mode mix measured by the mixed-mode DCB tests. The inset shows the Si/SiO_2/Hospbest/low-k(NGk1.9)/Hospbest film stack with the film thicknesses, where Hospbest is a siloxane-based hybrid material.

2.3 Mechanics of Cohesive and Interfacial Fracture in Thin Films

Integration of low-k and ultralow-k dielectrics in advanced interconnects has posed significant challenges for reliability issues due to compromised mechanical properties. Two types of failure modes have been commonly observed: cohesive fracture of the dielectrics [36–38] and interfacial delamination [39, 40]. The former pertains to the brittleness of low-k materials, and the latter manifests as a result of poor adhesion between low-k and surrounding materials. This section briefly reviews the mechanics underlying fracture and delamination in thin films with applications for integrated Cu/low-k interconnects.

In a generic thin-film structure with an elastic film on an elastic substrate, the mismatch in the elastic properties between the film and the substrate plays a critical role in the mechanical behavior and can be described by using two Dundurs' parameters [23]:

$$\alpha = \frac{\overline{E}_f - \overline{E}_s}{\overline{E}_f + \overline{E}_s} \quad \text{and} \quad \beta = \frac{\overline{E}_f\left(1-\nu_f\right)\left(1-2\nu_s\right) - \overline{E}_s\left(1-\nu_s\right)\left(1-2\nu_f\right)}{2\left(1-\nu_f\right)\left(1-\nu_s\right)\left(\overline{E}_f + \overline{E}_s\right)} \tag{2.6}$$

where $\overline{E} = E/\left(1-\nu^2\right)$ is the plane-strain modulus (N/m^2) and ν is Poisson's ratio, with the subscripts f and s for the film and substrate, respectively. When the film and the substrate have identical elastic moduli, we have $\alpha = \beta = 0$, while $\alpha > 0$ for a stiff film on a relatively compliant substrate (e.g., a SiN cap layer on low-k dielectrics) and $\alpha < 0$ for a compliant film on a relatively stiff substrate (e.g., a low-k film on a Si substrate). The role of β is often considered secondary compared to that of α and sometimes ignored for simplicity.

2.3.1 Channel Cracking

A tensile stress in an elastic film can cause cohesive fracture by channel cracking. Unlike a freestanding sheet, fracture of the film bonded to a substrate is constrained. As a result, the crack arrests at a certain depth from the film surface (often at or close to the film/substrate interface) and propagates in a direction parallel to the surface, forming a "channel crack," as illustrated in Figure 2.11 [23, 41]. Figure 2.12(a) shows an array of parallel channel cracks, and Figure 2.12(b) shows the cross section in the wake of a channel crack [42].

For an elastic thin film bonded to an elastic substrate, the energy release rate for steady-state growth of a channel crack takes form [23, 41]:

$$G_{ss} = Z(\alpha, \beta)\frac{\sigma_f^2 h_f}{\overline{E}_f} \tag{2.7}$$

where σ_f is the tensile stress in the film, h_f is the film thickness, and the dimensionless coefficient Z depends on the elastic mismatch between the film and the substrate. At steady state, the energy release rate is independent of the channel length. The value of Z represents the constraint effect on channel cracking due to the substrate and can be determined using a two-dimensional (2D) model [41, 43],

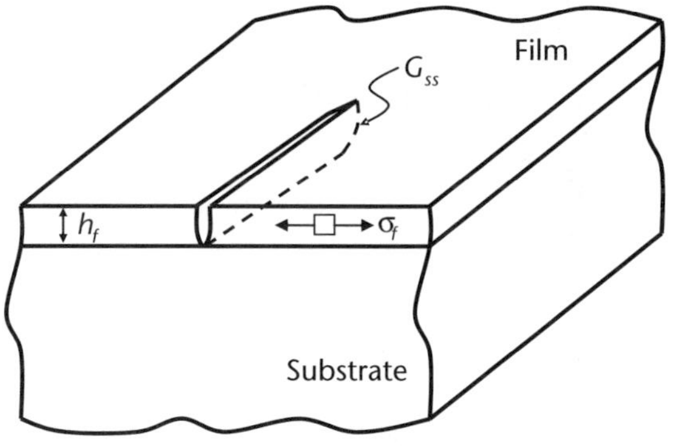

Figure 2.11 Illustration of a channel crack.

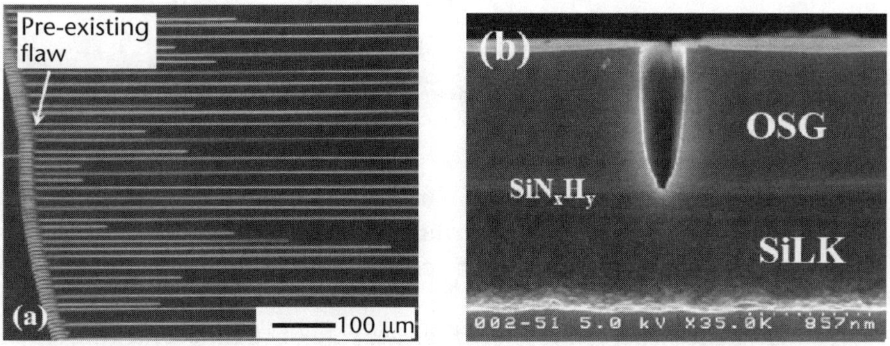

Figure 2.12 Top view (a) and cross-sectional view (b) of channel cracks in thin film stacks of low-k materials [42].

which is plotted in Figure 2.13 as a function of α. When the film and the substrate have identical elastic moduli, $Z = 1.976$. It deceases slightly for a compliant film on a relatively stiff substrate ($\alpha < 0$). A more compliant substrate, on the other hand, provides less constraint, and Z increases. For very compliant substrates (e.g., a SiN cap layer on low-k dielectrics), Z increases rapidly, with $Z > 30$ for $\alpha > 0.99$.

A three-dimensional analysis showed that the steady state is reached when the length of a channel crack exceeds two to three times the film thickness [44]. When the substrate material is more compliant than the film, however, the crack length to achieve the steady state can be significantly longer [45]. With all the subtleties aside, the steady-state energy release rate for channel cracking offers a robust measure for the reliability of thin-film structures, which has also been used for experimental measurements of cohesive fracture toughness of dielectric thin films [27] and crack-driving forces in integrated low-k interconnects [42]. Recently, channel cracking has been investigated in more complex integrated structures with low-k materials, such as multilevel patterned film structures [37] and stacked buffer layers [40].

In addition to the elastic constraint effect, the roles of interface debonding, substrate cracking, and substrate plasticity on film cracking have been studied [45–49]. As shown by Tsui et al. [38], while a brittle film cracks with no delamination on a

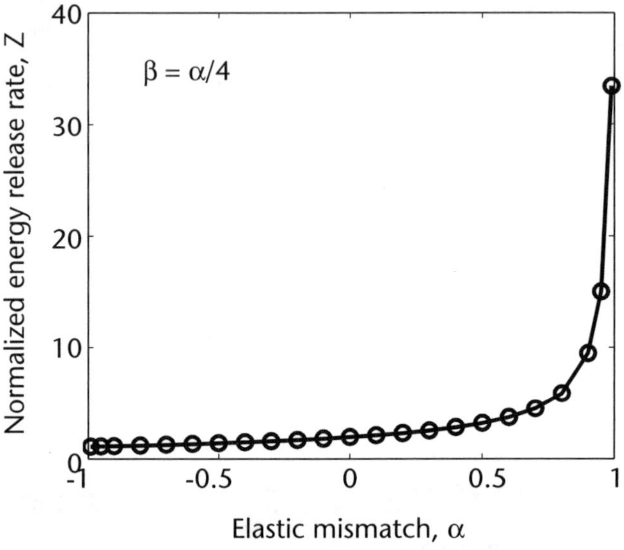

Figure 2.13 Normalized energy release rate for steady state channel crack growth.

stiff substrate, interfacial delamination was observed when the film lies on a more compliant buffer layer. Furthermore, the constraint effect can be significantly relaxed over time if the substrate creeps [50, 51], leading to higher energy release rates.

When the steady-state energy release rate of channel cracking reaches or exceeds the cohesive fracture toughness of the film, fast crack growth in the film is expected. In the subcritical regime ($G < G_c$), however, slow growth of channel cracks in thin films may be facilitated by environmental effects or thermal cycles. The consequence of slow crack growth can be critical for the long-term reliability and lifetime of devices. Several mechanisms for the slow growth of channel cracks in thin films have been studied, including environmentally assisted cracking [36, 38], creep-modulated cracking [50–53], and ratcheting-induced cracking [54, 55].

2.3.2 Interfacial Delamination

Integration of diverse materials relies on interfacial integrity. Typically, an interfacial crack nucleates from a site of stress concentration such as a free edge of the film or a geometric or material junction in a patterned structure. Under tension, a channel crack in a film may lead to delamination from the root of the channel [47]. Under compression, buckling of the film can drive propagation of buckle-delamination blisters (e.g., telephone cord blisters) [23].

Due to asymmetry in the elastic moduli with respect to a bimaterial interface, propagation of an interfacial crack occurs in general under mixed-mode conditions. As a result, the fracture toughness of an interface is necessarily expressed as a function of the mode mix. However, the stress field around an interfacial crack tip in general cannot be decoupled into pure mode I (opening) and mode II (shearing) fields, due to the oscillatory singularity at the crack tip [56, 57]. For a two-dimensional interfacial crack between two isotropic elastic solids joined along the x-axis, as illustrated in Figure 2.14, the normal and shear tractions on the interface directly ahead of the crack tip are given by [23]

$$\sigma_{yy} = \frac{K_1 \cos(\varepsilon \ln r) - K_2 \sin(\varepsilon \ln r)}{\sqrt{2\pi r}}, \sigma_{xy} = \frac{K_1 \sin(\varepsilon \ln r) + K_2 \cos(\varepsilon \ln r)}{\sqrt{2\pi r}} \tag{2.8}$$

where r is the distance from the crack tip, and ε is the index of oscillatory singularity depending on the second Dundurs' parameter,

$$\varepsilon = \frac{1}{2\pi} \ln\left(\frac{1-\beta}{1+\beta}\right) \tag{2.9}$$

The stress-intensity factors, K_1 and K_2, are the real and imaginary parts of the complex interfacial stress-intensity factor, $K = K_1 + iK_2$.

When $\varepsilon = 0$, the interfacial crack-tip stress field reduces to the homogeneous crack-tip field with tractions, $\sigma_{yy} = \frac{K_1}{\sqrt{2\pi r}}$ and $\sigma_{xy} = \frac{K_2}{\sqrt{2\pi r}}$, where K_1 and K_2 are the conventional mode I and mode II stress-intensity factors. In this case, the ratio of the shear traction to the normal traction is simply K_2 / K_1, which defines the mode mix. When $\varepsilon \neq 0$, however, the mode mix as a measure of the proportion of mode II to mode I requires specification of a length quantity since the ratio of the shear traction to the normal traction varies with the distance to the crack tip. As suggested by Rice [57], an arbitrary length scale (l) may be used to define a phase angle of the mode mix for interfacial delamination, namely,

$$\psi = \tan^{-1}\left[\left(\frac{\sigma_{xy}}{\sigma_{yy}}\right)_{x=l}\right] = \tan^{-1}\left[\frac{\operatorname{Im}\left(Kl^{i\varepsilon}\right)}{\operatorname{Re}\left(Kl^{i\varepsilon}\right)}\right] \tag{2.10}$$

The choice of the length l can be based on the specimen geometry, such as the film thickness, or on a material length scale, such as the plastic zone size at the crack tip. Different choices will lead to different phase angles. A simple transformation

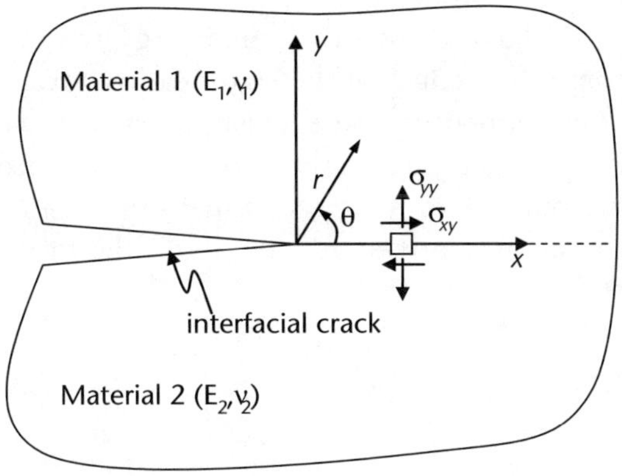

Figure 2.14 Geometry and convention for an interfacial crack.

rule was noted by Rice [57] that transforms the phase angle defined by one length scale to another, namely,

$$\psi_2 = \psi_1 + \varepsilon \ln(l_2 / l_1) \tag{2.11}$$

where ψ_1 and ψ_2 are the phase angles associated with lengths l_1 and l_2, respectively. Therefore, so long as a length scale is clearly presented for the definition of the phase angle, experimental data for the mode-dependent interface toughness can be unambiguously interpreted for general applications (i.e., $\Gamma(\psi_1, l_1) = \Gamma(\psi_2, l_2)$).

The energy release rate for a crack advancing along an interface is related to the interfacial stress-intensity factors by [23]

$$G = \frac{1-\beta^2}{E^*}\left(K_1^2 + K_2^2\right) \tag{2.12}$$

where $E^* = 2\left(\overline{E}_1^{-1} + \overline{E}_2^{-1}\right)^{-1}$. The criterion for interfacial delamination can then be stated as $G = \Gamma(\psi, l)$, where the same choice of the length l has to be used in the definition of the phase angle for the interface toughness and in the calculation of the phase angle for the specific problem along with the energy release rate G. For 3D problems, a mode III term must be added into the energy release rate, and another phase angle may be defined for the 3D mode mix.

For delamination of an elastic thin film from a thick elastic substrate under the plane-strain condition, a steady state is reached when the interfacial crack length is much greater than the film thickness. The energy release rate for the steady-state delamination is independent of the crack length:

$$G_{ss}^d = \frac{\sigma_f^2 h_f}{2\overline{E}_f} \tag{2.13}$$

Taking the film thickness as the length scale ($l = h_f$), the phase angle of mode mix at the steady state depends on the elastic mismatch as a function of the Dundurs' parameters (i.e., $\psi_{ss} = \omega(\alpha, \beta)$). This function was determined numerically and tabulated by Suo and Hutchinson [58]. When the film and the substrate have identical elastic moduli, $\psi_{ss} = \omega(0,0) = 52.1°$.

Yu et al. [59] have shown that the energy release rate for an interfacial crack emanating from a free edge can be significantly lower than the steady-state energy release rate. Consequently, there exists a barrier for the onset of delamination, which depends on the materials and geometry near the edge. For interfacial delamination from the root of a channel crack [46, 47], the energy release rate approaches the same steady-state value but follows a power law at the short crack limit [60]:

$$G_d \sim \left(\frac{d}{h_f}\right)^{1-2\lambda} \tag{2.14}$$

where d is the crack length, and λ depends on the elastic mismatch determined by

$$\cos \lambda\pi = \frac{2(\alpha-\beta)}{1-\beta}(1-\lambda)^2 - \frac{\alpha-\beta^2}{1-\beta^2} \tag{2.15}$$

As shown in Figure 2.15, the energy release rate approaches zero as $d / h_f \to 0$ when $\alpha < 0$ (compliant film on stiff substrate). Thus, there exists a barrier for the onset of delamination. On the other hand, when $\alpha > 0$ (stiff film on compliant substrate), the energy release rate approaches infinity as $d / h_f \to 0$, suggesting that interfacial delamination always occurs concomitantly with channel cracking. In Cu/low-k interconnects, the low-k dielectric is usually more compliant compared to the surrounding materials. Therefore, channel cracking of low-k dielectrics is typically not accompanied by interfacial delamination. However, when a more compliant buffer layer is added adjacent to the low-k film, interfacial delamination can occur concomitantly with channel cracking of the low-k film [38]. Moreover, a relatively stiff cap layer (e.g., SiN) is often deposited on top of the low-k film. Channel cracking of the cap layer could be significantly enhanced by interfacial delamination.

The energy release rate and mode mix of interfacial delamination in more complex integrated structures are commonly calculated for device reliability analysis. Here, finite-element-based models are typically constructed to compute the stress-intensity factors or energy release rates of interfacial cracks literally introduced into the model. Nied [61] presented a review focusing on applications in electronic packaging. Liu et al. [39] analyzed delamination in patterned interconnect structures. As one of the emerging reliability concerns for advanced interconnects and packaging technology, the impacts of chip-package interactions on interfacial delamination have been investigated by multilevel finite element models, which will be discussed in the next section.

The experimental techniques to measure interface toughness as the critical energy release rate ($\Gamma = G_c$) for fast fracture have been discussed in Section 2.2.2. In addition, interfacial cracks are often susceptible to environmentally assisted crack growth in the subcritical regime ($G < G_c$) [25, 27, 28, 31, 62, 63]. The kinetics of

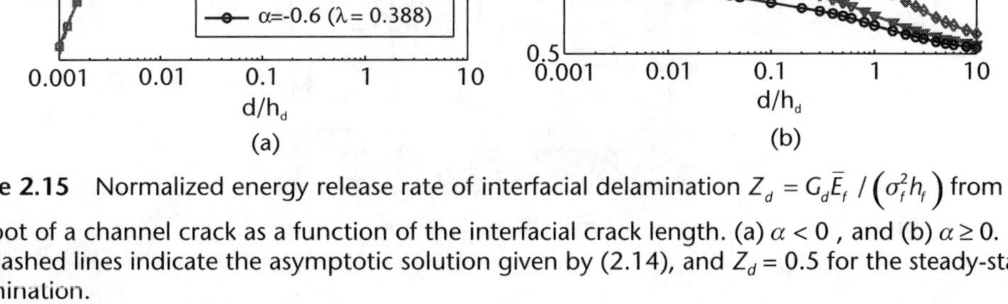

Figure 2.15 Normalized energy release rate of interfacial delamination $Z_d = G_d \bar{E}_f / (\sigma_f^2 h_f)$ from the root of a channel crack as a function of the interfacial crack length. (a) $\alpha < 0$, and (b) $\alpha \geq 0$. The dashed lines indicate the asymptotic solution given by (2.14), and $Z_d = 0.5$ for the steady-state delamination.

subcritical interfacial delamination have been understood as controlled by stress-dependent chemical reactions in stage I and by mass transport of environmental species (e.g., water molecules) to the crack tip in stage II [31]. Recently, by combining the kinetics of subcritical cracking and water diffusion, Tsui et al. [40] proposed a model to predict degradation of adhesion in thin-film stacks as a function of exposure time to water and found good agreement with experimental data for film stacks containing a low-k dielectric material.

2.4 Modeling of Chip-Packaging Interactions

Finite element analysis (FEA) is commonly used to evaluate the thermomechanical deformation and stress distributions in electronic packages and their impact on reliability. For stand-alone silicon chips, the modeling results show that thermal stresses in the Cu lines depends on the aspect ratio (i.e., the width versus height ratio) and the degree of confinement from the dielectric materials as well as the barrier and cap layers (Figure 2.16). For an aspect ratio greater than 1, the stress state is triaxial and behaves almost linear elastically under thermal cycling [64]. Wafer processing can induce additional residual stresses in the interconnect structures, which has also been investigated using FEA [65]. The general behavior is in quantitative agreement with results from X-ray diffraction measurements [64, 66]. After the silicon die is assembled into a flip-chip package, the package deformation can increase the thermomechanical stresses in the interconnect structures. Modeling the packaging effect on the thermal stress of the interconnect structure is challenging due to the large difference in the dimensions of the packaging and interconnect structures. For this reason, researchers from Motorola first introduced a multilevel submodeling technique to evaluate the energy release rate for interfaces in the interconnect structure after assembling the die into a flip-chip package [12, 13]. This technique bridges the gap between the packaging and wafer levels. The energy release rates for various

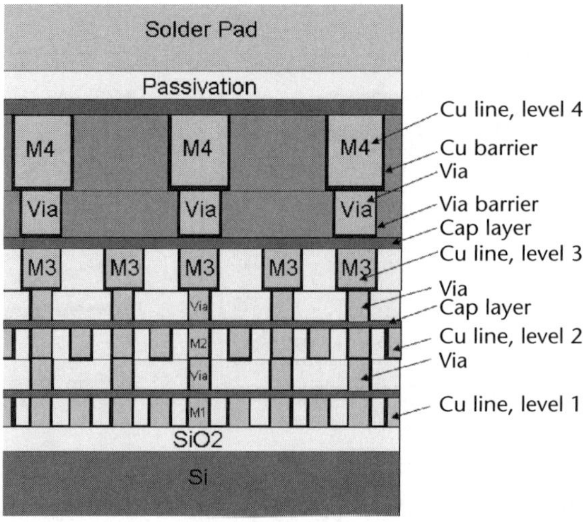

Figure 2.16 Cu/low-k structure schematics.

interconnect interfaces during packaging assembly were calculated using 2D FEA models. However, a flip-chip package is a complicated 3D structure that cannot be properly represented using a 2D model. We developed, therefore, a 3D FEA model based on a four-level submodeling technique to investigate the packaging effect on interconnect reliability, particularly focusing on the effects of low-k dielectrics and other materials used to form the Cu interconnect structures [14, 17].

2.4.1 Multilevel Submodeling Technique

Level 1. Starting from the package level, the thermomechanical deformation for the flip-chip package is first investigated. At this package level, a quarter-section of the package is modeled using the symmetry condition as illustrated in Figure 2.17(a). No interconnect structure detail was considered at this level because its thickness is too small compared to the whole package. Simulation results for this package-level model are verified with experimental results obtained from moiré interferometry.

Level 2. From the simulation results for the package-level modeling, the most critical solder bump is identified. A submodel focusing on the critical solder bump region with much finer meshes is developed, as shown in Figure 2.17(b). The built-in cut boundary technique in ANSYS [67] is used for submodeling. At this

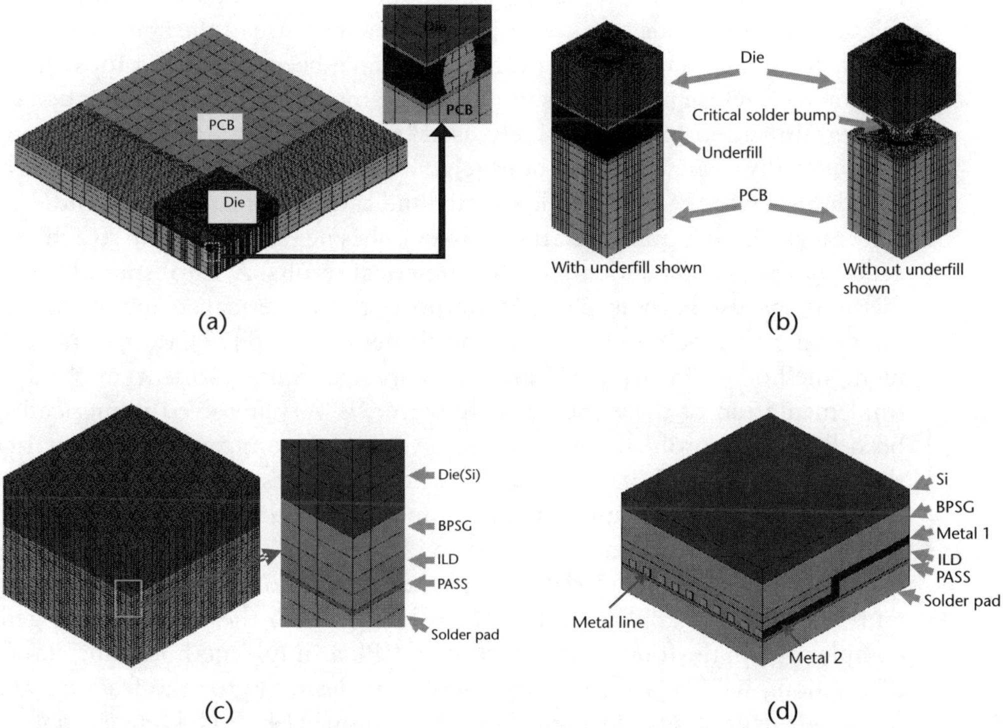

Figure 2.17 Illustration of four-level sub-modeling: (a) package level; (b) critical solder level; (c) die-solder interface level; and (d) detailed interconnect level.

submodel level, a uniform interlevel dielectric (ILD) layer at the die surface is considered, but still no detailed interconnect structure is included.

Level 3. Based on the level 2 submodeling results, a large peeling stress is found at the die-solder interface. At the critical die-solder interface region with the highest peeling stress, a submodel is created using the cut boundary technique, as shown in Figure 2.17(c). This submodel focuses on the die-solder interface region (a small region of level 2) containing a portion of the die, the ILD layer, and a portion of the solder bump. Still only a uniform ILD layer at the die surface is considered at this level, and no detailed interconnect structure is included.

Level 4. This submodel zooms in further from the level 3 model, focusing on the die-solder interface region as shown in Figure 2.17(d). Here, a detailed 3-D interconnect structure is included. An interconnect with two metal levels and vias is considered first, and effects of multilevel stacks are discussed in Section 2.6. The submodel is set up accordingly, and a crack with a fixed length is introduced along several interfaces of interest. The energy release rate and mode mix for each crack are determined using a modified virtual crack closure technique as discussed in the next section.

2.4.2 Modified Virtual Crack Closure Method

To investigate the impact of CPI on the reliability of low-k interconnect and packaging structures, interfacial cracks are introduced into the models, and both the energy release rates and mode mix are calculated as a measure of the crack-driving force for interfacial delamination. Several methods have been developed for calculating the interfacial fracture parameters within the framework of finite element analysis. The J-integral method has been widely used [68–70] and is a standard option in some commercially available FEA codes (e.g., ABAQUS [71]). This method is capable of calculating both the energy release rate and the mode mix for 2-D and 3-D interfacial cracks, but it requires relatively fine meshes near the crack tip to achieve convergence and path independence of the numerical results. A set of special finite element methods has also been developed to improve the numerical accuracy without requiring fine meshes, including the singular element method [72], the extended finite element method (XFEM) [73], and an enriched finite element method [74, 75]. Implementation of these methods, however, is very involved numerically and has been limited to problems with relatively simple geometry and material combinations. Alternatively, Liu et al. [19, 39] calculated stress-intensity factors by comparing the crack surface displacement to the analytical crack-tip solution, from which both the energy release rate and mode mix were determined. This approach requires very fine meshes near the crack tip for the accuracy of the displacement calculation and is not readily applicable to 3D problems. With the material and geometrical complexities in the four-level modeling of CPI, a simple method using standard FEA codes along with relatively coarse meshes is desirable for the fracture analysis. A modified virtual crack closure (MVCC) technique [14, 76] has emerged to meet such a need and is described as follows.

As illustrated in Figure 2.18, the MVCC method calculates the components of the energy release rate corresponding to the three basic fracture modes (I, II, and III) separately. With the local stress-strain and displacement distributions obtained by the finite element modeling, both the energy release rate and the mode mix for the interfacial cracks can be calculated accordingly. For the eight-node solid elements shown in Figure 2.18, the three energy release rate components G_I, G_{II} and G_{III} can be obtained as

$$\begin{aligned} G_I &= \sum_i F_z^{(i_1)} \delta_z^{(i_2)} / (2\Delta A) \\ G_{II} &= \sum_i F_x^{(i_1)} \delta_x^{(i_2)} / (2\Delta A) \\ G_{III} &= \sum_i F_y^{(i_1)} \delta_y^{(i_2)} / (2\Delta A) \end{aligned} \tag{2.16}$$

where $F_x^{(i_1)}$, $F_y^{(i_1)}$, and $F_z^{(i_1)}$ are nodal forces at node i_1 along the x-, y-, and z-directions, respectively, and $\delta_x^{(i_2)}$, $\delta_y^{(i_2)}$, and $\delta_z^{(i_2)}$ are relative displacements between node i_2 and node i_3 in the x-, y-, and z-directions, respectively. Note that, for simplicity, only one element set is shown along the crack front direction (y-direction). The total energy release rate is then

$$G = G_I + G_{II} + G_{III} \tag{2.17}$$

and the phase angles of mode mix may be expressed as

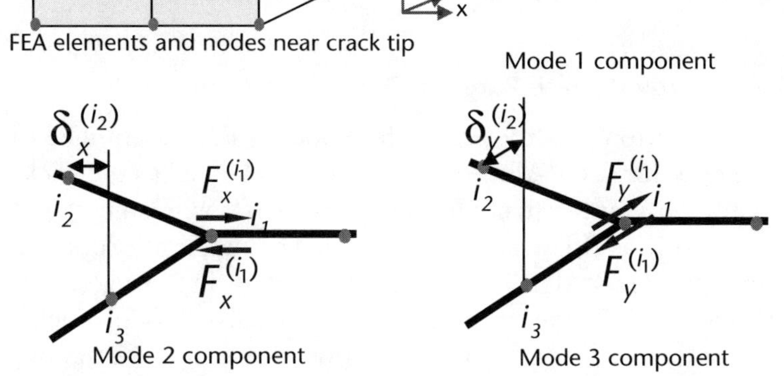

Figure 2.18 Illustration of the modified virtual crack closure (MVCC) technique.

$$\begin{aligned} \psi &= \arctan\left[\left(G_{II}/G_I\right)^{1/2}\right] \\ \varphi &= \arctan\left[\left(G_{III}/G_I\right)^{1/2}\right] \end{aligned} \tag{2.18}$$

The criterion for interfacial delamination can thus be established by comparing the total energy release rate to the experimentally measured mode-dependent interface toughness [i.e., $G = \Gamma(\psi, \varphi)$].

While the original virtual crack closure technique (VCCT) was proposed for cracks in homogeneous materials [77–79], it has been shown that care must be exercised in applying the technique for interfacial cracks [79–83]. As noted by Krueger [79], due to the oscillatory singularity at the interfacial crack tip, the calculated energy release rate and mode mix may depend on the element size at the crack tip. It was suggested that the element size shall be chosen to be small enough to assure a converged solution by the finite element model but also large enough to avoid oscillating results for the energy release rate. Furthermore, as discussed in Section 2.3.2, mode I and mode II in general cannot be separated for interfacial cracks (except for cases with $\beta = 0$). The separation of the energy release rate components in (2.16) is therefore dependent on the element size, as is the definition of the phase angles in (2.18). The total energy release rate on the other hand was found to be less sensitive to the element size [80, 81]. Several approaches have been suggested to extract consistent phase angles of mode mix independently of the element size using the VCCT [82, 83], following the standard definition in (2.10). For simplicity, the phase angles defined in (2.18) are used in the subsequent discussions.

2.4.3 Package-Level Deformation

The FEA results for the package-level modeling can be verified using results from moiré interferometry. Since the thermal load used in the moiré measurement was from 102°C to 22°C, we applied the same thermal load (102°C to 22°C) in the package-level modeling in order to compare the moiré and FEA results. Figure 2.19 shows the *z*-displacement (package warpage) distribution along the die centerline (line A-A in Figure 2.3). The FEA and moiré results are found to be in good agreement. Detailed moiré results can be found in [22].

2.4.4 Energy Release Rate for Stand-Alone Chips

After verification with moiré interferometry, FEA was applied to evaluate the energy release rates for stand-alone wafer structures as well as the packaging effect. Both Al and Cu interconnect structure with tetraethyl orthosilicate (TEOS) and a spin-on polymer SiLK as ILD were investigated. The material properties used in the modeling analysis are listed in Table 2.1. All materials in the wafer structure were assumed to be linear elastic except at the package level, where plasticity was considered for solder materials. To calculate the energy release rate, a crack was introduced at several relevant interfaces, as shown in Figure 2.20. The crack has a rectangular shape with a fixed length of 1.5 μm along the metal line direction and a width of 0.5 μm,

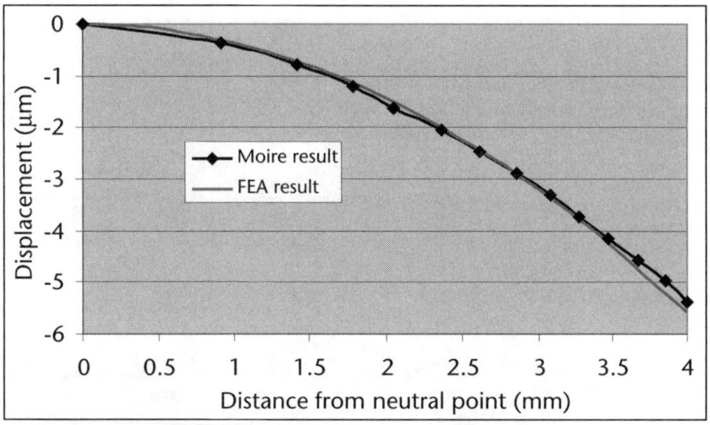

Figure 2.19 Comparison of FEA and Moiré results of thermal deformation for the flip-chip package in Figure 2.3.

Table 2.1 Mechanical Properties of Interconnect Materials [18, 22]

Material	*E (GPa)*	ν	α *(ppm/°C)*
Si	162.7	0.28	2.6
Al	72	0.36	24
Cu	122	0.35	17
TEOS (k = 4.2)	66	0.18	0.57
SiLK (k = 2.62)	2.45	0.35	66
MSQ (k = 2.7)	7	0.35	18
CVD-OSG (k = 3.0)	17	0.35	8
Porous MSQ-A (k < 2.3)	2	0.35	10
Porous MSQ-B (k < 2.3)	5	0.35	10
Porous MSQ-C (k ~ 2.3)	10	0.35	12
Porous MSQ-D (k ~ 2.3)	15	0.35	18
Porous MSQ-E (k ~ 2.3)	10	0.35	6
Porous MSQ-F (k ~ 2.3)	10	0.35	18

the same as the metal line width and thickness. In general, the energy release rate depends on the number of the metal levels and the crack dimension used in the calculation. In the following discussion, a fixed crack size in a two-level structure is used in order to simplify the CPI computation in the study of the material and processing effects. This point should be kept in mind as the crack-driving force is compared, particularly when the CPI study is extended to four-level interconnect structures with a different crack size to study scaling and ultralow-k effects in Section 2.6.

The energy release rates for interfacial fracture along the six interfaces shown in Figure 2.20 were first calculated for the stand-alone chip subjected to a thermal load of 400°C to 25°C, typical for wafer processing. The results summarized in Figure 2.21 show that the energy release rates for all the interfaces in Al/TEOS and Cu/TEOS structures are generally small, less than 1 J/m^2. The Cu/SiLK structure has the highest energy release rates for the two vertical cracks along the SiLK/barrier sidewall (crack 2) and along the barrier/Cu interfaces (crack 3), both exceeding 1

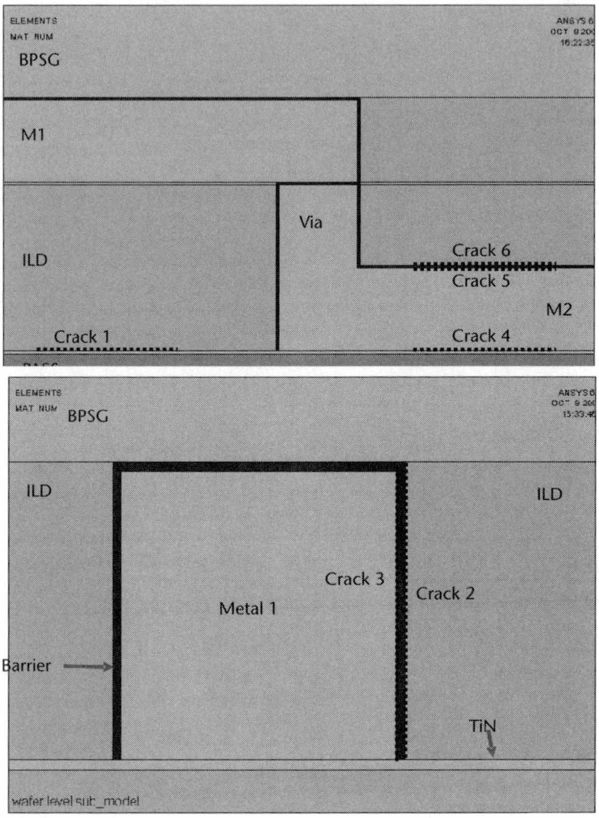

Figure 2.20 Cracks introduced along interfaces of interest.

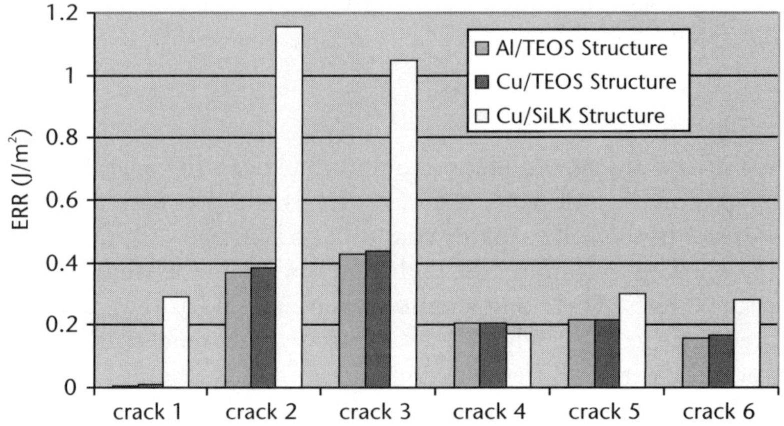

Figure 2.21 Energy release rates for interfaces in interconnect structures in stand-alone chip before packaging assembly (from 400°C to 25°C).

J/m^2. The fracture mode for these two cracks is almost pure mode I, indicating that for the stand-alone chip, the tensile stresses driving crack formation act primarily on the vertical interfaces due to the large CTEs of the low-k ILDs in comparison to the CTEs of the silicon substrate and metal lines. Compared to the critical energy release rates for low-k interfaces obtained from experiments (usually about 4 to 5 J/m^2 [84]), these values are considerably lower. Hence, interfacial delamination in Cu/low-k interconnect structures during wafer processing is not expected to be a serious problem, although the result does not rule out the possibility of delamination due to subcritical crack growth.

2.5 Energy Release Rate Under Chip-Package Interactions

2.5.1 Effect of Low-k Dielectrics

The energy release rates induced by CPI were evaluated using the four-step multilevel submodel. A stress-free state was assumed at –55°C for the flip-chip package, and the crack-driving force was obtained at 125°C to simulate a test condition of –55°C to 125°C. The package used has the same dimensions as the one used for moiré measurements, which has an organic substrate with a die size of 8 × 7 mm^2 and lead-free solders (95.5 Sn/3.8 Ag/0.7 Cu). The critical solder bump with the highest thermal stress is the outermost one at the die corner. The interconnect structure located at this critical solder bump–die interface was investigated. The results are given in Figure 2.22, which reveals a small CPI effect for Al/TEOS and Cu/TEOS structures. In contrast, the effect is large for the Cu/SiLK structure with the crack-driving force G reaching 16 J/m^2. Interestingly, the interfaces parallel to the die surface (cracks 1, 4, 5, and 6) are more prone to delamination, instead of the vertical interfaces 2 and 3 as is the case for the stand-alone chip. For these parallel interfaces, the mode mix is close to being pure mode I, although for the Cu/passivation interface, both mode I and III components are present. As compared with the results for the stand-alone wafers and after packaging, not only is a large increase in the crack-driving force evident due to chip-package interactions but the interfaces most prone to delamination also change to those parallel to the die surface. This indicates that the crack-driving force becomes dominated by thermal

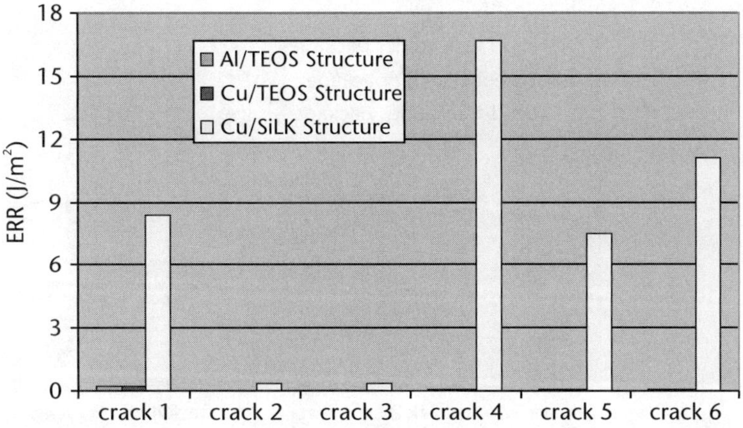

Figure 2.22 Energy release rates for interfaces in on-chip interconnect structures after assembly into packages with Pb-free solders.

stresses imposed by the package deformation where the package warpage has the most significant effect for the parallel interfaces.

These results indicate that the delamination induced by CPI occurs near the outermost solder bumps under mostly a mode I condition. As the crack propagates, both the energy release rate and the mode mix at the crack tip vary. The crack follows a path that maximizes G / Γ, the ratio between the energy release rate and the fracture toughness. Depending on the local material combination and wiring geometry, the crack may zigzag through the interconnect structure toward the lower Cu levels with weaker low-k dielectrics. As the crack propagates, the energy release rate will increase while the phase angle changes to mixed mode, depending on the local wiring geometry. The crack-propagation problem in a multilayer interconnect network is complex and will be further discussed in Section 2.6.

2.5.2 Effect of Solder Materials and Die Attach Process

As the semiconductor industry shifts from Pb-based solders to Pb-free solders, the effects of solder materials on CPI and low-k interconnect reliability become of interest. The energy release rates for the six interfaces are compared in Figure 2.23 for high-lead (95 Pb/5 Sn), eutectic lead alloy (62 Sn/36 Pb/2 Ag), and lead-free solder (95.5 Sn/3.8 Ag/0.7 Cu). The material properties used in these calculations are listed in Table 2.2. The mismatch in CTE between the lead-free solder and underfill is larger than that between the high-lead or eutectic solder and underfill. The Young's modulus of the lead-free solder is also larger than the high-lead and eutectic solders. Thus, larger thermal stresses are induced at the die surface for the lead-free solder package as compared to the high-lead and eutectic solder packages, resulting in the highest driving force for interconnect delamination in lead-free packages.

The processing step with the highest thermal load in flip-chip package assembly is the die attach before underfilling the package. The solder reflow occurs at a temperature higher than the solder melting point, and afterwards the package structure is cooled down to room temperature. Without the underfill serving as a stress buffer, the thermal mismatch between the die and the substrate can generate a large thermal

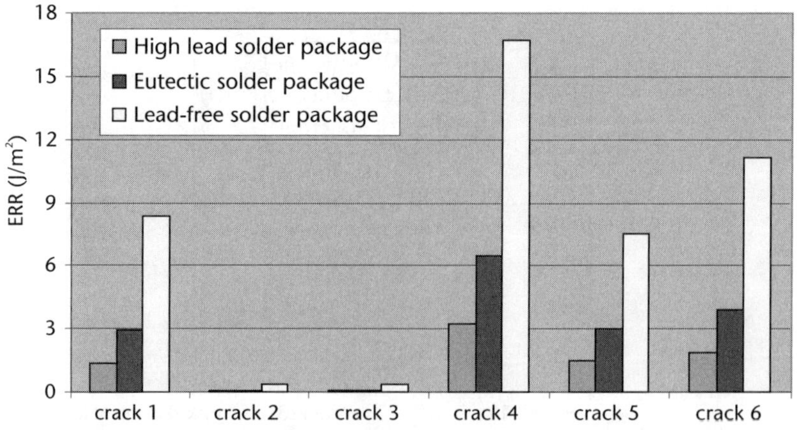

Figure 2.23 Energy release rates for Cu/SiLK interconnect structures in high-lead, eutectic solder and lead-free solder packages after underfilling.

Table 2.2 Material Properties for High-Lead, Eutectic Lead, and Lead-Free Solders [22] (Modulus Values Are a Function of Temperature *T*)

Solder Material	*E (GPa)*	ν	α *(ppm/°C)*
Eutectic	$75.84 - 0.152 \times T$	0.35	24.5
High lead	$39.22 - 0.063 \times T$	0.35	29.7
Lead free	$88.53 - 0.142 \times T$	0.40	16.5
Underfill	6.23	0.40	40.6
Organic substrate	Anisotropic elastic		16 (in plane) 84 (out of plane)

stress at the solder-die interface near the die corner, driving interfacial delamination. The CPI effect of the die-attach step for low-k structure was investigated for Cu/SiLK and Cu/MSQ structures for different solder materials. Here, the study was again performed for the high-lead, eutectic lead, and lead-free solders with different reflow cycles: 160°C to 25°C for eutectic solder, 250°C to 25°C for lead-free solder, and 300°C to 25°C for high-lead solder. The substrate in the package was organic with a die size of 8×7 mm^2, and the study assumed that the high-lead solder could be assembled onto an organic substrate in order to compare these solders on the same substrate. The results are summarized in Figure 2.24(a) for Cu/SiLK chips assembled on an organic substrate. The eutectic solder package has the lowest crack-driving force for interfacial delamination due to its lowest reflow temperature. In contrast, the lead-free solder package is most critical due to the high reflow temperature and the high Young's modulus of the lead-free solder material. For the high-lead solder, although it has the highest reflow temperature yet the lowest Young's modulus, the crack-driving force is lower than that for the lead-free solder package. For comparison, the results for the Cu/MSQ structure with eutectic and lead-free solders are shown in Figure 2.24(b). The energy release rate for the Cu/MSQ structure is generally about a factor of three lower than that of the Cu/SiLK structure. This can be attributed to the threefold higher Young's modulus of the MSQ dielectrics, indicating that the mechanical property of the low-k is an important factor to consider for the packaging effect.

Comparing Figure 2.23 and Figure 2.24(a), it is clear that the crack-driving force in the Cu/SiLK structure during the die-attach process is generally larger than that in an underfilled package during thermal cycling from –55°C to 125°C. This indicates that the die-attach process with a larger thermal load is a more critical step than thermal cycling in driving critical interfacial delamination in Cu/low-k structures.

2.5.3 Effect of Low-k Material Properties

To investigate the effect of dielectric properties, we first compare the CPI for a CVD-OSG (k = 3.0) [9] with an MSQ [10] and a spin-on polymer SiLK [7] to investigate how better material properties can improve interconnect reliability. Both MSQ and SiLK are fully dense with k ~ 2.7. The energy release rates were computed using the two-level interconnect structure with cracks 1 to 6, and the results are plotted in Figure 2.25. Among the dielectric materials, the energy release rates

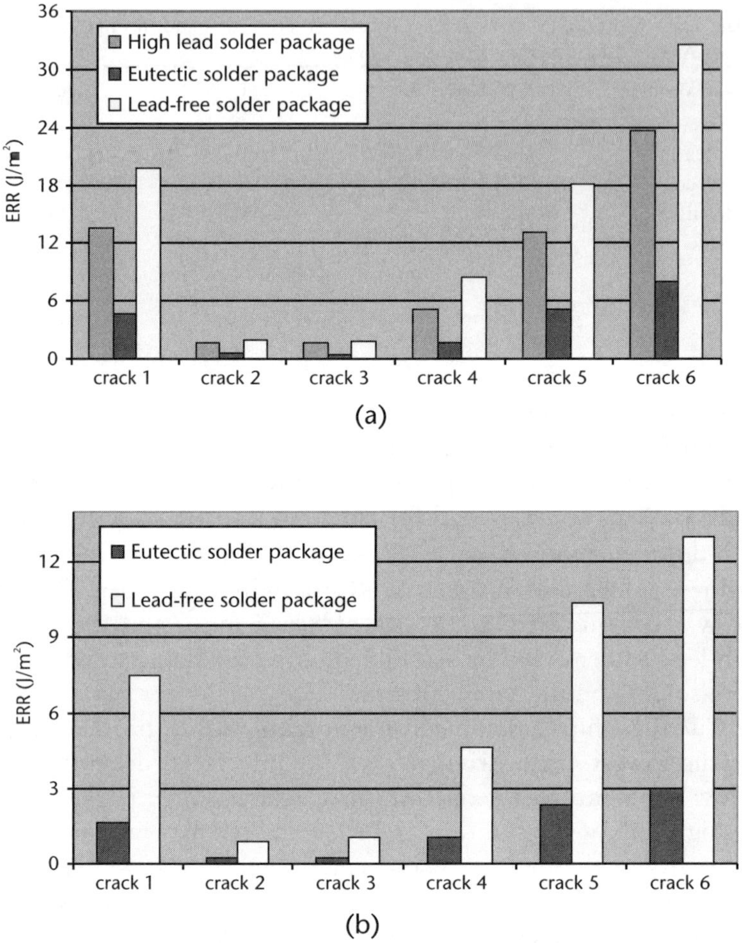

Figure 2.24 Energy release rates for interconnect interfaces in (a) Cu/SiLK and (b) Cu/MSQ structures in die attach process.

(ERRs) are the lowest for CVD-OSG, which has the highest Young's modulus (E). For the spin-on polymer, which has the lowest E, the ERR values for cracks 1 and 6 are about six times higher than those of CVD-OSG. This indicates that the on-chip interconnect fabricated with spin-on polymer needs about six times more adhesion strength at the interfaces of cracks 1 and 6 in order to obtain a mechanical reliability equivalent to interconnects fabricated using CVD-OSG.

Next, the study is extended to several porous MSQ materials (A to D) [11], which are being developed for interconnect structures of the 65 nm node and beyond. These porous low-k materials have k < 2.3 but with different thermomechanical properties, which are listed in Table 2.1. The results are plotted in Figure 2.26(a), which shows a good correlation between ERR and E. Comparing porous MSQ-D (k ~ 2.3) with fully dense CVD-OSG (k = 3.0), both with similar mechanical properties, their ERR values are similar. Interestingly, for the porous MSQ-E and the MSQ-F, even though they have very different CTE but similar E, their ERR values are about the same, too, as shown in Figure 2.26(b). Overall, there seems to be little effect due to the CTE of the low-k materials. In contrast, the ERR

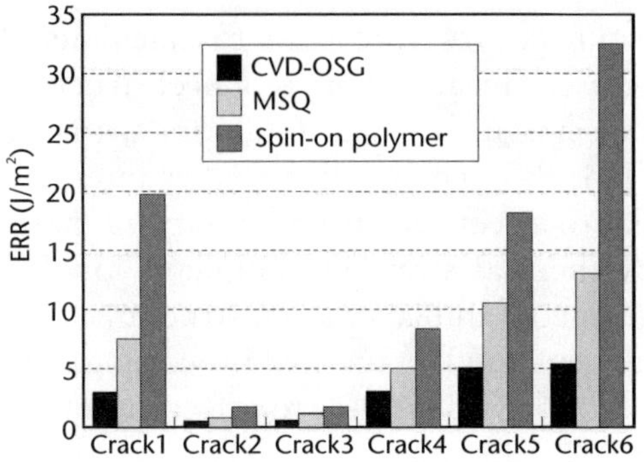

Figure 2.25 Comparison of ERR for low-k dielectrics of CVD-OSG, MSQ and a spin-on polymer. The cracks are the same as shown in Figure 2.20.

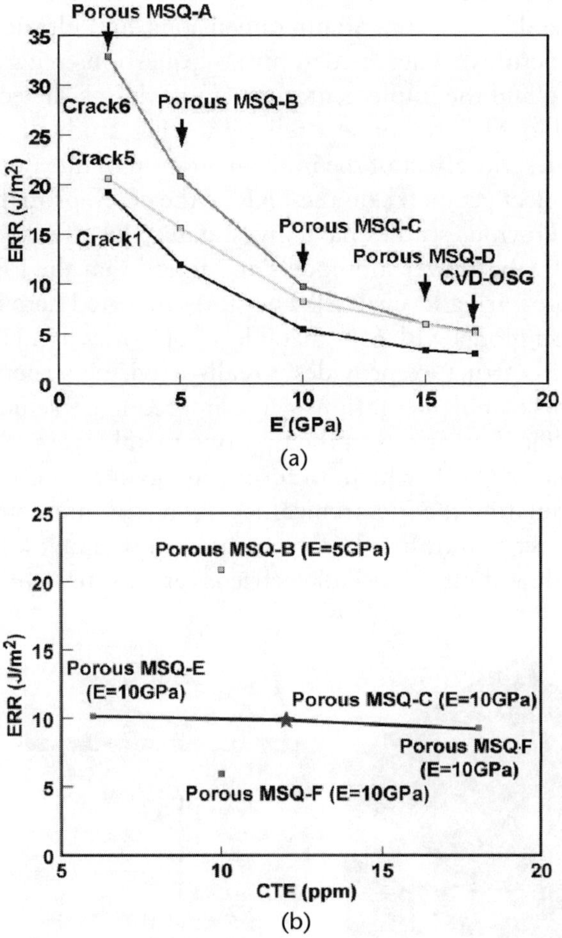

Figure 2.26 (a) Values of ERR as a function of Young's modulus for low-k dielectrics; and (b) Values of ERR as a function of CTE for low-k dielectrics.

increases considerably with decreasing E. Therefore, for low-k dielectrics, increasing E seems to be effective for improving the mechanical reliability.

The interconnect structure used to calculate ERR in this study is a simple two-layer structure. The actual interconnect structure for low-k chips for the 65 nm technology node has more than 11 layers with complex geometry and material combinations [84, 85]. There will be other interesting and important factors contributing to ERR to affect package reliability. Of particular interest is channel cracking induced by thermal stress in compliant low-k layers, which depends on the interconnect geometry and layer stack structure. There will also be the effect due to residual stresses generated by thermal processing during chip fabrication, which can superimpose onto the CPI stresses to affect the ERR driving force [42, 65].

2.6 Effect of Interconnect Scaling and Ultralow-k Integration

The scaling of interconnect structures has led to highly complex architectures with over 10 metal layers, sub–50 nm dimensions, and ultralow-k dielectrics (ultimately, air gap structures). There are important questions regarding the effect of interconnect scaling and the implementation of ultralow-k dielectric on chip-package interaction and low-k interconnect reliability. The study of the scaling effect is focused on two issues: the effect of the implementation of ultralow-k dielectric and the effect of interconnect geometry on the ERR as the crack propagates through the Cu/low-k structure. Previous studies have investigated the effect of increasing stacking layers based on 2D multilevel submodels and found that the ERR increases with the addition of more wiring levels [12]. The study reported here is based on a 3D multilevel interconnect model with four metal levels, as shown in Figure 2.27. We found that a four-level 3D structure provides a realistic wiring structure to analyze the effect of porous low-k implementation in the interconnect structure. In this structure, the pitch and line dimensions in the first two metal layers (M1 and M2) are doubled in the third layer (M3), which are doubled again in the fourth layer (M4), approximately simulating the hierarchical layers in real interconnect structures.

The effect of ultralow-k implementation was investigated using different stacking of low-k and ultralow-k dielectric layers. In this study, we are interested to find

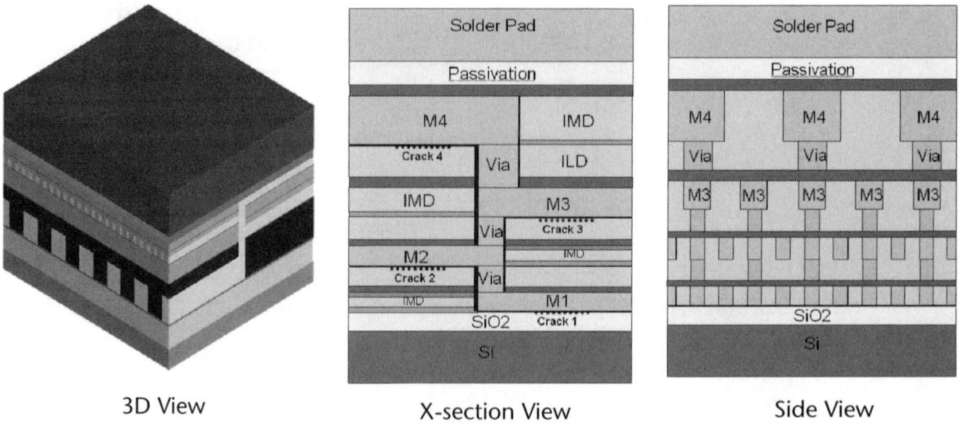

Figure 2.27 FEA model of 3D 4-layer interconnect.

out whether different combinations of low-k and ultralow-k dielectrics in selective metal layers could improve mechanical reliability without sacrificing electrical performance (RC delay). Energy release rates were calculated for horizontal cracks placed at each metal level at the interface between the etch stop/passivation (ESL) and the low-k dielectric, which is known to be one of the interfaces most prone to delamination [12–14]. Each crack has a width of 0.1 μm and a length of 2 μm extending in the multiple wiring directions, as shown in Figure 2.27. Results of the ERRs of the interfacial cracks in the four-level interconnect models with three different ILD combinations are summarized in Figure 2.28. The first model [Figure 2.28(a)] uses ultralow-k materials in all layers for which the interfacial crack at the uppermost level (crack 4) has the largest ERR. This is to be expected since the uppermost level is the thickest, being four times larger than M1 in thickness, and thus the maximum crack-driving force. In the second model [Figure 2.28(b)], SiO_2 is used to replace ULK at level 4. In this case, the high E of SiO_2 significantly reduces the ERR of crack 4 but raises the ERRs in the other three interfaces. This reflects the effect on the crack-driving force of the elastic mismatch between SiO_2 and the ULK layer as discussed in Section 2.3.2. In this structure, the ERR is highest for crack 3 in the M3 level, which is thicker than M1 and M2. In model 3 [Figure 2.28(c)], a fully dense low-k CVD-OSG is used at level 3, which has a higher E than the ULK. Consequently, the ERR of crack 3 is reduced, and the effect of elastic mismatch shifts the largest ERR to crack 2 in the M2 level with a magnitude comparable to that of crack 3 in model 2. This set of results indicates that the ultralow-k interface at the upper-

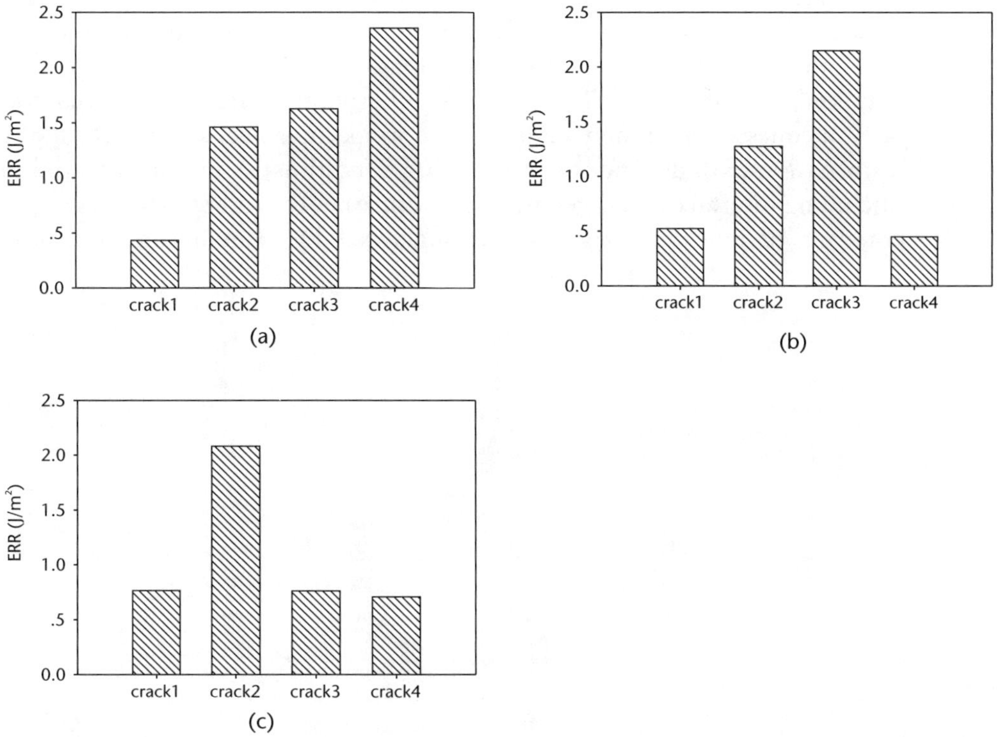

Figure 2.28 CPI-induced energy release rates for four-level interconnect models with different combinations of interlevel dielectrics: (a) ULK in all levels, (b) SiO_2 in the M4 level and ULK in others, and (c) SiO_2 in M4 and CVD OSG in M3 and ULK in others.

most level is the most critical, and the multilevel stacking structure has to be optimized in order to minimize the CPI effect on ULK interconnect reliability.

As shown in Figure 2.29, the calculated energy release rates increase dramatically from low-k (OSG and porous MSQ) to ultralow-k ILDs, especially for cracks 2 and 3. This trend is consistent with the results from the two-level interconnect model, which has shown increasing ERRs with decreasing ILD modulus (Figure 2.26). However, the magnitudes of the ERRs in the four-level model are considerably lower that those obtained from the two-level model, possibly due to denser metal lines providing stronger constraint on the cracks. Since ultralow-k materials are required for 45 nm technology and beyond, this result indicates that CPI will be a major concern due to the weak mechanical properties of the ultralow-k materials.

As a crack propagates in a multilevel interconnect structure, both the energy release rate and the mode mix at the crack tip vary. As illustrated by a two-dimensional model in Figure 2.30, as the crack grows from right to left along one interface, the energy release rate oscillates as a function of the crack length. When the crack tip is located close to the left corner of a metal line, the energy release rate peaks due to the peeling stress concentration. The magnitude of the peak increases with the crack length but seems to saturate toward a steady state. The phase angle of mode mix oscillates as well, but within a relatively small range. Apparently, the local material combinations and geometry complicate the stress field near the crack tip and thus the crack propagation along the interface. As a conservative design rule, the maximum energy release rate must be kept below the interface toughness at the corresponding mode mix to avoid interfacial delamination.

A crack propagation in a real interconnect structure due to CPI is shown in Figure 2.2. Apparently, the crack does not always propagate along one interface. Depending on the local material combination and geometry, an interfacial crack may kink out of the interface, causing cohesive fracture of low-k materials. Similarly, a cohesive crack may deflect into a weak interface. The selection of the crack propagation path depends on the loading conditions as well as material properties (including interfaces) and geometrical features in the interconnect structure. A general rule of crack propagation, as suggested by Hutchinson and Suo [23] for

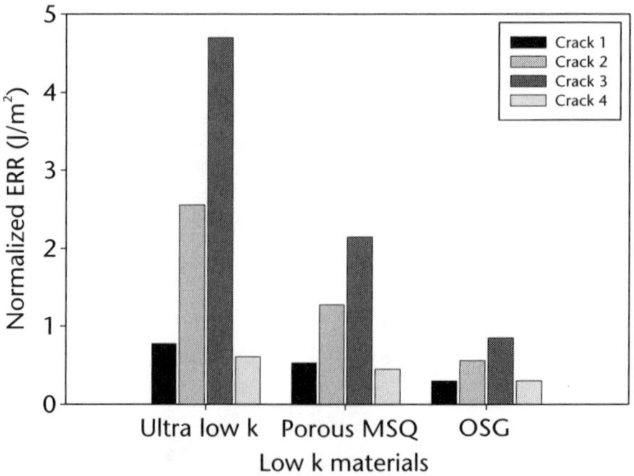

Figure 2.29 Comparison of CPI-induced energy releases rates in the four-level interconnects with low-k and ultralow-k ILDs.

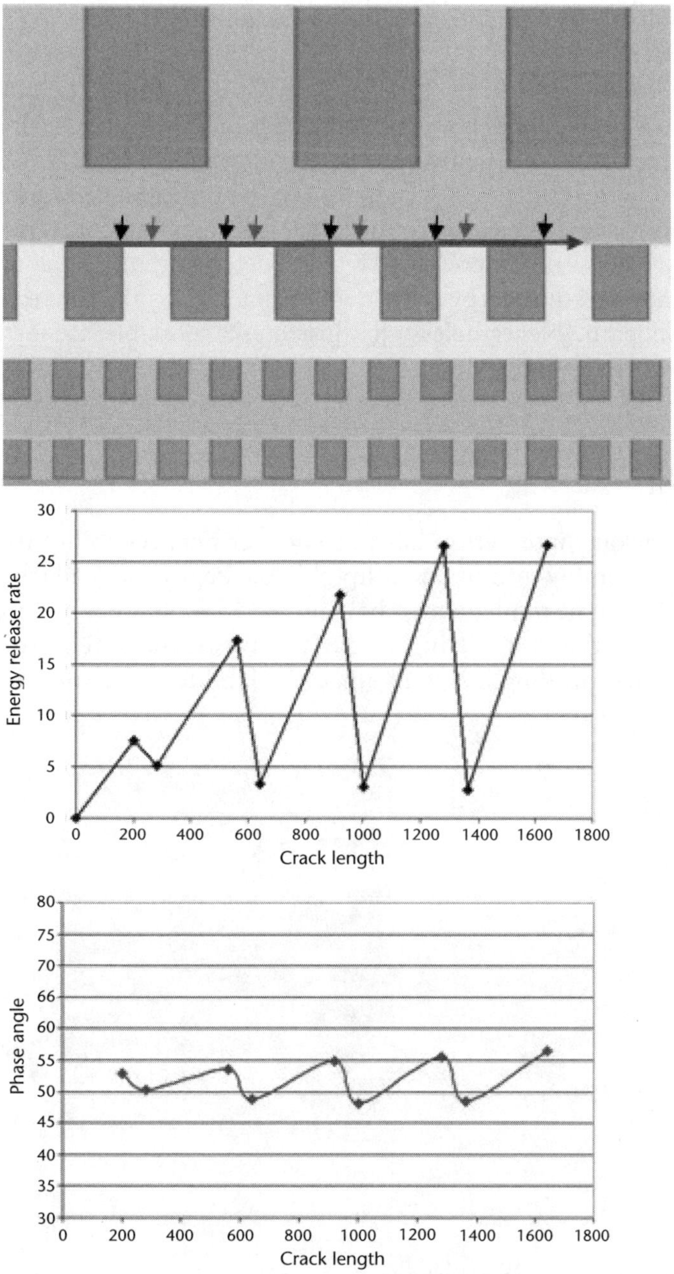

Figure 2.30 Crack length dependence of the energy release rate and mode mix of an interfacial crack.

anisotropic materials and composites, may be stated as follows: a crack propagates along a path that maximizes G / Γ, the ratio between the energy release rate and the fracture toughness. While cohesive fracture in an isotropic material typically follows a path of mode I ($\psi = 0$), the mode mix along an interfacial path varies, as does the interfacial fracture toughness. Therefore, the crack propagation not only seeks a path with the largest energy release rate but also favors a path with the lowest fracture toughness, either interfacial or cohesive. Due to the complexity of the materials and structures, modeling of crack propagation in multilevel interconnects has not been well developed. Experiments have shown that cracks often propagate from upper levels to lower levels, eventually causing failure by die cracking. Figure 2.31 depicts a simple model of crack propagation in a multilevel interconnect due to CPI. The crack initiates at an upper-level interface, which has been shown to have a higher energy release rate compared to the same crack in a lower-level interface. As the crack propagates toward the lower levels and the total crack length increases, the energy release rate increases. Without detailed data of the interface toughness, the calculation of the energy release rate alone is not sufficient to predict the crack propagation path. Nevertheless, it illustrates a possible scenario in consistency with experimental observations.

2.7 Summary

Chip-package interaction has become a critical reliability issue for Cu/low-k chips during assembly into organic flip-chip packages, particularly for ultralow-k porous dielectrics to be implemented beyond the 65 nm node. In this chapter, we review the experimental and modeling studies to investigate the chip-package interaction and its impact on low-k interconnect reliability. The problem is explored using

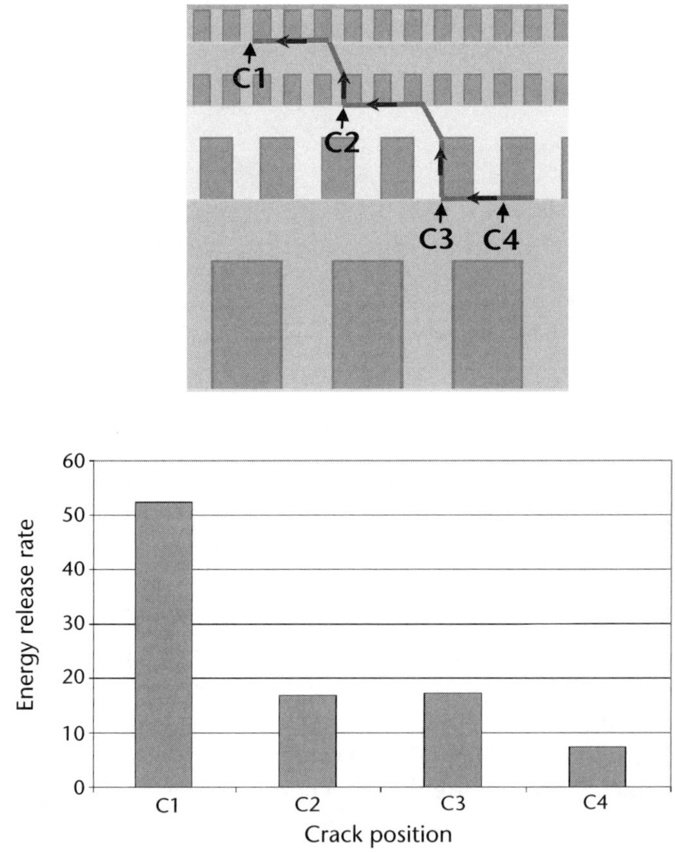

Figure 2.31 CPI induced crack propagation in a multilevel interconnect.

high-resolution moiré interferometry and multilevel submodeling, and its origin is traced to the large thermal stress induced by package deformation to drive crack propagation and the weak thermomechanical properties of the low-k dielectric material. The nature of interfacial delamination and crack growth in multilayered dielectric structures was discussed based on fracture mechanics. The chip-package interaction was investigated using 3D finite element analysis (FEA) based on a multilevel submodeling approach. The packaging-induced crack-driving force for relevant interfaces in Cu/low-k structures was deduced. The die-attach process was found to be a critical step, and the energy release rate was found to depend on the solder, underfill, and low-k material properties. Implementation of lead-free solder and ultralow-k material poses great threats to the Cu interconnect reliability by increasing the low-k delamination driving force. Finally, the effect of scaling and crack propagation in multiple Cu/dielectric line structures was investigated. Crack propagation was found to be a complex phenomenon depending on the local material combinations and geometry, which control the stress field near the crack tip and thus the crack propagation along the interface.

Recent efforts from within the industry and universities have significantly advanced the present understanding of chip-package interaction and its reliability impact on Cu/low-k interconnects. Many questions remain, and a major challenge in microelectronics packaging is to prevent cracks initiated at the edge of a chip from propagating into the functional area of the chip under thermomechanical loadings during packaging processes and service. The use of low-k and ultralow-k dielectrics in the interconnects presents even more of a challenge due to chip-package interactions and the significantly lower toughness of the low-k materials. One approach to preventing propagation of the edge cracks is to incorporate patterned metal structures around the perimeter of a chip as a crack-stop structure [19]. If designed properly, the metal structures can increase the fracture toughness along the path of crack propagation. A four-point-bend experiment has been used to determine the effective toughness of crack-stop structures [35]. The optimal design of crack-stop structures requires better understanding of crack propagation under the influence of chip-package interactions.

Acknowledgments

The authors are grateful for the financial support of their research from the Semiconductor Research Corporation, the Fujitsu Laboratories of America, and the United Microelectronics Corporation. They also gratefully acknowledge fruitful discussions with many colleagues, including R. Rosenberg, M. W. Lane, T. M. Shaw, and X. H. Liu from IBM; E. Zschech and C. Zhai from AMD; J. H. Zhao and D. Davis from TI; G. T. Wang and J. He from Intel; C. J. Uchibori and T. Nakamura from Fijitsu Laboratories; Z. Suo from Harvard; H. Nied from Lehigh; and R. Dauskardt from Stanford.

References

[1] See www.intel.com/technology/architecture-silicon/2billion.htm?iid = tech_arch+ body_ 2b.

[2] Edestein, D., et al., "Full Copper Wiring in a Sub-0.25 μm CMOS ULSI Technology," *IEEE Int. Electron Devices Conf.*, December 7–10, 1997, pp. 773–776.

[3] Venkatesan, S., et al., "A High Performance 1.8V, 0.20 μm CMOS Technology with Copper Metallization," *IEEE Int. Electron Devices Conf.*, December 7–10, 1997, pp. 769–772.

[4] Ingerly, D., et al., "Low-k Interconnect Stack with Thick Metal 9 Redistribution Layer and Cu Die Bump for 45 nm High Volume Manufacturing," *International Interconnect Technology Conference*, June 1–4, 2008, Burlingame, CA, pp. 216–218.

[5] Chae, S. H., et al., "Electromigration Statistics and Damage Evolution for Pb-Free Solder Joints with Cu and Ni UBM in Plastic Flip-Chip Packages," *J. Materials Science: Materials Electronics*, Vol. 18, 2007, pp. 247–258.

[6] Shang, J. K., et al., "Mechanical Fatigue of Sn-Rich Pb-Free Solder Alloys," *J. Materials Science: Materials Electronics*, Vol. 18, 2007, pp. 211–227.

[7] Ho, P. S., et al., "Reliability Issues for Flip-Chip Packages," *Microelectronics Reliability*, Vol. 44, No. 5, 2004, pp. 719–737.

[8] Suo, Z., "Reliability of Interconnect Structures," *Interfacial and Nanoscale Failure of Comprehensive Structural Integrity*, Vol. 8, 2003, pp. 265–324.

[9] Wu, T. Y., Y. Tsukada, and W. T. Chen, "Materials and Mechanics Issues in Flip-Chip Organic Packaging," *Proc. Electronic Comp. Technology Conf.*, May 28–31, 1996, Orlando, FL, pp. 524–534.

[10] Dai, X., et al., "In-situ Moiré Interferometry Study of Thermomechanical Deformation in Glob-Top Encapsulated Chip-on-Board Packaging," *Experimental/Numerical Mechanics in Electronic Packaging*, Vol. 1, 1997, p. 15.

[11] Miller, Mikel R., et al., "Analysis of Flip-Chip Packages Using High Resolution Moiré Interferometry," *Proc. 49th Electronic Components and Technology Conference,* June 1–4, 1999, San Diego, CA, pp. 979–986.

[12] Mercado, L., C. Goldberg, and S.-M. Kuo, "A Simulation Method for Predicting Packaging Mechanical Reliability with Low-k Dielectrics," *International Interconnect Technology Conference,* June 3–5, 2002, Burlingame, CA, pp. 119–121.

[13] Mercado, L., et al., "Analysis of Flip-Chip Packaging Challenges on Copper Low-k Interconnects," *Proc 53rd Electronic Components and Technology Conference,* May 27–30, 2003, New Orleans, LA, pp. 1784–1790.

[14] Wang, G. T., et al., "Packaging Effects on Reliability of Cu/Low-k Interconnects," *IEEE Trans. Device and Materials Reliability*, Vol. 3, 2003, pp. 119–128.

[15] Zhao, J. H., B. Wilkerson, and T. Uehling, "Stress-Induced Phenomena in Metallization," *7th International Workshop, AIP Conf. Proc.*, Vol. 714, 2004, pp. 52–61.

[16] Landers, W., D. Edelstein, and L. Clevenger, "Chip-to-Package Interaction for a 90 nm Cu/PECVD Low-k Technology," *Proc. International Interconnect Technology Conference*, June, 7–9, 2004, Burlingame, CA, pp. 108–110.

[17] Wang, G., P. S. Ho, and S. Groothuis, "Chip-Packaging Interaction: A Critical Concern for Cu/Low-k Packaging," *Microelectronics Reliability,* Vol. 45, 2005, pp. 1079–1093.

[18] Uchibori, C. J., et al., "Effects of Chip-Package Interaction on Mechanical Reliability of Cu Interconnects for 65 nm Technology Node and Beyond," *International Interconnect Technology Conference,* June 11–13, 2006, Burlingame, CA, pp. 196–198.

[19] Liu, X. H., et al., "Chip-Package Interaction Modeling of Ultra Low-k/Copper Back End of Line," *International Interconnect Technology Conference*, June 3–6, 2007, Burlingame, CA.

[20] Post, D., D. B. Han, and P. Ifju, *High Sensitivity Moiré: Experimental Analysis for Mechanics and Materials,* New York and Berlin: Springer-Verlag, 1994.

[21] Guo, Y., et al., "Solder Ball Connect (SBC) Assemblies under Thermal Loading: I. Deformation Measurement via Moiré Interferometry, and Its Interpretation," *IBM J. Research Development*, Vol. 37, 1993, pp. 635–647.

[22] Wang, G. T., PhD thesis, "Thermal Deformation of Electronic Packages and Packaging Effect on Reliability for Copper/low-k Interconnect Structures," University of Texas, Austin, 2004.

[23] Hutchinson, J. W., and Z. Suo, "Mixed-Mode Cracking in Layered Materials," *Advances in Applied Mechanics,* Vol. 29, 2002, pp. 63–191.

[24] Volinsky, A. A., N. R. Moody, and W. W. Gerberich, "Interfacial Toughness Measurements for Thin Films on Substrates," *Acta Materialia*, Vol. 50, 2002, 441–466.

[25] Kook, S.-Y., and R. H. Dauskardt, "Moisture-Assisted Subcritical Debonding of a Polymer/Metal Interface," *J. Appl. Phys.*, Vol. 91, 2002, pp. 1293–1303.

[26] Suo, Z., and J. W. Hutchinson, "Sandwich Specimens for Measuring Interface Crack Toughness," *Mater. Sci. Eng.*, A107, 1989, pp. 135–143.

[27] Ma, Q., et al., "A Four-Point Bending Technique for Studying Subcritical Crack Growth in Thin Films and at Interfaces," *J. Mater. Res.* Vol. 12, 1997, pp. 840–845.

[28] Dauskardt, R. H., et al., "Adhesion and Debonding of Multi-Layer Thin Film Structures," *Eng. Fract. Mech.*, Vol. 61, 1998, pp. 141–162.

[29] Liechti, K. M., and T. Freda, "On the Use of Laminated Beams for the Determination of Pure and Mixed-Mode Fracture Properties of Structural Adhesives," *J. Adhesion*, Vol. 28, 1989, pp. 145–169.

[30] Fernlund, G., and J. K. Spelt, "Mixed-Mode Fracture Characterization of Adhesive Joints," *Composites Science and Technology*, Vol. 50, 1994, pp. 441–449.

[31] Lane, M., "Interface Fracture," *Annu. Rev. Mater. Res.*, Vol. 33, 2003, pp. 29–54.

[32] Lane, M. W., et al., "Adhesion and Reliability of Copper Interconnects with Ta and TaN Barrier Layers," *J. Mater. Res.*, Vol. 15, 2000, pp. 203–211.

[33] Lane, M. W., E. G. Liniger, and J. R. Lloyd, "Relationship between Interfacial Adhesion and Electromigration in Cu Metallization," *J. Appl. Phys.*, Vol. 93, 2003, pp. 1417–1421.

[34] Lloyd, J. R., et al., "Electromigration and Adhesion," *IEEE Trans. Device and Materials Reliability*, Vol. 5, 2005, pp. 113–118.

[35] Shaw, T. M., et al., "Experimental Determination of the Toughness of Crack Stop Structures," *IEEE International Interconnect Technology Conference*, June 3–6, 2007, Burlingame, CA.

[36] Cook, R. F., and E. G. Liniger, "Stress-Corrosion Cracking of Low-Dielectric-Constant Spin-on-Glass Thin Films," *J. Electrochemical Soc.*, Vol. 146, 1999, pp. 4439–4448.

[37] Liu, X. H., et al., "Low-k BEOL Mechanical Modeling," *Proc. Advanced Metallization Conference*, October 19–21, 2004, San Diego, CA, pp. 361–367.

[38] Tsui, T. Y., A. J. McKerrow, and J. J. Vlassak, "Constraint Effects on Thin Film Channel Cracking Behavior," *J. Mater. Res.*, Vol. 20, 2005, pp. 2266–2273.

[39] Liu, X. H., et al., "Delamination in Patterned Films," *Int. J. Solids Struct.*, Vol. 44, No. 6, 2007, pp. 1706–1718.

[40] Tsui, T. Y., A. J. McKerrow, and J. J. Vlassak, "The Effect of Water Diffusion on the Adhesion of Organosilicate Glass Film Stacks," *J. Mech. Phys. Solids*, Vol. 54, 2006, pp. 887–903.

[41] Beuth, J. L., "Cracking of Thin Bonded Films in Residual Tension," *Int. J. Solids Struct.*, Vol. 29, 1992, pp. 63–191.

[42] He, J., G. Xu, and Z. Suo, "Experimental Determination of Crack Driving Forces in Integrated Structures," *Proc. 7th Int. Workshop on Stress-Induced Phenomena in Metallization*, edited by P. S. Ho et al., New York: American Institute of Physics, 2004, pp. 3–14.

[43] Huang, R., et al., "Channel Cracking of Thin Films with the Extended Finite Element Method," *Eng. Frac. Mech.*, Vol. 70, 2003, pp. 2513–2526.

[44] Nakamura, T., and S. M. Kamath, "Three-Dimensional Effects in Thin-Film Fracture Mechanics," *Mech. Mater.*, Vol. 13, 1992, pp. 67–77.

[45] Ambrico, J. M., and M. R. Begley, "The Role of Initial Flaw Size, Elastic Compliance and Plasticity in Channel Cracking of Thin Films," *Thin Solid Films* Vol. 419, 2002, pp. 144–153.

[46] Ye, T., Z. Suo, and A. G. Evans, "Thin Film Cracking and the Roles of Substrate and Interface," *Int. J. Solids Struct.*, Vol. 29, 1992, pp. 2639–2648.

[47] Pang, Y., and R. Huang, "Influence of Interfacial Delamination on Channel Cracking of Brittle Thin Films," in *Materials, Processes, Integration and Reliability in Advanced Interconnects for Micro- and Nanoelectronics*, edited by Q. Lin et al., Warrendale, PA: Materials Research Society, 2007, B06–04.

[48] Hu, M. S., and A. G. Evans, "The Cracking and Decohesion of Thin Films on Ductile Substrates," *Acta Metall.*, Vol. 37, 1989, pp. 917–925.

[49] Beuth, J. L., and N. W. Klingbeil, "Cracking of Thin Films Bonded to Elastic-Plastic Substrates," *J. Mech. Phys. Solids*, Vol. 44, 1996, pp. 1411–1428.

[50] Huang, R., J. H. Prevost, and Z. Suo, "Loss of Constraint on Fracture in Thin Film Structures Due to Creep," *Acta Materialia*, Vol. 50, 2002, pp. 4137–4148.

[51] Suo, Z., J. H. Prevost, and J. Liang, "Kinetics of Crack Initiation and Growth in Organic-Containing Integrated Structures," *J. Mech. Phys. Solids*, Vol. 51, 2003, pp. 2169–2190.

[52] Liang, J., et al., "Thin Film Cracking Modulated by Underlayer Creep," *Experimental Mechanics*, Vol. 43, 2003, pp. 269–279.

[53] Liang, J., et al., "Time-Dependent Crack Behavior in an Integrated Structure," *Int. J. Fracture*, Vol. 125, 2004, pp. 335–348.

[54] Huang, M., et al., "Thin Film Cracking and Ratcheting Caused by Temperature Cycling," *J. Mater. Res.*, Vol. 15, 2000, pp. 1239–1242.

[55] Huang, M., Z. Suo, and Q. Ma, "Plastic Ratcheting Induced Cracks in Thin Film Structures," *J. Mech. Phys. Solids*, Vol. 50, 2002, pp. 1079–1098.

[56] Williams, M. L., "The Stress around a Fault or Crack in Dissimilar Media," *Bull. Seismol. Soc. Am.*, Vol. 49, 1959, pp. 199–204.

[57] Rice, J. R., "Elastic Fracture Mechanics Concepts for Interfacial Cracks," *J. Appl. Mech.*, Vol. 55, 1988, pp. 98–103.

[58] Suo, Z., and J. W. Hutchinson, "Interface Crack between Two Elastic Layers," *Int. J. Fracture*, Vol. 43, 1990, pp. 1–18.

[59] Yu, H. H., M. Y. He, and J. W. Hutchinson, "Edge Effects in Thin Film Delamination," *Acta Mater.*, Vol. 49, 2001, pp. 93–107.

[60] He, M. Y., and J. W. Hutchinson, "Crack Deflection at an Interface between Dissimilar Elastic Materials," *Int. J. Solids Struct.*, Vol. 25, 1989, pp. 1053–1067.

[61] Nied, H. F., "Mechanics of Interface Fracture with Applications in Electronic Packaging," *IEEE Transactions on Device and Materials Reliability*, Vol. 3, 2003, pp. 129–143.

[62] Lane, M. W., X. H. Liu, and T. M. Shaw, "Environmental Effects on Cracking and Delamination of Dielectric Films," *IEEE Trans. Device Mater. Reliability*, Vol. 4, 2004, pp. 142–147.

[63] Vlassak, J. J., Y. Lin, and T. Y. Tsui, "Fracture of Organosilicate Glass Thin Films: Environmental Effects," *Mater. Sci. Eng.*, A391, 2005, pp. 159–174.

[64] Rhee, S. H., Y. Du, and P. S. Ho, "Thermal Stress Characteristics of Cu/Oxide and Cu/Low-k Submicron Interconnect Structures," *J. Applied Physics*, Vol. 93, No. 7, 2003, pp. 3926–3933.

[65] Wang, G. T., et al., "Investigation of Residual Stress in Wafer-Level Interconnect Structures Induced by Wafer Processing," *Proc. Electronic Components Technology Conf.*, May 31–June 1, 2006, San Diego, CA, pp. 344–349.

[66] Gan, D. W., S. Yoon, and P. S. Ho, "Effects of Passivation Layer on Stress Relaxation in Cu Line Structures," *IEEE International Interconnect Technology Conference,* June 2–4, 2003, Burlingame, CA.

[67] ANSYS Advanced Guide Manual, chapter 9, in ANSYS Version 9.0 Documentation, ANSYS, Inc., 2006.

[68] Shih, C. F., and R. J. Asaro, "Elastic-Plastic Analysis of Cracks on Bimaterial Interfaces: Part I-Small Scale Yielding," *J. Appl. Mech.*, Vol. 55, 1988, pp. 299–316.

[69] Nakamura, T., "Three-Dimensional Stress Fields of Elastic Interface Cracks," *J. Appl. Mech.*, Vol. 58, 1991, pp. 939–946.

[70] Begley, M. R., and J. M. Ambrico, "Delamination of Thin Films from Two-dimensional Interface Flaws at Corners and Edges," *Int. J. Fracture*, Vol. 112, 2001, pp. 205–222.

[71] ABAQUS Theory Manual, Section 2.16, in ABAQUS Version 6.6 Documentation, ABAQUS, Inc., 2006.

[72] Hughes, T. J. R., and M. Stern, "Techniques for Developing Special Finite Element Shape Functions with Particular References to Singularities," *Int. J. Numerical Methods in Engineering*, Vol. 15, 1980, pp. 733–751.

[73] Sukumar, N., et al., "Partition of Unity Enrichment for Bimaterial Interface Cracks," *Int. J. Numerical Methods in Engineering*, Vol. 59, 2004, pp. 1075–1102.

[74] Ayhan, A. O., and H. F. Nied, "Finite Element Analysis of Interface Cracking in Semiconductor Packages," *IEEE Trans. Components and Packaging Technology*, Vol. 22, 1999, pp. 503–511.

[75] Ayhan, A. O., A. C. Kaya, and H. F. Nied, "Analysis of Three-dimensional Interface Cracks Using Enriched Finite Elements," *Int. J. Fracture*, Vol. 142, 2006, pp. 255–276.

[76] Bucholz, F. G., R. Sistla, and T. Krishnamurthy, "2D and 3D Applications of the Improved and Generalized Modified Crack Closure Integral Method," in *Computational Mechanics'88*, (eds.) S. N. Atluri and G. Yagawa, New York: Springer-Verlag, 1988.

[77] Rybicki, E. F., and M. F. Kanninen, "A Finite Element Calculation of Stress Intensity Factors by a Modified Crack Closure Integral," *Eng. Frac. Mech.*, Vol. 9, 1977, pp. 931–938.

[78] Shivakumar, K. N., P. W. Tan, and J. C. Newman Jr., "A Virtual Crack-Closure Technique for Calculating Stress Intensity Factors for Cracked Three-dimensional Bodies," *Int. J. Fracture*, Vol. 36, 1988, pp. 43–50.

[79] Krueger, R., *The Virtual Crack Closure Technique: History, Approach and Applications*, NASA/CR-2002-211628, 2002.

[80] Sun, C. T., and C. J. Jih, "On Strain Energy Release Rates for Interfacial Cracks in Bimaterial Media," *Eng. Frac. Mech.*, Vol. 28, 1987, pp. 13–20.

[81] Raju, I. S., J. H. Crews, and M. A. Aminpour, "Convergence of Strain Energy Release Rate Components for Edge Delaminated Composite Materials," *Eng. Frac. Mech.*, Vol. 30, 1988, pp. 383–396.

[82] Chow, T. W., and S. N. Atluri, "Finite Element Calculation of Stress Intensity Factors for Interfacial Crack Using Virtual Crack Closure Integral," *Comput. Mech.*, Vol. 16, 1995, pp. 417–425.

[83] Agrawal, A., and A. M. Karlsson, "Obtaining Mode Mixity for a Bimaterial Interface Crack Using the Virtual Crack Closure Technique," *Int. J. Fracture*, Vol. 141, 2006, pp. 75–98.

[84] Scherban, T., et al., "Stress-Induced Phenomena in Metallization," *AIP Conference Proc.*, Vol. 817, 2006, pp. 741–759.

[85] *International Technology Roadmap for Semiconductors*, 2007 ed., San Jose, CA: Semiconductor Industry Association, available at www.itrs.net/Links/2007ITRS/ Home2007.htm.

CHAPTER 3

Mechanically Compliant I/O Interconnects and Packaging

Suresh K. Sitaraman and Karan Kacker

3.1 Introduction

As is pointed out throughout this book, off-chip interconnects must be compatible and must scale with advances in the semiconductor industry. Conventional off-chip interconnects include wire bonding, tape automated bonding (TAB), and C4 bumps. Wire bonding is widely used. However, it is inherently incapable of addressing the high I/O count, fine-pitch off-chip interconnect requirements because it is not area array. TAB is an improvement over wire bonding in the sense that it supports gang bonding (bonding of multiple wires simultaneously). However, it is more costly and suffers from the same drawbacks as wire bonding in its inability to support an area array of interconnects. Flip-chips with area-array solder bumps are being increasingly used today due to their advantages: higher I/O density, shorter leads, lower inductance, improved frequency response, better noise control, smaller package footprint, and lower profile [1]. Epoxy-based underfills are often used in such flip-chip assemblies to accommodate the coefficient thermal expansion (CTE) mismatch among different materials (e.g., silicon IC on an organic substrate) and to enhance the solder joint reliability against thermomechanical fatigue failure [2, 3]. However, additional underfill process steps, material and processing costs, reworkability, delamination, and cracking are some of the concerns with the use of underfills. Also, as the pitch size and chip-substrate gap decrease, the cost and the difficulties associated with underfill dispensing will increase dramatically [4, 5]. An approach similar to the flip-chip process based on copper bumps with underfills is also being pursued. Such an approach has been adopted by Intel Corporation [6] as well as by other companies. Copper bumps/columns, when compared to solder bumps, have improved resistance to electromigration due to lower joule heating, the ability to support higher current densities, and more uniform current distribution. Similar to area-array flip-chip solder bumps, copper bumps possess advantages such as higher I/O density, shorter leads, lower inductance, improved frequency response, better noise control, smaller package footprint, and lower profile. Gold bumps [7] are formed using a process similar to the wire-bonding process, and the chips with gold bumps are flipped and assembled on substrates. A modified wire-bonding process is used to form the gold bumps. This allows for a flexible and robust bump-formation process that supports die-level bonding. However, the bonding process is sequential in nature. Unlike conventional wire bonds, gold bumps support an area array of interconnects. Also, gold bumps have a lower resistivity than solder and are a lead-free solution. However, the assembly of gold bumps is challenging, as it requires a substrate with high planarity and a high temperature and pressure assembly process or the use of ultrasonic excitation.

In recent years, the semiconductor industry began incorporating low-k ($k < 3.0$ [8]) interlayer dielectrics into the multilayer on-chip interconnect network in order to reduce on-chip interconnect capacitance, which reduces RC delay and crosstalk. For example, IBM's Cell multiprocessor utilizes low-k interlayer dielectrics [9]. When such integrated circuits (ICs) are assembled on organic substrates, the chip-to-substrate interconnects or the input/output (I/O) interconnects, however, are subjected to extensive differential displacement due to the CTE mismatch between the die and the substrate under thermal excursions. The I/O interconnects, especially stiff solder bumps, could crack or delaminate the low-k dielectric material in the die, as described in Chapter 2. Hence, it is desirable that off-chip interconnects are designed to reduce the stresses introduced into the die due to the CTE mismatch between the substrate and the die.

One way to reduce such die stresses and to enhance the interconnect reliability without using an underfill material will be through the use of a compliant structure for an off-chip interconnect. Such a compliant interconnect will allow the interconnect to deform and hence accommodate the CTE mismatch between the silicon die and the organic substrate. Such an interconnect can be referred to as a "compliant interconnect" and is the focus of this chapter. A compliant interconnect decouples the die from the substrate and does not require an underfill material. This is in contrast to the solder bump approach, which requires an underfill material to couple the die to the substrate to ensure interconnect reliability. Elimination of the underfill material will facilitate the assembly of compliant interconnects with a pitch of 100 μm or less and will also make the assembly reworkable. Also, compliant interconnects exert minimal force on the die pads and therefore will not crack or delaminate the low-k dielectric material on the die. However, for a compliant interconnect to be viable, it must also be easy to fabricate and assemble using existing infrastructure, scalable in pitch, and preferably fabricated at the wafer-level. In addition it must meet the electrical, thermal, and mechanical requirements for next generation microsystems.

This chapter provides an overview of compliant interconnects. The requirements imposed on compliant interconnects are first discussed (Section 3.2). A description of past and current compliant interconnect technologies is then provided (Section 3.3). The design of compliant interconnects is then explored (Section 3.4). First, the design constraints imposed on compliant interconnects are described (Section 3.4.1). Then, G-Helix interconnects are considered to highlight the trade-offs between the electrical and mechanical performance of compliant interconnects. Subsequently, the reliability evaluation of compliant interconnects is described (Section 3.5) from the perspective of the thermomechanical reliability of the interconnect and the impact of the interconnect on low-k dielectrics employed in the die. The assembly of compliant interconnects is then discussed with case studies on the Sea of Leads (SoL) and G-Helix interconnects (Section 3.6). A description of an integrative approach toward designing compliant interconnects, which envisions employing varied off-chip interconnect geometries within a single package, is then provided (Section 3.7).

3.2 Compliant I/O Requirements

A viable compliant interconnect technology should meet the following requirements and meet them simultaneously.

1. *Mechanical reliability:* The interconnect must have sufficient compliance to not delaminate or crack low-k dielectric. For a 20 × 20 mm silicon die with a low-k material having a fracture toughness of 0.02 MPa $m^{1/2}$ [10], it can be shown that the required in-plane compliance will be 3.0 to 5.0 mm/N so as not to fracture the low-k material. Similarly, considering a substrate nonplanarity of 20 μm, it can be shown that the out-of-plane compliance should also be of similar magnitude [11]. This is roughly two orders of magnitude greater than today's solder bump compliance. A compliant interconnect must also have sufficient thermomechanical reliability to pass standard qualification tests without the use of an underfill material.
2. *Low electrical parasitics:* Interconnects should meet the high digital-speed and high data-rate (5–20 Gbps/channel) requirements for future ICs. These are discussed in greater detail in Chapters 5 and 6.
3. *Cost-effective fabrication process:* To be cost-effective, it is preferable that the proposed interconnects be batch fabricated at the wafer level (large-area fabrication), not the IC level. The fabrication and assembly of interconnects should be easily integrated into existing infrastructure, and the processes should be repeatable with a good yield. Also, the interconnects should be reworkable; therefore, it is preferable not to use underfill.
4. *Fine pitch and scalability with IC feature size:* As IC feature size scales from 20 nm in 2009 to 7 nm in 2019, the first-level interconnect pitch goes down from 130 μm in 2007 to 95 μm in 2019 (area-array configuration) [12]. Hence, a compliant interconnect that addresses today's needs must be scalable to address future pitch requirements.

Keeping in perspective these requirements, an overview of various compliant interconnect technologies is provided in the next section.

3.3 Overview of Various Compliant Interconnect Technologies

3.3.1 FormFactor's MOST

FormFactor had developed a compliant interconnect element called MicroSpring based on wire bonding. MicroSpring was first used in probe card applications. FormFactor extended this technology to realize a wafer-level package (WLP) that utilized the MicroSpring interconnect as a first-level interconnect. This application of the MicroSpring interconnect was called MicroSpring Contact on Silicon Technology (MOST) [13]. MOST utilizes a modified wire bonder to realize freestanding compliant interconnects on a silicon wafer. Once formed, the wire bond is plated with an alloy with a finish layer of gold. The interconnect shape is determined by controlling the motion of the wire bonder. The interconnects are assembled either by socketing or soldering. Though a successful technology for probe card applications, MOST has had limited success as a first-level interconnect. This can be attributed to the MOST fabrication process. The serial nature of the fabrication process is not viable for large I/O counts. Also, such a fabrication process is unable to achieve fine I/O pitches.

3.3.2 Tessera's μBGA and WAVE

Tessera's μBGA [14] is a compliant chip scale package (CSP) based on flexible circuit technology. The μBGA package uses a patterned polyimide flexible circuit with metal leads. The flexible circuit is attached to the die using an elastomer, which is a few mils thick. Leads attached along the edge of the flexible circuit are then thermosonically bonded in a sequential manner to the die pads present along the periphery of the chip. The polyimide flexible circuit has traces that fan in from the metal leads to pads on the flexible circuit layer. The package can then be bumped at the pads and assembled using standard surface-mount technology (SMT) techniques. Compliance between the chip and the substrate is provided by the metal leads and the low-modulus elastomer. The leads and the elastomer combine to take up the CTE mismatch between the die and the substrate. Hence, no underfill is required for the solder bumps. Two drawbacks of the μBGA package are the low compliance of the leads and the use of a sequential bonding process.

Tessera introduced its second generation of compliant packages, Wide Area Vertical Expansion (WAVE) [15, 16], to overcome some of the limitations of the μBGA package. The WAVE package utilizes leads with greater compliance than those used in the μBGA package. This enables greater reliability and an ability to address larger die sizes as compared to a μBGA package. Also, the WAVE package allows for gang bonding of the leads to the die, enabling a larger number of I/Os and more flexibility in die pad location. The WAVE package is again based on polyimide flexible circuits but utilizes a different fabrication process compared to the μBGA package. A two-layer polyimide flexible circuit is used with leads fabricated on it. Special attention is paid to the lead design to increase lead slack and hence improve reliability. The leads on the flexible circuit have a suitable bonding material deposited at their ends. The leads are then attached to die pads on the silicon and hence partially released from the surface of the flexible circuit. An elastomeric material is then injected between the flexible circuit and the die, vertically raising the leads attached between the flexible circuit and the die. The compliant leads, along with the elastomeric material, provide the necessary compliance to decouple the die from the substrate. Similar to the μBGA package, the flexible circuit can now be bumped and then assembled using standard SMT processes. WAVE represents an improvement over μBGA by batch processing interconnects that have a higher compliance. However, the encapsulating elastomer constrains the motion of the compliant lead, limiting its compliance and consequently its ability to accommodate the CTE mismatch between the die and the substrate.

3.3.3 Floating Pad Technology

Floating pad technology (FPT) enables pads on a device to move freely in all three directions (x, y, and z) [17] and hence be compliant. This is achieved by fabricating pads on a "micropocket." A schematic representation of FPT is shown in Figure 3.1. To realize the micropocket, a photodefinable compliant layer is spun on the wafer/chip carrier. Openings are defined at locations where the micropockets are desired. A polymer film is then attached, covering the micropocket. Pads, along with suitable routing, are defined on the polymer film on top of the micropocket. The pads also include an annular ring, which sits on the compliant polymer film outside of the micropocket area. The pads connect to the annular ring through thin metal lines, which allow the pad to remain compliant. The pads are bumped, and the solder bumps now sit on a compliant micropocket, which improves their reliability. Complex fabrication process and inability to scale interconnect pitch to small dimensions are some of the limitations with this technology.

3.3.4 Sea of Leads

Sea of Leads (SoL)–compliant interconnects evolved from the Compliant Wafer Level Package (CWLP) developed at Georgia Tech [18]. CWLP was based on batch fabrication of interconnects at the wafer level. Such an approach allowed for a potentially low-cost interconnect solution that could support a high interconnect density. However, the CWLP interconnects were fabricated on a polymer layer, which reduced the compliance of the interconnects. Sea of Leads (SoL) was developed to address some of the limitations of CWLP. In one implementation of SoL (Figure 3.2), leads were fabricated with embedded air gaps to increase their compliance [19]. The air gap was realized by defining a sacrificial material in the regions where an air gap was desired. An overcoat polymer material was spun on top of the sacrificial material, and the sacrificial material was then thermally decomposed and diffused through the overcoat polymer, thereby realizing the air gap. The intercon-

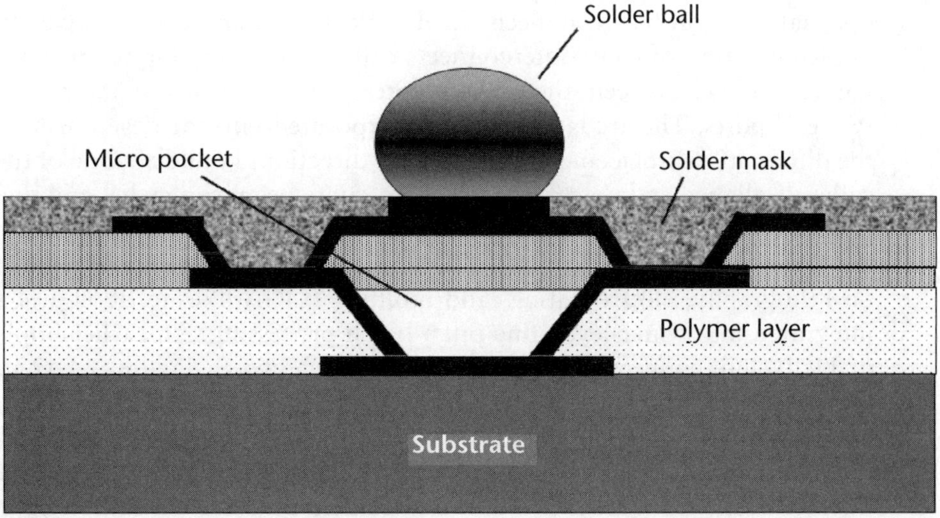

Figure 3.1 FPT cross section [17].

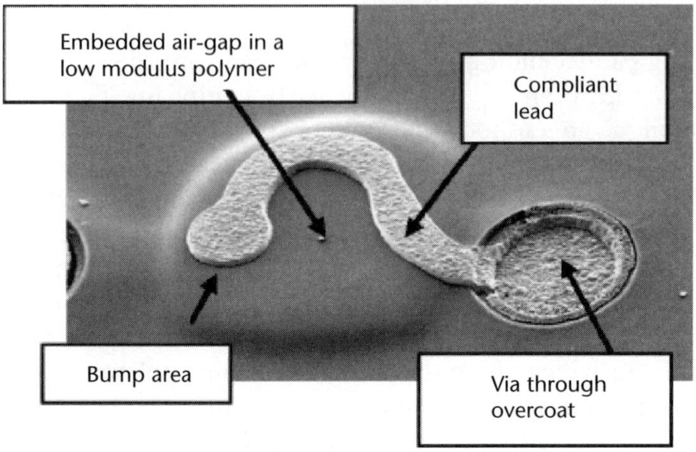

Figure 3.2 Sea of Leads with embedded air gap [20].

nects were then fabricated on top of the air gap. In another implementation, SoL interconnects were realized with some part of the interconnect not being adhered to the underlying polymer layer [20]. This improved the compliance of the interconnects; hence, these interconnects were called "slippery" SoL. A third implementation of SoL utilizes a sacrificial layer to realize interconnects that are freestanding and, hence, have higher compliance [21].

3.3.5 Helix Interconnects

Helix interconnects are a lithography-based, electroplated, compliant interconnect that can be fabricated at the wafer level. Developed at Georgia Tech, Helix interconnects are freestanding, which enables them to have a high compliance. A design-of-simulations approach, followed by an optimization process, was utilized to design the Helix interconnects. This allows the Helix interconnects to achieve an optimal trade off between mechanical performance and electrical performance. An implementation of Helix interconnects called "G-Helix" interconnects is shown in Figure 3.3 [22]. As seen, the G-Helix interconnects consist of an arcuate beam and two end posts. The arcuate beam is incorporated into the design to accommodate the differential displacement in the planar directions (x and z). One of the two vertically off-aligned end posts connects the arcuate beam to the die, and the other connects the arcuate beam to the substrate. The vertical end posts provide compliance in the three orthogonal directions. The fabrication of G-Helix interconnects is based on lithography, electroplating, and molding (LIGA-like) technologies and can be integrated into wafer-level, fine-pitch batch processing. An earlier implementation of Helix interconnects, called "β-Helix," is shown in Figure 3.4 [23]. Although β-Helix interconnects have improved mechanical performance over G-Helix interconnects, their poor electrical performance, combined with costly and time-consuming process steps, make the β-Helix interconnects not viable. G-Helix interconnects show great promise and can be used for consumer, computer, and other applications.

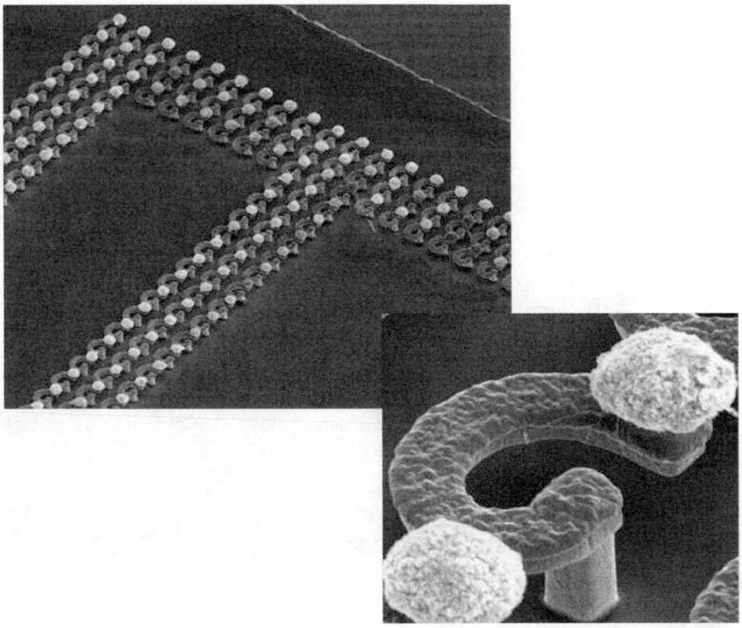

Figure 3.3 100 μm pitch G-Helix interconnects fabricated on a silicon wafer [11].

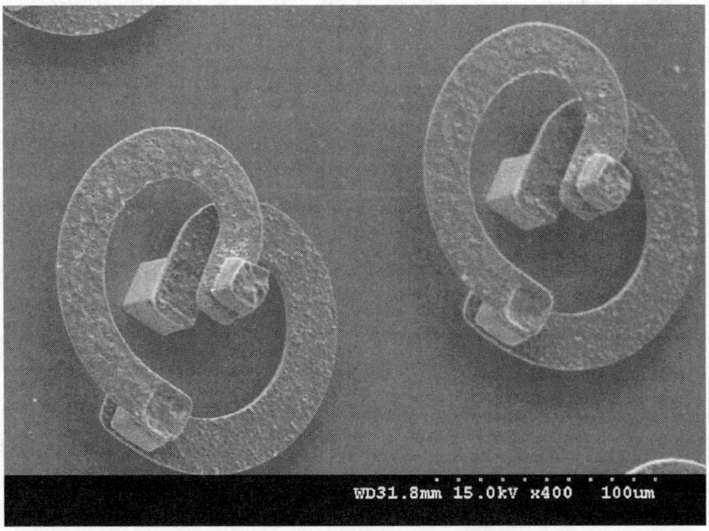

Figure 3.4 β-Helix interconnects fabricated on a silicon wafer [23].

3.3.6 Stress-Engineered Interconnects

A consortium comprising Xerox Palo Alto Research Center, Georgia Tech, and NanoNexus had developed linear and J-like stress-engineered compliant interconnects. These interconnects are fabricated using dc sputtering to realize intrinsically stressed patterns that curl off the surface of the wafer and are shown in Figure 3.5 [24–28]. To fabricate them, the interconnect metal is sputter-deposited at a low

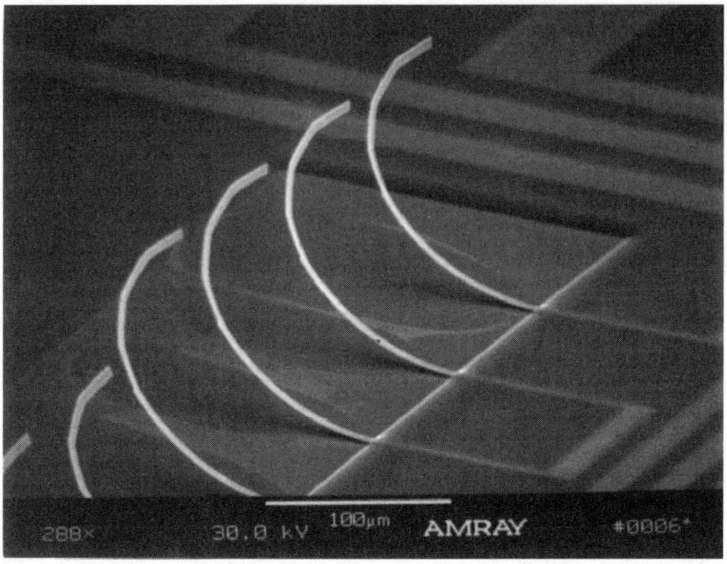

Figure 3.5 J-spring stress-engineered interconnects fabricated on a silicon wafer [24].

argon pressure on a patterned sacrificial layer. By changing the argon pressure, a stress gradient can be introduced into the metal. Once the sacrificial layer supporting the metal is removed, the intrinsic stress causes the metal to curl up in the regions where the sacrificial layer was present. The interconnects are anchored to the die at locations where the sacrificial layer was not present. Such an approach allows for batch fabrication of the interconnects. Assembly of the interconnect can be through contact and may not require solder [29]. The interconnect is sufficiently compliant to not require an underfill and can support very fine pitch interconnects (up to 6 μm). However, the use of a nonstandard sputtering fabrication process is a drawback with this technology. Also, the fabrication process results in interconnects that are not uniform across the wafer. To address this, replacing the sputtering process with an electroplating process has been explored [29].

3.3.7 Sea of Polymer Pillars

Sea of Polymer Pillars (SoPP) utilizes mechanically compliant, high-aspect-ratio polymer pins as electrical and optical interconnects [30]. The use of the polymer material allows the interconnects to have high compliance. The polymer material allows for transmission of optical signals, unlike any of the other compliant interconnects described. Also the polymer pins improve optical efficiency over free-space optical interconnects as optical transmission through air is avoided. The technology has been shown to be compatible with solder bumps and electrical compliant leads (Sea of Leads). In an alternate configuration, a metal film can be coated on the polymer pins to provide electrical interconnection simultaneously [31, 32]. In this manner, the same interconnect can transmit both optical and electrical signals. These interconnects are fabricated using wafer-level batch processes. Extending SoPP to a "trimodal" wafer-level package is also being explored [33]. In this configuration, the polymer pins perform a third function as fluidic I/Os. To achieve this hollow

I/Os are fabricated that transport a cooling fluid to support on-chip cooling. These are shown in Figure 3.6. The ability of these interconnects to perform three functions simultaneously makes them a promising interconnect technology, details of which are discussed in Chapter 11.

3.3.8 Elastic-Bump on Silicon Technology

Elastic-bump on Silicon Technology (ELASTec) is based on a resilient polymer bump with a metal lead plated on it [34]. The polymer bump provides the compliance, and the metal lead provides the electrical connection. A wafer-level packaging approach is adopted. The polymer (silicone) is printed on the wafer, and the metal leads are defined through lithography. The metal leads are soldered onto the printed circuit board (PCB). No underfill material is utilized. ELASTec has been demonstrated to pass a number of standard reliability tests [35]. However, ELASTec was developed as a first-level interconnect for memory applications characterized by a low I/O count. To date, it would appear that this technology would be unable to satisfy the finer pitch requirements of other applications due to limitations of the fabrication process adopted.

3.4 Design and Analysis of Compliant Interconnects

3.4.1 Design Constraints

As described in Section 3.2, there are four primary requirements for compliant interconnects: mechanical performance, electrical performance, fine pitch, and cost-effective fabrication. The fabrication process imposes an overall constraint on the realizable compliant interconnect designs. Typically, the interconnect design is constrained to planar structures. A second constraint imposed on the interconnect design is the pitch required, which limits its size. The pitch required also determines

Figure 3.6 Optical I/Os fabricated adjacent to fluidic I/Os [33].

if the fabrication process can be utilized as certain fabrication techniques are not amenable to interconnects at a fine pitch. The remaining two constraints, electrical performance and mechanical performance, represent the fundamental functions of a compliant interconnect: a compliant mechanical structure that conducts electrical signals while accommodating the CTE mismatch between the substrate and the die.

The mechanical function has two interrelated aspects: sufficient compliance and mechanical reliability. Compliance measures the amount a structure deforms per unit of the applied force. If one assumes the interconnect to be fixed at one end, the directional compliance can be obtained by applying forces (F) in the orthogonal x-, y-, and z-directions at the other end. From the resulting displacements u_x, u_y, and u_z, the directional compliance C_x ($= u_x/F_x$), C_y ($= u_y/F_y$), and C_z ($= u_z/F_z$) can be obtained. A freestanding first-level interconnect deforms due to the CTE mismatch between the silicon die and the organic substrate. For such a displacement-controlled load, compliance determines the force transmitted by the interconnect. The higher the compliance, the lower the force transmitted by the interconnect and the lower the stress on the die. This decreases the probability that the interconnect will crack or delaminate the low-k dielectric. In addition, the out-of-plane compliance is beneficial for the purposes of testing and accommodating substrate nonplanarity. However, when we consider the reliability of the interconnect, the compliance of the interconnect is not by itself a sufficient metric. Compliance determines the amount of energy stored in the interconnect as it deforms. For a displacement-controlled load, the higher the compliance, the lower the amount of energy stored in the interconnect, which increases interconnect reliability. However, the manner in which the energy is distributed within the interconnect is also important. If more energy is concentrated in particular regions of the interconnect, those regions will fail first. Hence, the shape of the interconnect is important as it determines the manner in which energy is distributed within it. Hence, a compliant design is a necessary though insufficient condition. The design must also realize an interconnect that has sufficient reliability. Based on reliability concerns (low-k dielectric cracking), the compliant interconnect should have an in-plane compliance in the range of 3 to 5 mm/N [11]. This is greater than two orders of magnitude when compared to conventional solder bumps. Such a high compliance is required as the interconnects accommodate the CTE mismatch between the die and the substrate without the use of an underfill material. Besides mechanical compliance, the interconnects should pass standard qualification tests like thermal cycling, shock and vibrations tests, the Highly Accelerated Stress Test (HAST), and so forth.

Another key characteristic of compliant interconnects, which influences their mechanical functionality, is that they are typically not rotationally symmetric like solder joints. Consequently, the in-plane compliance of the interconnects differs depending upon the direction in which the displacement is applied. Ideally, we would like the interconnects to experience a displacement in a direction in which their compliance is greatest. For a die assembled on a substrate, the center of the die represents the neutral point (i.e., the point at which the die does not move relative to the substrate when a thermal load is applied). A line drawn from the center of the die to the compliant interconnect is the direction in which the compliant interconnect deforms. It is hence desirable to orient the interconnect in such a manner that this line is in the direction in which the interconnect compliance is the greatest. This is illustrated in Figure 3.7. Such an approach allows maximal use of the interconnect compliance.

In terms of electrical performance, three key performance parameters for compliant interconnects are their resistance, capacitance, and inductance. A primary concern with compliant interconnects is their inductance. Compliant interconnects generally have a relatively high inductance. This is a direct consequence of their compliant design, which typically results in structures with long metal lines and consequently a high inductance. For interconnects at a 100 μm pitch, solder joints have an inductance of approximately 20 to 25 pH [36]. It would be desirable to bring the inductance of the interconnect as close as possible to this value. The resistance of compliant interconnects is typically not as much of a concern as it is relatively small when compared to the resistance of the traces on the substrate and die. However, it is desirable to reduce resistance, especially when joule heating is considered. Regarding capacitance, at Gbps a target value of 0.1 pF or below is desirable [36].

3.5 Case Study on Trade-Offs in Electrical/Mechanical Characteristics of Compliant Interconnects

To illustrate the trade-offs between the mechanical and electrical performance of compliant interconnects, we will consider the specific case of G-Helix interconnects [22]. As shown in Figure 3.8, the G-Helix interconnect consists of an arcuate beam and two end posts. The arcuate beam is incorporated into the design to accommodate the differential displacement in the planar directions (x and z). The two end posts connect the arcuate beam to the die and to the substrate. The G-Helix design is described by four geometry parameters: the beam width (W), the beam thickness (T), the mean radius of the arcuate beam (R), and post height (H).

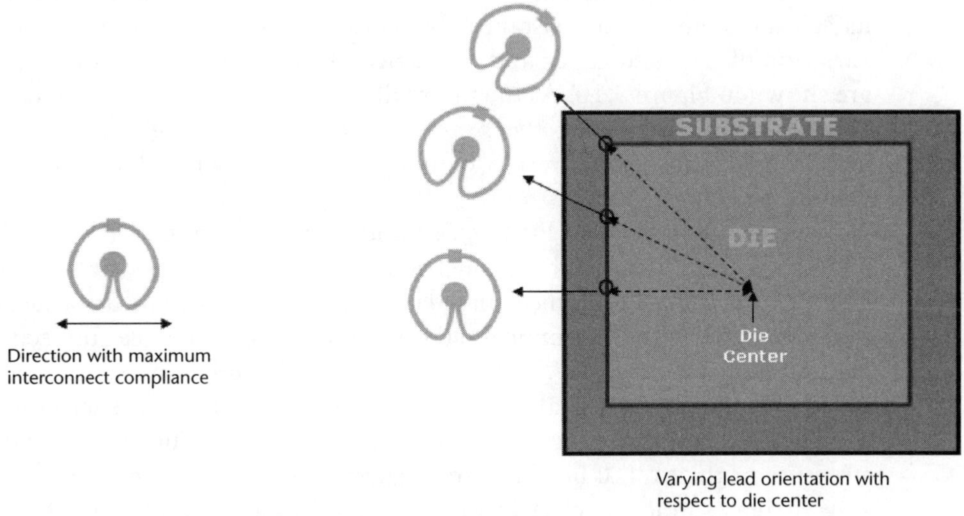

Figure 3.7 Compliant interconnect orientation.

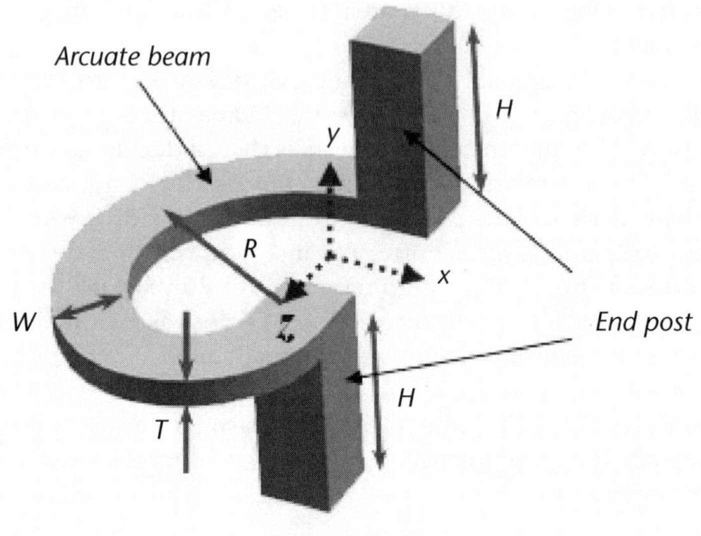

Figure 3.8 Schematic of G-Helix interconnect [22].

The mechanical performance is characterized in terms of the directional compliance in three orthogonal directions. The directional compliance can be obtained by the procedure described in Section 3.4.1. Analytical solutions and FEA-based models were used to determine the compliance of the G-Helix interconnect and its variation with the G-Helix geometry. The results are shown in Figure 3.9. As the figure illustrates, increasing the mean radius of the arcuate beam, decreasing the beam thickness, and decreasing the beam width results in an increase in the compliance of the interconnect in all three directions. Increasing the post height increases the compliance in the in-plane directions (x and z) but does not change the out-of-plane compliance.

The electrical performance of the G-Helix interconnect is described in terms of its self-inductance and resistance. Numerical models were used to determine the variation of self-inductance and resistance of the G-Helix interconnect. The results are shown in Figure 3.10. As the figure illustrates, increasing the mean radius of the arcuate beam, decreasing the beam thickness, decreasing the beam width, and increasing the post height results in an increase in the self-inductance and resistance of the interconnect.

The results shown in Figures 3.9 and 3.10 are summarized in Table 3.1. An interesting trend is seen: the geometric parameters have opposing effects on desirable electrical and thermomechanical parameters. In other words, when a geometric parameter is changed to improve the mechanical compliance, the self-inductance and the resistance increase. For example, when the interconnect thickness or width is decreased, the mechanical compliance increases; however, the self-inductance and the electrical resistance increase. Similarly, when the radius of the G-Helix arcuate structure or the overall height of the structure is increased, the mechanical compliance increases, and the electrical parasitics also increase. In general for compliant interconnects, improvements in mechanical performance come at the expense of electrical performance and vice versa. The optimum trade-off between mechanical

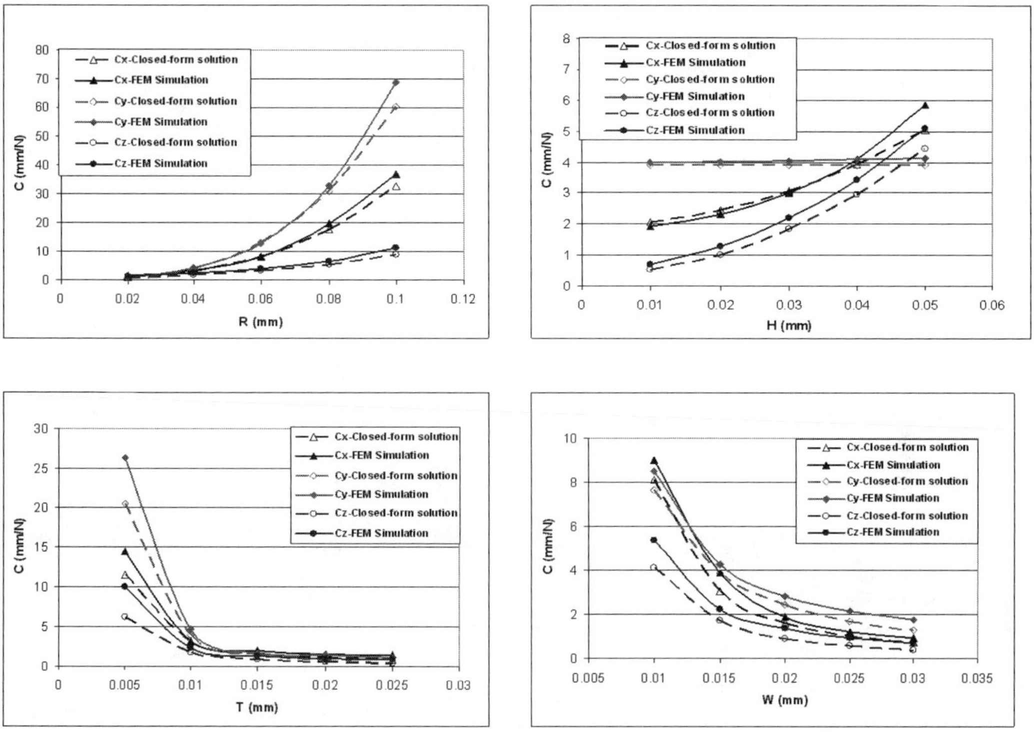

Figure 3.9 Effect of geometry parameters on directional compliance of G-Helix interconnect (baseline parameters: R = 40 μm, H = 30 μm, W = 15 μm, T = 10 μm) [22].

and electrical performance is determined by the specific application the compliant interconnect serves. A design optimization is a suitable technique to determine this and has been performed for stress-engineered interconnects in [37].

An alternate avenue to overcome this trade-off between mechanical and electrical performance is to innovate on the design of the compliant interconnect. To achieve this, a design concept that advocates the use of multiple electrical paths as part of a single compliant interconnect has been proposed [38]. It has been shown to improve the mechanical performance of the interconnect without compromising the electrical performance. Stated differently, using multiple-path compliant interconnects, if one keeps the mechanical compliance the same as a single-path interconnect, the electrical performance of the multiple-path interconnect will be superior to that of the single-path interconnect. An additional advantage of such an approach is that allows for a redundant interconnect design (i.e., the interconnect can continue functioning even if some of the electrical paths fail).

3.6 Reliability Evaluation of Compliant Interconnects

3.6.1 Thermomechanical Reliability Modeling

Compliant interconnects represent a paradigm shift from conventional solder bumps. Prior knowledge of the reliability of these interconnects is limited. A build-and-test approach by itself would not be a feasible procedure to adopt for

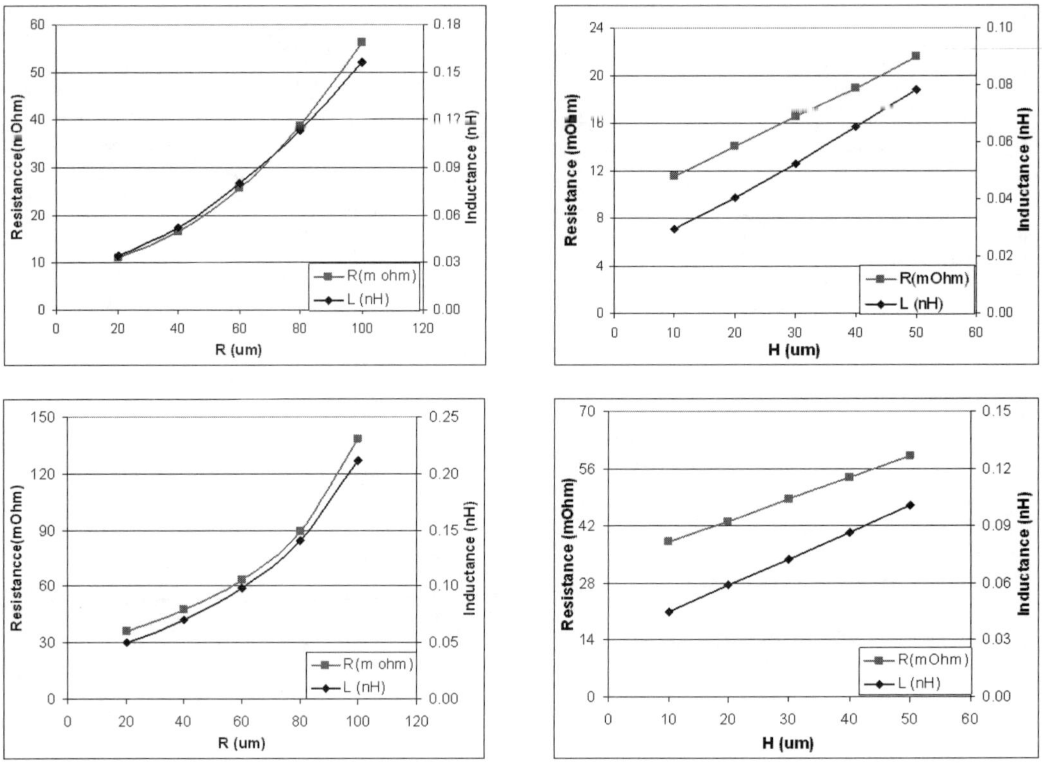

Figure 3.10 Effect of geometry parameters on self-inductance and resistance of G-Helix interconnect (baseline parameters: R = 40 μm, H = 30 μm, W = 15 μm, T = 10 μm) [22].

Table 3.1 Effect of Geometry Parameters on Electrical and Mechanical Performance of G-Helix Interconnect

Geometry Parameter	*Compliance*			*Electrical Parasitics*	
	x	*y*	*z*	*R*	L_{self}
↑*R*	↑	↑	↑	↑	↑
↑*H*	↑	↑	↔	↑	↑
↑*T*	↓	↓	↓	↓	↓
↑*W*	↓	↓	↓	↓	↓

compliant interconnects. On the other hand, developing models to assess the reliability of the interconnects allows for a quick assessment of different interconnect geometries, loading conditions, and other important factors.

A popular approach is to develop finite element models representing the interconnects as part of an electronic package. Three types of geometry models are commonly adopted to model electronic packages: (1) 2D models, (2) 2.5D models, and (3) 3D models. The choice of the model is based on the accuracy desired, the results desired from the model, and the limitations imposed by computing resources. 2D models, computationally the "cheapest," represent a cross section of the package. However, most compliant interconnects are 3D structures whose geometry cannot be adequately captured by a single cross section. Hence, 2D models are typically not used to model compliant interconnects. 2.5D models, also referred to as generalized plane displacement (GPD) models, represent a compromise between 2D and 3D models. These models are computationally more intensive than 2D models but can capture the 3D interconnect geometry and are hence more accurate. Compared to 3D models, they are computationally less intensive. In a 2.5D model, a predetermined width is modeled with 3D elements representing a strip of the package. The width of the strip is typically equal to the pitch of the interconnects. The remaining package geometry is approximated by applying appropriate boundary conditions. In a 3D model, no geometric assumptions are made, and the complete geometry of the package is represented. A quarter or one-eighth symmetry model with appropriate symmetry boundary conditions can be used. In addition, to reduce computation time, beam elements can be utilized to model the compliant interconnects. However, beam elements are unable to provide detailed stress-strain contours in the compliant interconnect. Hence, a combination of beam and solid elements is utilized, with solid elements used to represent the critical interconnects. Typically, for compliant interconnects, the outermost interconnect would fail first; therefore, it is critical and hence would be represented using solid elements.

Material models are also needed to capture the behavior of materials comprising the modeled package. The choice of material model is governed by the behavior of the material under the given loading conditions. For example, under typical accelerated thermal cycling (ATC) tests, silicon would be modeled as a linear elastic material, copper would be modeled using an elastic-plastic constitutive model, and solder would be modeled using an appropriate constitutive model (creep model with plasticity or a viscoplasticity model such as Anand's model) that captures the creep behavior of solder.

Once the geometry and the material models are created and the boundary conditions are applied, the thermal loading conditions are then applied to the model to determine the thermally induced stress-strain distribution in the geometry. The process temperature profile (assembly and cooldown) is initially applied. Typically, the solder reflow temperature is taken as the stress-free temperature. Subsequently, the ATC profile is applied. All the ATC cycles are normally not simulated. This is because the stress-strain profile stabilizes after a few thermal cycles with regard to the damage parameter; hence, the subsequent cycles do not need to be modeled. The results from the last modeled thermal cycle are used to calculate the appropriate damage parameter.

Under field-use conditions or under thermal cycling, the compliant interconnects experience repeated thermomechanical loads due to the CTE mismatch between the substrate and die and will fatigue-fail. In addition to thermomechanical loads, the compliant interconnects experience stresses due to mechanical loads such as vibration and shock. We will focus on thermomechanical loads. Also, the fatigue failure of compliant interconnects is typically in the low-cycle fatigue regime and determines the life-prediction model used. In general, variations of a Coffin-Manson equation are utilized to predict the low-cycle fatigue life of metals. The damage metric utilized is either strain based or energy based. The general form of the Coffin-Manson equation [39] follows:

$$N_f = A(\Delta\varepsilon_{in})^m \tag{3.1}$$

where N_f is the number of cycles to failure, $\Delta\varepsilon_{in}$ is the inelastic strain range, and m and A are constants. In this form of the Coffin-Manson equation, the inelastic strain range is utilized as the damage metric. Other damage metrics utilized include accumulated inelastic strain, accumulated creep strain, strain energy density, creep strain energy, and other variations. When an energy-based criterion is utilized in the form given by (3.2), it is known as a Morrow equation [39]:

$$N_f = B(\Delta W)^n \tag{3.2}$$

where N_f is the number of cycles to failure, ΔW is the energy-based criterion, and n and B are constants. The relationship drawn between the number of cycles to failure and the damage metric is generally obtained through a regression analysis of experimentally obtained failure data. Using the model developed, the damage parameter is evaluated and the fatigue life of the interconnects predicted.

The modeling of G-Helix interconnects to assess their thermomechanical reliability using a GPD model is described in [40]. In the paper, the organic substrate was modeled orthotropic and temperature dependent. The copper compliant interconnects were modeled elastic-plastic to capture potential plastic deformation in the interconnect. The solder alloy was modeled elastic-viscoplastic to capture the creep deformation. The model assumed the solder melting temperature as the stress-free temperature. The loading condition represented the model being cooled down to room temperature which followed by simulating the thermal cycles. The stress-strain hysteresis loop had stabilized in the simulations after three thermal cycles. Figure 3.11 shows equivalent plastic strain distribution at room temperature after three thermal cycles. Plastic strain range in the compliant interconnects was used as the damage metric and evaluated over the third thermal cycle. This was then used to determine the number of cycles to fatigue failure. The simulations indicated that the G-Helix compliant interconnects at a 100 μm pitch have the potential to exceed 1,000 thermal cycles between 0°C to 100°C on a 20 × 20 mm die on an organic substrate. These results suggest that compliant interconnects have sufficient reliability without using any underfill when used as interconnects between a silicon die and an organic substrate.

3.7 Compliant Interconnects and Low-k Dielectrics

The FEA models developed to determine the thermomechanical reliability of compliant interconnects can also be utilized to gauge their impact on dice that employ low-k dielectrics. For modeling low-k dielectrics, two approaches can be adopted. In the first approach, the low-k dielectric material is not represented in the FEA model. Instead, the stress induced in the silicon and at the die-interconnect interface by the

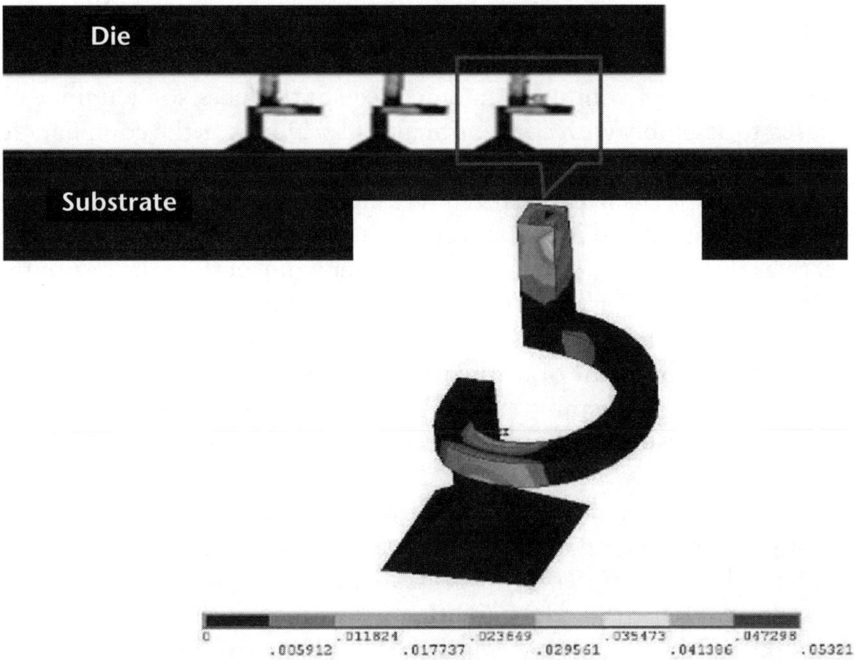

Figure 3.11 Equivalent strain distribution in G-Helix interconnect as predicted by finite element model [40].

compliant interconnects can be used as a metric to determine the probability of low-k dielectric cracking. Alternatively, in the second approach, for greater accuracy and at the expense of a more complicated model with increased computational time, the low-k dielectric material can be explicitly modeled. A fracture mechanics approach is generally applied in this case, and a global-local modeling approach must be implemented.

The benefits of utilizing compliant interconnects for a die employing low-k dielectric material are illustrated through numerical models developed in [11]. In the models developed, the low-k material was not explicitly modeled. The model considers a 20 × 20 mm die (600 μm thick) assembled on an organic substrate (800 μm thick) with G-Helix compliant interconnects. The stresses introduced into the die by the interconnects, as well as the interfacial/peel stresses, were utilized as metrics to measure the probability of cracking and delamination in the low-k dielectric material. It was observed that the stresses introduced into the die by the compliant interconnects, when the assembly is cooled down from reflow to –55°C, are less than 5 MPa. On the other hand, for flip-chip on organic board assemblies with underfills, simulations with identical die-substrate dimensions indicate that the die stresses will be of the order of 140 MPa. The standoff height for the solder bumps was 60 μm. The underfill was modeled as a linear elastic material with an elastic modulus of 7.8 GPa, Poisson's ratio of 0.33, and CTE of 28 ppm. The simulations demonstrate that the die stresses induced by the compliant interconnects are at least an order of magnitude lower than the die stresses for an equivalent flip-chip assembly; hence, the compliant interconnects are not likely to crack or delaminate low-k dielectric material.

3.8 Assembly of Compliant Interconnects

The assembly of compliant interconnects introduces some unique challenges compared to assembly with conventional solder bumps as the compliant leads can move. Assembly of compliant interconnects is typically done with solder. A challenge with using solder is ensuring localized wetting of the metal compliant interconnect. In the case of encapsulated interconnects such as Tessera's WAVE, this is not of great concern as the encapsulant protects the solder from wetting the rest of the interconnect. However, in the case of exposed metal interconnects, such as the Helix and SoL interconnects, the solder should not wet the complete interconnect. This would restrict the movement of the interconnect, impairing its ability to be compliant and preventing high assembly yield. Another novel aspect of compliant interconnects, especially freestanding interconnects, is their compliance in the out-of-plane direction. This allows the interconnects to displace in the out-of-plane direction and hence take up the nonplanarity of the substrate and uneven height of the solder bumps. However, to take advantage of this, an out-of-plane force needs to be applied during the assembly process. For conventional solder bump assembly, this is not done. To better understand the significance of these factors and other parameters that affect assembly yield, the assembly of SoL and G-Helix interconnects are considered.

3.9 Case Studies: Assembly of Sea of Leads and G-Helix Interconnects

In the case of SoL interconnects, two alternate approaches [21, 41] were adopted to ensure localized wetting of the interconnects. In [41], the SoL interconnects were plated with a layer of nickel (Ni), which was then oxidized using a O_2-rich plasma in a reactive ion etch (RIE) tool, resulting in the formation of a nonwetting layer of Ni oxide. In a localized region, this oxide layer was removed and solder plated. This ensured that the leads wetted the interconnect only in the region where solder was plated. This is shown in Figure 3.12. The choice of flux was also found to be critical in this case. Flux is needed during the assembly process to clean the solder in order to ensure good wetting. However, it was found that an aggressive flux would attack the Ni oxide layer and cause the solder to wick the interconnect. Hence, a milder organic flux was found to be more appropriate as it had sufficient activity to clean the solder but did not attack the Ni oxide layer. In [21], an alternate approach was adopted to contain the wetting of the solder to the tip of the interconnect. A polymer solder dam was fabricated at the tip of the interconnect. Subsequently, solder was plated into this dam, and the wetting was localized. This is shown in Figure 3.13. This approach is more robust but comes at the expense of an increased number of fabrication steps.

The assembly of G-Helix interconnects is now considered to highlight some of the other challenges associated with assembling a compliant interconnect with a high standoff height (~70 μm) [11]. The G-Helix assembly was performed for a die size of 20 × 20 mm. The interconnects were in a three-row peripheral array at a 100 μm pitch. For the case of G-Helix interconnects, localized wetting was achieved by plating a layer of nickel and gold (Au) on the tip of the interconnect. Ni serves as a

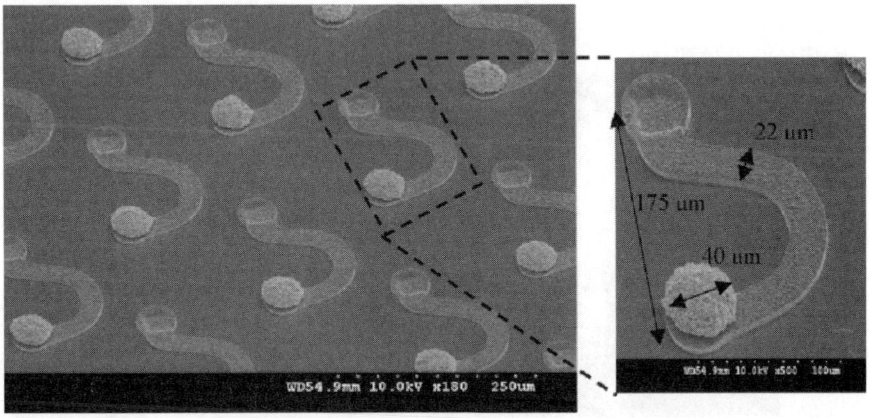

Figure 3.12 SoL interconnect with solder reflowed at interconnect tip [41].

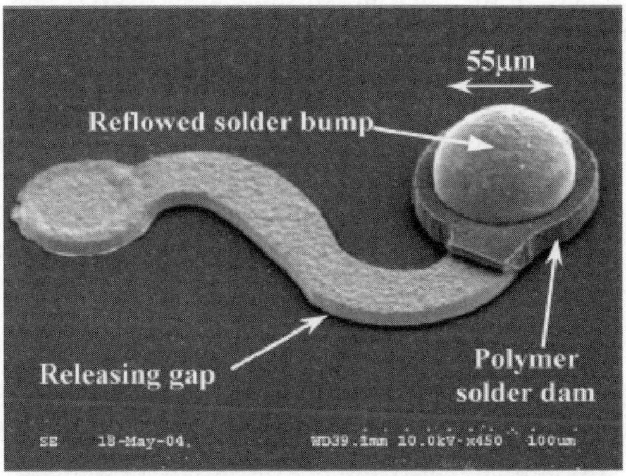

Figure 3.13 SoL interconnect with polymer dam containing the solder [21].

barrier metal, and Au protects the Ni from oxidizing and serves as a wetting layer for the solder. The G-Helix interconnects have a nonwetting copper oxide layer that covers its surface except at the tip of the interconnect. In this manner, the solder only wets the tip, as shown in Figure 3.14 depicting an assembled G-Helix interconnect.

During the assembly development process of G-Helix interconnects, other critical parameters were identified as follows:

1. *Flux volume:* Sufficient flux must be dispensed to reduce the oxides on the surface of the solder and to provide a deoxidized surface for the solder to wet. Excessive flux, however, was found to prevent the solder from wetting the interconnects. Therefore, an optimum flux volume needed to be determined.
2. *Compressive force profile:* During the assembly development process, it was seen that the yield was extremely low when there was a low force (100g)

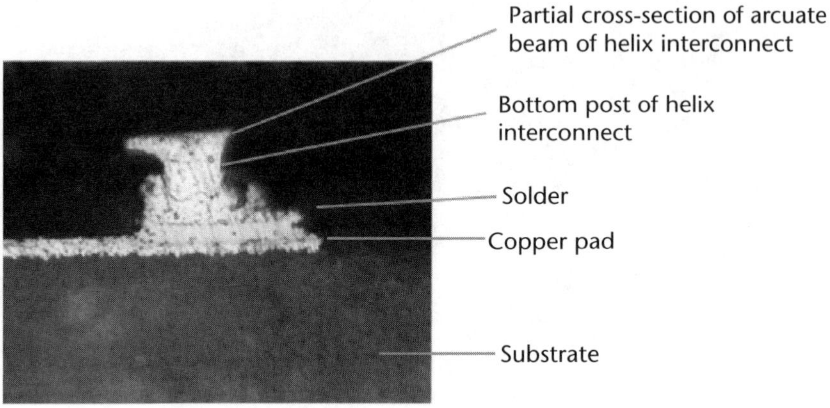

Figure 3.14 Cross section of assembled G-Helix interconnect [11].

applied on the backside of the die. However, it was also observed that if a large force (greater than 350g) was applied, it would excessively deform the G-Helix interconnects, causing the arcuate beam to contact the neighboring pad on the substrate. This in turn would cause the solder from the neighboring pad to wick onto the arcuate beam during reflow, resulting in misalignment. Therefore, through process development, a compressive force of 250g was found to be appropriate to get a good assembly yield. Such a force allowed all of the interconnects to make contact with the substrate pads and to overcome the nonplanarity of the substrate without contacting the neighboring pad on the substrate.

3. *Temperature profile:* The temperature profile for solder reflow can be divided into four stages: preheat, thermal soak, reflow, and cooldown. Each of the four stages needs to be optimized as it impacts assembly yield as well as the subsequent reliability of the solder joint.

The optimized force and temperature profile obtained for G-Helix interconnects is shown in Figure 3.15. As seen, after the preheat stage, a minimal amount of force is applied to bring the die with G-Helix interconnects into contact with the substrate that has solder plated on it. This is followed by a thermal soak stage, at the end of which the determined optimized force to overcome the nonplanarity is applied. This force is maintained during the subsequent solder reflow stage. This is followed by the cooldown stage, at the beginning of which the applied force is released.

3.10 Integrative Solution

A number of compliant interconnects use lithographic techniques to define their geometry, hence providing excellent opportunities for cost-effective I/O customization based on electrical, thermal, and mechanical requirements. As seen in Section 3.5, for the case of G-Helix interconnects, changing the geometric parameters of the interconnects has opposing effects on desirable electrical and thermomechanical parameters. In other words, when a geometric parameter is changed to improve the

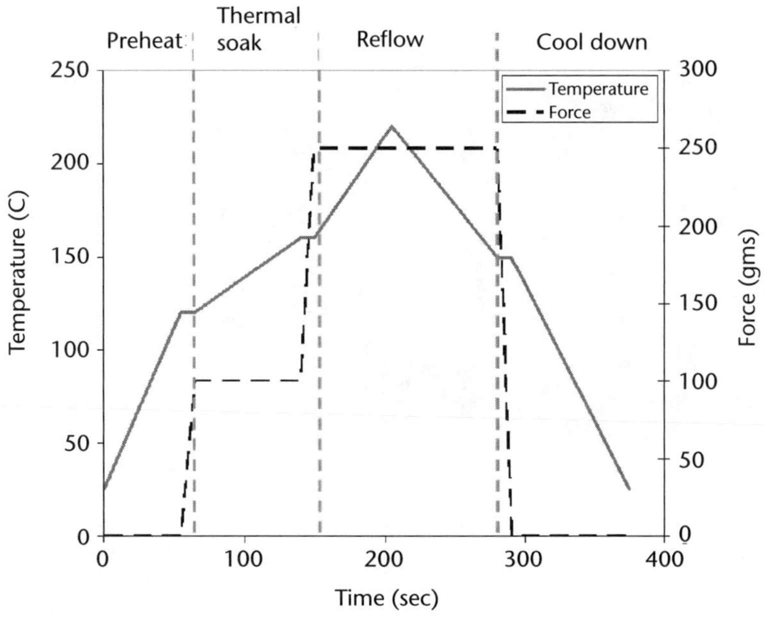

Figure 3.15 Assembly force and temperature profile for G-Helix interconnects.

mechanical compliance, the self-inductance and the resistance increase. This would hold true for most other compliant interconnects. Thus, by using different dimensions and different geometries for the interconnects, their thermomechanical and electrical performance can be optimized without compromising the thermomechanical reliability of the interconnects. Also, if the interconnects are defined through a lithographic process, different interconnect geometries can be fabricated without an increase in the number of processing steps.

In general, the interconnects near the center of the die need not have a high mechanical compliance as the differential displacement between the die and the substrate due to CTE mismatch is low near the center of the die. Thus, the interconnects at the center of the die can be fabricated in the shape of a column structure, while the interconnects near the edge of the die can be fabricated with a more compliant structure [42]. The column structures have lower electrical parasitics associated with them as compared to the compliant interconnects. As these columns are located near the center of the die where the CTE-induced differential thermal expansion is low, these columns will neither fatigue-fail nor exert excessive force on the low-k dielectric to crack or to delaminate. The interconnects away from the center of the die can be fabricated with increasing magnitude of compliance as one traverses to the corner/edge of the die. Typically, near the corner of the die, the CTE-induced differential thermal deformation is high; therefore, higher compliance is needed to reduce the force induced on the die pads by the interconnect. These compliant interconnects toward the edge of the die can be used as signal interconnects. Between the column interconnects near the center and highly compliant interconnects near the edges of the die, the interconnects in the middle region can be designed with intermediate compliance. This is illustrated for the case of G-Helix interconnects in Figure 3.16 [11], which shows such a configuration of intercon-

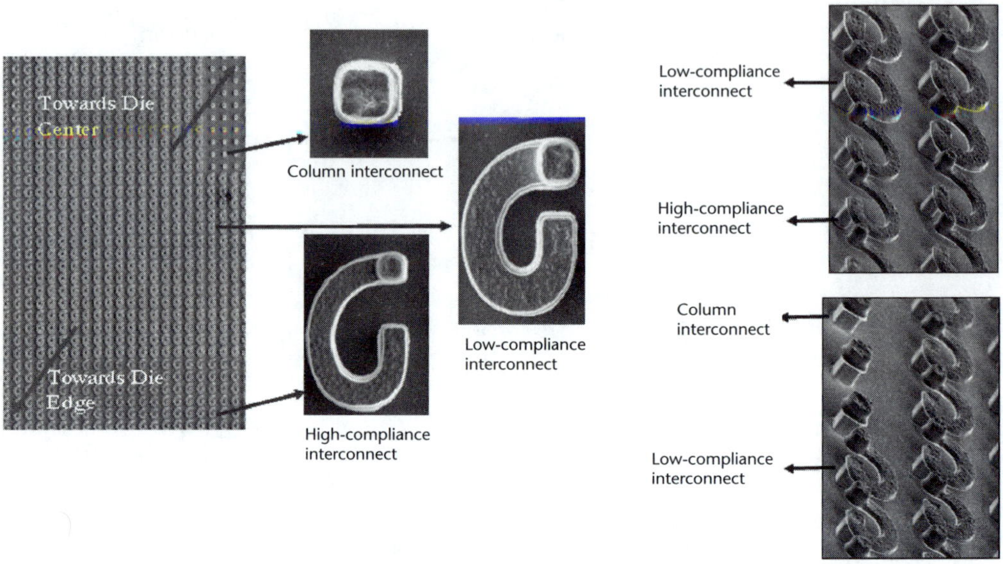

Figure 3.16 Varying G-Helix interconnect geometries fabricated on an individual chip [11].

nects fabricated on a silicon wafer. Alternatively, only compliant interconnects of varying dimensions can be employed across a single chip as described in [43].

Finite element models developed in [42] demonstrate the advantage of using such an approach. The models developed represent three different packages. In the first package (package 1), column interconnects with a square cross section were populated throughout the die in a similar 100 × 100 area-array configuration. In the second package (package 2), identical high-compliance G-Helix interconnects were populated throughout the die in a 100 × 100 area-array configuration. In the third package (package 3), the center of the die was populated with column interconnects, the peripheral rows were populated with high-compliance G-Helix interconnects, and the area in between was populated with low-compliance Helix interconnects. The interconnects formed a 100 × 100 array with the columns near the die center forming a 20 × 20 area array, the low-compliance interconnects forming the intermediate 15 rows, and the high-compliance interconnects forming the outer 25 rows. Package 3 was fabricated on a bare silicon die and is shown in Figure 3.16. Using the models developed, the thermomechanical reliability levels of packages 2 and 3 were found to be nearly equivalent. The predicted life of package 1, however, was much lower because the column interconnects are not compliant and cannot accommodate the CTE mismatch at the outermost locations. Although package 2 is good from a thermomechanical reliability perspective, it is not recommended from an electrical performance viewpoint due to the higher electrical parasitics associated with the high-compliance interconnects. Therefore, package 3, which uses a heterogeneous combination of interconnects, represents a judicious tradeoff between electrical parasitics and mechanical reliability. Hence, a heterogeneous array of interconnects appears to provide a balanced combination of mechanical and electrical performance without compromising the thermomechanical reliability.

3.11 Summary

A pressing need exists to develop new first-level interconnect technologies. Utilizing compliant structures as first-level interconnects appears to be a promising approach. Conventionally practiced solder bumps use an underfill material to couple the die to the substrate to ensure interconnect reliability. Compliant interconnects on the other hand decouple the die from the substrate to ensure interconnect reliability. Simulations demonstrate that the die stresses induced by the compliant interconnects are at least an order of magnitude lower than the die stresses for an equivalent flip-chip assembly; hence, the compliant interconnects are not likely to crack or delaminate low-k dielectric material [11]. Also, they eliminate the need to use an underfill material to accommodate the CTE mismatch between the die and the organic substrate. Apart from removing an additional manufacturing step and enabling fine pitch interconnects, the elimination of the underfill allows the interconnects to be reworkable. Additionally, the vertical compliance of compliant interconnects allows them to accommodate the nonplanarity of the substrates.

Numerous compliant interconnect technologies have been developed, each with its own set of limitations. Two significant barriers to the implementation of compliant interconnects are their fabrication cost and electrical performance. Lithography-enabled wafer-level batch fabrication of compliant interconnects appears to be the most promising approach to realizing these interconnects cost-effectively. A lithography-based approach also allows for scaling of these interconnects with technology nodes and hence enables increasingly finer pitches to be addressed. In terms of electrical performance, improvements can be made with respect to the design of the interconnects. Simulating the behavior of the interconnects and design optimization are powerful aids for achieving this. Adopting a system-level view toward compliant interconnects, as described in Section 3.7, is an additional approach to improving the electrical performance without compromising the mechanical reliability. Also, the use of column interconnects in conjunction with compliant interconnects provides for high enough rigidity against potential vibration or drop induced damage on the compliant interconnects. Additional work needs to be performed with regard to the assembly and reliability of these interconnects. Novel attachment methods, such as using conductive adhesives or thermocompression bonding, could be used in the future. Extending the use of compliant interconnects beyond electrical interconnects to optical interconnects represents another avenue for development [30]. In summary, compliant interconnects clearly have a number of advantages and represent a viable interconnect technology that can address the needs of the industry over the next decade. However, additional work needs to be performed to enable their commercial implementation.

References

[1] Lau, J. H., *Flip Chip Technologies*, New York: McGraw-Hill, 1996.

[2] Tummala, R. R., *Fundamentals of Microsystems Packaging*, New York: McGraw-Hill, 2001.

[3] Viswanadham, P., and P. Singh, *Failure Modes and Mechanisms in Electronic Packages*, New York: Chapman & Hall, 1998.

[4] Chi Shih, C., A. Oscilowski, and R. C. Bracken, "Future Challenges in Electronics Packaging," *Circuits and Devices Magazine, IEEE*, Vol. 14, 1998, pp. 45–54.

[5] Ghaffarian, R., "Chip-Scale Package Assembly Reliability," in *Chip Scale Magazine*, 1998.

[6] DeBonis, T., "Getting the Lead Out," May 5,2008, http://download.intel.com/pressroom/kits/45nm/leadfree/lf_presentation.pdf.

[7] Jordan, J., "Gold Stud Bump in Flip-Chip Applications," *Proc. Electronics Manufacturing Technology Symposium*, 2002, pp. 110–114.

[8] Miller, R. D., "In Search of Low-k Dielectrics," *Science*, Vol. 286, 1999, p. 421.

[9] Kahle, J. A., et al., "Introduction to the Cell Multiprocessor," *IBM J. Research and Development*, Vol. 49, 2005, pp. 589–604.

[10] Vella, J. B., et al., "Mechanical Properties and Fracture Toughness of Organo-Silicate Glass (OSG) Low-k Dielectric Thin Films for Microelectronic Applications," *Int. J. Fracture*, Vol. 120, 2003, pp. 487–499.

[11] Kacker, K., G. C. Lo, and S. K. Sitaraman, "Low-K Dielectric Compatible Wafer-Level Compliant Chip-to-Substrate Interconnects," *IEEE Transactions on Advanced Packaging* (see also *IEEE Transactions on Components, Packaging and Manufacturing Technology, Part B: Advanced Packaging*), Vol. 31, 2008, pp. 22–32.

[12] ITRS, "ITRS 2007 Roadmap—Assembly and Packaging," May 2, 2008, www.itrs.net/Links/2007ITRS/2007_Chapters/2007_Assembly.pdf.

[13] Novitsky, J., and D. Pedersen, "FormFactor Introduces an Integrated Process for Wafer-Level Packaging, Burn-In Test, and Module Level Assembly," *Proc. International Symposium on Advanced Packaging Materials. Processes, Properties and Interfaces*, Braselton, Georgia, 1999, pp. 226–231.

[14] DiStefano, T., and J. Fjelstad, "A Compliant Chip-Size Packaging Technology," in *Flip Chip Technologies*, (ed.) J. H. Lau, New York: McGraw-Hill, 1996, pp. 387–413.

[15] Fjelstad, J., "WAVE Technology for Wafer Level Packaging of ICs," *Proc. 2nd Electronics Packaging Technology Conference*, Singapore, 1998, pp. 214–218.

[16] Young-Gon, K., et al., "Wide Area Vertical Expansion (WAVE) Package Design for High Speed Application: Reliability and Performance," *Proc. 51st Electronic Components and Technology Conference*, Orlando, FL, 2001, pp. 54–62.

[17] Fillion, R. A., et al., "On-Wafer Process for Stress-Free Area Array Floating Pads," *Proc. 2001 International Symposium on Microelectronics*, Baltimore, MD, 2001, pp. 100–105.

[18] Patel, C. S., et al., "Low Cost High Density Compliant Wafer Level Package," *Proc. 2000 HD International Conference on High-Density Interconnect and Systems Packaging*, Denver, CO, 2000, pp. 261–268.

[19] Reed, H. A., et al., "Compliant Wafer Level Package (CWLP) with Embedded Air-Gaps for Sea of Leads (SoL) Interconnections," *Proc. IEEE 2001 International Interconnect Technology Conference*, Burlingame, CA, 2001, pp. 151–153.

[20] Bakir, M. S., et al., "Sea of Leads (SoL) Ultrahigh Density Wafer-Level Chip Input/Output Interconnections for Gigascale Integration (GSI)," *IEEE Transactions on Electron Devices*, Vol. 50, 2003, pp. 2039–2048.

[21] Dang, B., et al., "Sea-of-Leads MEMS I/O Interconnects for Low-k IC Packaging," *J. Microelectromechanical Systems*, Vol. 15, 2006, pp. 523–530.

[22] Zhu, Q., L. Ma, and S. K. Sitaraman, "Development of G-Helix Structure as Off-Chip Interconnect," *Transactions of the ASME, J. Electronic Packaging*, Vol. 126, 2004, pp. 237–246.

[23] Zhu, Q., L. Ma, and S. K. Sitaraman, "β-Helix: A Lithography-Based Compliant Off-Chip Interconnect," *IEEE Transactions on Components and Packaging Technologies*, Vol. 26, 2003, pp. 582–590.

[24] Ma, L., Q. Zhu, and S. K. Sitaraman, "Mechanical and Electrical Study of Linear Spring and J-Spring," *Proceeding of 2002 ASME-IMECE*, New Orleans, LA, 2002, pp. 387–394.

[25] Smith, D. L., and A. S. Alimonda, "A New Flip-Chip Technology for High-Density Packaging," *Proc. 46th Electronic Components and Technology Conference*, Orlando, FL, 1996, pp. 1069–1073.

[26] Smith, D. L., et al., "Flip-Chip Bonding on 6 um Pitch Using Thin-Film MicroSpring Technology," *Proc. 48th Electronic Components and Technology Conference*, Seattle, WA, 1998, pp. 325–329.

[27] Ma, L., Q. Zhu, and S. K. Sitaraman, "Contact Reliability of Innovative Compliant Interconnects for Next Generation Electronic Packaging," *Proc. 2003 ASME-IMECE*, Washington, DC, 2003, pp. 9–17.

[28] Ma, L., et al., "Compliant Cantilevered Spring Interconnects for Flip-Chip Packaging," *Proc. 51st Electronic Components and Technology Conference*, Orlando, FL, 2001, pp. 761–766.

[29] Chow, E. M., et al., "Solder-Free Pressure Contact Micro-Springs in High-Density Flip-Chip Packages," *Proc. 55th Electronic Components and Technology Conference*, Lake Buena Vista, FL, 2005, pp. 1119–1126.

[30] Bakir, M. S., et al., "Electrical and Optical Chip I/O Interconnections for Gigascale Systems," *IEEE Transactions on Electron Devices*, Vol. 54, 2007, pp. 2426–2437.

[31] Bakir, M. S., et al., "Mechanically Flexible Chip-to-Substrate Optical Interconnections Using Optical Pillars," *IEEE Transactions on Advanced Packaging* (see also *IEEE Transactions on Components, Packaging and Manufacturing Technology, Part B: Advanced Packaging*), Vol. 31, 2008, pp. 143–153.

[32] Bakir, M. S., et al., "Dual-Mode Electrical-Optical Flip-Chip I/O Interconnects and a Compatible Probe Substrate for Wafer-Level Testing," *Proc. 56th Electronic Components and Technology Conference*, San Diego, California, 2006, p. 8.

[33] Bakir, M. S., et al., "'Trimodal' Wafer-Level Package: Fully Compatible Electrical, Optical, and Fluidic Chip I/O Interconnects," *Proc. 57th Electronic Components and Technology Conference*, May 29, 2007 to June 1, 2007, Reno, NV, pp. 585–592.

[34] Dudek, R., et al., "Thermo-Mechanical Design of Resilient Contact Systems for Wafer Level Packaging," *Proc. EuroSimE 2006: Thermal, Mechanical and Multi-Physics Simulation and Experiments in Micro-Electronics and Micro-Systems*, Como, Italy, 2006, pp. 1–7.

[35] Dudek, R., et al., "Thermomechanical Design for Reliability of WLPs with Compliant Interconnects," *Proc. 7th Electronics Packaging Technology Conference*, Singapore, 2005, pp. 328–334.

[36] Kim, W., et al., "Electrical Design of Wafer Level Package on Board for Gigabit Data Transmission," *Proc. 5th Electronics Packaging Technology Conference*, Singapore, 2003, pp. 150–159.

[37] Klein, K. M., and S. K. Sitaraman, "Compliant Stress-Engineered Interconnects for Next Generation Packaging," Anaheim, California, 2004, pp. 219–226.

[38] Kacker, K., T. Sokol, and S. K. Sitaraman, "FlexConnects: A Cost-Effective Implementation of Compliant Chip-to-Substrate Interconnects," *Proc. 57th Electronic Components and Technology Conference*, May 29, 2007 to June 1, 2007, Reno, NV, pp. 1678–1684.

[39] Suresh, S., *Fatigue of Materials*, 2nd ed., Cambridge, UK: Cambridge University Press, 1998.

[40] Lo, G., and S. K. Sitaraman, "G-Helix: Lithography-Based Wafer-Level Compliant Chip-to-Substrate Interconnects," *Proc. 54th Electronic Components and Technology Conference*, Vol. 1, 2004, pp. 320–325.

[41] Bakir, M. S., et al., "Sea of Leads Compliant I/O Interconnect Process Integration for the Ultimate Enabling of Chips with Low-k Interlayer Dielectrics," *IEEE Transactions on Advanced Packaging*, Vol. 28, 2005, pp. 488–494.

[42] Kacker, K., et al., "A Heterogeneous Array of Off-Chip Interconnects for Optimum Mechanical and Electrical Performance," *Transactions of the ASME, J. Electronic Packaging*, Vol. 129, 2007, pp. 460–468.

[43] Bakir, M. S., et al., "Sea of Leads Ultra High-Density Compliant Wafer-Level Packaging Technology," *Proc. 52nd Electronic Components and Technology Conference*, May 28–31, 2002, San Diego, CA, pp. 1087–1094.

CHAPTER 4

Power Delivery to Silicon

Tanay Karnik, Peter Hazucha, Gerhard Schrom, Fabrice Paillet, Kaladhar Radhakrishnan

4.1 Overview of Power Delivery

Gordon Moore postulated that the number of transistors in an integrated circuit would double approximately every 18 months. This law, popularly known as Moore's law, has served as the guiding principle for the semiconductor industry. There is an aggressive effort across the industry to stay on this technology treadmill for at least the next decade. Back in 1971, the first processor manufactured by Intel had about 2,300 transistors, ran at a frequency of 740 kHz, and dissipated less than 1W. By way of comparison, the latest processor released by Intel, the Core 2 Duo, has close to 300 million transistors, runs at a frequency of 3.8 GHz, and dissipates 120W.

Every time a computer user runs an application, the transistors inside the processor go to work, with each transistor drawing a tiny amount of current for every clock cycle. Now, imagine a case in which the transistor is switching a few billion times per second like today's microprocessors. What happens when there are 100 million transistors switching three billion times per second? This can result in a large current demand at the die. So, when the computer transitions from an idle mode to a high-power mode, there is a sudden jump in the processor's current consumption. Any time this happens, there is a voltage drop associated with the current spike. This is not unlike what you see in older homes when a high-power appliance turns on and the associated voltage drop causes the light bulbs to dim (also known as "brownout"). In a similar fashion, when the processor draws a large amount of current, it will induce a voltage drop in the die power supply. The primary objective of power delivery design is to minimize this voltage drop [1].

4.1.1 Importance of Power Delivery

During each clock cycle, different portions of the silicon logic will need to communicate with each other. The time it takes for this to happen is a function of how much time it takes a signal to go from the input to the output of a transistor gate (gate delay), as well as the time it takes for the signal to travel from one gate to another (interconnect or RC delay). The RC delay can be reduced by using a low-k material to reduce the capacitance between metal lines or by reducing the transistor junction temperature, which will help reduce the metal line resistance. The gate delay is a function of the gate size as well as the voltage available to the processor.

When this voltage drops, there is an increase in the gate delay, which impacts the amount of time it takes for circuit blocks to communicate with each other. This in turn limits the maximum operating speed of the microprocessor. For the Core 2 Duo microprocessor, every millivolt of voltage drop translates to a ~3 MHz impact in the maximum operating frequency. For example, a poor power delivery design that increases the power-supply voltage drop by 70 mV could reduce the core frequency by up to 200 MHz, resulting in a substantial loss of revenue. In addition to lowering the maximum operating frequency of the core, excessive voltage drop can also corrupt the data that is stored in the memory cells, resulting in what are known as "soft errors."

4.2 Power Delivery Trends

The number of transistors on a microprocessor chip has been increasing at an exponential rate. At the same time, these transistors have been switching more quickly to improve performance. These two trends combine to drive up the current consumed by microprocessors. Even though a part of this increase is offset by the reduction in the voltage levels and the transistor size, microprocessor current consumption has still been increasing at an exponential rate over the last two decades, as shown in Figure 4.1. The brief respite in the current scaling in the mid-1980s can be attributed to the switch from n-channel metal-oxide-semiconductor field effect transistor (NMOS) to complimentary metal-oxide-semiconductor field effect transistor (CMOS) technology.

As the dimensions on die get smaller to accommodate the increasing device density, the die voltage levels have been scaling down to meet oxide reliability constraints. Figure 4.2 shows the silicon feature size as a function of time. From the figure, we can see that the feature size has been scaling by a factor of ~0.7 every 2 years. This corresponds to a doubling of the device density during the same period in

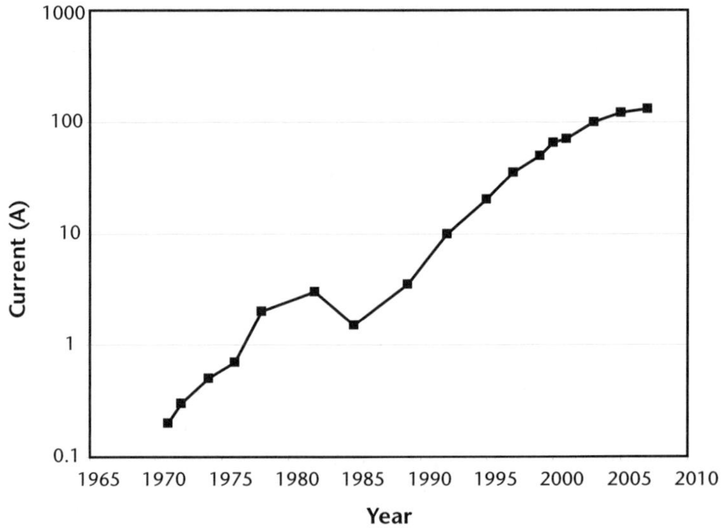

Figure 4.1 Microprocessor current consumption trend.

accordance with Moore's law. As the device dimensions have continued to get smaller, the gate oxide thickness has gone from about 100 nm back in the 1970s to close to 1 nm in today's fabrication process. In order to comply with the oxide reliability requirements, the die voltage has been scaling down as well, as Figure 4.2 shows. The lowered operating voltage drives a lowered noise requirement. This trend coupled with the increasing current yields a power delivery impedance target that is fast approaching submilliohm levels. This is a far cry from some of the early microprocessors, which had a target impedance of several ohms. Back in the early days, power was delivered through a handful of pins and routed to the die using traces on the package and wire bond interconnects. Compare this with the latest Core 2 Duo processor where power is delivered through hundreds of power and ground (P/G) pins and routed to the die through thick copper planes and thousands of P/G bumps [2]. In addition to these changes, which help reduce the dc resistance, the package decoupling capacitor technology has evolved over the years as well. For example, the early microprocessors used two-terminal capacitors to address their decoupling needs. However, since these capacitors have a large parasitic inductance associated with them, they lost their effectiveness at high frequencies. In an effort to combat this, interdigitated capacitors (IDC) were introduced as a more effective alternative. The IDC has eight or ten alternating power and ground terminals, which minimizes the parasitic inductance of the capacitor.

An unfortunate by-product of the reduction in the device dimensions is the increase in leakage power. Today's transistors conduct current even when they are turned off, and this current is referred to as leakage current. Figure 4.3 plots the growth trends of leakage current and active current [2]. While the active current is used to boost the performance of the processor, the leakage current adds nothing to the processor performance and exacerbates the thermal problem. Even as recently as the 1990s, leakage current was a negligible fraction of the total microprocessor current. However, from the curves, we can see that the growth rate for leakage cur-

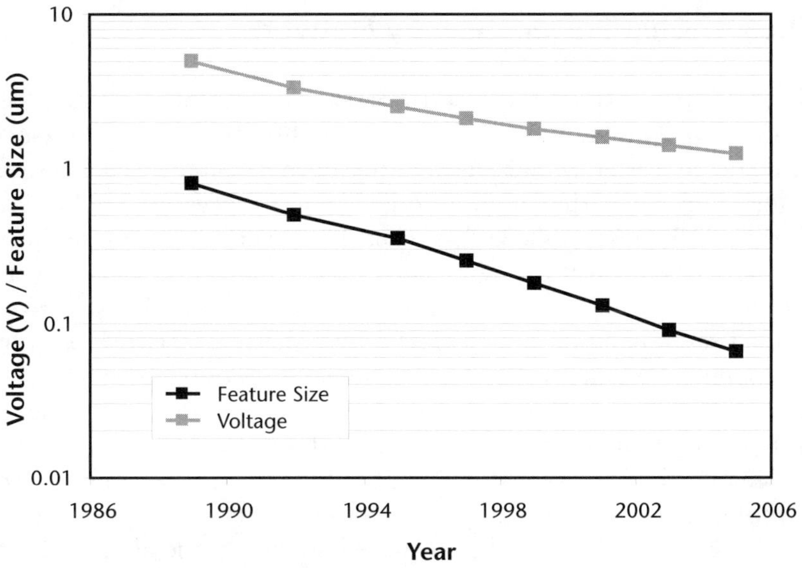

Figure 4.2 Microprocessor voltage/feature size scaling.

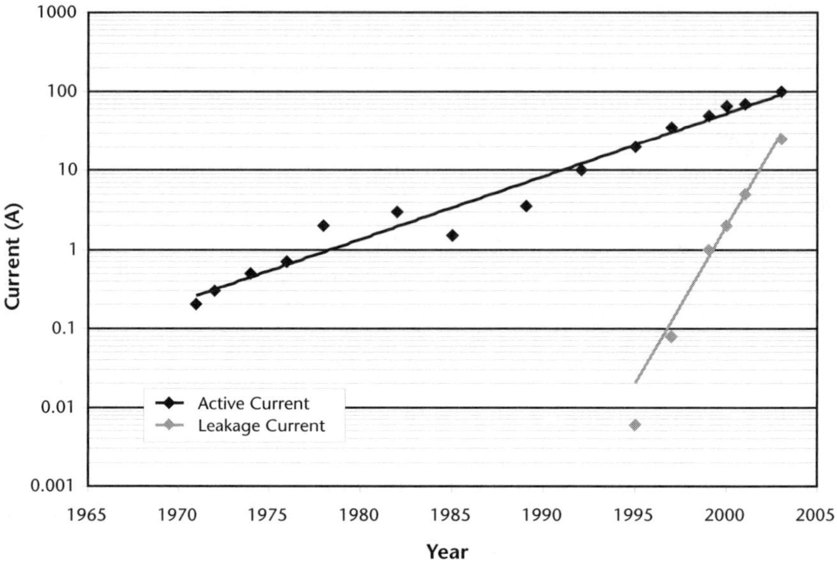

Figure 4.3 Active versus leakage current growths.

rent is much higher than that for active power, and if left unchecked, the former will soon exceed the latter. One way to combat the leakage issue is by slowing the frequency growth. With a reduced emphasis on the processor frequency, the process parameters can be tweaked to reduce leakage current at the expense of transistor switching speed. With frequency no longer being the primary knob for improving the processor performance, system architects have turned to other avenues in an effort to improve the overall performance. One example of this is the switch to multiple cores. By adding an extra logic core and reducing the switching frequency, the processor can get a performance boost without a significant power penalty.

4.3 The Off-Chip Power Delivery Network

Power from the wall outlet is first converted into a 12V dc supply by the power-supply unit. The 12V output from the power-supply unit is then sent through a dc-dc converter, which delivers the output voltage that is requested by the die. Most dc-dc converters used in today's microprocessors are located on the motherboard (MB) close to the CPU socket. However, the response time of today's dc-dc converter is not fast enough to meet the processor demands. This issue is usually addressed by using multiple stages of decoupling capacitors. Figure 4.4 shows a simplified representation of a typical power delivery network with multiple stages of decoupling capacitors [3]. The first stage of decoupling, as seen from the die, is the on-die capacitance. Due to the proximity of the on-die capacitance to the load, it can usually respond immediately to any load request. However, the amount of capacitance that can be designed in the die is limited and is of the order of a few hundred nanofarads [1]. The next stage of decoupling is typically present on the package. The package capacitors can either be placed on the topside of the package adjacent to the die or on the backside of the package under the die shadow as shown in Figure 4.5.

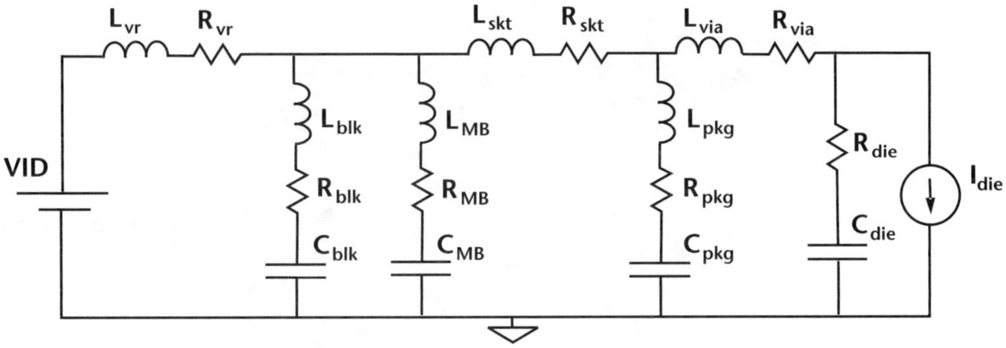

Figure 4.4 Typical power delivery network.

Dieside Capacitors on Package **Backside Capacitors on Package**

Figure 4.5 Microprocessor package capacitors.

The amount of capacitance present on the package is usually of the order of a few tens of microfarads. Unlike the on-die capacitance, the package capacitors have a finite inductance associated with them due to the inductance in the path through the package, as well as the parasitic inductance within the capacitor itself. The final stage of decoupling is present on the MB and typically comprises two types of capacitors: electrolytic capacitors and ceramic capacitors to meet the low-frequency and mid-frequency decoupling requirements, respectively [2]. Figure 4.6 shows a typical MB with the two types of capacitors highlighted.

4.3.1 Voltage Droops and Resonances on the Power Delivery Network

Any sudden change in the processor current is initially handled by the charge stored in the on-die decoupling capacitance. However, due to the relatively low value of the on-die capacitance, it can deliver charge to the processor only for a few nanoseconds before the voltage droops (first droop). The voltage continues to droop until the charge from the next stage of capacitors on the package can kick in to supply the processor current and replenish the charge on the on-die capacitance. After a few

Figure 4.6 Microprocessor motherboard.

tens of nanoseconds, the package capacitors will run out of charge as well, resulting in the second voltage droop, which continues until the MB capacitors can respond. Eventually, the MB capacitors run out of charge as well, causing a third voltage droop until the voltage regulator can respond. All three droops can be seen in Figure 4.7, which shows a scope capture of the die voltage measured on a Core 2 Duo processor in response to a step load [2].

While it is easier to visualize the droops in the time domain, more insight can be gained by looking at the impedance profile of the power delivery network in the frequency domain. Figure 4.8 shows the measured impedance profile of the power delivery network for a Core 2 Duo processor [2]. For every droop in the time

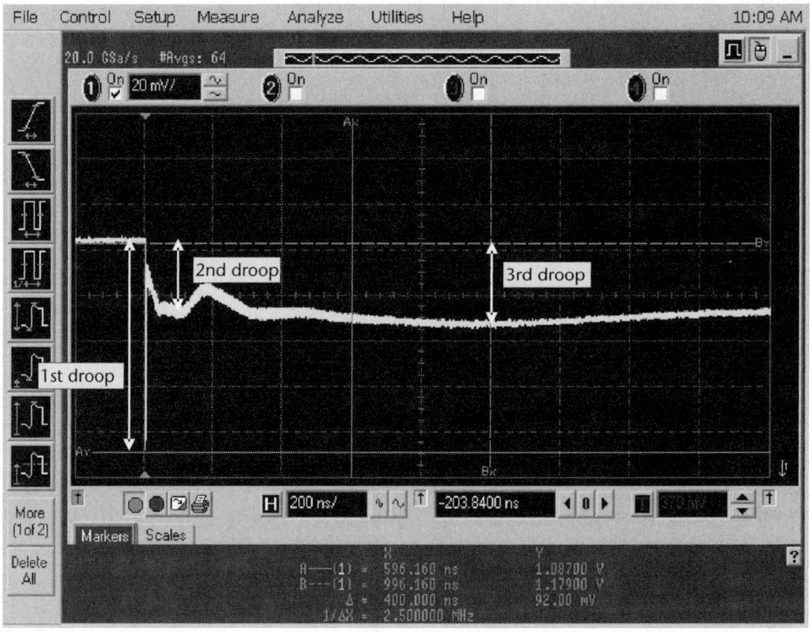

Figure 4.7 Oscilloscope capture of different voltage droops.

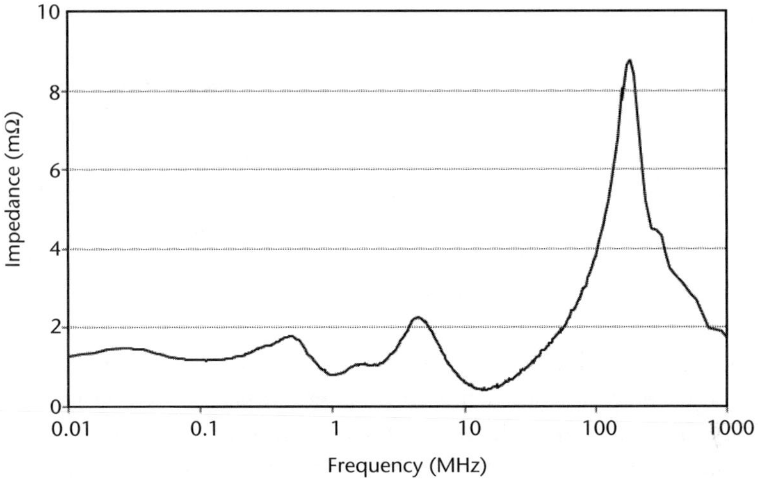

Figure 4.8 Impedance profile of power delivery network.

domain, there is a corresponding resonant peak in the frequency domain. For example, there is a direct correlation between first droop and the high-frequency resonant peak at around 200 MHz. This resonant peak can be attributed to the tank circuit that is created by the on-die capacitance and the inductance to the package caps. Potential ways to reduce first droop are by increasing the on-die capacitance or reducing the inductance to the package caps. Ironically, the first droop peak can also be reduced by increasing leakage current, which serves to dampen the resonance. Moving down the frequency axis, we see a smaller mid-frequency resonance at around 5 MHz, which corresponds to the second droop in the time domain. This resonant peak is caused by the package capacitance and the inductance to the next stage of decoupling on the MB. The simplest way to minimize second droop is by using package capacitors with large capacitance values. Finally, the resonant peak that corresponds to the third droop in the time domain is a relatively small bump at around the 600 kHz. This is caused by the interaction between the MB capacitors and the response time of the dc-dc converter. Potential ways to reduce third droop are by increasing the capacitance on the MB or using faster dc-dc converters with a higher bandwidth.

4.3.2 Current-Carrying Capability

One area of power delivery that is getting more attention lately is the current-carrying capability of the various elements of the power delivery network. Since today's high end processors can carry currents in excess of 100A, the power delivery designer needs to worry about having enough margin in the current-carrying capability of the bumps, the traces and vias in a package, and the pins in the socket. There are two kinds of failure mechanisms associated with excessive current. The first failure mechanism is that caused by electromigration and is primarily an issue for the on-die metal layers as well as the flip-chip solder bumps.

Electromigration failures are caused by mass electron movement across a relatively small area of cross section, such as the ones seen in on-die interconnects and solder bumps. Excessive current through these bumps or interconnects will cause significant voiding, as shown in Figure 4.9. The second failure mechanism associated with excessive current is due to joule heating and is more common among the package, socket, and MB interconnects [4]. If a processor is drawing up to 100A of current, even if the resistance in the path is as low as 1 mΩ, the power dissipated in the path can be as high as 10W. In the absence of a good thermal solution, the dissipated power will result in an increase in temperature. This increase in temperature is accompanied by an increase in resistance, which in turn will further increase the power dissipated. In some extreme case, the power dissipation and temperature rise can be high enough to result in thermal runaway, which will result in an instantaneous failure. A more likely, but still undesirable, scenario is that the temperature will stabilize at a value that is higher than the part can tolerate to meet its quality requirements. For example, if the package is subjected to a high temperature for an extended period of time, it could induce some sort of mechanical failure, which could eventually cause the part to fail.

4.4 dc-dc Converter

As explained in the previous section, meeting Vcc variation bounds in the presence of large current transients requires a prohibitively large amount of on-die decap. Alternatively, the voltage regulator needs to operate at a higher frequency, which may affect conversion efficiency [4]. We present the importance of a near-load dc-dc converter in this chapter.

4.4.1 Motivation for dc-dc Converter

Switching voltage regulators are widely used for microprocessor power delivery. The typical regulators accept high (~12V) input voltage and convert it to low

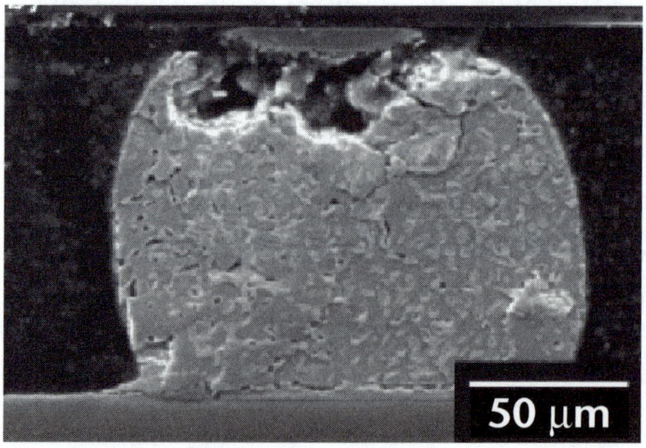

Figure 4.9 Bump voiding due to joule heating.

(~1.2V) die voltage with high efficiency. They are placed on the motherboard due to the large inductor and capacitor requirements dictated by the high conversion ratio and low switching frequency [2]. The response time is large, and the power delivery impedance needs to be very low to supply the processor's high current demands.

Figure 4.10 illustrates a near-load dc-dc converter inserted between the main voltage regulator module (VRM) and the microprocessor load. The near-load dc-dc converter reduces the VRM output current (I_{ext}) and allows an increase of the impedance (Z_{ext}). For a conversion ratio of N:1 and conversion efficiency of η, the VRM current is reduced by a factor of $N\eta$. With a converter-added droop of 5%, the decoupling requirement is reduced by a factor of $0.5N^2\eta$. These reductions directly translate into the reduction of losses in the power delivery network and the system component cost and size [4].

4.4.2 Modeling

Figure 4.11 shows a basic buck-type dc-dc converter. The transistors M1 and M2, which form the so-called bridge, switch the bridge output V_x between V_{in} and ground such that the average V_x, and therefore the output voltage V_{out}, are essentially the same as the reference voltage V_{ref}. A feedback circuit consisting of an amplifier and a network R_{fb} and C_{fb} (i.e., a type 1 compensator) controls a pulse-width modulator, which adjusts the duty cycle D accordingly (i.e., the percentage of time during which M2 is turned on). Since the average voltage across the inductor is essentially zero, the output voltage is $V_{out} = D_{Vin}$. The current through the inductor will increase and decrease as V_x is switched between V_{in} and ground, and a capacitance C_{decap} at the output is used to decouple the ripple current I_r from the load current I_L [see Figure 4.12(a)]. The decoupling capacitance is also needed to maintain the output voltage during sudden load current changes since the inductor limits the rate of current increase to $(V_{in} - V_{out}) / L$.

The efficiency of the dc-dc converter $\eta = P_{out} / P_{in} = P_{out} / (P_{out} + P_{loss})$ depends on the power loss due to parasitic capacitances, resistances, and leakage currents. Figure 4.12(b) shows a model of the dc-dc converter, where the bridge consists of two ideal switches, S1 and S2, and the transistor parasitics are captured by an effective bridge resistance R_b, bridge capacitance C_b, and leakage current I_{lkg}. The inductor parasitics are modeled by the wire resistance R_i and by the eddy current resistance R_e.

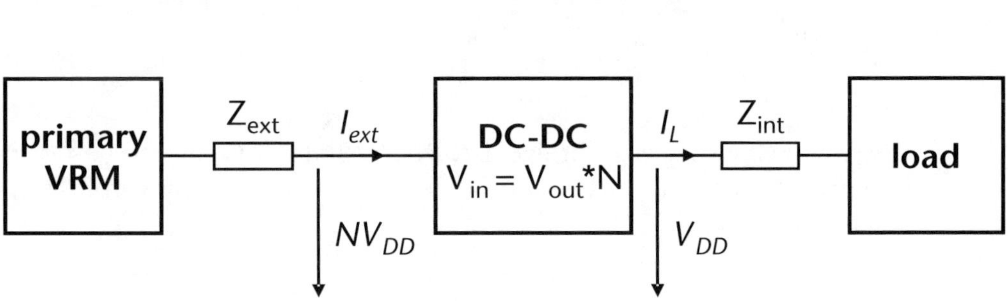

Figure 4.10 Inserting a dc-dc converter near the load.

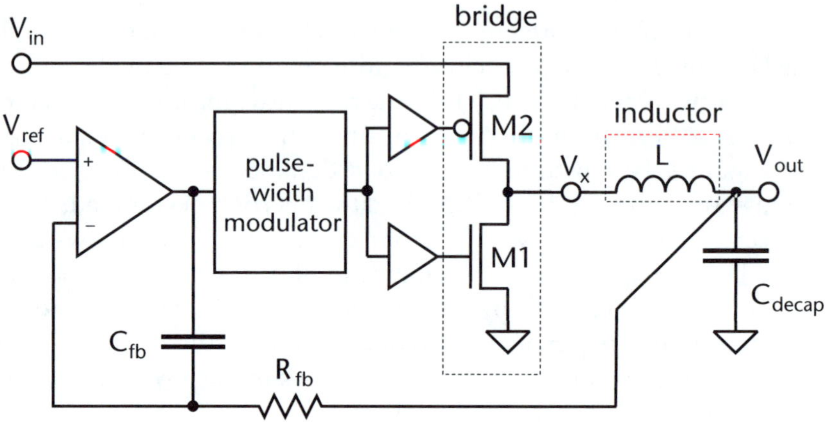

Figure 4.11 Basic buck-type dc-dc converter.

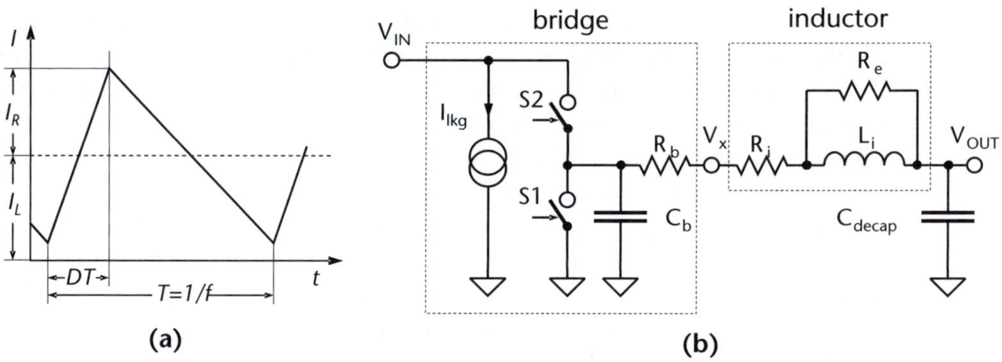

Figure 4.12 (a) Inductor current waveform, and (b) dc-dc converter power train model including bridge and inductor parasitics.

Unlike discrete-component voltage regulator designs, integrated voltage regulators can be easily optimized by changing the NMOS and PMOS transistor sizes W_N and W_P such that

$$C_b = W_N C_N + W_P C_P = WC_0; \quad C_0 = \frac{C_N + \alpha C_P}{1+\alpha}$$

$$R_b = \frac{R_N}{W_N}(1-D) + \frac{R_P}{W_P}D; \quad R_b = \frac{R_0}{W}; \quad R_0 = (1+\alpha)\left[(1-D)R_N + \frac{DR_P}{\alpha}\right]$$

where $W = W_N + W_P$ is the bridge size, R_N, R_P and C_N, C_P are the on-resistance and effective switched capacitance of the NMOS and PMOS transistors, respectively, and

$$\alpha = \frac{W_P}{W_N} = \sqrt{\frac{DR_P C_N}{(1-D)R_N C_P}}$$

is the optimal *P*-to-*N* width ratio. Notice that irrespective of the bridge size, $R_b C_b = R_0 C_0$, which is a key figure of merit for the transistor technology.

Similarly, the inductor can be optimized within a given volume or height constraint such that $\tau_i = L_i / R_i$ and $\tau_e = L_i / R_e$, which are the inductor technology figures of merit. For a given inductance L_i and frequency f, the inductor ripple current then is given as

$$I_R = \frac{V_{in} D(1-D)}{2fL_i}$$

and the inductor resistance and effective root-mean-square (RMS) current are

$$R_i = \frac{V_{in} D(1-D)}{2f\tau_i I_R}, \quad I_{rms}^2 = \left(I_L^2 + \frac{I_R^2}{3}\right)$$

The three most important power-loss components are the capacitive switching loss in the bridge and the resistive loss in the bridge and in the inductor:

$$P_{loss} = P_{cap} + P_{res} + P_{ind}$$

where

$$P_{cap} = C_b V_{in}^2 f = W C_0 V_{in}^2 f$$

$$P_{res} = R_b I_{rms}^2 \frac{R_0}{W}\left(I_L^2 + \frac{I_R^2}{3}\right)$$

$$P_{ind} = R_i I_{rms}^2 = \frac{V_{in} D(1-D)}{2f\tau_i I_R}\left(I_L^2 + \frac{I_R^2}{3}\right)$$

To find the optimum design (i.e., to minimize the power loss), we set the derivatives with respect to the three design variables—ripple current, bridge size, and frequency to zero—and after some manipulation we obtain

$$I_R = I_L$$

$$W = R_0 \frac{I_L}{V_{in}} \cdot \sqrt[3]{\frac{8\tau_i}{3R_0 C_0 D(1-D)}}$$

$$f = \sqrt[3]{\frac{D^2(1-D)^2}{3R_0 C_0 \tau_i^2}}$$

Remarkably, all three components $P_{cap} = P_{res} = P_{ind}$ are equal, and the total power loss is

$$P_{loss} = V_{in} I_L \cdot \sqrt[3]{24 \frac{R_0 C_0}{\tau_i} D(1-D)}$$

which directly links the optimal dc-dc converter design and efficiency to the key technology parameters R_0, C_0, and τ_i. Other power-loss contributions (e.g., from transistor off-state leakage) can be minimized separately and are typically small compared to the main components. A more detailed analysis, which also accounts for transistor off-state leakage as well as skin effect and eddy currents in the inductor and the impact of routing resistance, is given in [5].

4.4.3 Circuits

One of the biggest advantages of building dc-dc converters using high-speed digital CMOS processes is that switching frequencies can be as high as several hundred megahertz, which allows for the use of very small inductors and capacitors and therefore a drastic reduction in the area and/or volume required for power conversion; it also enables very fast response times. Since on-die component count and signal routing do not affect the cost of an integrated dc-dc converter, complex multiphase designs become economically feasible. A higher number of interleaved phases can effectively be used to reduce the switching noise on both input and output and to reduce the required amount of decoupling capacitance.

Figure 4.13 shows the block diagram of a 100 MHz, eight-phase, integrated dc-dc converter designed in a 90 nm CMOS process [6]. The bridges use a cascode topology, and the PMOS drivers are controlled by level shifters in order to support a higher input voltage of $V_{in} = 2V_{\max} = 2.4V$ without the need for specialized high-voltage devices. The cascode center rail is held in place by a linear regulator [7], which has to supply only the difference of the NMOS and PMOS driver supply currents. A 1.2V shunt regulator supplies the controller circuit (i.e., the eight-phase pulse-width modulator, the feedback circuit, and the seven-bit digitally controlled reference voltage generator). The bridges drive eight discrete 0402 size 1.9 nH air core inductors placed on the bottom of the package underneath the die to minimize routing resistance. This integrated dc-dc converter is designed for 12A load current and occupies only 10 mm^2 on a 4.5 × 5.5 mm chip, which also contains other test circuits.

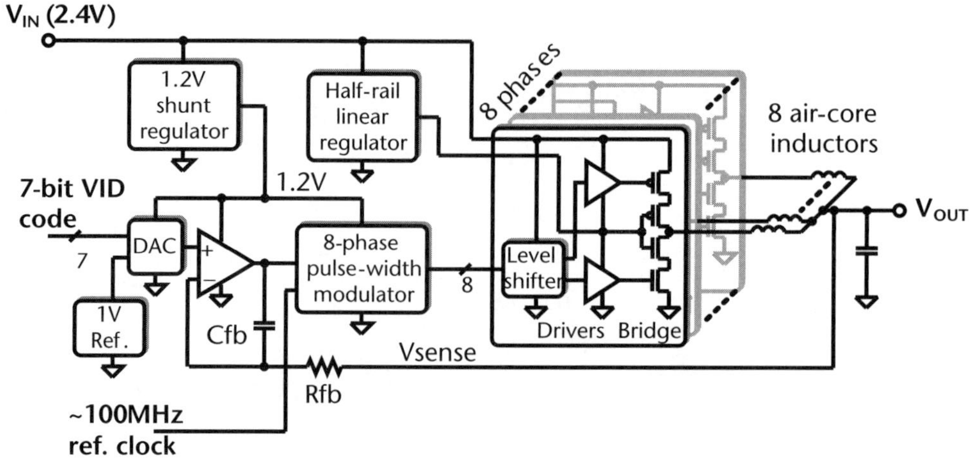

Figure 4.13 Eight-phase integrated dc-dc converter.

4.4.4 Measurements

Figures 4.14 and 4.15 show measurement results of two high-frequency near-load dc-dc converters. Figure 4.14(a) shows the measured efficiency of the synchronous near-load buck converter detailed in [6] for two different inductor values and two output voltages. The efficiency peaks at 10A to 12A (the rated load of this converter) with 79.3% and 85%. The optimum converter switching frequency depends on the inductor used: 60 MHz with the larger inductor (1.9 nH) and 80 to 100 MHz for the smaller 0.8 nH inductor such that the ripple current does not change much between the two. Figure 4.14(b) shows the transient response of the same converter when a load current is switched between 5A and 10A with a rise time of less than 100 ps. Thanks to the high switching frequency and high controller loop bandwidth, the droop recovery time is only about 100 ns. The first droop (very sharp 100 mV spike) is largely determined by the decoupling capacitor ESL (parasitic series inductance), whereas the second droop is only 30 mV because of the short 50 ns

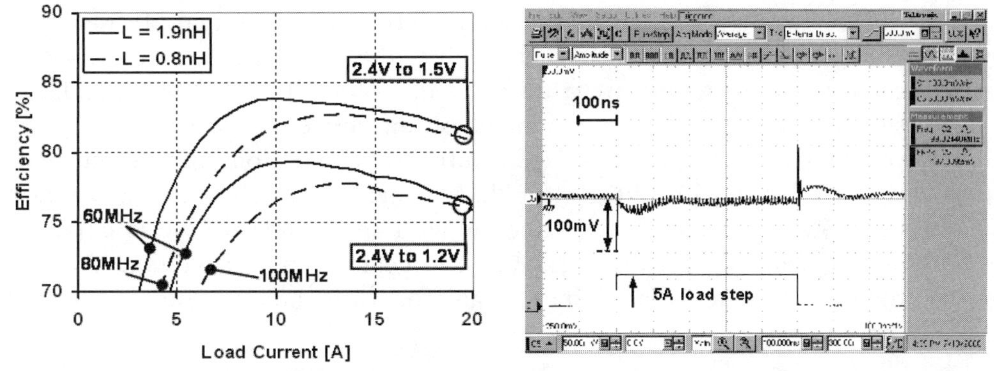

(a) (b)

Figure 4.14 (a) Measured efficiency as a function of load current for two inductor values and output voltages. (b) Measured transient output voltage droop in presence of a 5A load step with higher than 50 A/ns slew rate.

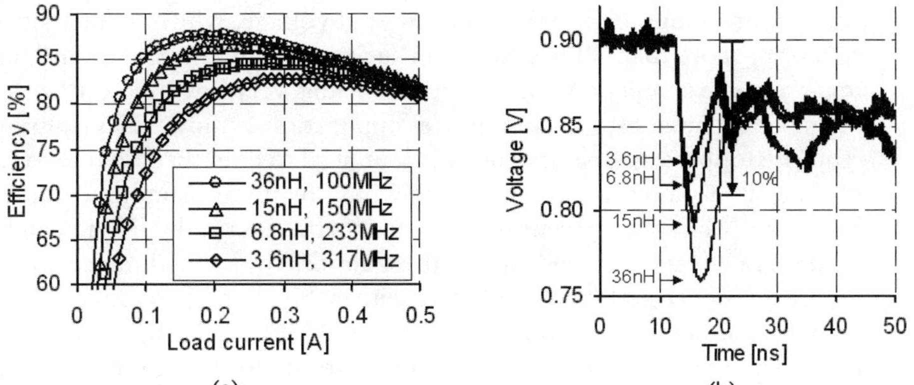

(a) (b)

Figure 4.15 (a) Measured efficiency as a function of load current for various inductor values for 1.4V to 1.1V conversion. (b) Measured transient output voltage droop for four inductor values in presence of a 150 mA load step (50% rated load) with 100 ps slew rate.

response time. Figure 4.15(a) shows the measured efficiency of a second switching converter with a hysteretic controller detailed in [8] in the presence of different inductor values. The same trend is visible. Optimum switching frequency decreases as the inductance increases. Figure 4.15(b) shows the transient response of the converter in the presence of a 50% instantaneous load change for various inductor values. The decoupling capacitors are integrated on die and exhibit negligible ESL. The response time of the converter is limited by the delay in the controller and the current slew rate in the inductors. It is dominated by the controller delay for small inductances (3.6 nH) and by the inductor current slew rate for larger inductance values (15 to 36 nH). In short, there is an apparent trade-off that higher inductance improves efficiency but degrades transient response and requires additional decoupling capacitance.

4.5 Linear Regulator

In a switching power converter, a device is switched on and off to chop a dc input voltage which is then filtered through a filter consisting of an inductor (L) and a capacitor (C) to generate a different dc voltage [4]. A linear regulator is much simpler. It uses a linear command on an active device (transistor) to deliver a constant output voltage independently from variations in load current or input voltage. No hard-to-integrate inductors are needed; hence, linear regulators can be easily integrated on the same die with any standard CMOS circuit. However, unlike switching converters, which can generate output voltages higher or lower than their input voltage, linear regulators only supply voltages lower than their input voltage.

4.5.1 Motivation

Multiple reasons can justify the selection of a linear regulator in place of a switching regulator, but both are often combined to take advantage of the best each can offer. Switching regulators can deliver large power and achieve 95% efficiency with input voltage three times higher than their output voltage, while an ideal linear regulator cannot be more than 33% efficient in such cases. On the other hand, linear regulators are much simpler circuits and can consume very low quiescent current, making them better suited for very low-power applications. A low quiescent power is a prerequisite for battery-operated devices that need to hold their charge for long periods of time at very low standby load. Switching regulators' higher complexity implies significantly larger quiescent power, even if efficiency at high load can be very high. Two extra advantages held by linear regulators over switching converters are their fast regulation capability combined with a high power-supply-rejection ratio in addition to their low output-voltage noise derived from the absence of switching noise. This makes them favorites for applications requiring low noise supplies, such as audio/RF, since switching converters' output always exhibits some residual voltage ripple. Not surprisingly, linear regulators are particularly popular in battery-operated devices (cell phone, PDA) where low dropout (LDO) linear regulators can compete in efficiency with switching converters, while providing low standby currents due to a low quiescent current. It is common to find a couple of switching converters with a dozen or so cascaded linear regulators inside cell phones.

Using a linear regulator to deliver a lower voltage to CMOS circuits with the objective to save overall power seems counterintuitive since the linear regulator losses increase linearly with the lowering of its output voltage. It actually makes sense if one considers the fact that the power dissipated by digital circuits like CPU cores increases/decreases with the square of the supply voltage, while leakage power varies exponentially with it. This means a linear reduction in the total power consumed is still achievable using a linear regulator while standby leakage current can be brought down to very low levels. To prove this point, let's consider the example of a low-power mobile CPU operating between 1V and 0.8V. At 1V the core consumes 1W. When the voltage drops by 20% to 0.8V, the core power is reduced by 36% to 0.64W. The efficiency of an ideal linear regulator is 80% for 1V to 0.8V conversion. This leads to a total power consumption of 0.8W (0.64/0.8) at 0.8V, a 20% power saving from operating at 1V, or 16% in the case of a nonideal linear regulator with 95% current efficiency. Such a solution can be easily implemented with low area overhead by integrating a linear regulator with the CPU core on the same die [9].

4.5.2 Modeling

The difference between the input and output voltage of a linear regulator is called the dropout voltage. It is one of four important parameters to consider when choosing between a switching converter and a linear regulator. The power efficiency of a linear regulator is at most $Eff = V_{out}/V_{in}$ for an ideal linear regulator. In practice, the linear regulator circuitry consumes a small amount of current for its biasing, even in the absence of load current. This current is referred to as the quiescent current of the linear regulator. Input current of the linear regulator is equal to the sum of the quiescent current I_q and the load current I_{out}. The second important parameter to evaluate performance of a linear regulator is the current efficiency, which is equal to $Eff_i = I_{out}/I_{in} = I_{out}/(I_{out} + I_q)$. Efficiency of a linear regulator is then $Eff = P_{out}/P_{in} = (I_{out} \times V_{out})/(I_{in} \times V_{in}) = Eff_i \times V_{out}/V_{in}$. The power-loss mechanism in a linear regulator is similar to that of a resistive divider. The transistor across which the dropout voltage is distributed acts as a controlled resistor (the top resistor of a resistive divider) and dissipates $V_{drop} \times I_{load}$ into heat. As a consequence, linear regulators are not recommended for applications where V_{in} is significantly higher than V_{out} since more energy would be wasted than delivered to the load. Despite the low efficiency of such a scenario, there are very low-power applications cases in which the low quiescent current of linear regulators can offset what would be an even poorer efficiency with switching converters due to their inherently higher quiescent current and design complexity.

The third aspect of a linear regulator's performance is the (transient) response time. Fast load regulation is important when supplying digital CMOS circuits such as CPU cores, with rapidly changing supply current. The response time, T_R, is found from the output decoupling C for a specified I_{MAX} and droop ΔV_{OUT}. $T_R = C \times \Delta V_{OUT}/I_{MAX}$. For the purpose of comparing regulators designed to different specifications of droop, decoupling C, quiescent current I_Q, and current rating I_{MAX}, a figure of merit (FOM) can be defined as $FOM = T_R \times I_Q/I_{MAX} = (C \times \Delta V_{OUT}/I_{MAX}) \times (I_Q/I_{MAX})$, which is a time constant.

The smaller FOM is, the better the regulator. For example, two identical regulators operating in parallel with two times higher I_{MAX}, I_Q, and C have the same FOM as each regulator operating stand-alone. Also, a reduction in the quiescent current by 50% and doubling of the capacitance does not affect FOM. Finally, the last element of performance of a linear regulator to consider is the power-supply noise-rejection ratio (PSRR). Depending on the linear regulator topology, the PSRR can be good or poor, especially at high-frequency. Let's review the respective performance of existing linear regulator topologies implemented in CMOS technology.

4.5.3 Circuits

Existing linear regulator topologies vary widely in their quiescent current, dc load regulation, transient response, decoupling capacitance, and silicon area requirements. Figure 4.16(a) is a linear regulator with source follower output driver. This topology achieves fast load regulation due to low output impedance of the source follower M_0. With sufficient voltage headroom (twice the transistor threshold voltage or more), M_0 will operate in saturation, which results in very good high-frequency power-supply noise rejection, but V_{GS} of M_0 can be small and result in a large silicon area requirement. Figure 4.16(b) features a gate overdrive to increase V_{GS} of M_0 to provide additional regulation headroom as well as to reduce silicon area [9, 10]. This topology uses replica biasing to further improve load regulation. Since the feedback loop does not observe output V_{OUT}, the load regulation is based only on the matching of transistor M_0's I-V characteristic [9, 11]. For a 90 mV droop, source conductance of transistor M_0 limits the ratio of the minimum to maximum output current to about 1:10. A source follower without replica biasing [10] does not have such problem, but load regulation is then limited by the bandwidth of the amplifier feedback loop. The topology in Figure 4.16(c) achieves low dropout

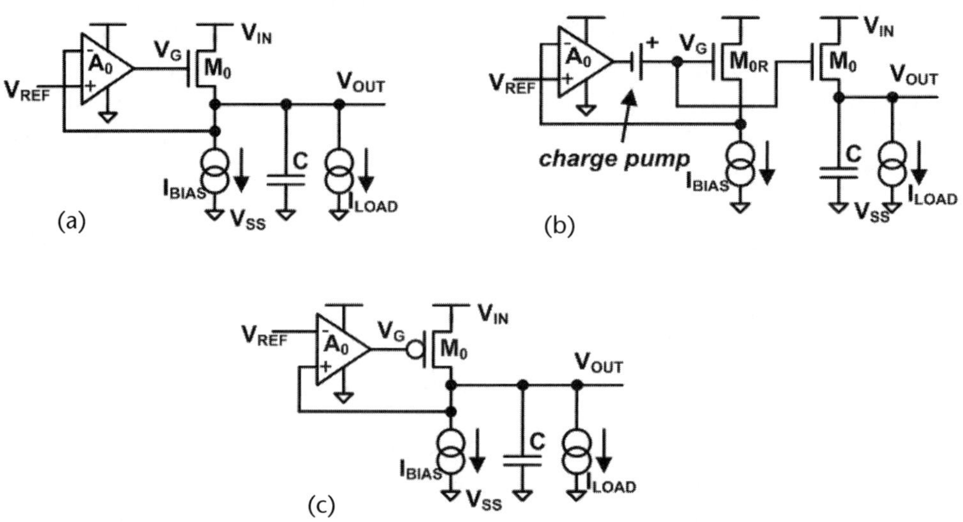

Figure 4.16 Linear regulator topologies in CMOS: (a) NMOS source follower; (b) replica-biased NMOS source follower with gate overdrive; and (c) PMOS common source topology.

voltage without a need for gate overdrive [12, 13]. Since M_0 can turn on without overdriving V_{GS}, the silicon area required for the output stage is much smaller than that for the regulator in Figure 4.16(a). This comes with a disadvantage of slow load regulation limited by the bandwidth of the amplifier feedback loop and poor power-supply noise rejection. None of the three topologies described above offers simultaneously low dropout voltage, small output droop, fast load regulation, and small silicon area. Let's look at two new topologies that fulfill these characteristics important for efficiently powering high-performance digital circuits such as CPU cores (Figures 4.17 and 4.18). The linear regulator topology in Figure 4.17 emulates a replica-biased source follower. The NMOS driver M_0 and its replica M_{0R} are replaced by unity-gain P-stage buffers PS_0 and PS_{0R} that utilize an internal PMOS driver transistor to achieve small dropout voltage and small area. Fast load regulation inherent to the source follower driver is not available in a common-source driver. Instead, load regulation is accomplished here by a fast, single-stage feedback loop within the P-stage that rapidly adjusts the gate voltage of transistor M_0 in the presence of a droop.

A PMOS output driver operating with large V_{GS} is preferable for small dropout voltage and small silicon area, but the load regulation in the circuit of Figure 4.16(c) was slow. The purpose of operational amplifier A_0 is to guarantee that V_{OUT} tracks

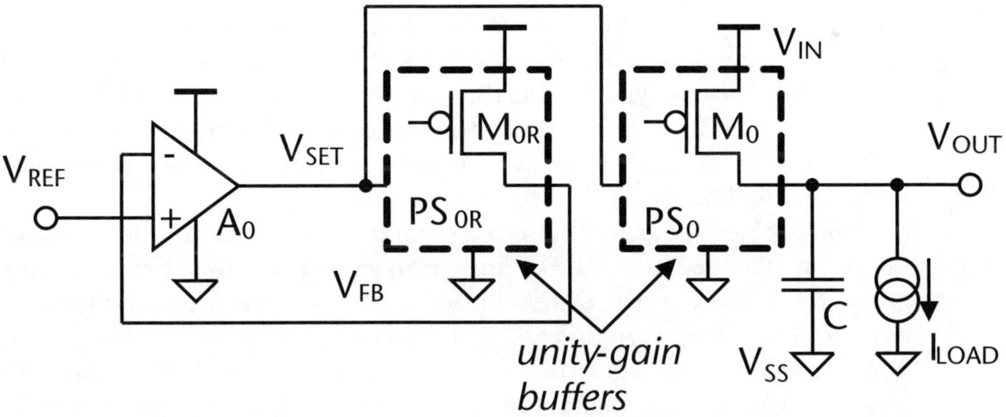

Figure 4.17 Low-dropout linear regulator topology with replica-biased common-source unity-gain buffer.

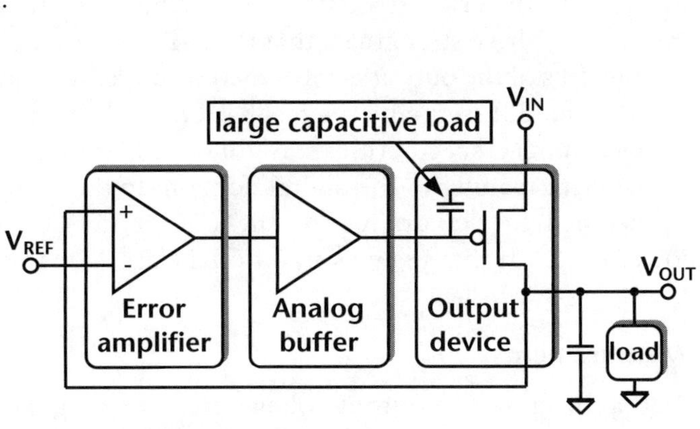

Figure 4.18 Conventional linear regulator topology.

V_{REF} across variations of process and temperature, which are slow events, and variations in V_{IN} and load current, which are fast. Tracking V_{REF} can therefore be separated into two independent problems. First, for certain nominal conditions of process, temperature, V_{IN}, and load current, the output voltage V_{OUT} should be equal to V_{REF}. Second, if any of the parameters deviates from the nominal conditions, the deviation of V_{OUT} from V_{REF} should be within specified limits (typically 10% P-P). To solve the first problem, the control loop must have a sufficient open-loop gain, typically >40 dB, but the bandwidth is not critical. The second problem requires a smaller gain, about 20 dB, but very high bandwidth is essential. Since the two problems have different requirements, they can be solved separately, as in the replica-biased source follower in Figure 4.16(b) and the linear regulator of Figure 4.17, where a slow control loop generates the gate bias voltage of M_0, while fast load regulation is accomplished by the device I-V characteristic.

Conventional linear regulator topologies can be summarized by the diagram in Figure 4.18. An error amplifier subtracts the feedback signal V_{OUT} from a reference voltage V_{REF}. The amplified error signal is buffered by an analog buffer with low output impedance, which drives the gate of the output device. The regulator output current is adjusted to meet the load current demand. This topology has several limitations. First, the same feedback loop is used for tracking V_{REF} and for responding to varying load demand. We have seen that this problem can be alleviated by using replica biasing with a fast local feedback loop for load regulation, as in Figure 4.17 [9]. Second, the response time depends on the slew rate of the analog buffer that drives the large output device. The slew rate of a class A buffer is directly proportional to its quiescent current, which limits the speed of a fast regulator with single-stage load regulation. A class AB buffer is more power efficient but tends to degrade the phase margin of the feedback loop, which leads to more aggressive compensation and lower bandwidth.

The problem of signal buffering from a small device to a large device is well known in digital circuits. The solution is to cascade multiple buffer stages that progressively increase in size. Digital buffers (e.g., inverters) are nearly perfect class AB circuits. They draw very little current when idle and provide large output current when switching. The regulator topology shown in Figure 4.19 leverages the power efficiency and speed of digital buffers. We call it a linear regulator with fast digital control [7]. The signal from the error amplifier is first translated by a simple analog/digital (A/D) converter structure into a thermometer-coded digital output. Digital buffers add drive strength so that the A/D converter can quickly turn on and off the parallel legs of the output device, thereby increasing or decreasing the current delivered to the load. Essentially the feedback goes through, and A/D/A conversion and the power-hungry, speed-critical task of driving the output device is executed in a fast and efficient manner in the digital domain. In the steady state, very little current is consumed in the driving of the output devices, which eliminates the speed-power trade-off that plagues traditional class A analog buffers.

4.5.4 Measurements

Figure 4.20(a, b) shows the output voltage of a linear regulator based on the topology of Figure 4.17 and that of a regulator based on the topology of Figure 4.19,

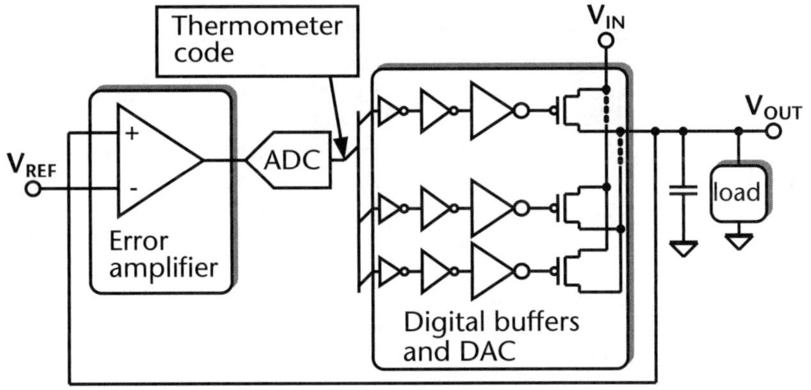

Figure 4.19 Linear regulator topology with digital control.

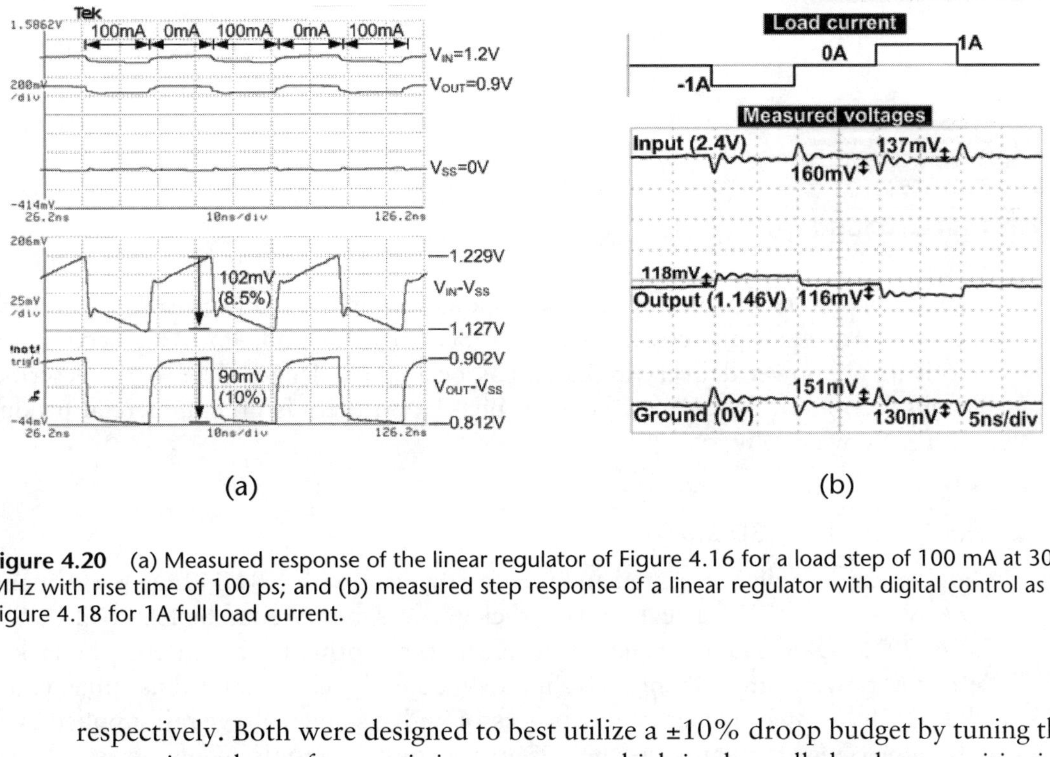

Figure 4.20 (a) Measured response of the linear regulator of Figure 4.16 for a load step of 100 mA at 30 MHz with rise time of 100 ps; and (b) measured step response of a linear regulator with digital control as in Figure 4.18 for 1A full load current.

respectively. Both were designed to best utilize a ±10% droop budget by tuning the output impedance for a resistive response, which is also called voltage positioning [14, 15]. Optimum droop response is achieved for a constant, resistive output impedance of the regulator across the full frequency range of the load current, including dc [14]. Counterintuitively, restoring the output voltage after a droop does not result in the minimum peak-to-peak variation. Voltage positioning is easily implemented in replica-biased designs by adjusting the gain of the load regulation loop so that the dc and ac droops are equal. To conclude this section, Table 4.1 compares the FOM performance of the linear regulators of Figure 4.17 [9] and Figure 4.19 [7] to previously published conventional linear regulator circuits.

Table 4.1 Figure of Merit Comparison of Linear Regulators

		[16] Figure 4.17	[7] Figure 4.19	[12] Figure 4.16(c)	[11] Figure 4.16(b)	[10] Figure 4.16(b)	[13] Figure 4.16(c)	[9] Figure 4.17
Year		2005	2006	1998	1998	2001	2003	2004
Technology		180 nm	90 nm	2 μm	0.5 μm	0.6 μm	0.6 μm	0.09 μm
Input voltage	V_{IN}	3.6V	2.4V	1.2	5	2	1.5	1.2
Output voltage	V_{OUT}	1.8V	1.2V	0.9	3.3	1.8	1.3	0.9
Output current	I_{OUT}	0.04A	1A	0.05	0.3	4.0	0.1	0.1
Output droop	ΔV_{OUT}	0.18V	0.12	0.019	0.4	0.22	0.13	0.09
Quiescent current	$I_Q[mA]$	4 mA	25.7 mA	0.23	0.75	0.2	0.038	6
Decoupling capacitance	C_{DECAP}	0.05 nF	2.4 nF	4.7 μF	0.18 nF	100 μF	10 μF	0.6 nF
Current efficiency		90.9%	97.5%	99.5%	99.8%	99.995%	99.96%	94.3%
Response time		225 ps	288 ps	1.8 μs	240 ps	8 μs	2 μs	540 ps
Figure of merit	*FOM*	22.5 ps	7.4 ps	8,200 ps	6 ps	280 ps	4,900 ps	32 ps

4.6 Power Delivery for 3D

We discussed the need for a fast-response multi-Vcc power supply in high-performance microprocessor systems in Sections 4.4 and 4.5. This section will motivate the power delivery in 3D and associated challenges. The works described in this section are in the experimental phase, and there is no commercial product with 3D power delivery.

4.6.1 Needs for 3D Stack

Intel researchers demonstrated an 80-core processor with 1.01 TFLOP performance at 62W power. [17] The peak power efficiency was measured to be 19.4 GFLOPS/W at 400 GFLOPS. The efficiency drops at high performance because of the lack of effective power management with a multi-Vcc supply with fast response time. Figure 4.21(a) shows an Intel Core Quad processor with a single voltage rail supplied by an off-chip voltage regulator. As the activity and voltage requirements across all cores are not the same, there is a significant energy loss incurred by this solution. Figure 4.21(b) illustrates a near-term solution where off-chip regulators will be delivering power to individual cores. The response time of an off-chip voltage regulator will continue to be a major bottleneck for effective power management. A regulator die attached on the same package as the processor in 2D fashion, called multichip packaging (MCP), provides a shorter response time but not enough independent Vcc rails.

Figure 4.21(c) shows the 3D power delivery solution with an scanning electron micrograph (SEM) image of a 3D IC, in which a regulator die is sandwiched

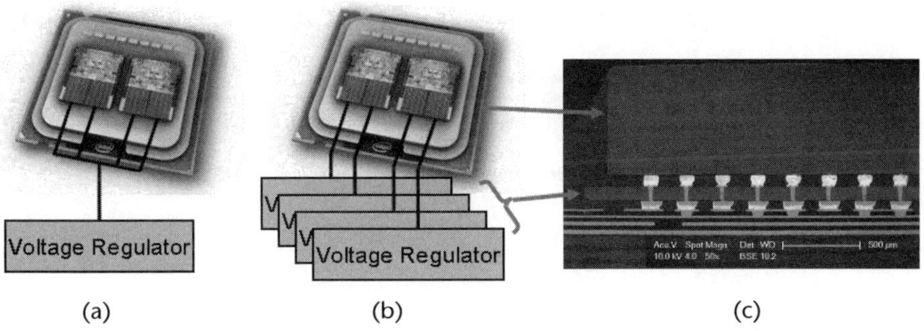

Figure 4.21 Multiple Supply Rail Power Delivery: (a) single Vcc; (b) multiple Vcc; and (c) 3D-stacked multiple Vcc.

between the processor and its package. Figure 4.22 describes a schematic of a 3D IC with two voltage regulators only for simplicity. It can be observed that the die-to-die connectivity is very dense with a direct bump attachment. It provides the possibility of many independent voltage supplies. The regulator die can be processed on a heterogeneous technology with high-voltage devices and low-loss interconnects. The I/O signals and Vss are fed to the top die through the sandwiched regulator die using a process technology called through-wafer-vias (TWVs) or through-silicon vias (TSVs). Control signals of the regulator are only supplied to the bottom die using TSVs. High efficiency requirements of the regulator impose a very low resistance specification on TSVs. L_1 and L_2 are the inductors for dc-dc conversion. As shown in Figure 4.23, there is significant power loss inside the die current traces when the current has to flow multiple times vertically through vias and metal layers in the 2D power delivery solution. The unidirectional current flow in 3D ICs alleviates this

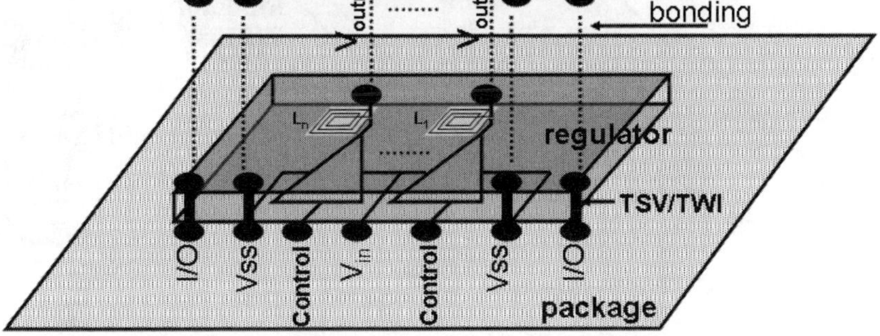

Figure 4.22 Multiple Vcc 3D power delivery.

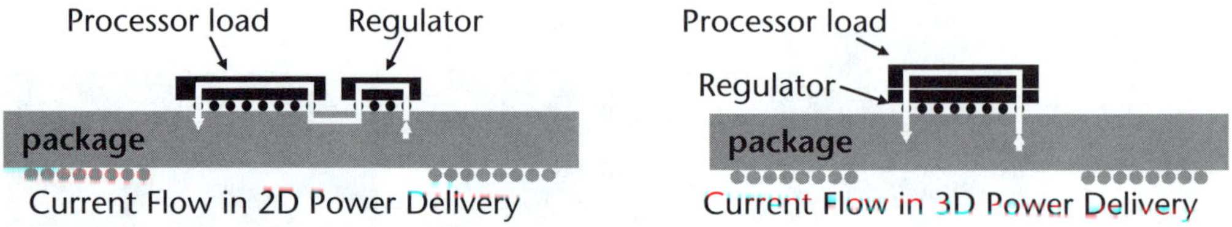

Figure 4.23 Current delivery.

problem. Additionally, the processor on top maintains the same heat-removal solution as a processor without a 3D regulator. Optimal 3D via allocation for power delivery and thermal dissipation was presented in [18].

4.6.2 3D-Stacked DC-DC Converter and Passives

Figure 4.24 shows various 3D ICs designed for 3D power delivery. Figure 4.24(a) shows a 3D solution without the need for TSVs [19]. The two dice are bonded by direct bump attachment, and the bottom die receives external power and I/Os through wire bonds. This solution saves TSV processing cost but is applicable to low-power products. The researchers proposed to have active circuits on the bottom die and passive L and C components on the top die. The same researchers proposed an alternative glass substrate L solution in the same paper as described in Figure 4.24(b). In [20], the researchers implemented a flip-chip bonded logic IC with multiple 3D-stacked, wire-bonded memory chips. The work can be extended to power delivery as drawn in Figure 4.24(c). The real 3D regulator implementation with

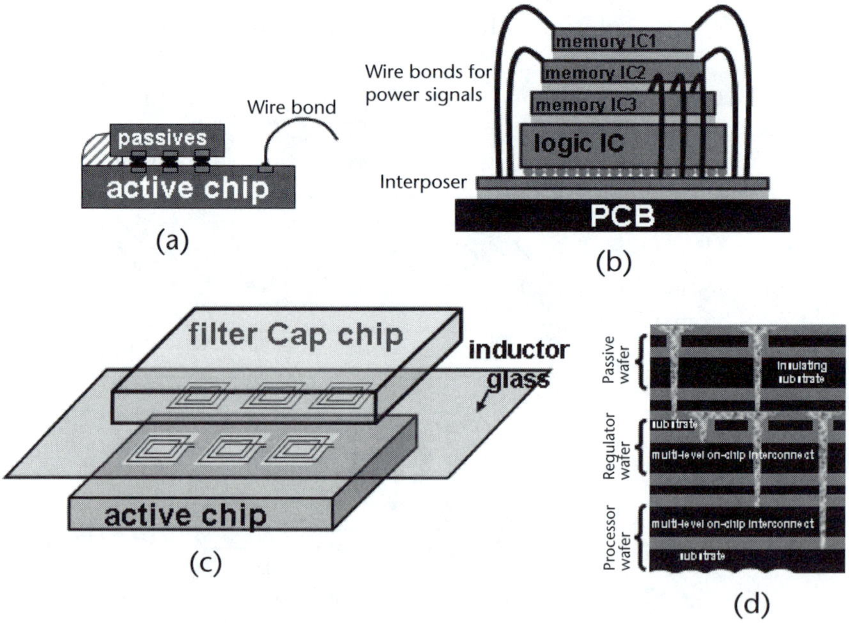

Figure 4.24 (a–d) 3D stacking approaches.

TSVs was demonstrated in [21] and is described in Figure 4.24(d). This implementation includes a logic chip at the bottom, a regulation circuits chip in the middle, and a chip with passive components on top. The current delivery was not unidirectional as described in Figure 4.23, and the efficiency was only 64% for an 800 mA solution; however, according to our knowledge, this is the only measured implementation of a 3D voltage regulator. If the logic chip is close to the package and away from the heat sink, the solution will not be feasible for high-power microprocessors.

4.7 Conclusion

We have presented the power delivery challenges in current microprocessor systems and the current and power scaling trends. High transient currents will be required to be supplied with minimal voltage fluctuations to multiple cores running on multiple independent power supplies on a single microprocessor die. The future power delivery trends are very demanding, and near-load dc-dc converters will become a necessity to support power-hungry microprocessor platforms. We have discussed the modeling and circuits for high-speed switching and linear regulators and provided an overview of 3D ICs attempting to solve the future power delivery problems.

References

[1] Muhtaroglu, A., G. Taylor, and T., Rahal-Arabi, "On-Die Droop Detector for Analog Sensing of Power Supply Noise," *IEEE J. Solid-State Circuits*, Vol. 39, No. 4, April 2004, pp. 651–660.

[2] Aygun, K., et al., "Power Delivery for High-Performance Microprocessors," *Intel Technology Journal*, Vol. 9, No. 4, November 2005, pp. 273–284.

[3] Wong, K., et al., "Enhancing Microprocessor Immunity to Power Supply Noise with Clock-Data Compensation," *IEEE J. Solid-State Circuits*, Vol. 41, No. 4, April 2006, pp. 749–758.

[4] Schrom, G., et al., "Feasibility of Monolithic and 3D-Stacked DC-DC Converters for Microprocessors in 90 nm Technology Generation," *Proc. International Symposium on Low Power Electronic Design*, Newport, CA, August 2004, pp. 263–268.

[5] Schrom, G., et al., "Optimal Design of Monolithic Integrated DC-DC Converters," *Proc. International Conference on Integrated Circuit Design and Technology*, Padova, Italy, June 2006, pp. 65–67.

[6] Schrom, G., et al., "A 100 MHz Eight-Phase Buck Converter Delivering 12A in 25 mm^2 Using Air-Core Inductors," *IEEE Applied Power Electronics Conference*, Anaheim, CA, March 2007, pp. 727–730.

[7] Hazucha, P., et al., "High Voltage Tolerant Linear Regulator with Fast Digital Control for Biasing of Integrated DC-DC Converters," *IEEE J. Solid-State Circuits*, Vol. 42, January 2007, pp. 66–73.

[8] Hazucha, P., et al., "A 233-MHz 80%–87% Efficient Four-Phase DC-DC Converter Utilizing Air-Core Inductors on Package," *IEEE J. Solid-State Circuits*, Vol. 40, No. 4, April 2005, pp. 838–845.

[9] Hazucha, P., et al., "Area-Efficient Linear Regulator with Ultra-Fast Load Regulation," *IEEE J. Solid-State Circuits*, Vol. 40, April 2005, pp. 933–940.

[10] Bontempo, G., T. Signorelli, and F. Pulvirenti, "Low Supply Voltage, Low Quiescent Current, ULDO Linear Regulator," *IEEE International Conf. on Electronics, Circuits and Systems*, Malta, September 2001, pp. 409–412.

[11] Den Besten, G. W., and B. Nauta, "Embedded 5V-to-3.3V Voltage Regulator for Supplying Digital IC's in 3.3V Technology," *IEEE Journal of. Solid-State Circuits*, Vol. 33, July 1998, pp. 956–962.

[12] Rincon-Mora, G. A., and P. A. Allen, "A Low-Voltage, Low Quiescent Current, Low Drop-Out Regulator," *IEEE J. Solid-State Circuits*, Vol. 33, January 1998, pp. 36–44.

[13] Leung, K. N., and P. K. T. Mok, "A Capacitor-Free CMOS Low-Dropout Regulator with Damping-Factor-Control Frequency Compensation," *IEEE J. Solid-State Circuits*, Vol. 38, October 2003, pp. 1691–1702.

[14] Redl, R., B. P. Erisman, and Z. Zansky, "Optimizing the Load Transient Response of the Buck Converter," *IEEE Applied Power Electronics Conference and Exposition*, Anaheim, CA, February 1998, pp. 170–176.

[15] Waizman, A., and C. Y. Chung, "Resonant Free Power Network Design Using Extended Adaptive Voltage Positioning (EAVP) Methodology," *IEEE Transactions on Advanced Packaging*, Vol. 24, August 2001, pp. 236–244.

[16] Rajapandian, S., et al., "High-Tension Power Delivery—Operating 0.18 µm CMOS Digital Logic at 5.4V," *IEEE International Solid-State Circuits Conference*, San Francisco, CA, February 2005, pp. 298–299.

[17] Vangal, S., et al., "An 80-Tile 1.28 TFLOPS Network-on-Chip in 65 nm CMOS," *Proc. International Solid-State Circuits Conference,* San Francisco, CA, February 11–15, 2007, pp. 98–99.

[18] Hao, Y., J. Ho, and L. He, "Simultaneous Power and Thermal Integrity Driven via Stapling in 3D ICs," *Proc. International Conference on Computer-Aided Design,* San Jose, CA, November 5–9, 2006, pp. 802–808.

[19] Lee, H., et al., "Power Delivery Network Design for 3D SIP Integrated over Silicon Interposer Platform," *Proc. Electronic Components and Technology Conference,* Reno, NV, May 2007, pp. 1193–1198.

[20] Sun, J., et al., "3D Power Delivery for Microprocessors and High-Performance ASICs," *Proc. Applied Power Electronics Conference*, Anaheim, CA, 2007, pp. 127–133.

[21] Onizuka, K., et al., "Stacked-Chip Implementation of On-Chip Buck Converter for Power-Aware Distributed Power Supply Systems," *Proc. Asian Solid-State Circuits Conference,* Hangzhou, China, November, 2006, pp. 127–130.

CHAPTER 5

On-Chip Power Supply Noise Modeling for Gigascale 2D and 3D Systems

Gang Huang, Kaveh Shakeri, Azad Naeemi, Muhannad S. Bakir, and James D. Meindl

5.1 Introduction: Overview of the Power Delivery System

As presented in Chapter 4, power dissipation has historically been increasing in high-performance chips. The supply voltage and supply current for Intel microprocessors are shown in Figures 4.1 and 4.2. Over the past 20 years, the supply voltage has decreased fivefold, and the supply current has increased fiftyfold. The net result of these trends is that the supply current that flows through the power distribution network has been increasing. Unfortunately, the increase in supply current causes an increase in the on-chip power-supply noise. Moreover, since the chip supply voltage is decreasing, the logic on the integrated circuit (IC) becomes more sensitive to any voltage change (noise) on the supply voltage [1]. The higher supply noise and the higher sensitivity of the circuits to it have made the design of the power distribution network a very important and challenging task.

There are two main components of power-supply noise: IR-drop and ΔI noise. The former, IR-drop, results from the voltage drop due to the supply current's passing through the parasitic resistance of the power distribution network. The latter, ΔI noise, is caused by the change in supply current passing through inductance of the power delivery network, and it becomes important when a group of circuits switch simultaneously. ΔI noise consists of three distinct voltage droops [2], and they result from the interaction between the chip, package, and board. The three droops are illustrated in Figure 5.1 [2]. The third droop is related to the bulk capacitors on the board level and has a time duration of a few microseconds. The third droop influences all critical paths but can be readily minimized by using more board space for bulk capacitors. The second droop is caused by the resonance between the inductive traces on the motherboard and package decoupling capacitors (decap). The second droop has a time duration of a few hundred nanoseconds and impacts a significant number of critical paths. The first droop is caused by the package inductance and on-die capacitance with a resonance frequency in the range of tens of megahertz to several hundred megahertz (related to package-level component sizes and on-chip decap). Among the three droops, the first droop has the smallest duration (tens of nanoseconds) but the largest magnitude. Chip performance can be severely degraded when the first droop interacts with some critical paths. Because of its

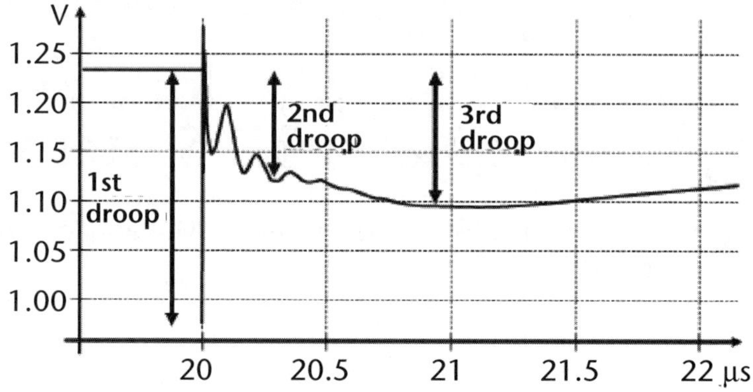

Figure 5.1 Simulated voltage droops [2].

severe impact on high-performance chips, the first droop is the main focus of this chapter.

Excessive power-supply noise can lead to severe degradation of chip performance and even logic failures. Thus, it is important to model and predict the performance of the power delivery network with the objective of minimizing supply noise. Different decisions need to be made in order to design an optimum system that includes the type of package, the power distribution network, and the size and number of the decoupling capacitors used in the power distribution network. To design an optimum power distribution system we need to understand the interaction between the components of the power distribution network. If problems associated with the design and implementation of a power distribution network are undetected early in the design cycle, they can become very costly to fix later. An overdesigned power distribution system would result in an expensive package and waste of the silicon and interconnect resources. An underdesigned system (if even an option) can lead to noise problems and difficulties regarding wire routing. Power-supply noise has traditionally been analyzed by extracting parasitic resistance and inductance with software tools and later simulating netlists with circuit simulators, such as SPICE. However, package and chip power delivery network models can be very large, and manipulating large networks by simulation is time-consuming and prolongs design cycles. As a result, compact and accurate physical models are needed for IR-drop and ΔI noise of the power distribution network. Such models would be critical in the early stages of design and can estimate the on-chip and off-chip resources needed for the power distribution network.

The outline of this chapter is as follows. Section 5.2 presents an overview of power distribution networks. Compact physical models for IR-drop are derived in Section 5.3. In Section 5.4, blockwise compact physical models are derived for the first droop ΔI noise for power-hungry blocks assuming uniform switching conditions. Analytical models are then introduced in Section 5.5 to extend blockwise models to general nonuniform switching conditions such as the hot spot case. To help identify challenges brought by 3D integration, models derived in Section 5.6 are also adapted in Section 5.4 to consider 3D chip stack. Finally, conclusions and future work are discussed in Section 5.7.

5.2 On-Chip Power Distribution Network

A microprocessor power distribution network (on-chip) typically employs a significant number of routing tracks that incorporate a large number of interconnects. The initial design and layout of the power distribution network must be done early in the design process and then gradually refined [3].

On-chip power distribution networks consist of global and local networks. Global power distribution networks carry the supply current and distribute power across the chip. Local networks deliver the supply current from global networks to the active devices. Global networks contribute most of the parasitics and are thus the main concern of this chapter. There are different methods for distributing power on the global wiring levels of a high-performance chip. The most common is to use a grid made of orthogonal interconnects routed on separate metal levels connected through vias [3]. Another method is to dedicate a whole metal level to power and another level to ground. This results in small on-chip power distribution parasitics and thus small voltage drop. This technique is relatively expensive and has been reported only in the Alpha 21264 microprocessor [4]. This chapter will focus mainly on grids.

Wire bond and flip-chip are the most common types of first-level chip interconnect. Wire bond is cheaper than flip-chip interconnections; however, the wire-bond interconnections cause a higher power-supply noise level in the power distribution network due to higher parasitics. In flip-chip technology, the parasitics are reduced by spreading the die pads along the surface of the chip and therefore reducing the noise. The development of gigascale integration (GSI) systems is not only driven by more efficient silicon real estate usage but also by more I/O pin counts. Hence, most of today's high-performance designs utilize flip-chip technology to provide the larger Input/Output bandwidth required. As such, the main focus of this chapter is supply noise when flip-chip technology is used.

Figure 5.2 illustrates power/ground (P/G) pads for a chip with a flip-chip package. The package supplies current through the power pads. The supplied current flows through power interconnects to the on-chip circuits, and then returns to the package through ground interconnects and ground pads.

5.3 Compact Physical Modeling of the IR-Drop

This section introduces compact physical IR-drop models for the flip-chip interconnects. These models are general and can be used for many kinds of chips and packages.

5.3.1 Partial Differential Equation for the IR-Drop of a Power Distribution Grid

In high-performance chips, the horizontal and vertical segments of the on-chip P/G grid are routed on different metal levels and connected through vias at the crossing points. The metal levels making the grid might have different thicknesses resulting

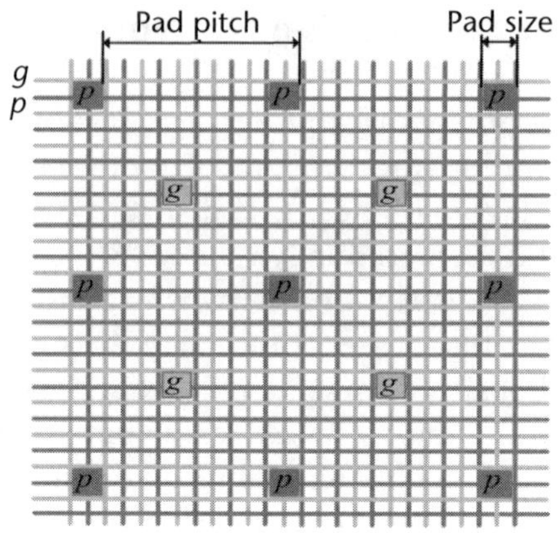

Figure 5.2 On-chip P/G grids and I/O pads in flip-chip packages.

in an *anisotropic* grid with different resistances in the x- and y-directions, as shown in Figure 5.3.

In Figure 5.3, each node of the power distribution grid is connected to the four neighboring nodes. A current source is placed at each node equal to the amount of current distributed to the chip by that node. Symbol J_0 is current per unit area distributed to the circuits by the grid, and R_{sx} and R_{sy} are the segment resistances in the x- and y-directions, respectively. The voltage drop due to via resistance is negligible as there are thousands of parallel vias in a grid. Therefore, it is neglected in these models. The IR-drop of a node on the grid is the voltage difference between that node and the voltage at the P/G I/O pad.

The double-grid structure in Figure 5.3 can be decoupled into two single-grid structures, as shown in Figure 5.4. For a single power grid (symmetrical for ground grid), the IR-drop of a point on the grid can be calculated from the IR-drop of the four neighboring points by using Kirchoff's current laws [5]:

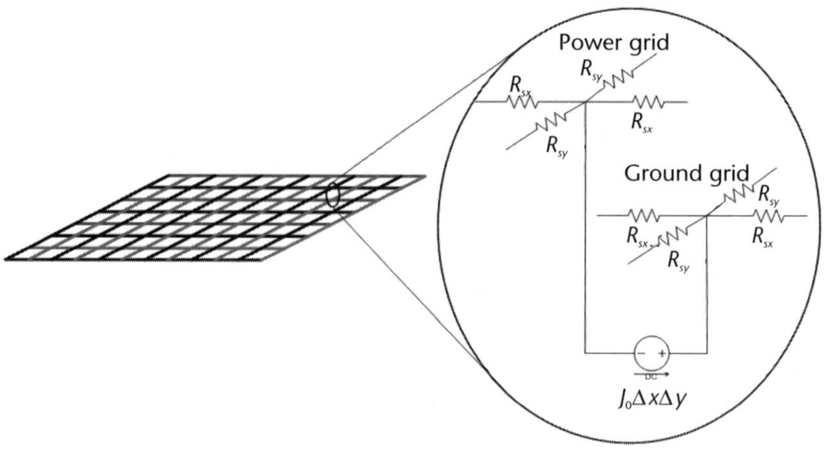

Figure 5.3 On-chip power distribution grid for IR-drop modeling.

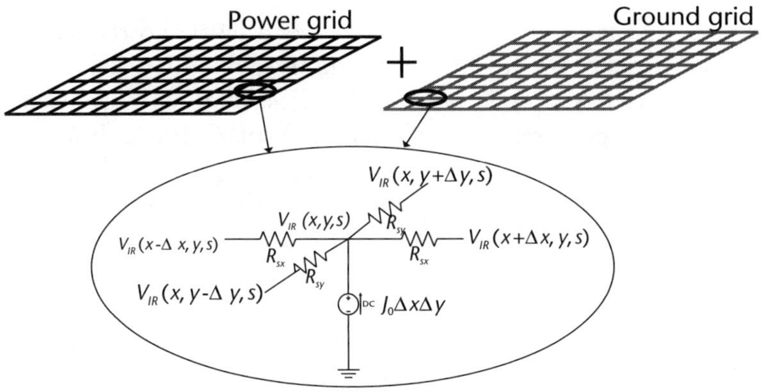

Figure 5.4 Double grid structure is decoupled into two single grids.

$$\frac{V_{IR}(x,y)-V_{IR}(x+\Delta x,y)}{R_{sx}\dfrac{\Delta x}{\Delta y}}+\frac{V_{IR}(x,y)-V_{IR}(x,y+\Delta y)}{R_{sy}\dfrac{\Delta y}{\Delta x}}+\frac{V_{IR}(x,y)-V_{IR}(x-\Delta x,y)}{R_{sx}\dfrac{\Delta x}{\Delta y}}$$
$$+\frac{V_{IR}(x,y)-V_{IR}(x,y-\Delta y)}{R_{sy}\dfrac{\Delta y}{\Delta x}}=-J_0\cdot\Delta x\cdot\Delta y \tag{5.1}$$

where $V_{IR}(x, y)$ is the IR-drop of a point at (x, y) on the P/G grid, J_0 is current per unit area distributed to the circuits by the grid ,and R_{sx} and R_{sy} are the segment resistances in x- and y- directions, respectively. The number of segments of the grid is usually large; therefore, the power distribution grid can be modeled as a continuous planar surface that distributes power across the chip. Replacing each of the neighboring voltages in (5.1) with their Taylor series and assuming that the grid is a continuous planar surface, (5.1) can be simplified as

$$\frac{1}{R_{sx}}\frac{\partial^2 V_{IR}(x,y)}{\partial x^2}+\frac{1}{R_{sy}}\frac{\partial^2 V_{IR}(x,y)}{\partial y^2}=J_0 \tag{5.2}$$

The IR-drop for different packages can be calculated by applying the appropriate boundary conditions to (5.2). For an *isotropic* grid, where the resistance in the x- and y-directions is equal ($R_{sx} = R_{sy} = R_s$), (5.2) can be simplified as

$$\nabla^2 V_{IR}(x,y)=R_s J_0 \tag{5.3}$$

which is the Poisson's equation.

In the following section, the boundary conditions defined by flip-chip interconnects is derived and then used to solve (5.3) to model the IR-drop.

5.3.2 IR-Drop of Isotropic Grid Flip-Chip Interconnects

In flip-chip technology, the package I/O pads are interconnected to the die I/O pads through metal bumps distributed across the chip surface. These bumps are connected to I/O pads located at the top metal level. To reduce the voltage drop in a high-performance processor, a large number of pads are allocated to the power distribution network; almost two-thirds of the total pads are used for power distribution [6]. These power and ground pads are spread throughout the surface of the chip to reduce voltage drop and loop inductance.

The chip is composed of macrocells, such as an ALU, clock circuits, and cache. The power grid IR-drop is calculated for each macrocell. The current density within each macrocell is assumed to be uniform (later in the chapter, hot spots are accounted for). Each macrocell is made of multiple cells, where each cell is defined as the area surrounded by four neighboring pads, as shown in Figure 5.5. Because of the uniform current density of each macrocell, the cells within that macrocell will have the same IR-drop. Hence, the partial differential equation needs to be solved only for one cell. Solving the partial differential equation with the boundary condition shown in Figure 5.6 results in the voltage drop [5] shown in Figure 5.7. The maximum IR-drop can be calculated from

$$V_{IR\max} = \frac{R_s I_{pad}}{2\pi} \ln\left(\frac{0.387a}{\alpha \cdot D_{pad}}\right) \tag{5.4}$$

Equation (5.4) gives the maximum IR-drop on the P/G grid as a function of on-chip and package parameters. The coefficient α is the pad shape parameter shown in Table 5.1. SPICE simulations show that (5.4) has less than a 5% error.

The IR-drop for a more general case with an anisotropic power distribution network and rectangular cells can be calculated using the same method:

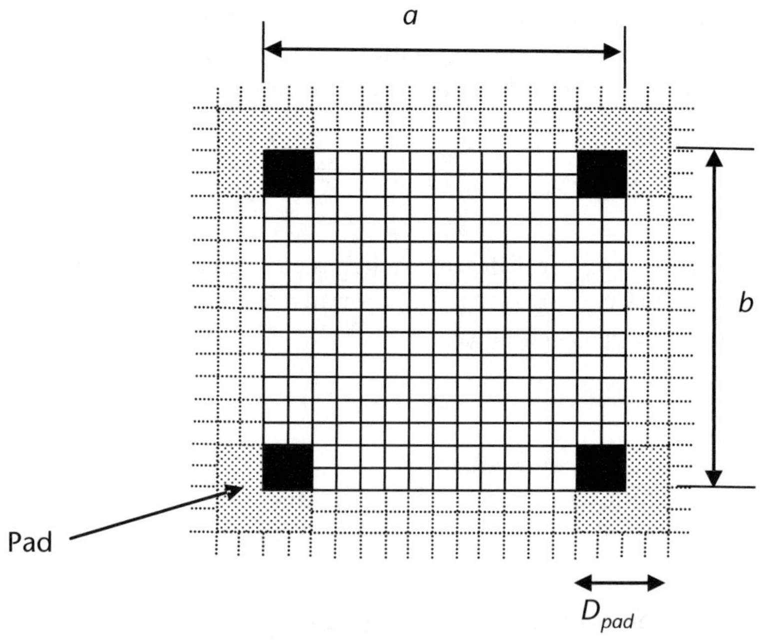

Figure 5.5 Grid between four neighboring pads. This area is called a cell.

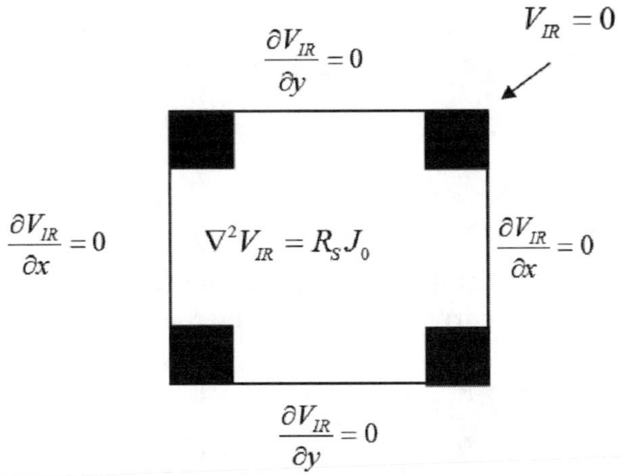

Figure 5.6 Boundary conditions for a cell. IR-drop is zero at each pad.

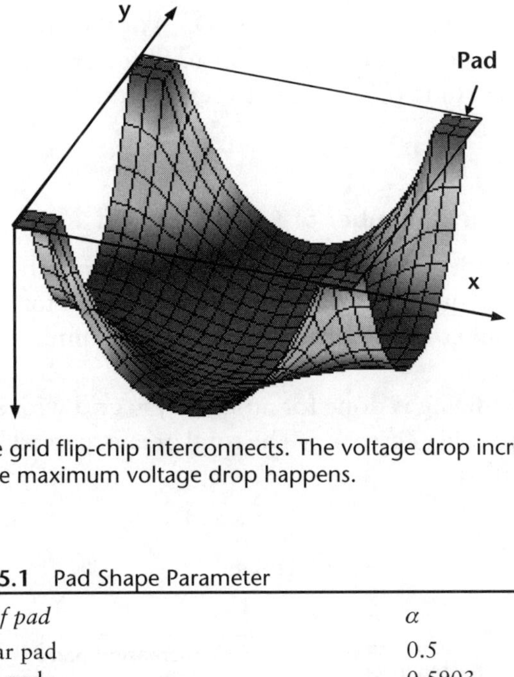

Figure 5.7 IR-drop on the grid flip-chip interconnects. The voltage drop increases toward the center of the cell, where the maximum voltage drop happens.

Table 5.1 Pad Shape Parameter

Kind of pad	α
Circular pad	0.5
Square pad	0.5903
Pad connected to a single node in the grid	0.2

$$V_{IR\max} = \frac{\rho I_{Total}}{2\pi n_{pg}} \sqrt{\frac{l_{segx} l_{segy}}{T_x T_y W_x W_y}} \ln\left(\frac{0.387\left(b\sqrt{T_x W_x l_{segx}} + a\sqrt{T_y W_y l_{segy}}\right)}{\alpha \cdot D_{pad}\left(\sqrt{T_x W_x l_{segx}} + \sqrt{T_y W_y l_{segy}}\right)} \right) \tag{5.5}$$

where I_{Total} is the total macrocell current, n_{pg} is the total number of P/G pads in a macrocell, W is the wire widths, T is thickness, and l_{seg} is length. Subscripts x and y represent the directions of the wires. This equation quantifies the trade-off between the power grid parameters (l_{segx}, l_{segy}, T_x, T_y, W_x, W_y) and pad parameters such as pad shape and size (α, D_{pad}), number of pads (n_{pg}), and the distance between pads (a, b).

5.3.3 Trade-Off Between the Number of Pads and Area Percentage of Top Metal Layers Used for Power Distribution

The IR-drop for flip-chip technology given in (5.5) can be rewritten as a function of the total macrocell current distributed by the pads:

$$V_{IR\max} = \frac{\sqrt{R_{segx}R_{segy}}\,I_{total}}{2\pi n_{pg}} \ln\left(\frac{0.387\left(a\cdot l_{segy}\sqrt{R_{segx}}+b\cdot l_{segx}\sqrt{R_{segy}}\right)}{\alpha\cdot D_{pad}\left(l_{segy}\sqrt{R_{segx}}+l_{segx}\sqrt{R_{segy}}\right)}\right) \quad (5.6)$$

The model shows the trade-off between the on-chip power distribution network (described by R_{segx}, R_{segy}, l_{segx}, l_{segy}) and package parameters (described by D_{pad}, n_{pg}, and α). Figure 5.8 shows this trade-off for a microprocessor in 2018. As shown in the figure, increasing the pad size (D_{pad}) or the number of pads (n_{pg}) reduces the percentage area of the top metal layer used for power distribution.

Another trade-off exists between the number of pads and is studied in the following section.

5.3.4 Size and Number of Pads Trade-Off

There is a trade-off between pad size and the number of pads. This section addresses the following question: Assuming a certain area for the total pads, is it better to have a large number of small pads or a small number of large pads to minimize the IR-drop?

The modeling is done for an isotropic grid with square pads; however, it can be extended to other cases too. The total area occupied by all pads, which is assumed to be constant, is

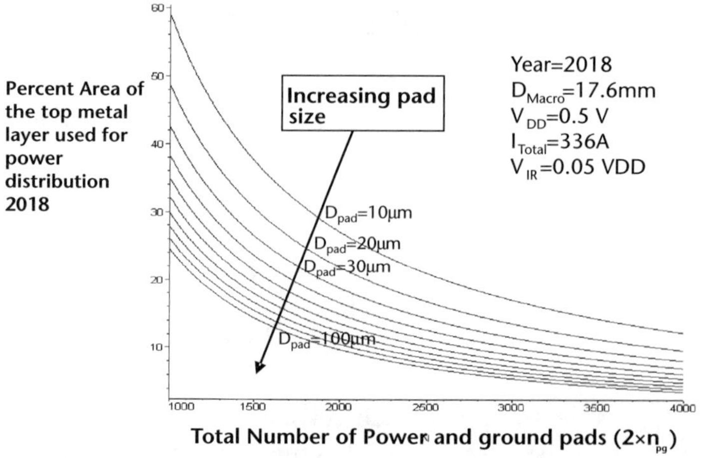

Figure 5.8 Trade-off between the number of pads and area percentage of top metal layers used for power distribution for different pad sizes.

$$A_{Tpad} = n_{pg} D_{pad}^2 \tag{5.7}$$

Assuming a constant total pad area, (5.5) can be rewritten as

$$V_{IR\max} = \frac{\rho l_{seg} I_{Total} D_{pad}^2}{2\pi WTA_{Tpad}} \ln\left(\frac{0.387 \cdot D_{Macro}}{\alpha\sqrt{A_{Tpad}}}\right) \tag{5.8}$$

All of the parameters in this equation are constant except for D_{pad}. This equation suggests reducing D_{pad} in order to reduce V_{IRmax}. In other words, the IR-drop is minimized by using a large number of small pads instead of a small number of large pads.

5.3.5 Optimum Placement of the Power and Ground Pads for an Anisotropic Grid for Minimum IR-Drop

Placement of the power and ground pads is important in reducing the IR-drop. This section derives the optimum placement of the power and ground pads, assuming a certain number of power and ground pads. The total number of pads dedicated to power and ground for a macrocell is assumed to be constant, resulting in a constant cell area.

$$a \cdot b = A_{Cell} \tag{5.9}$$

where a and b are the size of the cell in the x- and y-directions, and A_{Cell} is the cell area, which is constant. We want to find a and b so that the IR-drop is minimized for an anisotropic grid. In this case, the only variables are a and b; therefore, the maximum IR-drop can be written as

$$V_{IR\max} = K_1 \ln\left(K_2\left(a\sqrt{R_{sx}} + b\sqrt{R_{sy}}\right)\right) \tag{5.10}$$

where K_1 and K_2 are two constants. The minimum IR-drop happens when we have

$$a = b\sqrt{\frac{R_{sy}}{R_{sx}}} \tag{5.11}$$

This result suggests making a rectangular cell to minimize the IR-drop in an anisotropic grid. For an isotropic grid ($R_{sx} = R_{sy}$), the cells should be square to minimize the IR-drop.

5.4 Blockwise Compact Physical Models for ΔI Noise

This section derives a set of blockwise compact physical models for the first droop supply noise [7]. These models can be applied to a functional block with a large number of power and ground pads and give a quick snapshot of the power-supply noise for power-hungry blocks. These models can accurately capture the impact of package parameters as well as the distributed nature of on-chip power grid and decoupling capacitance.

5.4.1 Partial Differential Equation for Power Distribution Networks

As with the assumption made in Section 5.3 for a functional block with a large number of power and ground pads, the block can be divided into cells. Each of these cells is essentially a quarter of the cell used in the IR-drop model presented in previous section, and a cell is the identical square region between a pair of adjacent quarter power and ground pads. It can be assumed that no current passes normal to the cell borders. One cell is thus enough for the power-supply noise analysis, as shown in Figure 5.9.

The simplified circuit model of the power distribution network associated with a cell is shown in Figure 5.10. In this section, only the isotopic grids are considered. The segment resistance of the grid is still represented by R_s. Switching current density between a power grid node and the adjacent ground grid node is modeled as a current source $J(s)$ and represents the switching current density in the Laplace domain (to enable analysis of the current in a wide frequency range). The symbol C_d

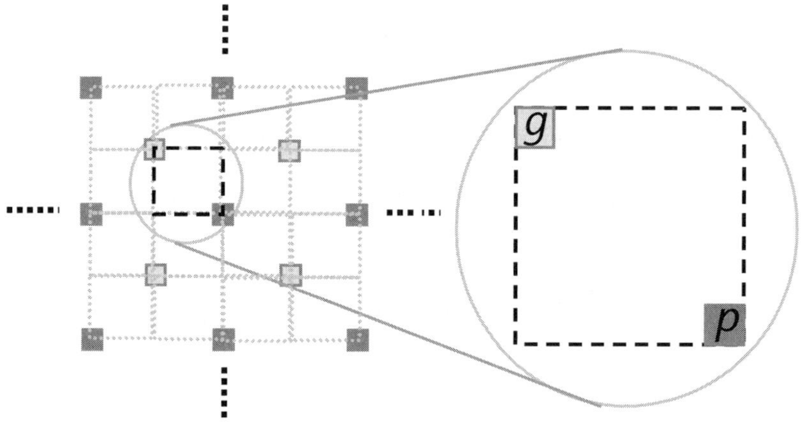

Figure 5.9 Division of power grid into independent cells.

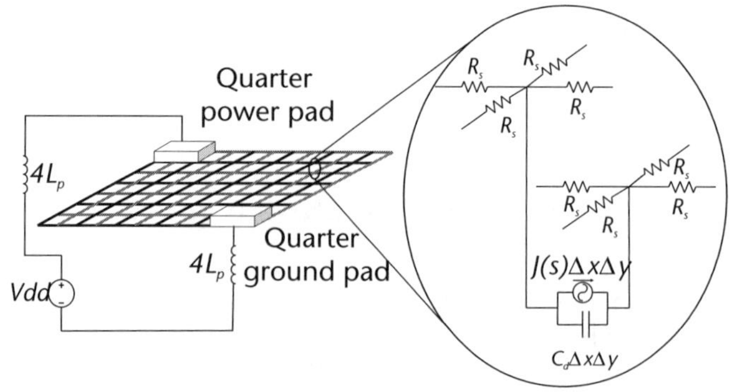

Figure 5.10 Simplified circuit model for ΔI noise.

denotes the decoupling capacitance (including both the intentionally added decoupling capacitors and the equivalent capacitance of the nonswitching transistors) per unit area. Finally, the symbol L_p represents the per-pad loop inductance of the package.

Because the on-chip inductive coupling between the power grid and ground grid is neglected, the double-grid structure can be decoupled into two individual grids, as shown in Figure 5.11. Assuming that the Laplace domain voltage of a given point (x, y) in a grid is $V(x, y, s)$, the voltage of this point can be calculated from the following partial differential equation using a method similar to the derivation for (5.3):

$$\nabla^2 V(x,y,s) = R_s J(s) + 2V(x,y,s)\cdot sR_sC_d + \Phi(x,y,s) \tag{5.12}$$

Equation (5.12) combines a Poisson's equation and a Helmholtz equation. $\Phi(x, y, s)$ is the source function of this differential equation and is added to take care of the voltage drop on L_p. As there is no current flowing through the cell boundaries, (5.12) should satisfy the following boundary conditions [7]:

$$\frac{\partial V(x,y,s)}{\partial y}\Big|_{x=0} = 0,\ \frac{\partial V(x,y,s)}{\partial y}\Big|_{x=a} = 0,\ \frac{\partial V(x,y,s)}{\partial x}\Big|_{y=0} = 0,\ \frac{\partial V(x,y,s)}{\partial x}\Big|_{y=a} = 0 \tag{5.13}$$

where a denotes the size of the square cell.

Equation (5.12) can be transformed into a pure Hemholtz equation and solved analytically by putting in the boundary condition of the second kind described by (5.13). The solution of $V(x, y, s)$ is

$$V(x,y,s) = -\frac{s\cdot\dfrac{J(s)}{2C_d} - \dfrac{J(s)\cdot R_s}{2C_d\cdot 4L_p}\left[G(x,y,0,0,s) - G(\alpha D_{pad},0,0,0,s)\right]}{s^2 + s\cdot\dfrac{R_s}{4L_p}G(\alpha D_{pad},0,0,0,s)} \tag{5.14}$$

where D_{pad} denotes the edge length of the quarter square pad, α is the pad shape parameter shown in Table 5.1, and $G(x, y, \xi, \eta, s)$ is the Green's function of a Helmholtz equation with the boundary condition for the second kind [8].

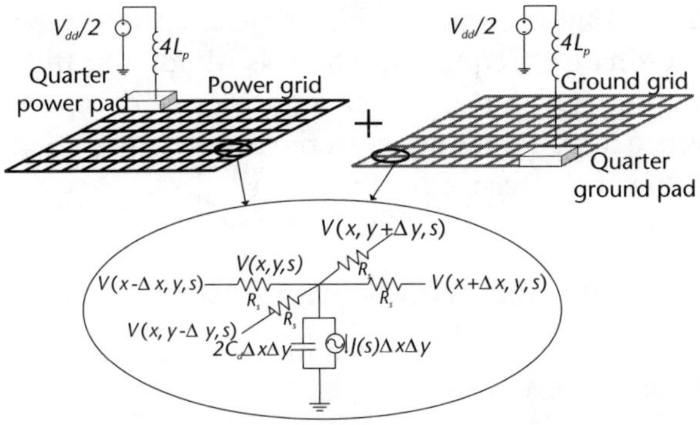

Figure 5.11 Differential model of a node for single P/G grid.

$V(x, y, s)$ can determine the frequency characteristics of the power noise at any location within a cell. By dividing $V(x, y, s)$ by the total switching current within a cell $J(s)a^2$, the transfer impedance of the power distribution network $Z(x, y, s)$ can be obtained:

$$Z(x,y,s) = -\frac{s\cdot\frac{1}{2C_d a^2} - \frac{R_s}{2C_d a^2\cdot 4L_p}\left[G(x,y,0,0,s) - G(\alpha D_{pad},0,0,0,s)\right]}{s^2 + s\cdot\frac{R_s}{4L_p}G(\alpha D_{pad},0,0,0,s)} \tag{5.15}$$

As the current source term is eliminated, $Z(x, y, s)$ incorporates the intrinsic impedance of a power distribution network.

Equation (5.15) can be simplified into a second-order transfer impedance function $Z_s(x, y, s)$:

$$Z_s(x,y,s) = -\frac{s\cdot\frac{1}{2C_d a^2} + \frac{z(x,y,0)}{2C_d a^2\cdot 4L_p}}{s^2 + s\cdot k\cdot\frac{Z(x,y,0)}{4L_p} + \frac{1}{2C_d a^2\cdot 4L_p}} \tag{5.16}$$

where

$$Z_s(x,y,0) = Z(x,y,0) = R_{IR}(x,y,0) \tag{5.17}$$

and

$$k = \frac{\sqrt{\left(\frac{4L_p}{2C_d a^2}\right)^2\left(\frac{1}{Z(x,y,0)}\right)^2 + \left(\frac{4L_p}{2C_d a^2}\right)}}{\left|Z(x,y,j2\pi f_{rf})\right|} \tag{5.18}$$

A comparison between (5.15), (5.16), and the results of SPICE simulation is performed to validate the proposed model and is shown in Figure 5.12. Good agreement can be observed from the figure.

Three corner points are selected in Figure 5.12(a), and it is noted from Figure 5.12(b) that the transfer impedance has a low-pass characteristic with only one peak resonance frequency. There is almost no difference between (5.15) and (5.16), and they both have less than a 4% error compared to the SPICE simulation. The difference in dc values between the three corner points results from the different IR-drop value at the respective locations.

5.4.2 Analytical Solution for Noise Transients

The current waveform induced by a function block is approximated by a ramp function, as shown in (5.19)

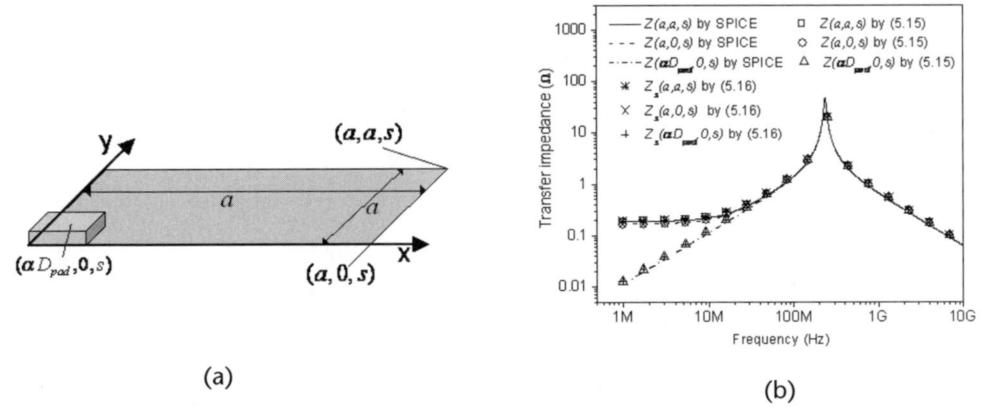

Figure 5.12 (a) Three corner points, and (b) transfer impedance comparison for three corner points between (5.15), (5.16), and SPICE simulation.

$$i(t) = \frac{I_p}{t_r}\left[t \cdot u(t) - (t - t_r)u(t - t_r)\right] \tag{5.19}$$

where I_p represents the peak current, and t_r is the rise time of the ramp. The Laplace transform of (5.19) is

$$I(s) = \frac{I_p}{t_r}\left(\frac{1}{s^2} - \frac{e^{-t_r \cdot s}}{s^2}\right) \tag{5.20}$$

We can also rewrite (5.16) as

$$Z_s(s) = \frac{K_1 s + K_0}{s^2 + 2Bs + \omega_{rf}^2} = K_1 \cdot \frac{s + \dfrac{K_0}{K_1}}{(s+B)^2 + \left(\omega_{rf}^2 - B^2\right)} \tag{5.21}$$

where

$$K_1 \equiv -\frac{1}{2C_d a^2}, K_0 \equiv -\frac{Z(x,y,0)}{2C_d a^2 \cdot 4L_p}, B \equiv \frac{k}{2} \cdot \frac{Z(x,y,0)}{4L_p}, \omega_{rf} \equiv \sqrt{\frac{1}{2C_d a^2 \cdot 4L_p}} \tag{5.22}$$

Using (5.20) and (5.21),

$$V(s) = I(s) \cdot Z_s(s) = K_1 \frac{I_p}{t_r} \cdot \frac{s + \dfrac{K_0}{K_1}}{s^2 \cdot \left[(s+B)^2 + \left(\omega_{rf}^2 - B^2\right)\right]} \cdot \left(1 - e^{-t_r \cdot s}\right) \tag{5.23}$$

The inverse Laplace transform of (5.23) represents the time-domain response of the power noise, and the transients can be divided into two parts. From $t = 0^+$ to $t = t_r$, the power noise transients can be written as $v_1(t)$:

$$v_1(t) = \frac{I_p}{t_r \cdot \omega_{rf}^2}\left(K_1 - \frac{2 \cdot K_0 \cdot B}{\omega_{rf}^2} + K_0 \cdot t\right) + \frac{I_p \cdot K_1 \sqrt{\left(\frac{K_0}{K_1} - B\right)^2 + \omega_{rf}^2 - B^2}}{t_r \cdot \omega_{rf}^2 \sqrt{\omega_{rf}^2 - B^2}} \cdot e^{-Bt} \cdot \sin\left(\sqrt{\omega_{rf}^2 - B^2} \cdot t + \phi\right) \tag{5.24}$$

where

$$\phi = \tan^{-1}\left(\frac{\sqrt{\omega_{rf}^2 - B^2}}{\frac{K_0}{K_1} - B}\right) + 2\tan^{-1}\left(\frac{\sqrt{\omega_{rf}^2 - B^2}}{B}\right) \tag{5.25}$$

Equation (5.24) is composed of a linear term and a sinusoidal term with exponential decay.

At $t > t_r$, the power noise transients can be written as $v_2(t)$:

$$v_2(t) = -I_p \cdot Z(x, y, 0) + \frac{I_p \cdot K_1 \sqrt{\left(\frac{K_0}{K_1} - B\right)^2 + \omega_{rf}^2 - B^2}}{t_r \cdot \omega_{rf}^2 \sqrt{\omega_{rf}^2 - B^2}} \times \sqrt{1 - 2 \cdot e^{Bt_r} \cos\left(\sqrt{\omega_{rf}^2 - B^2} \cdot t_r\right) + e^{2Bt_r}} \cdot e^{-Bt} \cdot \sin\left(\sqrt{\omega_{rf}^2 - B^2} \cdot t + \phi + \phi_0\right) \tag{5.26}$$

where

$$\phi_0 = \tan^{-1}\left(\frac{e^{Bt_r} \cdot \sin\left(\sqrt{\omega_{rf}^2 - B^2} \cdot t_r\right)}{1 - e^{Bt_r} \cdot \cos\left(\sqrt{\omega_{rf}^2 - B^2} \cdot t_r\right)}\right) \tag{5.27}$$

The first term in (5.26) is a constant dc value, which denotes the steady-state IR-drop. The second term is a sinusoidal function with an exponential decay.

As a result, the noise transient $v(t)$ can be written as the sum of $v_1(t)$ and $v_2(t)$:

$$v(t) = v_1(t) \cdot \left[u(t) - u(t - t_r)\right] + v_2(t) \cdot u(t - t_r) \tag{5.28}$$

Figure 5.13 illustrates the power noise transients of three corner points. Equation (5.28) matches SPICE simulations well and has less than a 5% error.

5.4.3 Analytical Solution of Peak Noise

The total noise $v_{total}(x, y, t)$ is equal to the sum of the noise produced by the power and ground grids, or

$$v_{total}(x, y, t) = v(x, y, t) + v(a - x, a - y, t) \tag{5.29}$$

Examining Figure 5.12(a), the minimum noise always occurs at the corner points $(0, 0)$ and (a, a), which is where the pads are located. The worst-case noise

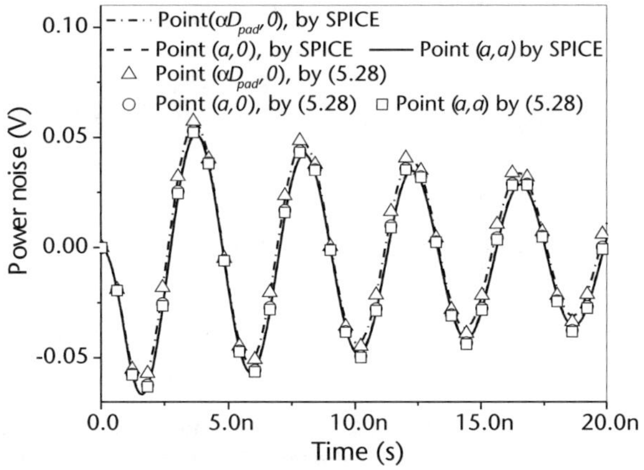

Figure 5.13 Power noise waveforms for three corner points: comparison between (5.28) and SPICE simulation.

occurs at the two remaining corner points, or $(0, a)$ and $(a, 0)$. For a single-grid network of two metal levels, the peak noise occurs when the sinusoidal function in (5.26) reaches its first peak value. The time at which this occurs, or the peak time t_p, can be solved by

$$\sin\left(\sqrt{\omega_{rf}^2 - B^2}\cdot t_p + \phi + \phi_0\right) = 1 \Rightarrow t_p = \frac{\frac{5}{2}\pi - \phi - \phi_0}{\sqrt{\omega_{rf}^2 - B^2}} \tag{5.30}$$

Consequently, the peak-noise value of a single-grid network is

$$V_{peak} = -I_p \cdot Z(x,y,0) + \frac{I_p \cdot K_1\sqrt{\left(\frac{K_0}{K_1} - B\right)^2 + \omega_{rf}^2 - B^2}}{t_r \cdot \omega_{rf}^2\sqrt{\omega_{rf}^2 - B^2}} \cdot\sqrt{1 - 2\cdot e^{Bt_r}\cdot\cos\left(\sqrt{\omega_{rf}^2 - B^2}\cdot t_r\right) + e^{2Bt_r}}\cdot e^{-Bt_p} \tag{5.31}$$

The total worst-case noise always occurs at points $(a, 0)$ and $(0, a)$. The total noise at $(a, 0)$ can be written as

$$v_{total}(a,0,t) = v(a,0,t) + v(0,a,t) = 2\cdot v(a,0,t) \tag{5.32}$$

and the worst-case peak noise for the double-grid network becomes equal to

$$V_{total-worstcase-peak} = 2\cdot V_{peak}(a,0) \tag{5.33}$$

SPICE simulations are performed on a pair of power and ground grids. These comparisons are illustrated in Figures 5.14 to 5.16. The worst-case peak noise can be greatly reduced by either adding more decoupling capacitors or decreasing the package-level inductance, as shown in Figures 5.14 and 5.15. Figure 5.16 shows

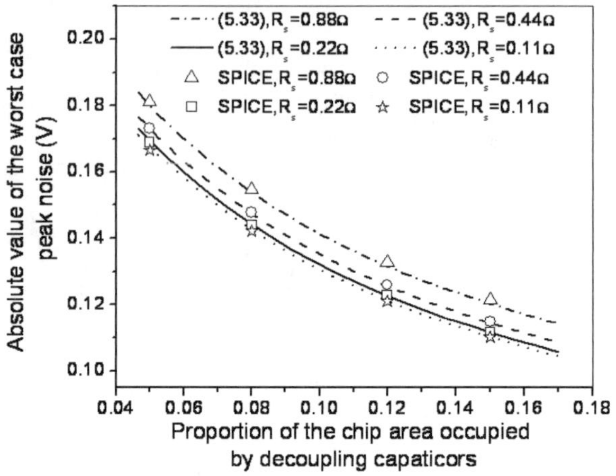

Figure 5.14 The worst-case peak noise as a function of the chip area occupied by decoupling capacitors. Comparison between (5.33) and SPICE simulation for a pair of grids.

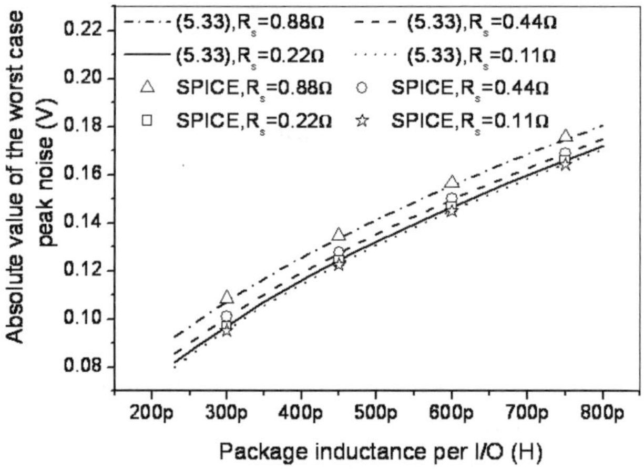

Figure 5.15 The worst-case peak noise as a function of L_p: comparison between (5.33) and SPICE simulation for a pair of grids.

that a higher I/O density can dramatically decrease the worst-case peak noise. In all plots, (5.33) has less than a 4% error compared to SPICE simulations.

It is observed that ΔI noise is sensitive to the amount of on-chip decoupling capacitance, package-level inductance, and the number of I/Os. Decoupling capacitor insertion is an effective way to reduce the noise level. However, the on-die area budget for decoupling capacitors can be limited. In this situation, package-level, high-density I/O solutions, such as Sea of Leads [9], can be used to suppress power noise. High-density chip I/Os can greatly reduce the loop inductance of power distribution networks, resulting in smaller noise. Larger numbers of I/Os can also reduce the IR-drop. From the above plots, it is clear that these compact physical models can be used to gain physical insight into the trade-offs between chip and package-level resources.

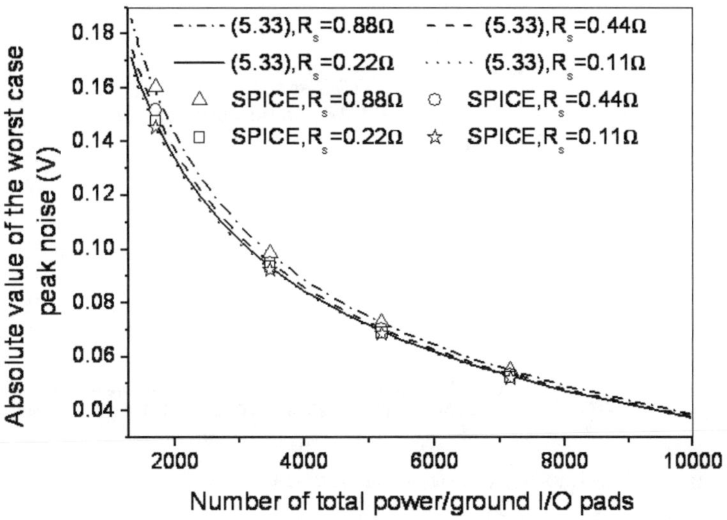

Figure 5.16 The worst-case peak noise changes as a function of the number of pads: comparison between (5.33) and SPICE simulation.

5.4.4 Technology Trends of Power-Supply Noise

The models can also be used to project the power noise trends of different generations of technology. In this section, the worst-case peak-noise value is calculated for a high-performance microprocessor unit (MPU) for each generation from the 65 nm node (year 2007) to the 18 nm node (year 2018) [6]. The values and scaling factors of each parameter for future generations are obtained as follows:

- The analysis is performed for a grid made of the top two metal levels.
- The total number of P/G pads, chip area, supply voltage, power dissipation, on-chip clock frequency, and equivalent oxide thickness (EOT) are selected based on the International Technology Roadmap for Semiconductors (ITRS) projections [6].
- For Intel microprocessors at the 180, 130, 90, and 65 nm nodes [10–13], metal thickness and signal wire pitch for the top two wiring levels do not scale with technology. The numbers for the 65 nm node are taken for each technology generation.
- Reducing the package-level inductance is associated with high costs. Therefore, we assume a constant L_p (0.5 nH) as a safe assumption [14, 15].

Figure 5.17 suggests that supply noise could reach 25% V_{dd} at the 18 nm node compared to 12% V_{dd} for current technologies if the ITRS scaling trends are followed. Excessive noise can cause severe difficulties for circuit designers, and new solutions to tackle this supply noise problem are needed in the future.

The importance of scaling package parameters such as the number of pads is also indicated in Figure 5.17. It can be seen that by increasing the pad number by 1.3-fold every generation, the supply noise can be kept well under control.

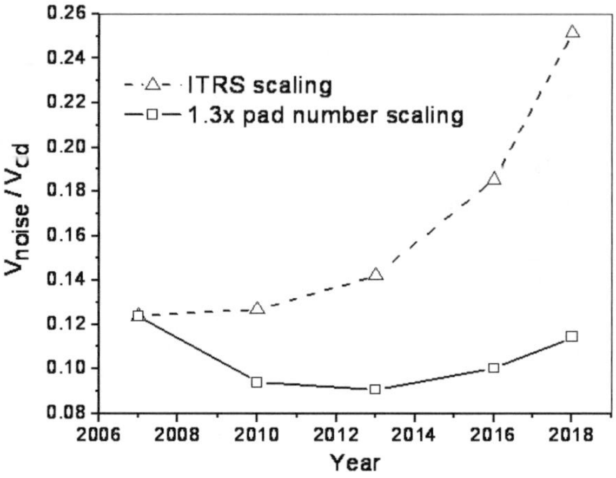

Figure 5.17 Technology trends of the worst-case peak noise.

5.5 Compact Physical Models for ΔI Noise Accounting for Hot Spots

Due to the increasing functional complexities of microprocessors, a nonuniform power-density distribution with local power densities greater than 300 W/cm^2 (hot spots) is not rare for today's high-performance chips [16]. These hot spots not only require advanced thermal solutions (to be covered in Chapters 9 to 11) but also challenge the design of power distribution systems. In this section, we extend the models presented in previous sections to more general cases by removing the assumption of uniform switching current conditions [17]. The new generalized analytical physical model enables quick recognition of the first droop noise for arbitrary functional block sizes and nonuniform current switching conditions.

5.5.1 Analytical Physical Model

A simplified circuit model that accounts for the hot spot is shown in Figure 5.18 and extends Figures 5.3 and 5.10. The symbols R_s, C_d, Δx, Δy, and L_p are the same as those used in Section 5.4.1. The current density for an active block, except for the central region of the block, is represented by $J(s)$ in the Laplace domain. In a high-performance chip, high local power dissipation can result in hot spots, as shown in the shaded region at the center of Figure 5.18, where $J_{hs}(s)$ denotes the current density inside the hot spot. The on-chip power distribution system consists of a power and a ground grid, and this double-grid structure can be decoupled into two individual grids. The main objective of this work is to model the power-supply noise caused by hot spots accurately. The single hot spot case is presented in this section, and the results can be extended to more hot spots by superposition. Hot spots are typically small compared to the chip area, and we consider a large square region that contains all the P/G pads carrying most of the supply current for a hot spot, as shown in Figure 5.19. Partial differential equation (5.34) can describe the frequency characteristics of the power noise $V(x, y, s)$ at each node in this square region:

$$\nabla^2 V(x,y,s) = R_s J(s) + 2V(x,y,s) \cdot sR_s C_d + \Phi(x,y,s) \qquad (5.34)$$

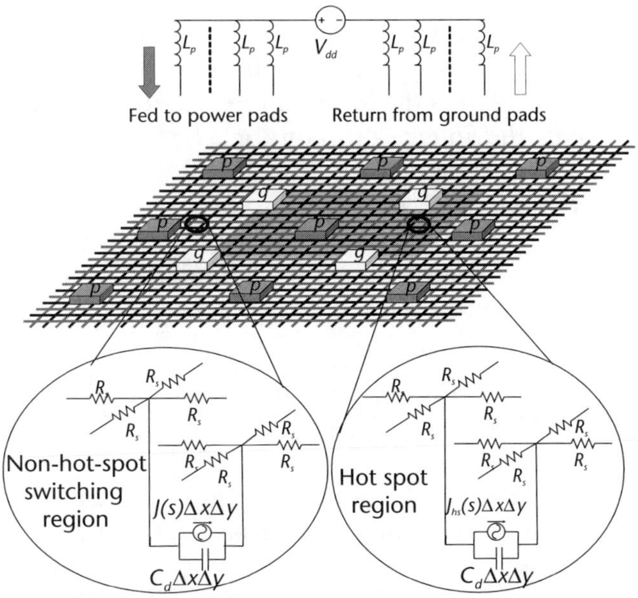

Figure 5.18 Simplified circuit model for GSI power distribution system with a hot spot.

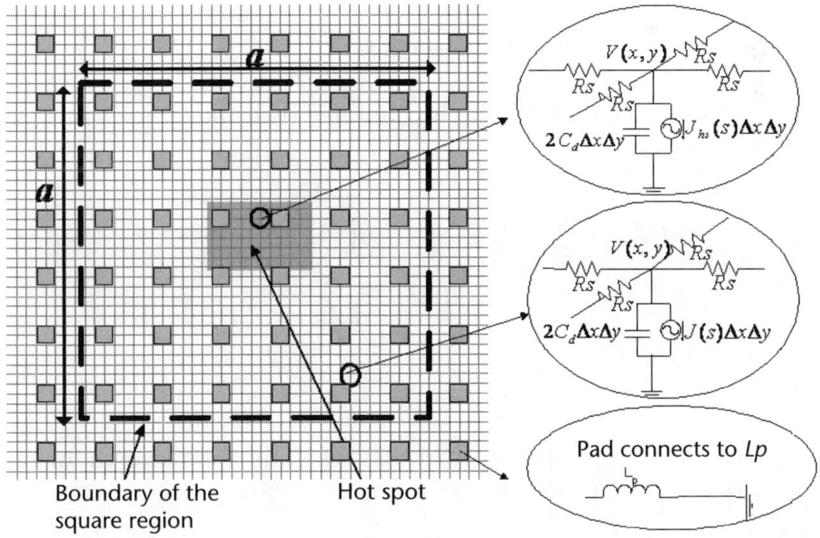

Figure 5.19 Circuit model for a single-grid structure and the square region allocated for the analysis.

where $\Phi(x, y, s)$ is the source function of this equation and can be written as

$$\begin{aligned}\Phi(x,y,s) = &-\sum_{i=1}^{M} R_s\left[J_{hs}(s) - J(s)\right]\cdot \Delta x \cdot \Delta y \cdot \delta\left(x - x_{spi}\right)\delta\left(y - y_{spi}\right) \\ &-\frac{R_s}{sL_p}\sum_{j=1}^{N} V_{padj}(s)\cdot \delta\left(x - x_{padj}\right)\delta\left(y - y_{padj}\right)\end{aligned} \tag{5.35}$$

The first term of $\Phi(x, y, s)$ represents the current sources associated with switching nodes in the hot spot region. M is the total number of nodes within the hot spot, and (x_{spi}, y_{spi}) represents the location of each node inside the hot spot. The second term represents the voltage drop on each L_p (package inductance associated with each pad). N is the total number of pads. $V_{padj}(s)$ and (x_{padj}, y_{padj}) denote the voltage and location of each pad, respectively.

Choosing a region large enough for the analysis would result in there being virtually no current flowing through the boundaries, which would produce boundary conditions of the second kind for (5.34) [8]:

$$\left.\frac{\partial V}{\partial y}\right|_{x=0} = 0, \quad \left.\frac{\partial V}{\partial y}\right|_{x=a} = 0, \quad \left.\frac{\partial V}{\partial x}\right|_{y=0} = 0, \quad \left.\frac{\partial V}{\partial x}\right|_{y=a} = 0 \tag{5.36}$$

where a denotes the size of the square region chosen for analysis. Equation (5.34) combines a Poisson's equation and a Helmholtz equation. If we let

$$V(x, y, s) = u(x, y, s) - \frac{J(s)}{2sC_d} \tag{5.37}$$

then (5.34) is modified into a pure Helmholtz equation:

$$\nabla^2 u(x, y, s) = 2u(x, y, s) \cdot sR_s C_d + \Phi(x, y, s) \tag{5.38}$$

The solution of (5.38) can be obtained by using Green's function $G(x, y, \xi, \eta, s)$. The solution is

$$\begin{aligned} u(x, y, s) = &\sum_{i=1}^{M} R_s [J_{hs}(s) - J(s)] \cdot \Delta x \cdot \Delta y \cdot G(x, y, x_{spi}, y_{spi}, s) \\ &- \frac{R_s}{sL_p} \sum_{j=1}^{N} \left[u_{padj}(s) - \frac{J(s)}{2sC_d} \right] \cdot G(x, y, x_{padj}, y_{padj}, s) \end{aligned} \tag{5.39}$$

However, in (5.39) $u_{padk}(s)$ $(k = 1..N)$ is still an unknown for each pad, and if we substitute $u_{padk}(s)$ $(k = 1..N)$ back into (5.39), we have

$$\begin{aligned} u_{pad1}(s) = &\sum_{i=1}^{M} R_s [J_{hs}(s) - J(s)] \cdot \Delta x \cdot \Delta y \cdot G(x_{pad1}, y_{pad1}, x_{spi}, y_{spi}, s) \\ &- \frac{R_s}{sL_p} \sum_{j=1}^{N} \left[u_{pad1}(s) - \frac{J(s)}{2sC_d} \right] \cdot G(x_{pad1}, y_{pad1}, x_{padj}, y_{padj}, s) \\ u_{padk}(s) = &\sum_{i=1}^{M} R_s [J_{hs}(s) - J(s)] \cdot \Delta x \cdot \Delta y \cdot G(x_{padk}, y_{padk}, x_{spi}, y_{spi}, s) \\ &- \frac{R_s}{sL_p} \sum_{j=1}^{N} \left[u_{padj}(s) - \frac{J(s)}{2sC_d} \right] \cdot G(x_{padk}, y_{padk}, x_{padj}, y_{padj}, s) \end{aligned} \tag{5.40}$$

$$u_{padN}(s) = \sum_{i=1}^{M} R_s \left[J_{hs}(s) - J(s)\right] \cdot \Delta x \cdot \Delta y \cdot G\left(x_{padN}, y_{padN}, x_{spi}, y_{spi}, s\right)$$
$$- \frac{R_s}{sL_p} \sum_{j=1}^{N} \left[u_{padj}(s) - \frac{J(s)}{2sC_d}\right] \cdot G\left(x_{padN}, y_{padN}, x_{padj}, y_{padj}, s\right)$$

Equation (5.40) includes N equations and N unknowns. The voltage $u_{padk}(s)$ ($k = 1..N$) associated with each pad can be solved from (5.40), and $u(x, y, s)$ and $V(x, y, s)$ can also be calculated accordingly.

5.5.2 Case Study

5.5.2.1 Configuration of the Functional Block and Hot Spot

A case study is performed for a functional block with a grid in the top two metal levels of a chip designed at the 45 nm node. The functional block contains a large number of pads (over 100 power and ground pads) and has a uniform current-density distribution except for a hot spot region with an extremely high current density. In this analysis, the switching functional block occupies a 3.75×2.5 mm^2 chip area and has an on-current density of 64 A/cm^2, which is the average current density given by the ITRS [6] for the 45 nm node. As shown in Figure 5.20, the hot spot region is assumed to have an on-current density of 400 A/cm^2, which is very common for chips nowadays [16]. This hot spot occupies a 0.39×0.39 mm^2 region located at the center of the switching block.

5.5.2.2 Comparison between the Physical Model and SPICE Simulations

In Figure 5.20, the double-grid structure needs to be divided into two single grids as previously noted. To apply the new model, a 6×6 pad region around the hot spot is

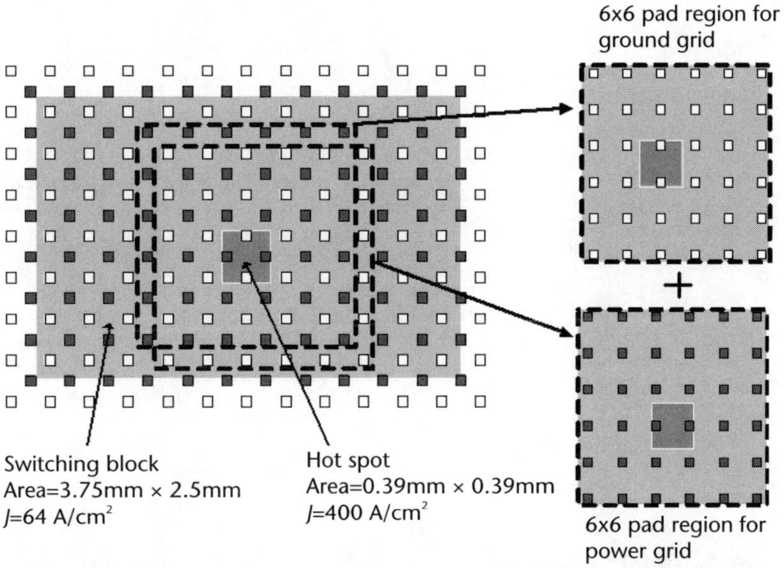

Figure 5.20 Illustration of the switching block, the hot spot, and the 6×6 pad regions allocated for the analysis.

selected for each grid. It is found that less than 1% of the total supply current consumed by the hot spot region flows through the pads outside the region; thus, a 6 × 6 pad region is sufficient for the analysis. Figure 5.21(a) illustrates the frequency-domain noise response at the center point of the hot spot. The results are also compared against SPICE simulations, and the new model shows less than a 1% error.

The total transient noise voltage at the center point of the hot spot is obtained and is represented by the solid line shown in Figure 5.21(b). Compared with the SPICE simulation results (square symbols), the peak-noise value has less than a 1% error.

To further understand the significance of this modified model in this case, it is necessary to look at the error from the blockwise model that ignores the nonuniform switching current caused by the hot spot. The average current density for the functional block is approximately 70 A/cm^2. By applying this average current, the transient noise response based on the blockwise models proposed in Section 5.4 is shown by the dashed line in Figure 5.21(b). The dash-dotted line is the noise response when we use the maximum current density within the functional block (400 A/cm^2) in the blockwise model. It is noted that if we neglect the nonuniformity of the current and

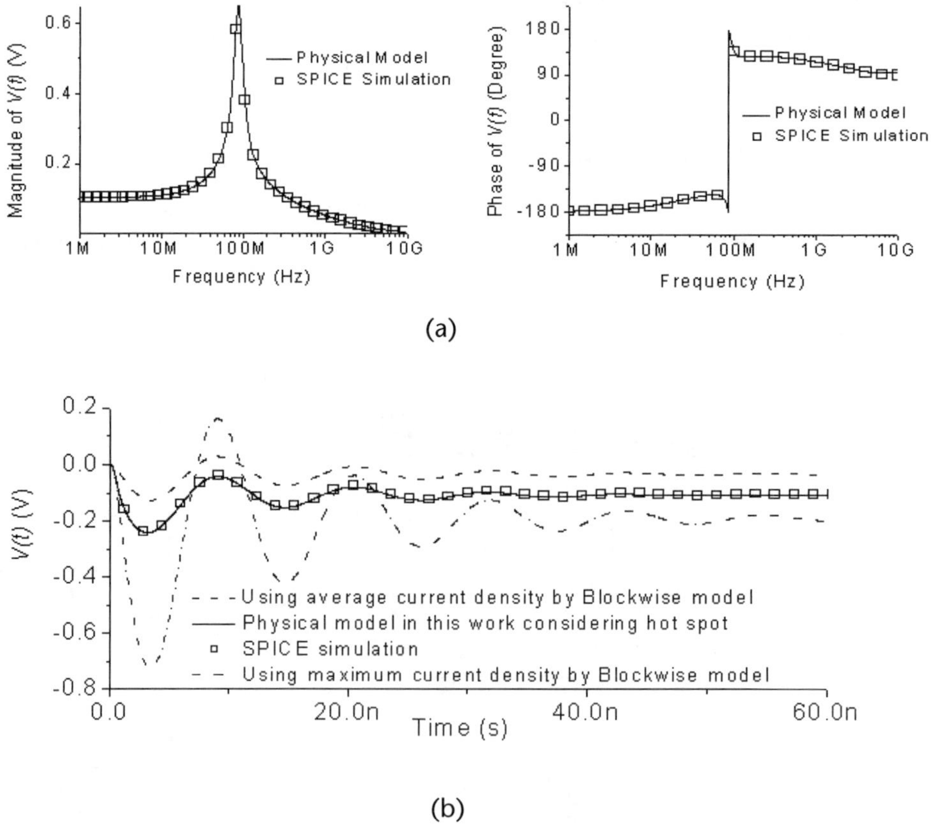

Figure 5.21 (a) Frequency-domain noise response for the center point of the hot spot (left: magnitude; right: phase), and (b) transient noise waveforms using SPICE simulation and different models.

use the average current density instead, we will underestimate the peak-noise value by 50%. If we use the maximum current density for the entire block to estimate noise, we will overestimate the peak-noise value voltage by three times.

5.5.2.3 Chip/Package Codesign and Solutions

To suppress supply noise to a safe level, we can either adopt an on-chip solution (adding more decoupling capacitors) or a package-level solution (adding more P/G pads). Decoupling capacitors are effective when the capacitance value is large enough and when the capacitors are close to the hot spot. Adding decoupling capacitors is costly for hot spots since the logic is already dense and the layout is already crowded. Decoupling capacitors also consume substantial gate leakage power. In this situation, package-level, high-density chip I/O techniques, such as Sea of Leads [9], can be an alternative option. The new physical model can help designers to identify the noise levels of hot spots, calculate how many more pads are needed, and fulfill chip/package codesign.

Adding more P/G pads locally can be quite effective in lowering the power-supply noise of the case studied in Section 5.4. To investigate this point, three cases are compared: (1) the number of pads in the hot spot region is the same as in the low-power region of the block, and either (2) 4 extra pads, or (3) 12 extra pads are utilized in the hot spot region as illustrated in Figure 5.22(a). As the peak noise changes almost linearly with the increase of current density within the hot spot, as shown in Figure 5.22(a), adding more pads can always provide more I/O

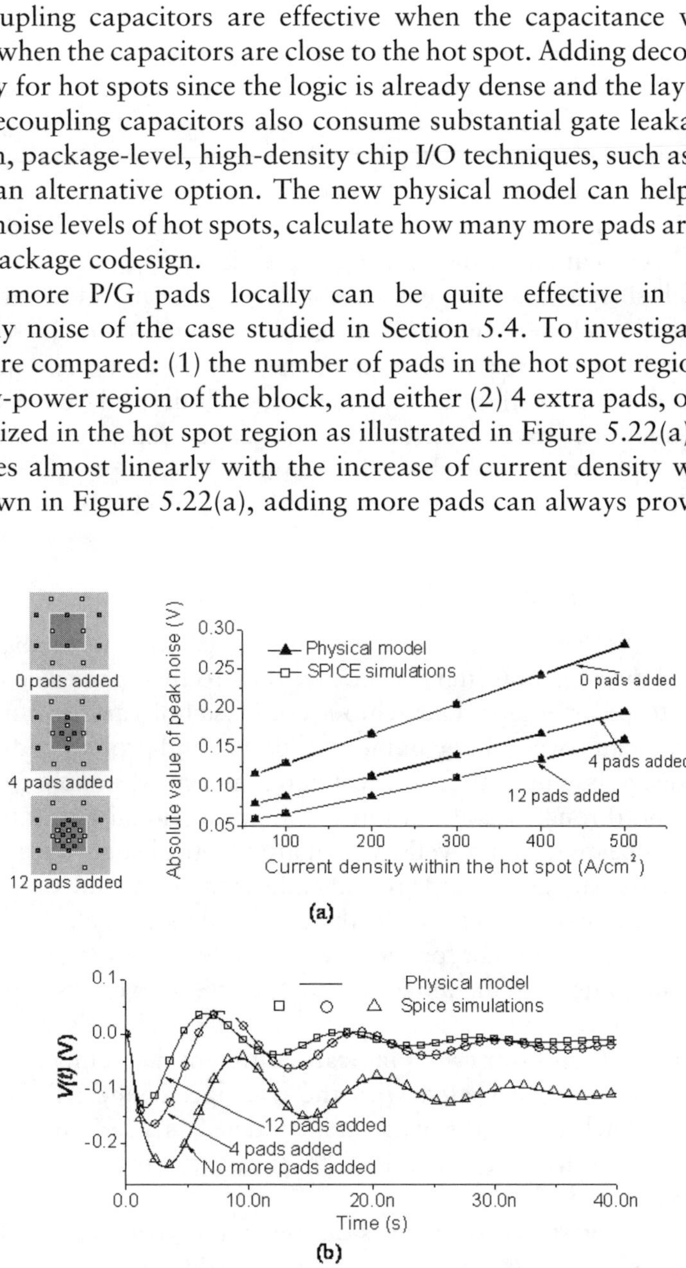

Figure 5.22 (a) Configurations of added pads within the hot spot and peak noise for different pad allocation schemes when the current density of a hot the spot changes, and (b) noise waveforms for different pad allocation schemes with a hot spot current density of 400 A/cm^2.

paths for the switching current and therefore reduce the peak noise. For example, for the hot spot current density of 400 A/cm^2, the peak noise is approximately 240 mV [Figure 5.22(b)]. The peak noise can be reduced to 165 mV (by 30%) by adding 4 pads and to 130 mV (by 45%) by adding 12 pads into the hot spot region.

5.6 Analytical Physical Model Incorporating the Impact of 3D Integration

Opportunities and challenges for 3D system integration are discussed in Chapters 13 to 15. 3D nanosystems can provide enormous advantages in achieving multifunctional integration, improving system speed, and reducing power consumption for future generations of ICs [18]. However, stacking multiple high-performance dice may result in severe thermal (discussed in Chapters 10 and 11) and power-integrity problems. Using flip-chip technology for 3D chip stacking causes the supply current to flow through the inductive solder bumps and narrow through-silicon vias that may exhibit large parasitic inductance. This may potentially lead to a large ΔI noise if stacked chips switch simultaneously. Thus, the power distribution networks in 3D systems need to be accurately modeled and carefully designed. In this section, an analytical model is derived from a set of partial differential equations that describe the frequency-dependent characteristics of the power-supply noise in each stack of chips to obtain physical insight into the rather complex power delivery networks in 3D systems [19].

5.6.1 Model Description

In 3D stacked systems, power is fed from the package through power I/O bumps distributed over the bottommost die and then to the upper dice using through-silicon vias and solder bumps. Each chip is composed of various functional blocks whose footprint can cover a large number of power and ground pads. Power-supply noise is modeled assuming that the switching current, decoupling capacitance distributions, and through-via allocation within a functional block are uniform. The footprint can be divided into cells, which are identical square regions between a pair of adjacent quarter power and ground pads. It can be assumed that no current passes in the normal direction relative to the cell borders in each die. Under these assumptions, one cell is enough for the power-supply noise analysis.

A simplified circuit model to analyze the power distribution network of 3D systems is shown in Figure 5.23. Each stacked chip is the same as the circuit model in Section 5.4.1, and subscript i indicates different die number. Also, L_p is the per-pad loop inductance associated with the package, connected to the bottommost die (layer 1). Each silicon through-via is modeled as a serially interconnected inductor L_{via} and resistor R_{via} (this includes the parasitics of the solder bumps when they are used between dice).

The whole structure consists of power and ground grids that can be decoupled into two single grids. The following partial differential equation describes the frequency characteristics of the power-supply noise $V_i(x, y, s)$ for each node in this region for stacked layer i:

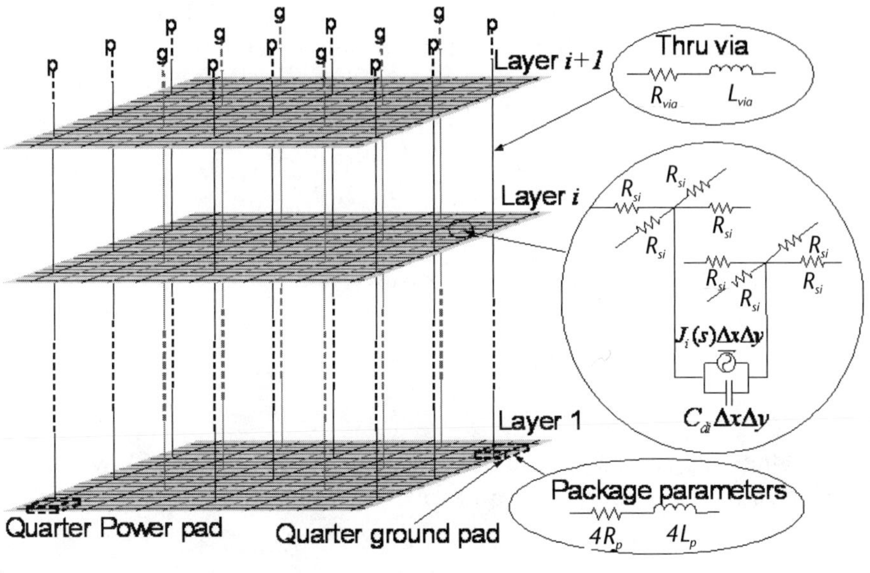

Figure 5.23 Simplified circuit model for 3D stacked system.

$$\nabla^2 V_i(x,y,s) = R_{si} J_i(s) + 2V_i(x,y,s)\cdot sR_{si}C_{di} + \Phi_i(x,y,s) \tag{5.41}$$

where $\Phi_i(x, y, s)$ is the source function of the partial differential equation for layer i (except for layer 1) and can be written as

$$\Phi_i(x,y,s) = R_{si}\cdot\sum_{k=1}^{N_{via}}\left(\frac{V_{(i-1)viak} - V_{iviak}}{sL_{viak} + R_{viak}} - \frac{V_{iviak} - V_{(i+1)viak}}{sL_{viak} + R_{viak}}\right)\cdot\delta(x - x_{viak})\delta(y - y_{viak}) \tag{5.42}$$

Equation (5.42) is derived to account for the discontinuity caused by through-wafer vias in a die, i. N_{via} denotes the total number of vias in each die, and V_{iviak} is the voltage of via k connected to layer i. Moreover, the source functions are used to make mathematical connections between layer $(i - 1)$, layer i, and layer $(i + 1)$. The source function for layer 1 is written as

$$\Phi_1(x,y,s) = \frac{R_{s1}}{4sL_p}V_{pad}(s)\cdot\delta(x)\delta(y) + R_{s1}\cdot\sum_{k=1}^{N_{via}}\left(-\frac{V_{1viak} - V_{2viak}}{sL_{viak} + R_{viak}}\right)\cdot\delta(x - x_{viak})\delta(y - y_{viak}) \tag{5.43}$$

In (5.43), the first term accounts for the contribution of the package inductance, where V_{pad} is defined as the voltage of the P/G pad in layer 1. As with (5.12), no current flows normal relative to the cell boundaries, and each partial differential equation should satisfy the boundary conditions of the second kind.

Equation (5.41) can be transformed into a pure Hemholtz equation and solved analytically. The supply noise at layer i and layer 1 is

$$V_i(x,y,s) = R_{si} \cdot \sum_{k=1}^{N_{via}} \left(\frac{V_{(i-1)viak} - V_{iviak}}{sL_{viak} + R_{viak}} - \frac{V_{iviak} - V_{(i+1)viak}}{sL_{viak} + R_{viak}} \right) \cdot G(x,y,x_{viak},y_{viak},s) - \frac{J_i(s)}{2sC_{di}} \quad (5.44)$$

$$V_1(x,y,s) = -\frac{R_{s1}}{4sL_p} V_{pad}(s) \cdot G(x,y,0,0,s) + R_{s1} \cdot \sum_{k=1}^{N_{via}} \left(-\frac{V_{1viak} - V_{2viak}}{sL_{viak} + R_{viak}} \right) \cdot G(x,y,x_{viak},y_{viak},s) - \frac{J_i(s)}{2sC_{d1}} \quad (5.45)$$

Since, V_{pad} and V_{iviak} are unknowns in (5.44) and (5.45), we can substitute them back into (5.44) and (5.45) and solve for them.

5.6.2 Model Validation

A comparison between the physical model and SPICE simulations is shown in Figure 5.24.

In Figure 5.24(a), the die with gray shade denotes the die that is switching, and the arrow points to the die for which we want to examine the supply noise. The

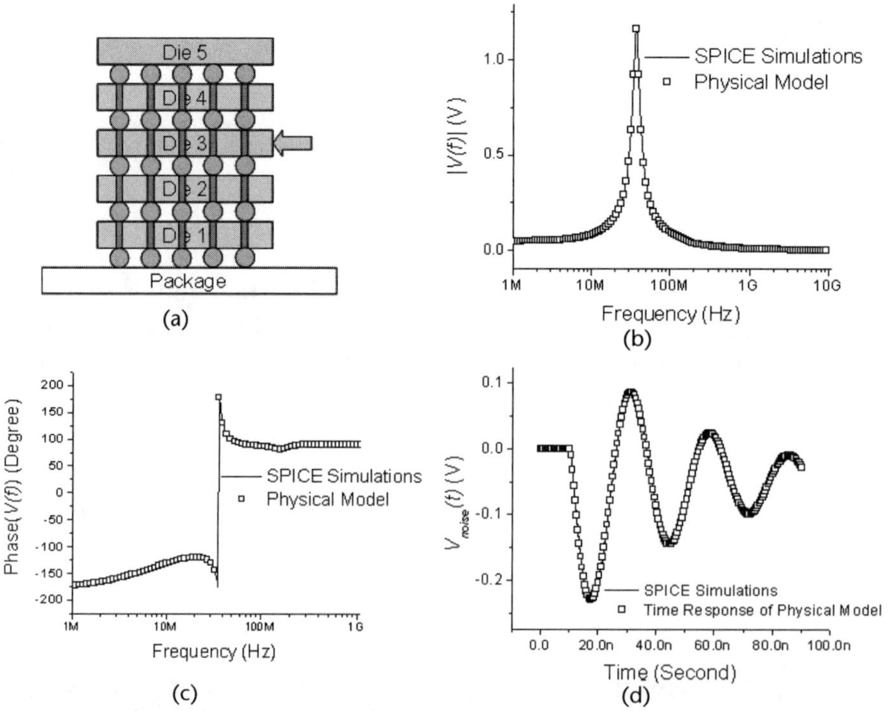

Figure 5.24 Comparison between the physical model and SPICE simulations. (a) Five dice stacking structure; (b) Magnitude response in frequency domain; (c) Phase response in frequency domain; (d) Noise waveform in time domain.

worst-case noise, which is the main concern in digital systems, normally occurs at the corners of the grid cell (furthest from P/G pads). This is similar to the previous findings in the case of a single chip. Figure 5.24(b, c) illustrates the frequency-domain response for the worst-case noise of the third die. The results are also compared against SPICE simulations, and the new model shows less than a 4% error. The transient supply noise of the worst-case scenario is also obtained and is represented by the solid line in Figure 5.24(d). Comparing with SPICE simulation (square dots), the peak-noise value has less than a 4% error.

5.6.3 Design Implication for 3D Integration

5.6.3.1 All Dice Switching Simultaneously

If only one die is switching, the noise is smaller than in the single-chip case (considered in Section 5.4.1), because the switching layer can use the decap of nonswitching layers in the 3D stack. Normally, the activities of the two blocks with the same footprints are highly correlated because an important purpose of 3D integration is to put the blocks that communicate most as close to each other as possible. Therefore, we must consider the worst-case scenario when all the dice are switching, as shown in Figure 5.25. If we increase the total number of dice and examine the noise levels in the topmost and bottommost dice, we can see that when all dice are switching, the noise produced in a 3D integrated system is unacceptable when compared to a single-chip case. This is especially true for the topmost layer, where the noise level changes dramatically (180 mV for the single-die case as opposed to 790 mV for the 10 dice case). Even for the bottommost layer, we need to identify methods for suppressing the noise! Traditionally, to suppress the noise to a safe level, we can either add more decoupling capacitors in a logic chip or add more P/G pads. In 3D systems, a power-integrity problem arises from the third dimension, and we can also push the solutions into the third dimension. In the following sections, new design methodologies will be presented in a "3D" way to tackle the 3D problem.

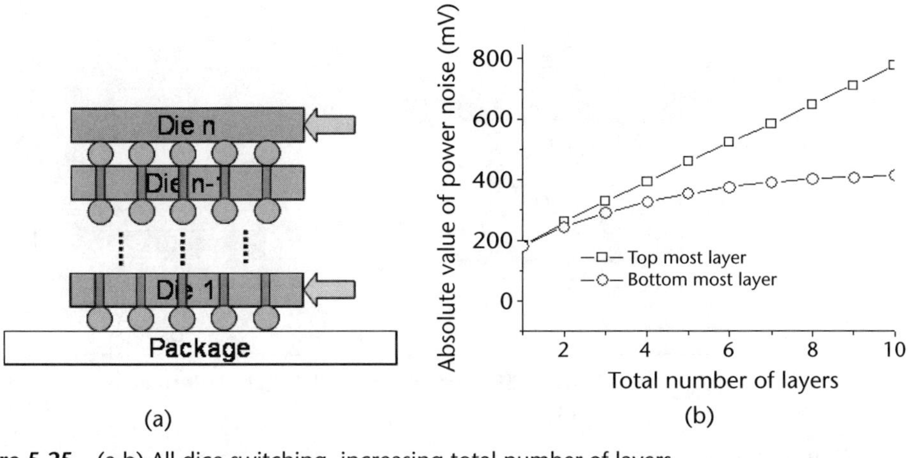

Figure 5.25 (a,b) All dice switching, increasing total number of layers.

5.6.3.2 "Decap Die"

If we can use a whole die as decap (100% area is occupied by decap) and stack the "decap die" with other dice, the noise can be suppressed to some extent. For example, if the same setup as discussed in previous sections is adopted and four dice with one decap die are stacked together, putting the decap die on the top can result in a 36% reduction in the worst-case peak noise (256 mV compared to 400 mV). Putting the decap die at the bottom of the stack can result in a 22% reduction (312 mV compared to 400 mV). Although improvements result from the decap die, we still need to add more decap dice to achieve the noise level of a single die (182 mV).

Figure 5.26(b–d) illustrates the case of different schemes for using two decap dice. By putting the two decap dice on the top, we can suppress the noise to the level of a single chip. It can be seen that putting the decap dice on the top is the best scheme to suppress the noise of the fourth die.

Instead of adding a decap die, it will be more efficient if high-k material is used between the power and ground planes (on-chip). Finally, it should be emphasized that cooling also presents challenges to 3D integration (to be discussed in Chapters 10 and 11), and the newly developed microfluid cooling technique can potentially alleviate this cooling problem.

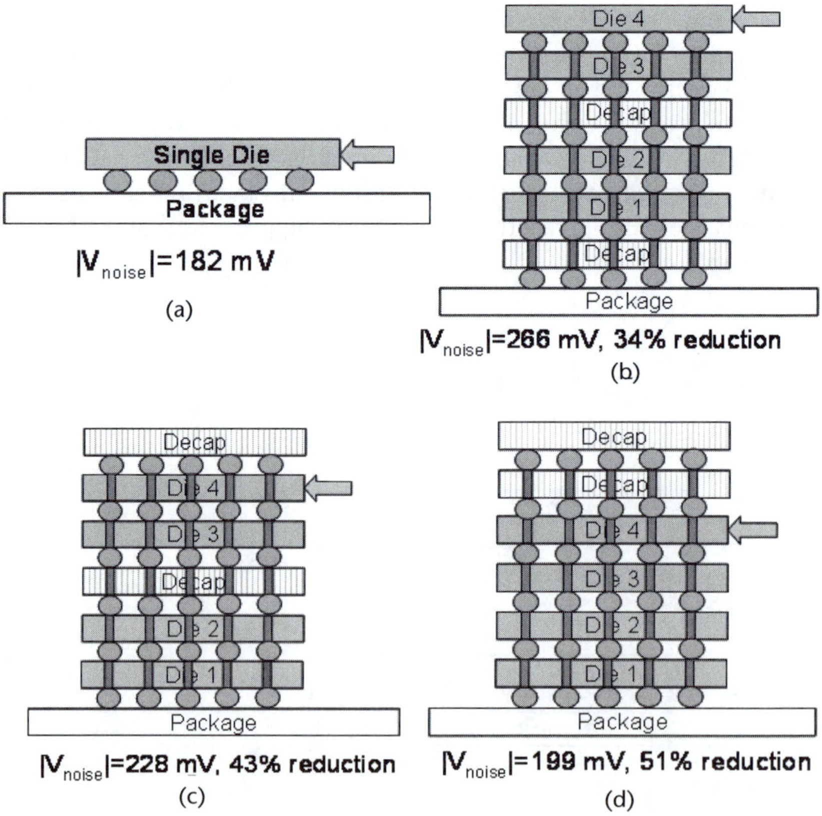

Figure 5.26 (a–d) Effect of adding decap dice when all dice are switching.

5.6.3.3 Through-Silicon Vias

Another possible solution is to put more through-silicon vias (TSVs). To examine the effect of increasing the number of TSVs, in the first case, the total number of P/G I/Os is fixed as 2,048. As Figure 5.27(a) shows, we cannot gain much benefit by solely increasing the number of TSVs. In the second case, the number of both P/G pads and TSVs in each layer is increased. This causes the power noise to decrease greatly and even reach the level of a single chip, as shown in Figure 5.27(b). The two cases show that the bottleneck is due to power and ground I/Os as they play a critical role in determining the power noise. The inductance of the package is the dominant part throughout the whole power delivery path for the first droop noise. Therefore, the power-integrity problem needs an I/O solution that can provide high-density interconnection without sacrificing the mechanical attributes needed for reliability.

5.7 Conclusion

The aggressive scaling of CMOS integrated circuits makes the design of power distribution networks a serious challenge. This is because the supply voltages, thus the circuit noise margins, are decreasing, while the supply current and clock frequency are increasing, which increases the power-supply noise. Excessive power-supply noise can lead to severe degradation of chip performance and even logic failure. Therefore, power-supply noise modeling and power-integrity validation are of great significance in GSI system designs. Accurate and compact physical models for the IR-drop and ΔI noise have been derived for power-hungry circuit blocks, hot spots, and 3D chip stacks in this chapter. Such models will be invaluable to designers in the early stages of the design to estimate accurately the on-chip and package-level resources need for the power distribution. The models have less than a 5% error compared to SPICE simulations. An analytical physical models are also derived to predict the first droop of power-supply noise when hot spots are accounted for. The

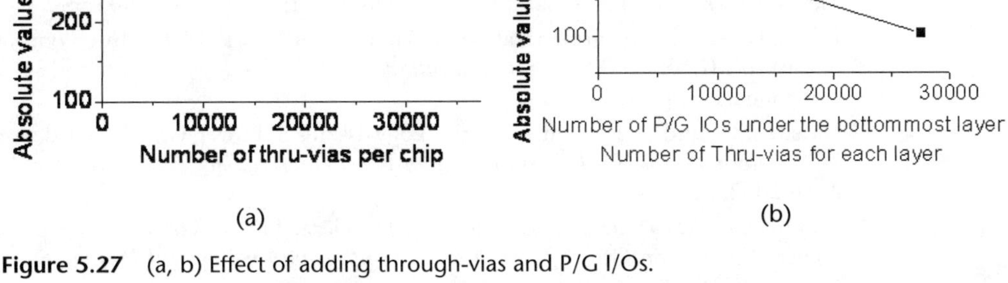

Figure 5.27 (a, b) Effect of adding through-vias and P/G I/Os.

model specifically addresses the nonuniformity problem for the power-density distribution brought by hot spots. The model gives less than a 1% error compared with SPICE simulations. The blockwise models were also extended to a 3D stack of chips and can be used to estimate accurately the first droop of the power-supply noise as a function of the number of through-silicon vias, chip P/G pads, and chip-level interconnect and decoupling capacitor resources.

References

[1] Swaminathan, M., and E. Engin, *Power Integrity: Modeling and Design for Semiconductor and Systems*, 1st ed., Upper Saddle River, NJ: Prentice Hall, 2007.

[2] Wong, K. L., et al., "Enhancing Microprocessor Immunity to Power Supply Noise with Clock-Data Compensation," *IEEE J. Solid-State Circuits*, Vol. 41, No. 4, April 2006, pp. 749–758.

[3] Dharchoudhury, A., et al., "Design and Analysis of Power Distribution Networks in PowerPC Microprocessors," *Design Automation Conference*, San Francisco, CA, June 15–19, 1998, pp. 738–743.

[4] Gowan, M. K., L. L. Biro, and D. B. Jackson, "Power Considerations in the Design of the Alpha 21264 Microprocessor," *Design Automation Conference*, San Francisco, CA, June 15–19, 1998, pp. 726–731.

[5] Shakeri, K., and J. D. Meindl, "Compact IR-Drop Models for Chip/Package C-Design of Gigascale Integration (GSI)," *IEEE Transaction on Electron Devices*, June 2005, Vol. 52, Issue 6, June 2005, pp. 1087–1096 .

[6] Semiconductor Industry Association, "International Technology Roadmap for Semiconductors (ITRS)," 2007, http://www.itrs.net/.

[7] Huang, G., et al., "Compact Physical Models for Power Supply Noise and Chip/Package Co-Design of Gigascale Integration," *Electronic Component and Technology Conference*, Reno, Nevada, June 2007, pp. 1659–1666.

[8] Polyanin, A. D., *Handbook of Linear Partial Differential Equations for Engineers and Scientists*, Boca Raton, FL, Chapman & Hall/CRC Press, 2002.

[9] Bakir, M. S., et al., "Sea of Leads (SoL) Ultrahigh Density Wafer Level Chip Input/Output Interconnections," *IEEE Transactions on Electron Devices*, Vol. 50, No. 10, October 2003, pp. 2039–2048.

[10] Bai, P., et al., "A 65 nm Logic Technology Featuring 35 nm Gate Lengths, Enhanced Channel Strain, 8 Cu Interconnect Layers, Low-k ILD and 0.57 μm 2 SRAM Cell," *International Electron Device Meeting Technical Digest*, San Francisco, CA, November 2004, pp. 657–660.

[11] Yang, S., et al, "A High Performance 180 nm Generation Logic Technology," *Proc. International Electron Device Meeting*, San Francisco, CA, December 1998, pp. 197–200.

[12] Tyagi, S., et al, "A 130 nm Generation Logic Technology Featuring 70 nm Transistors, Dual Vt Transistors and 6 Layers of. Cu Interconnects," *International Electron Device Meeting Technical Digest*, San Francisco, California, December 2000, pp. 567–570.

[13] Jan, C. H., et al., "90 nm Generation, 300 mm Wafer Low k ILD/Cu Interconnect Technology," *Proc. IEEE 2003 International Interconnect Technology Conference*, San Francisco, CA, June 2003, pp. 15–17.

[14] Nassif, S. R., and O. Fakhouri, "Technology Trends in Power-Grid-Induced Noise," *Proc. 2002 International Workshop on System-Level Interconnect Prediction*, San Diego, CA, April 2002, pp. 55–59.

[15] Muramatsu, A., M. Hashimoto, and H. Onodera, "Effects of On-Chip Inductance on Power Distribution Grid," *International Symposium on Physical Design*, San Francisco, CA, April 2005, pp. 63–69.

[16] Prakash, M., "Cooling Challenges for Silicon Integrated Circuits," *Intersociety Conference on Thermal and Thermomechanical Phenomena in Electronic Systems*, Las Vegas, Nevada, May 2004, pp. 705–706.

[17] Huang, G., et al., "Physical Model for Power Supply Noise and Chip/Package Co-Design in Gigascale Systems with the Consideration of Hot Spots," *IEEE Custom Integrated Circuits Conference*, San Jose, CA, September 2007, pp. 1659–1666.

[18] Banerjee, K., et al., "3D ICs: A Novel Chip Design for Improving Deep-Submicrometer Interconnect Performance and Systems-on-Chip Integration," *Proc. IEEE*, Vol. 89, No. 5, May 2001, pp. 602–633.

[19] Huang, G., et al., "Power Delivery for 3D Chip Stacks: Physical Modeling and Design Implication," *IEEE 16th Conference on Electrical Performance and Electronic Packaging*, Atlanta, Georgia, October 2007, pp. 205–208.

CHAPTER 6

Off-Chip Signaling

John C. Eble III

6.1 Historical Overview of Off-Chip Communication

As the raw computational power of a silicon chip continues to grow exponentially, an ever-greater requirement exists to sustain the input/output (I/O) needs of this marvelous engine such that a well-balanced system is realized. Figure 6.1 captures the six orders of magnitude increase in total off-chip I/O bandwidth throughout the microprocessor stages of evolution: simple multicycle, pipelined, superscalar, multithreaded, and now multicore.

While it is well known that chip performance tends to double every 18 months (Moore's law corollary), the off-chip bandwidth also doubles at a rate of every 2 years, relying on increases in signal pin count and per-pin bandwidth. Looking to the future, the International Technology Roadmap for Semiconductors (ITRS) [1] provides predictions of these two components for microprocessors and application-specific integrated circuits (ASICs) and for different pin classes [Figure 6.2(a)]. Assuming that a quarter of pins operate at the highest speed using differential signaling, that a quarter of pins are cost-performance pins, and that the remaining are low-cost pins, an extrapolation of off-chip bandwidth [Figure 6.2(b)] can be made

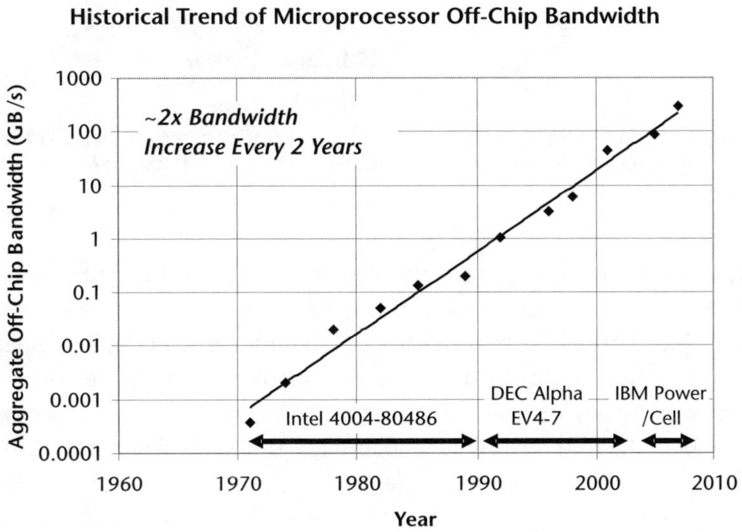

Figure 6.1 Microprocessor off-chip bandwidth over the last 30 or more years.

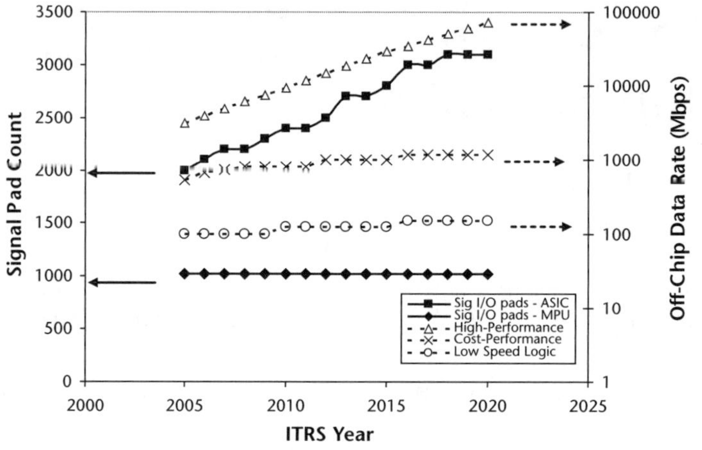

(a)

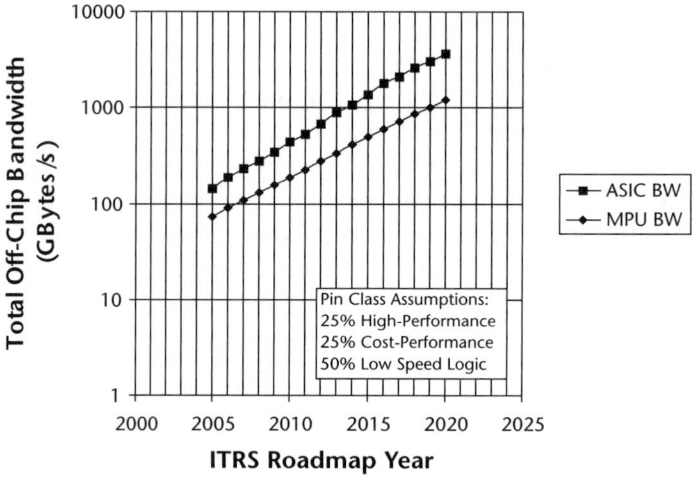

(b)

Figure 6.2 (a) ITRS projections of signal pin count and per-pin bandwidth, and (b) extrapolation of off-chip microprocessor and ASIC bandwidth based on ITRS projections of signal pin count and per-pin bandwidth.

that is fairly consistent with the historical picture in Figure 6.1. The fractions proposed above are based on assuming 128 bidirectional, differential high-speed ports in 2008 and an MPU bandwidth consistent with the Cell processor [2] in 2006.

Two primary application areas continue to push the total bandwidth requirements of a highly integrated single die. The first area is "processing units," whether they be dedicated graphics processing units (GPUs), moderate-count, multicore CPUs, or throughput-centric designs that more aggressively scale the number of processing elements [2, 3]. The system bandwidth needs of such dice are normally dominated by a memory interface (either directly to onboard dynamic random access memory (DRAM) chips, a buffer chip, or a "southbridge" chip) and a chip-to-chip interface to one or more other processing elements to create a simple coprocessor arrangement or multi-CPU network. Over the first 20 years or so of the microprocessor's development, bandwidth needs were driven by the increased computation rate of a single thread of execution due to the dramatic increase in clock frequency afforded by CMOS scaling and architectural advances that increased the instructions per cycle. Future bandwidth needs will be driven by the increased number of cores or processing elements integrated into a single die and the number of independent threads that can be run. Amdahl's balanced system rule of thumb [4] states a system needs a bit of I/O per second per instruction per second; therefore, these throughput-centric processors will require immense bandwidth.

The second area is switch-fabric ASICs that coordinate the flow of packets in advanced network and telecommunications equipment. A state-of-the-art die could have 200 differential, bidirectional ports (800 signal pins), each operating at 6.25 gigabits per second (Gbps), resulting in an astounding total I/O bandwidth of 312.5 Gbps. The ITRS identifies the networking and communications space as the application driver of bandwidth and predicts a fourfold bandwidth increase every 3 to 4 years [1]. A different characteristic of this space is that communication occurs through a backplane rather than just a single board or substrate. Backplane environments are complex and challenging as routes between chips are longer (through backplane and two line cards), routes go through high-density connectors, and the signal path may contain many impedance discontinuities and vias.

Overall system performance is not only dependent on the I/O bandwidth but also on the latency to obtain needed data. However, many of the architectural decisions and innovations over the last 20 years have been made to hide or minimize the impact of memory latency, so this tends to favor maximizing bandwidth over minimizing latency. Latency in a memory interface is primarily dominated by the internal access times of the DRAM core. Also, as cycle times scale faster than the reduction in connection length, the time-of-flight across wires can become appreciable. It is roughly one cycle per inch to travel when the on-chip frequency is 5 GHz. Latency remains important in direct chip-to-chip communication. In a multi-CPU network, the latency is heavily dependent on the number of hops; therefore, the latency from an in port to an out port must be minimized. When the granularity of data needed is large, latency becomes a minor concern. Many signaling techniques may take a slight hit on latency in order to get large gains in overall bandwidth.

Off-chip bandwidth needs have so far been met by a combination of packaging advances, better channels and signal integrity, and, to a great extent, sophisticated pin electronics. I/O circuits have the distinct advantage of reaping the benefits of silicon scaling, whose circuit-level benefits can be succinctly captured using the fanout-of-4 (FO4) delay metric [5]. The FO4 is the propagation delay through a canonical CMOS circuit loaded by a capacitance four times its own input capacitance. The bandwidth (BW) of a particular signaling implementation in terms of this scaling metric can always be expressed as

$$BW = 1/(k \times FO4) \tag{6.1}$$

where k is a constant indicating how aggressive an implementation is. The inverse of BW is called the unit interval (UI). Figure 6.3 shows that the constant k remains within a fairly constant range (UI equals one to eight FO4 delays) over a survey of signaling solutions across a number of process generations.

Not only can the pin drivers, receivers, and clocking circuits run more quickly through technology scaling, but more functionality can be fit into a fixed area or

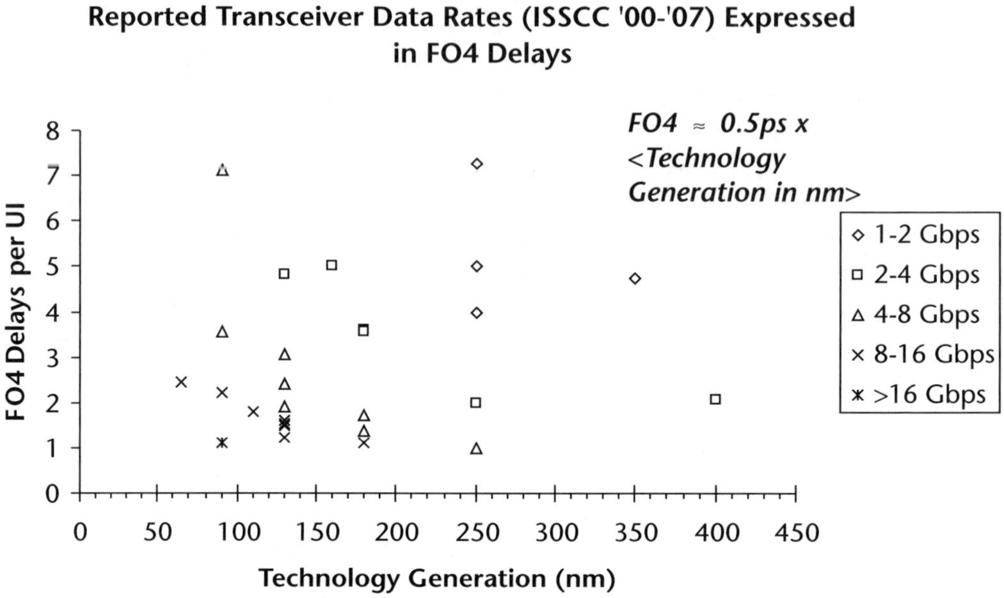

Figure 6.3 Reported transceiver signaling rates in terms of FO4 delays.

power budget. I/O circuits were originally generic and essentially just full-swing CMOS buffers. A complicated pad type would have been bidirectional with a tristateable transmitter and a receiver that was essentially an inverter with its threshold set by the process and the *P* to *N* ratio. A parallel set of such buffers would be used to transmit a word of information from one chip to another synchronized with two external clocks similar to an on-chip synchronous path. It would take several round-trip delays of "ringing the line up" before reliable latching of the data could occur. This simple implementation, sufficient for the bandwidth needs of a prior generation, has progressed steadily over many generations to a profoundly complex system that: (1) terminates both ends of the lines for incident-wave switching; (2) optimally designs an I/O interface for a class rather than one general solution; (3) prefers differential over single-ended signaling to maximize signal-to-noise ratio (SNR), minimize interference from other channels, and greatly reduce self-generated simultaneous switching output (SSO) noise by using constant current output structures; (4) counteracts the effects of signal-integrity effects such as reflections and intersymbol interference (ISI) through signal conditioning and advanced receivers; and (5) optimizes link timing on a per-bit basis and then resynchronizes and aligns across a parallel word at a lower-frequency parallel clock. Fortunately, even with this complexity, efficiency metrics such as milliwatts per gigabit per second (mW/Gbps) and micrometers squared per gigabit per second (μm^2/Gbps) have continued to scale through process and moderate signaling-voltage scaling. Signaling-voltage levels have not scaled as quickly as core-voltage values because of interoperability requirements and a belief that reducing voltage margins can be risky. In order to simultaneously meet power and bandwidth requirements of the ITRS roadmap, research such as that shown in Figure 6.4 [6] must continue to drive down (mW/Gbps).

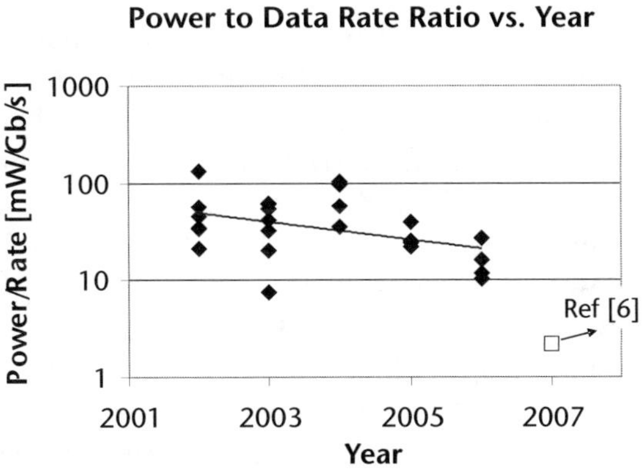

Figure 6.4 Historical power efficiency trend in mW/Gbps and recent research to make significant gains in efficiency. (*From:* [6]. © IEEE 2007. Reprinted with permission.)

6.2 Challenges in Achieving High-Bandwidth Off-Chip Electrical Communication

This section identifies and attempts to quantify the key challenges of continuing to scale off-chip electrical bandwidth to meet projected needs. The challenge areas are continued scaling of high-speed I/O count per die, on-chip I/O capacitance, reflections, losses, noise, jitter, and electrical route matching.

6.2.1 System-on-a-Chip Impact of Large-Scale I/O Integration

Total die bandwidth is the product of signal pin count and bandwidth/pin. While most of the challenges are with respect to bandwidth/pin, it is equally important to continue the scaling of high-speed I/O signal count. Scaling count is not only a challenge from the perspective of packaging technology and escaping between packaging levels (Section 6.6), but also from the die standpoint of integrating large numbers of high-speed I/Os. The die-side challenges include keeping the power and area to a reasonable fraction of the die totals and ensuring that the analog portions of the I/O cell work within specification across a large number of replications in the face of transistor variation.

The total I/O power of a system-on-a-chip ($P_{I/O}$), expressed in (6.2), is a function of pin count (N), bandwidth per pin (BW) in units of Gbps, and power efficiency (E) in units of mW/Gbps. A particular system-on-a-chip (SoC) will have k different classes of pins, each with their own unique characteristics represented by the subscript j.

$$P_{I/O} = \sum_{j=1..k} \left(N_j \times BW_j \times E_j \right) \tag{6.2}$$

For example, the first-generation Cell processor I/O count is dominated by a multibyte parallel I/O and dual-channel memory interface. These two classes of I/Os sum to approximately 14W (assuming peak bandwidth and an efficiency of 21.6 mW/Gbps [7]), which is a significant portion, possibly 15% to 20% of the estimated 80W [8], of its power budget. SoC performance scaling requires that BW and pin count increase, so the only way to manage power is through scaling power efficiency. Figure 6.4 indicates that the power efficiency has scaled down from ~100 mW/Gbps to ~10 mW/Gbps and below over the last 6 years. Following the trend line shown, the end-of-the roadmap figure of merit would be ~1.1 mW/Gbps or ~1.1 pJ energy dissipation per bit transferred. Since the maximum power dissipation is completely flat starting in 2008 and it is necessary for the I/O to stay at a fixed percentage of the power budget (~14% for ASIC based on a 20 mW/Gbps assumption in 2005), the power efficiency scales inversely proportionally to the off-chip bandwidths projected in Figure 6.2(b). Based on this requirement, the 2020 target would be 0.93 mW/Gbps, which is not that far off the historical trend.

A differential I/O cell's area can be expressed in number of flip-chip bumps. An aggressively designed transmitter or receiver differential unidirectional link will consume four bumps worth of area (two signals, power, and ground). Because of the aggressive data rates, current and future I/Os will almost always need to be designed local to the bump such that capacitance is minimized. Since the bump pitch scales at a rate slower than the minimum feature size, this is not an issue for the digital portions of the design and allows for more functionality. Analog portions of the design may not scale as aggressively, and some portions, like electrostatic discharge (ESD) protection, may not scale at all! The area for I/O signal ESD protection as well as power-supply shunts can be 5% to 10% of the total I/O area footprint in current-generation technology designs.

High-speed I/O macrocells incorporate a number of analog building blocks that are especially sensitive to transistor variation. Each block needs to be designed to a specification (e.g., sampler sensitivity) such that the overall link works reliably. Transistor variation will cause the block's performance from instance to instance to be a random draw from a probability density function, which is assumed to be Gaussian. If a component is given a portion of the chip yield budget in terms of the failure rate due to that component, fr_{comp}, and the number of such components on a chip is n_{comp}, then the number of standard deviations, σ_{mult}, over which the design must still meet specification is given by

$$\sigma_{mult} = \sqrt{2} \times erf^{-1}\left[\left(1 - fr_{comp}\right)^{1/n_{comp}}\right] \tag{6.3}$$

This multiplier will increase with greater component count, while the effects of transistor variation are expected to increase as a fundamental problem. Techniques to solve this problem include increasing device length and widths, while accepting area and power increases, calibrating circuits at power-up and possibly periodically, and doing as much processing as possible in the digital domain.

6.2.2 Pad Capacitance: On-Chip Low-Pass Filters

The signal pad capacitance, or C_i, is present at both ends of a point-to-point link and they act as poles in the system that fundamentally limit the bandwidth of the overall channel. Figure 6.5 shows an idealized view of the signaling system and results for an ideal transmission line with a state-of-the-art C_i value of ~800 fF.

The 12-inch channel shows resonances in its transfer function because of the nonideal termination at high frequencies. Equalization techniques can deal with these bandwidth limitations, shown more clearly in the 0-inch case, to a degree.

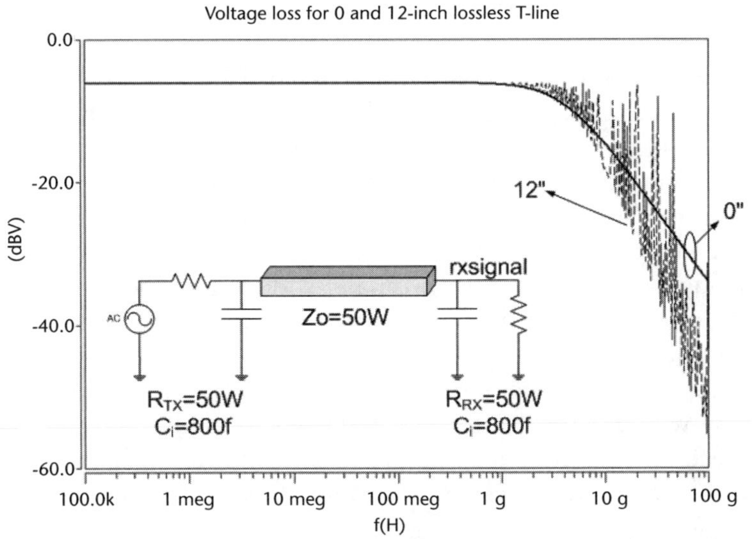

Figure 6.5 Ideal channel model with C_i bandwidth-limiting components and its voltage response.

With an aggressive equalization value of 20 dB, bandwidths of 80 Gbps are possible with this ideal channel. In order to achieve 72 Gbps at the end of roadmap, C_i values will need to continue to scale, which will be a significant challenge because of the issues now discussed.

Signal pad capacitance includes not only the physical landing pad but also the interconnection network down to the active driver/receiver transistors and ESD, the capacitance of the active circuitry, and the capacitance of ESD structures. The scaling of these capacitances does not follow traditional scaling rules.

The capacitance of the physical landing pad is dependent on the size of the landing pad (more determined by C4 bumping packaging technology as opposed to wafer-scale processing) and the distance to the "ground plane" beneath it. This distance is a function of the technology metal stack-up and the highest metal level the particular design uses in this region.

The interconnection network from the signal bump pad to the active components must be sized not only to withstand the large current surges and provide a near-zero impedance path to clamping devices of an ESD event (for example, 2A peak for a 2,000 KV HBM event [9]) but also to be electromigration tolerant during normal signaling modes as well as during fault conditions, where a signal is shorted to a supply or left floating due to a fault in the environment. Because of these hefty metallization requirements, this capacitance component can dominate and scales slowly proportional to changes in required ESD zap currents [10] and the fault currents, which are a function of the I/O supply voltage.

The capacitance of the ESD protection components—which shunt the large ESD zap current away, while keeping the voltage rise low enough not to damage oxides connected to the pad or cause a second breakdown in transistors whose drain is connected to the pad—also scale slowly as the currents do not scale, while the maximum voltage rise permitted in fact scales down. Research on gated structures

[11] that present a lower capacitance when powered on in a system appears promising.

Finally, the capacitance of the actual driver or receiver and any auxiliary circuits that monitor link integrity must be included. The driver's drain area, thus capacitance, is ultimately determined by the amount of signaling current drawn. If signaling current does not scale, this component will stay roughly constant with process generation as microamps per micrometer remain relatively constant. The receiver capacitance is primarily determined by the size of the input devices of the first amplifier stage. These devices are normally wide to provide gain at the bit-rate frequency and tend to be longer than minimum length to improve matching and reduce the offset of the amplifier.

For the state-of-the-art C_i value of ~800 fF used above, an approximate breakdown would be 25% for the ESD device and wiring, 25% for the transmitter/receiver/termination circuitry, 20% for the actual bump, 20% for redistribution from the bump to active circuitry, and 10% for the auxiliary circuit load.

6.2.3 Reflections Due to Impedance Discontinuities and Stubs

In a typical chip-to-chip interface, many boundary conditions exist where the impedance is not continuous and matched. At the transmitter and receiver ends of the interface, on-die passive and/or active devices are used to terminate the signaling channel. Because of component variations, voltage and temperature dependencies, and parasitic capacitances (Section 6.2.2), the channel is not matched perfectly across the frequency range of interest. Additional discontinuities will occur when changing levels of packaging hierarchy. For instance, the transmission lines in a package have some tolerance that does not necessarily match perfectly the board on one side and the silicon on the other. When transitioning from an outer layer to an inner layer in a package/board substrate using plated-through-hole vias, stubs are introduced that appear capacitive. These sources of reflections can cause resonances in the system that introduce significant notches in the transfer function (Section 6.3) of the channel.

Within an incident-wave switching system, any energy that is reflected will not make it to the receiver during the bit time. Furthermore, reflections cause energy unrelated to the bit being sent to interfere (Section 6.3). This lowers the signal-to-noise ratio at the receiver and limits the maximum bandwidth at which signaling can occur. For this reason, high-speed links are almost exclusively terminated at both ends to absorb reflections. Reflections are deterministic; therefore, techniques like receiver decision feedback equalization (DFE) and feed-forward equalization (FFE) can be used to actively cancel reflections to a degree.

6.2.4 Dielectric and Skin-Effect Loss and Resulting Intersymbol Interference

A lossless transmission line can be represented as cascade of inductors and capacitors. Any length of transmission line will have some dc resistance, but this amount is usually insignificant and has no bearing on signaling as the cross-sectional area is quite large. However, as the frequency of a signal increases, the electrons flowing through the conductive material tend to concentrate themselves further and further to the surface, or "skin," of the media, thereby decreasing the effective cross-sectional area. This physical phenomenon is called the "skin-effect," and a conductive media can be characterized by its skin depth, which is the depth into the wire at which the current density decays to $1/e$ of the current density at the surface. The skin depth is dependent on the material properties, such as resistivity and permeability, as well as the inverse square root of the current switching frequency. Therefore, this is a frequency-dependent resistive term in the transmission line equation leading to attenuation and dispersion of the signal proportional to the square root of frequency. In the gigahertz region, the skin depth begins to approach the root-mean-square value of the surface roughness of copper foils used in PCB manufacturing, and the attenuation increases even more quickly with frequency [12, 13].

The second effect causing frequency-dependent loss is energy dissipation in the insulating dielectric between the conductor and return path(s), which can be represented by a shunt conductance. Dielectric loss is described by the loss tangent or dissipation factor of a particular material, which is the ratio of the real (resistive power loss) and imaginary (capacitive reactance) portions of the capacitor's impedance. This attenuation is directly proportional to frequency and becomes the dominant loss mechanism when approaching 1 GHz [13].

Since these losses are frequency dependent, they deteriorate the signal during rising and falling transitions in which its highest-frequency components are present. An edge is smeared across a much longer period of time, at some point extending the energy into one or more successive unit intervals. The energy that interferes with successive bits is termed intersymbol interference (ISI). The degree of ISI present in a signaling system can be quantified by considering the single-bit response of the system (Section 6.3). These losses are deterministic and can therefore be canceled to some degree using equalization techniques. These effects will continue to be a significant challenge as the frequency increases.

6.2.5 Interference and Noise

The analog voltage launched by the transmitter and received at the far end must be discriminated at each unit interval to a digital value in the presence of both proportional-to-signal and fixed interference (deterministic) and noise (random) sources [14]. Bandwidth limitations and reflected energy have been discussed explicitly and can be considered a proportional interference that reduces the signal-to-noise ratio.

Self-generated power-supply noise, due to time-varying current demands through an inductive package, can affect the actual voltages transmitted as well as interfere with critical timing and receiver circuits. The output drivers typically consume the most current, so this has been given the name simultaneous switching output (SSO) noise. In single-ended terminated links, simultaneous switching inputs (SSIs) can also produce local noise [15]. This problem has existed for decades, and various solutions have been developed, including using differential, constant current-steering drivers and differential signaling, staggering driver turn-on, and coding data spatially or temporally, such as using data-bus inversion (DBI) [16] to shape the current waveforms. Additionally, the problem can be addressed by decreasing the effective inductance of the packaging system by increasing the number of balls/bond wires and providing decoupling capacitance throughout the silicon/packaging hierarchy. These noises are especially important in pseudo-differential signaling systems where a single-ended data line is compared to a fixed reference. In this case, the noise transfer functions to the signal and the reference are not identical; therefore, this common noise is converted to a differential noise source. The power integrity on-chip and its impact on ever decreasing signaling margins will continue to be a challenge.

A second category of interference is crosstalk, where signaling channels in proximity capacitively couple energy to one another in a direct fashion or inductively couple energy through return path loops [14, 17, 18]. Forward crosstalk in the direction of wave propagation results in far-end crosstalk (FEXT) at the sensitive receivers of a unidirectional bus. In a stripline configuration, where the dielectric is uniform, the capacitive and inductive components cancel, and FEXT is cancelled. Reverse crosstalk in the opposite direction of signal propagation results in near-end crosstalk (NEXT) at the transmitters of a unidirectional bus. The capacitive and inductive components sum, which makes source termination imperative. FEXT and NEXT components increase with frequency, and their energy can approach that of the signal at extreme frequencies. Increasing the spacing or use of guard traces at the expense of density will reduce crosstalk. As mentioned, good stripline stack-ups on both board and package go a long way toward mitigating these issues. Another unfortunate side effect of crosstalk is that the noise components are at a frequency quite near the signal frequencies of interest. Therefore, linear amplification in that frequency band can either be ineffective or worsen the signal-to-noise ratio.

While interference is bounded and generated by other portions of the system, noise is unbounded and a result of fundamental thermal and flicker noise within the transistors making up the analog circuits in the I/O cell. Noise can affect the precise timing circuits in the silicon (Section 6.2.6), resulting in unbounded jitter, as well as the receiving sampler that makes the final binary decision after any amplifying/equalizing prestages. Noise effects must be carefully considered in the design of these classes of circuits and can fundamentally limit the time and voltage resolution of future signaling systems.

6.2.6 Timing and Jitter

The effects discussed in the previous sections impact not only voltage margins but also the time at which the voltage difference is maximal. Transmitter jitter, the dynamic timing uncertainty of an edge with respect to a perfect phase reference, must be tightly controlled so that the beginning eye (Section 6.3) is clean. It is then the job of the clocking system to ensure that the data received is sampled at the optimal point—near the middle of a unit interval.

Early high-speed chip-to-chip interfaces tried to extend the on-chip synchronous clocking methodology to neighboring chips such that skews between transmitter and receiver clocks were controlled using balanced clock distribution and delay locked loops [19, 20]. This became impractical with the scaling of data rates and number of chips involved in a communication network. The next generation of clocking technologies shipped an explicit clock bundled and synchronous with some number of data signals ("clock forwarded" or "source synchronous" systems) that to first order removed the path delay between chips from the timing equation. This technology has advanced from single-data rate (SDR) clocks with no active delay adjustment to dual-data rate (DDR) clocks with active clock placement and possible calibration. This type of clocking is still used (e.g., DDR, GDDR, HyperTransport) in a number of chip-to-chip standards and has the advantage that jitter up to some frequency is common across clock and data. Two timing components that still limit this clocking architecture are offsets in the centering of the clock and skew between the clock path and the associated data bits. Many links and standards now define a data-training sequence that allows positioning of the sampling clock where the eye opening is maximal. This can either be done once at startup or periodically to track system environment variations, with the drawback of having to take down the link. Furthermore, a set of independent clocks for each bit can be generated, each with its own offset control [7]. Once periodic data training is introduced, it becomes possible to dispense with the explicit forwarded clock. A clean clock source, with jitter superior to that of the forwarded clock, at the receiver is generated a clock whose phase can be adjusted to optimally sample the data.

State-of-the-art clocking solutions require data coding/scrambling such that timing information can be directly extracted from the data stream of a single link. By ensuring that the data stream has a high enough edge density (probability that two successive bits are different), a phase detector can compare an internal recovered clock to this timing information and make updates (Section 6.5.5). This has the advantage of continuous tracking of the data timing. This topology requires clean clock sources, possibly with PPM differences and/or SSC modulation, on both ends of the link, and a clock and data recovery (CDR) unit to continuously adjust the receiver clock.

To satisfy ever decreasing timing budgets, electrical link standards specify a stringent requirement on the total transmit jitter (TJ) at a specific bit-error rate (BER). The total jitter [21] is further divided into a random jitter component that is statistically unbounded, characterized by its standard deviation σ, and a deterministic component that is bounded and caused by physical phenomena such as power-supply noise, nonideal circuit components, and data-dependent jitter through the lossy medium. The required σ, in UI or picoseconds, can be found with the following equation:

$$\sigma = \frac{\left(f_{rj} \times TJ\right)}{\left[2 \times \sqrt{2} \times erf^{-1}(1 - BER)\right]} \tag{6.4}$$

where f_{rj} is the fraction of the total jitter budget devoted to random jitter. At current line rates of 6.25 Gbps, the mean time between failures (MTBF) is about 2.67 minutes when targeting a BER of 1e–12. Assuming $f_{rj} \times TJ$ is 150 mUI, this requires a sigma of 10.5 mUI or 1.7 ps RMS jitter. An end-of-the-roadmap system with constant MTBF and random jitter budget would be 10.1 mUI or 0.14 ps RMS jitter. About 25% to 35% of the timing budget is granted to the transmitter side, and the remaining is split between the channel and the receiver clocking/recovery system. All components of timing error will need to scale proportionally for electrical bandwidth to scale.

6.2.7 Route Matching

A practical challenge is the matching of channel lengths through the package/board hierarchy across a synchronous or electromagnetically coupled bundle of wires. The timing aspect of this challenge can be solved with the techniques introduced previously that allow per-bit calibration of sampling times or adoption of an encoded/serial link topology. However, differential signaling systems, which are used almost exclusively in high-bandwidth links, require matching of the two halves of the pair. Assuming a 100 mUI requirement and a propagation velocity of 7 ps/mm, the matching target for a 6.25 Gbps system is 2.3 mm and scales down to 0.09 mm for a 72 Gbps system.

6.3 Electrical Channel Analysis

For compact modeling, simulation efficiency and accuracy, and ease of extraction, the electrical signal-integrity community has adopted techniques from the RF/microwave disciplines that treat a channel component as a black box linear network. Scattering parameters (S-parameters) describe the electrical behavior of linear networks and are related to y- and z-parameters, but they are defined with respect to a fixed characteristic impedance (i.e., 50Ω environment). A two-port S-parameter network (Figure 6.6) contains four elements that describe the incident and reflected power wave from each port. Port 1 is typically the input port, and port 2 is the output port. The "a" terminals represent incoming waves, and "b" terminals represent outgoing waves of the model.

S-parameters are always frequency dependent (i.e., a value is valid only at a single frequency) such that all phenomena discussed in Section 6.2 that pose a challenge to signaling are captured, and models can be simply cascaded. The S11 parameter is the input port voltage reflection coefficient and is equivalent to gamma defined in transmission line theory. Input return loss is often specified in high-speed signaling standards and is expressed as $20\log_{10}(S11)$. The S21 is the forward voltage gain, and the magnitude is often plotted in decibels to show the loss versus frequency of transmission channels. The S12 component represents reverse isolation, and S22 is the output port voltage reflection coefficient, which is also the basis for output return loss. The S-parameters of linear, passive channel components, such as a package

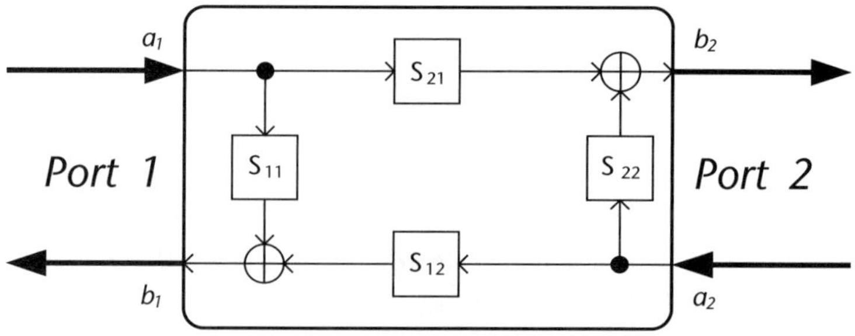

$$b_1 = S_{11}a_1 + S_{12}a_2$$
$$b_2 = S_{21}a_1 + S_{22}a_2$$

Figure 6.6 Two-port S-parameter network block and characteristic equations.

routes, board traces, connectors, sockets, cables, and SMAs, can be measured with a vector network analyzer. Multiport networks allow the characterization of coupling and crosstalk between signals and power planes.

Given an end-to-end channel whose characteristics are captured with scattering parameters, a quantitative and qualitative assessment of the channel can be derived by a number of means. The first is to view the S21 (transmitted energy) in the frequency domain as in Figure 6.7.

At low frequencies below 10 MHz, no loss is experienced in either case. For an example target bit rate of 20 Gbps, the Nyquist rate of the signal will be at 10 GHz. For the clean channel case, the loss at 10 GHz is –6.2 dB, while the loss for the loaded channel is –12.2 dB. The loaded channel also shows resonances indicating impedance discontinuities and thus reflections. The S21 plot can give a quick indication of the bandwidth limitations of the channel and what equalization methods would need to be employed. A single I/O cell may need to signal over a large family of such S21 curves. Other frequency-domain plots of interest are the return loss (logarithmic magnitude of S11) and the FEXT and NEXT, which capture crosstalk power, in terms of decibels, versus frequency.

For time-domain analysis, data-sequence simulations can be run through the channel using ideal transmitter and receiver models. The received waveforms are then folded over each other at a period equal to the bit time to form an eye diagram. The eye diagram is a simple yet powerful analysis tool as it captures the "eye opening" in terms of voltage and timing margin. A receiver mask can be defined that cap-

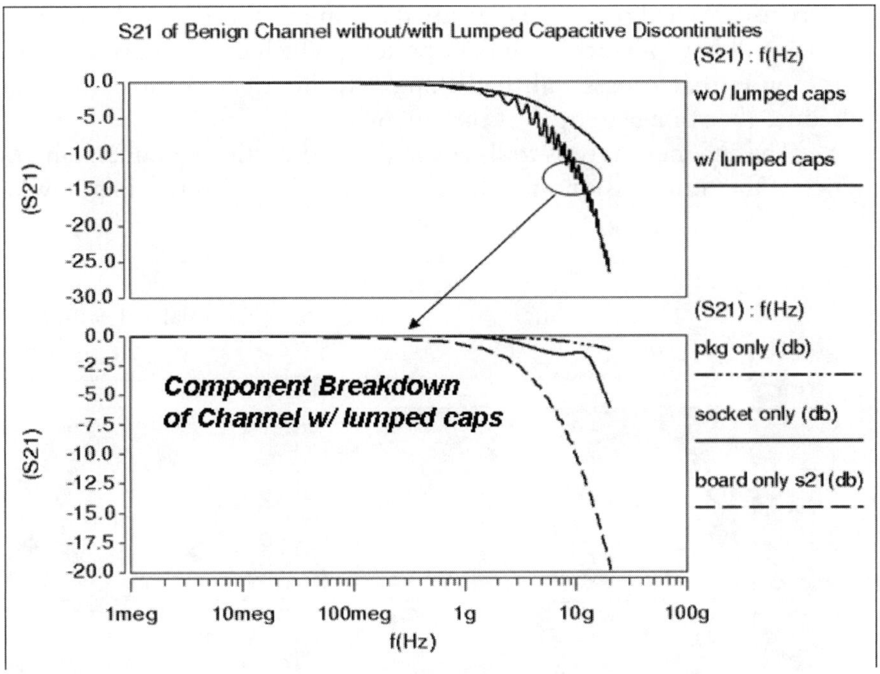

Figure 6.7 S21 plot of a benign environment including bond wire, 13 mm of package trace, socket with pogo pin connectors, and ~4-inch FR4 printed circuit board trace. Adding a 500 fF discontinuity at the pogo pin and 1 pF of load capacitance alters the S21 significantly. A breakdown of the individual components is provided in the second panel.

tures the sensitivity, aperture time, and timing noise and a pass/fail determination can be made. Besides design-time analysis, eye diagrams are extremely useful diagnostic tools during characterization of silicon (Figure 6.8). Jitter is defined as the uncertainty in the edge crossings at the boundaries between successive UIs.

The eye diagram simulation can be enhanced to use actual transmitter and receiver models, possibly with equalization. An eye that was completely closed can be "opened" with equalization. A time-domain simulation approach is straightforward, but the length of simulation time to obtain statistical information is unreasonable for typical BERs. Statistical frequency-domain approaches to build up eye diagrams that can predict BER, as a function not only of the passive channel but also of noise sources and nonidealities in the silicon, have been created [22, 23].

A third useful analysis technique is the time-domain single-bit response, which can be measured at end of the channel after the transmitter has launched a positive step followed by a negative step one bit time later. In a channel with negligible reflections but a significant amount of attenuation, this lone one or zero pattern represented by the single-bit response will be the worst in terms of voltage margin. This response also clearly shows the amount of ISI introduced by the channel as the square pulse introduced at the beginning of the line is bandwidth-limited by the channel and distributes its energy over multiple bit times. If the main cursor is chosen to be the time of maximum voltage amplitude, the amount of precursor (–N × UI) and postcursor (+N × UI) ISI can be calculated from the response as shown in Figure 6.9.

The single-bit response also shows the effects of reflections or echoes that can occur many bit times later than the original cursor. With enough history, these can also be cancelled out since they are deterministic. A channel with significant reflections will have a worst-case data pattern, which can be derived from the single-bit response, that causes "all bad things" to interfere constructively and maximally reduce the voltage margin at the cursor.

The channel analysis tools covered in this section to understand the characteristics of the transmission medium are invaluable in determining how much equaliza-

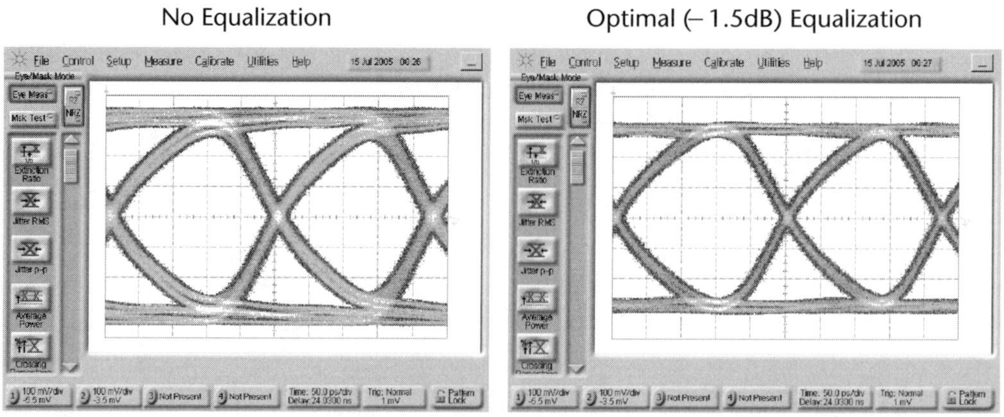

Figure 6.8 4.25 Gbps eye diagrams, with no equalization and optimal equalization, for the representative benign channel of Figure 6.7.

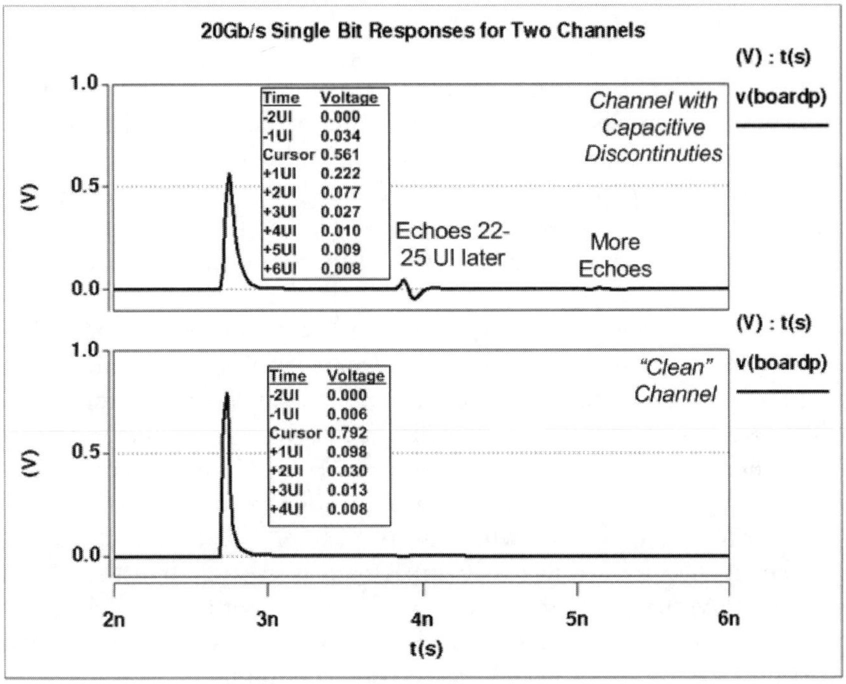

Figure 6.9 Single-ended single-bit (1V, 50 ps pulse) response of two channels introduced in Figure 6.7 showing attenuation, ISI, and reflections.

tion and possibly crosstalk cancellation is required to meet a certain BER specification, as well as in making an a priori prediction of BER.

6.4 Electrical Signaling Techniques

Before introducing circuit techniques for high-speed off-chip electrical communication, the physical-layer specification of a unidirectional electrical interconnection is now discussed.

6.4.1 Analog Line Representation

The most basic specification is how digital information is converted into an analog electrical waveform at the transmitter and then discerned by a receiver at the far end. The de facto industry standard is nonreturn-to-zero (NRZ) signaling that encodes binary information in a high and low voltage or current amplitude. This is also referred to as 2-PAM (pulse amplitude modulation) and is not unlike the signaling within a CMOS chip. Each symbol sent during a unit interval, T_{symbol}, is equivalent to a single bit; therefore, the bandwidth of such a system expressed in bits per second is $1/T_{symbol}$ and is exactly equal to the baud rate expressed in symbols per second. The primary frequency component, or Nyquist frequency, is always half the symbol rate. The main strengths of this technique are its conceptual and implementational simplicity, long legacy and compatibility, and ease of interoperability and testing with available equipment. An alternative 2-PAM scheme is return-to-zero (RZ) signaling in which, after each bit is sent, the line is driven to an intermediate state for an equivalent period of time. A practical example of such a scheme is Manchester encoding or biphase modulation, which encodes a "high" bit as a rising edge at the center of the bit time and a "low" bit as a falling edge. The strength of this encoding is that clock recovery is vastly simplified because of the abundance of edges in the data stream. The weakness is that the frequency component goes up to the bit rate and therefore suffers increased attenuation over NRZ.

A 2-PAM NRZ signaling system can be expanded to an N-PAM system, where the signal takes on N discrete voltage/current amplitudes, and the signal is discerned using $N - 1$ decision thresholds [24–26]. A single symbol then encodes $\log_2 N$ bits. The great advantage of this technique in lossy transmission systems is that the Nyquist frequency reduces by a factor of $1/\log_2 N$ for the same effective bit rate as in the 2-PAM case. However, the signal-to-noise ratio (SNR) of the system is decreased by a factor of $1/(N - 1)$, which is –6 dB for $N = 3$ and –9.5 dB for $N = 4$. To first order, if the increase in loss from A to B is greater than the SNR reduction, then multi-PAM makes sense. However, this argument is not as clear when NRZ equalization is considered [27]. Multi-PAM systems also suffer from interoperability problems and lack of test/characterization equipment.

A third signaling approach is duo-binary coding [28, 29], which uses three levels to reduce the required channel bandwidth. Its transmitter is normal NRZ and depends on the low-pass effects of the channel to create open, triangular eyes above and below the zero crossing point.

The techniques discussed so far are all baseband techniques where the frequency components extend from dc up to the maximum frequency components that arise during voltage/current transitions. An analog multitone system [30] chops the channel bandwidth into a small number of subchannels and signals over each particular frequency range. This technique can be extremely effective when a channel may have severe notches at intermediate frequencies.

6.4.2 Data Coding and AC/DC Coupling

Data-coding techniques can be used to constrain the frequency spectra of a random series of data. Dc-balanced coding or block coding, such as 8b/10b [31] with 25% bandwidth overhead, ensures that over a small window of data an even number of ones and zeros is transmitted. This also limits the maximum run length to some number of bits, five in the case of 8b/10b. Scrambling data with a pseudorandom bit sequence will also limit the run length to a statistically probable amount but is not necessarily dc balanced. Ensuring a sufficient number of edges is required for signaling systems that depend on CDRs to extract a sampling clock as well as for systems that attempt to do continuous-time timing calibration. A dc-balanced coding allows the use of ac coupling between transmitter and receiver. This enables interoperability between chips with different termination voltages or with significant ground shift, and it removes the effect of dc offsets in the transmitter and receiver that can eat into single-ended budgets. It also allows the receiver to independently set the common mode to an optimal point. Ac coupling normally requires an extra external component per signal pin, although signaling rates are becoming such that they can be built on-chip for a modest area penalty. This decision affects the equivalent load the driver sees, thus the output common mode and bias conditions of the transmitter.

6.4.3 Single-Ended Versus Differential Signaling

Two canonical approaches to communicating a binary 2-PAM signal exist. In one approach, the voltage/current is sent on a single physical wire and is compared to a fixed reference on the receiving side. Return currents flow through a shared ground plane, and any noise on the signal (or in the return path) that is not common with the fixed receive reference reduces the signal-to-noise ratio. This approach is very pin efficient as there is a single chip bump/pad, a single package escape and trace, a single pin, and a single board escape and route to the companion chip, along with a shared power plane that does double duty as a power delivery system. However, this approach is susceptible to noise (both external interference like crosstalk and self-generated noise like SSO). This approach remains the dominant signaling system in communication to commodity DRAMs [32] and some direct chip-to-chip interfaces [33]. An alternative approach is to utilize a pair of wires to send a voltage/current along with its complement such that the signal is self-referenced and the two received signals can be differentially compared. This approach is much more noise immune because of the tight coupling between the two wires (it has less crosstalk and balanced signaling currents, so little to no SSO). Its only drawback is the addition of a second pad/bump, route, and so forth, to transmit a single bit. To achieve the same signal pin efficiency of single-ended system, the achievable bit rate must be twice that of the single-ended system. This simple formula can be misleading as a single-ended system will need to have more power/ground bumps to keep its inductance low. In a differential signaling system, even (common) and odd (differential) modes of wave propagation must be well terminated, and the receiver should have sufficient common-mode rejection. Determining the best solution is application specific and often determined by backwards-compatibility requirements. However, differential signaling systems do a better job of harnessing the full potential of silicon scaling and enabling high-bandwidth interfaces as they more directly remove the external limitations discussed in Section 6.2 that could limit SoC performance.

6.4.4 Termination

Contemporary high-bandwidth systems terminate both ends of the channel. Driver termination becomes necessary when significant reflected or reverse crosstalk energy makes it back to the transmitter, which can then reflect back into the channel toward the receiver. Looking into the transmitter, it must be matched impedance (low S22) in the high state, low state, and during transitions, and it needs to be relatively flat across the frequency range of interest and the voltage swing range. This termination will be to one or more low-impedance supplies or a node that is heavily bypassed to a supply. Receiver termination is a little more straightforward as there are no active networks driving the line. A dc-coupled receiver will need to be terminated in such a way that is compatible with the driving structure; that is, the receiver termination network directly affects the transmitter's bias conditions. An ac-coupled receiver has more flexibility and has no impact on the dc bias conditions of the transmitter. In differential systems, the differential mode of propagation can be terminated by simply attaching 100Ω across the receiver inputs. To terminate the common mode of propagation, the terminator can be split and the middle node heavily bypassed to a power supply, or a hybrid network can be constructed.

6.4.5 Voltage Mode Versus Current Mode and Signal Levels/Swing

The final specification needed to fully define a signaling system is whether the transmitter encodes binary values using voltages or currents and the magnitude difference of the two states. To transmit voltages, resistive switches, ideally matched to the characteristic impedance of the line, alternatively connect the line to a low and high voltage reference. In most cases, the low voltage is simply ground, and the high voltage is an I/O voltage supply rail. This I/O voltage supply rail, VDDIO, has historically lagged the scaling of the core's VDD voltage. The reasons for this are interoperability with multiple generations of standards and noise-margin concerns. Assuming an impedance matched system, the VDDIO voltage alone sets the signal swing. VDDIO along with the termination voltage set the VOH and VOL levels seen at the receiver. VDDIO along with the characteristic impedance of the environment, Z, set the current draw, thus the power required to achieve that signal swing. Table 6.1 provides these single-ended dc-coupled values for different receiver termination options.

Current-mode signaling allows signal swings and power to be independent of VDDIO and is quite suitable for differential signaling and equalization. One or more high-impedance current sources whose current can be steered with switches or whose current can be switched on and off are used. A unipolar current-mode transmitter will use zero current and $+I$ for low and high, respectively. A bipolar current-mode transmitter will use $-I$ and $+I$ for low and high. Since the current source is high impedance, a termination resistor must be provided. The current choice is independent of Z and VDDIO; therefore, the signal swing can be independently set to save power or satisfy other constraints. For the same signal swing, current-mode signaling consumes more power than voltage-mode signaling.

6.4.6 Taxonomy of Examples

Based on the application space and expected channel environment, the above signaling decisions can be made, and the voltage levels then become well defined.

Table 6.1 Fundamental Equations for Canonical Voltage-Mode Driver (Z Pullup to VDDIO and Z Pulldown to GND) for Three Different Rx Termination Cases

Rx Termination	*VOH*	*VOL*	*IOH*	*IOL*	*Average Power*
Z to GND	0.5 VDDIO	0	VDDIO/2Z	0	$VDDIO^2/4Z$
Z to VDDIO	VDDIO	0.5 VDDIO	0	VDDIO/2Z	$VDDIO^2/4Z$
Z to VDDIO/2 (2Z to GND and 2Z to VDDIO)	0.75 VDDIO	0.25 VDDIO	3 VDDIO/5Z	3 VDDIO/5Z	$9\ VDDIO^2/25Z$

Table 6.2 presents a number of popular signaling standards as an example of the wide design space that exists.

6.5 Circuit Techniques for I/O Drivers/Receivers and Timing Recovery

This section will review circuit implementations and topologies in high-volume CMOS technology to realize the functions and concepts that have been introduced earlier to achieve high-bandwidth electrical off-chip signaling systems.

6.5.1 Transmitter and Bit-Rate Drivers

A general high-speed I/O macro will have a wider, slower parallel interface to the chip core and a narrow, faster interface to the pins that can stress the technology limits. The transmitter path to the pin includes a final, clocked multiplexor that combines some number of lower-speed datapaths, a predriver block that buffers up and possibly preconditions the input signal into the final driving stage, and the final driver itself that drives high and low current/voltage symbols onto the line and matches the characteristic impedance. The multiplexor can be embedded into the driver such that the function is accomplished at the pin. However, this tends to make the final stage more complicated, increasing the input capacitance of the final driver as well as the output capacitance on the pin itself, which can limit the bandwidth as mentioned in Section 6.2.2. Common degrees of multiplexing include one

Table 6.2 Taxonomy of Signaling Standards and Their Electrical Characteristics

Signaling Type	*Standards/ Products*	*Voltage/ Current Mode*	*ac/dc Coupled*	*SE/ Differential*	*Tx Termination*	*Rx Termination*	*Voltage Swings*	*Common Mode (TX or both if dc)*
SSTL	DDR2/3	Voltage	dc	SE	25Ω–50Ω series	Center-tapped VDDIO/2 termination	0.5–0.67 × VDDIO	VDDIO/2
PODL	GDDR3/4	Voltage	dc	SE	40Ω series pulldown; 60Ω series pullup	60Ω to VDDIO	0.6 × VDDIO	0.7 × VDDIO
RSL	RDRAM	Current	dc	SE	None	28Ω to VDDIO	800 mV	VDDIO–400 mV
GTL	Intel processor bus	Current	dc	SE	50Ω to VDDIO = 1.2V	50Ω to VDDIO = 1.2V	1V	VDDIO–500 mV
Bipolar CML	LVDS, Hyper-Transport	Current	dc	Differential	None	50Ω to virtual ground	350 mV	1.25V
Unipolar CML	PCI-E, SATA, FibreChannel, XAUI	Current	ac	Differential	50Ω to VDDIO	50Ω to Rx choice	150 mV–1V	VDDIO–V_{swing}
Unipolar CML	CEI-6 SR, XDR	Current	dc	Differential	50Ω to VDDIO	50Ω to VDDIO	150 mV–1V	VDDIO-$V_{swing}/2$

(full-rate clocking), two (half-rate clocking), and four (quarter-rate clocking). Increasing degrees of multiplexing has the potential of reducing power and the clock frequency of datapath and control circuits through parallelism; however, the number of clock phases needed to retime the data increases from one to two to four. Half-rate clocking architectures are a good trade-off but require active duty cycle control of the dual-data rate clock. Quarter-rate clocking architectures require even more complex quadrature correction circuits that ensure the four UIs of the clock period are evenly spaced. The overall path must be designed very carefully such that no ISI is introduced on-chip due to intermediate nodes not fully switching during a bit time. Also, the path length from the last clocking point should be minimized as to reduce the susceptibility to supply noise and thus the deterministic/bounded jitter on the output. A near ideal solution from the jitter perspective is to use the final 2:1 output multiplexor as part of the capacitive tank load of an LC oscillator [34].

The final driver design connected to the pad has many constraints, which must be simultaneously considered (Table 6.3).

Two specific transmitter designs are considered here. The first is a classic unipolar current-mode logic (CML) equalizing driver [27]. Each of the n taps can sink its current, I_n, from the TP or TN output through the differential pair switch, and the currents are summed at the line to create 2^n unique currents. Swing control is set by the baseline current source magnitudes, and the relative currents set the tap weights (Section 6.5.4.1). Matched impedance is set by the poly load resistor. The second is an ultralow-power voltage-mode design [6] that does not require Tx side equaliza-

Table 6.3 Transmitter Design Constraints and Their Impact

Constraint	*Impact*
Meet reliability limits under both normal and fault conditions	Consider thick oxide transistors; wires must be sized to handle worst-case EM and thus more C_i
Minimize crossover currents to reduce power dissipation and supply noise (voltage mode)	Driver topology and required predriver timing/waveform accuracy
Match impedance (S22 specification)	Calibrated terminator/voltage-mode driver switches; make-before-break operation; need low C_i
Keep current source transistor in saturation (current mode)	Must allocate voltage and design current tail to have enough headroom
Maintain low power	Minimize input capacitance of final driver since signals transitioning at bit rate; consider efficiency of driver and fraction of current delivered to load
Maximize on-chip bandwidth so equalization is not wasted	Low C_i; ensure no ISI introduced into predriver
Ensure output swings are predictable across PVT and meet tight min/max specs	Active calibration that compares a replica transmitter to an off-chip precision resistor and/or a bandgap voltage
Maintain low jitter	Keep supply noise sensitivity low through circuit choices and minimizing path length from last clock point
Meet ESD and latchup rules	Area impact of guard rings, spacing to other circuits, ESD circuits; C_i impact of ESD
Meet min/max edge rates	Active edge rate control if process fast compared to signaling rate
Maintain common-mode noise (differential)	Symmetric rise and fall times and low skew

tion and whose swing is set by a voltage, V_s, regulator (Figure 6.10). Matched impedance is ensured by the continuously calibrated predriver stage swing (V_r regulator) that ensures the pull-up/pull-down nmos transistors have the correct equivalent impedance.

6.5.2 Receiver and Bit-Rate Analog Front End

The receive path from the pin usually begins with a bit-rate, analog, signal-conditioning circuit block. The main purposes of this first stage could be to: (1) provide some buffering and programmable gain (AGC) either to attenuate large swings or amplify small swings before hitting the samplers; (2) tightly control the common mode, possibly shifting it to a better spot, for the sampler circuit, while rejecting common-mode noise; and (3) provide some frequency selectivity in order to equalize the channel and remove ISI effects (Section 6.5.4). The first stage of demultiplexing is achieved by having multiple samplers and independent even and odd datapaths in the case of half-rate clocking. The sampler must have high gain and be regenerative to resolve fully to digital levels within a bit time, maintain a small aperture window, and achieve low offset. An example sampler design (Figure 6.11) is a modified StrongARM latch that has a precharge phase followed by an evaluate phase. Auxillary transistors are included for precharging (P4, P5), equilibrating the sense nodes (P1), and desensitizing the input after evaluation (N6), as well as for offset calibration.

At high bit rates, full amplification can take additional stages and power. As the sensitivity requirements increase with shrinking UI, thermal noise of the sampler

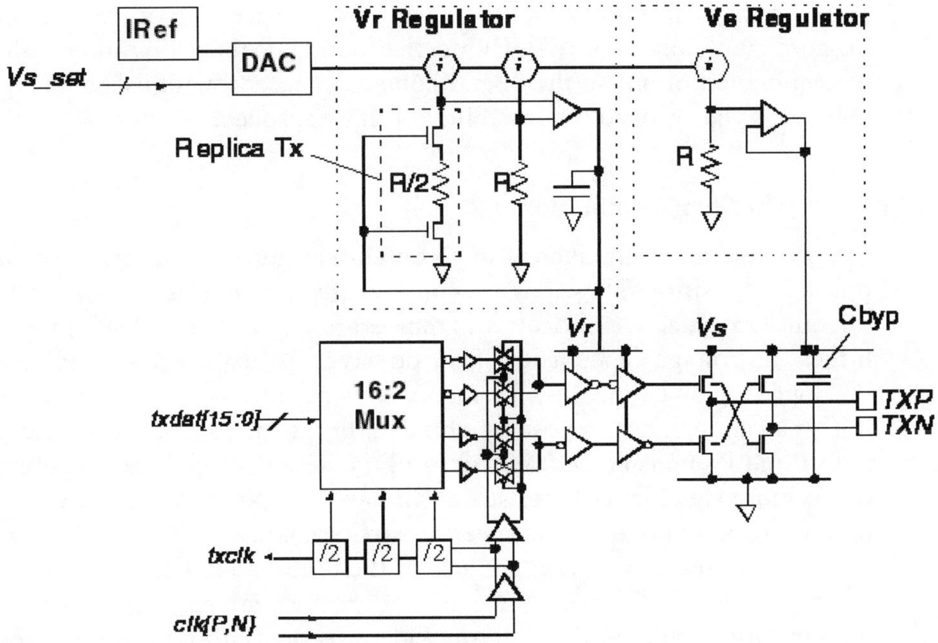

Figure 6.10 Ultra-low power voltage-mode driver with aggressive swing control and continuous, analog impedance control. (*From:* [6]. © IEEE 2007. Reprinted with permission.)

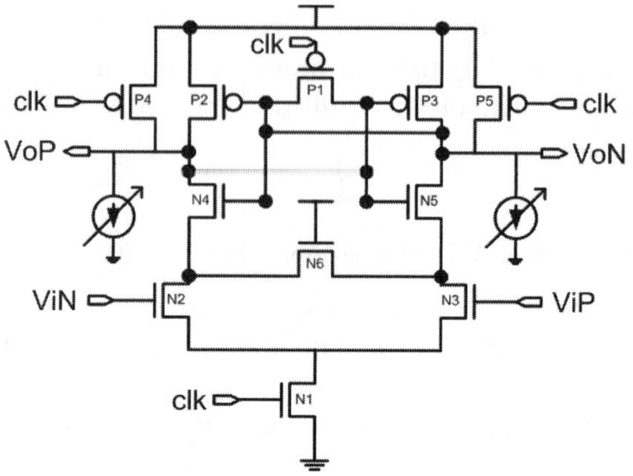

Figure 6.11 Example sampler design based on a clocked precharge/evaluate sense amplifier that includes offset compensating current sources.

transistors can become a significant and unbounded source of noise that sets a lower bound on the minimum swing that can reliably be detected. Besides the even and odd data samplers, a CDR-based electrical link (Section 6.5.5) will need to have edge samplers that are positioned 90° from the data samplers. An additional sampler, with programmable voltage and timing offset, can be used to scan the incoming eye over a period of many cycles to get a quality measure [25]. Furthermore, these results can be fed back to adaptation engines (Section 6.5.4).

It has been a common belief that speed, power, and area constraints require the front-end processing/signal conditioning to be done solely in the analog domain. A final binary detection with the sampler then converts to the digital domain. An alternative approach would be to have a high-speed, moderate-resolution analog-to-digital converter (ADC) at the pin and then do all equalization and discrimination of bits in the digital domain. A recently published backplane serdes macro [35] has proved the feasibility of this approach.

6.5.3 On-Chip Termination

The workhorse circuit element of an on-chip terminator is a precision, nonsilicided polysilicon resistor. These passive components can have as good as ±10% to ±15% tolerance range across process and temperature in a digital/ASIC process and have little to no voltage dependency. They do have a parasitic capacitance component to the doped region below the polysilicon, but since the poly is over the field oxide/STI, it is small enough not to pose a significant problem. Some processes will offer both an N+ and P+ nonsilicided resistor, and the designer will need to determine which one is more suitable. Poly resistors will have a maximum current density specified because of reliability and/or sheet-resistivity change with time. This requirement tends to constrain the minimum size of the terminator. For example, a terminator designed to handle 20 mA of current in a process with maximum current density of 0.25 mA/μm would need an 80 μm wide resistor. With typical sheet resistivities of ~400Ω/μm, the length of a 50Ω termination resistor would be 10 μm.

The resistor tolerance is usually sufficient to meet the dc S11/reflection coefficient targets for a particular standard or link budget. However, if multiple termination impedances need to be supported, if the terminator needs to be turned off, or if the link requires smaller low-frequency S11, the terminator can be segmented into portions consisting of a series resistor and a switchable transistor in the triode region. Terminators can be calibrated to an external resistor, but one must be careful not to increase the C_i to the point that the high-frequency S11 near Nyquist suffers at the expense of lower-frequency S11.

6.5.4 Equalization

Equalization is the process of approximating a filter that has the inverse transfer function of the lossy channel such that the frequency response of the two cascaded systems is flat across the frequency range of interest, which in most signaling systems is from dc to the Nyquist rate (half the baud rate). A digital filter operates at discrete time units (each UI or fraction thereof), while an analog filter is continuous. Equalizing the channel mitigates the effects of ISI and can "open the eye" at the decision point of the receiver.

6.5.4.1 Transmitter Equalization

At the transmitter, a finite impulse response (FIR) filter can be efficiently implemented [36] as the transmitter has full digital knowledge of the bit stream, and the filter's length only needs to be as long as the number of bit times over which a pulse sent is smeared due to the low-pass effects of the channel. The general difference equation describing this filter is

$$y[n] = a_0 b_{tx}[n] + a_1 b_{tx}[n-1] + \dots a_N b_{tx}[n-N] \tag{6.5}$$

Each term in (6.5) is called a tap. This is referred to as an $N + 1$ tap filter, a_0 is the main tap coefficient, $a_{1..N}$ are the postcursor tap coefficients, $b_{tx}[n]$ is the current bit (taking on values of 1 and –1 for logic high and low, respectively) to be transmitted, $b_{tx}[n-1] \dots b_{tx}[n-N]$ are the previous transmitted bits whose energies will be present at the end of the channel, and $y[n]$ is the voltage/current waveform launched at the transmitter end at time n. In extreme cases of pulse stretching, there may be precursor ISI (nonzero energy at one-bit time earlier than the main tap arriving) present. The filter above can be extended to cancel this ISI by introducing a precursor tap term:

$$y[n] = a_{-1} b_{tx}[n+1] + a_0 b_{tx}[n] + a_1 b_{tx}[n-1] + \dots a_N b_{tx}[n-N] \tag{6.6}$$

Differential current-mode transmitters (Figure 6.12) lend themselves to a straightforward implementation of the above filter while maintaining constant current, fixed impedance, and fixed common mode. Programmable current sources for each of the $N + 1$ taps can be implemented as current DACs and then summed at the pad. The tap magnitude sets the digital value going into the DAC, the tap sign either inverts or passes the bit using an XOR gate, and that result drives the current-steering switches at the bit rate. The dynamic range and resolution of each

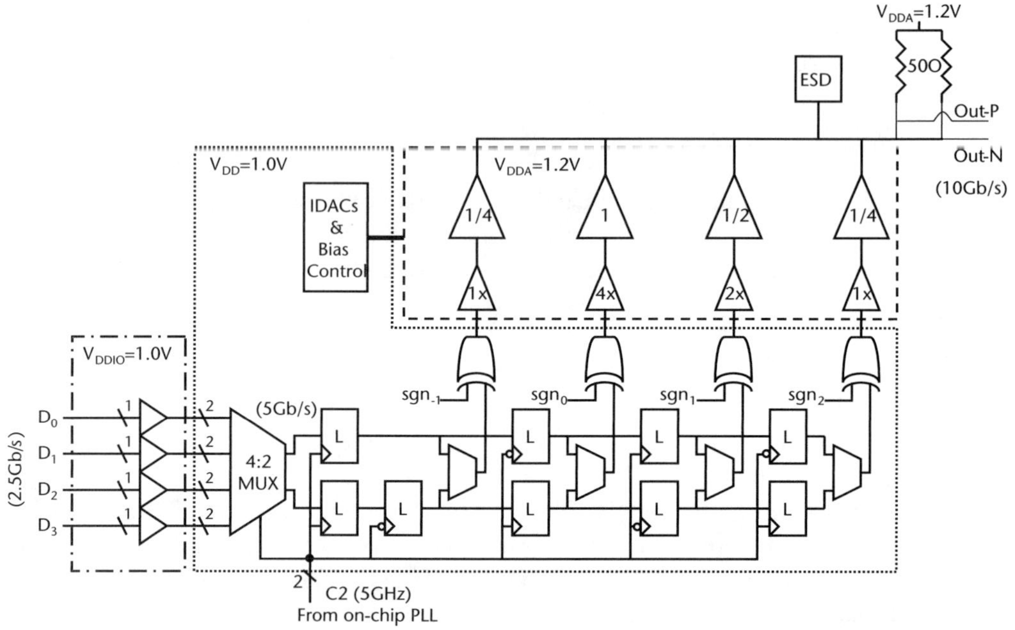

Figure 6.12 Implementation of Tx equalization using a differential CML driver with one precursor, one main, and two postcursor taps. (*From:* [27]. © 2007 IEEE. Reprinted with permission.)

DAC can be optimized for each tap and the class of targeted channels. An alternative implementation, which minimizes the pad complexity at the expense of digital logic complexity, is to have one single current DAC at the pad. The digital code driven into the DAC is a function of the current bit, previous bits and possibly a future bit, and the equalization coefficients. An example implementation, which supports the widest possible range of equalization levels and efficiently reuses all unit current cells, is an N-bit DAC that can output 2^N current levels from I_{swing} to $-I_{swing}$. The current resolution or step size of the DAC is then $I_{swing}/2^{N-1}$. Normalizing the equalization coefficients to I_{swing}, the absolute sum of the coefficients must equal one. Therefore, each coefficient is discretized into normalized units of $1/2^{N-1}$ or $N - 1$ bits of resolution. Many possible implementations besides these two are possible that would trade off equalization range, resolution, C_i, power, and area.

Transmitter equalization is often referred to as "preemphasis" or, possibly more accurately, as "de-emphasis." There will be a data pattern (likely a lone zero or one) where the coefficients will all add in one direction to give the full positive or negative I_{swing}. There will be a second data pattern (likely a long string of ones or zeros) where the postcursor coefficients will all subtract from the main tap, thereby, deemphasizing that transition. For example, in a simple two-tap filter where the main tap coefficient is positive and the first postcursor tap is negative with value α, a normalized value of one will be transmitted when the current bit is high and the previous bit was low, but a normalized value of $1 - 2\alpha$ will be transmitted when the current bit is high and the previous bit was high. This amount of deemphasis is sometimes referred to in terms of $20\log_{10}(1/(1 - 2\alpha))$ dB of equalization. This deemphasis does in fact, for a fixed transmitter power, reduce the SNR ratio available at the receiver in order to achieve the most open eye.

The strengths of Tx equalization are that it is done mostly in the digital domain, the coefficients are set with a DAC using ratios of currents and are therefore fairly insensitive to PVT variation, and the effects of equalization are highly observable with an oscilloscope. Its main drawback is that setting the tap coefficients to achieve the best equalization and most open eye at the end of the channel is a difficult problem. In most cases there is no exact solution, but a closed-form solution exists that minimizes the square of the residual error [37]. However, without knowing the channel a priori, this is not practical. Solutions have been invented that use back-channel communication from receiver to transmitter to set these coefficients in a continuous and adaptive manner by observing the eye and using an LMS algorithm and objective function, to either minimize timing or voltage uncertainty [38].

6.5.4.2 Receiver Equalization

The cascading of two systems suggests that the same sort of digital FIR filter could be implemented on the receiver side, referred to as an analog discrete-time linear equalizer [39]. The challenge here is that the governing expression, given in (6.7), is in terms of instantaneous analog voltages at discrete bit times, which implies the use of sample and hold circuits and an overall increase in analog circuit complexity:

$$y_{eq}[n] = a_0 y[n] + a_1 y[n-1] + \ldots + a_N y[n-N] \tag{6.7}$$

An alternative solution often implemented to achieve linear equalization on the receive side is to use a continuous time filter. To approximate the inverse transfer function of the channel, this filter is in the form of a high-pass filter with a fixed dc gain, and it is typically just a second-order system. An entirely passive system could be built by summing an attenuated version of the dc path, through a simple resistive divider, with a high-pass filter path consisting of a series capacitor and a load resistor. The RC time constant of the high-pass filter sets the –3 dB frequency. The difference in decibels between the Nyquist gain and the dc gain is referred to as the amount of equalization, similar to the terminology used for transmitter equalization. The weakness of this approach is that the gain at the Nyquist will always be at or below unity, and the filter values are set by passive components with wide variation ranges.

An active linear equalizer, possibly with multiple stages, can be created that offers greater-than-unity gain at the Nyquist frequency [40]. The general circuit implementation is to start with a common source differential amplifier and then add source degeneration to reduce the dc gain. By placing frequency-dependent impedance (capacitor) in parallel with the real resistance, the amount of degeneration, thus gain reduction, becomes a function of frequency. The actual and equivalent circuits along with the equations governing the gain are captured in Figure 6.13.

The dc gain, which also defines the equalization boost, is solely a function of the degeneration resistor. The degeneration capacitor and resistor set the zero at which the gain starts to diverge from the dc gain up to the high-frequency gain. Poles further out in frequency (set by the load resistor and capacitor) will then roll off the gain, hopefully after the Nyquist frequency. The key parameters of the linear equalizer (gain, gain boost, 3 dB point) can be placed under programmable control by

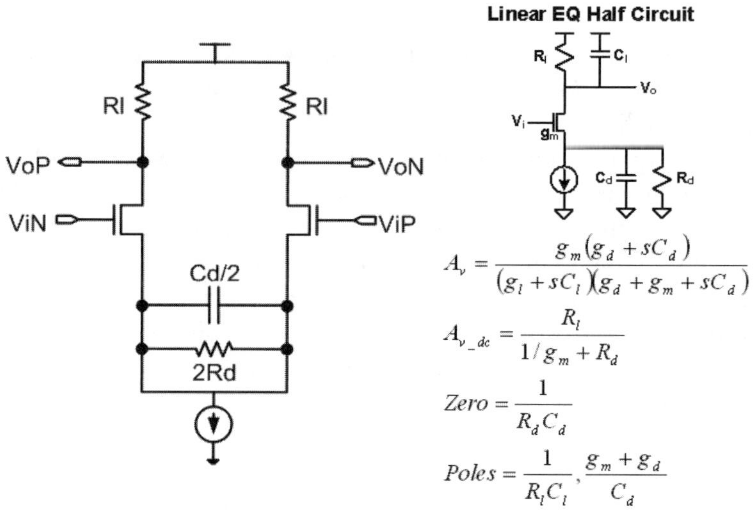

$$A_v = \frac{g_m(g_d + sC_d)}{(g_l + sC_l)(g_d + g_m + sC_d)}$$

$$A_{v\_dc} = \frac{R_l}{1/g_m + R_d}$$

$$Zero = \frac{1}{R_d C_d}$$

$$Poles = \frac{1}{R_l C_l}, \frac{g_m + g_d}{C_d}$$

Figure 6.13 Circuit-level implementation of linear receive equalization and the fundamental circuit analysis equations.

having switchable capacitor and resistor components. A drawback of this approach is that interference at the receiver, such as FEXT, within the passband of the filter will also be amplified. Continuous time filtering also does not address reflections.

A nonlinear approach to receive equalization can fundamentally improve the SNR at the sampler. Decision feedback equalization (DFE) uses the binary decisions made during previous bit times, whose residual energies are still present in the channel, to affect the current signal being presented to the sampler. This type of equalization is called nonlinear since binary, noiseless slicer outputs are used to affect the incoming signal; this same property makes it more noise immune from crosstalk as the incoming analog signal is not directly amplified. The equation describing this equalization method is below. The instantaneous analog voltage at bit interval time n is $y[n]$; the ISI compensation terms follow and are a summation of N products of the previous bits received ($b_{rx}[n-1]$ to $b_{rx}[n-N]$) and their weighting coefficients (d_1 to d_N). This equalized analog voltage is then sliced to a digital value using a sampler. This digital value will then be used during the subsequent N bit periods.

$$y_{eq}[n] = y[n] + d_1 b_{rx}[n-1] + \ldots + d_N b_{rx}[n-N] \tag{6.8}$$

$$b_{rx}[n] = \text{sgn}\left(y_{eq}[n]\right) \tag{6.9}$$

DFE is effective in cancelling reflections (echoes) that may occur in the first N bit intervals due to close-in impedance mismatches. The key implementation challenge to DFE is the tight feedback that must occur between resolving the bit at time n and then using that result at time $n + 1$. In that UI (160 ps for a 6.25 Gbps data rate), the analog value must be resolved to a digital value, the digital value is then used to change a switch, and then the analog voltage must settle prior to the next sampling point. The timing inequality for this path is expressed in (6.10). Some designs have successfully accomplished this first-tap DFE feedback [41].

$$T_{ck-q} + T_{dfesettle} + T_{setup} \leq T_{UI} \tag{6.10}$$

To address this critical path problem, an alternative implementation was proposed to handle the first DFE tap by "loop-unrolling" and recasting the system [42]. The first tap is removed from the feedback system, and two parallel paths are added in the feed-forward direction. The samplers in these two paths are skewed (either in the sampler or with a preamplifier that adjusts the offset) to have a positive and negative offset equivalent to the first tap coefficient d_1. The result of the proper path can then be selected in the next clock cycle with increased timing margin. This general approach has been called partial-response DFE and proven in silicon [43].

A third alternative implementation of DFE is to digitize the incoming signal using a baud-rate ADC and then performing the equalization using digital processing techniques [35]. Performing the filtering in the digital domain allows scaling with process, more robust production-level test, and flexibility and configurability. The direct conversion to digital also allows FFE to be implemented with no analog delay elements.

6.5.4.3 Equalization Summary

A number of 90 to 130 nm designs operating in the 4 to 10 Gbps range have converged on a Tx FFE and Rx DFE architecture [27, 41, 43–45]. The FFE usually handles precursor taps and the DFE postcursor taps. As much equalization as possible is put in the receiver since it does not amplify channel noise (crosstalk) and can be made adaptive to the specific channel environment and changes in voltage and temperature. Equalization continues to be an active area of research, and as bit rates and technology scale, the best solutions will change.

6.5.5 Clocking and CDR Systems

The most critical circuit components in a high bit-rate I/O macro are the high-speed transmit and receive clocks, which must meet very stringent jitter specifications to ensure low BER. Delivery of such a high-speed clock through a package and across many links is not practical, so a lower-speed reference clock from a clean source such as an external oscillator is normally used. An on-chip phase-locked loop (PLL) is then used to multiply this reference clock. This particular circuit is one of the most critical of any electrical link and has been the intense focus of many textbooks and edited collections [46–48]. PLLs implemented for link integration tend to be third-order charge pump systems built out of a phase-frequency detector (PFD) charge pump, loop filter, voltage-controlled oscillator (VCO), and feedback divider building blocks (Figure 6.14).

A key design parameter of a PLL is its loop bandwidth. From the perspective of the reference clock, the PLL will track phase-noise components less than the loop bandwidth and reject components higher than the loop bandwidth, essentially acting as a low-pass filter. Noise components originating in the oscillator below this frequency, either through random thermal or flicker noise components in the transistors or noise coupled through the power supply, are rejected, whereas components higher than this frequency pass through to the clock output. Therefore, given

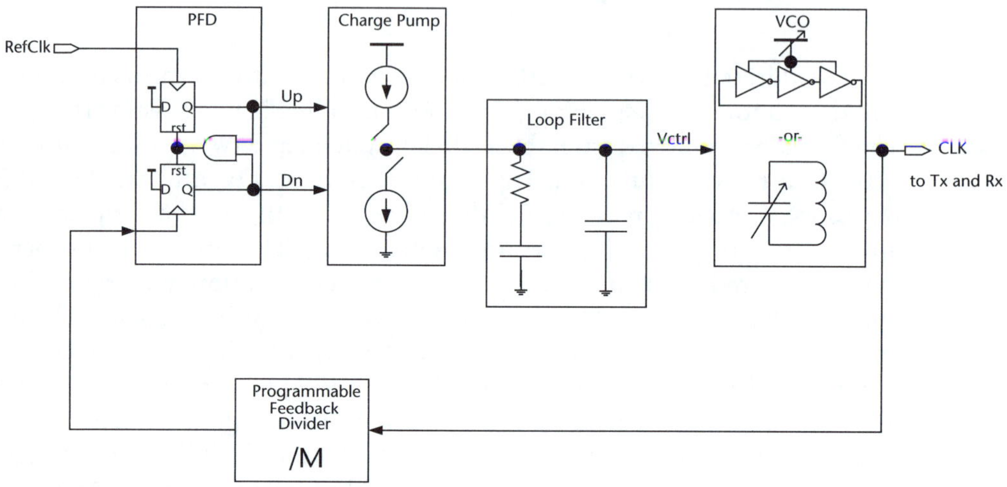

Figure 6.14 Block diagram of a typical PLL used in an I/O interface.

a clean reference clock, the bandwidth should be set as high as possible; however, it can be no higher than the update rate (reference clock frequency divided by any predivider) divided by about 10 to ensure stability and maintain linear system assumptions [46]. A single implementation will often need to work across a number of different data rates and reference clock frequencies, so the PLL will require a wide tuning range and a loop bandwidth that scales accordingly. The exact circuit architecture of such a PLL has changed with time. Maneatis first published a wide-tuning-range PLL that maintained the same loop bandwidth by using self-biased techniques [49]. The VCO uses current-mode stages with symmetric loads. As supply voltage decreases, maintaining enough headroom for these circuits presents a challenge. A ring-based approach using inverters running off a regulated supply can be more scalable, and its bandwidth can also be made adaptive [50]. The gain, or K_{vco}, of CMOS-buffer-based PLLs is typically very large and therefore very sensitive to power-supply noise. The availability of thick, low-resistance, copper, upper-level interconnects and CAD analysis tools has enabled the robust and predictable design of on-chip inductors. Therefore, LC tank oscillators have become the most desirable VCO structures and moderate to high Qs have been achievable to produce clocks with superior phase-noise characteristics. The tuning range of an LC oscillator is necessarily small. Therefore, to cover a wide range of data rates requires a combination of possibly multiple high-frequency oscillators and programmable dividers.

The use of an LC oscillator can greatly reduce the random component of jitter, but deterministic jitter, mostly due to power-supply noise, remains a challenge. The supply sensitivity of the circuits in the clocking path and the overall path delay must be considered.

Receiver-side clocking has the additional requirement of actively placing the sampling clock at the optimal point within a UI; therefore, the phase of the synthesized high-speed, low-jitter clock must be adjustable across the entire phase space (unit circle) of the fundamental clock period. The phase-detector portion of a CDR unit determines the phase of the incoming data against the phase of the internal sampling clock. A linear phase detector will output an analog signal proportional to the phase difference. A digital or "bang-bang" phase detector will determine for each transition received whether the internal clock is "early" or "late." An Alexander phase-detector system [51] oversamples the data stream by placing an extra sampler, the edge sampler, 0.5 UI offset from the primary data sampler right at the data-edge crossing point (Figure 6.15). If two successive data samples are different, then an edge is identified. The edge sample will either resolve to match the first data bit or the second, providing an indication of whether the sampling clock is early or late.

This has become the method of choice in high-integration electrical interfaces as it simply requires another instance of existing hardware and operates entirely in the digital domain. The deserialized data and edge samples can then be shipped to the digital filter component of the CDR that determines whether the phase needs to be adjusted and to what new value. The raw data and edge samples are converted into an early/late count that is then presented to a digital filter. There are many linear and nonlinear implementations of this filter possible, and the filtering choices can affect the maximum PPM that can be reliably tracked and the overall jitter tolerance of the system [52].

The final component of the CDR is the phase-adjustment circuitry. The new clock with optimal phase is synthesized using "phase mixing," which is fundamentally derived from the weighted sum of two sine waves with a fixed phase offset. In (6.11) and (6.12), a 90° offset is used:

$$A\sin(\omega t) + (1 - A)\sin(\omega t + 90) = A\sin(\omega t) + (1 - A)\cos(\omega t) = 0 \tag{6.11}$$

$$\omega t = 90 + \arctan[A/(1 - A)] \tag{6.12}$$

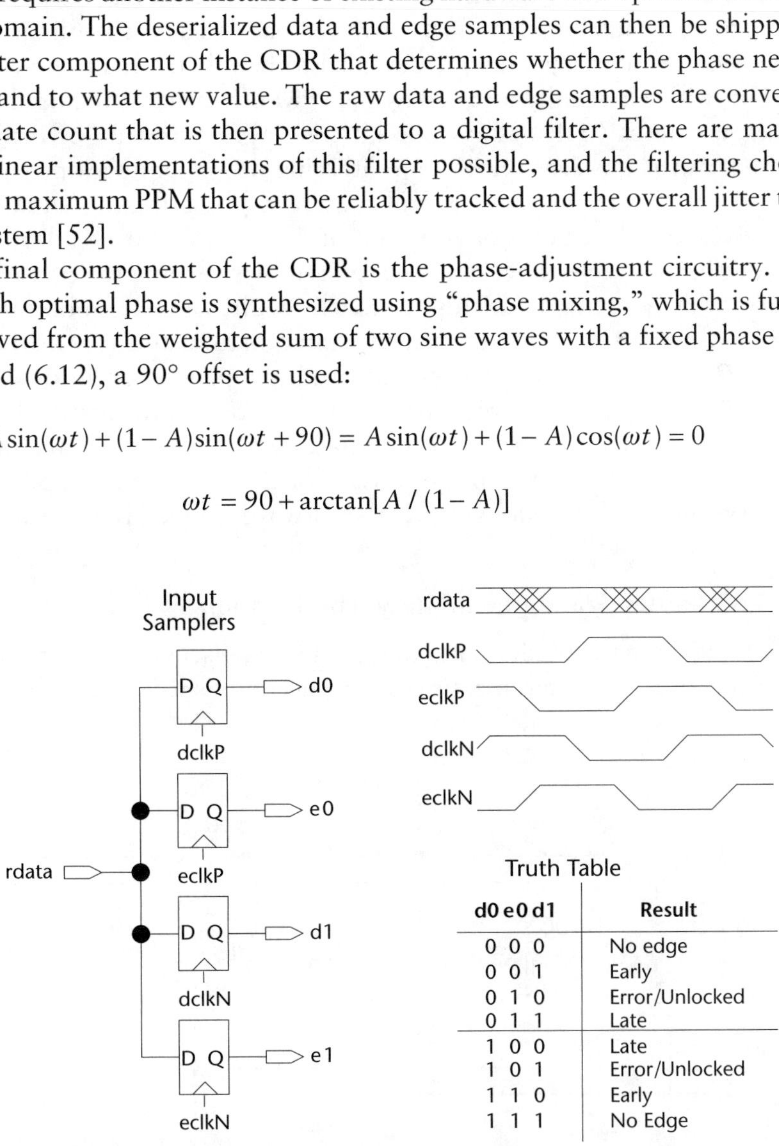

d0 e0 d1	Result
0 0 0	No edge
0 0 1	Early
0 1 0	Error/Unlocked
0 1 1	Late
1 0 0	Late
1 0 1	Error/Unlocked
1 1 0	Early
1 1 1	No Edge

Figure 6.15 Canonical Alexander phase-detector system with timing diagrams and truth table defining the result provided to the CDR based on three successive samples.

As A is varied from zero to one in a linear fashion, the phase change follows the path of the arctan function, which has a slight S-shaped curve with its largest integral linearity at the 22.5° and 67.5° points. In order to do phase mixing, multiple, evenly spaced phases of the clock (four-eight) are required. These phases can either be tapped off directly from a multistage VCO of the PLL, they can be generated from a delay-locked loop (DLL) that is driven by a two-phase clock, or a higher frequency clock can be divided down to generate the needed phases. The circuit implementation of such a phase mixer is normally accomplished using current-mode mixing (Figure 6.16). Two adjacent phases of a multiphase clock are first chosen. These phases then control a differential pair that converts the varying clock voltage to a varying differential current. The magnitude of the differential current is varied by the phase interpolator setting. The two differential currents are then summed together at a load resistor and converted back to a clock voltage.

The phase mixer must be designed to handle the maximum update rate of the CDR, which is usually dictated by the PPM offset that is required to be tracked. In a mesochronous CDR system, the differential nonlinearity of the current mixing is important as this determines your dithering size around the optimal sampling point.

Systems without a CDR still require a phase mixing circuit to fine tune the receiver sampling point on a per-bit or byte basis. During a training sequence, the width of the received eye can be explored by sweeping the phase mixer setting and then finding a midpoint. This approach requires a phase mixer with excellent integral linearity. The source to the phase interpolator can either be the forwarded clock directly (through a DLL to generate multiple phases) or an on-chip locked replica clock tracking any low-frequency phase variation of the forwarded clock but rejecting high-frequency phase information. If tracking temperature variation and other time-dependent effects is not necessary or the phase can be updated periodically in the system at a high enough rate, the forwarded clock then becomes superfluous.

6.5.6 Serdes, Framing, and Resynchronization

Building upon the pin level muxing/demuxing discussed in Sections 6.5.1 and 6.5.2, serialization and deserialization can be accomplished efficiently using either a shift

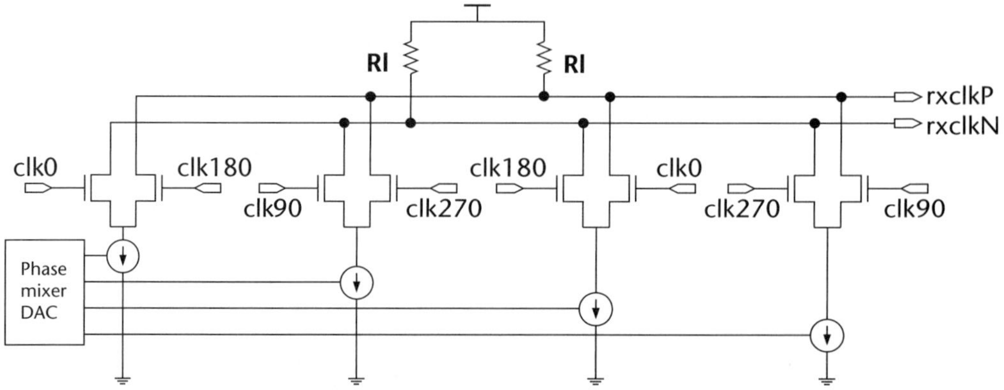

Figure 6.16 Current-mode phase-mixing circuit where the phase of the output clock is controlled by a DAC that partitions a fixed current amount to two adjacent phases to achieve the desired phase mixing.

register approach that accumulates bits at the high-speed clock and then unloads into the parallel domain or a tree-based approach that doubles the parallel width and halves the frequency with each stage until the desired parallel width is reached. Both these approaches require the synthesis of lower rate clocks with predictable/controllable phase offsets.

Receive framing is the process of realigning the parallel word to a byte boundary. This is normally accomplished either with training sequences of known data or through the use of primitives in the code space. The clocks can be actively managed to achieve alignment, or small latency can be added to the system with the addition of a barrel shifter stage.

Synchronization between the SoC clock domain and that of the high-speed internal PLL clocks of the electrical link must be accomplished on both the transmitter and receiver sides. On the transmitter side, the transmitter can provide a clock for the SoC (and let it manage the clock domains), it can actively servo its internal phase to the SoC's parallel data clock (concerns about output jitter), or it can use a low-latency FIFO to cross from the low-speed parallel clock domain to the high-speed clock domain (concerns about managing phase wander). On the receiver side, there is less flexibility, and the options are either providing the recovered parallel clock to the SoC or using a FIFO.

These lower-speed datapath stages are all digital, and the primary goals are low power, robustness, and ease of integration into SoCs.

6.6 Packaging Impact on Off-Chip Communication

From the chip I/O design perspective, an electrical link connects to an encapsulating package either through wire-bonding pads or C4 flip-chip bumps. These two options affect the I/O cell floor plan and the on-chip signal parasitics. Wire-bond pads are constrained to the perimeter of the chip; therefore, this option is used more for low-cost, low-integration parts and does not push the limits of aggregate system bandwidth. The high-speed circuitry/ESD is pushed near the edge as much as possible such that routing to the wire-bond pad is minimized. The width of an I/O cell is constrained by the number of pads dedicated to it (i.e., two signal pads and two supply pads for a width of four pads), and the height is determined by the complexity of the interface. The wire-bond pad capacitance itself can be made quite low if the wire-bond pad is only in upper metal and void of metals underneath, although some packaging/silicon technologies now allow circuit-under-pad (CUP), which decreases die size possibly at the cost of extra parasitics. C4 bumps (first-level interconnection) are generally area-arrayed across the die and therefore provide a signal count proportional to the die area as opposed to the perimeter; therefore, this is the choice for high-bandwidth systems. The high-speed circuitry/ESD is located underneath the signal bumps as much as possible to provide a mostly vertical connection out to the package. The I/O cell size is now dictated in both the width and height based on the number of signal/supply bumps reserved. The signal bump capacitance can be quite significant if there is dense power gridding on the metallization levels just below it. In the past, dual wire-bond/flip-chip designs were accommodated through the use of a redistribution layer, which was an extra routing layer that connected a wire-bond interface over to a C4 bump. For lower speeds, this is a good solution, but the parasitics of such a route will limit the signal bandwidth and increase the S11 at high frequencies.

The wire-bond and C4 options also affect the electrical propagation path to the package route. The bonding finger can have a significant amount of inductance and capacitance and is more prone to crosstalk. The inductance can actually be good for signal paths and reflections as it can cancel out some of the on-chip capacitance, but it is quite detrimental to the power-supply connections that are feeding the circuitry needed to achieve these high bandwidths. The C4 in general is electrically superior.

In chip-to-chip systems where two chips can be placed in close proximity to one another on a board (e.g., game systems), a significant fraction of the path distance between the transmitter bump and receiver bump is actually in the packages. This distance is proportional to the package size itself, which is a function of the total ball count, packaging technology, and the ball-to-ball spacing (second-level interconnection). Current ball spacings are typically 1 mm with 0.8 mm on the horizon.

An electrical package usually dedicates some number of layers for signal routing such that bumps can fan out to each ball. A package is normally built from a core that provides stability as well as at least the first two metallization layers. Further layers are then built up on both sides from this core. For example, a 3-2-3 packaging stack-up would have a total of eight layers, two of which were part of the original core and then three built up on top and bottom (Figure 6.17).

The signal integrity of these layers is critical to achieving high bandwidths. Since the distance is nontrivial, the frequency-dependent attenuation must be modeled and kept to a minimum. The characteristic impedance of the routes and vias between layers should closely match that of the board and the silicon to minimize reflections. These reflections are especially troublesome as they occur near the transmitter and have not been attenuated by the full path traversal, and they can cause

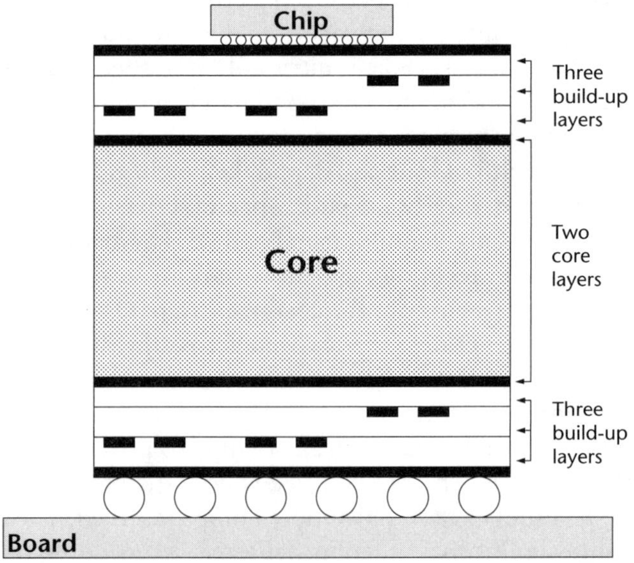

Figure 6.17 Typical 3-2-3 ball grid array (BGA) stack-up. Cooling interconnect network is not shown in the schematic.

resonances in the S21. Finally, the package design has a huge impact on crosstalk. Similar to a board stack-up, the signal-routing layers can be stripline, which is superior for crosstalk, or microstrip. Also, the number of routing layers will dictate how dense signal wiring must occur.

A C4 area array can accommodate a very large number of signals. For example, with an 180 μm bump pitch, a 15 × 15 mm die size will have on the order of 7,000 bumps, of which ~25% could be signals. The diameter of the actual C4 solder ball connecting this bump to the package, as well as the vias down to other routing layers, will be less than the pitch to accommodate routing between C4 bumps/vias on the signal layers of the package. The scale of the ball grid array (BGA) ball field is possibly a factor of three to four larger than the actual die, and most if not all the BGA balls under the die will be power-supply connections, so there is a critical perimeter around the die through which all signals need to escape on the various routing layers (Figure 6.18).

Practically, this means that for a given number of routing layers on the package, there is a maximum number of C4 bumps that can be escaped per unit perimeter of the die. Also, having many routing layers on the package means going through vias, including possibly core vias, that can be quite capacitive and have poor signal integrity. Scaling design rules and packaging technology to ensure this is not a bottleneck will continue to be a challenge.

This same escape-routing argument can be made for the next level of packaging hierarchy: from the BGA balls landing on the board and out the perimeter of the packaged part. Given two chips adjacent to one another on a board, the maximum cross-sectional bandwidth will be limited by either the number of board trace routes that can be made in a three-dimensional slice between the chips or by the ability to escape the number of signals out one side of the packaged parts perimeter. Board technology will need to keep pace with shrinking BGA ball pitch.

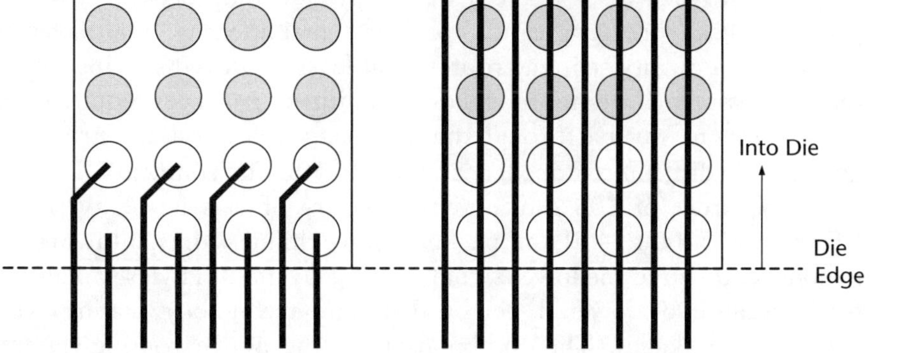

Figure 6.18 Example escape routing of eight differential pairs using two packaging layers.

Chip/package codesign and even chip/package/board codesign will grow in importance as electrical signaling is pushed to its limits.

6.7 New Interconnect Structures, Materials, and Packages

Most of the frequency-dependent attenuation discussed in Section 6.2.4 comes from traces between chips on motherboards, plug-in cards, and across backplanes. The high-volume manufacturing/cost-performance solution is to use printed circuit boards (at least four layers) with 0.5 to 2 oz. electroplated (corresponding to thicknesses of 0.675 to 2.7 mils) copper on an FR4 substrate with plated through hole (PTH) vias through the entire board thickness. Advanced board materials focus on decreasing the loss tangent of the dielectric materials used to create the insulating substrate. For FR4, dielectric loss becomes more significant to the overall attenuation than skin effect at just 750 MHz [13]. The dielectric loss tangent of FR4 (0.02 to 0.03) is simply too large to support data rates approaching 10 Gbps in backplane systems with significant trace length. Table 6.4 identifies PCB materials that have been quoted in the literature, their dielectric loss tangent, and the S21 loss per inch at 20 GHz [13, 53].

This table shows that materials and PCB processing can reduce the loss tangent by an order of magnitude or more and that the loss for a 12-inch channel can be reduced from –30 dB down to –5 dB. Therefore, PCBs need not limit bandwidths if cost can be controlled. In addition to materials, advanced PCB manufacturing steps can continue the scaling of electrical signaling, once again at the challenge of increased cost. Processes to make multilayer boards more cost competitive with standard four-layer boards would allow for stripline routing and crosstalk reduction. Manufacturing processes that aid signal integrity include blind (connecting inner layer to outer layer) or buried (connecting two inner layers) vias, which remove via stubs that produce capacitive discontinuities. This is accomplished either through controlled-depth drilling or by predrilling individual PCB layers before lamination. An alternative to these more costly solutions is to counterbore PTH vias by back-drilling the plated copper to reduce the length of the via and thus the stub and lumped capacitance. Recent work considering the backplane and electrical requirements to scale bandwidths to 25 Gbps proposed a roadmap to getting there, which is reproduced in Table 6.5 [13].

In short-reach links where distances are less than backplane applications, the package has a more significant impact on signal integrity. Multilayer packages can provide high-quality stripline routes or at least improved spacing on signal lines to reduce crosstalk. Current "thick-core" multilayer packages require roughly half the signal routes to traverse through the core using large vias that introduce capacitance and also are difficult to impedance-match to 50Ω environments. This core thickness is on the order of 800 μm, and the density of vias is set by the aspect ratio. "Thin-core" (200 to 400 μm) packages have thinner cores, allowing for better vias to connect down to the lower signal-routing layers. PTH vias can also have smaller pitch because of lowered height. Ultimate technology scaling would lead to "coreless" packages. The combination of the above routing/via improvements with finer BGA ball pitches should constrain the total distance routed, minimize

Table 6.4 Potential Improvements in Dielectric Loss and Overall Attenuation Through the Use of Advanced Materials Technology in PCB Manufacturing

Materials Technology	*Common/ Brand Name*	*Products*	*Loss Tanget*	*Loss (dB)/Inch at 20 GHz*
Epoxy resin with woven glass reinforcement	Generic FR4	Mainstream consumer and computing products	0.025	–2.5
Silicon	Multichip module	High-end processor solutions	0.015	–2.4
Thermoplastic modified epoxies	Low-loss FR4, Nelco4000-13, Nelco4000-13 SI	High-performance backplanes	0.01	–1.1
Hydrocarbon ceramic woven glass	Rogers 4350/4003	Micro-/millimeter wave	0.005	–0.7
Polytetrafluorothylene (PTFE) resin	Teflon, Duroid	Micro-/millimeter wave	0.001	–0.4

Table 6.5 Projection of PCB Materials, Manufacturing Processes, System Physical Architecture, and Pin Electronics Needed to Achieve Future Data Rate Nodes

Data Rate, Gbps	*2.5–3.125*	*5.0–6.25*	*10.0–12.5*	*20.0–25.0*
Trends				
Dielectric material	FR4	FR4	FR4/lower-loss material	Lowest-loss material
Via length reduction (counterboring)	No	Potentially	Yes for backplanes	Yes for line/switch cards
Board interconnect architecture	Backplane	Backplane	Backplane/midplane	Midplane
Tx eq. taps	1 or 2	≥2	>2	>2
Rx linear eq.	No	Potentially	Yes	Yes
Rx DFE taps	No	Potentially	Yes	Yes
Rx prDFE or duo binary	No	Potentially	Yes	Yes
Tx/Rx capacitive loading	1 to 2 pF	1 to 1.5 pF	<1 pF	<1 pF

(*Source:* R. Kollipara, Rambus Inc. Reprinted with permission from paper presented at DesignCon 2006, Santa Clara, California)

crosstalk, and minimize impedance mismatches causing reflections. A second class of packaging solutions that attempts to remove the PCB board altogether is called system-in-package (SiP), where multiple bare or packaged die communicating are stacked vertically or side by side in an multichip module (MCM) configuration. This approach has been most successful in cell phones, where the SiP advantages of system footprint and integration of disparate technologies have been proven. SiPs will have signaling advantages in terms of performance, power, and chip-to-chip density as the interconnect distance is shorter due to the vertical connectivity. SiPs are a simple, cost-effective form of 3D integration. True 3D integration of silicon substrates with through wafer vias remains an active area of research and could produce the most compelling system-in-package solutions and enable the full capability of silicon scaling.

Even with advanced packages and board materials and manufacturing, there are still impedance tolerances across the die-package and package-board interfaces giving rise to reflections. To extend the lifetime of copper-based chip-to-chip connections, researchers have proposed low-loss ribbon cables direct-attached to the package substrate, using flex-circuit technologies, that claim a threefold improvement of raw bandwidth over FR4-based systems [54, 55]. This solution bypasses the conventional microprocessor socket, vertical interconnect to board, and motherboard, yielding an environment with no via stubs, fewer discontinuities, and less loss. One such flex-circuit material is liquid crystal polymer (LCP), which has a loss tangent between 0.002 and 0.006. Challenges that remain are defining a separable connection point for upgrades or replacements and scaling to high pin counts. Current proven solutions [54] are limited to about 10 differential pairs on one package side, while conventional solutions can provide 100 to 150 differential pairs. The strategy is to segregate the most critical signals to a more easily controlled and optimized signal path, possibly through the top of the package. Beyond performance enhancement, power reduction is possible due to smaller swings and simpler pin electronics.

While most of the chapter has focused on the circuit ends of a link, there is a tremendous community of engineers solving the key interconnect challenges in the path from circuit to circuit, and the continued scaling of copper-based electrical signaling looks promising.

6.8 Conclusion

Copper-based electrical signaling continues to be the dominant mode of off-chip communication within consumer-based high-bandwidth systems, home/office computing platforms, and backplane-based servers/networks; however, signal count and per-signal bandwidth will need to continue aggressive scaling from several gigabits per second to nearly 100 Gbps over the next 10 to 15 years so as not to become the limiter of semiconductor scaling. The challenges are many: the power/area of large-scale integration of high-speed I/O macrocells; bandwidth limitations on-chip such as C_i; signal-integrity concerns such as attenuation, crosstalk, reflections, and ISI; power integrity, supply noise, and managing circuit sensitivity; and scaling of random and deterministic jitter components. Improvements and hard work on multiple fronts will maintain the momentum of electrical-based interconnects in the face of these challenges: (1) following Moore's law to enable the continued scaling of FO4 delay, as well as of the power (mW/Gbps) of increased pin electronics fitting in the area of a C4 bump, (2) clever circuits and architectures enabled by the ever-freer transistors in this C4 footprint and proven, yet evolving, statistical analysis techniques; (3) evolutionary and revolutionary packaging technologies to interconnect chips that improve signal and power integrity; and (4) lower-loss materials and potentially segregated paths for the critical interfaces of a high-bandwidth chip. Paradigm shifting/revolutionary changes such as directly integrated optics continue to be forestalled by these concerted efforts.

References

[1] International Technology Roadmap for Semiconductors (ITRS), 2006, www.itrs.net.

[2] Gschwind, M., et al., "Synergistic Processing in Cell's Multicore Architecture," *IEEE Micro*, Vol. 26, No. 2, 2006, pp. 10–24.

[3] Vangal, S., et al., "An 80-Tile 1.28 TFLOPs Network-on-Chip in 65 nm CMOS," *ISSCC Dig. Tech. Papers*, San Francisco, CA, February 11–15, 2007, pp. 98–99.

[4] Bell, G., J. Gray, and A. Szalay, "Petascale Computational Systems," *IEEE Computer*, Vol. 39, No. 1, 2006, pp. 110–112.

[5] Horowitz, M., C.-K. K. Yang, and S. Sidiropoulos, "High-Speed Electrical Signaling: Overview and Limitations," *IEEE Micro*, Vol. 18, No. 1, 1998, pp. 12–24.

[6] Poulton, J., et al., "A 14 mW 6.25 Gb/s Transceiver in 90 nm CMOS," *IEEE J. Solid-State Circuits,* Vol. 42, No. 12, 2007, pp. 2745–2757.

[7] Chang, K., et al., "Clocking and Circuit Design for a Parallel I/O on a First-Generation CELL Processor," *ISSCC Dig. Tech. Papers*, San Francisco, CA, February 6–10, 2005, pp. 526–527.

[8] Krewell, K., "Cell Moves into the Limelight," *Microprocessor Report Newsletter*, February 14, 2005, pp. 1–9.

[9] JEDEC Solid State Technology Association, "JESD 22-A114D: Electrostatic Discharge (ESD) Sensitivity Testing Human Body Model (HBM)," October 2007, www.jedec.org.

[10] Industry Council on ESD Target Levels, "White Paper 1: A Case for Lowering Component Level HBM/MM ESD Specifications and Requirements," October 2007, www.esdtargets.blogspot.com.

[11] Salman, A. A., et al., "Field Effect Diode (FED): A Novel Device for ESD Protection in Deep Sub-Micron SOI Technologies," *International Electron Devices Meeting*, San Francisco, California, December 11–13, 2006, pp. 1–4.

[12] Deutsch, A., et al., "Prediction of Losses Caused by Roughness of Metallization in Printed-Circuit Boards," *IEEE Trans. Advanced Packaging*, Vol. 30, No. 2, 2007, pp. 279–287.

[13] Kollipara, R., et al., "Practical Design Considerations for 10 to 25 Gbps Copper Backplane Serial Links," *DesignCon 2006*, Santa Clara, California, February 6–9, 2006.

[14] Daly, W. J., and J. W. Poulton, *Digital Systems Engineering*, New York: Cambridge University Press, 1998.

[15] Brox, M., et al., "A 2 Gb/s/pin 512 Mb Graphics DRAM with Noise-Reduction Techniques," *ISSCC Dig. Tech. Papers*, San Francisco, CA, February 5–9, 2006, pp. 158–159.

[16] Stan, M. R., and W. P. Burleson, "Bus-Invert Coding for Low Power I/O," *IEEE Transactions on VLSI*, Vol. 4, No. 1, 1995, pp. 49–58.

[17] Bakoglu, H. B., *Circuits, Interconnections, and Packaging for VLSI*, Reading, MA: Addison-Wesley Publishing Company, 1990.

[18] Johnson, H., and M. Graham, *High-Speed Digital Design: A Handbook of Black Magic*, Upper Saddle River, NJ: Prentice Hall, 1993.

[19] Johnson, M. G., and E. L. Hudson, "A Variable Delay Line Phase Locked Loop for CPU-Coprocessor Synchronization," *ISSCC Dig. Tech. Papers*, San Francisco, CA, February 17–19, 1988, pp. 142–143.

[20] Watson, R., and R. Iknaian, "Clock-Buffer Chip with Multiple-Target Automatic Skew Compensation," *ISSCC Digest of Technical Papers*, San Francisco, CA, February 15–17, 1995, pp. 106–107.

[21] Technical Committee T11, "Fibre Channel—Methodologies for Signal Quality Specification—MSQS," October 2007, www.t11.org.

[22] Stojanovic, V., and M. A. Horowitz, "Modeling and Analysis of High-Speed Links," *Proc. Custom Integrated Circuits Conference*, San Jose, CA, September 21–24, 2003, pp. 589–594.

[23] Sanders, A., M. Resso, and J. D'Ambrosia, "Channel Compliance Testing Utilizing Novel Statistical Eye Methodology," *DesignCon West 2004*, Santa Clara, CA, February 2–5, 2004.

[24] Farjad-Rad, R., et al., "A 0.3-um CMOS 8 Gb/s 4-PAM Serial Link Transceiver," *IEEE J. Solid-State Circuits*, Vol. 35, No. 5, 2000, pp. 757–764.

[25] Zerbe, J., et al., "1.6 Gb/s/pin 4-PAM Signaling and Circuits for a Multidrop Bus," *IEEE J. Solid-State Circuits*, Vol. 36, No. 5, 2001, pp. 752–760.

[26] Stonick, J. T., et al., "An Adaptive PAM-4 5-Gb/s Backplane Transceiver in 0.25-um CMOS," *IEEE J. Solid-State Circuits*, Vol. 38, No. 3, 2003, pp. 436–443.

[27] Bulzacchelli, J. F., et al., "A 10-Gb/s 5-tap DFE/4-tap FFE Transceiver in 90-nm CMOS Technology," *IEEE J. Solid-State Circuits*, Vol. 41, No. 12, 2006, pp. 2885–2898.

[28] Lender, A., "The Duobinary Technique for High-Speed Data Transmission," *IEEE Transactions on Communication and Electronics*, Vol. 82, No. 5, 1963, pp. 214–218.

[29] Götz, H., and J. Sinsky, "The Duobinary Format—A New Application for an Idea Published Long-Ago," *Euro DesignCon 2005*, Munich, Germany, October 24–27, 2005.

[30] Amirkhany, A., et al., "A 24 Gb/s Software Programmable Multi-Channel Transmitter," *IEEE Symposium on VLSI Circuits*, Kyoto, Japan, June 14–16, 2007, pp. 38–39.

[31] Widmer, A. X., and P. A. Franaszek, "A DC-Balanced, Partitioned-Block, 8B/10B Transmission Code," *IBM J. Res. Dev.*, Vol. 27, No. 5, 1983, pp. 440–451.

[32] JEDEC Solid State Technology Association, "JESD 79–3A: DDR3 SDRAM Standard," October 2007, www.jedec.org.

[33] Dreps, D., "The 3rd Generation of IBM's Elastic Interface on POWER6," *Hot Chips 19*, Palo Alto, CA, August 19–21, 2007.

[34] Chiang, P., et al., "A 20-Gb/s 0.13-um CMOS Serial Link Transmitter Using an LC-PLL to Directly Drive the Output Multiplexer," *IEEE J. Solid-State Circuits*, Vol. 40, No. 4, 2005, pp. 1004–1011.

[35] Harwood, M., et al., "A 12.5 Gb/s SerDes in 65 nm CMOS Using a Baud-Rate ADC with Digital Receiver Equalization and Clock Recovery," *ISSCC Dig. Tech. Papers*, San Francisco, CA, February 11–15, 2007, pp. 436–437.

[36] Dally, W. J., and J. Poulton, "Transmitter Equalization for 4-Gbps Signaling," *IEEE Micro*, Vol. 17, No. 1, 1997, pp. 48–56.

[37] Lee, Edward M. J., "An Efficient I/O and Clock Recovery Design for Terabit Integrated Circuits," PhD thesis, Stanford University, August 2001.

[38] Ho, A., et al., "Common-Mode Backchannel Signaling System for Differential High-Speed Links," *IEEE Symposium on VLSI Circuits*, Honolulu, Hawaii, June 17–19, 2004, pp. 352–355.

[39] Jaussi, J. E., et al., "8 Gb/s Source-Syncrhonous I/O Link with Adaptive Receiver Equalization, Offset Cancellation and Clock Deskew," *IEEE J. Solid-State Circuits*, Vol. 40, No. 1, 2005, pp. 80–88.

[40] Farjad-Rad, R., et al., "0.622–8.0 Gbps 150 mW Serial IO Macrocell with Fully Flexible Preemphasis and Equalization," *IEEE Symposium on VLSI Circuits*, Kyoto, Japan, June 12–14, 2003, pp. 63–66.

[41] Payne, R., et al., "A 6.25 Gb/s Binary Adaptive DFE with First Post-Cursor Tap Cancellation for Serial Backplane Communications," *ISSCC Dig. Tech. Papers*, San Francisco, CA, February 6–10, 2005, pp. 68–69.

[42] Kasturia, S., and J. H. Winters, "Techniques for High-Speed Implementation of Non-Linear Cancellation," *IEEE J. Select. Areas Communications*, Vol. 9, No. 6, 1991, pp. 711–717.

[43] Stojanovic, V., et al., "Autonomous Dual-Mode (PAM2/4) Serial Link Transceiver with Adaptive Equalization and Data Recovery," *IEEE J. Solid-State Circuits*, Vol. 40, No. 4, 2005, pp. 1012–1026.

[44] Balan, V., et al., "A 4.8–6.4 Gb/s Serial Link for Backplane Applications Using Decision Feedback Equalization," *IEEE J. Solid-State Circuits*, Vol. 40, No. 11, 2005, pp. 1957–1967.

[45] Beukema, T., et al., "A 6.4-Gb/s CMOS SerDes Core with Feed-Forward and Decision-Feedback Equalization," *IEEE J. Solid-State Circuits*, Vol. 40, No. 12, 2005, pp. 2633–2645.

[46] Gardner, F. M., *Phaselock Techniques*, New York: John Wiley & Sons, 1979.

[47] Best, R. E., *Phase-Locked Loops: Design, Simulation, and Applications*, New York: McGraw-Hill, 1999.

[48] Razavi, B. (ed.), *Phase-Locking in High-Performance Systems: From Devices to Architectures*, Hoboken, NJ: John Wiley & Sons, 2003.

[49] Maneatis, J. G., "Low-Jitter Process-Independent DLL and PLL Based on Self-Biased Techniques," *IEEE J. Solid-State Circuits*, Vol. 31, No. 11, 1996, pp. 1723–1732.

[50] Sidiropoulos, S., et al., "Adaptive Bandwidth DLLs and PLLs Using Regulated Supply CMOS Buffers," *IEEE Symposium on VLSI Circuits*, Honolulu, HI, June 15–17, 2000, pp. 124–127.

[51] Alexander, J. D. H., "Clock Recovery from Random Binary Signals," *Electronics Letters*, Vol. 11, No. 22, 1975, pp. 541–542.

[52] Lee, H., et al., "Improving CDR Performance via Estimation," *ISSCC Dig. Tech. Papers*, San Francisco, CA, February 5–9, 2006, pp. 332–333.

[53] McMorrow, S., and R. Weiss, "Feasibility of 40 to 50 Gbps NRZ Interconnect Design for Terabit Backplanes," *DesignCon East*, Worcester, MA, September 19–21, 2005.

[54] Braunisch, H., et al., "Flex-Circuit Chip-to-Chip Interconnects," *Electronic Components and Technology Conference Proceedings*, San Diego, CA, May 30 to June 2, 2006, pp. 1853–1859.

[55] Grundy, K., et al., "Designing Scalable 10G Backplane Interconnect Systems Utilizing Advanced Verification Methodologies," *DesignCon 2006*, Santa Clara, CA, February 6–9, 2006.

CHAPTER 7

Optical Interconnects for Chip-to-Chip Signaling

Alan F. Benner and Oluwafemi O. Ogunsola

7.1 Introduction

Optical interconnects, or optical links, have been regularly considered for decades as a solution to increasing problems in interconnect bandwidth, distance, and density requirements. Despite continuing difficulties with cost—the set of things that can be demonstrated with optical technology is far larger than the set of things that can be cost-competitively deployed against electrical counterparts—there has been a steadily increasing use of optical interconnects as system performance requirements have increased, particularly in large-scale systems [1]. As common I/O bit rates for silicon chips have approached or exceeded the super high frequency (SHF) range (3 to 30 GHz, 10 to 1 cm wavelength), the use of optical signaling for chip-to-chip communications between silicon chips is starting to be quite viable versus copper.

With the advent of small, power-efficient semiconductor lasers and low-loss optical fibers in the late 1970s and early 1980s, optical interconnects became feasible but were only economically viable for long-distance links (e.g., >1 km), where the cost and bulkiness of RF cables made optical fiber transmission a lower-cost alternative. Since then, silicon CMOS technology has steadily improved at a remarkable rate due to scaling [2]. Interconnect technology, measured in bandwidth-distance product (BDP), has also steadily improved, but at a slower rate, particularly for off-chip signaling [2], leading to a steady need for innovations in system design and architecture to accommodate the changing ratios between communications and computational capabilities. Many of the inventions in system design, such as caches, vector processors, very large instruction word (VLIW), parallel processing, and chip multiprocessing or multicore processors, have become practical because of the growing inability of processing cores to gather and transport data at a rate commensurate with their ability to process it. The migration from electrical to optical interconnect, while long resisted for good cost-performance reasons, is starting to directly impact significant aspects of system design, particularly in high-end systems, as designers work to maintain overall design balance by improving the system interconnect matched to the rate of improvement in silicon processing capabilities.

As shown in Figure 7.1, links of different lengths, widths, and formats are required to satisfy the communications requirements for systems of various sizes;

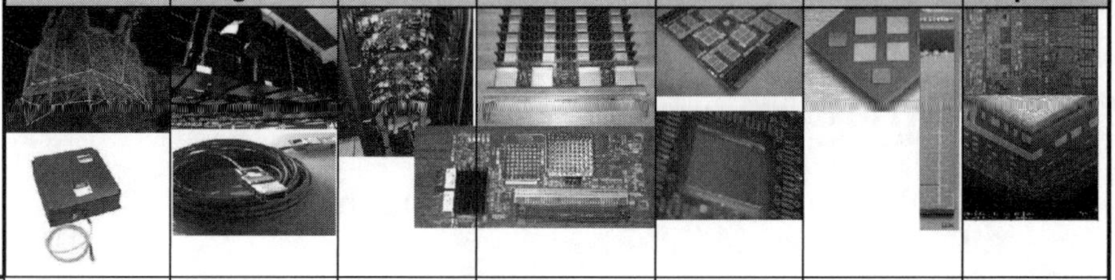

	MAN & WAN	Cables – Long	Cables – Short	Card-to-Card	Intra-Card	Intra-Module	Intra-chip
Length	Multi-km	10m - 300 m	1 m - 10 m	0.3 m - 1 m	0.1 m - 0.3 m	5 mm - 100 mm	0 mm - 20 mm
Typical # lines per link	1	1 - 10s	1 - 10s	1 - 100s	1 - 100s	1 - 100s	1 - 100s
Use of optics	Since 80s	Since 90s	Now	2005-2010 with effort	2010 - 2015	Probably after 2015	Later

Figure 7.1 Interconnect hierarchy showing the use of electrical and optical technology at various levels of the systems.

optical technology has become common for the longer-distance and higher-speed links with steady migration toward shorter-distance links over time [1].

Optical components (e.g., laser sources), photodetectors, and signal-routing and transmission components (e.g., fibers, waveguides, splitters and amplifiers) have been available at density, power-efficiency, and performance levels that meet the needs of systems designers using advanced silicon technology. More recently (since the late 1990s), the integration of silicon technology with optical interconnect capabilities has started to become practical, with microphotonic components such as waveguides being well developed [3, 4] and a variety of different options for detectors and even sources being demonstrated [5].

Several key factors have contributed to the practicality of integrating photonics technology in standard or near-standard silicon processing and CMOS-based designs. One is that the dimensional accuracy of lithography in silicon processing progressed past the threshold needed for working with light. Typical semiconductor laser light, with wavelengths in the range of 850 to 1,550 nm in free space (~575 to 1,055 nm in SiO_2, ~244 to 450 nm in Si), could only be managed in silicon devices when the accuracy of lithography was reduced to ~1/5 to 1/10 of the wavelength, which happened as standard silicon technology made the 130 to 90 to 65 nm transition in technology node between 2001 and 2007 [6]. A further factor is the increased ability to exploit different materials. Where silicon processing in the 1970s used perhaps a dozen elements from the periodic table, current processing uses well over 50. Procedures have been established to prevent cross-contamination and assure mechanical integrity, which is critical for use of materials like germanium (Ge) and indium phosphide (InP), which have different and better optical properties than silicon (Si) or silicon dioxide (SiO_2).

With these progressions in technology, integration of optical devices and components onto silicon chips for chip-to-chip communications has progressed to the level of commercial deployment, with steady improvement in performance and cost-effectiveness. Devices with promise even for on-chip communications (with both transmitter and receiver of a link on the same chip) have started to approach feasibility, although the competition against electrical interconnect is much stiffer for these links [7]. More practically, the integration of optical components on the same first-level packages as silicon chips, removing the need for electrical I/O between the first-level packages and printed circuit boards, is rapidly approaching commercial competitiveness and is likely to dramatically improve the BDP and power efficiency of chip-to-chip communications in the 2009 to 2012 and beyond time frame. In the end, optical interconnect technology has to prove its worth against the next-best alternative, the electrical interconnect, which continues to improve at a high rate. The confluence of increasing cost-competitiveness, silicon-compatible devices and integrations schemes, and the need for high-performance I/O chip bandwidth due to technology scaling serve to motivate the use of optics. The focus of this chapter, therefore, is to provide the reader with insight into why optical interconnects are of interest, how they compare with their electrical counterparts, and how the fundamental requirements of an optical link can be used to achieve high-bandwidth chip-to-chip and board-to-board interconnection.

7.2 Why Optical Interconnects?

At a fundamental level, electrical communication through wires involves the transport of energy through the medium of induced voltage differences that are transported by displacements of electrons in conductors. Optical communication involves emission, transport, and absorption of photons.[1] The transport of photons versus displacement of electrons produces different effects, reflected in attenuation or distortion of the signals as a function of frequency and distance. The question then becomes, do these fundamental differences make optics advantageous over electrical interconnects for high-speed chip-to-chip and board-to-board communication? To answer that question, we first need an understanding of the problem that optics is being proposed to solve. As electrical interconnects are covered in great detail elsewhere in this book, a summary is provided below.

7.2.1 The Semiconductor Industry's Electrical Interconnect Problem

For four decades, the semiconductor industry has delivered chips with greater transistor densities, higher clock speeds, and greater I/O bandwidth. This achievement is the result of scaling down the minimum feature sizes and reducing the cost of production [6]. While scaling transistor dimensions reduces their intrinsic switching delay [2], it increases the intrinsic on-chip interconnect delay by increasing the effective resistance [2, 8]. As is shown in [9], on-chip wires' delay operates in the resistance-capacitance (RC) region. Thus, higher resistance produces longer delay. To

1. Note that RF transmission—microwave, radio, wireless, and so forth—shares much of the same physics as optical communication and many of the same advantages. However, the huge difference in wavelength—meters or centimeters for RF versus micrometers for optical—implies a completely different set of components for the generation, transmission, and detection of the two types of communications.

illustrate this effect, [2] shows that the intrinsic delay of a 1-mm-long interconnect fabricated using 35 nm technology will be 100 times longer than that of a transistor fabricated in the same technology. As scaling moves interconnect dimensions closer to those of the mean free path of electrons, surface scattering effects are introduced. Also, at high frequencies, skin depth is on the order of the wire dimensions; thus, the effective cross-sectional area of the conductor decreases [9]. Both effects further increase effective resistance.

Wires that are thicker than 1 μm operate in the resistance-inductance-capacitance (RLC) delay region [9]. These interconnects correspond to board-level wires. At high frequencies, they are limited by time-of-flight and not increased resistance. The board-level problem is then a bandwidth issue [10] as signal speed is limited by material losses [11]. The authors of [11, 12] show that above 1 GHz, dielectric loss is greater than skin-effect loss for copper traces on FR4. These losses diminish bandwidth density. As an example, [11] shows that a $203 \times 17.8\ \mu$m copper trace that is 40 cm long suffers a 20 dB loss at 10 GHz on FR4.

To summarize, increased effective resistance on-chip and frequency-dependent loss on the board constitute the electrical interconnect problem. The former increases on-chip interconnect delay, while the latter reduces board-level bandwidth. While optical interconnects may indeed migrate on-chip, their need to resolve interconnect delay is not yet compelling. To this end, on-chip optics may be restricted to the monolithic integration of vertical cavity surface emitting lasers (VCSELs) or photodetectors (PDs) in the near future to enable chip-to-chip and, ultimately, board-to-board communication to address the board-level bandwidth requirements.

Historically, when the demand for low-loss, high-speed, and high-bandwidth data for long-distance applications has grown, the telecommunications industry has replaced electrical interconnects with their optical counterparts. The seminal case illustrating the effect of such a crossover is the transition from thousands of kilometers of electrical transatlantic transmission cables, TAT-7, to single-mode optical fiber, TAT-8, in the late 1980s [13, 14]. At that time, the number of circuits serviceable by TAT-7 was 4,000, while TAT-8 serviced 40,000 [13]. This change from coaxial copper cables was done to accommodate higher bandwidth; it also yielded greater connectivity at a lower cost—switching to fiber versus inserting amplifiers. As of 1993, the equivalent cost of a transatlantic cable circuit had decreased from $6 million in 1958 to $4,000 [13], and by 1995, the cost had dropped further to $150 [14]. As a result of this transition, optical interconnects are ubiquitous in global telecommunications. From a hardware perspective, they facilitated the growth of the Internet. This example demonstrates how switching to optics can increase bandwidth and connectivity, lower cost, and enable high-performance applications. It is in an attempt to reap these same kinds of demonstrated benefits that optical interconnects are being investigated in the chip semiconductor world.

7.2.2 The Optical Interconnect Solution

While it is clear from history that optical interconnects outperform electrical ones over thousands of meters, their benefit for lengths that are less than a meter—corresponding to board-level interconnects—has been a source of significant debate since the early 1980s [15–18]. Since the idea was first published in 1984 [15, 16], years of work on this subject have followed. The accumulated rationale and challenges for optical interconnects were recently summarized in [16]. The fundamental question is, if optical interconnects can be made in a compatible manner, will they be a direct answer to the I/O bandwidth requirements for the industry?

Though there are many arguments presented in [16], the most compelling of these is bandwidth. The authors of [17] have compared the aggregate bandwidth of equivalently dimensioned board-level optical waveguides and electrical wires over a range of transmission rates. A partition length was derived for which optical bandwidth is always greater. As technology scales, this length decreases from about 17 cm at 5 GHz to below 1 cm at 30 GHz [17]. Hence, the bandwidth advantage of optical interconnects can be reaped for board-level dimensions. In addition, an optical waveguide allows wavelength-division multiplexing (WDM), transporting data on multiple wavelengths simultaneously through the same waveguide, so the ratio of bandwidth per area of transmission can (with WDM on single-mode waveguides) be improved over electrical interconnects by one or two orders of magnitude for on-chip interconnects and at least three orders of magnitude for chip-to-chip links.

A further advantage for optical interconnects is signal density, as shown in Figure 7.2, which shows pictorially and to scale the relative sizes of on-chip wiring, single-mode waveguides in glass fibers, multimode waveguides, and differential pairs of striplines in printed circuit board wiring. It is quite clear that optical waveguides are dramatically denser than circuit board wiring. Even multimode waveguides provide a nearly 16-fold improvement in interconnection density; that is, a single optical waveguide layer could conceivably replace roughly 16 signaling layers, or nearly 30 total layers of PCB material. Note, though, that the comparison versus on-chip wiring is quite different: a multimode waveguide is dramatically less dense than on-chip electrical interconnect wiring. This difference in density between

Figure 7.2 Interconnect density comparison between on-chip wires, optical waveguides, and printed circuit board electrical differential pairs.

on-chip and chip-to-chip interconnect density will have implications for what types of optical interconnect technology are practical.

At the system level, the electrical interconnect problem can ultimately limit the aggregate performance, especially for links such as CPU/DRAM, where the bit rate per channel is limited by other factors, such as chip circuit design [19]. For example, source-synchronous interconnects are used to improve chip-to-chip interconnect bandwidth by reading and writing data on both edges of a clock cycle [20]. In this way, pin I/O bandwidth can be doubled. These interconnects introduce new design-analysis challenges as latency is not easily predictable. This is because poorly terminated lines increase settling time; hence, intersymbol data interference occurs [20]. Solving this predictability issue requires added I/O electrical packaging performance, which serves to drive up cost. Optical I/O and accompanying board-level interconnects can help reduce this bottleneck for connection to printed circuit boards and between boards.

Ideally, all communication in a processing system would be done across short distances of homogeneous transmission media. However, processing practically often requires transmission over tens of centimeters or more—a current processor is capable of effectively using multiple DRAM chips worth of data. For example, in high-performance computing (HPC) systems, a ratio of 1 GB capacity per peak GFLOP per second is common such that a 10 to 100 GFLOP/s processor chip would be matched with 10 to 100 GB of memory. This is the equivalent of between 12 and 120 DRAM chips. Communication between the processor and the 120th DRAM chip can occur over a significant distance as they will not reside on a single board, limiting the memory system performance [21]. These architectural designs beg for optics to achieve high-speed and high-bandwidth board-to-board interconnection. If that happens, it will elucidate an even more compelling need for board-level optics; otherwise, we stop short of completely solving the bandwidth problem. By terminating optical wires at some chip close to the microprocessor or the board's edge, bandwidth-limited electrical wires still deliver the data to the processor. Such a scenario would be akin to the *last-mile bottleneck* currently facing the telecommunications industry. This bottleneck refers to the fact that high-speed and high-bandwidth optical fiber terminates close to the home, but lower-bandwidth coaxial cables are still being used for home delivery. This limits the residential services that can be provided [22, 23]. Optical fiber is proposed. Hence, optical I/O cannot be treated simply at one level. It must span the I/O, board-level, and board-to-board hierarchy.

System-level packaging and performance constraints determine the performance and type of interconnection capabilities required. There is a clear bandwidth advantage for optical interconnects over electrical interconnects at dimensions of interest for the semiconductor industry. For optics to be used, the key challenges are therefore cost competitiveness and architectural implementation.

7.3 Cost-Distance Comparison of Electrical and Optical Links

Figure 7.3 is a representative diagram illustrating the final user-level cost of links, as a function of distance, using both electrical and optical transmission technology.

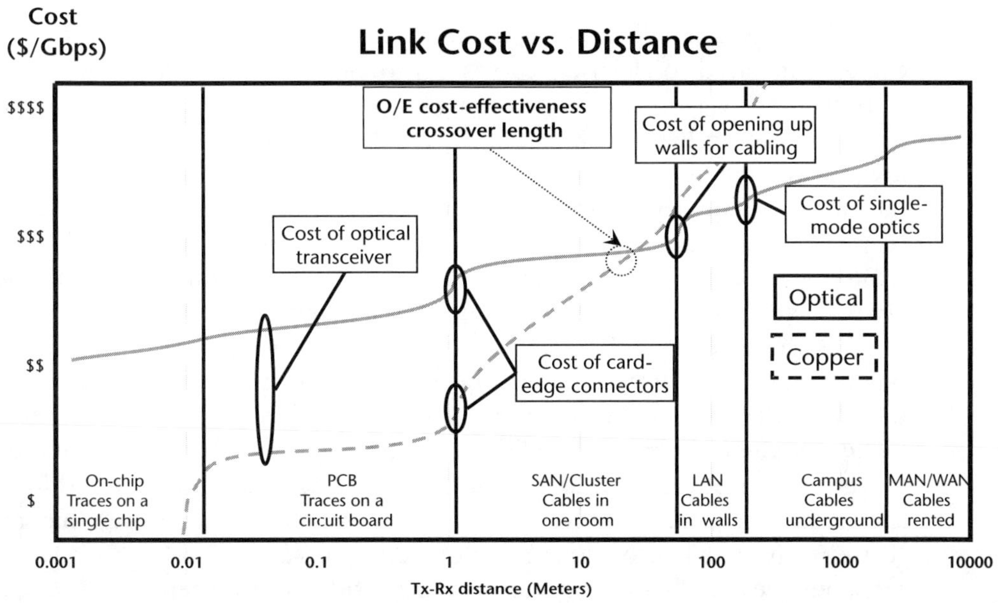

Figure 7.3 Link cost versus distance plot showing the factors determining effective optical/electrical crossover length.

The general trend of the graph—that longer links have higher cost per gigabit per second—is obvious and intuitive since longer cables cost more, and more sophisticated and expensive circuitry is required for transmission over longer, higher-loss cables. The particular ways in which the cost scales with distance are perhaps not, however, quite so intuitive and have strong implications for how systems are designed.

It should be noted that costs do not scale linearly with distance as there are threshold lengths at which the cost dramatically jumps discontinuously. These occur at distances corresponding to natural packaging limitations. For example, links shorter than ~1 cm can stay on a single silicon chip and be extremely inexpensive. Links longer than the size of a chip require the addition of first- and second-level packaging, disruptively increasing the link cost as compared to on-chip links. However, the cost of a link across a circuit board is largely independent of the length of the link (although more expensive preemphasis and equalization circuitry may be needed, especially at high bit rates, for longer traces). There is another discontinuity as the link lengths exceed the ~0.8m dimensions of a circuit board; thus, cables, connectors, and receptacles are needed. Further discontinuities occur as link lengths exceed the size of a single room, or a single building, or a single campus, as shown in the figure.

Figure 7.3 also shows the qualitative difference between copper and optical links. For short links, the optical links are dramatically more expensive since they require other components (e.g., lasers and photodetectors) that are not needed for electrical links. The electrical drive and receiver circuitry is similar in cost for both electrical and optical links. The slope of the optical links, at various distances, is dramatically lower than the slope for electrical links since optical fiber is relatively inexpensive compared to high-performance electrical cable that is needed to support high-speed signaling, especially for long distances. Also, the low loss of optical cables, compared to electrical cables, allows less-complex preemphasis and equalization circuitry, compared to the complex circuits needed for longer electrical links. The key point in Figure 7.3 is that due to the extra cost of an optical link and the lower incremental cost for longer optical links, there is an "O/E crossover length" (i.e., a critical link length analogous to the bandwidth partition length) at which the cost of optical and electrical links are the same. Shorter link lengths as of today remain less expensive to implement using electrical transmission, and longer links are less expensive to implement using optical transmission.

The distances and characteristics of the links in Figure 7.3 correspond to one operating at a moderate 2.5 Gbps bit rate. Comparative behavior at several other representative bit rates is shown in Figure 7.4. As would be expected, the costs are higher for higher-bit-rate links. For electrical interconnects, the slope of the cost/distance curves are steeper at higher bit rates since complex circuitry and high-quality cables are needed. However, there are exceptions. At lower bit rates, the cost-effectiveness falls off. If the cost of a 0.6 Gbps and a 2.4 Gbps transmitter were the same, the 2.4 Gbps transmitter would provide fourfold better cost per gigabit per second. As of this writing, the "sweet spot" for transmission bit rate, providing best performance for least overhead in circuits and wires, is in the range of 2.4 to 5 Gbps. For optical interconnects, since the cables support 40 Gbps signaling just as well as lower-bit-rate signaling, the slope is lower for higher bit rates Once optical cabling is installed, it makes sense to transport through it the highest bit rate that the transmitting and receiving equipment can cost-effectively support.

In summary, Figures 7.3 and 7.4 represent empirical data gathered over time by studying the data communications industry. They have been assembled to deliver these key points: (1) there are O/E cost-effectiveness crossover lengths, and (2) those lengths are shorter for higher bit rates. This means that optics begins to make financial sense for chip-to-chip and board-to-board interconnection from a cost perspec-

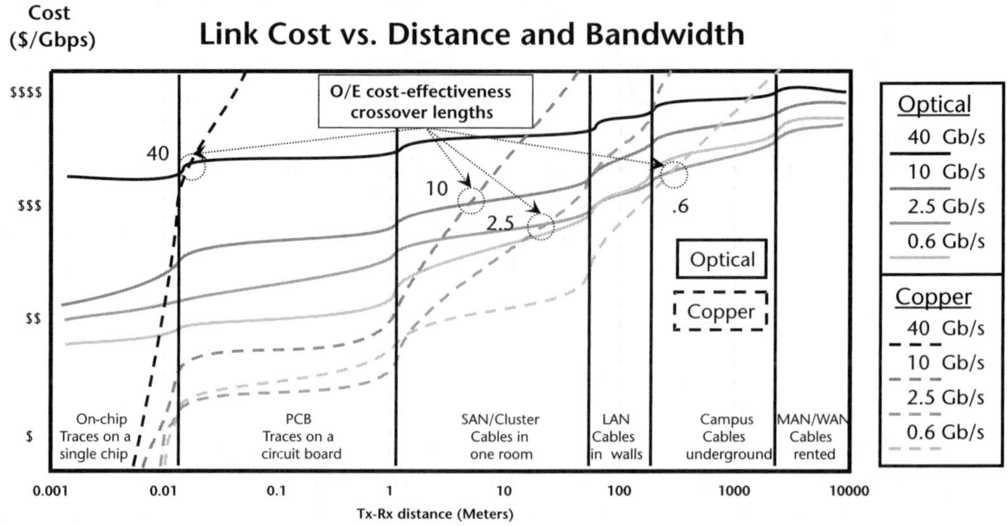

Figure 7.4 Link cost versus distance and bandwidth, showing dependency of O/E crossover length on link bit rate.

tive, and silicon technology drives higher I/O bandwidth. Since the electrical and optical technologies have historically had similar rates of cost/performance improvement, these O/E crossover lengths have been surprisingly constant over time. Also surprisingly, if power dissipation per gigabit per second is measured instead of cost per gigabit per second, a very similar set of behaviors is observed, as discussed in [24]. The net result for system designers and technology builders is that even if electrical links improve in cost/performance similarly to optical link cost/performance, there will still be a steady evolution toward a larger fraction of the links in a system being optical. It will only be a few years, as common bit rates go to 10 Gbps and above, before the only electrical signals in a system will be those that stay on a chip or circuit board, and board-to-board signals will use optical transmission.

7.4 Chip-Based Optical Interconnects

We now discuss what the optical interconnect design may ultimately evolve towards to satisfy bandwidth requirements.

7.4.1 The Optical Interconnect System

The idea of using optical interconnects in computer systems dates back to the late 1980s and early 1990s [25–29]. At the most basic level, an optical link consists of at least one optical source, constituent parts for routing, and at least one detector.

A review of the proposals for optical-interconnect-based systems cited above [25–29] identifies the following implementation scheme: (1) bring optical fibers to a board, (2) couple and route optical data between the fibers and a board-level waveguide routing network (WRN), and (3) couple light between the WRN and a CMOS driver chip with III-V vertical cavity surface emitting lasers (VCSELs) or detectors bonded on top. The driver chip communicates electrically with the microprocessor.

Admittedly, this kind of scheme provides better bandwidth than is possible today and can serve as a cost-effective approach to introducing optics into computer systems. It stops short, however, of completely solving the bandwidth problem. The reason is that by terminating the high-bandwidth optical wires at a chip close to the microprocessor, lossy electrical wires still deliver the high-speed data to the processor. This would be akin to the aforementioned last-mile bottleneck. For optics truly to solve the electrical bandwidth problem, light must originate and terminate at the CMOS microprocessor. This requires the monolithic integration of lasers, modulators, and/or detectors on silicon chips. These tasks represent critical challenges as projected by the ITRS [6]; however, on-going work on the monolithic fabrication and integration of detectors [30], modulators [31], and emitters [32] on silicon CMOS shows promise. These constituent components are discussed in Chapter 6 and are only mentioned here as needed for optical links.

Assuming such capabilities, a future optical bus configuration would now: (1) bring optical fibers to a board, (2) couple and route optical power between the fibers and the WRN, and (3) couple light between the WRN and the microprocessor. Simple point-to-point and multipoint systems are illustrated in Figures 7.5 and 7.6, respectively. The figures show polymer pillars (also called polymer pins), which

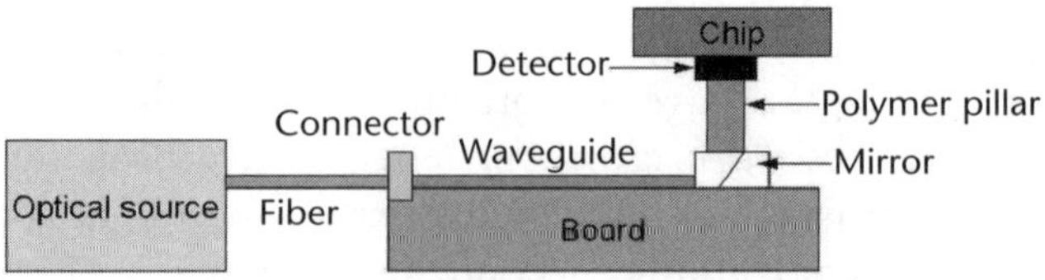

Figure 7.5 Schematic of a simple point-to-point optical interconnect link.

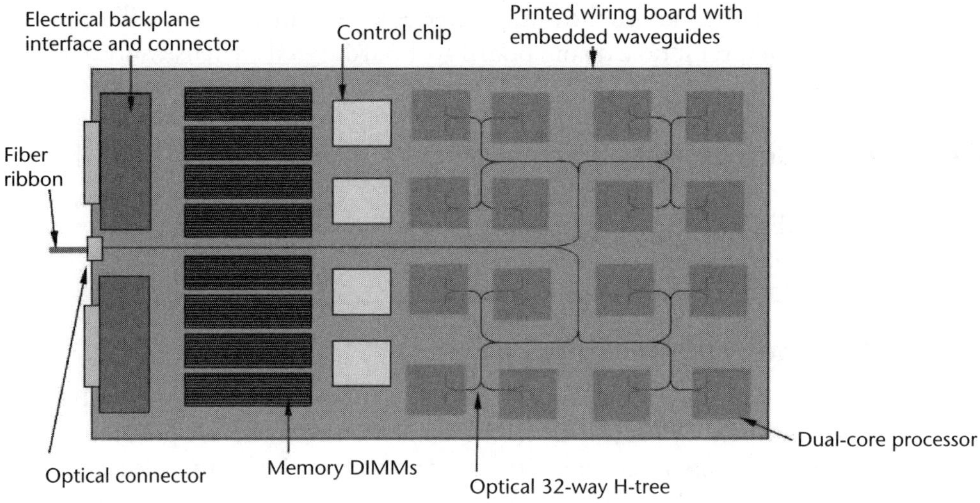

Figure 7.6 A 32-way optical board-level architecture with 16 mirror-enabled polymer pillar dual-core processor chips, control chips, dual in-line memory modules (DIMMs), electrical backplane interface and connectors, embedded waveguide H-tree, and an optical connector.

are discussed electrically and thermally elsewhere in this book. Their optical performance will be discussed shortly. To enable that, each of the three steps in the optical bus are now discussed.

7.4.2 Bringing Optical Fibers to a Board

The task here is to connect multiple boards using optical fiber. There are two ways by which optical fibers can be brought to a board as defined by what happens at its edge: termination at transmitters/receivers or at a WRN. Literature examples of transmitter/receiver-terminating fiber-to-board configurations include the Optoelectronic Technology Consortium (OETC) [33], Parallel Inter-Board Optical Interconnect Technology (ParaBIT) [34, 35], and the Multiwavelength Assemblies for Ubiquitous Interconnects (MAUI) modules [36]. The OETC module has 32 channels operating at 500 Mbps for a total of 16 Gbps; the ParaBIT bus operates at 1.25 Gbps over 48 channels for a total of 60 Gbps; and the first-generation MAUI bus operates at 5 Gbps over 48 channels for a total of 240 Gbps aggregate bandwidth. This level of bandwidth is immense. As has been stated, termination at the board edge may help introduce optics, but light must ultimately terminate at the microprocessor and not with lossy electrical wires. Thus, at the board edge, the fiber should couple light into a board-level WRN. Implementations for fiber-to-WRN coupling vary from using compact nanotapers to couple light between silicon waveguides and fiber [37] to the use of standard fiber connectors from the side [38] and top [39] of a board's edge. Any of these can be used for fiber-to-WRN depending on the system.

7.4.3 Waveguide Routing Network

The task here is to route light between chips and between chips and the board edge. The medium for guiding the light can be polymer waveguides or optical fibers; however, polymer waveguides are discussed because they lend themselves readily to denser networks due to their small dimensions. There have been successful demonstrations of optical interconnection using free-space signal transmission. For example, the Optical Transpose Interconnection System (OTIS) project [40] and an optoelectronic crossbar switch containing a terabits per second free-space optical interconnect [41] have both been demonstrated as research projects. A high-speed, large-scale IP router from Chiaro Networks using an optical phased-array structure for signal routing did reach commercial development [42]. However, in practice, free-space interconnects for applications that do not require imaging, display, or other inherently 3D operations, while technically possible, are uncompetitive against electrical interconnect for short distances and against waveguides like optical fibers for longer-distance scales. In data-processing environments, where fans blow and components grow and shrink with temperature changes by more than 30 μm, guided wave configurations maximize flexibility in packaging to use for cooling and power delivery. As such, free-space schemes are not addressed here but have been mentioned for completeness.

Fundamental to any optical interconnect work involving waveguides is the existence of a high-optical-quality polymer material. Optical-quality refers to the low-loss coefficient (decibels per centimeter) of a material at a given wavelength. There are innumerable polymers for optical waveguiding. An example polymer that is emerging in the literature for making optical waveguides [43, 44] is SU-8 2000 (SU-8) from MicroChem [45–47]. Looking briefly at this material helps identify the relevant characteristics. SU-8 2000 is an epoxy-based polymer that was invented at IBM as a high-aspect-ratio, negative-tone photoresist [45, 46]. MicroChem modified its solvent to provide better adhesion. Patterns developed in SU-8 have straight sidewalls, excellent thermal stability, good adhesion, and good chemical resistance [45]. Table 7.1 compares the loss coefficients of SU-8 waveguides measured using the destructive cutback method as reported in the literature. The table serves to confirm that SU-8 is a good example of an inherently high-quality optical material, while also illustrating the impact of different waveguide fabrication techniques. The optical properties of SU-8 summarized here serve simply to provide guidance on evaluating any optical material for waveguiding.

While the inherent nature of the waveguide films is critical, another factor that influences the optical quality of a waveguide is the surface of each interface. As shown in Table 7.1, when an SU-8 waveguide has air as a cladding, the surface roughness effects are more pronounced due to the high index contrast. As such,

Table 7.1 SU-8 Waveguide Loss-Coefficient Comparison

Fabrication Technique	*Wavelength*	*Waveguide Dimensions*	*Upper/ Lower Cladding*	*Loss Coefficient*
Photolithography	635 nm	125 × 20 μm ($W \times H$)	Air/2 μm SiO_2	1.42 dB/cm*
Molding [48]	850 nm	50 × 50 μm ($W \times H$)	127 μm Topas/Topas	0.6 dB/cm
E-beam writing [49]	1,310 nm	1.8 × 4–8 μm ($W \times H$)	4 μm NOA61/4 μm of NOA61	0.22 dB/cm 0.41 dB/cm*
E-beam writing [49]	1,550 nm	1.8 × 4–8 μm ($W \times H$)	4 μm NOA61/4 μm of NOA61	0.48 dB/cm 0.49 dB/cm*

Note: * represents waveguide configurations in which air is the upper cladding.

these waveguides exhibit more loss. The interface problem is particularly bad for waveguides on FR4 due to the natural undulations on the surface of the board [50]; hence, a thick cladding layer is required. Using this approach, [51] has demonstrated 12.5 Gbps data transmission in 50 × 50 μm waveguides that were 1m long on FR4. It is argued that modal dispersion, the major bandwidth limiter in multimode waveguides, is not an issue at this transmission rate or length [51].

Finally, WRNs can be as simple as a point-to-point link or any configuration that successfully implements the desired distribution. Two example WRNs that benefit system operation are star couplers for broadcasting data [52] and H-trees for clock distribution [53]. A start coupler is an $N \times N$ passive optical device that takes data from one of the N inputs and distributes it to each of the N outputs simultaneously. An H-tree is as illustrated in Figure 7.6.

7.4.4 Chip-Board Coupling

The task here is to couple light in and out of the WRN at its termination(s) beneath chips. The solutions for out-of-plane coupling in waveguides are mirrors and grating couplers [54]. A mirror reflects light that is incident upon it, while a grating coupler diffracts the light due to its periodic or quasiperiodic modulation in refractive index recorded at the surface of a material or in its volume [54, 55]. An ideal chip-board coupling solution must be a packaging one that actively integrates these passive devices with the chip's I/O. It must also be compatible with chip assembly and thermal management requirements. Figure 7.7 shows different chip-board optical I/O interconnect configurations. Figure 7.7(a, b) shows quasi-free-space examples as light is not confined between the chip and the board. The impact of optical confinement will become clear when the vertical waveguide depicted in Figure 7.7(c) is discussed.

The most commonly reported chip-board configurations bypass integration with microprocessors by bonding III-V optoelectronic arrays and their drivers onto an intermediate interposer or directly onto a board with embedded mirror-terminated waveguides [53, 56–59]. These configurations may represent the earliest introduction of optical interconnects as they do not change CMOS fabrication. While monolithic integration remains the ultimate goal, an approximation with a similar effect is hybrid integration. In hybrid integration, III-V VCSEL and

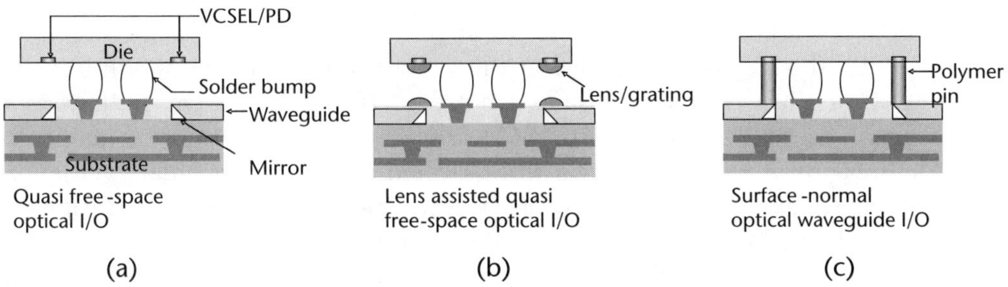

Figure 7.7 Different chip-board optical I/O interconnect configurations where (a) shows a configuration in which mirror-terminated board-level waveguides couple light in/out of chips with monolithically integrated VCSELs/PDs, (b) is the same as (a) but with improved coupling due to the presence of focusing elements like a lens or grating, and (c) represents a fully guided surface-normal coupling configuration where light is optically confined between the chip and board, which provides the highest coupling and packing efficiency.

detector arrays are bonded directly onto the microprocessor chip. The microprocessor is then bonded onto an interposer, and this assembly is bonded to a board with embedded mirror-terminated waveguides [60–63]. These configurations may represent another way of introducing optical interconnects as they approach monolithic integration but still separate the III-V and CMOS fabrication. Another promising approach reported in the literature is the direct-chip-attachment of different-material chips on a single board using compliant interconnects; this is called polylithic integration [54, 64]. The compliant interconnects are *S*-like structures called Sea of Leads [65]. They are batch-fabricated on chips at the wafer level to account for the CTE mismatch between chips and boards, as well as to provide electrical interconnection. In this configuration, they further serve to align volume-grating-coupler-terminated waveguides on the chip and board for optical interconnection. This work represents the first wafer-level configuration aimed at providing an integrated optical and electrical packaging solution for delivering optical interconnects in future GSI chips.

While all the aforementioned chip-board solutions are schematically different, they can be categorized as quasi-free-space because light is not confined in the vertical direction as illustrated in Figure 7.7(a, b). The schemes represented by Figure 7.7(b) realize this fact and try to assist with alignment tolerances due to the CTE mismatch by using lenses and/or grating couplers. For example, 230-μm diameter microlenses with 780 μm focal lengths are used to collimate and focus light between substrates in [55] to achieve an 82% theoretical coupling efficiency, and 300-μm–volume grating couplers are used to transfer light between substrates—on-chip mirror-terminated waveguides couple light to small active devices [54, 55]. Hence, confinement in the vertical direction helps minimize active device dimensions, and the compensation for the CTE mismatch is imperative. The polymer pillar represents a fully guided chip-board optical interconnection solution [48, 49], and its potential to change the way processors communicate was recently recognized [66].

A polymer pillar is a compliant cylindrical structure designed to provide optical, electrical, and thermal I/O interconnection between a chip and a substrate. As stated, this structure is addressed in multiple chapters of this book; only the optical interconnection application is treated here. Pillars were originally photolithographically defined at the wafer level in Avatrel 2000P from Promerus LLC; in this way, their densities can exceed 10^5 pillars per cm^2 [49]. More recently, they have been fabricated using SU-8 2000 due to its higher optical quality [67, 68].

Figure 7.8 shows an array of fabricated SU-8 polymer pillars that are 50 μm in diameter and 300 μm tall, for an aspect ratio of 7:1. Pillar fabrication is not discussed here but is treated in great detail in [67].

Being polymers, the pillars' compliant nature accommodates the CTE mismatch between substrates. Optical interconnection is provided because the polymer structure is a vertical waveguide [69]. The pillars are batch-fabricated on the chip side, and the sockets for their attachment are batch-fabricated on the board side. Attachment is executed by flip-chip-bonding the chip directly into the sockets and holding them in place by solder for electrical interconnects and/or some optical adhesive for the optical pillars, as depicted in Figure 7.9.

Figure 7.8 SEM image of an array of 50 × 300 μm ($D \times H$) SU-8 2100 polymer pillars on a 250-μm pitch. This represents a 7:1 aspect ratio.

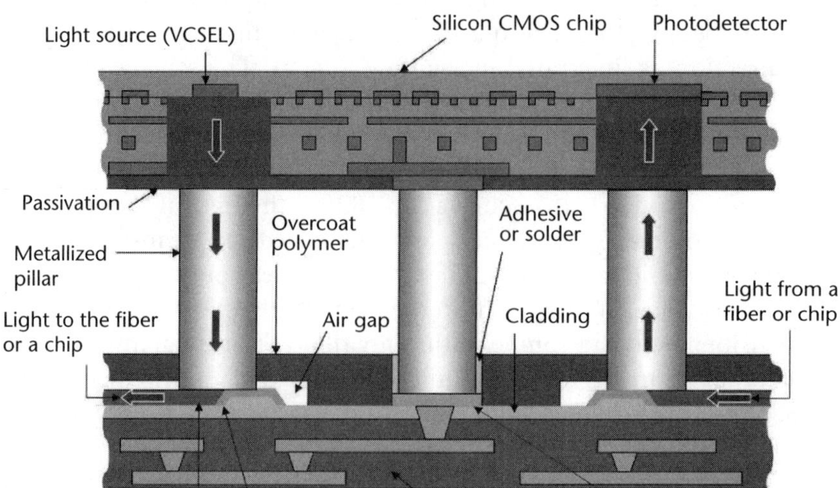

Figure 7.9 Mirror-enabled polymer pillar I/O interconnection for GSI chips with photodetectors and VCSELs. The board-level waveguides are embedded in air for high index contrast.

Optoelectronic devices are generally placed directly above their corresponding coupling elements as shown in Figure 7.9. This means that surface-normal coupling is needed. While grating couplers can be designed to achieve this, their fabrication technology limits how small they are. Mirrors integrated with waveguides and pillars are discussed here as they are the ubiquitous solution in the majority of the aforementioned approaches in the literature. From geometrical optics, mirrors produce reflected rays with angles of reflection equal to the angle of incidence. Thus, a 45° angle mirror is the intuitive choice for a mirror to couple light out of a waveguide at 90°. Fabrication of 45° angle mirrors is difficult and can lead to rough mirror surfaces that are off-angle [58]. To maximize the light incident on an optical device, these fabrication tolerances must be considered. Furthermore, the spread of the outcoupled light in the region between the substrate and the chip needs to be taken into account, together with device dimensions. Large lenses can rectify these problems as already stated, but their size can limit device densities. Since a polymer pillar has a high refractive index, light coupled into it will be confined along its height. This spatial confinement is advantageous because: (1) pillars can be fabricated with cross-sectional dimensions approaching those of the chip-level devices such that device size dictates density, (2) cross-coupling between devices can be minimized, and (3) intentional and unintentional deviations from 45° mirrors can be accounted for while still producing an out-of-plane 90° bending of the light.

Figure 7.8 already showed an array of pillar with cross-sectional diameters of 30 μm as compared to the 230 μm lens described in [60]. Figure 7.10 shows an array of pillars of different diameters fabricated atop waveguides, terminating at a metalized, anisotropically etched silicon sidewall that serves as a mirror. The pillars are 40 and 70 μm in diameter. For a reference to how these are fabricated, please see

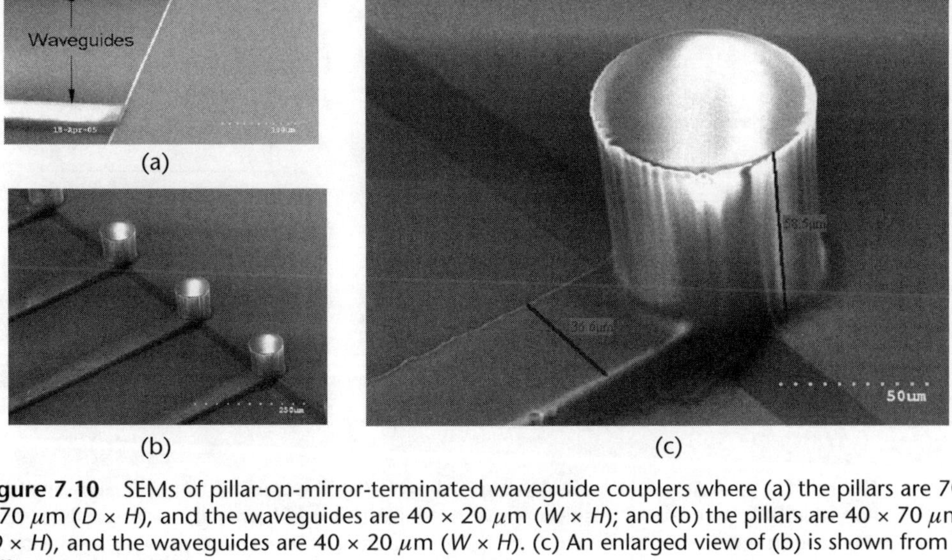

Figure 7.10 SEMs of pillar-on-mirror-terminated waveguide couplers where (a) the pillars are 70 × 70 μm ($D \times H$), and the waveguides are 40 × 20 μm ($W \times H$); and (b) the pillars are 40 × 70 μm ($D \times H$), and the waveguides are 40 × 20 μm ($W \times H$). (c) An enlarged view of (b) is shown from a different perspective.

[67]. As is clear from the images, these pillars can be as small as desired since they are photolithographically defined; thus, devices limit density.

With respect to cross-coupling between adjacent devices, the authors of [70] experimentally compared the relative transmitted optical intensity of a 50 × 150 μm optical pillar and a 50 μm optical aperture. Figure 7.11(a) illustrates the experimental setup used to characterize the surface-normal optical coupling efficiency of the pillars and apertures. In this measurement, the light source with the 12° divergence angle was used. The fiber was scanned in the x-axis and in the y-axis across the endface of the pillar and across the surface of the aperture (at a z-axis distance equal to the pillar's height). The transmitted optical power was measured with a Si detector as a function of the fiber (light source) position in the lateral direction. The results are shown in Figure 7.11(b, c).

The transmitted intensities are normalized to the maximum transmission at the center of the aperture without a pillar. The x- and y-axis scans are essentially equal due to the radial symmetry of the light source and the pillars. The difference between the coupling efficiency of the two measurements (using data from the x-axis scan) is plotted in Figure 7.11(c). The data clearly demonstrates that at the 0 μm displacement position, the optical pillar enhances the coupling efficiency by approximately 2 dB when compared to direct coupling into the aperture. At distances of ±25 μm away from the center, the optical-coupling improvement due to the pillar exceeds 4 dB. The 4 dB coupling improvement is significantly larger than the 0.23 dB excess loss of the pillars, which clearly demonstrates the benefits of the pillars. Note that the profile of the relative intensity curve of the optical pillar is almost flat across the entire endface of the pillar and abruptly drops beyond the edges of the pillar (x = ±25

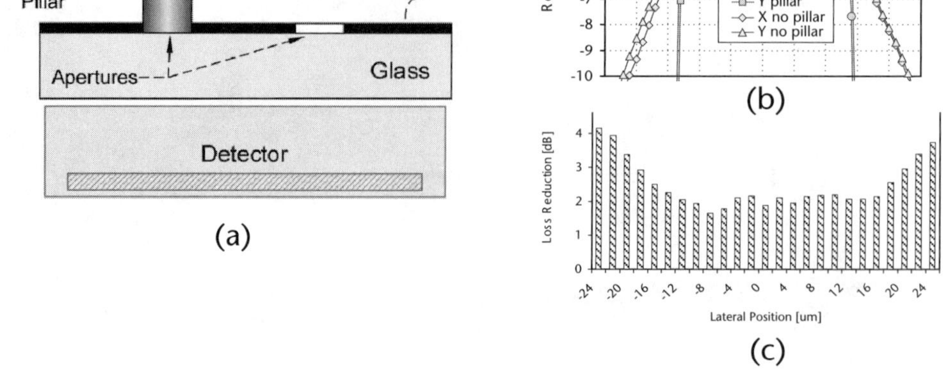

Figure 7.11 (a) An experimental setup is shown. (b) The transmitted optical intensity as a function of light source lateral position above the pillar (50 × 150 μm) and aperture is measured using the experimental setup shown in (a). (c) The reduction in the coupling loss due the optical pillars ranges from 2 to 4 dB [70].

μm). On the other hand, the intensity curve of the aperture resembles an inverse parabola. This signifies the importance of alignment. Any misalignment in the lateral direction would cause a fast roll-off in the intensity. Even with perfect alignment during assembly, any lateral misalignment between the mirror and the detector due to either CTE mismatch or other factors may reduce the coupling efficiency and limit the achievable bandwidth. This demonstrates that the optical crosstalk between adjacent I/Os is eliminated through the use of the optical pillars.

With respect to manufacturing tolerances and integration with a passive device such as a mirror, extensive work has been done to characterize the input and output coupling of the mirror-enabled polymer pillars shown in Figure 7.10 [67, 71]. Figure 7.9 shows one potential configuration where a silicon chip is assumed to have monolithically integrated detectors and VCSELs. As depicted in the figure, mirror-terminated waveguides on the board are used to couple light into and out of the chip. The waveguides are embedded in air to provide a high index contrast. The effect of the pillar is to spatially confine the light between the chip and board. Since the pillar confines light, the mirror angle does not have to be 45° to attain a 90° bending of the light. Figure 7.12 shows two-dimensional finite difference time domain (FDTD) transverse electric (TE) simulations. It shows that light can be coupled in and out of a pillar with good efficiency using non-45 degree mirror-terminated waveguide. This analysis was done using FULLWAVE from Rsoft [72].

The data shows that a 54.74° mirror is used to couple light into and out of a pillar. The waveguide-to-pillar and pillar-to-waveguide coupling efficiencies are 89.3% and 87.1%, respectively. The waveguide-to-pillar efficiency compares well with that of 82% efficiency for a total internal reflection mirror–terminated multimode waveguide beneath a 230-μm diameter lens at a focal length of 780 μm [60]. If the mirror angle is changed to 45°, with no other changes, the wave-

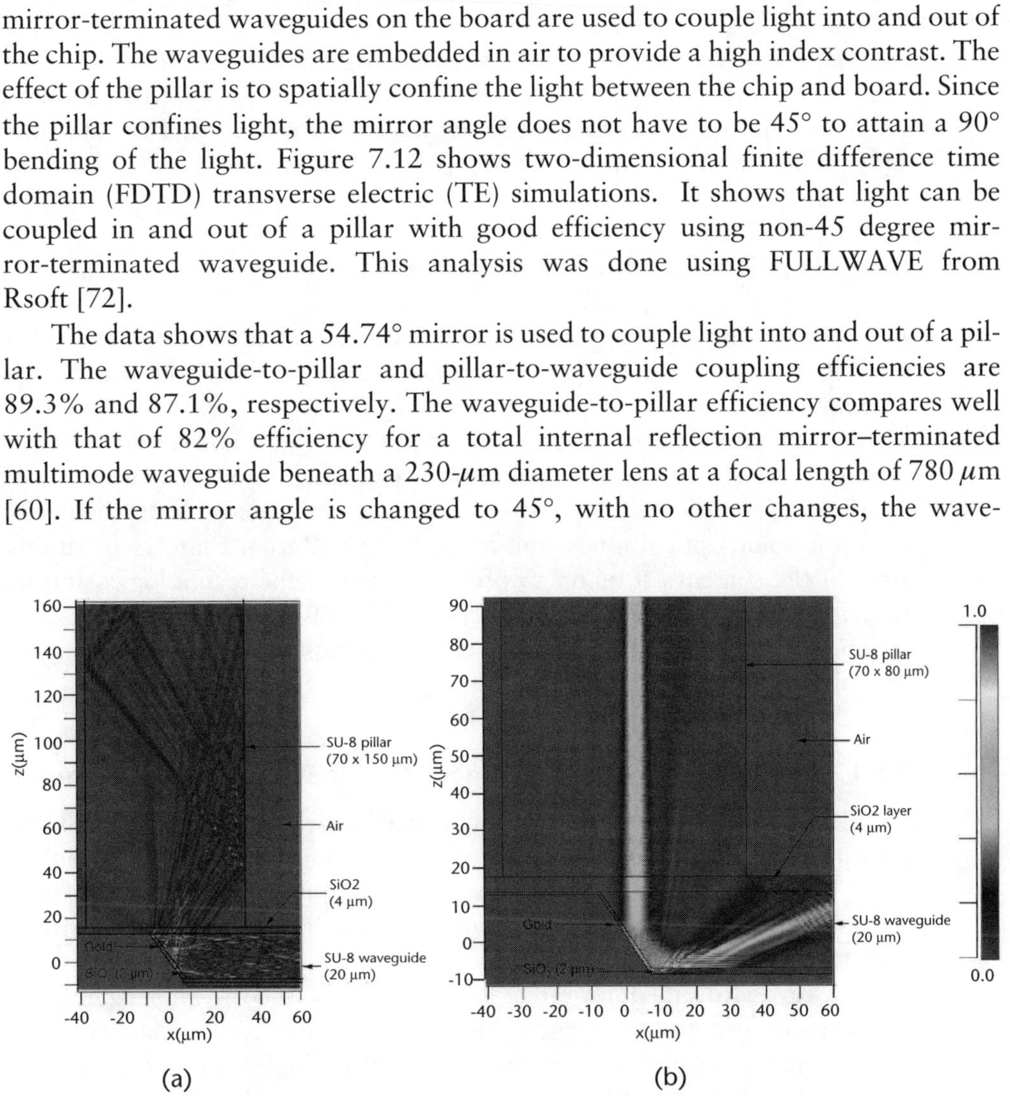

Figure 7.12 Time-averaged TE polarization result for (a) waveguide-to-pillar and (b) pillar-to-waveguide coupling when a 4-μm high SiO_2 layer is introduced between an SU-8 pillar located directly on top of an SU-8 mirror-terminated waveguide. The mirror angle is 54.7°.

guide-to-pillar and pillar-to-waveguide coupling efficiencies are now 89.3% and 89.3%, respectively. Hence, there is no inherent disadvantage to using a non-45° mirror to perform a 90° bending of light when integrated with a polymer pillar. Polymer pillars are therefore flexible to substrate technology such that silicon, FR4 with imprinted silicon mirrors, and FR4 with a conventional 45° mirror can be used, and unwanted manufacturing angular deviations due to the fabrication of a 45° mirror can be accounted for. A complete design, fabrication, theoretical analysis, and optical testing of chip-to-board, board-to-chip, and chip-to-chip interconnection facilitated by mirror-enabled polymer pillars is presented here [67] for further study.

At this point, the description of the optical interconnect system is complete. The key to remember is that high-bandwidth I/O, board-level wiring, and board-to-board interconnection is needed. If optics is able to penetrate at one level, it must suffuse to the others in order to truly solve the bandwidth problems associated with tracking the advances of semiconductor scaling. To this end, requirements and literature examples for achieving each of these I/O, board-level wiring, and board-to-board interconnects optically have been provided. The data shows that a fully guided solution may be the best one.

7.5 Summary, Issues, and Future Directions

In reviewing the various research and development activities described here, it is clear that the future directions for optical interconnect for the various application areas have not been clearly defined yet. Many technological options have yet to prove themselves relative to other alternatives (e.g., direct laser modulation or continuous wave lasers with external modulators, single-mode versus multimode transmission, 1,550 versus 850 nm wavelengths) before large-scale deployment of silicon microphotonic optical interconnects can occur. Rather than try to summarize the state of the current art or try to predict which of the technologies and techniques described here will prove most successful, it seems more useful to highlight a set of questions that come up related to silicon photonics and optical interconnect, such as the following.

7.5.1 Which Links Will Use Multimode Versus Single-Mode Transmission?

It is clear that the longest (>300m) high-speed links will have to be with single-mode transmission due to bandwidth limitations from modal dispersion in multimode fiber. It is also clear that the shortest link, on-chip links, must also be single mode due to the large size of multimode waveguides relative to the typical size of transistors, storage cells, and other devices in CMOS chips.

However, for medium-range links (between 0.01m and 100m), multimode transmission provides the least expensive and most mechanically robust transmission. A set of single-mode transmission technologies (sources, connectors, and waveguides) that could fill this gap cost-effectively would eliminate the need for separate multimode transmission components, but such a technology does not yet appear likely.

7.5.2 Which Wavelengths Will Be Used for Which Types of Links?

Data transmission at the 1,550-nm range is widely deployed for telecom and WDM links. However, components in this wavelength range have tended to be much more expensive than, for example, 850 nm VCSELs or edge-emitting lasers at even shorter wavelengths used for optical storage (e.g., CDs and DVDs). Shorter wavelengths have commonly been used for shorter links, but, again, the clarity of silicon waveguides at long wavelengths and the advantage of longer wavelengths for patterning larger structures have encouraged the use of 1,550-nm range wavelengths for the very shortest on-chip components. A cost-effective, long-wave laser, particularly in vertically emitting arrays, would seem to be widely useful, but the commercial impact of large markets for alternative designs and the inertia of already tested solutions lead one to expect a variety of wavelengths to be used for at least several years. Also, the use of polymer waveguides (e.g., for optical flex circuits or optical PCBs) will impact wavelength selection since shorter wavelengths have lower attenuation in all currently planned waveguide polymers.

7.5.3 How Important Will WDM Be Versus Multiple Separate Waveguides?

One of the often-listed advantages of optical transmission is that waveguide bandwidth is so high (tens of terabits per second), which means, since electronic circuitry cannot modulate signals so quickly, that multiple logical channels at different wavelengths can be transported across a single waveguide simultaneously. This wavelength-division multiplexing allows dramatic reduction in the cost of waveguides and allows effects (e.g., routing by wavelength in array waveguide gratings) that are not possible in multimode or single-wavelength waveguides. However, the use of WDM has its own overheads and complications associated with multiple separate and distinct source designs for different wavelengths, the difficulty of achieving sufficient wavelength stability across fabrication and temperature ranges, and the cost of wavelength multiplexing and demultiplexing components relative to the aggregated cost of extra waveguides to carry multiple channels. To date, WDM has been worthwhile for those links where the cost of the optical fibers has been higher than the cost of these more complex sources and other WDM components—which has meant only for links of ~10 km and more. However, as optical interconnect moves progress onto chips, where the competition for space is more intense, WDM may make more sense.

7.5.4 How Much Power and Cost Advantage Is to Be Gained by On-Chip Integration of Optical Interconnects Versus Integration of Other Components?

The competition for space inside or on the surface of silicon chips is very fierce: each component or circuit integrated on a chip must prove its value in comparison to the universe of other components or circuits that could be integrated in that same space, both on a cost/performance basis and a design complexity basis. As an example, optical transceivers could be incorporated onto a processor chip to accelerate the communications with another processing device; alternatively, since the optical transceivers tend to be fairly large, the same chip area could be used for pulling the functions of that processing device on-chip itself. These trade-offs (what circuits and functions go on-chip? what goes on-package? what goes on the same PCB? what goes in other PCBs?) are part of the overall system design space that optical interconnects enter as they become more practical.

7.5.5 How Much Optics Is On-Chip Versus On-Package?

As the review here demonstrates, all of the key technologies for optical interconnect (sources, detectors, waveguides, splitters, and so forth) are possible in silicon chips or in silicon chips modified to incorporate other materials (e.g., III-Vs, Ni/NiO_2/Ni, SiO_2) in optically useful ways. However, the components built this way have fairly fundamental penalties as compared to components built specifically in optimized materials systems. There is a lot of advantage yet to be gained for optical interconnect through more advanced packaging and closer integration of silicon processing and optical communications technology at the package level before going to the extent of incorporating optics directly onto silicon chips.

7.6 Summary

These questions, at the level of both research feasibility and commercial competitive viability against competing alternatives, have yet to be answered. As these questions are addressed by many groups of dedicated researchers and developers over the next few years, the most promising approaches for optical interconnects in the various applications will be clearer.

References

[1] Benner, A., et al., "Exploitation of Optical Interconnects in Future Server Architectures," *IBM J. Res. Dev.*, Vol. 49, 2005, pp. 755–775.

[2] Meindl, J. D., et al., "Interconnect Opportunities for Gigascale Integration," *IBM J. Res. Dev.*, Vol. 46, March/May 2002, pp. 245–263.

[3] Fitzgerald, E. A., and L. C. Kimerling, "Silicon-Based Microphotonics and Integrated Optoelectronics," *MRS Bulletin*, Vol. 23, No. 4, April 1998, pp. 39–47.

[4] Kimerling, L. C., "Devices for Silicon Microphotonic Interconnection: Photonic Crystals, Waveguides and Silicon Optoelectronics," *57th Annual Device Research Conference Digest,* June 28–30, 1999, Santa Barbara, CA, pp. 108–111.

[5] Lipson, M., "Guiding, Modulating, and Emitting Light on Silicon—Challenges and Opportunities," *J. Lightwave Technology*, Vol. 23, No. 12, December 2005, pp. 4222–4238.

[6] International Technology Roadmap for Semiconductors at www.itrs.net/home.html.

[7] Haurylau, M., et. al., "On-Chip Optical Interconnect Roadmap: Challenges and Critical Directions," *IEEE. J. Selected Topics, Quantum Electronics*, Vol. 12, No. 6, November/December 2006, pp. 1699–1705.

[8] Davis, J. A., et al., "Interconnect Limits on Gigascale Integration (GSI) in the 21st Century," *Proc. IEEE*, Vol. 89, March 2001, pp. 305–324.

[9] Sarvari, R., and J. D. Meindl, "On the Study of Anomalous Skin Effect for GSI Interconnections," *Proc. IITC*, 2003, Burlingame, CA, pp. 42–44.

[10] Naeemi, A., et al., "Optical and Electrical Interconnect Partition Length Based on Chip-to-Chip Bandwidth Maximization," *IEEE Photon. Technol. Lett.*, Vol. 16, April 2004, pp. 1221–1223.

[11] Dawei, H., et al., "Optical Interconnects: Out of the Box Forever?" *IEEE J. Sel. Topics in Quant. Electron.*, Vol. 9, March–April 2003, p. 614.

[12] Svensson, C., and G. H. Dermer, "Time Domain Modeling of Lossy Interconnects," *IEEE Trans. Adv. Packag.*, Vol. 24, May 2001, pp. 191–196.

[13] Davis, J. H., N. F. Dinn, and W. E. Falconer, "Technologies for Global Communications," *IEEE Commun. Mag.*, Vol. 30, 1992, pp. 35–43.

[14] Ayres, R. U., and E. Williams, "The Digital Economy: Where Do We Stand?" *Technological Forecasting and Social Change*, Vol. 71, May 2004, pp. 315–339.

[15] Goodman, J. W., et al., "Optical Interconnections for VLSI Systems," *Proc. IEEE*, Vol. 72, July 1984, pp. 850–866.

[16] Miller, D. A. B., "Rationale and Challenges for Optical Interconnects to Electronic Chips," *Proc. IEEE*, Vol. 88, June 2000, pp. 728–749.

[17] Naeemi, A., et al., "Optical and Electrical Interconnect Partition Length Based on Chip-to-Chip Bandwidth Maximization," *IEEE Photon. Technol. Lett.*, Vol. 16, April 2004, pp. 1221–1223.

[18] Horowitz, M., C.-K. K. Yang, and S. Sidiropoulos, "High-Speed Electrical Signaling: Overview and Limitations," *IEEE Micro*, Vol. 18, January–February 1998, pp. 12–24.

[19] Katayama, Y., and A. Okazaki, "Optical Interconnect Opportunities for Future Server Memory Systems," *High Performance Computer Architecture 2007, IEEE 13th International Symposium*, February 10–14, 2007, Phoenix, AZ, pp. 46–50.

[20] Collins, H. A., and R. E. Nikel, "DDR-SDRAM, High Speed, Source Synchronous Interfaces Create Design Challenges," *EDN* (U.S. edition), Vol. 44, September 2, 1999, pp. 63–72.

[21] National Energy Research Supercomputing Systems, science-driven systems overview, at http://cbcg.nersc.gov/nusers/systems.

[22] Kettler, D., H. Kafka, and D. Spears, "Driving Fiber to the Home," *IEEE Commun. Mag.*, Vol. 38, November 2000, pp. 106–110.

[23] Green, P. E., Jr., "Fiber to the Home: The Next Big Broadband Thing," *IEEE Commun. Mag.*, Vol. 42, September 2004, pp. 100–106.

[24] Cho, H., P. Kapur, and K. C. Saraswat, "Power Comparison between High-Speed Electrical and Optical Interconnects for Interchip Communication," *J. Lightwave Technology*, Vol. 22, No. 9, September 2004, pp. 2021–2033.

[25] Dowd, P. W., "High Performance Interprocessor Communication through Optical Wavelength Division Multiple Access Channels," *Comp. Architecture News*, Vol. 19, May 1991, pp. 96–105.

[26] Ghose, K., R. K. Horsell, and N. K. Singhvi, "Hybrid Multiprocessing Using WDM Optical Fiber Interconnections," *Proc. MPPOI*, 1994, Cancun, Mexico, pp. 182–196.

[27] Collet, J. H., W. Hlayhel, and D. Litalze, "Parallel Optical Interconnects May Reduce the Communication Bottleneck in Symmetric Multiprocessors," *Appl. Optics*, Vol. 40, July 2001, pp. 3371–3378.

[28] Louri, A., and A. K. Kodi, "SYMNET: An Optical Interconnection Network for Scalable High-Performance Symmetric Multiprocessors," *Appl. Optics*, Vol. 42, June 2003, pp. 3407–3417.

[29] Kodi, A. K., and A. Louri, "RAPID: Reconfigurable and Scalable All-Photonic Interconnect for Distributed Shared Memory Multiprocessors," *IEEE J. Lightwave Technol.*, Vol. 22, September 2004, pp. 2101–2110.

[30] Chi On, C., A. K. Okyay, and K. C. Saraswat, "Effective Dark Current Suppression with Asymmetric MSM Photodetectors in Group IV Semiconductors," *IEEE Photon. Technol. Lett.*, Vol. 15, November 2003, pp. 1585–1587.

[31] Liu, A., et al., "A High-Speed Silicon Optical Modulator Based on a Metal-Oxide-Semiconductor Capacitor," *Nature*, Vol. 427, February 2004, pp. 615–618.

[32] Groenert, M. E., et al., "Monolithic Integration of Room-Temperature CW GaAs/AlGaAs Lasers on Si Substrates via Relaxed Graded GeSi Buffer Layers," *J. Appl. Phys.*, Vol. 93, January 2003, pp. 362–367.

[33] Wong, Y.-M., et al., "Technology Development of a High-Density 32-Channel 16-Gb/s Optical Data Link for Optical Interconnection Applications for the Optoelectronic Technology Consortium (OETC)," *IEEE J. Lightwave Technol.*, Vol. 13, June 1995, pp. 995–1016.

[34] Katsura, K., et al., "Packaging for a 40-Channel Parallel Optical Interconnection Module with an Over 25-Gb/s Throughput," *Proc. ECTC*, 1998, Seattle, WA, pp. 755–761.

[35] Usui, M., et al., "ParaBIT-1: 60-Gb/s-Throughput Parallel Optical Interconnect Module," *Proc. ECTC*, 2000, Las Vegas, NV, pp. 1252–1258.

[36] Lemoff, B. E., et al., "MAUI: Enabling Fiber-to-the-Processor with Parallel Multiwavelength Optical Interconnects," *IEEE J. Lightwave Technol.*, Vol. 22, September 2004, p. 2043.

[37] Almeida, V. R., R. R. Panepucci, and M. Lipson, "Nanotaper for Compact Mode Conversion," *Optics Lett.*, Vol. 28, August 1, 2003, pp. 1302–1304.

[38] Li, Y., et al., "Multigigabits per Second Board-Level Clock Distribution Schemes Using Laminated End-Tapered Fiber Bundles," *IEEE Photon. Technol. Lett.*, Vol. 10, June 1998, pp. 884–886.

[39] Van Steenberge, G., et al., "MT-Compatible Laser-Ablated Interconnections for Optical Printed Circuit Boards," *IEEE J. Lightwave Technol.*, Vol. 22, September 2004, pp. 2083–2090.

[40] Zane, F., et al., "Scalable Network Architectures Using the Optical Transpose Interconnection System (OTIS)," *Proc. Second International Conference on Massively Parallel Processing Using Optical Interconnections* (MPPOI'96), 1996, Maui, HI, pp. 114–121.

[41] Walker, A. C., et al., "Operation of an Optoelectronic Crossbar Switch Containing a Terabit-per-Second Free-Space Optical Interconnect," *IEEE Quantum Electronics*, Vol. 41, No. 7, July 2005, pp. 1024–1036.

[42] McDermott, T., and T. Brewer, "Large-Scale IP Router Using a High-Speed Optical Switch Element," *J. Optical Networking*, Vol. 2, No. 7, 2003, pp. 229–240.

[43] Choi, C., et al., "Flexible Optical Waveguide Film Fabrications and Optoelectronic Devices Integration for Fully Embedded Board-Level Optical Interconnects," *IEEE J. Lightwave Technol.*, Vol. 22, September 2004, pp. 2168–2176.

[44] Wong, W. H., J. Zhou, and E. Y. B. Pun, "Low-Loss Polymeric Optical Waveguides Using Electron-Beam Direct Writing," *Appl. Phys. Lett.*, Vol. 78, April 2001, pp. 2110–2112.

[45] LaBianca, N. C., and J. D. Gelorme, "High Aspect Ratio Resist for Thick Film Applications," *Proc. SPIE*, Vol. 2438, 1995, pp. 846–852.

[46] Lorenz, H., et al., "SU-8: A Low-Cost Negative Resist for MEMS," *J. Micromechanics and Microengineering*, Vol. 7, September 1997, pp. 121–124.

[47] MicroChem, Inc., at www.microchem.com.

[48] Bakir, M. S., et al., "Sea of Polymer Pillars: Compliant Wafer-Level Electrical-Optical Chip I/O Interconnections," *IEEE Photon. Technol. Lett.*, Vol. 15, November 2003, pp. 1567–1569.

[49] Bakir, M. S., and J. D. Meindl, "Sea of Polymer Pillars Electrical and Optical Chip I/O Interconnections for Gigascale Integration," *IEEE Trans. Electron Devices*, Vol. 51, July 2004, pp. 1069–1077.

[50] Chang, G.-K., et al., "Chip-to-Chip Optoelectronics SOP on Organic Boards or Packages," *IEEE Trans. Adv. Packag.*, Vol. 27, May 2004, pp. 386–397.

[51] Bona, G. L., et al., "Characterization of Parallel Optical-Interconnect Waveguides Integrated on a Printed Circuit Board," *Proc. SPIE*, Vol. 5453, 2004, pp. 134–141.

[52] Israel, D., et al., "Comparison of Different Polymeric Multimode Star Couplers for Backplane Optical Interconnect," *IEEE J. Lightwave Technol.*, Vol. 13, June 1995, pp. 1057–1064.

[53] Chen, R. T., et al., "Fully Embedded Board-Level Guided-Wave Optoelectronic Interconnects," *Proc. IEEE*, Vol. 88, 2000, pp. 780–793.

[54] Schultz, S. M., E. N. Glytsis, and T. K. Gaylord, "Design, Fabrication, and Performance of Preferential-Order Volume Grating Waveguide Couplers," *Appl. Optics*, Vol. 39, March 2000, pp. 1223–1232.

[55] Mulé, A. V., "Volume Grating Coupler-Based Optical Interconnect Technologies for Polylithic Gigascale Integration," PhD thesis, Georgia Institute of Technology, May 2004.

[56] Sadler, D. J., et al., "Optical Reflectivity of Micromachined {111}-Oriented Silicon Mirrors for Optical Input-Output Couplers," *J. Micromechanics and Microengineering*, Vol. 7, December 1997, pp. 263–269.

[57] Rho, B. S., et al., "PCB-Compatible Optical Interconnection Using 45 Degrees–Ended Connection Rods and Via-Holed Waveguides," *IEEE J. Lightwave Technol.*, Vol. 22, September 2004, pp. 2128–2134.

[58] Cho, H. S., et al., "Compact Packaging of Optical and Electronic Components for On-Board Optical Interconnects," *IEEE Trans. Adv. Packag.*, Vol. 28, February 2005, pp. 114–120.

[59] Cho, M. H., et al., "High-Coupling-Efficiency Optical Interconnection Using a 90 Degrees Bent Fiber Array Connector in Optical Printed Circuit Boards," *IEEE Photon. Technol. Lett.*, Vol. 17, March 2005, pp. 690–692.

[60] Ishii, Y., and Y. Arai, "Large-Tolerant 'OptoBump' Interface for Interchip Optical Interconnections," *Electronics and Communications in Japan, Part 2 (Electronics)*, Vol. 86, 2003, pp. 1–8.

[61] Mikawa, T., et al., "Implementation of Active Interposer for High-Speed and Low-Cost Chip Level Optical Interconnects," *IEEE J. Sel. Topics Quantum Electron.*, Vol. 9, March–April 2003, pp. 452–459.

[62] Tooley, F., et al., "Optically Written Polymers Used as Optical Interconnects and for Hybridisation," *Optical Materials*, Vol. 17, June–July 2001, pp. 235–241.

[63] Patel, C. S., et al., "Silicon Carrier with Deep Through-Via, Fine Pitch Wiring, and Through Cavity for Parallel Optical Transceiver," *Proc. ECTC*, May 31–June 3, 2005, Orlando, FL, pp. 1318–1324.

[64] Mulé, A. V., "Volume Grating Coupler-Based Optical Interconnect Technologies for Polylithic Gigascale Integration," PhD thesis, Georgia Institute of Technology, May 2004.

[65] Bakir, M. S., et al., "Sea of Leads (SoL) Ultrahigh Density Wafer-Level Chip Input/Output Interconnections for Gigascale Integration (GSI)," *IEEE Trans. Electron Devices*, Vol. 50, October 2003, pp. 2039–2048.

[66] Gibbs, W. W., "Computing at the Speed of Light," *Scientific American*, November 2004, pp. 80–87.

[67] Ogunsola, O. O., "Prospects for Mirror-Enabled Polymer Pillar I/O Optical Interconnects for Gigascale Integration," PhD thesis, Georgia Institute of Technology, December 2006.

[68] Bakir, M., et al., "Trimodal' Wafer-Level Package: Fully Compatible Electrical, Optical, and Fluidic Chip I/O Interconnects," *Proc. Electronic Components and Technol. Conf.*, May 29–June 1, 2007, Reno, NV, pp. 585–592.

[69] Bakir, M. S., et al., "Optical Transmission of Polymer Pillars for Chip I/O Optical Interconnections," *IEEE Photonics Technology Letters*, Vol. 16, January 2004, pp. 117–19.

[70] Bakir, M., et al., "Mechanically Flexible Chip-to-Substrate Optical Interconnections Using Optical Pillars," *IEEE Trans. Adv. Packaging*, Vol. 31, No. 1, 2008, pp. 143–153.

[71] Ogunsola, O. O., et al., "Chip-Level Waveguide-Mirror-Pillar Optical Interconnect Structure," *IEEE Photon. Technol. Lett.*, Vol.18, No.15, August 2006, pp. 1672–1674.

[72] See Rsoft Design Group at www.rsoftdesign.com.

CHAPTER 8

Monolithic Optical Interconnects

Eugene A. Fitzgerald, Carl L. Dohrman, and Michael J. Mori

As with other generations of integration, monolithic optical interconnection is the ultimate solution for creating a plethora of integrated electronic/photonic systems with the highest functionality at the lowest possible cost. Silicon transistor technology has revealed the benefits of ultraminiaturization: extreme low cost, maximum performance, and creation of new applications and markets. Interestingly, the arguments against monolithic integration are identical to the arguments that have been made against silicon technology for more than 40 years. The reason is that, initially, integrating two optimized, separate components creates a sacrifice in component performance and cost. All too often, it is forgotten that an integrated technology is initially more expensive, and, at a component level, the performance is decreased. However, if a market-in-need is identified and an integration technology exists, volume will increase, performance of the components and the system will improve, and cost will lower. Assuming high volume, the ultimate yield in an integrated technology can be much lower than the hybrid technology; cost will always be lowered due to the fact that interconnects higher than the chip-level are increasingly expensive, by orders of magnitude.

In this review, we concentrate our efforts on the potential integration technologies that may allow for monolithic optical interconnects, when and if an appropriate first market is identified. However, the boundaries of silicon manufacturing are paramount; this infrastructure must be the manufacturing infrastructure for monolithic optical interconnects to achieve silicon-like scalability and integration with silicon complementary metal-oxide semiconductor (CMOS) electronics.

It is assumed in most of the research that a telecom or datacom application is the likely driver; therefore, that may impose some additional boundaries in choosing research direction. This constraint is certainly likely, although we must keep in mind the nature of innovation and remember that the technology might seed itself in a simple application outside of these more obvious targets for integrated monolithic electronic/photonics. However, since much research work is defined along these paths, we break down our discussion into integrated components that can achieve these goals. The components of an integrated optical link would be a light emitter, possibly a modulator, a waveguide, and a detector. We briefly attempt to review the work for integrating these components on silicon, although the area is heavily researched, and we are forced to do an abbreviated job and surely cannot review all work in this area.

First, the requirements of and principles behind optical emitters are discussed, and several materials approaches (III-V, Group IV), along with corresponding foreseeable obstacles, are covered. Next, we discuss optical modulators and their application in optoelectronic integration. In the third section, we discuss the operation and integration of optical detectors, again exploring promising options with both III-V and Group IV materials. The final component of optical integration is presented in the fourth section on waveguides. We finish with a brief discussion of the potential for commercialization of the discussed technologies.

8.1 Optical Sources on Si

The realization of an optoelectronic integrated circuit (OEIC) will require a bright and efficient photon source, preferably laser as opposed to LED. This source could be directly driven or modulated externally, but with a more flexible design, fewer additional components are needed, and OEIC design is simplified. The current research focus is on creating an efficient, small, electrically driven laser, which can be manufactured on the wafer scale in a method compatible with current CMOS process technology. The poor light-emitting qualities of silicon and the incompatibility of the best-known light-emitting devices with silicon remain the main limiters of silicon photonic systems. Traditionally, semiconductor light emitters utilize direct band-to-band transitions of carriers, which ultimately combine an electron-hole pair across the band gap to emit a photon of corresponding energy. Silicon's indirect band gap separates free electrons and holes in momentum space, making radiative recombination less likely. Since the radiative lifetime of Si is long (milliseconds) and the nonradiative lifetime is relatively short (nanoseconds), the internal quantum efficiency of Si emitters is very low (about 10^{-6}), many orders of magnitude lower than the efficiency of typical direct band gap semiconductor devices, such as those based on GaAs and InP. Fabrication of light emitters integrated on a Si wafer may be possible through a variety of mechanisms, and the approaches stem from a wide range of physical principles, from coaxing band-to-band emission from native Si, to using nonlinear optical properties of Si, to introducing luminescent impurities to the Si lattice, to hybridizing Si with other luminescent materials systems.

8.1.1 Interband Emission: III-V Sources

Perhaps the most promising approach to integrating laser sources on Si is through integration of III-V materials directly onto the Si wafer. Decades of telecommunication technology have provided advanced designs for InP-based III-V diode lasers that emit around 1.3 and 1.55 μm. Using photons near these colors makes sense for a Si OEIC; silicon's transparency at these wavelengths makes it a natural choice for the core material of waveguides, and with its high refractive index contrast, Si/SiO_2 waveguides have excellent potential for a convenient and high-performance waveguiding system. But this color choice also requires development of integrated photodetectors using new materials. Figure 8.1 shows absorption as a function of wavelength for a variety of common semiconductors and indicates that at the telecommunication wavelengths of 1.3 and 1.55 μm, germanium's reasonable absorption makes it a reasonable material of choice for photodetectors. Instead, shifting the light source into the yellow-green region of the visible spectrum (500 to 600 nm)

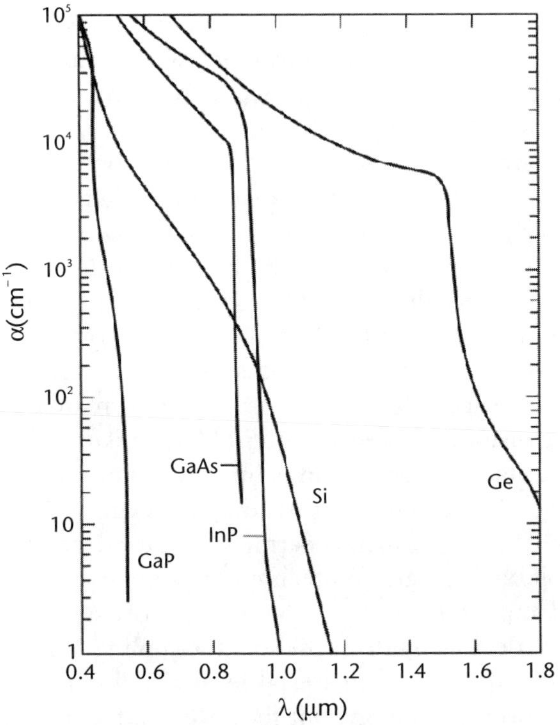

Figure 8.1 Absorption coefficient as a function of wavelength for a variety of common semiconductors. Absorption coefficient is a key parameter for photodetector materials, the choice of which has a dramatic effect on all elements of the OEIC. Note that Ge is the only material shown with appreciable absorption near 1.3 and 1.55 μm.

would enable an OEIC with Si photodetectors, but sources amenable to integration are not yet readily available.

The III-V approach for on-Si light sources requires not a focus on how to provide luminescence (as Si-based sources do) but a focus on bringing known photonic solutions onto the Si wafer by bringing together traditionally incompatible materials, while controlling lattice defects, preserving CMOS process integrity, and facilitating efficient exchange of photons between the III-V and IV systems in a low-cost and inherently manufacturable process. In the following sections, we focus on integration challenges and approaches currently being taken toward demonstrating integrated, interband photon sources.

8.1.1.1 Issues for Integration on Si

The three basic routes to integration of III-V on Si are monolithic (heteroepitaxy), wafer-scale hybridization, and more conventional hybridization of smaller components, like dice or individual devices. Heteroepitaxy is the deposition of III-V layers directly onto a Si wafer. This is made difficult due to the large lattice mismatch of GaAs or InP on Si, the complexity of engineering the heterovalent on homovalent crystal types, and incompatible growth processes. The growth of high-quality material [with permissibly low threading dislocation density (TDD)] requires engineering of the lattice to confine the dislocations that bridge the mismatch to layers away from the active device region. Much effort has been placed on optimizing growth processes that bridge all lattice mismatch with a simple buffer layer design such as in a two-step growth process. This two-step process was discovered in the 1980s [1] and has recently become popular again as researchers attempt to find shortcuts and deposit high mismatched films directly on silicon [2, 3], despite historical data continuing to show that threading dislocation densities below 10^7 cm^{-2} cannot be achieved. In this two-step method, a thin, low-temperature GaAs layer is deposited directly on Si, then annealed at high temperature to improve crystalline quality. These techniques result in high TDD layers of about 10^9 cm^{-2}, which degrade performance of optical devices. TDD can be lowered to about mid to low 10^7 cm^{-2} at best with cyclic thermal annealing processes. Alternately, dislocation engineering techniques are very successful at bridging lattice mismatch without prohibitive increases in threading dislocation density [4, 5]. Here, misfit dislocations are confined to mismatched interfaces where strain is introduced gradually in compositionally graded SiGe layers on Si, keeping the threading dislocation density at a maximum value of around 2×10^{-6} cm^{-2}, which is permissible for fabrication of lasers [5]. More details of heteroepitaxial integration techniques will be discussed later.

Hybrid approaches on the other hand allow separate processing of III-V epitaxy and CMOS devices with subsequent coupling and further processing. Wafer-bonding techniques at the material level are closer to monolithic integration than any other hybrid technology, but they are still far from achieving the device density and design freedom that true monolithic integration provides. The bonding of dissimilar wafer materials strives to join them mechanically through room-temperature formation of Van der Waals forces or formation of covalent bonds at elevated temperatures. Thermal budgets of bonding techniques tend to be limited to avoid many types of problems. These include thermal expansion mismatch (which results in crack formation in extreme cases), chemical interdiffusion across the bonding interface, dopant diffusion in prefabricated devices in either of the bonded wafers, and propagation of mismatch defects from the bonded interface. Sometimes the use of a thin adhesive layer, such as the thermosetting polymer divinylsiloxane-benzocyclobutene (DVS-BCB), is reported to improve the strength of the bonded interface while lowering the necessary processing temperature and relaxing the need for complicated chemical mechanical polishing technologies [6–8] (see Figure 8.2). But excellent results have also been obtained using no adhesive and low temperatures when the bonding surfaces are very well controlled [9]. If an adhesive is to be used, its choice is quite important; qualities to consider include optical transparency, planarization properties, curing temperature, tendency to outgas during cure, and glass transition temperature. Regardless of bonding method, proper surface preparation is critical for effective bonding and often involves a sequence of solvent cleans, chemical etches, and oxygen plasma treatments interspersed with careful visual inspections.

Differences in available wafer sizes make wafer bonding less attractive technically and economically, but schemes such as die-to-wafer boding are employed to allow integration of III-V on a limited area of the host Si wafer. Due to the potentially low thermal budget for bonding processes, mismatch defects are confined to the bonding interface and do not impinge either the Si or III-V device layers around the interface. One severe drawback of this method is the difficulty of precisely plac-

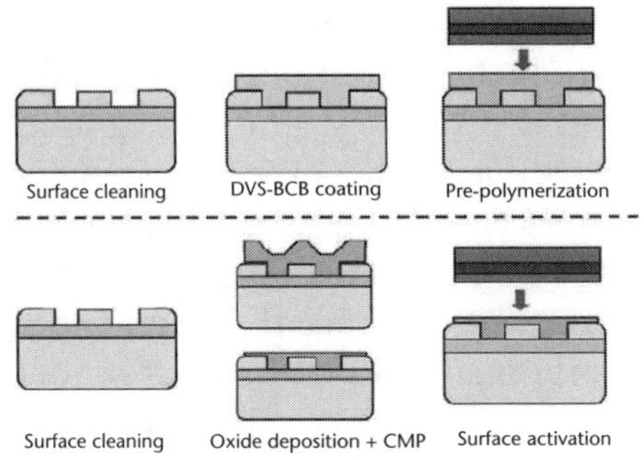

Outline of the DVS-BCB adhesive die-to-wafer bonding process (top) and the molecular adhesion die-to-wafer bonding process (bottom).

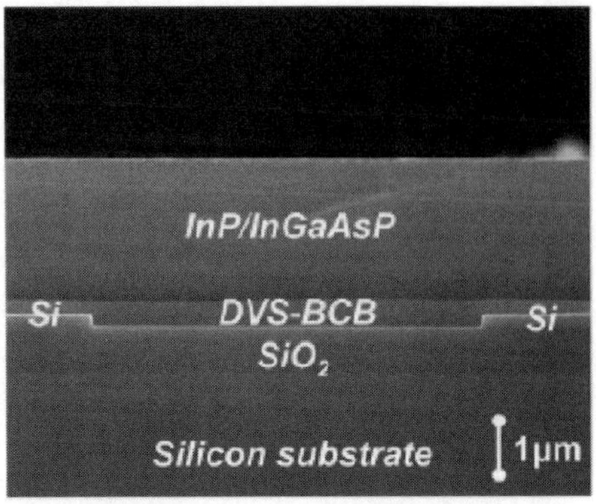

Scanning electron microscopy (SEM) cross-section of a DVS-BCB adhesive bonding interface on which the SOI waveguides can clearly be seen, together with the planarizing DVS-BCB bonding layer and the bonded InP/InGaAsP epitaxial layer stack.

Figure 8.2 Schematic comparison of adhesive and molecular die-to-wafer bonding processes and scanning electron micrograph (SEM) of an adhesive-bonded interface showing the excellent planarizing properties of DVS-BCB without benefit of a CMP step [7].

ing optical devices such as lasers on the Si host platform. Tight confinement of the optical mode to the high-index-contrast Si/SiO_2 waveguides and to the III-V laser requires precise alignment of the laser onto the waveguide. It is difficult to envision any cost-effective manufacturing process whereby more than a few lasers are aligned on each Si wafer [10]; thus, fundamental limits on the complexity of an OEIC built with this method exist. Despite this, recent progress has been made at the University of California, Santa Barbara (UCSB), Intel, and Ghent University to relax the alignment tolerances and shows strong promise. In these schemes, the exact position of the III-V laser is set by processing after the bonding step, making the laser self-aligned. In principle, this allows much faster pick and place methods to be implemented; we will discuss experimental results further below. Though successful bonded devices have been demonstrated, this method is not ideal due to practical problems in scaling up to large batch production.

More conventional hybrid technology such as flip-chip bonding continues to evolve and is another possible route for III-V integration until monolithic integration is widely available. Flip-chip bonding is a hybrid approach in which dice of prefabricated III-V devices are soldered in place on a host wafer. The density of devices is limited due to the relatively large size of the solder bumps used, and unlike the mask alignment steps used in typical photolithography (where the masks contain alignment features facilitating exact placement), precise alignment of the die on the host wafers is complicated and time-consuming. Flip-chip bonding is not a wafer-scale process; bringing together two separately fully processed dice prohibitively escalates the cost of any large-scale integration scheme. Because this technology is fairly developed and cannot reach densities of devices that push the monolithic interconnection frontier, we will not address this technology further.

In both monolithic and hybrid approaches, thermal mismatch between the silicon wafer (α ~2.6×10^{-6}/°C) and III-V layers (α ~5×10^{-6}/°C) must be considered. For the highest optical interconnection densities, close integration at the material level must be achieved between silicon and the photonic material, for example,

GaAs. Either by limiting the thickness of the III-V layers to several microns or by limiting all process steps to below ~300°C, cracking and delamination can be avoided. The onset of cracking was studied in detail in [11] in which the authors show that there are reasonable limits to the thickness of the integrated devices and that the thermal mismatch problem is tenable (with a processing temperature differential of 700°C the critical cracking thickness of GaAs is still greater than 2 μm). Engineering of thermal stress is also possible; by intentionally incorporating lattice mismatch equal in magnitude but of the reverse sense than that expected through thermal mismatch, one can counteract thermal residual strain at room temperature. Even so, low temperature processes are preferred not only to avoid the mismatch problem altogether but also to prevent undesirable interdiffusion between III-V and IV layers.

8.1.1.2 Hybrid Technology Progress

To date, using both hybrid and monolithic techniques, only a handful of electrically pumped lasers on Si have been demonstrated, and to our knowledge, no lasers have been integrated with CMOS electronics. We first focus on recent progress in hybrid approaches to integration on Si.

At Ghent University, Roelkens et al. have recently demonstrated both Fabry-Perot and Microdisk InP-based lasers bonded to silicon-on-insulator (SOI) [7, 12]. Here, the fabrication of Fabry-Perot lasers was motivated by the need for high optical-power output (milliwatts) at the expense of large footprint and power consumption for an optical transceiver application such as fiber-to-the-home. Microdisk lasers are motivated by the need for a lower-power source (tens of microwatts) for an intrachip optical interconnect application. Much work was focused on the coupling of light out of the laser structures and into waveguides. Next to the 500-μm-long Fabry-Perot laser structure, a polymer waveguide is butt-coupled to a tapered SOI waveguide such that the optical mode is gradually passed into the Si wire (see Figure 8.3). Almost 1 mW of multimode optical power coupled to the Si wire was demonstrated. Although this structure requires a dedicated coupling structure, this has advantages for high-density integration as it couples light into relatively thin Si waveguides. Also, due to the low thermal conductivity of the bonding layers (SiO_2 and DVS-BCB) continuous wave (cw) operation of these lasers was deteriorated. By including a heat sink design where the p contact was extended to reach the Si substrate, cw properties were improved. Further thermal optimization will be needed for this device design.

Another very recent breakthrough in continuous wave, electrically pumped lasers on Si was developed by teams at UCSB and Intel. Here, researchers created the first electrically pumped hybrid evanescent laser on Si by using an AlInGaAs/InP structure emitting at about 1,577 nm. One key accomplishment in this study was the demonstration of evanescent coupling to bring laser emission onto the Si platform. In this design, the unpatterned III-V wafer dice are bonded to a relatively thick (690 nm) Si rib waveguide structure already fabricated on the Si wafer. The surface of the laser structure is intimately bonded, without an adhesive layer and with a low temperature process (<300°C), to the Si waveguide. By designing the dimensions of the waveguide properly, the optical mode is controlled to fall primarily in the wave-

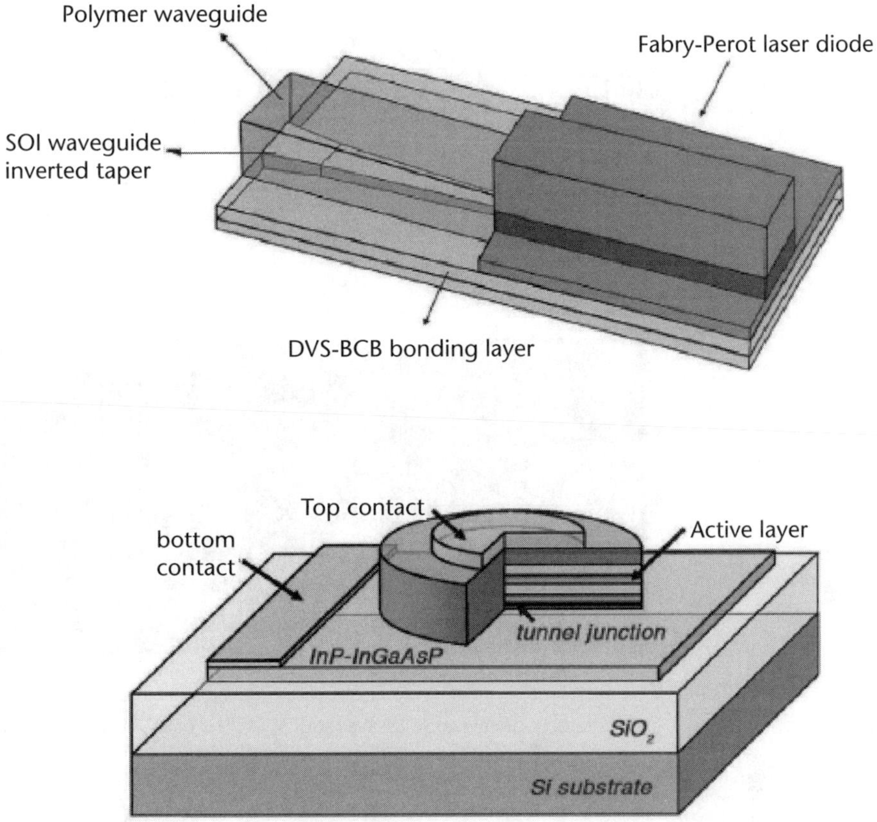

Figure 8.3 Schematic drawings of two successful hybrid approaches to integrated laser sources on Si. The Fabry-Perot design is larger in footprint but could provide higher powers than the smaller, low-power, low-threshold Microdisk design. (*Source:* [7, 12].)

guide (see Figure 8.4). One major advantage of this process is that it removes the need for precise alignment in the bonding step. The position of the Si waveguide and subsequent electrical isolation provided by H+ implants [13] define the gain region of the laser. Similar techniques were used to demonstrate a laser operating at 1.31 μm as well, showing the flexibility and broad applicability of the method [14]. The authors claim that this process could provide thousands of lasers to the host wafer in a single bonding step, thus solving many scale-up problems. However, we assume the reference to scale-up refers to a limitation of die-level scale-up, as the high-volume InP and Si wafer diameters are not the same.

One problem with the UCSB-Intel design is the requirement to cleave and polish the Si hybrid waveguide structure to define the laser cavity. The team addresses this difficulty with a subsequent waveguide racetrack design that requires no facet definition to create the cavity. The authors also state that single wavelength sources are possible on this platform by shrinking the cavity length to about 50 μm. This solution requires building lasers with larger footprints but still small enough that many thousands can be placed on a typically sized die. Much work remains to be done with these laser designs. Although the output power of the racetrack laser is 29 mW (cw), which is sufficient for many optical interconnect applications [16], lasing was

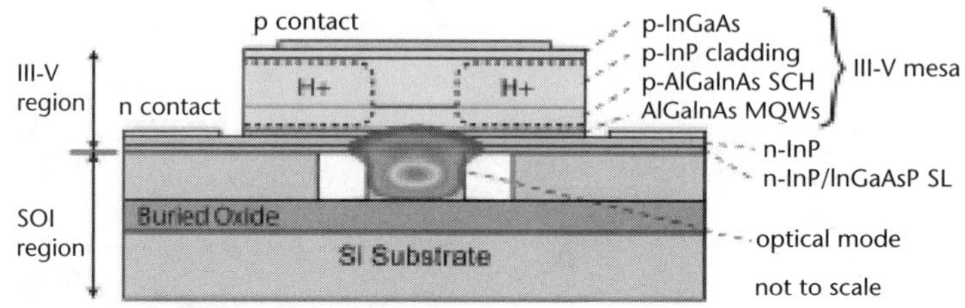

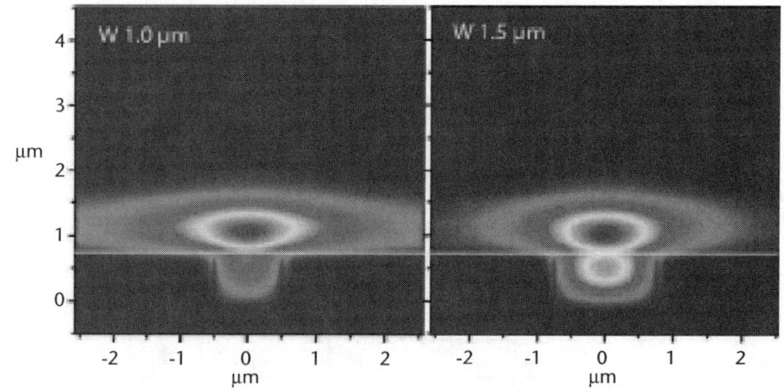

Calculated optical mode for waveguide widths of 2.5 μm and 3 μm.

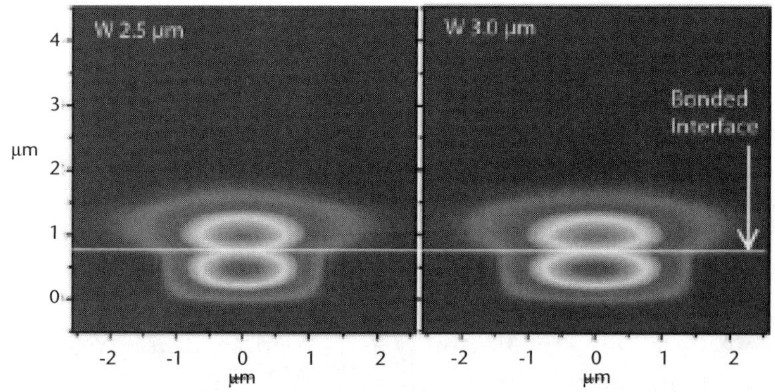

Figure 8.4 Schematic representation of the bonded evanescent design. Here the optical mode of the emitted photons is designed to lie primarily in the Si rib waveguide. As shown by the mode calculations, varying the rib waveguide width controls the overlap of the modes from the III-V material with the Si waveguide. (*Source:* [10] for mode, [15] for schematic.)

only observed up to 60°C, and the operation temperature of microprocessors can be well upward of this. Similarly, the faceted laser design only lased (cw) up to the relatively low temperature of 40°C and had a maximum output power of 1.8 mW. Certainly, thermal management of the optoelectronic device will be very important.

Lastly, we point out one fundamental problem with the design of the evanescently coupled waveguide laser: since the waveguide is designed to confine most of the optical mode (>95%) outside of the gain/quantum well (QW)region, these lasers have relatively weak amplification. This can, in theory, be counteracted through the use of waveguide engineering to physically separate gain regions from coupling regions of the device [17]. In [17], Yariv proposes a design called the "Supermode Si/III-V Hybrid Laser" in which the width of the Si waveguide (as described above) is varied to push the optical mode either up into the III-V slab for amplification or down into the Si waveguide for optical coupling. With very few simple modifications to the geometry of the UCSB-Intel system (etching the III-V slab into a mesa waveguide and adding a 50 nm Silica layer between the Si and III-V), Yariv calculates that by simply changing the Si rib width (from 0.75 to 1.35 μm) rather than keeping the Si rib width constant (1.1 μm), the confinement factor in the QW region in the amplifying stage can be increased from 0.067 to 0.268 and the confinement fraction in the Si waveguide in the transport/coupling region can be increased modestly from 0.757 to 0.892. Thus, with simple modifications, the hybrid laser performance can be further improved. It should be noted that this modification adds SiO_2 between the III-V and Si layers, and it is not clear if such a thin layer might affect the thermal properties of the structure to its detriment during cw operation.

The UCSB-Intel group has further demonstrated a mode-locked laser (MLL) using their hybrid technique [18]. MLLs are useful for producing short pulses from a wide optical spectrum. When combined with optical modulators, these are useful for optical clock signal generation, optical time-division multiplexing, wavelength-division multiplexing, and optical code-division multiple access. This device was operated at repetition rates of up to 40 GHz and was both actively and passively mode locked. Currently, the requirement for separate cleaving and polishing steps to define the cavity mirrors prevents this device from being integrated with other optical devices. Future designs may include etched mirrors, ring cavities, or DBRs will allow for integration.

8.1.1.3 Monolithic Technology Progress

Up to this point we have focused on hybrid fabrication of III-V on Si devices. While there are distinct advantages to implementing the proven technology of decades of telecommunications sources, we also pointed out that die-to-wafer bonding of III-V epitaxial structures to a Si wafer has associated scale-up difficulties. As an alternative, III-V layers could be grown directly on Si wafers in a wafer-scale process, removing the need to bond multiple dice to each Si wafer. This approach has strong advantages but brings up many concerns as well. With a III-V epitaxy step in the CMOS process, (1) wafer size difference between the typical III-V and Si platforms can be ignored, (2) III-V substrates that are more expensive than Si on a per-area basis are not needed, (3) growth of the "optical" layer on Si could be a natural addition to the Si CMOS process, and (4) selective area epitaxy methods (which have been well proven in III-Vs) could be implemented. Insertion of a III-V growth step into the CMOS manufacturing process requires overcoming many difficulties: (1) bridging the lattice mismatch between Si and typical III-V emitters, (2) dealing with thermal mismatch between Si and III-V, (3) growing heterovalent materials (III-V) on homovalent (Si) materials while avoiding antiphase defects (misregistry of the

III-V bonds), (4) processing of III-V materials in a CMOS fab (where material contamination concerns are paramount), and (5) managing and balancing the thermal budgets of both the Si CMOS circuits and III-V optical devices. While the difficulties are large and the paths to monolithic integration unclear, the benefits to successful heteroepitaxial integration are immense.

Of primary importance for the fabrication of optical devices is III-V material quality. Several demonstrations of III-V lasers directly grown on Si via high-quality graded buffers have been made [4, 19]. Although other recent progress has demonstrated metamorphic buffers beyond the Ge lattice constant to InP [20] and beyond [21], we will focus here on better-developed technology at the Ge/GaAs lattice constant. In this work, lattice mismatch is alleviated gradually through epitaxial layers with increasing amounts of Ge in SiGe, building in and relaxing strain. This technique spreads misfit dislocations through the thickness of the epitaxial film and assures that efficient dislocation glide can take place, minimizing the nucleation of excessive populations of threading dislocations. This method requires somewhat thick films (10 μm, typically, to bridge Si to Ge/GaAs), which are disadvantageous in themselves but give the highest quality of integrated mismatched material available (~10^6 cm^{-2} TDD). As a result, both III-As and III-P lasers (emitting at 858 and 680 nm, respectively) at the GaAs lattice constant were realized, underlining the high quality of material that can be obtained by this method.

Heterovalent-on-homovalent growth tends to lead to antiphase domains and boundaries. This is the mismatch of domains initiated on a substrate with far-spaced single atomic steps; III-V material that nucleates on each side of a single step will suffer misregistry of bonds upon coalescence at the step edge [Figure 8.5(a)] [22]. These III-III and V-V bonds, called antiphase boundaries (APBs), create nonradiative defect centers in optical devices and increase leakage current in diodes. APBs can be avoided or dealt with through two complementary techniques, both of which feature growth on offcut substrates. Offcut introduces a high density of step edges that tend to coalesce into double atomic steps during a high-temperature anneal prior to III-V growth. III-V growth on a double atomic step does not result in an antiphase defect [see Figure 8.5(b)]. Growth on a substrate with a high density of single steps (provided by the wafer offcut) will contain APBs, but their high density allows them to self-annihilate as shown in Figure 8.5(c).

More practical aspects of monolithic integration are also being addressed by the team at MIT with creation of a platform for practical integration of Si and III-V epitaxial devices [23, 24]. The thickness of a SiGe graded buffer grown directly on a Si wafer is prohibitive for III-V device integration (interconnecting III-V devices that are 10 μm higher than the surrounding Si CMOS circuits would be difficult). The platform, called Silicon on Lattice Engineered Substrate (SOLES), consists of an SOI structure wafer bonded on top of a Ge/SiGe/Si high-quality virtual substrate. This design enables coplanar fabrication of Si CMOS (in the top SOI) with III-V structures grown in wells etched to reach the underlying Ge virtual substrate (see Figure 8.6). Visible LEDs were successfully demonstrated on the SOLES platform by Chilukuri et al. (see Figure 8.7). Here AlInGaP/GaAs encapsulated in Si was used to demonstrate CMOS-compatible integration of III-V optical materials. Current work in the group also focuses on thinning the SiGe graded buffer while maintaining material quality and relaxation.

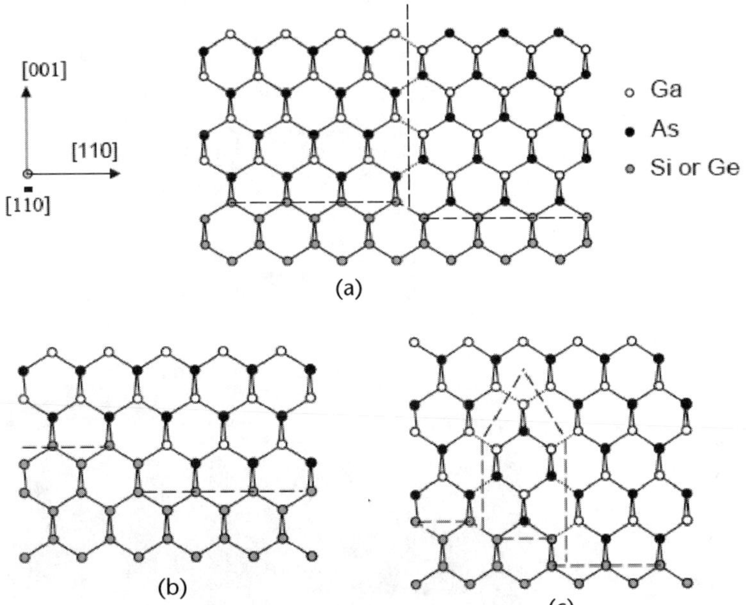

Figure 8.5 [110] projection of III-V on Si growth. (a) At a single atomic step edge, the III-V material will suffer a misregistry of bonds resulting in an APB. (b) The presence of a double atomic step prevents the formation of an APB. (c) A high density of single steps allows self-annihilation of APBs. (*Source:* [22].)

Other epitaxial integration techniques are also possible. Epitaxial lateral overgrowth (ELO) has been used to achieve 1.5-μm–thick, low-threading-dislocation-density InP on Si, using a single 0.5 μm GaAs buffer layer [25]. The process uses a Si_3N_4 mask with openings to the underlying offcut Si substrate. ELO was shown to result in high-quality InP with very low strain, due to thermal expansion mismatch between the InP and mask materials.

III-V quantum dot (QD) lasers have many desirable attributes and are being developed on the GaAs lattice constant for integration on Si. Recent advances in QD lasers have made them excellent candidates for an OEIC source. Attractive, demonstrated parameters have included very low current threshold (J_{th} = 100 A/cm^2), large output power (=14W), temperature-invariant operation, large small-signal modulation bandwidth ($f_{-3\ \mathrm{dB}}$ = 24.5 GHz), and near-zero chirp and α parameters [26]. The wavelength of III-As QD lasers is relatively short, at around 1 to 1.3 μm.

The team at University of Michigan, Ann Arbor, in [27] grew QD devices directly on Si using only an metalorganic chemical vapor deposition (MOCVD) GaAs buffer layer of about 2 μm. They claim that although the overall dislocation density of this structure is high ($2 - 5 \times 10^7$ cm^{-2}), the stress fields present from the strained QDs deflect threading dislocations away, and since higher stress fields are more effective, light-emitting properties of large, highly strained QDs are improved over those of QDs with less strain. The performance of the QD lasers on Si (e.g., J_{th} = 1,100 A/cm^2) is worse than that on GaAs, and work is being done to improve the devices by incorporating a SiGe compositionally graded buffer. The team further studied an integration scheme that included growth of a QD laser coupled to a QW

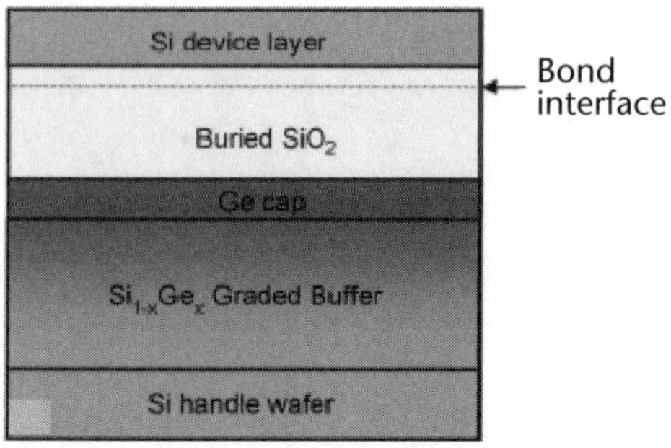

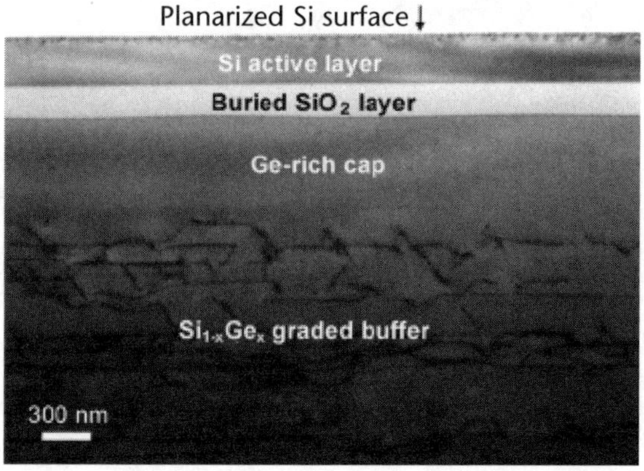

TEM ,micrograph of SOLES structure, with residual exfoliation damage at surface due to incomplete CMP removal. Bond interface is within the buried SiO_2 layer but is not visible because it does not produce contrast in TEM.

Figure 8.6 Schematic and transmission electron micrograph (TEM) of the SOLES structure with minimal exfoliation damage at the surface due to incomplete CMP. SOLES allows coplanar integration of highest-quality heteroepitaxially integrated materials. (*Source:* [23].)

modulator (QD density is typically too low to enable an effective QD-based modulator), but the process required focused ion beam (FIB) etching and subsequent regrowth of the modulator structure, making the device purely experimental [27]. Nonetheless, QD lasers' excellent performance attributes make them a quite attractive prospect for a Si OEIC.

A high-quality GaAs-on-Si solution opens the possibility of any GaAs-based photonic solution to be epitaxially integrated on Si, but still other routes to epitaxial photon sources on Si exist, including lattice-matched solutions [28]. Here Kunert et al. use pseudomorphically strained GaNAsP/GaP for infrared emission. While the material is in the early stages of research, the authors claim that integration on Si instead of GaP will be straightforward, and they have also published lasing results for the system [29]. Additionally, epitaxial integration on Si of smaller lattice constant materials (III-Ns) has been studied. Most studies focus on GaN-based materials on (111) Si substrates [30]. The addition of rare-earth ions in GaN heteroepitaxially grown on Si has also been investigated for the application of a wide range of color emission, from ultraviolet through visible into IR [31] and has resulted in a red laser on Si.

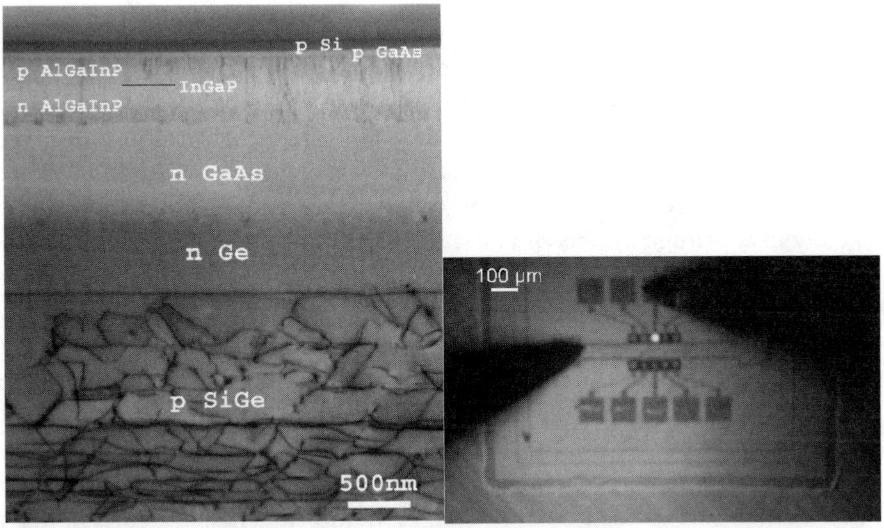

Figure 8.7 The SOLES platform was used to demonstrate visible light emission from AlInGaP/GaAs LED arrays in a practical CMOS-integrated scheme. A TEM micrograph shows the III-V material that was grown in wells etched through the SOI structure to expose the underlying high-quality Ge-on-Si graded structure. In the optical micrograph of the electrically probed LED, crosshatch from the underlying graded structure is clearly visible through the device Si layer. (*Source:* [24].)

8.1.2 Native Si and Impurity-Based Luminescence

Many methods have been applied to enhance Si's native light-emitting properties [32]. Insertion of dislocation loops into Si diodes modifies local band structure and helps to confine carriers (0.01% to 0.1% internal quantum efficiency) [33]. Texturing the surface of ultrapure Si diodes suppresses nonradiative processes, yielding relatively high internal quantum efficiencies of 1% [34]. Nanostructured devices have shown stimulated emission but very low efficiencies of 0.013% [35]. The discovery of strong visible light emission of porous Si [36] (which is brittle, weak, and difficult to work with) has led to the development of nanocrystalline Si (Si-nc) embedded in SiO_x. This technique improves the optical and mechanical properties of porous Si, largely through the benefits of the strong Si-O bond. Unfortunately, the oxide matrix that holds the Si-nc is also a good electrical insulator, making it difficult to electrically pump devices. These problems are being addressed through study of silicon nitride–embedded Si-nc as well.

Impurity-based schemes, where light-emitting rare-earth dopants are added to nanocrystalline Si, enable strong optically and electrically pumped luminescence. In particular, the use of erbium (Er) ions allows emission at 1.54 μm with IQEs claimed to be upwards of 10% [37]. We focus here on work in Si-nc.

8.1.2.1 Nanocrystalline Silicon: Si-nc

Si-nc layers are typically fabricated through plasma-enhanced CVD (PECVD) [38], Si+ implantation [39], low-pressure chemical vapor deposition [40], coevaporation [41] or sputter deposition [42], and annealing of Si-rich SiO_x films ($x < 2$). Annealing separates the phases of Si and SiO_2, while Si crystallinity (amorphous to nanocrystalline) is controlled through anneal temperature and time. A typical device is illustrated in Figure 8.8. Varying the atomic fraction of Si in the films affects the mean separation and radius of the Si phase.

For layers deposited on a p-type substrate, fabrication of an n-type polysilicon electrical contact creates a metal-oxide semiconductor (MOS) device capable of electroluminescence (EL). Large voltages are needed for effective excitation, typically in the range of tens of volts, but by thinning the active layer and optimizing the Si-nc, operating biases as low as –4V have been obtained [43]. Threshold voltage varies widely based on the Si concentration in the SiO_x film. EL near 850 nm is attributed to electron-hole recombination in the Si, and its wavelength varies with the mean radius of the nanograins. Smaller crystallites exhibit larger band gap and shorter wavelength emission. Current through the active layer depends strongly on the structure of the Si nanoclusters. For example, samples with high Si fraction and low anneal temperature (amorphous clusters) consist of highly interconnected Si structures surrounded by SiO_2 rich in Si and thus display current densities as much as five orders of magnitude greater than well-separated Si-nc samples for the same bias. EL brightness is strongly dependent on injected current, and despite their high conductivity, amorphous active layers are about 10 times dimmer than crystalline layers for a given current injection. Lower EL intensity, together with the much higher conductivity, gives the amorphous nanograins similar power efficiencies as nanocrystalline samples. In fact, radiative lifetimes in the amorphous active layers are faster than in nanocrystalline layers; it is the presence of relatively stronger nonradiative processes that lowers the overall brightness of the amorphous layers. Presumably then, amorphous nanograins remain a potentially promising technology for Si-integrated EL.

Investigation on silicon nitride–embedded Si-nc is also underway. Here, the motivation is to move away from wide band gap, resistive silica in favor of another CMOS-compatible, smaller band gap material, such as the Si-nc matrix. Visible emitting devices with 0.005% efficiency [44] and near-infrared structures have been demonstrated [45] with this system.

Note that Si-nc work is converging on a device structure that is very similar to older electroluminescent display devices. In these devices, an insulator is evaporated with a rare-earth dopant—essentially, electroluminescent displays are phosphors in

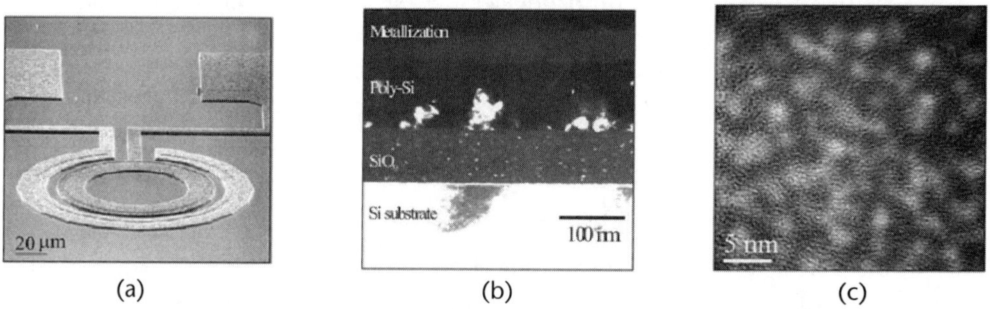

Figure 8.8 Several electron micrographs of Si nanostructures: (a) SEM image of a device, (b) dark-field TEM cross section of device showing Si nanocrystals in SiO_x film, and (c) high-resolution image of amorphous Si nanoclusters (bright) in SiO_2 matrix (dark). (*Source:* [32].)

thin-film form so that high voltages can electrically inject carriers through the insulator, exciting the rare-earth elements, and subsequently the rare earth will emit light. Full displays have been commercialized. However, reliability is the ultimate problem for any such device. Thus, Si-nc appears to be headed along a path on which much research, development, and commercialization has already been attempted and failed. We remain optimistic that Si-nc researchers can fully investigate the failure mechanisms of electroluminescent display devices and head Si-nc research in a direction to avoid a similar fate.

8.1.2.2 Impurity-Doped Silicon

The characteristic wavelength of emission of Si-nc devices can be shifted toward the more technologically significant wavelength of 1.54 μm by implanting erbium ions in the nanostructured layer (Si-nc's act as efficient energy sensitizers to rare-earth ions). After annealing to repair implant damage and activate the Er ions, these samples behave similarly to the Er-free structures, except that the wavelength of emission is shifted [46–48]. The mechanism of energy transfer to the erbium ions is typically explained as a Förster-Dexter nonradiative energy coupling. Due to their similar design, these devices have many of the same limitations as the Si-nc described above, but they are more attractive given the wavelength of emission.

8.1.3 Nonlinear Optical Properties of Si: Raman Emission

Recently much interest has been generated by the demonstration of silicon Raman-based lasers [49]. Raman scattering is routinely used for light stimulation in optical fiber but requires fiber lengths of kilometers. On the wafer, silicon's large Raman gain coefficient allows laser designs to be incorporated at the chip scale, but the resulting devices are still large (approximately millimeters) in comparison to conventional semiconductor lasers. In the Si-Raman laser, the nonlinear optical properties of Si are used to convert the wavelength of a pump laser (from off-chip) to a signal wavelength (on-chip). When optically excited by a well-defined energy (purely or nearly monochromatic source), faint sidebands can be observed at frequencies spaced above and below that of the pump energy. Atomic vibrational modes of a crystal define these energies (Stokes and anti-Stokes transitions) by which excited photons differ from the pump photons. Stimulated emission is possible by forming a cavity where the signal wavelength resonates and the pump wavelength excites carriers. Since the material properties of the medium define the level of the Stokes shifts (energy change of the pump), theoretically a very wide spectrum of outputs is possible by varying the color of the pump, from infrared into the ultraviolet.

Si-Raman lasers use no exotic dopants or materials, so they are fully CMOS compatible. This technique has the inherent drawback of requiring a separate laser source and optics (increasing system complexity), which must be linked into the chip (increasing manufacturing complexity). The gain bandwidth of these lasers is relatively small and only supports several WDM channels. Some flexibility in the design seems to be available by varying the pump wavelength or by adding multiple pump wavelengths (to tune output spectrum). Incorporation of Ge structures may introduce a possibility to tune characteristic Stokes shifts (the shift is a function of SiGe composition), broaden the beam spectrum (graded SiGe), or increase the number of Raman active phonon modes and output wavelengths (SiGe superlattice). Overall, it seems that much improvement is needed, not only in terms of the lasing efficiency (demonstrated threshold was 9W of optical pump power) and bandwidth of the devices but in terms of practical aspects like size, support infrastructure complexity, and optical coupling onto the chip, before Si-Raman lasers can be included in a large-scale, integrated optoelectronic circuit.

8.1.4 Future Photon Source Technologies

Up to this point, we have elucidated several enabling technologies for the Si CMOS OEIC, focusing on approaches with the most advanced and promising results. Two approaches that are also interesting, but in the early stages of research, remain to be discussed.

The use of Ge in Si photonic structures is intriguing because of its natural compatibility with Si CMOS processes. Being a Group IV element, there is the perception that Ge is not a serious CMOS contaminant compared to III-V materials (in reality, Ge requires more careful attention as GeO is very volatile). In fact, Ge has already been in use for several years in the production of modern microprocessors to provide strain in metal-oxide-semiconductor field-effect transistor (MOSFET) channels to increase minority carrier mobility and thus performance characteristics. While Ge was traditionally held as an indirect material and not considered for optical source applications, recent years have seen increasing interest in coercing Ge to behave as an optical material. Much work has been done to create Ge-based photodetectors for Si photonics (discussion to follow), and this is being expanded to include Ge as a gain material for a Ge-based laser [50].

The basic premise of this study is based on the observations that the indirect (L) conduction band valley is only 136 meV below the direct (Γ) valley and that tensile strain in Ge reduces this difference. According to deformation potential theory, a strain of 2% will make Ge direct, but it will also decrease the direct band gap from 0.8 eV to about 0.5 eV. While this may be useful for long wavelength devices (~2.5 μm), shorter wavelengths are desired for Si integration technologies, and, coincidentally, germanium's native direct band gap of 0.8 eV corresponds to 1,550 nm light emission. A relatively modest tensile strain of 0.25% (easily achieved through thermal expansion mismatch between Si and Ge) will reduce the energy difference between direct and indirect valleys. By introducing relatively large n-type doping in the Ge (7.6×10^{19} cm^{-3}), the authors propose to fill the remaining portion of the indirect valley below the direct with carriers such that injected electrons populate the direct valley. Furthermore, tensile strain splits the light and heavy hole valence bands, bringing the light hole band closer to the direct valley. This increases optical gain with injected carrier density more quickly due to the lower density of states in the valence band. The authors believe that despite the large free carrier absorption caused by heavy n-type doping and short (100 ns) defect-limited carrier lifetime, a net gain of about 400 cm^{-1} and a lasing threshold current density of about 6 kA cm^{-2} will be attainable. We additionally point out that it may be possible to reduce the density of threading dislocations in this design through the use of bonding or metamorphic graded buffer techniques to extend the minority carrier lifetime in the device, thus lowering the lasing threshold and extending device lifespan.

Another technology under development that may prove important to optoelectronic integration efforts is a SiGe quantum cascade laser such as that theorized in [51] and also by others. In this study, the prospect of an intersubband device utilizing a cascade of transitions in germanium's conduction band L valley is analyzed. Although such a device could probably be designed only for operation in the far infrared, it would also naturally integrate with other Si-based devices. Given that so few Si-based lasers have been demonstrated, any SiGe-based optical device using CMOS-compatible materials is truly an intriguing prospect.

8.1.5 Fundamental Questions: Localized Luminescence and Reliability

In pursuing a relatively important, but fundamental, goal like monolithic integration on silicon, connecting the underlying physics to the overall goals in the context of previous work is crucial. It is often difficult to determine what research paths could be most viable. In such an exercise, we should try to keep in mind accumulating trends, especially broad trends that may give insight into paths that may have a low chance of succeeding. After researchers have created the seeds in many different areas, it is reasonable to ask if potential trends come to light. Two areas deserving careful thinking are the nature of localized luminescence and reliability.

Broadly, research has lined up into two categories for creating light emitters on silicon: interband semiconductor transitions, like those in III-V technology, or localized luminescence, like the injection of rare earths into semiconductors, the formation of nanocrystals, the formation of intermediate defect states, and so forth. We may speculate beforehand that these localized phenomena suffer immediately from efficient electrical injection or, at a minimum, from a low density of states. By definition, using some barrier to create a localized luminescent center that may be efficient in optical emission brings into question whether such a structure can ever have high electrical-to-optical conversion efficiency. After decades of research, it seems that experimental data supports this early hypothesis. It seems that semiconductor bands remain the most efficient electrooptical converters because the energy gap is between conducting electron states; a photon is immediately interacting with "conducting rails" that are conducting and connected to contacts. The trend suggests that semiconductors, like III-Vs, should be used for light sources and that performing lattice-mismatch engineering to create true monolithic integration on silicon is a promising path. Some attributes of the discussed technologies are shown in Table 8.1.

For any light emitter to have prospects, it must follow the test path:

1. Can it emit light from electrical injection?
2. Can it emit light *efficiently* from electrical injection?
3. Can the material be integrated on silicon and retain efficient light emission?
4. Is it reliable?
5. Can it be manufactured in a silicon fabrication facility?

Table 8.1 Summary of Some of the Potentials and Challenges Offered by Several of the Integrated Optical Sources Discussed

Integrated Optical Source	*Potential*	*Challenges*
III-V monolithic	Mature optical technology Best electron-to-photon conversion efficiency Production volume very scalable Excellent material quality possible	Integration: lattice mismatch, CMOS incompatibility, thermal mismatch
III-V hybrid	Mature optical technology Best electron-to-photon conversion efficiency Best material quality	Integration: CMOS incompatibility, thermal mismatch Limited scalability in production volume Photon coupling
Si-Raman	CMOS compatible	External optical pump required Large device size Poor performance
Si-nc	CMOS compatible	Poor performance: low efficiency, high voltages required
Ge	CMOS compatible	Innovative design unproven

Much of the work has apparently been in category 1 due to an unquantified interpretation of category 5, with very little work occurring in categories 2 to 4. However, we wonder here whether an effort to turn silicon or another inefficient light emitter into something that can transgress to stage 5 will have a convenient lattice and few defects such that it can be reliable (e.g., by the time it is converted into something that could work, it looks like a pretty bad semiconductor). So far, the research appears to suggest that steps 2 and 4 are the most difficult for modified silicon. It is interesting to note that III-V materials are at stage 4, and from modern investigations, III-V material is likely to already be accepted in silicon-fabrication facilities if needed, due to the modern incorporation of many new materials into the silicon-fabrication facility. In other words, material compatibility appears to depend on need and demand and is not a physical problem.

This current summary suggests that III-V integration is likely the first path to commercial monolithic optical integration. Breakthroughs with any other solution will compete in the future with an installed base of III-V/Si expertise (assuming any monolithic technology is commercialized).

8.2 Optical Modulators and Resonators on Si

Optical modulators and resonators comprise an essential aspect of many nanophotonic CMOS circuits. Modulators make it possible to encode digital information into an optical carrier wave, which relieves the burden of having to directly modulate the optical light source. This enables faster data-transmission rates since higher modulation rates can be achieved with less distortion by using a separate modulator than by direct modulation of the light source. In addition, use of modulators could potentially reduce the number of on-chip optical light sources required, and it also opens the possibility of using an exclusively off-chip light source, which can then be coupled into the OEIC (although the typical assumption is that this can be done much more easily and more cost-effectively than is actually the case).

Modulator performance is evaluated using a number of criteria. Some of the primary ones to consider include

- *Insertion loss:* This is the amount of loss a light signal experiences in passing through the device in its off state. It is given by

$$\left(P_{in} - P_{out}\left(V_{off}\right)\right) / P_{in} \tag{8.1}$$

  where P_{in} is the signal input power, P_{out} is the output power, and V_{off} is the voltage corresponding to the off state of the device.
- *Contrast ratio:* This is the ratio between the output optical intensity in the device off state, $P_{out}(V_{off})$, and the on state, $P_{out}(V_{on})$. Alternatively, the modulation depth is also sometimes used to describe this characteristic. It is defined as follows:

$$M = \frac{P_{out}\left(V_{off}\right) - P_{out}\left(V_{on}\right)}{P_{out}\left(V_{off}\right)} \times 100\% \tag{8.2}$$

- *Bandwidth* (Δf).
- *Turn-on voltage* (V_{on}).
- *Power consumption:* For most of the devices currently being proposed, the main source of power consumption is in the capacitive charging and discharging of the device as it is turned on and off (a nonideal component is the leakage current of the device).
- *Device size.*

Most schemes for light modulation in CMOS-compatible devices can be grouped into two broad categories. The first consists of electroabsorption modulators (EAMs) in which the absorption of the device can be controlled by application of a voltage. The second category consists of devices in which the applied voltage is used to control the refractive index instead of the absorption. By using this refractive index control in an interferometer arrangement, this phenomenon can be used to produce amplitude modulation in the carrier wave. These two categories are explained in the sections below.

8.2.1 Electroabsorption Modulators

In electroabsorption modulators, a voltage is applied to the device, producing an electric field in the active region, which increases the optical absorption in this region. Electroabsorption in semiconductor devices is caused by the Franz-Keldysh effect in bulk layers and the quantum-confined Stark effect (QCSE) in quantum-sized heterostructures. Both of these effects utilize the slanted band edges produced by the electric field across the device. In bulk layers, application of an electric field causes the ground-state electron and hole wavefunctions to form an electron standing wave outside the band gap with an evanescent tail inside the band gap. Overlapping evanescent tails enable absorption inside the band gap, thus lowering ("redshifting") the onset of absorption. In the QCSE, the electron ground-state energy is lowered and the hole ground-state energy is raised by an applied electric field, thus decreasing the transition energy. In addition, the QCSE prevents the field-induced separation of excitons by spatially confining the electron and holes together within the quantum well. In bulk layers, the electrons and holes are separated by the electric field, thus inhibiting exciton formation, which has a blueshifting effect that partially counteracts the Franz-Keldysh effect [52]. Thus, QW-based structures utilizing the QCSE can attain a much more significant degree of electroabsorption than one can obtain with bulk samples.

EAMs composed of III-V materials are the most mature of current semiconductor modulator designs. They have found use in fiber-optic telecommunication systems, where they have proven especially effective when monolithically integrated with a semiconductor laser [53]. Representative electroabsorption spectra of a few AlGaAs/GaAs heterostructures are shown in Figure 8.9 [54], illustrating the kind of strong electroabsorption that can be achieved with this material system. Integrated laser-modulators are commercially available with transmission rates in the tens of gigahertz.

In addition to III-V-based modulators, SiGe-based EAMs have also been an active area of research for optoelectronic-CMOS integration. The research community is pursuing these devices because, unlike III-V devices, they would not require

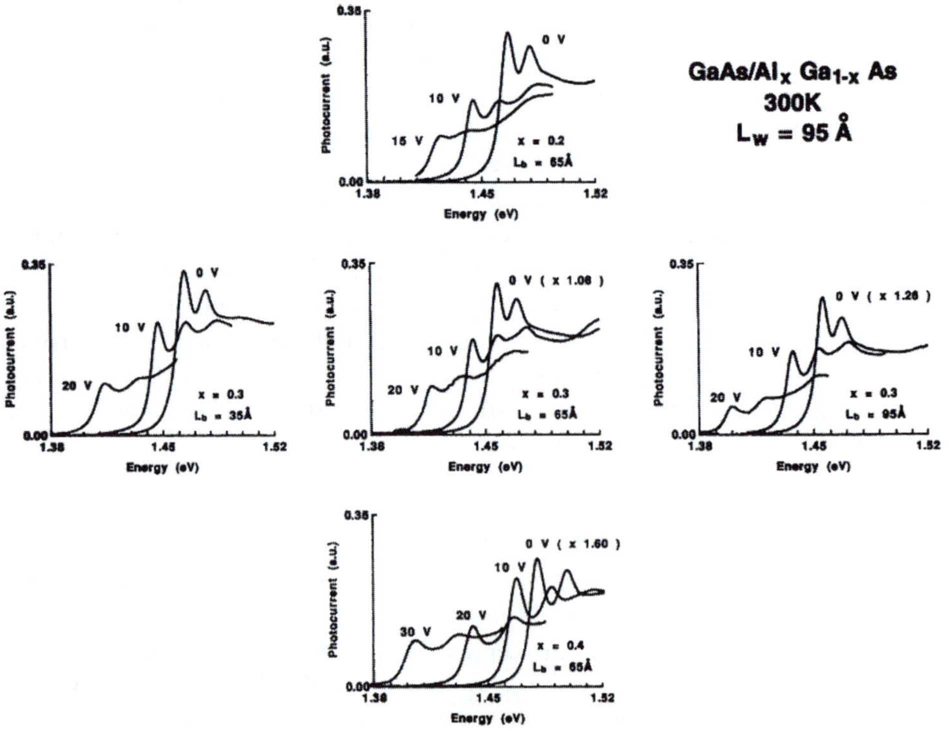

Figure 8.9 Electroabsorption spectra for various AlGaAs/GaAs multiquantum well pin structures. The quantum well barrier width increases horizontally across the figure, while the quantum well barrier height increases vertically down the figure. (*Source:* Courtesy of [54].)

the addition of new materials to the standard CMOS materials set (SiGe is an increasingly ubiquitous material for Si CMOS devices). However, the primary disadvantage of SiGe compared to III-Vs is its indirect band gap, which causes the optical absorption to be relatively low and to have a broad edge, thus limiting the modulation depth of the device. Electroabsorption in bulk Si was evaluated for modulator potential by [55], but this work only considered electroabsorption in terms of its effect on electrorefraction via the Kramers-Kroenig relations.

The first known report of electroabsorption phenomena in SiGe heterostructures is [56]. The authors reported the measurement of an electroabsorption redshift in a $Si_{1-x}Ge_x$/Si multiquantum well (MQW) heterostructure on a $Si_{1-y}Ge_y$ substrate, which has a type II band alignment. This result demonstrated that electroabsorption was possible, but the large width of the absorption onset would make it very difficult to design an EAM with a large modulation depth using this heterostructure.

In a study of strained SiGe-Si multiquantum well (MQW) structures with type I band offsets, [57] found experimentally that the structure did not undergo a redshift in its absorption spectrum when an electric field was applied. They attributed this to a field-driven decrease in the exciton binding energy in this structure, which creates a blueshift in the absorption spectrum that roughly cancels the redshift in the spectrum produced by the QCSE. This decrease in exciton binding energy is in turn caused by the relatively small conduction band offset in the strained SiGe-Si MQW structure. Further study of type I SiGe-Si MQW structures by [58] revealed a more complicated behavior in which absorption experienced a redshift at low electric fields and a blueshift at higher fields. At high electric fields, the electron is not confined by the small conduction band offset of the SiGe quantum well, leading to a drop in exciton binding energy and thus a blueshift in absorption. However, below a "critical" electric field, the electron stays confined in the SiGe quantum well, and the heterostructure undergoes a redshift in absorption similar to the QCSE observed in III-V materials (but to a much lesser extent since the electric field must remain below the critical value).

An EAM design using a type I SiGe/Si MQW structure as an EAM was investigated theoretically by [59] and then tested experimentally [60]. This design utilized a blueshift in the structure with applied electric field similar to that described by [58] at high fields. This device achieved modulation, but with a low contrast ratio and very high insertion loss due to the high absorption by the device in the off state.

While prior attempts at modulation by direct electroabsorption in SiGe structures have not demonstrated high performance, the work of [61] has demonstrated strong electroabsorption in SiGe structures in a remarkable way. Instead of using the indirect interband transition as the physical mechanism for absorption, the device relies on the direct transition from the valence band to the Γ-valley in the conduction band. The MQW structure employed consists of compressively strained Ge and tensile-strained $Si_{0.15}Ge_{0.85}$ on a $Si_{0.1}Ge_{0.9}$ relaxed layer, which has large band offsets in the Γ-valley to provide quantum confinement for the QCSE. The band-edge diagram of the device is depicted in Figure 8.10. The direct transition in Ge lies at a slightly higher energy (~0.8 eV) than the lowest indirect transition (~0.6 eV, using the L-valley in the conduction band). Thus, when operating at a photon energy near 0.8 eV, indirect absorption occurs in both the off state and on state of

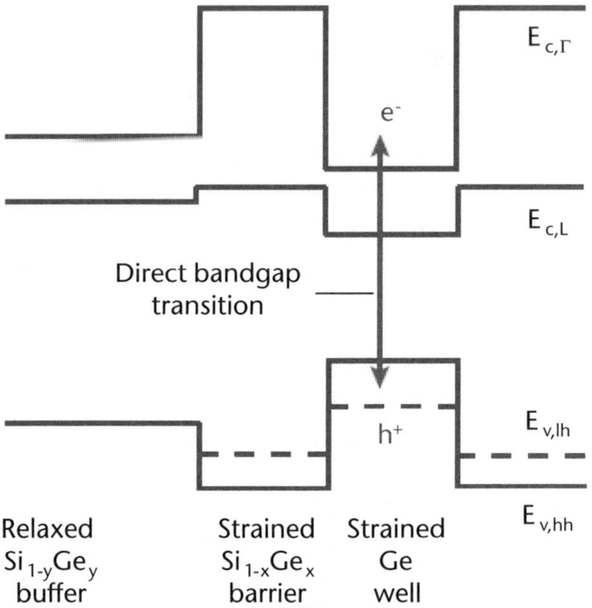

Figure 8.10 Band-edge diagram of SiGe MQW structure exhibiting QCSE [61].

the device; however, the contribution of the indirect transition to absorption is small enough that the Stark shift in the direct transition is clearly visible. The absorption versus photon energy of the device at various applied voltages is shown in Figure 8.11. In this figure, one can clearly see the redshift of the exciton absorption peak as the applied reverse bias is increased. This device is capable of providing much higher modulation depths than other SiGe-based EAMs, and the proximity of the band gap of Ge to the important telecommunications wavelength of 1.55 μm makes it even more attractive as a practical device (although attainment of 1.55 μm operation required substrate heating to 90°C [62]). Thus, this device is by far the most promising of the SiGe-based EAMs proposed.

Another strategy is to use strain to modify the absorption behavior of Ge. Tensile strain in Ge causes the reduction of the energy of the Γ-valley in the conduction band, bringing it closer to the L-valley. This change effectively makes the Ge more like a direct gap semiconductor, which will produce a sharper absorption onset. Recent results [63] have shown that application of an electric field of 70 kV/cm to a tensile Ge p-i-n diode changes the absorption of 1,647 nm radiation from 57 to 230 cm^{-1} (as shown in Figure 8.12), a change that could be used to fabricate a 100 μm EAM with a contrast ratio of 7.5 dB and insertion loss of 2.5 dB.

One conclusion is that Ge or Ge-rich devices offer the best promise. However, since Ge is nearly lattice matched to GaAs, one realizes that high-quality Ge on silicon can result in high-quality GaAs on Si. Thus, wider band gap, direct band GaAs devices are generally preferred over Ge devices, forcing a reconsideration of how desired or useful a Ge-based device is.

With this in mind, a number of III-V modulators have been monolithically integrated on Si. The work of [64] demonstrated the presence of the QCSE in AlGaAs/GaAs MQW structures on Si. The QCSE was distinctly visible despite the

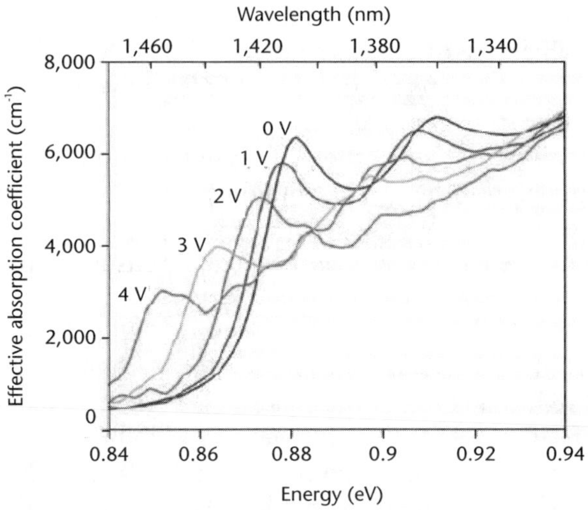

Figure 8.11 Absorption versus wavelength for SiGe QCSE EAM. (*Source:* [61].)

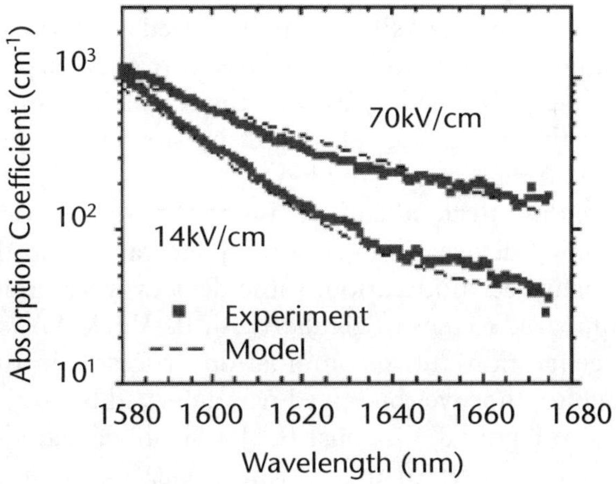

Figure 8.12 Electroabsorption spectra from a tensile Ge p-i-n diode utilizing the Franz-Keldysh effect [63].

fact that GaAs growth was initiated directly on Si, a technique that is known to generate high levels of threading dislocations, which degrade device performance. Compositionally graded $Si_{1-x}Ge_x$ buffers have been shown to enable growth of GaAs-based devices (such as lasers) on Si with device performance similar to identical devices grown on GaAs. The use of a $Si_{1-x}Ge_x$ buffer layer to integrate this EAM device on Si would be expected to yield a dramatic improvement in performance.

8.2.2 Phase-Modulation Devices

Phase-modulation devices comprise the other main category of modulator devices with potential application to nanophotonic CMOS OEICs. These devices function by altering the refractive index of the active region, which can be harnessed to produce an amplitude modulation of the carrier wave. This can be accomplished in various ways, which are discussed below.

Mach-Zehnder interferometers (MZIs) are one of the most common device designs used to convert phase modulation of a signal into amplitude modulation. The MZI temporarily splits the optical signal into two paths: one path being a passive waveguide, the other path containing the active region in which refractive index can be controlled by a voltage. The signal in the active region can be tuned in or out of phase with the passive waveguide signal, thus using the interference to modulate the signal.

In long-haul telecommunications systems, MZI modulators using the nonlinear optic material $LiNbO_3$ are the predominant choice because of the low insertion losses and reduced chirp of these devices, as well as their better modulation depth compared to III-V EAMs [52]. However, as $LiNbO_3$ is not a material used in CMOS processing, it is not being considered for CMOS OEIC applications, and the research community is focusing on devices using materials with proven CMOS compatibility. Also, a nonsemiconductor material has more material dissimilarities, such as larger lattice mismatch and thermal mismatch.

Of the standard CMOS semiconductors, Si and SiGe do not exhibit a linear electrooptic effect due to their centrosymmetric crystal structures. However, despite this, the refractive index of silicon can be varied by increasing the population of electrons and/or holes, thus increasing the free carrier absorption, which also results in a change in refractive index by virtue of the Kramers-Kroenig relations. This effect is referred to as the free carrier plasma dispersion effect, and its magnitude in silicon was investigated theoretically by [55].

To utilize this effect, silicon MZI modulators were developed using p-i-n diode or three-terminal devices to inject or deplete carriers in the active region [65, 66]. While they achieved modulation, these devices were limited to speeds of up to 20 MHz (as compared to tens of gigahertz for III-V EAMs) as they were rate limited by the carrier generation and recombination processes in the active region. A major breakthrough to improve this speed was achieved by [67]. Their device, depicted in cross section in Figure 8.13, consists of a Si rib waveguide with an MOS capacitor positioned at the center of the optical mode. This change in structure enabled a major enhancement in speed since the charging and discharging of the MOS capacitor is a much faster process. As a result, this device performed at speeds of over 1 GHz with further improvements possible [68].

While the MZI is a fairly simple way to convert phase modulation to amplitude modulation, it requires long device lengths to harness the refractive index contrast produced by a Si p-i-n diode. The device of [67] requires a length of about 8 mm, which is much too long to be practical for most on-chip CMOS OEIC applications. However, other modulator geometries can utilize the free carrier plasma dispersion effect more efficiently, allowing smaller device footprints. For instance, the device of [69], shown in Figure 8.14, uses a ring resonator geometry in combination with the free carrier plasma dispersion effect in silicon to achieve modulation. The transmission of the ring resonator structure is greatly reduced at resonant optical wavelengths, which occur when the ring circumference equals an integer number of wavelengths. The resonant wavelength can be tuned by applying a voltage to a Si p-i-n diode to modulate the free carrier population in the ring resonator. The high

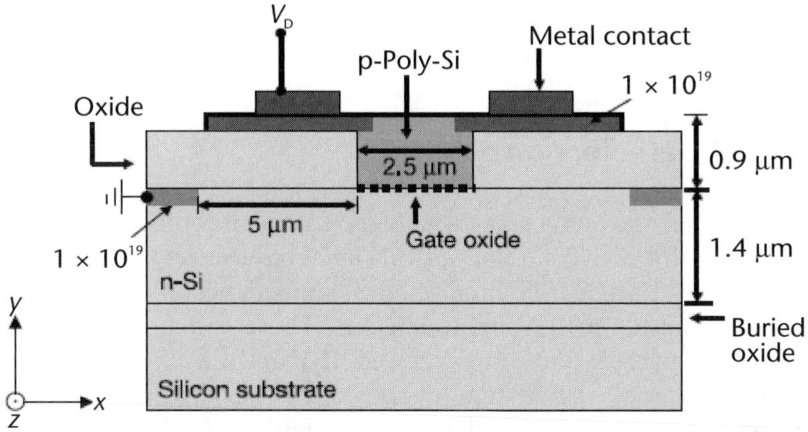

Figure 8.13 Cross-sectional schematic of MOS capacitor used for modulation of the free carrier population in the MZI modulator of [67].

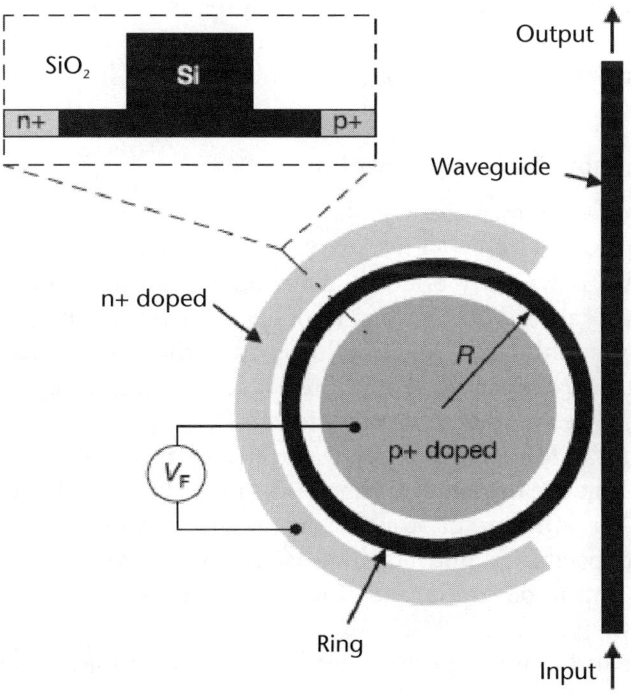

Figure 8.14 Ring resonator EAM design of [69]. Like the MZI modulator, the ring resonator design utilizes the free carrier plasma dispersion effect, but with significantly reduced footprint.

sensitivity of this geometry to variation in refractive index has enabled a modulator to be produced with a diameter of 12 μm, a dramatic improvement over the MZI modulator of [67]. Static operation of the device of [69] exhibited a 15 dB modulation depth at a 1,573.9 nm operation wavelength using only a 0.3V change in bias.

Dynamic testing achieved modulation speeds of 1.5 Gbps while maintaining a modulation depth of near 7 dB.

8.3 Optical Detectors on Si

To complete the optoelectronic integrated circuit toolbox, a photodetector (PD) will be needed for conversion of optical signal to electrical. At present, optical links are composed of Si CMOS circuitry (transimpedance amplifier, preamplifier, and so forth) alongside a III-V photodetector. These systems are integrated with hybrid packaging of the separately processed III-V and CMOS components. Not only is the performance of the system degraded by packaging parasitics and wire-bond crosstalk, but scale-up for large-scale integration is cost prohibitive. In efforts to make the PD a Si CMOS compatible technology, we must either select a new material for the PD portion of the link or devise a way to bring III-V to the Si CMOS fabrication facility. This quest is very similar to finding a photon source; in fact, many methods of generating photons can be reversed for photon absorption instead. Unfortunately, while some material systems are promising sources, they seem to offer much less when operated as detectors. We will discuss some of these limitations and outline promising research below.

8.3.1 Photodetector Principles

The key requirements of the photodetector are to convert optical to electrical signal efficiently, preferably with an explicitly designed method for coupling in light at low loss. This is typically done in semiconductors through generation of photocurrent such as in a pin PD: photons impinge on a reverse-biased pin junction where the active material has a band gap lower than the photon energy. Excited carrier pairs are swept apart by the provided electric field and thus create an electrical current (signal). The speed at which the detector can operate is determined by the length of time it takes for carriers to reach the electrical contacts or by the capacitance of the device that creates an RC time constant. Thick intrinsic regions allow sufficient length for efficient photon absorption (therefore high responsivity) and reduce device capacitance, but they increase the time of transit of carriers. Hence, detector design must be optimized for trade-offs in transit time, capacitance, and responsivity.

The technique used to couple light into the photodetector is quite important. In the vertically incident PD designs typical for discretely assembled systems, light absorption is parallel to the direction of current extraction. As discussed above, this places limits on the design since, for optimum responsivity, a thick intrinsic region is needed, but the large distance for carriers to travel lowers available bandwidth. If the direction of light propagation through the device is instead perpendicular to the current extraction, as in a waveguide-coupled device, then the parameters of responsivity and bandwidth are decoupled. This requires much more complex design of the device as the method for inserting the light must be explicitly designed with assumptions about the integrated waveguiding method. In a final OEIC design, PDs will most likely be waveguide coupled, so many research groups are demonstrating these devices now, showing optimization of the coupling geometry [6, 70]. The reader is referred to [71] for further discussion.

The 1.1 eV band gap Si makes it transparent to near-infrared light; thus, Si is a good near-infrared waveguide core material, but it cannot be used for detectors in this wavelength range. In optoelectronic integration schemes where the photon color is higher energy (visible), Si p-i-n detectors would be an excellent choice for high-performance, simple, hence low-cost detectors. As discussed above, tactics for integration of visible emitters have been investigated, primarily to exploit the existing technology for Si detectors. We will not focus on Si p-i-n detectors, given their well-developed technology, but instead on solving problems with integrating near-infrared detectors on the Si wafer.

In order to detect light at 1.3 and 1.55 μm, we must seek to integrate other materials on the Si wafer. Because of its compatibility with Si, a compelling approach is to look toward Ge to decrease the band gap of Si. Approaches to be discussed below include mostly relaxed bulk Ge layers or highly strained (Si)Ge QDs or QWs.

In our photon source discussion above, we pointed out that schemes for tweaking Si to do the optical work traditionally left to III-Vs comes up short in terms of efficiency; the same remains true in our search for a Si detector. Er-doped Si p-n junctions have the possibility of acting as detectors but are extremely inefficient (10^{-4}) [72] and require devices on the order of centimeters in length [73]. Similarly, the low dimensionality of Si-nc could allow generation of photocurrent, but their small absorption cross section and difficult current extraction would also make them very inefficient. For these reasons, we will not discuss these tactics further.

While SiGe materials may prove to be quite useful for near-infrared photodetection, III-V detector technology is the most advanced. InGaAs PDs have unequalled performance, including low dark current, high speed, and high sensitivity. If an OEIC's optical source is also to be of III-V materials, it might be natural to include III-V PDs as well. Integration of the III-V PD material involves the same challenges and opportunities listed above.

8.3.2 Highly Strained Group IV–Based Designs

Unlike bulk Ge device layers, SiGe QWs and QDs can be grown pseudomorphically on Si substrates, but they typically result in PDs with low absorption efficiencies, low responsivities, and very large device sizes. The use of $Si_{0.65}Ge_{0.45}$-strained QWs in [74] established that while SiGe PDs for 1.3 μm operation were possible, internal efficiencies were quite low at 11%. Similarly, [75] demonstrates $Si_{0.5}Ge_{0.5}$ MQW structures for detection at 1.55 μm, but only low responsivities of 0.1 A/W were obtained for relatively large 240 μm devices.

Interest in QD device designs for photon sources has been high in recent years, and QD-based absorbers have also been investigated. For near-infrared applications, Ge self-assembled islands on Si would constitute a convenient system, but because of their low volume density, absorption is relatively weak. El kurdi et al. [108] demonstrated large (3–7 mm) PDs using Ge QDs embedded in a pin structure with responsivities of only 0.025 A/W and 0.00025 A/W for 1.3 and 1.55 μm operation, respectively. Elfving et al. [76] used a more complex QW/QD hybrid design to increase responsivity of a QD-based detector. Here, Ge QD detection layers were paired with $Si_{0.79}Ge_{0.21}$ QW conduction layers and fabricated in a MOSFET-type, three-terminal design. Thus, applied gate voltages aided photocurrent extraction and resulted in responsivities of 0.35 and 0.03 A/W at 1.3 and 1.55 μm, respectively.

8.3.3 Mostly Relaxed Bulk Ge–Based Designs

Perhaps the most promising results for a Group IV–based PD have been attained with pure-Ge devices. The absorption length (α^{-1}) of $Si_{1-x}Ge_x$ at several important wavelengths is shown in Figure 8.15 [77]. From this figure, it is apparent that the addition of modest amounts of Ge to Si does not have a terribly strong impact on absorption efficiency; therefore, to fabricate sensitive PDs, high-Ge SiGe or pure-Ge devices with high material quality will be needed. This will additionally leverage the higher mobility, and therefore lower-voltage-requirement advantages, that Ge has to offer. Using high-Ge SiGe for PD applications is not without disadvantages: at a wavelength of 1.55 μm, Ge is still hampered by a long absorption length of about 10 μm (based on absorption coefficient in Ge); Ge is more difficult to passivate than Si; its small band gap leads to higher dark currents; its low melting point has implications for processing incompatibility with Si, including higher dopant diffusivities; and the 4.2% lattice mismatch between Ge and Si make *direct* epitaxial growth with low defect density extremely difficult.

Growth of Ge on Si is possible through a variety of means. Pseudomorphic Ge on Si is only possible in layer thicknesses below the critical thickness of about 10 to 20 nm [78]. Instead, as previously described, the use of SiGe compositionally graded buffers permits thick relaxed layers of monolithically integrated pure Ge of highest quality [79]. Dark current is increased by the presence of threading dislocations [80] and can be minimized by reducing TDD. One MIT study used SiGe buffers to demonstrate a very low dark current (2 pA/μm^2) [81], while another recent study showed improvement of dark current by a factor of 10 achieved by using two relatively thin buffer layers [82] (see Figure 8.16). Alternatively, a thin, low-temperature-grown Ge

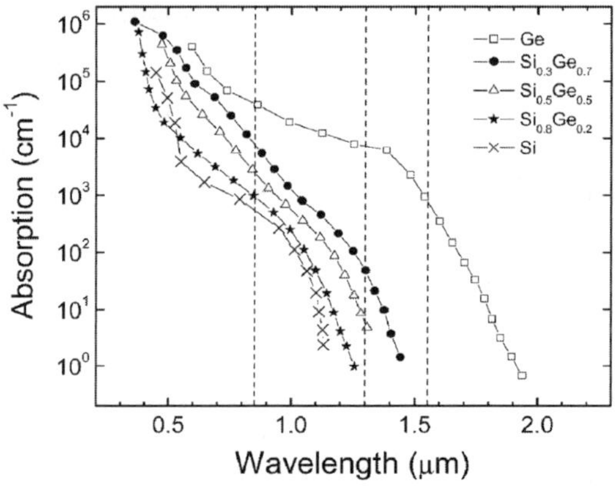

Figure 8.15 Absorption in SiGe as a function of composition and wavelength. Technologically important wavelengths (850 nm, 1.3 μm, and 1.55 μm) are marked with dashed lines and emphasize the large fraction of Ge in SiGe that is needed to make effective PDs at the longer wavelengths. (*Source:* [77].)

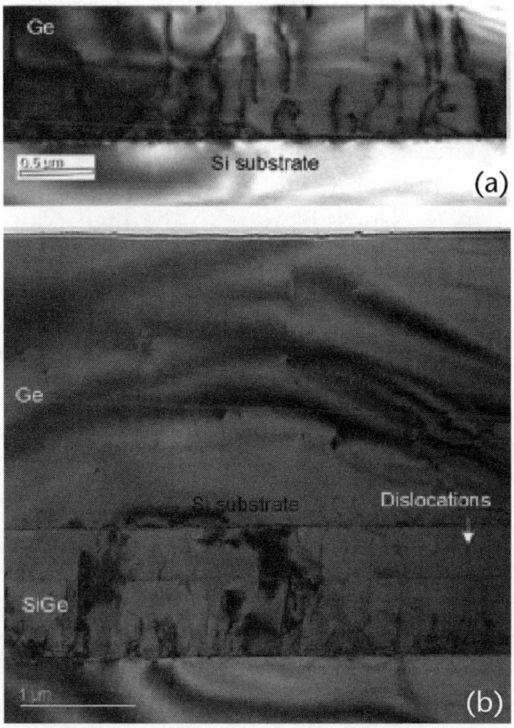

Figure 8.16 TEM comparison of direct Ge on Si growth and Ge on Si with several relatively thin SiGe buffer layers. (a) A very high density of threading dislocations is visible, indicating a likely TDD of about 10^9 cm^{-2}. (b) This density is much reduced, but the presence of dislocations in the cap layer of this cross section micrograph indicates a probable TDD of about 10^8 cm^{-2}. (*Source:* [82].)

buffer layer directly on Si followed by a high-temperature, thick Ge layer results in threading defect densities of about 10^9 cm^{-2}. After cyclic thermal annealing, this can be reduced to 10^7 to 10^8 cm^{-2} TDD [77, 83], which may be low enough for effective PDs at room temperature as the band gap of Ge is small and Ge devices are therefore intrinsically leaky at room temperature (thus, a high level of dislocations may not affect such an intrinsically large leakage). Due to its relative simplicity, this method is most often used in demonstration of Ge PDs for Si CMOS integration.

Recent results on Ge-based PDs have shown impressive progress. At IBM, Koester et al. [77] presented low dark-current devices in a vertically illuminated, lateral p-i-n PD. The efficiency is relatively high at 52% for λ = 895 nm, and the responsivity is 0.38 A/W. Additionally, the group demonstrated –3 dB bandwidths of 27 GHz with a dark current of 24 nA and efficiency of 46%. While this performance is excellent, it has been achieved for short wavelength operation. Improved performance at 1.3 μm can be brought about by adding an antireflective coating to increase efficiency. Or by optimizing the Ge, higher absorption may be possible; the authors state that diffusion of Si from the SOI underneath the Ge effectively makes part of the device SiGe, hurting long wavelength absorption. By reducing the volume of Si available (thickness of the SOI layer) and minimizing thermal budget, 1.3 μm absorption would be improved.

In a very recent study at MIT [84], Ge p-i-n photodetectors were monolithically integrated below silicon oxynitride and silicon nitride waveguides (see Figure 8.17), resulting in high responsivity and speed with low dark current, $f_{-3\text{ dB}}$ = 7.5 GHz, 1.08 A/W at 1.55 μm and <1 μA. The authors also state that the device was RC limited by the relatively large size of the contact pads used and, if integrated with a transimpedance amplifier, would have been capable of much higher bandwidth. The high responsivity was possible due to the high coupling efficiency of the design, which is capable of >90% insertion without limiting the intrinsic layer thickness. Light is inserted into the device in a direction perpendicular to carrier extraction; hence, long absorption lengths of about 10 μm are possible without large intrinsic layer thickness. The waveguide-coupled design of the device also solves the problem of absorption roll-off at long wavelengths, which is much stronger in vertically illuminated designs. These devices meet the performance requirements of high-speed receiver designs but have not been demonstrated with SOI waveguides.

8.3.4 III-V-Based Designs

In our discussion above regarding photon sources, we covered methods of integrating III-V material on Si through heteroepitaxy and hybrid integration techniques. These methods are applicable for III-V-based PDs as well. We focus here on the opportunities offered through integration of III-V-based PDs over Group IV.

InGaAs PDs are unchallenged and dominate in optical receiver design. Their unique combination of high-performance parameters includes low dark current, high speed, and high sensitivity [6]. Bonded InGaAs devices have long shown excellent parameters for integration. In 1997, Hawkins et al. reported in [85] vertically illuminated InGaAs avalanche photodetectors (APDs) integrated on Si with a gain-bandwidth product of 300 GHz for optical-fiber transmission applications. Devices with a 1 μm active layer showed responsivities of 0.57 A/W, internal quantum efficiencies of 60% (compared to the 69% expected for the design), and maximum bandwidth of 13 GHz at 1.31 μm. It should be noted that although photocurrent traverses the InGaAs/Si-bonded interface in this design, this does not

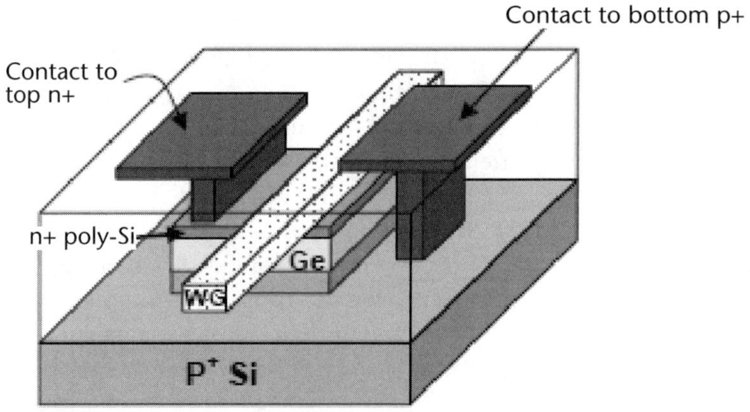

Figure 8.17 Schematic drawing of a waveguided, integrated PD design resulting in high responsivity and speed. (*Source:* [84].)

detrimentally affect the device. Instead, the authors attribute the small shortcoming of the quantum efficiency to diffusion of the p-type dopant (Zn) during the 650°C, 10 minute bonding step, which created an unintentionally thick absorbing layer at the back of the device. Additional progress on bonded InGaAs/Si photodetectors was show by Levine et al. in [86]. In their results on the performance of the diodes in 1.55 μm illumination, the authors quote 95% to 100% internal quantum efficiency and a 3 dB bandwidth of 20 GHz, along with maximum ultralow dark currents of 180 pA when reverse biased as much as 10V.

Lastly, we point out the results for integrated PDs from both the Ghent and Intel-UCSB teams. In these demonstrations, no additional processing steps were desired, so the photon source material is the same as the photodetector material. In [12], Roelkens et al. used an identical InGaAsP/InP structure as that used for a laser to demonstrate a 50 μm long waveguided PD with a 0.23 A/W responsivity at 1.555 μm. Similarly, Park et al. also demonstrated a QW-based, waveguided PD alongside the UCSB-Intel integrated laser discussed above [87]. The authors state that the internal responsivity (excluding losses for coupling light into and out of the platform) is 1.1 A/W with an internal quantum efficiency of 90%. The PD responds over the entire 1.5 μm range and shows dark currents of less than 100 nA at a reverse bias of 2V. The bandwidth of this PD was lower than desirable at 467 MHz due to the large device size and therefore high capacitance. The authors propose many improvements to reduce the capacitance, for example, through shrinking the size and choosing different materialsay. This is calculated to result in devices with 10 GHz bandwidth and 60% internal quantum efficiency.

All of these results show that not only have III-V/Si devices held a unique position as the strongest overall players over the years but also that Group IV devices have made great strides recently. Both technologies hold exciting prospects for future optoelectronic integration.

8.4 CMOS-Compatible Optical Waveguides

Waveguides are a crucial component to any OEIC as they enable optical signal transport from one device to another. Si-based waveguides are essentially nanophotonic-sized analogues to optical fibers used in telecommunications. The underlying physics behind waveguides is the same as that of optical fibers, but many of the design considerations for waveguides are markedly different.

In addition to traditional designs for dielectric and semiconductor waveguides, other schemes for nanophotonic waveguiding include use of photonic crystals and plasmonics. Photonic crystals use a periodic refractive index contrast to create a photonic band gap, thus inhibiting the passage of light through the device [88]. By introducing controlled defects into the photonic crystal, one can theoretically produce waveguides with exceptionally low loss and bend radii, as well as novel photonic components [89].

The field of plasmonics is attempting to use surface plasmons to propagate electromagnetic waves along the surface of nanoscale conductors [90]. This technique has the potential to direct photons on chip using structures much smaller than traditional waveguides, which are limited to feature sizes on the order of the wavelength of light. Both photonic crystals and plasmonics have promise for application in future generations of nanophotonic devices; however, as they are not likely to be used in the first generation of CMOS-nanophotonic OEICs, this section will concentrate on traditional waveguides. Readers interested in more information on these topics are directed to reviews dedicated to photonic crystals [89] and plasmonics [90].

8.4.1 Types of Waveguides and Basic Physical Principles

Waveguides utilize the well-known optical principle of total internal reflection to guide the optical signal. The physics of waveguides has been treated at length by other authors, both for silicon photonics [91] and in general [92]. When examining the performance of a waveguide system, the following are useful metrics to consider:

- Wavelength of operation;
- Loss (measured in decibels per unit length, the loss of a waveguide can generally come from three sources: material absorption, substrate leakage, and scattering due to surface roughness);
- In/out coupling efficiency;
- Minimum waveguide size;
- Minimum bend radius;
- Thermal expansion/thermal budget.

The current schemes under investigation for nanophotonic CMOS waveguiding can be roughly divided into two categories: on-silicon waveguides (fabricated on top of the Si substrate) and in-silicon waveguides (fabricated out of the Si substrate).

On-silicon waveguides consist mostly of CMOS-compatible dielectrics (e.g., SiO_2, Si_3N_4) that can be easily deposited on a Si wafer. Of these, SiO_2-based waveguides have already undergone much development for use in planar lightwave circuits (PLCs) for dense wavelength-division multiplexing (DWDM) systems [93, 94]. Thus, methods for fabrication of waveguides and a wide variety of passive devices, including splitters, directional couplers, and arrayed waveguide gratings, have already reached a high level of technological maturity through PLC development. Similarly, techniques for coupling light into and out of on-silicon waveguides from outside fibers have already been mastered through the development of PLCs.

However, the primary drawback of SiO_2-based waveguides is the small refractive index contrast (Δn) between core and cladding in such a system. The refractive index of SiO_2 can be manipulated through the addition of various dopants (this technique is used in optical fibers), but these index values do not stray far from the undoped value of 1.5, giving an index contrast typically on the order of $\Delta n \sim 0.01$. Thus, in order to maintain modal confinement, the waveguides must be large, typically on the order of microns or larger [93]. This large footprint makes these waveguides prohibitive for nanophotonic/CMOS applications. The minimum bend radius is also restricted to sizes on the order of millimeters, which would make its use in nanophotonic systems such as optical interconnects impossible.

Another on-silicon dielectric candidate is SiN. With a refractive index of 2.0 to 2.25 (depending on stoichiometry), this material provides a large enough Δn (using

SiO_2 as the cladding material) to enable waveguide widths below 1 μm with bend radii in the tens of microns [95]. SiN is widely used in microelectronics; thus, methods for depositing [96] and processing this material are well established. The work of [95] found that submicron-dimension waveguides could be fabricated on SiO_2 with minimum losses of 1 to 2 dB/cm at 1,620 nm, but the loss increased at shorter wavelengths due to absorption by N-H bonds from hydrogen impurities incorporated during the deposition process.

In addition to on-silicon methods, a great deal of work has investigated the use of in-silicon waveguides for CMOS nanophotonics. Silicon has a much higher refractive index (~3.5), making submicron waveguides possible [97]. SOI wafers are considered by some an ideal platform for single-crystalline in-silicon waveguides since the Si is already clad on the underside by the buried oxide, and channel waveguides can be formed by etching the Si down to the buried oxide [98]. In addition, the bend radius of Si waveguides on SiO_2 can be reduced down to microns; the work of [99] demonstrated losses below 0.1 dB for a 90° turn with a 1 μm bend radius.

One drawback of in-silicon waveguides is the higher losses experienced. Scattering due to roughness on the waveguide surface is a key contributor to loss. The roughness scattering of waveguides increases roughly proportionally with Δn^2, although the exact functional dependence is debated [97]. For silicon waveguides on SiO_2, the refractive index contrast is large ($\Delta n \sim 2$), far greater than the $\Delta n < 0.1$ usually used in PLCs or in SiN-based waveguides ($\Delta n = 0.5$ to 0.75). Thus, these waveguides can experience much higher values of loss compared to on-silicon methods.

Nanometer-scale sidewall roughness is generated by the etching methods used to define the waveguide (usually dry etching), and this roughness is generally on the order of a few nanometers [95]. To mitigate this effect, this roughness can be reduced by oxidizing the waveguides after etching. Thermal oxidation in the reaction rate-limited regime, followed by oxide removal using wet chemical etches, can serve to smooth the Si sidewalls. Figure 8.18 is an example of a Si sidewall smoothed

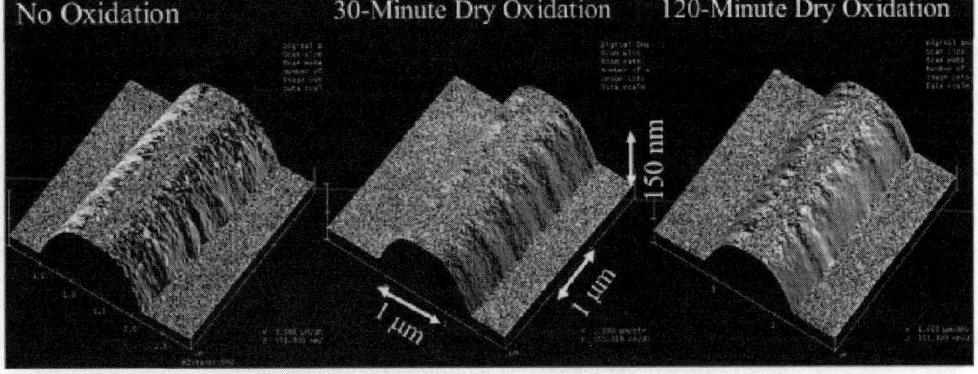

Figure 8.18 Demonstration of the effect of thermal oxidation to smooth the sidewalls of etched Si waveguides. Thermal oxidation in the reaction rate-limited regime can serve to smooth the Si sidewalls, reducing optical losses due to sidewall roughness scattering [95].

using this thermal oxidation technique. A variation on this concept, termed "wet chemical oxidation," involves repeated cycles of oxidation of and oxide removal from the silicon waveguide using wet chemicals at low temperatures [100]. This method typically removes on the order of 10 nm of waveguide material, while thermal oxidation methods remove on the order of 100 nm; thus, wet chemical oxidation is more efficient (in terms of material removed) than thermal oxidation and better preserves the original geometry of the waveguide as a result. The work of [100] showed that this process reduced sidewall-roughness losses of a proof-of-concept Si waveguide from 9.2 to 1.9 dB/cm.

Another problem associated with Si waveguides is the ability to couple them to fibers. The mode size of a standard fiber can be up to 250 times larger than that of a Si waveguide [101], and this size difference can lead to severe coupling losses. Fortunately, there are a number of techniques available to resolve this issue. One of the most popular methods is the inverse taper approach. In this method, the silicon waveguide is gradually tapered down to a size that can no longer support an optical mode [97]. A large single-mode dielectric waveguide is fabricated around this tapered region such that, as the light passes through the Si into the taper, its mode gradually expands into the dielectric, which can then easily be coupled to the fiber using standard PLC techniques. This is depicted in Figure 8.19. Coupling losses using this technique have been reported as low as 0.2 dB [102]. Another popular coupling method is the grating coupler. In this technique, a grating can be etched onto the sides of a silicon waveguide, enabling the light to couple at a 90° angle into the wafer [103]. This is depicted in Figure 8.20. After optimization, this method was shown to achieve a coupling loss of less than 1 dB over a 35 nm range using a 13 × 12 μm-sized grating [104].

In addition to monocrystalline silicon, it is also possible to use poly-Si as a waveguiding material. This has the advantage of being able to be deposited on top of other layers, making multiple levels of optical waveguiding possible. However, the presence of grain boundaries increases loss in the poly-Si by causing scattering from grain boundaries, increasing sidewall roughness at grain boundaries, and increasing

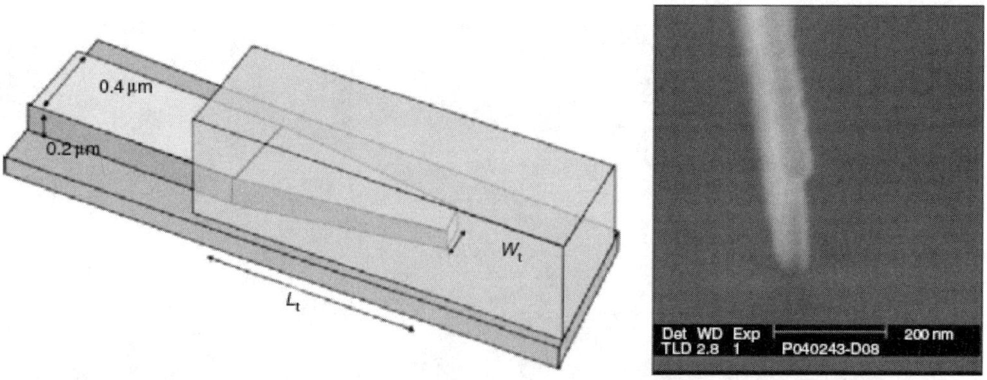

Figure 8.19 Example of the inverse taper approach for coupling light from a waveguide to an optical fiber. In this method, the silicon waveguide is gradually tapered down to a size that can no longer support an optical mode. A large, single-mode dielectric waveguide is fabricated around this tapered region such that, as the light passes through the Si into the taper, its mode gradually expands into the dielectric [97].

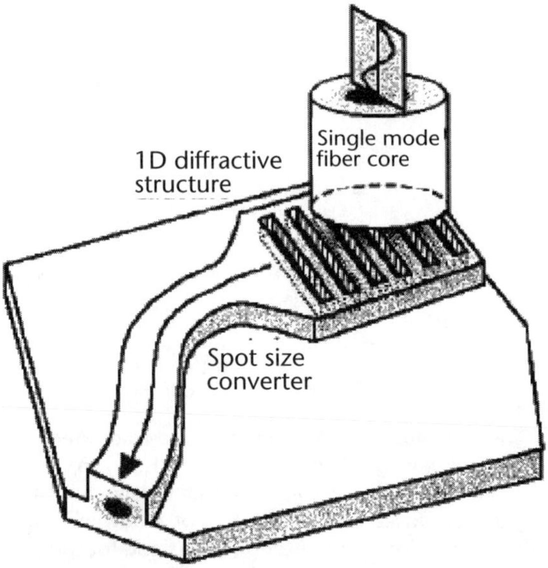

Figure 8.20 Schematic of a grating coupler used for coupling light between an optical fiber and an on-chip waveguide. In this technique, a grating can be etched onto the sides of a silicon waveguide, enabling the light to couple at a 90° angle into the wafer [103].

dangling bonds, which can contribute to scattering [95]. In addition, sidewall smoothing by oxidation is not possible with poly-Si because the different grain orientations will cause different oxidation rates, leading to a roughening effect instead of a smoothing effect. The results of [105, 106] showed that poly-Si waveguides could attain losses as low as 15 dB/cm.

In considering the different materials available for waveguide fabrication, it is important to recognize the trade-off between scattering losses and waveguide footprint (as measured by minimum bending radius) that is caused by Δn. This trade-off was expressed graphically by [107] and is shown in Figure 8.21. Thus, selection of a waveguiding material will require selecting a Δn that yields the best balance of scattering losses and footprint for the particular application.

Overall, the waveguide component of an optical interconnect presents the lowest technological barriers in the emitter-modulator-waveguide-detector sequence; therefore, it is also the most developed. The waveguide is also the least affected by reliability constraints and has few integration challenges.

8.5 Commercialization and Manufacturing

As we have mentionedk, the perceived drivers for monolithic optical interconnects are related primarily to reducing the size of current telecom and datacom systems. Although a driver for research investment, we need to imagine technological success in order to evaluate the actual usefulness of a particular solution emerging from research. For example, let us imagine that we are within a few years of creating an array of lasers, detectors, modulators, and waveguides on silicon substrates, com-

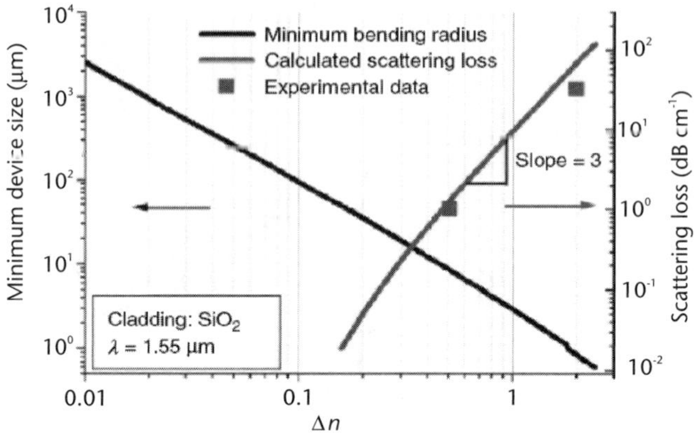

Figure 8.21 Plot of minimum bending radius and calculated scattering loss versus refractive index contrast (Δn) for waveguide materials at $\lambda = 1.55\ \mu$m, assuming SiO_2 cladding. This graph depicts the trade-off between scattering losses and waveguide footprint (as measured by minimum bending radius) that is caused by Δn. Selection of a waveguiding material will require selecting a Δn that yields the best balance of scattering losses and footprint for the particular application [107].

patible with silicon manufacturing, such that monolithic integrated systems can now be realistically envisioned. We are forced to imagine, then, microprocessors, memory, or other large, complex silicon dice with many reliable optical devices appearing quickly in a state-of-the-art, 300 mm silicon fabrication facility. Reasonable readers familiar with commercialization know the paradox we are entering. Even with a completely defined, successful monolithic technology in the laboratory, the envisioned "first application" is so complex in nature and in manufacturing that it appears unlikely that it can be rolled out in any rational and economic way.

Should we then say "nice research," but realize that maybe we should have thought of this before? Or is there another path? There are only two main paths. Either a consortia of very large semiconductor companies invests in a big way over a decade to make that successful research vision come to reality (unlikely), or a smaller market that needs optical integration can appear in which smaller silicon infrastructure (i.e., trailing edge) can be used to successfully penetrate the first market. This second path has been the more likely path for truly disruptive innovations; therefore, if integrated photonics on silicon is to be a large disruption, it will likely need to find a more humble beginning. The nature of this first application would be one in which the ability to add optical capability to CMOS is so advantageous that a trailing-edge CMOS technology is sufficient to create novel value at the integrated chip level. For example, a 0.18μ CMOS process running on a 150 or 200 mm silicon infrastructure is combined with LED or laser or detector arrays to create a disruption in some market space. Note that this market space will be much smaller than traditional large-die CMOS chip markets, like memory, processors, and the like. And the chip will be less complex as well. After successful penetration in this small market, resources gained from that success can fuel the next, more complex, larger monolithic optical CMOS integrated circuit. Eventually, the manufacturing environment will be large enough that more complex integrated circuits can be manufactured, and the integrated circuit drivers that researchers envisioned will at long last appear, in a real way, on the horizon.

References

[1] Lum, R. M., et al., "Improvements in the Heteroepitaxy of GaAs on Si," *Appl. Phys. Lett.*, Vol. 51, July 5, 1987, pp. 36–38.

[2] Hu, C. C., C. S. Sheu, and M. K. Lee, "The Fabrication of InGaP/Si Light Emitting Diode by Metalorganic Chemical Vapor Deposition," *Mater. Chem. Phys.*, Vol. 48, March 15, 1997, pp. 17–20.

[3] Akahori, K., et al., "Improvement of the MOCVD-Grown InGaP-on-Si towards High-Efficiency Solar Cell Application," *Solar Energy Mater. Solar Cells*, Vol. 66, February 2001, pp. 593–598.

[4] Currie, M. T., et al., "Controlling Threading Dislocation Densities in Ge on Si Using Graded SiGe Layers and Chemical-Mechanical Polishing," *Appl. Phys. Lett.*, Vol. 72, April 6, 1998, pp. 1718–1720.

[5] Groenert, M. E., et al., "Monolithic Integration of Room-Temperature cw GaAs/AlGaAs Lasers on Si Substrates via Relaxed Graded GeSi Buffer Layers," *J. Appl. Phys.*, Vol. 93, January 1, 2003, pp. 362–367.

[6] Brouckaert, J., et al., "Thin-Film III-V Photodetectors Integrated on Silicon-on-Insulator Photonic ICs," *J. Lightwave Technol.*, Vol. 25, April 2007, pp. 1053–1060.

[7] Roelkens, G., et al., "III-V/Si Photonics by Die-to-Wafer Bonding," *Mater. Today,* Vol. 10, July-August 2007, pp. 36–43.

[8] Tong, Q. Y., and U. M. Gosele, "Wafer Bonding and Layer Splitting for Microsystems," *Adv Mater,* Vol. 11, December 1, 1999, pp. 1409–1425.

[9] Kim, M. J., and R. W. Carpenter, "Heterogeneous Silicon Integration by Ultra-High Vacuum Wafer Bonding," *J. Electron. Mater.*, Vol. 32, August 2003, pp. 849–854.

[10] Fang, A. W., et al., "Hybrid Silicon Evanescent Devices," *Mater. Today,* Vol. 10, July-August 2007, pp. 28–35.

[11] Yang, V. K., et al., "Crack Formation in GaAs Heteroepitaxial Films on Si and SiGe Virtual Substrates," *J. Appl. Phys.*, Vol. 93, April 1, 2003, pp. 3859–3865.

[12] Roelkens, G., et al., "Laser Emission and Photodetection in an InP/InGaAsP Layer Integrated on and Coupled to a Silicon-on-Insulator Waveguide Circuit," *Opt. Express,* Vol. 14, September 4, 2006, pp. 8154–8159.

[13] Boudinov, H., H. H. Tan, and C. Jagadish, "Electrical Isolation of n-Type and p-Type InP Layers by Proton Bombardment," *J. Appl. Phys.*, Vol. 89, May 15, 2001, pp. 5343–5347.

[14] Chang, H. H., et al., "1310 nm Silicon Evanescent Laser," *Opt. Express,* Vol. 15, September 3, 2007, pp. 11466–11471.

[15] Fang, A. W., et al., "Electrically Pumped Hybrid AlGaInAs-Silicon Evanescent Laser," *Opt. Express,* Vol. 14, October 2, 2006, pp. 9203–9210.

[16] Kobrinsky, M. J., et al., "On-Chip Optical Interconnects," *Intel Technology Journal,* Vol. 8, 2004, pp 129–140.

[17] Yariv, A., and X. K. Sun, "Supermode Si/III-V Hybrid Lasers, Optical Amplifiers and Modulators: A Proposal and Analysis," *Opt. Express,* Vol. 15, July 23, 2007, pp. 9147–9151.

[18] Koch, B. R., et al., "Mode-Locked Silicon Evanescent Lasers," *Opt. Express,* Vol. 15, September 3, 2007, pp. 11225–11233.

[19] Kwon, O., et al., "Monolithic Integration of AlGaInP Laser Diodes on SiGe/Si Substrates by Molecular Beam Epitaxy," *J. Appl. Phys.,* Vol. 100, July 1, 2006, pp. 013103–013103.

[20] Quitoriano, N. J., and E. A. Fitzgerald, "Relaxed, High-Quality InP on GaAs by Using InGaAs and InGaP Graded Buffers to Avoid Phase Separation," *J. Appl. Phys.,* Vol. 102, August 1, 2007, pp. 033511–033511.

[21] Hudait, M. K., et al., "High-Quality InAsyP1-y Step-Graded Buffer by Molecular-Beam Epitaxy," *Appl. Phys. Lett.*, Vol. 82, May 12, 2003, pp. 3212–3214.

[22] Ting, S. M., "Monolithic Integration of III-V Semiconductor Materials and Devices with Silicon," 1999, PhD thesis, MIT.

[23] Dohrman, C. L., et al., "Fabrication of Silicon on Lattice-Engineered Substrate (SOLES) as a Platform for Monolithic Integration of CMOS and Optoelectronic Devices," *Materials Science and Engineering B (Solid-State Materials for Advanced Technology),* Vol. 135, December 15, 2006, pp. 235–237.

[24] Chilukuri, K., et al., "Monolithic CMOS-Compatible AlGaInP Visible LED Arrays on Silicon on Lattice-Engineered Substrates (SOLES)," *Semiconductor Science and Technology,* Vol. 22, February 2007, pp. 29–34.

[25] Sun, Y. T., K. Baskar, and S. Lourdudoss, "Thermal Strain in Indium Phosphide on Silicon Obtained by Epitaxial Lateral Overgrowth," *J. Appl. Phys.,* Vol. 94, August 15, 2003, pp. 2746–2748.

[26] Mi, Z., et al., "High Performance Self-Organized InGaAs Quantum Dot Lasers on Silicon," *J. Vac. Sci. Technol. B,* Vol. 24, May–June 2006, pp. 1519–1522.

[27] Yang, J., P. Bhattacharya, and Z. Wu, "Monolithic Integration of InGaAs-GaAs Quantum-Dot Laser and Quantum-Well Electroabsorption Modulator on Silicon," *IEEE Photonics Technol. Lett.,* Vol. 19, May–June 2007, pp. 747–749.

[28] Kunert, B., et al., "Luminescence Investigations of the GaP-Based Dilute Nitride Ga(NAsP) Material System," *J. Lumin.,* Vol. 121, December 2006, pp. 361–364.

[29] Kunert, B., et al., "Near Room Temperature Electrical Injection Lasing for Dilute Nitride Ga(NAsP)/GaP Quantum-Well Structures Grown by Metal Organic Vapour Phase Epitaxy," *Electron. Lett.,* Vol. 42, May 11, 2006, pp. 601–603.

[30] Yablonskii, G. P., et al., "Luminescence and Stimulated Emission from GaN on Silicon Substrates Heterostructures," *Phys. Stat. Sol. (A)—Appl. Res.,* Vol. 192, July 16, 2002, pp. 54–59.

[31] Park, J. H., and A. J. Steckl, "Demonstration of a Visible Laser on Silicon Using Eu-Doped GaN Thin Films," *J. Appl. Phys.,* Vol. 98, September 1, 2005, pp. 0561-8-1–056108-3.

[32] Iacona, F., et al., "Silicon-Based Light-Emitting Devices: Properties and Applications of Crystalline, Amorphous and Er-Doped Nanoclusters," *IEEE J. Sel. Top. Quantum Electron.,* Vol. 12, November–December 2006, pp. 1596–1606.

[33] Ng, W. L., et al., "An Efficient Room-Temperature Silicon-Based Light-Emitting Diode," *Nature,* Vol. 410, March 8, 2001, pp. 192–194.

[34] Green, M. A., et al., "Efficient Silicon Light-Emitting Diodes," *Nature,* Vol. 412, August 23, 2001, pp. 805–808.

[35] Chen, M. J., et al., "Stimulated Emission in a Nanostructured Silicon pn Junction Diode Using Current Injection," *Appl. Phys. Lett.,* Vol. 84, March 22, 2004, pp. 2163–2165.

[36] Canham, L. T., "Silicon Quantum Wire Array Fabrication by Electrochemical and Chemical Dissolution of Wafers," *Appl. Phys. Lett.,* Vol. 57, September 3, 1990, pp. 1046–1048.

[37] Castagna, M. E., et al., "Si-Based Materials and Devices for Light Emission in Silicon," *Physica E-Low-Dimensional Systems and Nanostructures,* Vol. 16, March 2003, pp. 547–553.

[38] Irrera, A., et al., "Electroluminescence Properties of Light Emitting Devices Based on Silicon Nanocrystals," *Physica E,* Vol. 16, March 2003, pp. 395–399.

[39] Lalic, N., and J. Linnros, "Light Emitting Diode Structure Based on Si Nanocrystals Formed by Implantation into Thermal Oxide," *J. Lumin.,* Vol. 80, December 1998, pp. 263–267.

[40] Photopoulos, P., and A. G. Nassiopoulou, "Room- and Low-Temperature Voltage Tunable Electroluminescence from a Single Layer of Silicon Quantum Dots in between Two Thin SiO_2 Layers," *Appl. Phys. Lett.,* Vol. 77, September 18, 2000, pp. 1816–1818.

[41] Jambois, O., et al., "Photoluminescence and Electroluminescence of Size-Controlled Silicon Nanocrystallites Embedded in SiO_2 Thin Films," *J. Appl. Phys.,* Vol. 98, August 15, 2005, pp. 046105–046105.

[42] Dal Negro, L., et al., "Light Emission from Silicon-Rich Nitride Nanostructures," *Appl. Phys. Lett.,* Vol. 88, May 1, 2006, pp. 183103–183103.

[43] Franzo, G., et al., "Electroluminescence of Silicon Nanocrystals in MOS Structures," *Appl. Phys. A—-Mater. Sci. Process.,* Vol. 74, January 2002, pp. 1–5.

[44] Cho, K. S., et al., "High Efficiency Visible Electroluminescence from Silicon Nanocrystals Embedded in Silicon Nitride Using a Transparent Doping Layer (vol 86, pg 071909, 2005)," *Appl. Phys. Lett.,* Vol. 88, May 15, 2006, pp. 209904–209904.

[45] Dal Negro, L., et al., "Spectrally Enhanced Light Emission from Aperiodic Photonic Structures," *Appl. Phys. Lett.,* Vol. 86, June 27, 2005, p. 261905-1.

[46] Michel, J., et al., "Impurity Enhancement of the 1.54-Mu-M Er3+ Luminescence in Silicon," *J. Appl. Phys.,* Vol. 70, September 1, 1991, pp. 2672–2678.

[47] Palm, J., et al., "Electroluminescence of Erbium-Doped Silicon," *Physical Review B,* Vol. 54, December 15, 1996, pp. 17603–17615.

[48] Michel, J., et al., "Erbium in Silicon," *Light Emission in Silicon: From Physics to Devices,* Vol. 49, 1998, pp. 111–156.

[49] Jalali, B., et al. "Raman-Based Silicon Photonics," *IEEE J. Sel. Top. Quantum Electron.,* Vol. 12, May–June 2006, pp. 412–421.

[50] Liu, J., et al., "Tensile-Strained, n-Type Ge as a Gain Medium for Monolithic Laser Integration on Si," *Opt. Express,* Vol. 15, September 3, 2007, pp. 11272–11277.

[51] Driscoll, K., and R. Paiella, "Silicon-Based Injection Lasers Using Electronic Intersubband Transitions in the L Valleys," *Appl. Phys. Lett.,* Vol. 89, November 6, 2006, pp. 191110–191110.

[52] Cunningham, J. E., "Recent Developments and Applications in Electroabsorption Semiconductor Modulators," *Materials Science and Engineering R: Reports,* Vol. R25, August 31, 1999, pp. 155–194.

[53] Brinkman, W. F., et al., "The Lasers behind the Communications Revolution," *Bell Labs Technical Journal,* Vol. 5, January–March 2000, pp. 150–167.

[54] Fox, A. M., et al., "Quantum Well Carrier Sweep Out: Relation to Electroabsorption and Exciton Saturation," *IEEE J. Quant. Electron.,* Vol. 27, 1991, pp. 2281–2295.

[55] Soref, R. A., and B. R. Bennett, "Electrooptical Effects in Silicon," *IEEE J. Quant. Electron.,* Vol. QE-23, January 1987, pp. 123–129.

[56] Park, J. S., R. P. G. Karunasiri, and K. L. Wang, "Observation of Large Stark Shift in Ge_xSi_{1-x}/Si Multiple Quantum Wells," *J. Vac. Sci. Technol. B,* Vol. 8, March 1990, pp. 217–220.

[57] Miyake, Y., et al., "Absence of Stark Shift in Strained $Si_{1-x}Ge_x$/Si Type-I Quantum Wells," *Appl. Phys. Lett.,* Vol. 68, April 8, 1996, pp. 2097–2099.

[58] Li, C., et al., "Observation of Quantum-Confined Stark Shifts in SiGe/Si Type-I Multiple Quantum Wells," *J. Appl. Phys.,* Vol. 87, June 1, 2000, pp. 8195–8197.

[59] Qasaimeh, O., S. Singh, and P. Bhattacharya, "Electroabsorption and Electrooptic Effect in SiGe-Si Quantum Wells: Realization of Low-Voltage Optical Modulators," *IEEE J. Quant. Electron.,* Vol. 33, 1997, pp. 1532–1536.

[60] Qasaimeh, O., and P. Bhattacharya, "SiGe-Si Quantum-Well Electroabsorption Modulators," *IEEE Photonics Technology Letters,* Vol. 10, June 1998, pp. 807–809.

[61] Yu-Hsuan Kuo, et al., "Strong Quantum-Confined Stark Effect in Germanium Quantum-Well Structures on Silicon," *Nature,* Vol. 437, October 27, 2005, pp. 1334–1336.

[62] Yu-Hsuan Kuo, et al., "Quantum-Confined Stark Effect in Ge/SiGe Quantum Wells on Si for Optical Modulators," *IEEE J. Selected Topics in Quantum Electronics,* Vol. 12, November 2006, pp. 1503–1513.

[63] Jongthammanurak, S., et al., "Large Electro-Optic Effect in Tensile Strained Ge-on-Si Films," *Appl. Phys. Lett.,* Vol. 89, October 16, 2006, p. 161115-1.

[64] Cunningham, J. E., et al., "Growth of GaAs Light Modulators on Si by Gas Source Molecular-Beam Epitaxy for 850 nm Optical Interconnects," *13th North American Molecular-Beam Epitaxy Conference,* 1994, pp. 1246–1250.

[65] Tang, C. K., and G. T. Reed "Highly Efficient Optical Phase Modulator in SOI Waveguides," *Electron. Lett.,* Vol. 31, 1995, pp. 451–452.

[66] Dainesi, P., "CMOS Compatible Fully Integrated Mach-Zehnder Interferometer in SOI Technology," *IEEE Photon. Technol. Lett.,* Vol. 12, 2000, pp. 660–662.

[67] Liu, A., et al., "A High-Speed Silicon Optical Modulator Based on a Metal-Oxide-Semiconductor Capacitor," *Nature,* Vol. 427, February 12, 2004, pp. 615–618.

[68] Liu, A., et al., "Scaling the Modulation Bandwidth and Phase Efficiency of a Silicon Optical Modulator," *IEEE J. Selected Topics in Quantum Electronics,* Vol. 11, March 2005, pp. 367–372.

[69] Xu, Q., et al., "Micrometre-Scale Silicon Electro-Optic Modulator," *Nature,* Vol. 435, May 19, 2005, pp. 325–327.

[70] Michel, J., et al., "Advances in Fully CMOS Integrated Photonic Devices," *Silicon Photonics II: 22–25 January 2007, San Jose, California, USA*, edited by Joel A. Kubby and Graham T. Reed. Bellingham, WA: SPIE, 2007.

[71] Bowers, J. E., and C. A. Burrus, "Ultrawide-Band Long-Wavelength p-i-n Photodetectors," *J. Lightw. Technol.,* Vol. LT-5, No. 10, October 1987, pp. 1339–1350.

[72] Coffa, S., G. Franzo, and F. Priolo, "High Efficiency and Fast Modulation of Er-Doped Light Emitting Si Diodes," *Appl. Phys. Lett.,* Vol. 69, September 30, 1996, pp. 2077–2079.

[73] Kik, P. G., et al., "Design and Performance of an Erbium-Doped Silicon Waveguide Detector Operating at 1.5 mu m," *J. Lightwave Technol.,* Vol. 20, May. 2002, pp. 834–839.

[74] Splett, A., et al., "Integration of Wave-Guides and Photodetectors in Sige for 1.3 Mu-M Operation," *IEEE Photonics Technol. Lett.,* Vol. 6, January 1994, pp. 59–61.

[75] Lafontaine, H., et al., "Growth of Undulating $Si_{0.5}Ge_{0.5}$ Layers for Photodetectors at Lambda = 1.55 mu m," *J. Appl. Phys.,* Vol. 86, August 1, 1999, pp. 1287–1291.

[76] Elfving, A., et al., "Three-Terminal Ge Dot/SiGe Quantum-Well Photodetectors for Near-Infrared Light Detection," *Appl. Phys. Lett.,* Vol. 89, August 21, 2006, pp. 083510-1–083510-3.

[77] Koester, S. J., et al., "Germanium-on-SOI Infrared Detectors for Integrated Photonic Applications," *IEEE J. Sel. Top. Quantum Electron.,* Vol. 12, November–December 2006, pp. 1489–1502.

[78] Hartmann, J. M., et al., "Reduced Pressure-Chemical Vapor Deposition of Ge Thick Layers on Si(001) for 1.3–1.55-mu m Photodetection," *J. Appl. Phys.,* Vol. 95, May 15, 2004, pp. 5905–5913.

[79] Fitzgerald, E. A., et al., "Totally Relaxed Ge_xSi_{1-x} Layers with Low Threading Dislocation Densities Grown on Si Substrates," *Appl. Phys. Lett.,* Vol. 59, August 12, 1991, pp. 811–813.

[80] Giovane, L. M., et al., "Correlation between Leakage Current Density and Threading Dislocation Density in SiGe p-i-n Diodes Grown on Relaxed Graded Buffer Layers," *Appl. Phys. Lett.,* Vol. 78, January 22, 2001, pp. 541–543.

[81] Samavedam, S. B., et al., "High-Quality Germanium Photodiodes Integrated on Silicon Substrates Using Optimized Relaxed Graded Buffers," *Appl. Phys. Lett.,* Vol. 73, October 12, 1998, pp. 2125–2127.

[82] Huang, Z. H., et al., "Effectiveness of SiGe Buffer Layers in Reducing Dark Currents of Ge-on-Si Photodetectors," *IEEE J. Quant. Electron.,* Vol. 43, March–April 2007, pp. 238–242.

[83] Luan, H. C., et al., "High-Quality Ge Epilayers on Si with Low Threading-Dislocation Densities," *Appl. Phys. Lett.,* Vol. 75, November 8, 1999, pp. 2909–2911.

[84] Ahn, D., et al., "High Performance, Waveguide Integrated Ge Photodetectors," *Opt. Express,* Vol. 15, April 2, 2007, pp. 3916–3921.

[85] Hawkins, A. R., et al., "High Gain-Bandwidth-Product Silicon Heterointerface Photodetector," *Appl. Phys. Lett.,* Vol. 70, January 20, 1997, pp. 303–305.

[86] Levine, B. F., et al., "Ultralow-Dark-Current Wafer-Bonded Si/InGaAs Photodetectors," *Appl. Phys. Lett.,* Vol. 75, October 4, 1999, pp. 2141–2143.

[87] Park, H., et al., "A Hybrid AlGaInAs-Silicon Evanescent Waveguide Photodetector," *Opt. Express,* Vol. 15, May 14, 2007, pp. 6044–6052.

[88] Joannopoulos, J. D., R. D. Meade, and J. N. Winn, *Photonic Crystals: Molding the Flow of Light.* Princeton, NJ: Princeton University Press, 1995, p. 137.

[89] Krauss, T. F., "Photonic Crystal Microcircuit Elements," in *Optical Interconnects: The Silicon Approach* 119, edited by L. Pavesi and G. Guillot, Berlin and New York: Springer, 2006, p. 381.

[90] Ozbay, E., "Plasmonics: Merging Photonics and Electronics at Nanoscale Dimensions," *Science*, Vol. 311, January 13, 2006, pp. 189–193.

[91] Reed, G. T., and A. P. Knights, *Silicon Photonics: An Introduction,* Chichester, UK: John Wiley, 2004, p. 255.

[92] Saleh, B. E. A., and M. C. Teich, *Fundamentals of Photonics*, New York: Wiley, 1991, p. 966.

[93] Miya, T., "Silica-Based Planar Lightwave Circuits: Passive and Thermally Active Devices," *IEEE J. Selected Topics in Quantum Electronics,* Vol. 6, January 2000, pp. 38–45.

[94] Doerr, C. R., and K. Okamoto, "Advances in Silica Planar Lightwave Circuits," *J. Lightwave Technol.*, Vol. 24, December 2006, pp. 4763–4789.

[95] Sparacin, D. K., and the Massachusetts Institute of Technology's Department of Materials Science and Engineering, *Process and Design Techniques for Low Loss Integrated Silicon Photonics*, Cambridge, MA: MIT, 2006, p. 260.

[96] Agnihotri, O. P., et al., "Advances in Low Temperature Processing of Silicon Nitride Based Dielectrics and Their Applications in Surface Passivation and Integrated Optical Devices," *Semiconductor Science and Technology*, Vol. 15, July 2000, pp. 29–40.

[97] Van Thourhout, D., W. Bogaerts, and P. Dunon, "Submicron Silicon Strip Waveguides," in *Optical Interconnects: The Silicon Approach* 119, edited by L. Pavesi and G. Guillot, Berlin and New York: Springer, 2006.

[98] Barkai, A., et al., "Integrated Silicon Photonics for Optical Networks [Invited]," *J. Optical Networking,* Vol. 6, January 2007, pp. 25–47.

[99] Vlasov, Y. A., and S. J. McNab, "Losses in Single-Mode Silicon-on-Insulator Strip Waveguides and Bends," *Optics Express,* Vol. 12, April 19, 2004.

[100] Sparacin, D. K., S. J. Spector, and L. C. Kimerling, "Silicon Waveguide Sidewall Smoothing by Wet Chemical Oxidation," *J. Lightwave Technol.,* Vol. 23, August 2005, pp. 2455–2461.

[101] Jalali, B., and S. Fathpour, "Silicon Photonics," *J. Lightwave Technol.,* Vol. 24, December 2006, pp. 4600–4615.

[102] McNab, S. J., N. Moll, and Y. A. Vlasov, "Ultra-Low Loss Photonic Integrated Circuit with Membrane-Type Photonic Crystal Waveguides," *Optics Express,* Vol. 11, November 3, 2003.

[103] Taillaert, D., et al., "An Out-of-Plane Grating Coupler for Efficient Butt-Coupling between Compact Planar Waveguides and Single-Mode Fibers," *IEEE J. Quant. Electron.,* Vol. 38, July 2002, pp. 949–955.

[104] Taillaert, D., P. Bienstman, and R. Baets, "Compact Efficient Broadband Grating Coupler for Silicon-on-Insulator Waveguides," *Opt. Lett.,* Vol. 29, December 1, 2004, p. 1.

[105] Foresi, J. S., et al., "Losses in Polycrystalline Silicon Waveguides," *Appl. Phys. Lett.,* Vol. 68, April 8, 1996, pp. 2052–2054.

[106] Agarwal, A. M., et al., "Low-Loss Polycrystalline Silicon Waveguides for Silicon Photonics," *J. Appl. Phys.*, Vol. 80, December 1, 1996, pp. 6120–6123.

[107] Wada, K., et al., "Si Microphotonics for Optical Interconnection," in *Optical Interconnects: The Silicon Approach* 119, edited by L. Pavesi and G. Guillot, Berlin and New York: Springer, 2006, p. 381.

[108] El kurdi, K., et al., "Silicon–on–Insulator Waveguide Photodetector with Ge/Si Self-Assembled Islands," *J. Appl. Phys.*, Vol. 92, August 15, 2002, pp. 1858-1861.

CHAPTER 9

Limits of Current Heat Removal Technologies and Opportunities

Yogendra Joshi, Andrei G. Fedorov, Xiaojin Wei, and Siva P. Gurrum

9.1 Introduction

Electronics thermal management spans over 10 decades of length scale from the semiconductor devices and interconnects (tens of nanometers) to data-center facilities (hundreds of meters). At the smallest scales, the device and interconnect feature dimensions in high-performance microprocessors will soon be in the mean free path regime (for electrons in copper at 300K, ~ 40 nm; for phonons in Si, ~300 nm). The active and leakage power dissipation components in the transistors have already been discussed in Chapters 1 and 4. These losses appear as heat dissipation, which has to be removed by effective thermal management techniques. An increasing contribution to the overall power loss and heat dissipation is the joule heating in the interconnects. Thermal transport in metal interconnects is due to flow of electrons. At these dimensions, electron scattering from the interfaces and grain boundaries affects the transport process, resulting in a lowering of effective thermal and electrical conductivity. It can be inferred that this reduction of thermal conductivity will substantially impact the joule heating within interconnects since the current densities continue to increase. The effective thermal and electrical conductivities of complex networks of lines and vias also deviate substantially from their continuum values. Furthermore, novel low-k dielectric materials with significantly lower thermal conductivity than silicon dioxide are being introduced to circumvent the interconnect delay problem. Im et al. [1] compiled the thermal-conductivity data and its trend as a function of dielectric constants as shown in Figure 9.1. Interconnect temperature rise at a given power level depends on several factors, among which the dielectric thickness and thermal conductivity are highly important. Increased temperatures result in decreased failure times by diffusion-driven mechanisms such as electromigration. It is therefore clear that thermal issues within the chip are becoming an increasing concern for maintaining performance and reliability.

As seen in Figure 9.2, the high volumetric heat-generation rates need to be transferred across a length-scale cascade from the chip through the package, server, rack or cabinet, data center, and finally to the environment. Increasing levels of heat dissipation are a challenge at each length scale. At the interfaces of the chip with the first-level package, *thermal interface materials* (TIMs) are a key concern. The spreading and rejection of heat to the air within a server or computer enclosure is via high-performance *heat spreaders* and *heat sinks*, respectively. Multiple server or

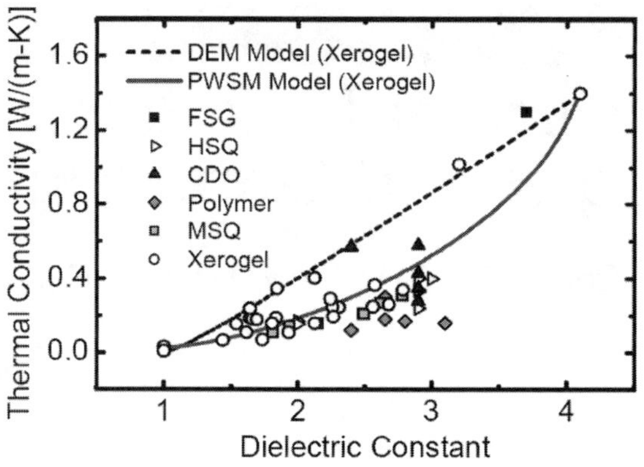

Figure 9.1 Correlation between thermal conductivity and dielectric constant. (*From:* [1]. © 2005 IEEE. Reprinted with permission.)

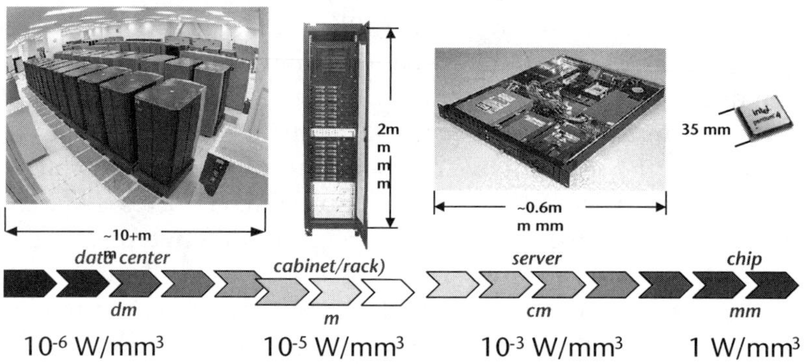

Figure 9.2 The length scale cascade involved in the thermal management of microsystems. Inefficiencies at any level degrade overall performance and energy efficiency. Typical current volumetric heat generation rates are also indicated.

computer modules are placed in standard-sized racks or cabinets. The trend is to provide increasing processing capability with each generation of servers. Architectures for server packaging are evolving to thin, vertically stacked units, called blades, that result in dramatic increases in volumetric heat-dissipation rates. With continuing projected increase in rack heat loads [2], direct air cooling at the rack level will be inadequate. Alternate cooling technologies are being explored to replace or augment air cooling, including single-phase liquid cooling, two-phase liquid cooling, and refrigeration. Several companies have already introduced solutions for rack heat loads in the range of 35 kW. Examples include air-to-liquid "rear-door" heat exchangers and vapor-compression refrigeration to cool the hot air prior to discharge into the data center.

In Section 9.2, we begin with a brief discussion of the heat removal at the data-center level to emphasize the ultimate bottleneck in the multiscale heat-removal hierarchy. The focus is then shifted in Section 9.3 to the chip and the first-level package. In Section 9.4, a thermal-resistance chain allows the identification of the various heat-removal bottlenecks. These components are then addressed, starting with *thermal interface materials* in Section 9.5. In Section 9.6, *heat spread-*

ers to reduce the effective heat flux are discussed. A general discussion of convective heat removal is presented in Section 9.7. *Heat sinks*, for the rejection of the heat to the ambient coolant are explored in Section 9.8. The current state-of-the-art capabilities of these devices is provided, along with projected limits. Finally, Section 9.9 is devoted to the design of microchannel heat sinks.

9.2 Thermal Problem at the Data-Center Level

Data centers house large amounts of high-performance data-processing, storage, and communications equipment within standard electronic cabinets or racks. These facilities are utilized by a broad range of end users, including Internet service providers, banks, stock exchanges, corporations, educational institutions, government installations, and research laboratories. Recent benchmarking studies by Lawrence Berkeley National Laboratories [3] show a doubling in data-center floor heat loads per unit area, from 25 W/ft^2 to 52 W/ft^2 from 2003 to 2005. This is consistent with the emerging trend toward volumetrically compact computing architectures, such as blade servers. Due to the relatively frequent upgrades in the computing equipment, both existing and new facilities are being subjected to these sharp increases in floor heat loading. In 2006, data centers in the United States consumed about 61 billion kilowatt-hours, or 1.5% of total U.S. electricity consumption, for a total electricity cost of about $4.5 billion [3]. This estimated level of electricity consumption is equivalent to the amount of electricity consumed by approximately 5.8 million average U.S. households and is estimated to be more than double the electricity that was consumed for this purpose in 2000. Such a sharp rise in energy consumption by data centers has prompted a directive by the U.S. Congress and a coordinated response by the various stakeholders, as detailed in [4]. A significant fraction of the energy costs associated with the operation of a typical data center can be ascribed to the cooling hardware. The ratio of the total input power to a data center to that consumed by information technology (IT) equipment has dropped from 1.95 to 1.63 between 2003 and 2005 for a number of benchmarked facilities [3]. Despite this, energy usage by cooling equipment continues to be a major concern.

Rejection of the heat from the data center involves computer-room air conditioning (CRAC or AC) units that deliver cold air to the racks arranged in alternating cold/hot aisles through perforated tiles placed over an underfloor plenum, as seen in Figure 9.3. The plenum depth ranges from around 25 cm to over 1m and is partially occupied by cabling and ductwork. Placement of racks in the alternate hot and cold aisles is meant to avoid the mixing of cold supply air with hot exhaust air from the racks. Several alternate air-delivery and -return configurations are employed, particularly when a raised floor arrangement or underfloor plenum is unavailable. These include through-the-ceiling delivery and return. Typical data centers with air-cooling systems have an average design cooling capacity of 3 kW per rack, with a maximum of 10 to 15 kW per rack, while the typical airflow rate supplied by the CRAC units to a single rack is approximately 0.094 to 0.24 m^3/s (200 to 500 CFM), with 0.47 m^3/s (1,000 CFM) being an upper bound, based on constraints such as blower acoustic noise. Inadequate airflow may cause recirculation and mixing of the hot discharge air with the chilled supply air before the air enters the racks. Such

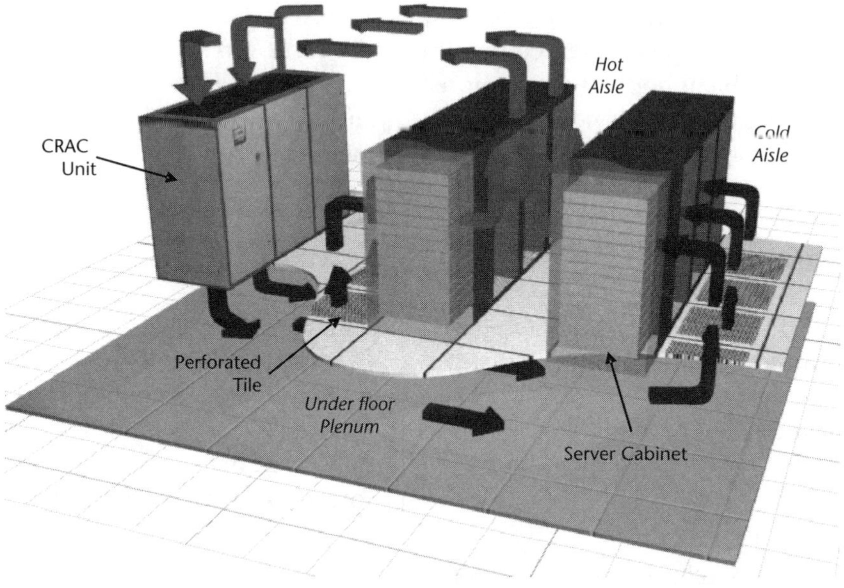

Figure 9.3 The most commonly employed raised floor hot-aisle/cold-aisle arrangement. Cold air from Computer Room Air-Conditioning (CRAC) units is discharged into a sub-floor plenum. It comes up through perforated floor tiles in the cold aisle and moves across the racks. The hot air discharged at the back of the racks moves up among the hot aisles and is returned to the CRAC units for heat rejection.

hot spots may cause chip temperatures to exceed the specified range. The airflow patterns within the delivery plenum are also very important for insuring adequate cooling air supply to the cabinets. Placement of the cold isles very close to the CRAC units has been found to result in reverse flows, causing the hot air from the data center to be supplied to the plenum. This is thought to be a result of the reduced static pressure near the CRAC discharge within the plenum due to the high air velocities. The hot return air from the cabinets exchanges heat typically with a chilled water loop within the CRAC units. This loop in turn rejects heat to a vapor-compression refrigeration system that rejects the heat ultimately to the environment via an air-cooled condenser or a cooling tower. Since the heat from the electronics as well as that rejected from the cooling hardware ultimately needs to be rejected to the environment, it is crucial to have the devices run most efficiently and provide only the necessary cooling.

9.3 Emerging Microprocessor Trends and Thermal Implications

Microprocessor evolution has gone through several design trends. These trends have appeared as higher clock speed, lower voltage scaling, and power-aware design in the past few decades. Temperature effects were considered second-order, and thermal design mainly appeared only at the end of the design flow. In recent years, emphasis on thermal design integrated with the device design and analysis has been growing rapidly. Temperature rise and its variation across the die now appears to be a first-order effect, directly affecting chip performance and power dissipation.

9.3.1 Influence of Temperature on Power Dissipation and Interconnect Performance

Power dissipation in a microprocessor is the sum of dynamic and static power. Dynamic power is associated with transistor switching, and static power mainly results from transistor leakage currents. In the past, static power was negligible when compared to dynamic power. Continuing with the present scaling as presented in the ITRS [4], it is expected that static or leakage power will account for a relatively large fraction of total power in future generations. Absolute temperature rise is then ever more critical since the leakage power increases exponentially with temperature.

In addition to absolute temperature rise, temperature variation across the die has become critical. Current chip architecture results in localized power densities, which result in temperature *hotspots* on the die. Temperature variation across the die can be as high as 50°C in a real microprocessor, as shown by Borkar et al. [5] in Figure 9.4. These variations can severely affect interconnect and device performance. Ajami et al. [6] performed a detailed analysis to show the importance of clock skew caused by nonuniform temperature profiles across interconnects. More recently, Sundaresan and Mahapatra [7] analyzed nonuniform joule heating during the transient event of signal propagation along the interconnect. They show that temperature will be higher at the sending end of the wire compared to the receiving end. The underlying silicon substrate temperature nonuniformity will add to this thermal gradient. In their full chip simulations, they observe a temperature gradient as high as 24°C across the bus line. This temperature gradient resulted in timing violations, with 2.27 violations per hundred bus references for the 130 nm node and 6.2 violation per hundred references for the 45 nm node on average.

9.3.2 Three-Dimensional Stacking and Integration

Three-dimensional integrated circuits (3D ICs) can alleviate the interconnect problem since they allow close proximity of different cells, leading to shorter interconnect lengths. This can reduce both delay and power dissipation over traditional 2D ICs. Hua et al. [8] achieved a 27% reduction in energy per operation and a 20% improvement in speed in a low-power Fast Fourier Transform test chip due to 3D

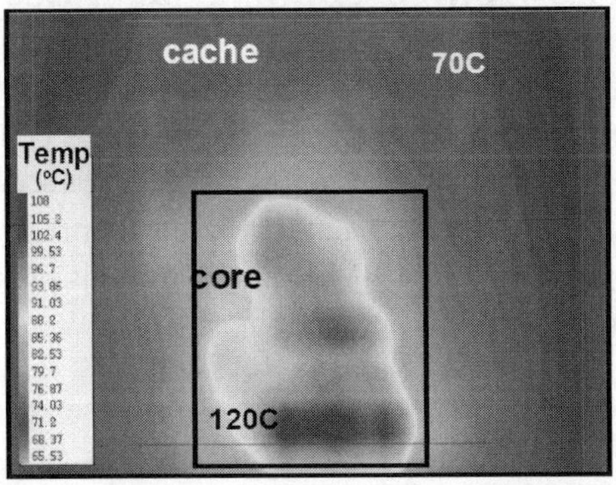

Figure 9.4 Temperature variation within the die of a microprocessor. (*From:* [5]. © 2003. Reprinted with permission.)

integration. They also observed that increasing thermal vias does not necessarily lead to lower temperature design due to routing congestion caused by them. Link et al. [9] investigated temperature rises in 3D stacking of two 2D processors. The 2D die layout closely matched the AMD Athlon 64 "Winchester" core processor. It was found that the power-density location vertically is not significant when compared to aerial power density. They considered several scenarios, including the case where dynamic power (DP) is reduced in a vertical-integration configuration. It is expected that DP will be smaller in vertical integration due to shorter wire length and capacitance. It was found that even with vertical integration, it is essential to avoid placing high-power-density blocks directly one over the other.

9.3.3 Multicore Design as the Next Exponential

Due to limitations associated with voltage scaling, clock-speed scaling, and design complexity, the industry is actively pursuing multicore architectures to continue the exponential improvement in performance (Parkhurst et al. [10]). The multicore approach deals efficiently with power and thermal issues and makes it suitable for applications where parallel processing can improve performance significantly. Pham et al. [11] implemented the first-generation CELL processor with a dual-threaded power processor element (PPE) and eight synergistic processor elements (SPE), which is capable of 10 simultaneous threads. Performance parameters in these architectures strongly depend on temperature profile, and a temperature-aware design is essential right from the start of the design flow. Several articles have appeared recently on the thermal design of multicore processors [12–14]. Due to the physical location of different modules, dynamic thermal management (DTM) appears essential. The devices have built-in temperature sensors that dynamically control clock frequency scaling, dynamic voltage and frequency scaling (DVFS), clock gating, and computing migration. As an example, the CELL processor includes 1 linear sensor for global temperature monitoring, and 10 local digital thermal sensors to monitor on-chip temperature variations. Large parameter fluctuations due to fabrication will certainly necessitate DTM in future multicore architectures.

9.4 The Thermal Resistance Chain: Challenges and Opportunities

The power dissipated in a chip passes through a chain of thermal resistances before ultimately being dissipated to the ambient air. A typical high-performance-package thermal management is shown in Figure 9.5(a) [15]. The die is flip-chipped onto a substrate, which is further attached to the printed wiring board (PWB) through a ball grid array (BGA) or pins. A spreader plate is attached to the backside of the die, which is in turn attached to a large heat sink.

9.4.1 Thermal Resistance Chain

The thermal-resistance path can be mainly divided into two parallel paths, one through the topside and the other through the bottom side, as shown in

Figure 9.5(b). Consider the topside chain. Silicon is highly conductive (thermal conductivity k = 148 W/mK) and results in a small temperature gradient across its thickness. The heat from the active side flows through the silicon to a spreader plate across a thermal interface material (TIM1). The spreader plate essentially spreads the heat to a larger area and further dumps the heat into the heat sink through another thermal interface material (TIM2). In the past several years, researchers both in industry and academia have actively pursued better thermal-interface-material design, heat sinks, and heat spreaders. These efforts have resulted in very small overall thermal-resistance values on the topside. On the bottom side, the power dissipated goes through flip-chip bumps and underfill layer to the substrate. The substrate in turn transfers some of the heat from the chip into the PWB, which is ultimately rejected to the ambient air. Efforts to minimize bottom-side thermal resistance have focused only on applications where a good topside thermal-resistance path is not available, such as molded packages without heat sink.

9.4.2 Challenges and Opportunities in the Thermal Resistance Chain

Ever-shrinking thermal budgets now demand a closer look at the bottom-side thermal-resistance chain, even in applications that have a much smaller topside thermal

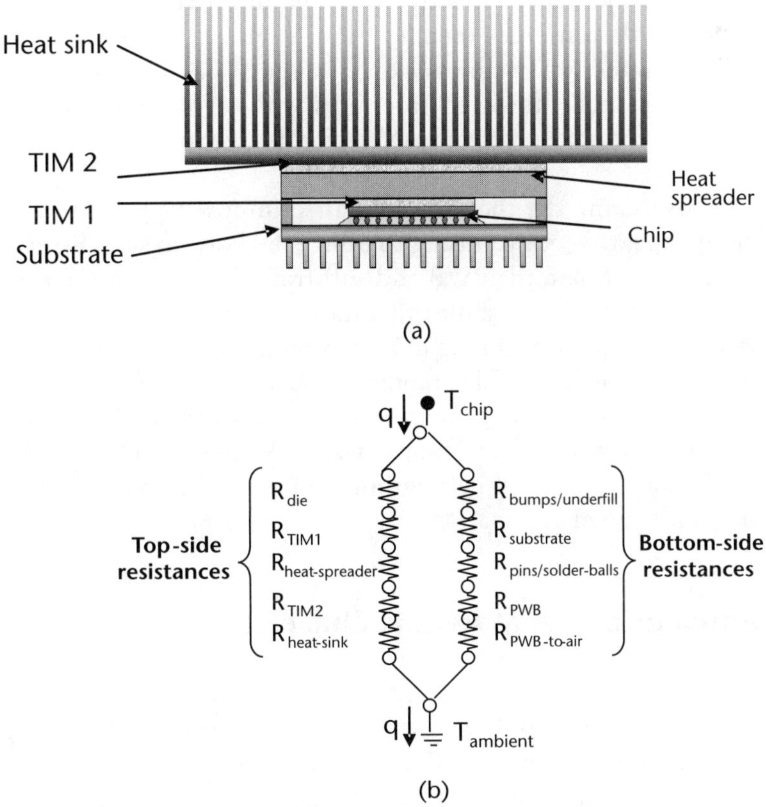

Figure 9.5 (a) A typical configuration for microprocessor thermal management [15]. (b) The top and bottom-side thermal resistance chain in parallel between the chip and ambient [15]. (*From:* [15]. © 2004 IEEE. Reprinted with permission.)

resistance. The first layer in this chain is the solder and underfill layer. Solder is highly conductive (thermal conductivity ~ 50 W/mK) and can thus be a good conduit for heat transfer through conduction. Underfill, which surrounds the solder bumps, has a significantly lower thermal conductivity (less than 1 W/mK). The overall thermal resistance of this layer can still be reduced to well below 0.05 C/W by increasing the bump density. The thermal conductivity of underfill materials can be improved by using different filler particle materials or particle size distribution. The next layer is the substrate, and its thermal performance depends largely on the metallization and vias within the layers. In general, ceramic packages have a lower thermal resistance, but recent simulations by Calmidi and Memis [16] show that dense core substrates can offer two times lower thermal resistance than the standard core counterparts and can in some cases exceed the thermal performance of ceramic packages. They consider a situation where an additional heat sink is attached to the backside of the PCB. It was shown that up to 73% of heat can be diverted to the board even with a 0.4 C/W heat sink on top of the package. They obtained an overall thermal-resistance reduction of 43% with this configuration. However, this is not an ideal solution because it reduces system packing efficiency.

PWB represents an important bottleneck in the bottom-side resistance chain. Typical cross-plane thermal-conductivity values are very low (~0.3 W/mK), but the in-plane conductivity can be much larger due to copper trace layers. Cross-plane thermal conductivity can be improved by including dummy thermal vias across the different metallization levels. In some cases, a metallic stud can be incorporated below the package in the PWB for better thermal performance. The metallic stud, however, interferes with signal lines and power planes in the PWB and may not be very attractive for microprocessors that have a large number of I/Os.

To overcome the thermal resistance imposed by PWB, through-the-substrate cooling appears very promising to utilize the bottom side. Essentially, a fluid loop is constructed through the PWB and substrate that collects heat from the die through either microchannels or some other means. Dang et al. [17] demonstrated a working device that has thermofluidic I/O interconnects from the die to the PWB, the details of which will be discussed Chapter 11. Schaper et al. [18] proposed integrating dice fabricated with various front-end technologies through flexible Cu posts. A fluid dam is also proposed using copper walls. Wego et al. [19] and Chason [20] demonstrated microfluidic channels within the PWB to enable efficient heat removal from space-constrained applications.

9.5 Thermal Interface Materials Challenges

Thermal interface material (TIM) plays a critical role in maintaining the thermal mechanical performance and reliability of the microelectronic packages. When applied between the inactive side of the die and the heat spreader (TIM1) or the heat spreader and the heat sink (TIM2), thermal interface material helps to fill the air gap between the otherwise mated solid surfaces. If a TIM is not used at either location, 95% to 99% of the interface will be filled by air, which is a poor thermal conductor. An implementation of the TIMs is illustrated in Figure 9.5. The effectiveness of the thermal interface material depends on the ability of the material to flow and fill the asperities in the solid surface, the bulk thermal conductivity of the material itself, and the achievable bond-line thickness (BLT), as shown in (9.1). Here BLT is the bond-line thickness of the TIM, and K_{TIM} is the bulk thermal conductivity. R_{CONT} is the contact resistance.

$$R_{TIM} = \frac{BLT}{K_{TIM}} + R_{CONT} \tag{9.1}$$

To minimize the thermal resistance due to the TIM, it is important to consider all three aspects of the material. For a typical polymeric matrix loaded with conductive filler particles, the thermal conductivity of the TIM is a function of the thermal conductivity of the matrix and the filler and the filler volume fraction, as described in Maxwell's model shown in (9.2). Here K_m is the thermal conductivity of the polymerix matrix, K_f is the filler thermal conductivity, and ϕ is the filler particle volume fraction.

$$K_{TIM} = \frac{K_f + 2K_m + 2\phi\left(K_f - K_m\right)}{K_f + 2K_m - \phi\left(K_f - K_m\right)} K_m \tag{9.2}$$

Prasher [21] reported a modified Bruggeman model for $K_f/K_m > 1$, which takes into account of the interface resistance between the filler particles and the polymer matrix. This model, as shown in (9.3), provides better agreement with measured data for high particle volume fraction.

$$K_{TIM} = \frac{1}{(1-\phi)^{3(1-\alpha)(1+2\alpha)}} K_m \tag{9.3}$$

Here α is given as

$$\alpha = \frac{R_b K_m}{d} \tag{9.4}$$

where R_b is the interface resistance between the filler particles and the polymer matrix, and d is the particle size. Both (9.2) and (9.3) suggest increased TIM thermal conductivity for increasing particle volume fraction. However, it is not always desirable to increase the particle volume fraction as the BLT will also increase. Prasher [21] proposed an empirical correlation for BLT:

$$BLT = 1.31 \times 10^{-4} \left(\frac{\tau_y}{P}\right)^{0.166} \tag{9.5}$$

where τ_y is the yield stress for the TIM and P is the applied pressure. Although not explicitly shown, the volume fraction influences the yield stress. The contact resistance between the TIM and the solid surface has not received much attention until recently.

As the TIM performance envelope is pushed further, the contact resistance becomes increasingly important. Characterization of this contact resistance is not well established. In general, the intercept of the curve of thermal resistance versus BLT is considered to be the contact resistance. Depending on the chemistry of the polymeric matrix and solid surface conditions, this number can fall within a wide range. It is not uncommon for the contact resistance to dominate the overall resistance for high-performance TIM materials. Understanding of the mechanisms of the contact resistance is still developing. Gowda et al. [22] reported several characterization techniques to identify micron and submicron voids at the interface between the interface material and the surfaces of the heat spreader and semiconductor device. Using SEM for cross-sectioned samples suffered from the artificial effects caused by the sample preparation. The nonintrusive computed tomography (CT) seems to be a better choice to provide some virtual section images of the interface. However, nonintrusive techniques such as CT and scanning acoustic microscopy (SAM) are unable to provide the spatial resolution needed for this kind of analysis. Hu et al. [23] used an IR camera to measure the temperature distribution at the cross section of a sample. It was reported that the gradient at the interface between the TIM and solid wall is higher at the bulk region, suggesting boundary thermal resistances at the TIM and solid wall interfaces.

9.5.1 State of the Art of Thermal Interface Materials

Thermal interface materials fall into several types, depending on their chemistry and physical appearance. Figure 9.6 shows typical ranges of thermal resistances for these different types of TIMs. As can be seen, TIM thermal resistances vary widely due to differences in manufacturing, void fraction, surface condition, and chemistry. In selecting thermal interface materials, thermal performance is clearly the key criterion. However, close attention has been paid to reliability, manufacturability, reworkability, and cost. Six types of TIMs will be discussed with emphasis on silicone gel and solder.

Greases are compliant thermal compounds based on silicone or hydrocarbon oils with solid filler particles for thermal enhancement. The advantages of greases are their high thermal conductivity and relative ease to compress to a thin bond line. Since no curing process is required, grease is often considered reworkable. One well-known problem with grease is the "pump-out" issues associated with gradual loss of grease materials due to repeated cycling actions of the package. At severe conditions, dry out can happen. As a result of the "pump-out" phenomena, thermal performance of the package degrades significantly [24]. Considering these characteristics, greases are often used as TIM2 materials for high-end desktop computers or low-mid-range servers [25]. For applications where silicone contamination is a concern, non-silicone-based thermal compounds can be applied. For instance, the ATC family of materials has been used extensively in high-end IBM servers [26].

An alternative to the greases is the elastomeric pad, which is polymerized silicone rubbers in the form of easy-to-handle solids. Elastomeric pads usually incorporate a woven fiber-optic carrier that contains filler particles. Despite their poor thermal performance, elastomeric pads have been often used as TIM2 materials that are preattached to the heat sink at one vendor before final assembly at another

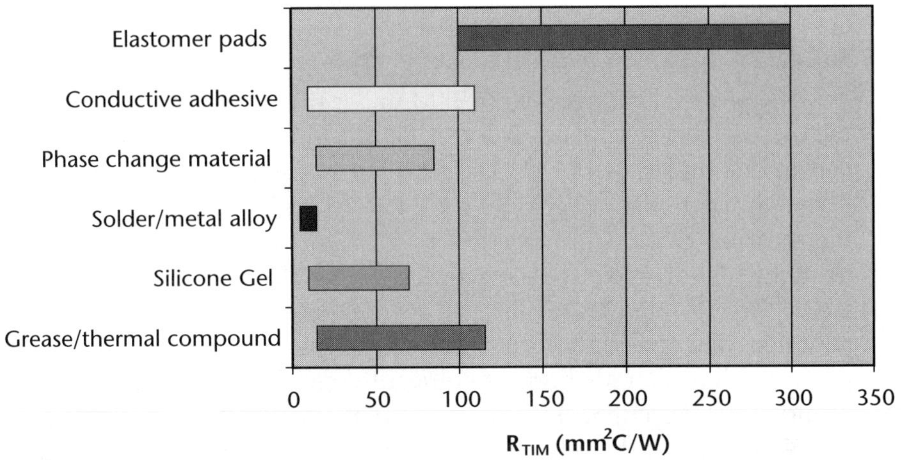

Figure 9.6 Typical thermal performance for the different category of the TIMs. Typical TIM bond line thickness is in the tens of microns.

vendor. One disadvantage of the elastomer pads is the high pressure required to achieve certain bond-line thicknesses. The low thermal performance limits the elastomer pads to low-thermal-requirement applications.

Phase-change material is a unique family of thermal interface materials that combines the thermal performance of grease with the ease of handling in solid pads. At room temperature, phase-change materials remain in a solid state and as such can be preattached to the back of the heat sink with a covering foil to shield the material. Upon melting, typically in the range of 50°C to 80°C, the liquid material flows to fill the small voids in the solid surface, thereby improving the heat-flow continuity. As a TIM2 material, phase-change material can provide noticeable mechanical coupling between the heat spreader and the heat sink, which becomes a concern for package and interconnect reliability under dynamic loading conditions, such as shock and vibration test [27].

Conductive adhesive materials are typically an epoxy matrix filled with silver particles. Very thin bond line can be achieved, and the thermal performance of the material can be very high (~10 mm^2C/W). Upon curing, due to the high modulus, epoxy adhesives provide rigid coupling between the surfaces such that CTE mismatch becomes a prohibitive problem. In applications where a CTE-matching heat spreader is used and flat ceramic substrate is used, conductive adhesive materials provide excellent and reliable thermal performance [28].

Silicone gel material has been the mainstream material, used primarily as TIM1 for microprocessor cooling. Like grease, gel material can easily be dispensed to the surface of the die. The filler particles, typically aluminum and alumina, enhance the bulk thermal conductivity (~4 W/m-K). Gel material requires curing to form a stable polymer network. Upon curing, the TIM is not susceptible to pump-out like grease. As the TIM1 material is in close proximity to the chip junction, it is extremely important to maintain the thermal performance and reliability at operating conditions. Extensive tests have been conducted for this type of material for both organic and ceramic packages. Significant degradation has been identified under accelerated reliability-stressing conditions [29]. Dal [29] observed that at 96 hours of highly accelerated stress testing (HAST), the TIM sample appears grainy and flat, suggesting loss of adhesion at the heat spreader surfaces. At 192 hours of HAST, the TIM becomes powdery and possibly brittle. There is apparent resin-filler separation.

The physical changes of the TIM under the HAST condition correlate well with the thermal measurement readouts. Significant degradation took place between the end-of-line and post-HAST. Dal [29] further analyzed the effect of moisture and temperature on the integrity of the TIM. At high temperatures of 125°C to 250°C, the presence of moisture causes a reversible reaction that breaks the Si-O bond to form the silanol group. This reaction is reversible in the absence of the moisture. The silanol group migrates to the surface, leading to the loss of hydrophobicity and adhesion. The hydrophobicity recovers gradually after removal of the stress. Temperature alone can also pose significant degradation to the silicone network. As a result, the TIM is hardened and becomes brittle. At temperatures above 180°C, decomposition of the polymer chains leads to a cyclic compound referred to as D3. Due to the change in the polymer chain, the TIM evolves from a tacky end-of-line state to the grainy and brittle state after stressing. The combined effects of moisture and temperature cause significant degradation in the thermal performance of the TIM. In field-application conditions, this can be exacerbated by the cycling condition, which imposes repeated tensile and shear stress on the TIM. For high-end applications, the degraded TIM material may not meet the requirements that were met at the end of line. It is thus necessary to introduce high-performance solder thermal interface materials.

Solders are not new to microelectronic packages as they frequently appear in the interconnects of the first and second levels. As a thermal material, solder has thermal conductivity that is orders of magnitude higher than the polymeric type of material. As pointed out previously, thermal conductivity is not the only criterion for selecting a TIM material. The reflow temperature of the solder has to be appropriate as the heat spreader attachment is typically the last stage of the first-level assembly process [30]. For instance, the same solder used for the c4 interconnects should not be considered for the TIM. It is also desirable to use a ductile solder at the CTE mismatching surfaces of the copper heat spreader and the silicon die. Considering these criteria, Hua et al. [30] selected a pool of candidate solders, including indium (In), 63Sn37Pb, 42Sn58Bi, 97In3Ag, and Sn. Thermal test vehicles were built using 12 mils of preforms of various solders. Thermal test after thermal cycling stress shows that pure indium and 63Sn37Pb are the most promising. Considering the lead-free requirement, only pure indium is chosen finally.

As described in Hua et al. [30] and Deppisch et al. [31], the thermal performance and reliability of the indium thermal interface depend on the thickness of the solder preform used, the Au metallization thickness, and the die sizes. Deppisch et al. [31] point out that thicker BLT helps to absorb the stress induced by the CTE mismatch, while thinner BLT increases the mechanical coupling between the lid and silicon. As a result, although thin BLT is desired from a pure thermal point of view, thicker BLT actually helps maintain the integrity of the thermal interface and should be used instead. The Au metallization thickness is also found to be very important to the reliability of the interface. Extremely thin Au thickness should be avoided as this jeopardizes the wetability of the solder to the surface. However, an increasingly thicker Au layer was found to compromise the integrity of the interface between the solder and the inter-metalic compound (IMC). The failure mode after thermal cycling seems to be different for small and large die sizes. A center-initiated fracture is seen only at small die sizes. On the other hand, corner failure is observed for both small and large die sizes. To understand the failure mechanism, SEM-EDS analysis is conducted for cross sections prepared by focused ion beam (FIB) to examine the IMC structures. Fractures have been observed at the interface between the AuIn2 and the bulk In. It is conjectured that the nodular Au-rich IMC structure acts as a pinch point that causes stress concentration. In summary, the thermal performance and reliability of the solder thermal interface material is sensitive to the BLT, metallization thickness and quality, and die sizes. It is necessary to exercise careful modeling and testing to explore the optimum process window.

9.5.2 Challenges and Opportunities

As indicated in the ITRS roadmap [105], the thermal budget for high-power devices continues to shrink. As shown in Figure 9.7, if the same silicone gel material is used, as the heat flux increases to 100 W/cm^2, thermal resistance due to the TIM will account for half of the total resistance. On the other hand, as more functionality is integrated into handheld devices, enhancement of the conductive heat transfer seems to be the only option, due to spatial constrictions. New material that has extremely high thermal conductivity, such as carbon nanotubes (CNT), may eventually receive more attention.

9.6 Conductive and Fluidic Thermal Spreaders: State of the Art

Heat spreaders are utilized in order to reduce the heat flux near the chip by effectively spreading the heat over a larger area. At the package level, metallic alloy heat slugs have been commonly employed, as seen in Figure 9.8. Requirements for

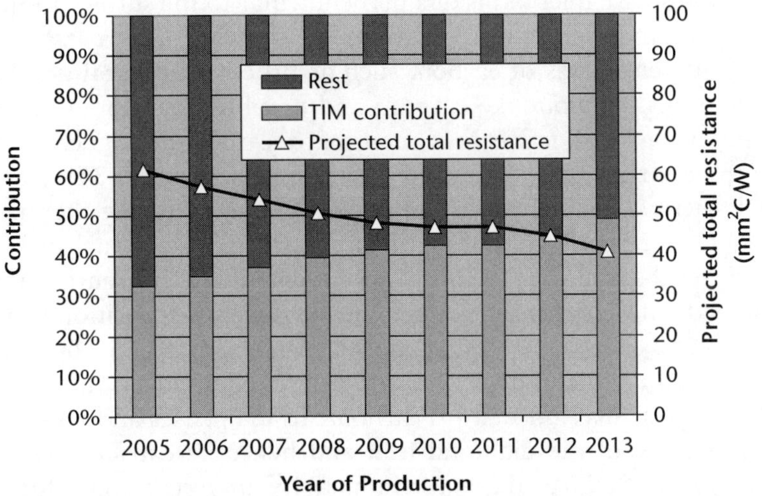

Figure 9.7 Contribution of the thermal interface material to the projected overall resistance (ITRS 2006 updates for cost performance [105]) of the package, assuming the fixed thermal interface material (at 20 mm^{2}C/W) is used for all applications. The ambient temperature is assumed at 45°C and the junction temperature at 85°C.

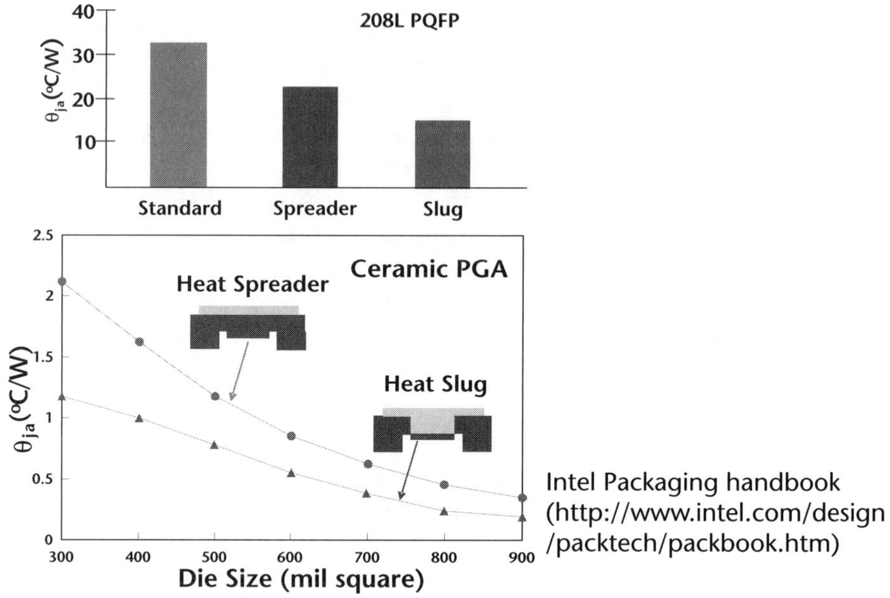

Figure 9.8 Package level heat spreaders and slugs are used to reduce the internal heat conduction thermal resistance. These gains are more significant for smaller die, due to the more concentrated heating. For the ceramic pin-grid array package shown above, the spreader/slug is Cu/Cu-Tungsten alloy attached to the package top.

spreaders include high thermal conductivity, low tailorable coefficient of thermal expansion, and low densities. Advanced composite materials with high thermal conductivity and matching thermal expansion coefficient to that of silicon are progressively being used to replace the traditional aluminum- and copper-based metal-plate heat spreaders (Table 9.1) [32, 33]. Recent attempts in improving the thermal performance of the spreader material have also focused on the development of silicon substrates with microwhiskers perpendicular to the surface [34]. Examples of these materials and their implementations are shown in Figure 9.9.

Different forms of carbon, such as processed natural graphite, carbon-carbon composites, diamondlike carbon, and graphite foam offer lots of possibilities for maximizing conductive heat transfer due to their higher thermal conductivity. Graphite foams have been used to fabricate heat sinks [35]. Natural diamond with high thermal conductivity (2,000 W/mK) and matching thermal expansion coefficient (1 to 2 ppm/K) has been used in bonding devices, such as laser diodes, to dissipate the thermal load ([36]). The advent of low-pressure synthesis of diamond, coupled with inexpensive, large, chemical-vapor-deposition processes, has made it possible to consider the use of diamond in electronics for heat removal [37]. However, continuous and defect-free growth of diamond is a difficult task. The presence of defects such as amorphous carbon/carbide phases and voids reduces the thermal conductivity of the diamond film and hence its capacity to be an effective heat spreader [38]. High-thermal-conductivity materials like aluminum nitride (370 W/mK) can be used to fill the voidlike regions formed during the growth of diamond, thereby making the multilayer AlN/diamond composite structure an effective heat spreader [39]. Unfortunately, the contamination of wafers during deposition of

Table 9.1 Material Properties of Composites Solids Used Instead of Solid Copper and Aluminum Metal Plates

Material	*Density(g/cm³)*	*Thermal Conductivity (W/mK)*	*CTE (ppm/C°)*
Silicon	2.3	151	4.2
Aluminum	2.7	238	23.6
AlN	3.3	170–200	4.5
Beryllia	3.9	250	7.6
Copper	8.9	398	17.8
Cu W (10–20% Cu)	15.7–17.0	180–200	6.5–8.3
Cu Mo (15–20% Mo)	10.0	160–170	7.0–8.0

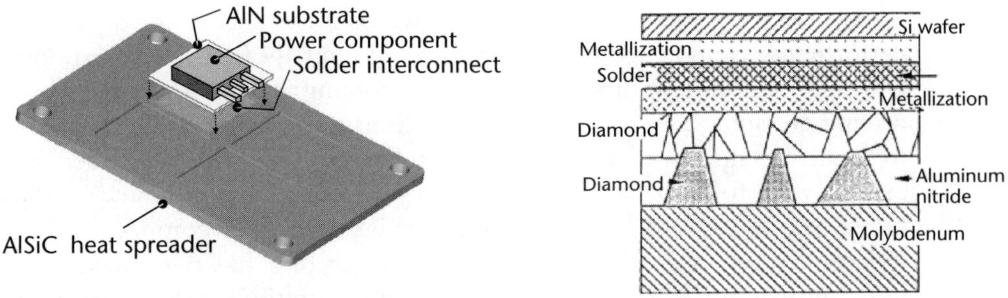

(a) Liquid cooled heat sink (Moores et al., 2001)

(b) Multilayer diamond heat spreader bonded to device wafer (Jagannadham, 1998)

Figure 9.9 Heat spreaders and sinks based on advanced composite materials. (a) A AlSiC based liquid cooled spreader with integral pin fins [33]. In (b), [42] a multilayer diamond spreader is attached to the device.

diamond has prevented the integration of diamond heat spreaders with silicon technology.

The use of liquid-vapor phase change to transport heat across the system has been proposed as an alternative to polycrystalline diamond or AlN ceramic heat spreaders, mainly due to the high cost of fabricating heat spreaders from these types of materials. Particular emphasis has been placed on the development of flat-plate heat pipes for cooling electronic systems in space-constrained applications. Successful demonstration of the use of flat heat pipes for cooling printed wiring boards [40] with heat fluxes up to 2 W/cm^2 has been achieved in the past.

There has been a considerable amount of research on micro heat pipe arrays and flat-plate micro heat pipes ([41, 42]). Flat heat pipes with a segmented vapor space machined on a silicon substrate have been suggested as an alternative to the conductive cooling of integrated circuits using diamond films [43] with the flat-plate heat pipe exhibiting a thermal conductivity approximately five times that of the silicon material over a wide range of power densities. The design of wick structures in flat-plate heat pipes has also become a focus of renewed interest. Experimental and theoretical analyses of the heat-transport capabilities of flat miniature heat pipes with trapezoidal and rectangular micro capillary grooves [44] and triangular microgrooves [45] have been carried out with the objective of increasing the fluid flow to the evaporator section. Heat pipe integrated with an aluminum plate has been utilized for CPU heat removal of 18W from notebook computers while maintaining the CPU below 85°C [46]. Selected examples of flat heat pipe–based heat spreaders are presented in Figure 9.10.

As seen in Figure 9.11 *for a peripherally, convectively cooled, thin heat spreader*, the conduction thermal resistance between the heat source and wall, $R_{th} = (T_{heater} - T_{wall})/Q$, for a given solid heat spreader of constant thermal conductivity sharply increases with reduction in thickness, due to the increase in spreading resistance. To meet a given thermal-resistance target using pure conduction, one needs to move from aluminum to copper to diamond, as the thickness of the spreader continues to reduce. As such, a thin spreader with a *variable, on-demand*, effective thermal conductivity is highly desirable.

A heat pipe offers some of the characteristics of such a spreader. However, its performance is, in general, subject to several limits [47]. The circulation rate of the working fluid in heat pipes for electronic cooling is usually limited by insufficient driving pressure. This so-called capillary limit restricts the application of heat pipes to moderate chip heat dissipation and relatively small heat spreader areas. There exists a need for thermal management devices that are not limited by the capillary limitation of conventional heat pipes, while remaining compact and orientation independent. This motivation led to the design of a flat two-phase heat spreader [48–50]. Figure 9.12 shows a schematic representation of this novel boiling-based device, which consists of an evaporator area in the middle and a hollow frame, called a pool belt, along its periphery. The heating is supplied at localized regions of

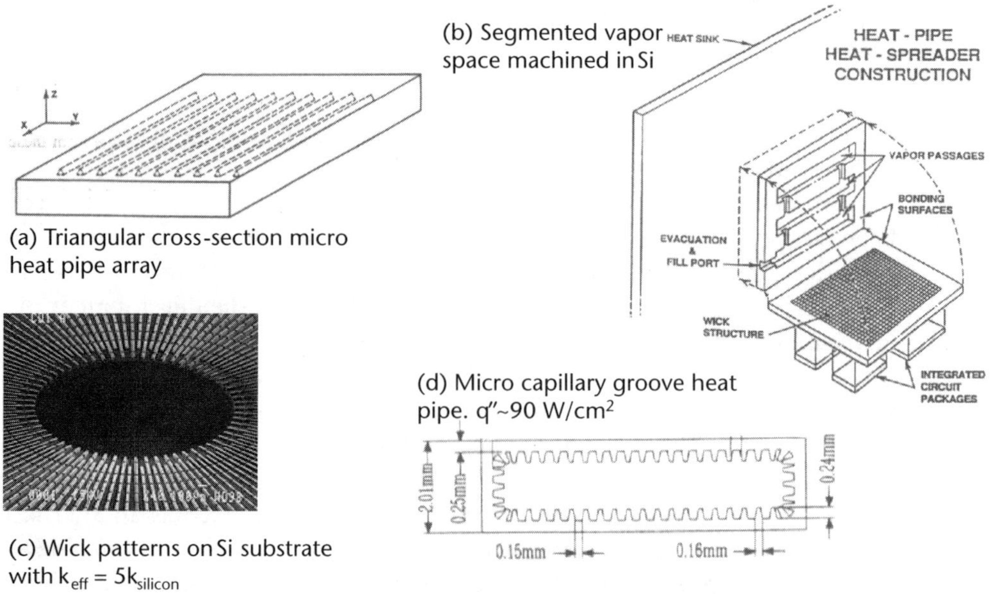

Figure 9.10 Heat pipe based flat heat spreaders. In (a) an array of triangular cross-section micro-heat pipes is utilized [45]. In (b) and (c) different patterns of micro-fabricated wicks are used to construct flat heat pipe based spreaders [46, 47]. In (d), the internal walls of a flat spreader cavity are lined with micro-grooves [48].

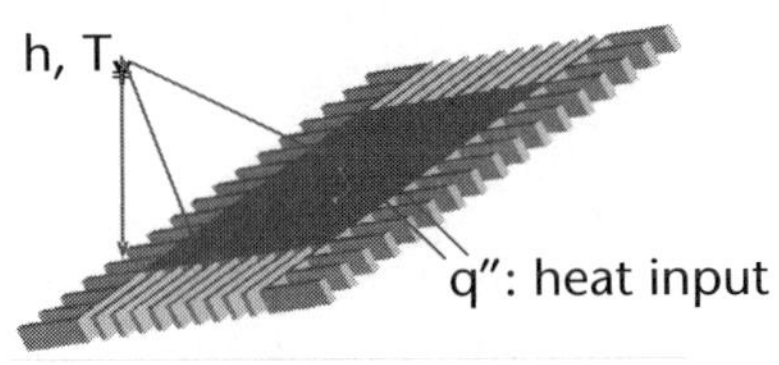

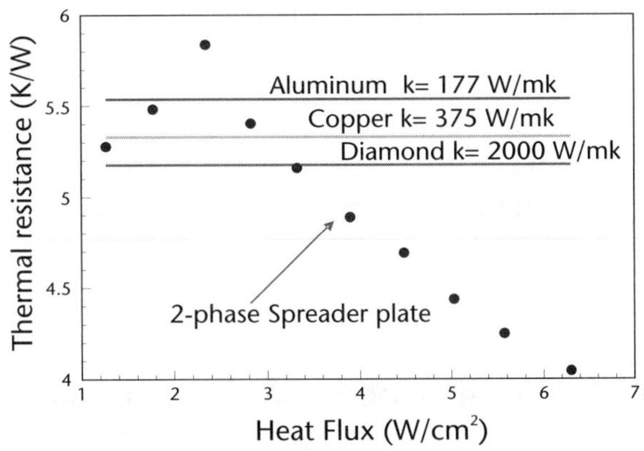

Figure 9.11 Aluminum prototype 4.5 mm thick two-phase heat spreader thermal performance comparison to various solid spreaders of identical dimensions, under external natural convection cooling. Note, higher heat fluxes can be achieved through reduction of air-side resistance.

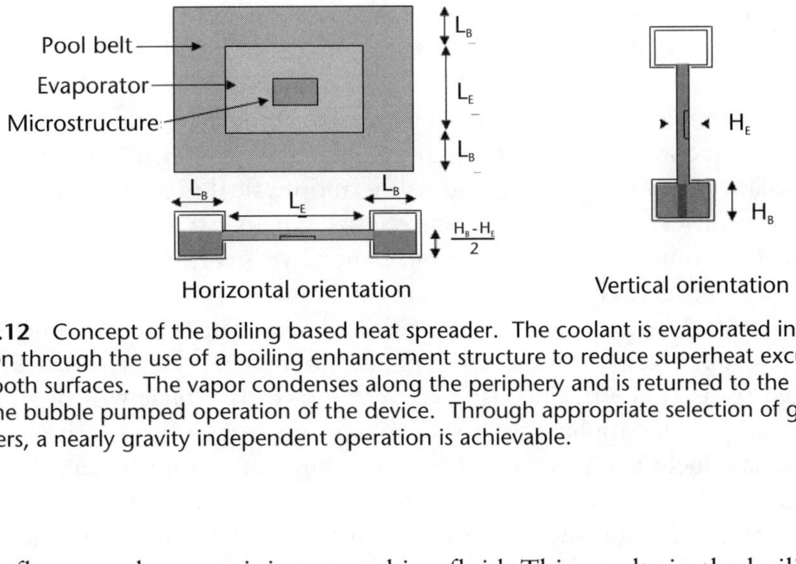

Figure 9.12 Concept of the boiling based heat spreader. The coolant is evaporated in the central region through the use of a boiling enhancement structure to reduce superheat excursion seen with smooth surfaces. The vapor condenses along the periphery and is returned to the center to enable the bubble pumped operation of the device. Through appropriate selection of geometrical parameters, a nearly gravity independent operation is achievable.

the thin flat spreader containing a working fluid. This results in the boiling of the fluid from the boiling-enhancement structure. Vigorous circulation of the liquid and vapor is maintained by the interconnected microfabricated network of microchannels within the structure. The bubbles move toward the finned periphery of the device, where they condense. Orientation-independent performance is achieved by ensuring that the evaporator section of the spreader remains flooded under all inclinations through satisfying the design constraint:

$$H_B / H_E = 2(1 + L_B / L_E) \tag{9.6}$$

Ample supply of liquid to the evaporator, coupled with known boiling-enhancement techniques, helps in overcoming the performance constraints of conventional heat pipes. Vapor and liquid trains are formed within the microchannels, and the differential in capillary effects across each vapor or liquid slug is thought to produce the driving force. Since these trains are different from the transport of vapor and liquid in a heat pipe or vapor chamber, it is expected that these devices are not constrained by the typical capillary limits. The performance of the spreader fabricated in aluminum was compared with a solid spreader of equivalent dimensions under similar external cooling conditions. This comparison is seen in Figure 9.13 and it shows the superior performance of the boiling-based heat spreader at the high heat fluxes. It is seen that the two-phase spreader resistance drops sharply as boiling is established. The overall value of the thermal resistance is even superior to that of diamond.

9.7 Heat-Transfer Coefficient for Various Cooling Technologies

Convective heat transfer plays a critical role in defining the total thermal resistance that needs to be overcome for the heat generated at the die to be ultimately rejected to the ambient environment. For a given junction-to-ambient temperature difference, the convective thermal resistance is often a dominant term in the overall resistance network and needs to be minimized, both for the package-level thermal management (e.g., air-cooled heat sinks) as well as in the case of an off-package heat exchanger (chiller or condenser) for liquid cooling and refrigeration. Thus, the range of heat-transfer coefficients that can be achieved in essence defines the performance envelope that any given thermal management device design could deliver. The magnitude of convective heat transfer coefficient depends on four factors: (1) the type of coolant (gas versus liquid) and its thermophysical properties, (2) the mechanism of heat transfer (single phase versus phase change), (3) the hydrodynamics of the coolant flow (internal versus external, natural versus forced, laminar versus turbulent, streaming versus jetting versus spraying) and the fluid velocity/flow rate, and (4) the design of the heat sink (e.g., utilization of heat transfer rate augmentation structures to disrupt thermal and hydrodynamic boundary layers and therefore increase the heat-transfer coefficient). An excellent, up-to-date review of different heat transfer techniques in application to electronic cooling with an emphasis on practical aspects, including heat sink design and implementation schemes, has been recently presented in the online *ElectronicsCooling* magazine [51].

Figure 9.13(a) shows the values of the intrinsic convective heat transfer coefficients for different fluids and heat transfer modes that have been estimated for typical operating conditions of heat sinks [52]. The range spans more than six orders of magnitude. Clearly, the higher the intrinsic heat transfer coefficient, the less surface area (and the smaller the heat sink) needed to dissipate any given heat load. The general trend is that the heat transfer coefficient increases from natural convection to force convection along the heated surface to jet impingement flow and peaks at the phase-change heat transfer, regardless of the nature of the coolant. Gases are the least potent coolants with the heat transfer coefficient ranging from between 1 and 5 W/m^2K for natural convection to between 100 and 150 W/m^2K for forced convec-

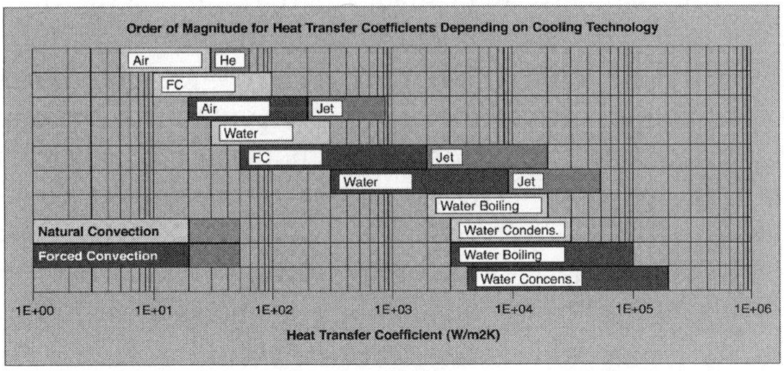

(a)

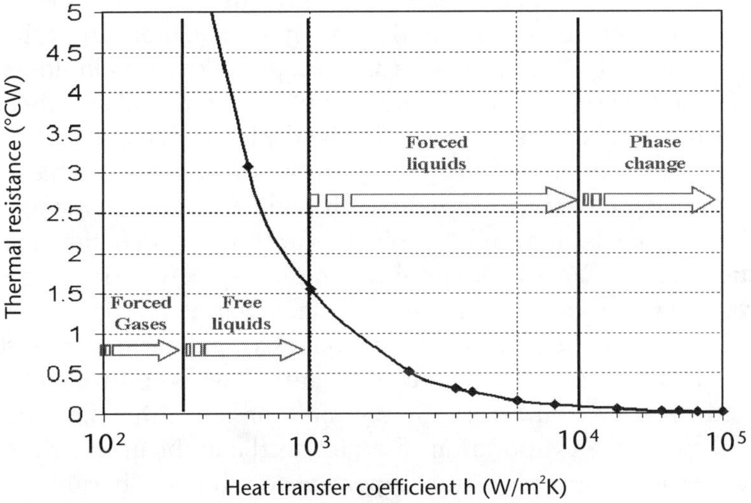

(b)

Figure 9.13 (a) Heat transfer coefficients for different cooling fluids in gas and liquid phase and for different modes of heat transfer. (*From:* [52]. © 1997. *Electronics Cooling.* Reprinted with permission.) (b) Die-to-ambient conduction-convection thermal resistance as function of the convective heat transfer coefficient (assuming a square 17.6m x 17.6 mm, 0.5 mm thick silicon die with k=148 W/mK, and neglecting all interface/contact resistances). (*From:* [55]. © 2004 IEEE. Reprinted with permission.)

tion due to their low thermal conductivity. The low heat capacity (product of density and specific heat) further reduces the cooling potential of gases due to their diminishing ability to store thermal energy dissipated from the package.

In principle, forced convection of liquids in microchannels can achieve the record high heat fluxes with a decrease in the hydraulic diameter of flow channels, as the heat transfer coefficient for fully developed laminar flow scales inversely with the channel diameter. This was amply demonstrated by Tuckerman and Pease [53] in their pioneering work on microchannel heat sinks. The effect is especially profound for the liquid coolants with high thermal conductivity, such as water or liquid metals. However, an excessive increase in the pumping power places a limit on how small the microchannels can be made (thereby limiting the value of practically realized heat transfer coefficients). Further improvement of performance with both gas and liquid cooling could be obtained by using jet impingement flow in which a stream of coolant is directed toward (normal or under a certain angle) the substrate, leading to significant reduction of the boundary-layer thickness and associated increase in the convective heat transfer coefficient. This extends the range of achievable values up to 1 kW/m^2K for air as a coolant and 10 to 50 kW/m^2K for liquids, remarkably with manageable pressure drops. One drawback of using jet impingement cooling for high power dissipation is an increased acoustic signature of the system, often resulting in unacceptable noise levels.

Finally, phase-change heat transfer is the most efficient mechanism of heat transfer due to an advantage offered by the significant latent heat of vaporization of liquids. There are two distinct ways to enable phase-change heat transfer: boiling and evaporation. The main difference between these two methods lies in the location at which phase-change occurs. In the case of boiling, it occurs at the bottom (heated) surface beneath the fluid layer—as a result, the key factors limiting the heat transfer rate are the rate of bubble nucleation and their removal/transport away from the heated surface. The latter is controlled by the hydrodynamics of the boiling process and imposes a limit on the maximum heat fluxes that can be achieved via boiling, the critical heat flux (CHF), beyond which boiling becomes unstable and ineffective. Despite the CHF limitation, the boiling heat transfer coefficient is remarkably high, ranging from 1 kW/m^2K for pool boiling to 100 kW/m^2K for convective boiling, with higher values obtained for water as a coolant.

In contrast, in the case of evaporation, phase change occurs at the free surface of the liquid film. As a result, the rate of heat transfer is controlled by two resistances: conduction/convection across the film and mass transfer (i.e., for saturated vapor removal) from the evaporation interface to the ambient environment. Since these resistances act in sequence, it is important to minimize both of them to be able to achieve a high heat transfer rate. The film conduction/convection resistance is controlled by the film thickness (the thinner the film is, the smaller the resistance), while the mass transfer resistance for removal of evaporated fluid is controlled by three factors: the mass transfer coefficient defined by velocity and flow mode), the relative humidity (dryness) of the sweeping gas blown over the film, and the saturation density of the liquid that is being evaporated. This leads to two very important observations [54]: (1) Fundamentally, evaporation may be a much more efficient method of heat removal as compared to boiling if certain conditions are met. Indeed, theoretically, if one can maintain a stable monolayer of liquid on the surface and blow fully dry, sweeping gas (e.g., air) at high velocity above this liquid monolayer, one can dissipate heat fluxes of the order of 1 MW/cm^2. (2) More volatile fluids, such as fluorocarbon dielectric liquids (e.g., FC-72), perform in a superior manner to water in evaporation cooling schemes, even though the thermophysical properties of water (i.e., thermal conductivity and latent heat of vaporization) are much better. However, the saturation density is much higher for FC-72 than for water.

An alternative way to display this information, which is more revealing for thermal design, is to plot the total conduction-convection resistance of an electronic package and the heat sink as a function of the convective heat transfer coefficient [55]. Figure 9.14(b) shows such a comparison for the silicon die subjected to differ-

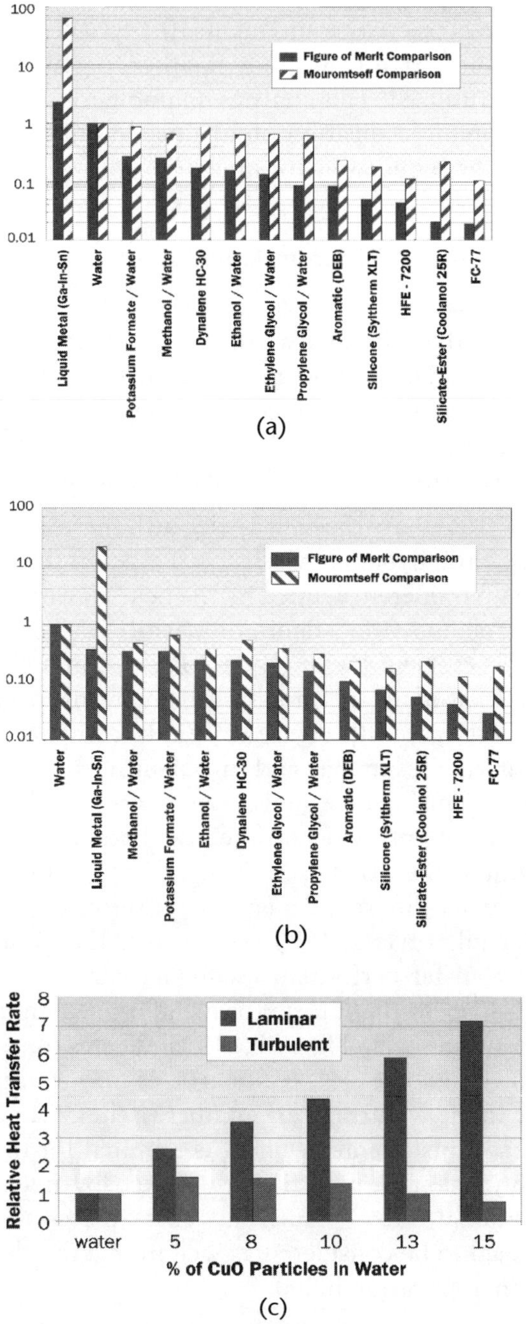

Figure 9.14 Liquid coolant comparison using the Mouromtseff Number and the thermohydraulic figure-of-merit (FOM) for (a) laminar flow, and (b) turbulent flow. (*From:* [56]. © 2006. *Electronics Cooling*. Reprinted with permission.) (c) Heat transfer enhancement (as compared to water alone) for a nanofluid containing copper oxide (CuO) nanoparticles suspended in water for fully developed internal laminar and turbulent flows (liquid temperature is 20° C). (*From:* [59]. © 2007. *Electronics Cooling*. Reprinted with permission.)

ent cooling methods, neglecting (for simplicity) all interfacial/contact resistances. The results clearly indicate that it is possible to remove heat load corresponding to the most severe projected requirements (0.14°C/W resistance for high-performance chips at the end of 2016 ITRS) only if either forced-liquid or phase-change convective cooling are utilized. This analysis unambiguously calls for the accelerated development and commercialization of advanced liquid-cooling technologies to sustain the growth of the semiconductor industry.

9.7.1 Comparison of Different Liquid Coolants

Considering its very high convective heat transfer coefficient, liquid cooling has recently become the subject of a growing body of academic and industrial research and development. The pertinent question is then, which liquid is a better coolant? Two criteria, which were designed to account for a desire to maximize thermal performance and minimize hydraulic pressure losses [56], have been proposed to answer this question: the Mouromtseff number (Mo) [57] $Mo = \rho^a k^b C_p^c / \mu^d$ (where ρ, k, C_p and μ are the density [kg/m^3], thermal conductivity [W/mK], specific heat [J/kgK], and dynamic viscosity [Pa.s] of the fluid, and (a,b,c,d) are the dimensionless parameters defined by the heat-transfer mode and appropriate correlations) and Yeh and Chu's figure of merit (FOM) FOM = Cp*h/P (where h is the heat transfer coefficient [W/m^{2}K], and P is the pumping power [W], which can be expressed in terms of the thermophysical and transport properties of the liquid using relevant correlations) [58]. Figure 9.14(a, b) compares different coolants using Mo and FOM criteria for laminar and turbulent flow, respectively. Clearly, both methods indicate that the high thermal conductivity liquids such as water and liquid metals are the best coolants, whereas the low thermal conductivity fluorocarbons (e.g., FC-77) are among the worst. Yet, the relative superiority score assigned to any given liquid is somewhat different depending on the criterion (Mo versus FOM) used for comparison. Kulkarni et al. [59] recently used the Mouromtseff number to evaluate relative heat transfer performance enhancement, which can be achieved by using nanofluids (a suspension of metal or metal oxide nanoparticles in the carrier fluid). The comparison shown in Figure 9.14(c) indicates that addition of even a small fraction (<6% in volume) of CuO nanoparticles into water leads to as much as a sevenfold increase in the Mouromtseff number in the case of laminar flow. On the other hand, the heat transfer enhancement is minimal for the turbulent flows, which are already highly chaotic and efficiently mixed by macroscopic eddy transport. Of course, additional factors such as cost, material compatibility, and system operability need to be considered as well in making the final selection of the liquid coolant for an application in hand.

9.7.2 Subambient Operation and Refrigeration

In addition to demands for improved power dissipation, lowering the junction temperature can result in a significant improvement in microprocessor reliability and performance. In recent work, Nowak [60] has shown that for the conventional ICs, latency improves with a decrease in temperature of $T^{-0.5}$ to $T^{-1.0}$ for the transistor and $T^{-1.5}$ for the interconnects, yielding $T^{-0.5}$ to $T^{-0.75}$ scaling for the overall increase in the microprocessor speed. Further, it has also been shown that a decrease in the leakage current enabled by the low-temperature operation at –100°C allows for achieving potentially a 4.3-fold enhancement in throughput of a CMOS chip, as compared to that at 85°C operation, when the chip is operated in a power-limited mode [61]. Different refrigeration technologies, ranging from mechanical to thermochemical to solid-state thermoelectric (TEC), are capable of delivering subambient cooling of microprocessors.

In general, mechanical (e.g., vapor-compression) refrigeration features the best performance from the power-consumption prospective [see Figure 9.15(a)] [62], whereas the solid-state TEC cooling is attractive from the simplicity and ease-of-integration viewpoint. Calculation of temperature at the cold end of a refrigeration cycle, where it interfaces with the chip, is similar to the conventional above-ambient cooling with the only difference being that the net heat flow to the ambient is equal to the sum of the chip heat dissipation and the work input to the

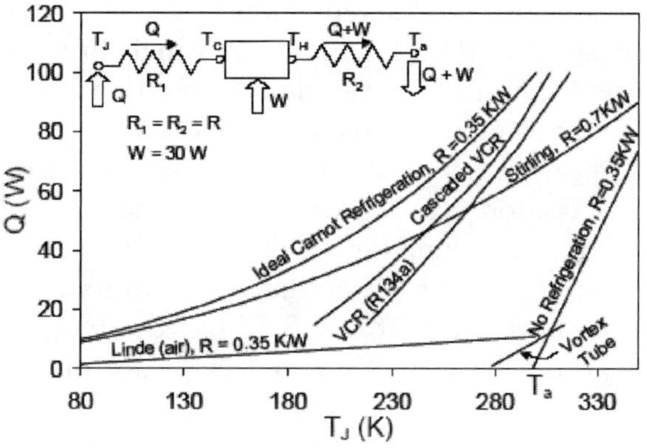

(a)

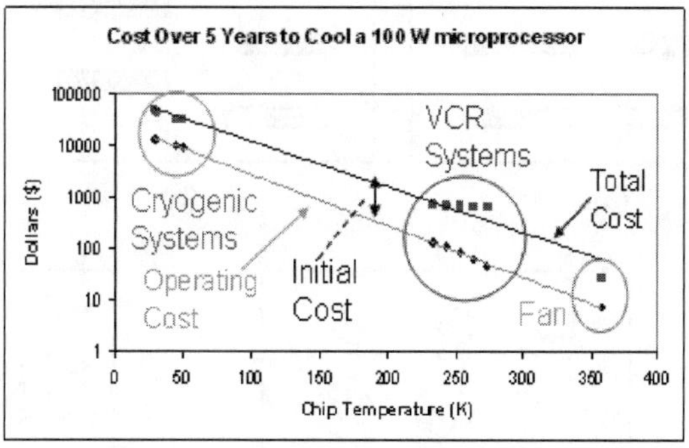

(b)

Figure 9.15 (a) Dependence of power that is possible to dissipate at the chip by various refrigeration methods as function of junction temperature: power input is taken constant across all technologies at 30 W and ambient temperature is set to 25°C; the resistanes to heat transfer at the evaporator R_1 (from the chip) and condenser R_2 (to the ambient) are taken to be the same and equal to 0.35 K/W. (*From*: [62]. © 2004. Reprinted with permission.) (b) The total cost of ownership of the cooling system, including the installation (initial) cost and the operating cost, as function of the chip temperature for a 100 W microprocessor during 5 years of operation.

refrigeration cycle. As a result, one can use a modified thermal resistance network analysis to determine the scaling law for the minimum achievable junction temperature T_j for both above-ambient and subambient (refrigeration) cooling of the microchip by using a single, simple equation $T_J = Q(R_1 + R_2) + T_a \times Q/(Q+W)$. Here, R_1 and R_2 (K/W) are thermal resistances of the evaporator/heat exchanger (interfaced with the chip) and condenser (interfaced with the ambient environment), respectively; Q (W) is the power dissipated at the chip; W (W) is the power input for driving the refrigeration cycle, and T_a (K) is the ambient temperature. Despite its simplicity, this equation is a very powerful tool for analyzing the effect of refrigeration on the cooling temperature. Indeed, in the case of no refrigeration, the power input is equal to zero ($W = 0$W), and the multiplier in front of the ambient temperature is maximum and equal to unity. Including the refrigeration system into the cooling loop (i.e., nonzero power input W) reduces the multiplier of the ambient temperature, in effect lowering the achievable junction temperature. The equation clearly shows that any sizable reduction of the junction temperature in the presence of large thermal loads (Q is large) is increasingly difficult, as it requires proportionally greater power input W to run the refrigeration system. The latter translates into a drastically increasing size and complexity of the refrigeration system (bigger compressors and heat exchangers) and associated escalation in the cost of installation (fixed) and operation/maintenance (mainly electricity). Figure 9.15(b) illustrates these cost trends as a function of the junction temperature for cooling a 100W microprocessor for 5 years of continuous operation.

Ultimately, in the case of any fluid-based refrigeration method, heat rejection occurs by heat transfer from a die to deeply subcooled fluid (gas or liquid) via sin-

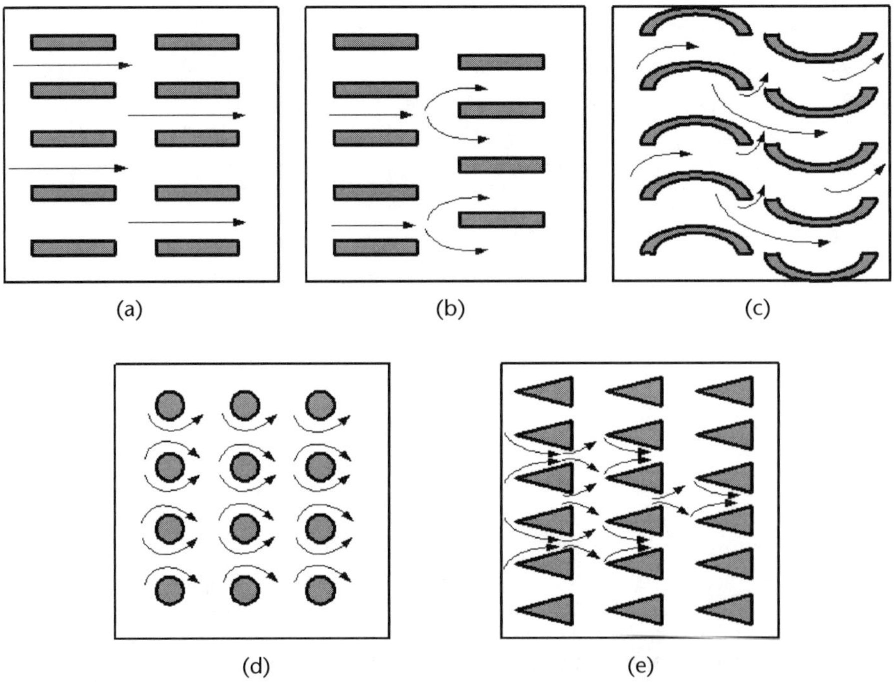

Figure 9.16 Different designs (view from the top) of heat sink structures with (a) straight fins, (b) offset fins, (c) wavy fins, (d) pin fins, and (e) louvered fins.

gle-phase or phase-change convective heat transfer. Thus, it is of fundamental interest and practical importance to investigate the heat-transfer properties and cooling rates, which can be achieved using commonly used refrigerants. Experiments with deeply subcooled liquids are challenging due to difficulties in controlling the heat leaks from the ambient environment into the test structure, resulting in an artificial decrease of measured heat transfer coefficient values as compared to the true performance. Only recently, Wadell et al. [63] performed a comprehensive investigation of heat transfer and pressure drop associated with convective boiling of a refrigerant R-508 at –80°C. Deep subambient operation was achieved using a two-stage vapor-compression refrigeration system, and different designs of the microstructured evaporator (from a baseline planar slit structure to microchannels to inline and staggered pin fin arrays) were tested. Experimental results indicate that very high effective heat transfer coefficients (ranging between 20 and 50 kW/m^2K) can be readily achieved while keeping the die at –80°C, which is on par with the values found in two-phase heat exchangers operated under the normal ambient conditions.

9.8 Air-Cooled Heat Sinks and Alternatives

The air-cooled heat sink has been and remains the main workhorse of the electronic cooling industry. The simplicity and low cost of operation, combined with the abundance and environmental friendliness of the coolant, makes air cooling uniquely appealing as a first-to-go approach to thermal management. To overcome the inferiority of air as a heat-transfer fluid, over the years the design of air-cooled heat sinks has evolved to a staggering level of sophistication with a main goal of providing the highest possible surface area for convective heat transfer in the smallest package and with the lowest possible pressure-drop (pumping-power) requirement. It should be noted that despite an increasing interest and push toward the adaptation of liquid cooling, the air-cooled heat sinks will never disappear entirely from the research landscape. This is simply because ultimate heat rejection to ambient environment, even in the case of liquid cooling or refrigeration, occurs at the liquid chiller/condenser with an air side of the heat exchanger often defining the overall system size and performance. Despite many similarities in analytical methodologies employed for performance analysis, design requirements for the package-level air-cooled heat sinks and those for the external air-cooled heat exchangers are rather different and require separate consideration.

It is instructive to provide an example of the current state of the art in air cooling. According to recent DARPA-released data, a state-of-the-art air-cooled heat sink dissipating 1 kW of power has a 4 inch × 4 inch footprint with 1 inch long fins, weighs ~0.3 kg, and features a thermal resistance of 0.2 K/W. It requires an air blower/fan that consumes 100W of power, weighs ~0.5 kg, and supplies 200 CFM (ft^3/min.) of airflow against a ~0.6 inch H_2O (~150 Pa) pressure drop of the heat sink and connecting manifolds. In this review, our focus is on package-level air-cooling, and the interested reader is referred to an excellent recent review [64] summarizing the state of the art and advances in air-cooled heat-exchanger technology.

Figure 9.16 shows a schematic of state-of-the-art air-cooled heat sinks with different fin structure designs, which can be produced using a variety of manufacturing methods, ranging from conventional machining to forging and extrusion to die-casting to fin folding [65]. Different manufacturing methods have their advantages and disadvantages; many are only applicable to heat sinks made of certain materials and are widely varied in the resulting cost of fabrication. The common theme for all manufacturing methods is difficulty and an increased cost associated with making high-density and high-aspect-ratio fins for high-performance heat sinks. This challenge gave rise to a new approach to heat sink design, known as "design for manufacturability," which aims to identify the heat sink structure and materials that result in achieving desired performance targets using readily available production capabilities at minimum manufacturing cost [65]. Design for manufacturability is just one aspect of a broader approach to the second-law (entropy) based thermoeconomic optimization of heat sinks [66], which aims not only at maximizing the thermal performance of the heat sink but also accounting for other important aspects of "sustainable" product development, including resource losses due to irreversibilities associated with heat sink production and operation [67].

Finally, to push the limits of air-cooled heat sinks, increased attention has recently been given to two important practical aspects of heat sink design and operation. First is an issue of the coolant bypass when air introduced into the heat sink avoids traveling through a finned (i.e., active heat-transfer) zone, but instead takes the path of minimal hydraulic resistance around and above the heat sink. This scenario has been recently evaluated by Simons [68], who showed that air bypass results in a rather dramatic increase in the heat sink thermal resistance, which more than doubles with an increase in the number of fins [see Figure 9.17(a)]. The second important design aspect concerns evaluation of heat sink performance normalized by the heat sink size [Figure 9.17(b)] and weight [Figure 9.17(c)]. An increase in dissipated heat loads translates into a need for greater heat transfer area, thus bigger and heavier heat sinks. The space limitations motivate a closer look at which fin structures are better suited to compact and lightweight thermal management systems. Recent analysis [69] showed that louvered fins significantly outperformed all other fin designs (straight fins, pin fins, wavy fins, and offset fins) when heat-transfer performance was compared based on the modified (normalized by the heat sink size and weight) evaluation metric [Figure 9.17(b, c)].

9.8.1 Fundamental Limits and Performance Models of Air-Cooled Heat Sinks

A question of what sets the fundamental performance limit of the air-cooled heat sink has been revisited in the literature a number of times, with the most recent comprehensive account coming from Rogers et al. [70]. A simple, yet instructive and elegant, approach to this problem was recently suggested by Nakayama [71], whose conclusions are briefly summarized here as being applicable to a wide range of electronic equipment. The analysis considered the upper bound of heat dissipation from a fan-cooled electronic box and found that availability of a fan capable of supplying the required volumetric flow rate is often a factor defining the limit of air cooling. Also, in most practical cases, the bottleneck to the heat flow is not the result of heat

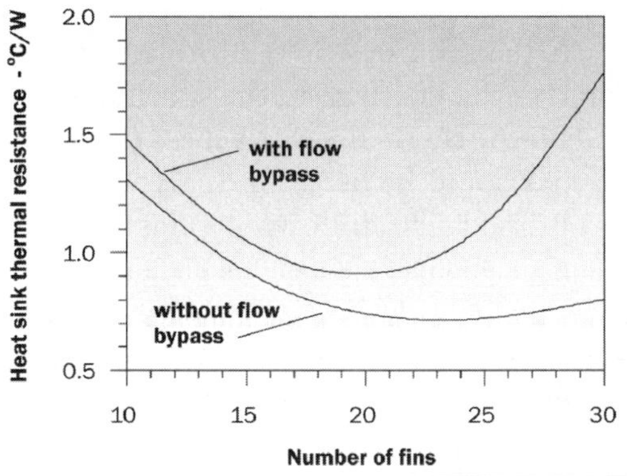

(a)

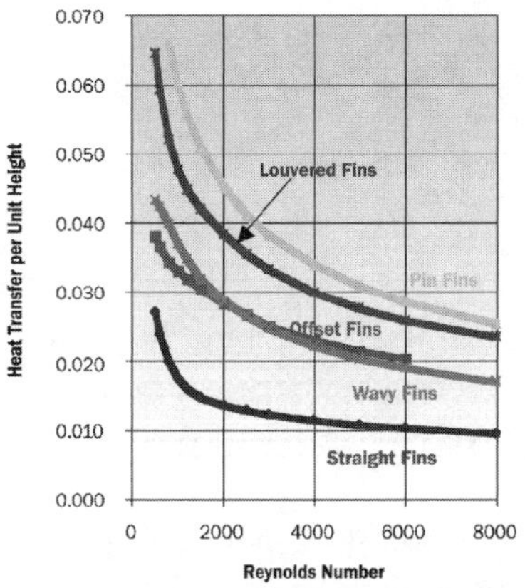

(b)

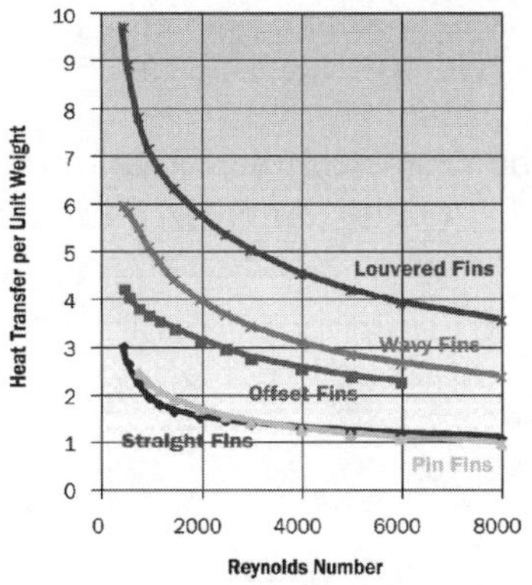

(c)

Figure 9.17 (a) Effect of removing air flow bypass on heat sink thermal resistance. (*From:* [68]. © 2004. Reprinted with permission.) Performance comparison of various heat sink/fin designs when normalized by heat sink size (b) and weight (c). (*From:* [69]. © 2004. *Electronics Cooling.* Reprinted with permission.)

spreading by conduction through the finned structure but due to the resistance to convective heat transfer from the finned surface. Thus, it is ultimately the magnitude of the air-convective heat-transfer coefficient (which approaches its practical maximum at around 200 kW/m^2K for conventional fans with acceptable noise levels [51, 52]) that sets the limit on the realizable performance of air-cooled heat sinks. For a given heat transfer coefficient, the specific power dissipation levels that can be attained depend on the thermal effectiveness of the finned structure of the heat sink. The procedure that could be used for analytical evaluation of air-cooled heat sink performance is outlined next, following a classic text by Kraus and Bar-Cohen [72].

Step 1. The analysis of the finned surface starts with a generalized fin equation for the local excess temperature $\theta(x) = T(x) - T_s$ (T_s is the surrounding air temperature), which is given, for a longitudinal fin [Figure 9.18(a)], by

$$\frac{d}{dx}\left\{\left[y_1(x) - y_2(x)\right]\frac{d\theta(x)}{dx}\right\} - \frac{h}{k}\theta(x)\left\{\left[1 + y_1'(x)^2\right]^{0.5} + \left[1 + y_2'(x)^2\right]^{0.5}\right\} = 0 \quad (9.7)$$

and, for a spine fin [Figure 9.18(b)], by

$$r\frac{d^2\theta(x)}{dx^2} + 2\frac{dr}{dx}\frac{d\theta(x)}{dx} - \frac{2h}{k}\left[1 + \left(\frac{dr}{dx}\right)^2\right]^{0.5} \theta(x) = 0 \quad (9.8)$$

All geometric parameters used in (9.7) and (9.8) are defined in Figure 9.18, and h and k are the convective heat-transfer coefficient and thermal conductivity of the fin material, respectively.

Step 2. Solving (9.7) and (9.8) yields expressions for the local temperature excess and rate of heat transfer in terms of the fin base temperature excess θ_b and the base (total) rate of heat transfer q_b:

$$\theta(x) = \theta_b \lambda_1(x) + q_b \lambda_2(x) \quad \text{and} \quad q(x) = kA(x)\left[\theta_b \lambda_1'(x) + q_b \lambda_2'(x)\right] \quad (9.9)$$

Here, $A(x)$ is the cross-sectional area variation of the fin, and specific expressions for $\lambda_1(x)$ and $\lambda_2(x)$ are summarized in Table 9.2 for different fin geometries.

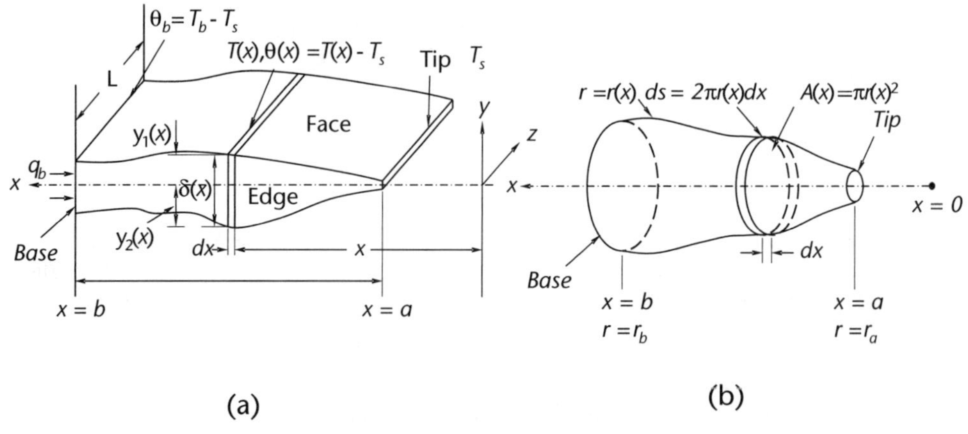

Figure 9.18 Schematic diagram and definition of the main parameters used for performance analysis of (a) longitudinal fins with an arbitrary profile and (b) generalized spine fins.

Table 9.2 Summary of $\lambda_1(x)$ and $\lambda_2(x)$ for Longitudinal Fins and Spines

Longitudinal Fins		
Rectangular profile	$\lambda_1(x) = \cosh m(b-x), \lambda_2(x) = \frac{1}{Y_0}\sinh m(b-x)$	$m = \sqrt{\frac{2h}{k\delta}}, \quad Y_0 = \sqrt{2hk\delta}L$
Rectangular profile with one face insulated	$\lambda_1(x) = \cosh m(b-x), \lambda_2(x) = \frac{1}{Y_0}\sinh m(b-x)$	$m = \sqrt{\frac{h}{k\delta}}, \quad Y_0 = \sqrt{hk\delta}L$
Trapezoidal profile	$\lambda_1(u) = u_b\left[K_1(u_b)I_0(u) + I_1(u_b)K_0(u)\right]$ $\lambda_1(u) = \frac{u_b}{Y_0}\left[K_0(u_b)I_0(u) - I_0(u_b)K_0(u)\right]$	$n = \sqrt{\frac{h}{k\sin\phi}}, \quad u(x) = 2n\sqrt{x}$ $Y_0 = 2k\delta_b Ln^2 / u_b, \quad u_b = 2n\sqrt{b}$
Half-trapezoidal profile	$\lambda_1(u) = u_b\left[K_1(u_b)I_0(u) + I_1(u_b)K_0(u)\right]$ $\lambda_1(u) = \frac{u_b}{Y_0}\left[K_0(u_b)I_0(u) - I_0(u_b)K_0(u)\right]$	$n = \sqrt{\frac{h(\sec\phi + 1)}{k\tan\phi}}, \quad u(x) = 2n\sqrt{x}$ $Y_0 = 2k\delta_b Ln^2 / u_b, \quad u_b = 2n\sqrt{b}$
Truncated, concave, parabolic profile	$\lambda_1(u) = \frac{1}{\alpha-\beta}\left[\alpha\left(\frac{x}{b}\right)^\beta - \beta\left(\frac{x}{b}\right)^\alpha\right]$ $\lambda_2(x) = \frac{1}{Y_0(\alpha-\beta)}\left[\left(\frac{x}{b}\right)^\alpha - \left(\frac{x}{b}\right)^\beta\right]$	$\alpha, \beta = \frac{-1 \pm \sqrt{1+4m^2b^2}}{2}$ $m = \sqrt{\frac{2h}{k\delta_b}}, \quad Y_0 = \frac{k\delta_b L}{b}$
Spines		
Cylindrical spine	$\lambda_1(x) = \cosh m(b-x), \lambda_2(x) = \frac{1}{Y_0}\sinh m(b-x)$	$m = \sqrt{\frac{4h}{k\delta}}, \quad Y_0 = \frac{\sqrt{\pi^2 hkd^3}}{2}$
Rectangular spine	$\lambda_1(x) = \cosh m(b-x), \lambda_2(x) = \frac{1}{Y_0}\sinh m(b-x)$	$m = \sqrt{2h(\delta_1+\delta_2)/k\delta_1\delta_2}$ $Y_0 = \sqrt{2hk\delta_1\delta_2(\delta_1+\delta_2)}$
Elliptical spine	$\lambda_1(x) = \cosh m(b-x)$ $\lambda_2(x) = \frac{1}{Y_0}\sinh m(b-x)$	$P = \pi(\delta_1+\delta_2)\left[1 + \frac{p^2}{4} + \frac{p^4}{64} + \frac{p^6}{256}\ldots\right]$ $m = \sqrt{\frac{hP}{kA}}, \quad Y_0 = kAm, \quad p = \frac{\delta_1 - \delta_2}{\delta_1 + \delta_2}$
Truncated, conical spine	$\lambda_1(w) = \frac{u_b^2}{u}\left[K_2(u_b)I_1(u) + I_2(u_b)K_1(u)\right]$ $\lambda_1(x) = \frac{8b}{Y_0 r_b w_b w}\left[K_1(u_b)I_1(u) - I_1(u_b)K_1(u)\right]$	$n = \sqrt{\frac{4h^2}{r_b^2k^2}\left(b^2 + r_b^2\right)}, \quad Y_0 = \frac{k\pi r_b}{n^2 b}$ $u_b = 2n\sqrt{b}, \quad u(x) = 2n\sqrt{x}$
Truncated, concave, parabolic spine	$\lambda_1(u)\frac{1}{\alpha-\beta}\left[\alpha\left(\frac{x}{b}\right)^\beta - \beta\left(\frac{x}{b}\right)^\alpha\right]$ $\lambda_2(u)\frac{1}{Y_0(\alpha-\beta)}\left[\left(\frac{x}{b}\right)^\alpha - \left(\frac{x}{b}\right)^\beta\right]$	$\alpha, \beta = \frac{-3 \pm \sqrt{9+4m^2b^2}}{2}$ $m = \sqrt{\frac{2h}{k\delta_b}}, \quad Y_0 = \frac{k\delta_b L}{b}$

Step 3. Using linear transformation, the temperature excess and heat transfer rate at the fin tip are determined using the conditions at the fin base:

$$\begin{bmatrix}\theta_a \\ q_a\end{bmatrix} = \mathbf{\Gamma}\begin{bmatrix}\theta_b \\ q_b\end{bmatrix} = \begin{bmatrix}\gamma_{11} & \gamma_{12} \\ \gamma_{21} & \gamma_{22}\end{bmatrix}\begin{bmatrix}\theta_b \\ q_b\end{bmatrix} = \begin{bmatrix}\lambda_1(a) & \lambda_2(a) \\ k(a)A(a)\lambda_1'(a) & k(a)A(a)\lambda_2'(a)\end{bmatrix}\begin{bmatrix}\theta_b \\ q_b\end{bmatrix} \tag{9.10}$$

where Γ is the matrix of the linear transformation, and γ_{ij} is the element of Γ.

Step 4. Input admittance, which is the ratio of the base (total) heat transfer rate to the base temperature excess, is computed:

$$Y_{in} \equiv \frac{q_b}{\theta_b} = \eta h S = \frac{\tanh mb}{mb} hS \tag{9.11}$$

where η, m, b, and S are fin efficiency, fin performance parameter, fin height, and the total convective heat transfer surface, respectively.

Step 5. Once thermal performance of an individual fin is determined using (9.7) to (9.11), the array of fins is assembled as follows:

a. *Cascade algorithm*:

$$\begin{bmatrix} \theta_{b2} \\ q_{b2} \end{bmatrix} = \mathbf{\Gamma}_2^{-1} \begin{bmatrix} \theta_{a2} \\ q_{a2} \end{bmatrix} = \mathbf{T}_2 \begin{bmatrix} \theta_{a2} \\ q_{a2} \end{bmatrix} = \mathbf{T}_2 \begin{bmatrix} \theta_{b1} \\ q_{b1} \end{bmatrix} = \mathbf{T}_2 \mathbf{T}_1 \begin{bmatrix} \theta_{a1} \\ q_{a1} \end{bmatrix} = \mathbf{T}_e \begin{bmatrix} \theta_{a1} \\ q_{a1} \end{bmatrix} \tag{9.12}$$

$$\mathbf{T}_e = \mathbf{T}_n \mathbf{T}_{n-1} \mathbf{T}_{n-2} \cdots \mathbf{T}_2 \mathbf{T}_1 \tag{9.13}$$

b. *Fins in cluster*:

$$q_{am} = q_{b1} + q_{b2} + \ldots + q_{bn}, \quad \theta_{am} = \theta_{a1} = \theta_{a2} = \ldots \theta_{an} \tag{9.14}$$

$$Y_c = \frac{q_{am}}{\theta_{am}} = \frac{q_{b1} + q_{b2} + \ldots + q_{bn}}{\theta_{am}} = \sum_{i=1}^{n} Y_i \text{ (Thermal admittance of the cluster)} \tag{9.15}$$

where Y_i is the input admittance of each fin in cluster, and Y_c is the cluster input admittance.

c. *Fins in parallel*:

$$\begin{bmatrix} q_b \\ q_a \end{bmatrix} = \mathbf{Y} \begin{bmatrix} \theta_b \\ \theta_a \end{bmatrix} = \begin{bmatrix} y_{11} & y_{12} \\ y_{21} & y_{22} \end{bmatrix} = \begin{bmatrix} \theta_b \\ \theta_a \end{bmatrix} = \begin{bmatrix} \tau_{22}/\tau_{12} & -(\tau_{11}\tau_{12} - \tau_{22}\tau_{21})/\tau_{12} \\ 1/\tau_{12} & -\tau_{11}/\tau_{12} \end{bmatrix} \begin{bmatrix} \theta_a \\ \theta_a \end{bmatrix} \tag{9.16}$$

$$\mathbf{Y}_e = \mathbf{Y}_1 + \mathbf{Y}_2 + \mathbf{Y}_3 + \ldots + \mathbf{Y}_n \text{ (Thermal admittance matrix in parallel)} \tag{9.17}$$

where τ_{ij} is an element of the matrix $\mathbf{T}$, $\mathbf{Y}_i$ is the input admittance matrix of each fin acting in parallel, and $\mathbf{Y}_e$ is the equivalent thermal admittance matrix that relates the total heat transfer rate from the array to the temperature excess at the base and the tip for fins acting in parallel.

9.8.2 Active Performance Augmentation for Air-Cooled Heat Sinks

Despite the technological maturity of air cooling, the art and science of air-cooled heat sink design continues to blossom with innovative ideas pushing the boundaries of the performance envelope to new heights. Fundamentally, once the junction-to-ambient temperature difference is fixed, there are only two available means to increase the power dissipation capabilities of a heat sink: increasing the surface area available for heat transfer and decreasing the total resistance to heat transfer. An increase in the heat transfer surface area leads to greater size and weight of the package, both of which are typically constrained by the specific application requirements. Therefore, efforts to reduce the total thermal resistance of the heat sink have gained increasing attention in the last decade as being universally useful across all application domains. The most promising avenues for innovation appear to be in exploring and exploiting various methods of active augmentation targeting the two performance limiting factors—the air throughput enabled by a fan and effective heat-transfer coefficients—as identified by Nakayama [71]. Several subjectively selected approaches are mentioned next, emphasizing the key general ideas that are being utilized to motivate further research.

Synthetic jets [73, 74] [Figure 9.19(a)] and piezoelectric fans [75, 76] [Figure 9.19(b)] are two recent examples of successful attacks on the problems of air throughput and the limited heat-storage capacity of cooling air. Specifically, significant performance improvements have been realized by increasing ambient "cold" air delivery, enhanced mixing, and "warm" air rejection using active (actuator-driven) devices: forced flexing of a perforated diaphragm in the case of the synthetic jet and piezo-driven flapping of a blade in the case of the piezoelectric fan. This is done in combination with the extended heat transfer surfaces provided by the conventional heat sinks, and with no increase in pressure-drop (pumping-power) penalty. Another recent, interesting idea for augmenting air throughput

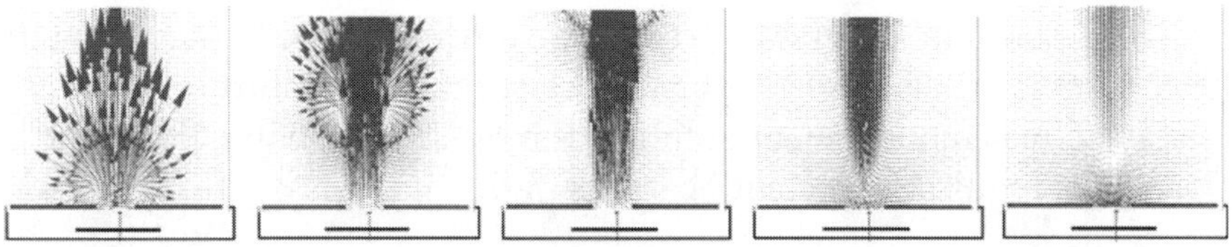

(a)

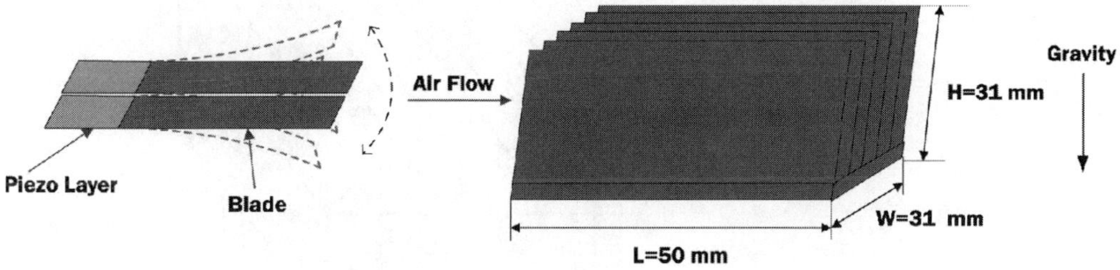

(b)

Figure 9.19 Examples of performance augmentation schemes for air-cooled heat sinks. (a) Synthetic jets, showing an increased mixing of air which results in a decrease of effective thermal boundary layer thickness and associated increase in the heat transfer coefficient. (*From:* [74]. © 2007. *Electronics Cooling*. Reprinted with permission.) (b) Piezoelectric fans are utilized for enhancing cold air throughput of the heat sink and therefore reduction in the total thermal resistance. (*From:* [76]. © 2007. *Electronics Cooling*. Reprinted with permission from.)

through the heat sink exploits gas ion generation by emission from field-enhancing nanostructures, resulting in a microscale ion-driven airflow [77].

In complimentary efforts, exciting advances are being made in developing means for enhancing the convective heat transfer coefficient of air cooling. In particular, new twists on the general idea of gas-assisted evaporative cooling [78] have recently been described. In one approach, called perspiration nanopatch, enhanced evaporation from a capillary-confined thin liquid film subjected to a high velocity dry gas (air) streaming is exploited [79], potentially allowing for dissipation of heat fluxes approaching 300 W/cm^2 [54]. In another approach, the surface of the heat sink is modified with a sorption material and is cyclically exposed to cold/dry or warm/wet air streams, resulting in thermochemical (desorption-based) enhancement of total dissipated heat fluxes as compared to air cooling alone [80]. Yet another approach to air-cooling enhancement takes advantage of mist-impingement cooling with evaporation of ultrafine droplets generated by a liquid atomization array and delivered to the active (fin) heat transfer surface of the heat sink by the primary airflow supplied by the fan [81]. The planar atomization device is placed on top of the heat sink, thus effectively forming a capping lid, preventing parasitic air bypass of the heat sink finned area. Finally, it has been recently shown that forced convective liquid cooling can be utilized in combination with air cooling in a hybrid heat sink configuration to provide for synergetic heat removal at different rates from different domains of the microprocessor with the possibility of internal regeneration of the liquid coolant via heat exchange with air [82].

In summary, continuous progress in the optimal design and manufacture of advanced finned structures, in combination with recent innovations in enhancing air throughput, mixing, and convective heat transfer coefficient using active flow control, provide the basis for optimism about the bright future of air-cooled heat sinks as a thermal management solution for rapidly evolving electronic equipment and products requiring efficient heat dissipation.

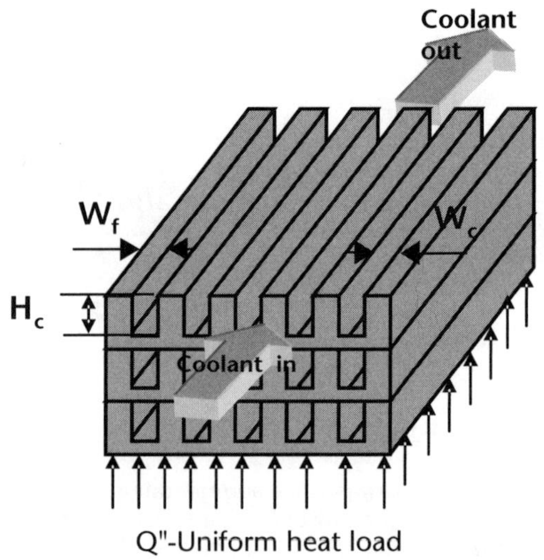

Figure 9.20 Three-dimensional stack of micro-channels. (*From:* [98]. © 2003 IEEE. Reprinted with permission.)

9.9 Microchannel Heat Sink Design

One of the promising liquid-cooling techniques is to attach a microchannel heat sink to, or directly fabricate microchannels on, the inactive side of the chip. Usually in a closed-loop arrangement, coolant such as water is pumped through the microchannels to take away the heat generation. In fully developed laminar flow, the nondimensional Nusselt number, Nu, is constant. Thus, the heat-transfer coefficient, h, is inversely proportional to the hydraulic diameter, as shown in (9.18), where d_H is the hydraulic diameter, and k is the thermal conductivity of the fluid:

$$Nu = \frac{h \cdot d_H}{k} = const \Rightarrow h \propto \frac{1}{d_H} \tag{9.18}$$

The benefit of microchannel heat sink is demonstrated in Tuckerman and Pease [53], who report a microchannel heat sink, consisting of parallel microflow passages 50 μm wide and 302 μm deep. Thermal resistance substantially lower than that of conventional heat sinks was demonstrated (0.09 Kcm^2/W with 2.14 bar pressure drop using water).

The successful design of microchannel heat sinks relies on understanding the fundamental characteristics of the flow and heat transfer inside microchannels. Following Tuckerman and Pease [53], several early studies [83–91] indicated that the flow and heat-transfer parameters deviate from the classical theory developed for macrosize channels, and the transition from laminar to turbulent flow occurs at a considerably smaller critical Reynolds number. However, there is no consensus on the trend of this deviation among different researchers. Some of the reported deviations may be attributed to experimental conditions, such as surface roughness and flow maldistribution. In contrast, results from more recent studies [92–97] appear to indicate that the pressure-drop and transport characteristics for macroscale channels also hold for the microchannels intended for microelectronics cooling. They are used consistently here, considering the practical channel sizes for microelectronic cooling.

9.9.1 Simple Model for Microchannel Heat Sink Design

This section describes the procedure for microchannel heat sink design [99]. A schematic of a stacked microchannel heat sink is shown in Figure 9.21. At each layer, a number of parallel microchannels are machined in the surface of a substrate (e.g., copper, silicon, or diamond). These layers are then bonded into a stacked micro heat sink. The benefits of using stacked microchannels are the significant reduction in pressure drop and temperature gradient when compared with a single-layered microchannel heat sink. It also allows tailored design of the fluid passage for potential spot cooling for highly nonuniform heat-density applications. Although the focus of this section is on stacked microchannels, the same procedure should be followed since the single-layered microchannel heat sink is simply a special case of the stacked microchannel heat sink.

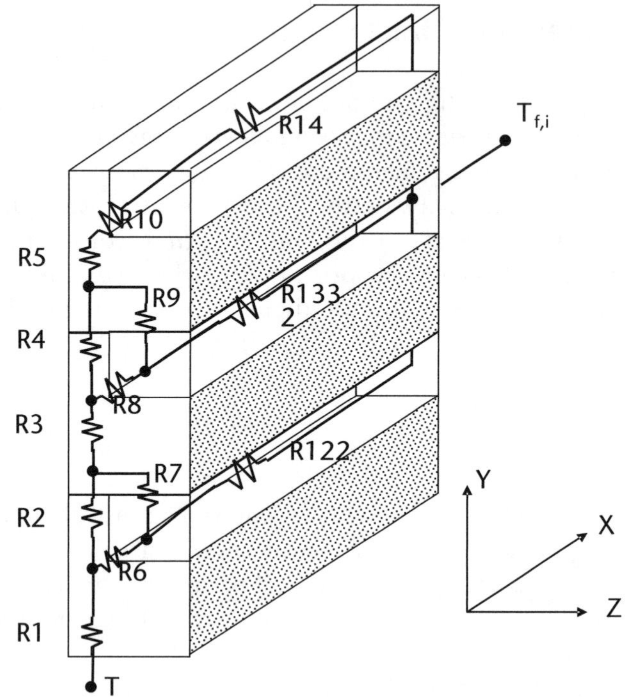

Figure 9.21 Thermal resistance network for a three-layered micro-channel stack. (From: [98]. © 2003 IEEE. Reprinted with permission.)

9.9.1.1 Correlations for Friction Factor and Nusselt Number [98]

The pressure drops associated with the stack structure include the contraction and expansion pressure drops in the inlet and outlet, respectively, the pressure drop due to 90° bend, and the flow friction loss. As pointed out by many researches [53, 99], the friction losses dominate the pressure drop in laminar flow in a rectangular duct. Therefore, only the friction loss is included here.

To consider the friction factor for developing laminar flow in a rectangular duct, a Churchill-Usagi asymptotic type of model shown in (9.19) to (9.21) is used [100]:

$$f_{app}\,\mathrm{Re}_{\sqrt{A_c}} = \left[\left(\frac{3.44}{\sqrt{y^+}}\right)^2 + \left(8\sqrt{\pi}G(\alpha)\right)^2\right]^{0.5} \tag{9.19}$$

$$G(\alpha) = \left[1.086957^{1-\alpha}\left(\sqrt{\alpha} - \alpha^{1.5}\right) + \alpha\right]^{-1} \tag{9.20}$$

$$y^+ = \frac{y}{\mathrm{Re}_{\sqrt{A_c}}\sqrt{A_c}} \tag{9.21}$$

In these equations, y is the length of the channel, and α is the aspect ratio or the inverse of the aspect ratio of the microchannel that is always smaller than one. It is noted here that the length scale for the Reynolds number and the dimensionless length is the square root of the channel cross-sectional area. Similarly, the Nusselt number for thermally developing conditions in a rectangular duct is given as [101]

$$Nu_{\sqrt{A_c}}\left(y^*\right) = \left[\left(\left(C1 \cdot C2\left(\frac{8\sqrt{\pi}G(\alpha)}{y^*}\right)^{\frac{1}{3}}\right)^5 + \left(C3\left(\frac{G(\alpha)}{\alpha^{\gamma}}\right)\right)^5\right)\right]^{0.2} \tag{9.22}$$

$$y^* = \frac{y}{\mathrm{Re}_{\sqrt{A_c}} \cdot \mathrm{Pr} \cdot \sqrt{A_c}} \tag{9.23}$$

where $C1 = 1$, for isoflux condition, $C2 = 0.501$, $C3 = 3.66$, and $\gamma = 0.1$.

9.9.1.2 Thermal Resistance Network [98]

A thermal-resistance analysis based on existing correlations for heat-transfer coefficients and friction factor in laminar channel flow was developed. Spreading and constriction resistance adapted from [99] are incorporated into the resistance network.

The thermal-resistance network for a three-layered microchannel stack is illustrated in Figure 9.13. All component resistances except R_2 and R_4 are listed in Table 9.3. To calculate these resistances, one-dimensional conduction is assumed. This assumption is reasonable as long as the aspect ratio is sufficiently large. In the base area, however, the aspect ratio is close to one. Two-dimensional heat conduction in that region is taken into account by considering constriction resistance for area contraction and spreading resistance for area expansion.
where

$$R_{cond} = \frac{t}{K_s WL} \tag{9.24}$$

$$R_{cont} = \frac{Ln\left[1/\sin\left(\frac{\pi}{2}\frac{1}{1+W_c/W_f}\right)\right]}{K_s WL} \tag{9.25}$$

$$R_{spread} = \frac{Ln\left[1/\sin\left(\frac{\pi}{2}\frac{1}{1+W_c/W_f}\right)\right]}{K_s WL} \tag{9.26}$$

$$R_{Conv,fin} = \frac{1}{h \cdot n \cdot L \cdot W_c + 2 \cdot h \cdot \eta_f \cdot n \cdot L \cdot H_c} \tag{9.27}$$

$$R_{Conv,c} = \frac{1}{h \cdot n \cdot L \cdot W_c} \tag{9.28}$$

Table 9.3 Component Resistances for the Network

$R_1 = R_{cond} + R_{cont}$	$R_6 = R_{conv,fin}$	$R_7 = R_{conv,c}$	$R_{12} = R_{bulk}$
$R_3 = R_{cond} + R_{cont} + R_{spread}$	$R_8 = R_{conv,fin}$	$R_9 = R_{conv,c}$	$R_{13} = R_{bulk}$
$R_5 = R_{cond} + R_{cont} + R_{spread}$	$R_{10} = R_{conv,fin}$		$R_{14} = R_{bulk}$

$$R_{bulk} = \frac{1}{\rho \cdot C_p \cdot v_m \cdot n \cdot H_c \cdot W_c} \tag{9.29}$$

The fin equation can be solved to get the two conduction resistances R_2 and R_4:

$$\frac{d^2\theta}{dx^2} = m^2 \cdot \theta, \quad 0 < x < H_c \tag{9.30}$$

where $\theta = T - T_f$, $m = \sqrt{2 \cdot h / k_s \cdot W_f}$.

Since the boundary condition for the fin equation is part of the solution, the solution procedure is iterative in nature. The simple model described in this section can be readily applied in an optimization algorithm to achieve optimum designs [98].

9.9.2 Conjugate Heat-Transfer and Sidewall Profile Effects

The simple model above provides a useful tool for first-order-type designs. A more detailed numerical model is needed to investigate the conjugate heat-transfer effects and the sidewall profile effects.

Conjugate heat transfer for two-layered microchannel with parallel and counterflow arrangements has been numerically modeled using a commercially available tool [102]. Half of the unit cell of the channel is considered in the model. Figure 9.22 displays the heat fluxes in kilowatts per square meter at the six solid-liquid interfaces for the parallel and counterflow conditions. For both cases, the flow rate is 1.38×10^{-6} m^3/s (83 mL/min). For parallel flow in Figure 9.22(a), the coolant flows along the positive x-direction in both layers. For counterflow in Figure 9.22(b), the coolant flows along the positive x-direction in the bottom layer and the negative x-direction in the top layer. For parallel flow, the heat flux at the bottom layer peaks at the entrances for each of the three interface walls. As the flow moves along x-axis, the boundary-layer thickness increases, and the heat flux decreases rapidly. The same trend was also reported in Fedorov and Viskanta [103] for a single-layered microchannel. However, the heat-flux pattern for the sidewall of the top layer is significantly different. Near the entrance, the heat flux drops initially because of the boundary-layer development. After a short distance, it starts increasing. Near the exit, it approaches a local maximum, after which the heat flux drops again.

For the counterflow configuration in Figure 9.22(b), the interesting feature is that near the outlets of the channels of both layers, the heat flux is negative such that heat is locally dissipated from the heated water to the surrounding cooler silicon. This is because the solid region near the outlets is also cooled by fresh coolant entering the inlets of the other layer. At low flow rates, the temperature of the heated fluid

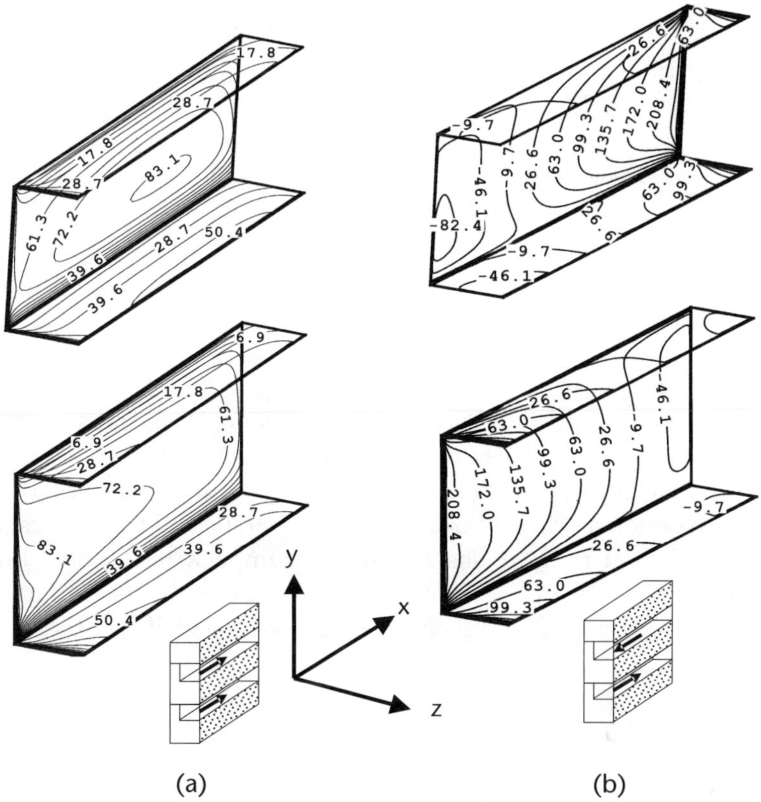

Figure 9.22 Heat flux at solid-liquid interfaces in kW/m^2 for a total flow rate of $1.38\text{x}10^{-6}$ m^3/s (83 ml/min). (a) Parallel flow, and (b) counter flow. Silicon is used for the microchannel while DI water is the working fluid. (*From:* [97]. © 2007. Reprinted with permission.)

is actually higher than the surrounding solid. An effective heat exchange between the hot fluid and cool fluid exists. This helps to smooth out the temperature gradient, but the effective heat-transfer area is also reduced. This has a negative effect on the overall thermal performance.

A second effect discussed here is the nonideality due to fabrication of the channel. Numerical modeling was conducted to look at the effects of the deviation from the vertical sidewall with a 5° sidewall slope [97]. Constant heat-flux condition was assumed at all four walls, a condition referred as H2 in Shah and London [104]. The longitudinal heat-transfer and Nusselt numbers are defined in (9.14) and (9.15). As shown in Table 9.4, compared with rectangular channels of the same flow area, the Nusselt number for a tapered channel is significantly lower. At fully developed condition, the degradation in the Nusselt number is about 60%. It is interesting that near the entrance, this degradation is not as significant because there the boundary-layer development just starts and the velocity and temperature profiles over the majority of the cross section are rather close to uniform.

$$\overline{h}_x = \frac{\overline{q_x''}}{\overline{T}_{W,x} - T_{m,x}} \tag{9.31}$$

Table 9.4 Local Nusselt Number for 5° Trapezoidal Channel and Rectangular Channel

$\frac{x}{d_H \cdot \mathrm{Re}\,\mathrm{Pr}}$	*0.09*	*0.18*	*0.45*	*0.85*	*Fully developed*
$Nu_{H2,x}$ rectangular	3.89	3.44	3.09	2.96	2.94
$Nu_{H2,x}$, 5° sidewall	2.7	2.09	1.56	1.35	1.21
Degradation in Nu	30%	40%	50%	54%	60%

$$Nu_{H2,x} = \frac{\bar{h}_x D_h}{k_f} \tag{9.32}$$

In summary, liquid-cooling microchannel heat sinks feature much enhanced heat-transfer coefficients due to reduced thermal boundary-layer thickness and increased heat-transfer area-to-volume ratio. Simple models described here can be used to conduct the first-order designs. Detailed 3D numerical models reveal that conjugate heat-transfer has a significant impact on the heat-flux distribution and overall efficiency of the heat-transfer, particularly for counterflow arrangement. Manufacturing defects such as tapered sidewall profile tend to reduce the heat-transfer efficiency as well.

9.10 Conclusion

In this chapter the thermal management challenges from the chip to the data-center level are introduced. The focus is then maintained on the state of the art of thermal management solutions available at the chip and package levels. Efficient heat removal requires consideration of the entire thermal-resistance chain and insuring that all resistances in this path are maintained as comparable and minimized simultaneously. This requires continuing attention to thermal interface materials, heat spreaders, and heat sinks in the current packaging schemes. The nonuniform heating within the die is becoming increasingly significant due to emerging trends such as continuing feature-size reduction and the move toward three-dimensional architectures. This may require incorporation of cooling ever-closer to the heat sources, in the form of direct liquid cooling, in order to be able to address the local hot spots. This will have a significant impact on the overall heat-removal chain. The use of substrate cooling is another scenario that will provide a new heat-removal path, altering the overall resistance chain significantly.

References

[1] Im, S., et al., "Scaling Analysis of Multilevel Interconnect Temperatures for High-Performance ICs," *IEEE Transactions on Electron Devices*, Vol. 52, No. 12, 2005, pp. 2710–2719.

[2] American Society of Heating, Refrigeration and Air-Conditioning Engineers (ASHRAE), *Datacom Equipment Power Trends and Cooling Applications*, Atlanta: ASHRAE, 2005.

[3] Greenberg, S., et al., "Best Practices for Data Centers: Lessons Learned from Benchmarking 22 Data Centers," *Proc. ACEEE Summer Study on Energy Efficiency in Buildings*, Asilomar, CA, August, 2006, pp. 76–87.

[4] Report to Congress on Server and Data Center Energy Efficiency Public Law 109-431, August 2, 2007, U.S. Environmental Protection Agency ENERGY STAR Program, pp. 67–87.

[5] Borkar, S., et al., "Parameter Variations and Impact on Circuits and Microarchitecture," *Proc. 40th Conference on Design Automation* (*DAC 2003*), Anaheim, CA, June 2–6, 2003, pp. 338–342.

[6] Ajami, A. H., K. Banerjee, and M. Pedram, "Modeling and Analysis of Nonuniform Substrate Temperature Effects on Global ULSI Interconnects," *IEEE Transactions on Computer-Aided Design of Integrated Circuits and Systems*, Vol. 24, No. 6, 2005, pp. 849–861.

[7] Sundaresan, K., and N. R. Mahapatra, "An Analysis of Timing Violations Due to Spatially Distributed Thermal Effects in Global Wires," *Proc. 44th Design Automation Conference* (*DAC 2007*), San Diego, California, June 4–8, 2007, pp. 515–520.

[8] Hua, H., et al., "Exploring Compromises among Timing, Power and Temperature in Three-Dimensional Integrated Circuits," *Proc. 43rd Design Automation Conference* (*DAC 2006*), San Francisco, California, June 24–28, 2006, pp. 997–1002.

[9] Link, G. M., and N. Vijaykrishnan, "Thermal Trends in Emerging Technologies," *Proc. 7th International Symposium on Quality Electronic Design* (*ISQED'06*), San Jose, CA, March 27–29, 2006, pp. 625–632.

[10] Parkhurst, J., J. Darringer, and B. Grundmann, "From Single Core to Multi-Core: Preparing for a New Exponential," *Proc. 2006 IEEE/ACM International Conference on Computer-Aided Design* (*ICCAD'06*), San Jose, CA, November 5–9, 2006, pp. 67–72.

[11] Pham, D., et al., "The Design Methodology and Implementation of a First-Generation CELL Processor: A Multi-Core SoC," *Proc. IEEE 2005 Custom Integrated Circuits Conference*, September 18–21, 2005, San Jose, CA, pp. 45–49.

[12] Mukherjee, R., and S. O. Memik, "Physical Aware Frequency Selection for Dynamic Thermal Management in Multi-Core Systems," *Proc. 2006 IEEE/ACM International Conference on Computer-Aided Design* (*ICCAD'06*), San Jose, CA, November 5–9, 2006, pp. 547–552.

[13] Monchiero, M., Canal, R., and A. Gonzalez, "Power/Performance/Thermal Design Space Exploration for Multicore Architectures," *IEEE Transactions on Parallel and Distributed Systems*, 2007, Vol. 19, No. 5, 2007, pp. 666–681.

[14] Rao, R., S. Vrudhula, and C. Chakrabarti, "Throughput of Multi-Core Processors under Thermal Constraints," *Proc. 2007 International Symposium on Low Power Electronics and Design* (*ISLPED'07*), Portland, OR, August 27–29, 2007, pp. 201–206.

[15] Gurrum, S., et al., "Thermal Issues in Next Generation Integrated Circuits," *IEEE Transactions on Device and Materials Reliability*, Vol. 4, No. 4, 2004, pp. 709–713.

[16] Calmidi, V., and I. Memis, "Thermal Enhancement of Systems Using Organic Flip-Chip Packages (FC-PBGA) with an Alternate Cooling Path through the Printed Wiring Board," *Proc. 56th Electronic Components and Technology Conference*, San Diego, CA, May 30–June 2, 2006, pp. 315–321.

[17] Dang, B., M. S. Bakir, and J. D. Meindl, "Integrated Thermal-Fluidic I/O Interconnects for an On-Chip Microchannel Heat Sink," *IEEE Electron Device Letters*, Vol. 27, No. 2, 2006, pp. 117–119.

[18] Schaper, L., et al., "Integrated System Development for 3-D VLSI," *Proc. 57th Electronic Components and Technology Conference*, Reno, NV, May 29–June 1, 2007, pp. 853–857.

[19] Wego, A., S. Richter, and L. Pagel, "Fluidic Microsystems Based on Printed Circuit Board Technology," *J. Micromechanics and Microengineering*, Vol. 11, 2001, pp. 528–531.

[20] Chason, M., "Organic Electronics for Large Area Electronic Devices," *Materials Research Society Symposium Proceedings*, Vol. 769, 2003, pp. H3.1.1–H3.1.9.

[21] Prasher, R. S., "Rheology Based Modeling and Design of Particle Laden Polymeric Thermal Interface Materials," *IEEE Transactions on Components and Packaging Technologies*, Vol. 28, No. 2, 2005, pp. 230–271.

[22] Gowda, A., et al., "Micron and Sub-Micron Scale Characterization of Interfaces in Thermal Interface Material Systems," *9th Intersociety Conference on Thermal and Thermomechanical Phenomena in Electronic Systems*, Vol. 2, 2004, pp. 556–563.

[23] Hu, X., L. Jiang, and K. E. Goodson, "Impact of Wall Region Particle Volume Fraction Distribution on Thermal Resistance of Particle Filled Thermal Interface Materials," *19th Annual IEEE Semiconductor Thermal Measurement and Management Symposium*, San Jose, CA, USA, March 11–13, 2003, pp. 106–111.

[24] Mahajan, R., C.-P. Chiu, and R. Prasher, "Thermal Interface Materials: A Brief Review of Design Characteristics and Materials," *Electronics Cooling*, Vol. 10, No. 1, February 2004.

[25] Stern, M. B., et al., "Evaluation of High Performance Thermal Greases for CPU Package Cooling Applications," *IEEE 21st Annual Semiconductor Thermal Measurement and Management Symposium*, San Jose, CA, USA, March 15–17, 2005, pp. 39–43.

[26] Singh, P., et al., "A Power, Packaging, and Cooling Overview of the IBM eServer z900," *IBM J. Research and Development*, Vol. 46, No.2, 2002, pp. 711–738.

[27] Stern, M., D. Kearns, and B. Ong, "Adhesion of Thermal Interface Materials for CPU Heat Sinks, an Overlooked Issue," *Electronics Cooling*, Vol. 13, No. 1, February 2007, pp. 28–29.

[28] Knickerbocker, J. U., et al., "An Advanced Multichip Module (MCM) for High-Performance UNIX Servers," *IBM J. Research and Development*, Vol. 46, No.6, 2002, pp. 779–804.

[29] Dal, S., "Degradation Mechanisms of Siloxane-Based Thermal Interface Materials under Reliability Stress Conditions," *IEEE 42nd Annual International Reliability Physics Symposium*, Phoenix, AZ, April 25–29, 2004. pp. 537–542.

[30] Hua, F., and C. Deppisch, "Solder as Thermal Interface Material for High Power Devices," *SMTA Journal,* Vol. 9, No. 1, 2006, pp. 21–26.

[31] Deppisch, C., et al., "The Material Optimization and Reliability Characterization of an Indium-Solder Thermal Interface Material for CPU Packaging," *JOM,* Vol. 58, No. 6, 2006, pp. 67–74.

[32] Saraswati, R., and F. J. Polese, "Aluminum Matrix Composite Heat Sinks for Microchips and Microcircuits," *Proc. SPIE—The International Society for Optical Engineering*, Vol. 3582, 1999, pp. 681–686.

[33] Moores, K. A., Y. K. Joshi, and G. H. Schiroky, "Thermal Characterization of a Liquid Cooled AlSiC Base Plate with Integral Pin Fins," *IEEE Transactions on Components and Packaging Technologies*, Vol. 24, No. 2, 2001, pp. 213–219.

[34] Hammel, E., et al., "Silicon Substrates with Microwhisker Structure—An Innovative Heat Spreader Technology for Power Electronics Application," *Proc. SPIE—The International Society for Optical Engineering*, Vol. 3906, 2001, pp. 474–479.

[35] Klett, J., et al., "Heat Exchangers Based on High Thermal Conductivity Foams," *Proc. 1st World Conference on Carbon*, Berlin, Germany, July 9–15, 2000, p. 244.

[36] Eisele, H., and G. I. Haddad, "GaAs TUNNETT Diodes on Diamond Heat Sinks for 100 GHz and Above," *Microwave Theory and Techniques*, Vol. 43, 1995, pp. 210–213.

[37] Fabis, P. M., and E. Windischmann, "Thermal Management Enhancement of GaAs Devices Using CVD Diamond Heat Spreaders in a Plastic Package Environment," *J. Electronic Packaging*, Vol. 122, 2000, pp. 92–97.

[38] Jagannadham, K., "Multilayer Diamond Heat Spreaders for Electronic Power Devices," *Solid State Electronics*, Vol. 42, 1998, pp. 2199–2208.

[39] Yoganand, S. N., et al., "Integrated AlN/Diamond Heat Spreaders for Silicon Device Processing," *J. Vacuum Science and Technology A*, Vol. 20, 2002, pp. 1974–1982.

[40] Basiulis, A., H. Tanzer, and S. McCabe, "Thermal Management of High Power PWBs through the Use of Heat Pipe Substrates," *6th Annual International Electronic Packaging Conference*, San Diego, California, November 1986, pp. 501–515.

[41] Peterson, G. P., "Modeling, Fabrication and Testing of Micro Heat Pipes: An Update," *Applied Mechanics Review*, Vol. 49, November 1996, pp. 175–183.

[42] Adkins, D. R., et al., "Silicon Heat Pipes for Cooling Electronics," *1st Annual Spacecraft Thermal Control Symposium*, Albuquerque, New Mexico, November 16–18, 1994.

[43] Benson, D. A., et al., "Micro-Machined Heat Pipes in Silicon MCM Substrates," *IEEE Multi-Chip Module Conference*, Santa Cruz, California, February 1996, pp. 127–129.

[44] Hopkins, R., A. Faghri, and D. Khrustalev, "Flat Miniature Heat Pipes with Micro Capillary Grooves," *J. Heat Transfer*, Vol. 121, 1999, pp. 102–109.

[45] Park, J. S., et al., "Flat Micro Heat Pipe Arrays for Cooling and Thermal Management at the Package," *Proc. SPIE—The International Society for Optical Engineering*, Vol. 4408, April 2001, pp. 424–429.

[46] Take, K., and R. L. Webb, "Thermal Performance of an Integrated Plate Heat Pipe with a Heat Spreader," *J. Electronic Packaging*, Vol. 123, 2001, pp. 189–195.

[47] Faghri, A., *Heat Pipe Science and Technology*, New York: Taylor & Francis, 1995.

[48] Murthy, S., Y. Joshi, and W. Nakayama, "Single Chamber Compact Two-Phase Heat Spreaders with Micro-Fabricated Boiling Enhancement Structures," *IEEE Trans. on Components and Packaging Technologies*, March 2002, Vol. 25, pp. 156–163.

[49] Murthy, S., Y. Joshi, and W. Nakayama, "Orientation Independent Two-Phase Heat Spreaders for Space Constrained Applications," *Microelectronics Journal,* Vol. 25, 2003, pp. 1187–1193.

[50] Murthy, S., Y. Joshi, and W. Nakayama, "Two-Phase Heat Spreaders Utilizing Microfabricated Boiling Enhancement Structures," *Heat Transfer Engineering*, Vol. 25, 2004, pp. 26–36.

[51] Lasance, C., and R. E. Simons, "Advances in High-Performance Cooling for Electronics," *Electronics Cooling*, Vol. 11, No. 4, November 2005.

[52] Lasance, C., "Technical Data Column," *Electronics Cooling*, Vol. 3, No. 1, January 1997.

[53] Tuckerman, D. B., and R. F. W. Pease, "High-Performance Heat Sinking for VLSI," *IEEE Electron Device Letters*, Vol. EDL-2, No. 5, 1981, pp. 126–129.

[54] Narayanan, S., A. Fedorov, and Y. Joshi, "Perspiration Nanopatch for Hot Spot Thermal Management," *Proc. InterPack 2007*, Vancouver, British Columbia, Canada, July 8–12, 2007, CD proceedings.

[55] Gurrum, S., et al., "Thermal Issues in Next Generation Integrated Circuits," *IEEE Transactions on Device and Materials Reliability*, Vol. 4, No. 4, 2004, pp. 709–713.

[56] Ellsworth, M. J., "Technical Brief: Comparing Liquid Coolants from Both a Thermal and Hydraulic Perspective," *Electronics Cooling*, Vol. 12, No. 3, August 2006.

[57] Simons, R. E., "Comparing Heat Transfer Rates of Liquid Coolants Using the Mouromtseff Number," *Electronics Cooling*, Vol. 12, No. 2, May 2006.

[58] Yeh, L.-T., and R. C. Chu, *Thermal Management of Electronic Equipment*, New York, NY: ASME Press, 2002, pp. 261–265.

[59] Kulkarni, D. P., P. K. Namburu, and D. K. Das, "Comparison of Heat Transfer Rates of Different Nanofluids on the Basis of the Mouromtseff Number," *Electronics Cooling*, Vol. 13, No. 3, August 2007.

[60] Nowak, E. J., "Maintaining the Benefits of CMOS scaling When Scaling Bogs Down," *IBM J. Res. Dev.*, Vol. 46, No. 2/3, pp. 169–180, 2002.

[61] Naeemi, A., et al., "The Urgency of Deep Sub-Ambient Cooling for Gigascale Integration," *Proc. Int. Conf. on Integrated Circuit Design and Technology ICICDT05*, Austin, Texas, May 9–11, 2005, CD Proceedings.

[62] Suman, S., Y. Joshi, and A. Fedorov, "Cryogenic/Sub-Ambient Cooling of Electronics: Revisited," *Proc. ITherm 2004*, Las Vegas, Nevada, June 1–14, 2004, CD Proceedings.

[63] Wadell, R., Y. Joshi, and A. Fedorov, "Experimental Investigation of Compact Evaporators for Ultralow Temperature Refrigeration of Microprocessors," *ASME J. Electronic Packaging*, Vol. 129, No. 3, 2007, pp. 291–299.

[64] Webb, R. L., and N.-H. Kim, "Advances in Air-Cooled Heat Exchanger Technology," *J. Enhanced Heat Transfer*, Vol. 14, No. 1, 2007.

[65] Iyengar, M., and A. Bar-Cohen, "Design for Manufacturability of SISE Parallel Plate Forced Convection Heat Sinks," *IEEE Transactions on Components and Packaging Technologies*, Vol. 24, No. 2, 2001, pp. 150–158.

[66] Poulikakos, D., and A. Bejan, "Fin Geometry for Minimum Entropy Generation in Forced Convection," *ASME J. Heat Transfer*, Vol. 104, 1982, pp. 616–623.

[67] Bar-Cohen, A., and M. Iyengar, "Design and Optimization of Air-Cooled Heat Sinks for Sustainable Development," *IEEE Transactions on Components and Packaging Technologies*, Vol. 25, No. 4, 2002, pp. 584–591.

[68] Simons, R. E., "Calculation Corner: Estimating the Effect of Flow Bypass on Parallel Plate-Fin Heat Sink Performance," *Electronics Cooling*, Vol. 10, No. 1, February 2004.

[69] Marthinuss, J., and G. Hall, "Air Cooled Compact Heat Exchanger Design for Electronic Cooling," *Electronics Cooling*, Vol. 10, No. 1, February 2004.

[70] Rodgers, P., V. Eveloy, and M. Pecht, "Limits of Air-Cooling: Status and Challenges," *Proc. 21st IEEE SEMI-THERM Symposium*, San Jose, California, March 14–17, 2005, pp. 116–124.

[71] Nakayama, W., "Exploring the Limits of Air Cooling," *Electronics Cooling*, Vol. 12, No. 3, August 2006.

[72] Kraus, A. D., and A. Bar-Cohen, *Design and Analysis of Heat Sinks,* New York: John Wiley and Sons, 1997.

[73] Allen, M. G., and A. Glezer, "Synthetic Jet Actuators for Cooling Heated Bodies and Environments," U.S. Parent 6123145, 2000.

[74] Mahalingam, R., et al., "Synthetic Jets for Forced Air Cooling of Electronics," *Electronics Cooling*, Vol. 13, No. 2, May 2007.

[75] Nelson, R. D., "Heat Exchanger Having Piezoelectric Fan Means," U.S. Patent 4923000, 1990.

[76] Sauciuc, I., "Piezo Actuators for Electronic Cooling," *Electronics Cooling*, Vol. 13, No. 1, February 2007.

[77] Schlitz, D. J., S. V. Garimella, and T. S. Fisher, "Ion-Driven Air Pump Device and Method," US Patent App., 2004.

[78] Bar-Cohen, A., G. Sherwood, and M. Hodes, "Gas-Assisted Evaporative Cooling of High Density Electronic Modules," *Proc. I-THERM IV: Concurrent Engineering and Thermal Phenomena*, Washington, DC, May 4–7, 1994, pp. 32–40.

[79] Fedorov, A. G., "Nano-Patch Thermal Management Devices, Methods, and Systems", U.S. Patent App. 11/748,540 (2007).

[80] Launay, S., A. G. Fedorov, and Y. Joshi, "Thermal Management Devices, Systems, and Methods," U.S. Patent App. 11/867,070, 2007.

[81] Fedorov, A., and J. M. Meacham, "Evaporation-Enhanced, Dynamically-Adaptive Air (Gas)-Cooled Heat Sink for Thermal Management of High Heat Dissipation Devices," *ITherm 2008*, Orlando, Florida, May 28–31, 2008, pp. 333–341.

[82] Green, C., A. Fedorov, and Y. Joshi, "Fluid-to-Fluid Spot-to-Spreader (F^2/S^2) Hybrid Heat Sink for Integrated Chip-Level and Hotspot-Level Thermal Management," *ITherm 2008*, Orlando, Florida, May 28–31, 2008, pp. 510–520.

[83] Wu, P., and W. A. Little, "Measurement of Friction Factors for the Flow of Gases in Very Fine Channels Used for Microminiature Joule-Thompson Refrigerators," *Cryogenics*, Vol. 23, No. 5, 1983, pp. 273–277.

[84] Pfahler, J., et al., "Gas and Liquid Flow in Small Channels," in *Micromechanical Sensors, Actuators and Systems*, edited by S. B. Choi et al., ASME DSC 32, New York: American Society of Mechanical Engineers, 1991, pp. 49–60.

[85] Peng, X. F., G. P. Peterson, and B. X. Wang, "Frictional Flow Characteristics of Water Flowing Through Rectangular Microchannels," *Experimental Heat Transfer*, Vol. 7, 1994, pp. 249–264.

[86] Peng, X. F., G. P. Peterson, and B. X. Wang, "Heat Transfer Characteristics of Water Flowing through Rectangular Microchannels," *Experimental Heat Transfer*, Vol. 7, 1994, pp. 265–283.

[87] Papautsky, I., et al., "Laminar Fluid Behavior in Microchannels Using Micropolar Fluid Theory," *Sensors and Actuators*, Vol. 73, 1999, pp. 101–108.

[88] Harms, T. M., M. J. Kazmierczak, and F. M. Gerner, "Developing Convective Heat Transfer in Deep Rectangular Microchannels," *Int. J. Heat and Fluid Flow*, Vol. 210, 1999, pp. 149–157.

[89] Mala, G. M., and D. Li, "Flow Characteristics of Water in Microtubes," *Int. J. Heat and Fluid Flow,* Vol. 20, 1999, pp. 142–148.

[90] Mala, G. M., et al., "Flow Characteristics of Water Through Microchannels Between Two Parallel Plates with Electrokinetic Effects," *Int. J. Heat and Fluid Flow,* Vol. 18, 1997, pp. 489–496.

[91] Tso, C. P., and S. P. Mahulikar, "The Use of the Brinkman Number for Single Phase Forced Convective Heat Transfer in Microchannels," *Int. J. Heat and Mass Transfer*, Vol. 41, No. 12, 1998, pp. 1759–1769.

[92] Xu, B., K. T. Ooi, and N. T. Wong, "Experimental Investigation of Flow Friction for Liquid Flow in Microchannels," *International Communications in Heat and Mass Transfer*, Vol. 27, 2000, pp. 1165–1176.

[93] Qu, W., and I. Mudawar, "Experimental and Numerical Study of Pressure Drop and Heat Transfer in a Single-Phase Micro-Channel Heat sink," *Int. J. Heat Mass Transfer*, Vol. 45, 2002, 2549–2565.

[94] Liu, D., and S. V. Garimella, "Investigation of Liquid Flow in Microchannels," *8th AIAA/ASME Joint Thermophysics and Heat Transfer Conference*, St. Louis, Missouri, Paper No. AIAA 2002-2776, 2002, pp. 1–10.

[95] Lee, P., and S. V. Garimella, "Experimental Investigation of Heat Transfer in Microchannels," *ASME Summer Heat Transfer Conference,* Las Vegas, Nevada, Paper No. HT2003-47293, 2003, pp. 1–7, 2003.

[96] Kohl, M. J., et al., "An Experimental Investigation of Microchannel Flow with Internal Pressure Measurements," *Int. J. Heat Mass Transfer*, Vol. 48, 2005, pp. 1518–1533.

[97] Wei, X., and Y. Joshi, "Experimental and Numerical Study of Sidewall Profile Effects on Flow and Heat transfer inside Microchannels," *Int. J. Heat Mass Transfer*, Vol. 50, 2007, pp. 4640–4651.

[98] Wei, X., and Y. Joshi, "Optimization Study of Stacked Micro-Channel Heat Sinks for Micro-Electronic Cooling," *IEEE Transactions on Components and Packaging Technologies*, Vol. 26, No. 1, 2003, pp. 55–61.

[99] Phillips, R. J., "Micro-Channel Heat Sinks," *Advances in Thermal Modeling of Electronic Components and Systems*, edited by A. Bar-Cohen and A. D. Kraus, Vol. 2, 1990, New York, ASME Press, pp. 109–184.

[100] Muzychka, Y. S., and M. M. Yovanovich, "Modeling Friction Factors in Non-Circular Ducts for Developing Laminar Flow," *2nd AIAA Theoretical Fluid Mechanics Meeting,* Paper No. 98-2492, Albuquerque, NM., June 15–18, 1998.

[101] Muzychka, Y. S., and M. M. Yovanovich, "Modeling Nusselt Numbers for Thermally Developing Laminar Flow in Non-Circular Ducts for," *7th AIAA/ASME Joint Thermophysics and Heat Transfer Conference,* Paper No. 98-2586, Albuquerque, NM., June 15–18, 1998.

[102] Wei, X., Y. Joshi, and M. K. Patterson, "Experimental and Numerical Study of a Stacked Microchannel Heat Sink for Liquid Cooling of Microelectronic Devices," *J. Heat Transfer*, Vol 129, No.10, October 2007, pp. 1432–1444.

[103] Fedorov, A. G., and R. Viskanta, "Three-Dimensional Conjugate Heat Transfer in the Micro-Channel Heat Sink for Electronic Packaging," *Int. J. Heat Mass Transfer*, Vol. 43, 2000, pp. 399–415.

[104] Shah, R. K., and A. L. London, *Laminar Flow Forced Convection in Ducts*, New York: Academic Press, 1978, p. 200.

[105] International Technology Roadmap for Semiconductors, 2005 and 2006 editions:, www.itrs.net/Links/2005ITRS/Home2005.htm, www.itrs.net/Links/2006Update/.

CHAPTER 10

Active Microfluidic Cooling of Integrated Circuits

Carlos H. Hidrovo and Kenneth E. Goodson

10.1 Introduction

The thermal management of high heat fluxes is a critical roadblock in the way of higher-performance microelectronics. The ongoing reduction in microtransistor size translates into heat fluxes comparable to those encountered in nuclear reactions and rocket nozzles, but under much severe temperature constraints. Although the average heat flux may remain in the vicinity of 100 to 300 W/cm^2, the peak heat flux at localized hot spots may approach or exceed 1 kW/cm^2. In order to handle these heat fluxes properly at lower operating temperatures, integrated cooling technologies that remove heat closer to the source are required. Microfluidic cooling in a heat sink built within or attached directly to the silicon chip is one route to address this problem. Three primary microfluidic technologies have the potential to accommodate this very large heat flux. These include microjet impingement, spray cooling, and microchannel heat sinks [1]. Microjet impingement relies on the use of a high-speed liquid jet that emerges from a nozzle to reduce thermal boundary-layer thicknesses and increase convection coefficients at the incident surface. Despite being one of the leading cooling technologies, microjet impingement has serious drawbacks in terms of fluid recovery for open systems and temperature uniformity. Therefore, intricate architectures using multiple microjets and specially designed outlet ports are needed. This leads to designs that are highly optimized for very specific thermal operating conditions and are therefore not robust to time and spatially varying thermal loads. Similar to microjet impingement, spray cooling relies on the impingement of liquid onto the heated surface. Spray cooling relies on a blanket of liquid droplets rather than a continuous liquid jet striking the heated source, providing better surface coverage and temperature uniformity. Furthermore, rather than relying on the convective transport capabilities of the thin liquid film formed when the droplets hit the surface, spray cooling is achieved through the evaporation of the aforementioned film. The latent heat associated with the phase change translates into lower surface-temperature requirements, another advantage over microjet impingement cooling. This makes spray cooling one of the leading contenders in the race to achieve heat-flux removal values exceeding 1 kW/cm^2. However, it has serious drawbacks in terms of the pressures required to achieve proper spray droplet breakup and the distribution needed to exploit the cooling potential of this technology fully.

Microchannel heat sinks have received much attention from the research community over nearly three decades. The seminal paper by Tuckerman and Pease [2] demonstrated a microchannel heat sink capable of removing 790 W/cm^2 with a temperature rise of 71°C using water at a flow rate of 516 mL/min and just over 2 atm of required pressure head. Their work has been followed by a number of studies of single- and two-phase microchannel flows, including those by Phillips et al. [3], Samalam [4], Peng and Wang [5], Bowers and Mudawar [6], Peles et al. [7], Lin et al. [8], Qu and Mudawar [9], and Zhang et al. [10], as well as several comprehensive review articles on the topic [11–14]. As the semiconductor industry and research community move toward the realization of 3D IC, packaging constraints become the primary consideration when choosing a cooling technology that can allow seamless integration with these novel architectures. Both microjet impingement and spray cooling require stacking of several fabrication layers for the creation of the nozzle structures and to achieve the required separation for proper jet and spray formation. For this reason, and given the maturity of this technology, this chapter focuses on microchannel heat sink cooling since it is at the forefront of implementation feasibility for 3D IC applications. In Section 10.2 we discuss the fundamentals of single-phase flow and convective heat transfer. We then proceed to delineate the most salient features of two-phase flow boiling, as they pertain to microchannel cooling, in Section 10.3. Section 10.4 is devoted to the topic of fluid flow and convection modeling in microchannels, paying particular attention to the two-phase flow boiling regime, where, for the most part, existing macroscale models are not applicable. The substantial pumping requirements associated with microchannel flow, in the context of cooling performance, are covered in Section 10.5. Intrinsically related to the large pressure drops associated with microchannel cooling is the design of optimal and novel microchannels network architectures that can lessen the effect of this detrimental characteristic. This is explored in Section 10.6, particularly from the perspective of 3D IC. Section 10.7 concludes the chapter with an outlook on future and promising technologies that might prove key in the successful development and commercialization of next generation microchannel cooling systems.

Before proceeding, a clarification note: we will employ the term *microchannel* to indistinguishable designate channels with characteristic cross-sectional length scales on the order of micrometers, regardless of geometry. It is not uncommon to associate the term microchannel instinctively with a rectangular cross section since this is the most prevalent type of microchannel geometry. However, depending on the fabrication procedures and processes, one can attain other types of cross-sectional geometries, including semicircular and triangular, just to mention two.

10.2 Single-Phase Flow Cooling

The original microchannel heat sink concept introduced by Tuckerman and Pease used single-phase liquid water. In this section we discuss the fundamentals behind single-phase flow convection. Continuum theory can be used to describe the flow of liquids, such as water and other commonly employed refrigerants (water is generally chosen as the working fluid in microchannel heat sink cooling systems due to its superb thermal properties and benign environmental characteristics), and their behavior in mini- and microchannels [15]. This is not necessarily the case for gases, where the Knudsen number (Kn, a dimensionless number defined as the ratio of the molecular mean free path to a representative physical length scale) can approach values above one depending on the microchannel characteristic dimension. However, for microchannels with values of the hydraulic diameter greater than 10 μm, air and water-vapor behavior can be treated by the classical continuum theory near atmospheric conditions. As such, we will omit treatment of rarefaction effects in single-phase gas flows and concentrate instead on liquid flows, only noting that for rarefied gas flows, the Nusselt number (Nu) and therefore the heat-transfer coefficient are reduced when compared to their continuum counterparts [16].

Any discussion of single-phase-flow forced-convection cooling needs to start with the fundamental fluid mechanics principles governing the flow. In what follows, we restrict our treatment to circular cross section microchannels. Although most cooling microchannels have rectangular cross sections, the fundamental flow physics and relevant concepts between the two geometries are unequivocally equivalent. The relative simplicity of the flow solutions for the circular cross section microchannel favors its use for the illustrative purposes of this section.

10.2.1 Laminar Flow Fundamentals

The simplest and most basic form of internal flow is that of incompressible, fully developed, steady-state, single-phase laminar flow through an axisymmetric uniform circular duct for a Newtonian fluid. Solution of the Navier-Stokes and continuity equations for this type of flow gives rise to the well-known Hagen-Poiseuille parabolic velocity profile (Figure 10.1).

$$u = -\frac{1}{4\mu}\frac{dp}{dx}\left(R^2 - r^2\right) \tag{10.1}$$

The existence of an implicit analytical solution for the velocity profile as a function of pressure gradient and duct geometry lends itself to a simple formulation relating pressure drop and flow rate in a circular tube, the quantities of interest for the designer. Integration of the velocity profile over the area of the duct gives rise to the following equation relating pressure gradient and volumetric flow rate:

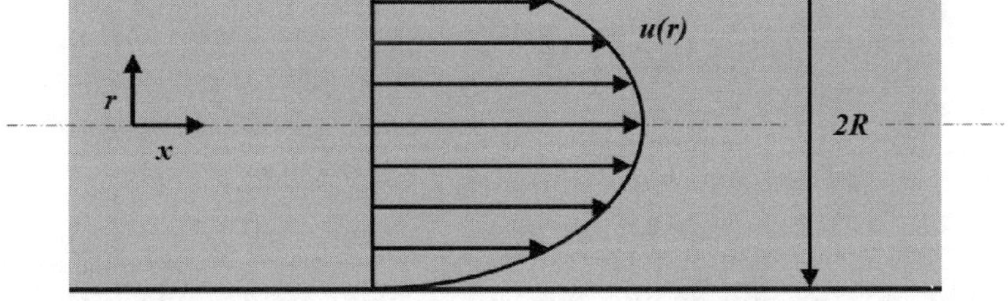

Figure 10.1 Parabolic velocity profile for fully developed laminar flow in a circular cross section tube. The pressure drop balances the shear stress at the wall, which is constant for fully developed flow and proportional to the velocity gradient at the wall.

$$-\frac{dp}{dx} = \frac{8\mu Q}{\pi R^4} = \frac{8\mu u_{avg}}{R^2} \quad (10.2)$$

For a duct of length L, we have

$$\Delta p = \frac{8\mu QL}{\pi R^4} = \frac{8\mu u_{avg} L}{R^2} \quad (10.3)$$

However, most internal flows do not lend themselves to implicit analytical velocity profiles and corresponding simple formulations for the relationship of pressure drop to volume flow rate. In these instances an alternative formulation more practical for design purposes must be used. One such approach employs the friction coefficient, or Fanning friction factor, which is defined as the ratio of the wall shear stress to the dynamic pressure:

$$C_f = \frac{\tau_w}{(1/2)\rho u_{avg}^2} \quad (10.4)$$

A force balance on an infinitesimal circular fluid element of length dx requires that the force due to pressure difference be opposed by an equal but opposite direction shear-stress force acting against the flow. This can be explained using

$$\pi R^2 dp = 2\pi R dx \tau_w \quad (10.5)$$

and

$$\frac{dp}{dx} = \frac{2\tau_w}{R} \quad (10.6)$$

Again, for a duct of length L, we have

$$\Delta p = \frac{2\tau_w L}{R} \quad (10.7)$$

Substituting (10.4) into (10.7), we have

$$\Delta p = \frac{2C_f (1/2)\rho u_{avg}^2 L}{R} = \frac{C_f \rho u_{avg}^2 L}{R} \quad (10.8)$$

Comparing (10.3) and (10.8), we have

$$C_f = \frac{8\mu}{\rho u_{avg} R} = \frac{16\mu}{\rho u_{avg} D} = \frac{16}{\mathrm{Re}_D} \quad (10.9)$$

Thus, the Fanning friction factor is inversely proportional to the flow Re through a proportionality constant, in this case 16, which is dependent on the flow-velocity profile. In the case of incompressible, fully developed, steady-state, single-phase laminar flows for Newtonian fluids, it is only dependent on flow-channel geometry. Thus, (10.9) can be generalized for any arbitrary channel geometry as

$$C_f = \frac{Po}{\mathrm{Re}_{D_h}} \tag{10.10}$$

where the circular tube diameter D has been replaced by the hydraulic diameter D_h, which is given by

$$D_h = \frac{4A_c}{P_w} \tag{10.11}$$

The constant C_fRe product is known as the Poiseuille number (Po), and as stated before, it is only a function of channel geometry. Po correlations and values for different channel geometries can be found in authoritative books, such as that from Shah and London [17] (for rectangular cross sections it ranges from 14.23, for a square duct with aspect ratio of one, to 24, in the limiting case of two infinite parallel plates or aspect ratio of zero). The generalized pressure-drop equation can now be stated as

$$\Delta p = \frac{2C_f \rho u_{avg}^2 L}{D_h} \tag{10.12}$$

It is important to explore the underlying implications that these pressure correlations have. Equations (10.10) and (10.12) can be rearranged to provide a general correlation between pressure drop and volumetric flow rate:

$$\Delta p = \frac{8Po\mu QL}{D_h^3 P_w} \tag{10.13}$$

If we compare (10.3) and (10.13), it is apparent that for a given fluid flow rate, the pressure drop is inversely proportional to the fourth power of the channel characteristic length scale (this is somewhat softened toward a third power inverse dependence for very high-aspect-ratio channels by the wetted perimeter term). This has detrimental implications for microchannel flow as the required pressure drops can be quite substantial.

10.2.2 Entrance Effects: Developing Flow and Sudden Contraction and Expansion

The previous section treatment is applicable only to fully developed flows where the velocity profiles remain constant and are not a function of axial position. In reality, internal flows require a finite length of duct before reaching fully developed conditions. As the flow enters the channel, the velocity profile is in constant change as the effects of the wall are propagated into the bulk of the flow (Figure 10.2). Under the assumption of uniform velocity conditions at the entrance, the hydrodynamic entrance length is given by

$$\frac{L_h}{D_h} = 0.05\,\text{Re} \tag{10.14}$$

Depending on flow conditions and overall channel length, the entrance effects can be quite substantial and need to be accounted for. This can be particularly true for microchannels, whose lengths can be relatively short to lessen the toll on pressure drop already imposed by their cross-sectional area, as discussed before. Pressure-drop effects due to the entrance can be accounted for using an apparent Fanning friction factor, which embodies both frictional losses in the entrance and fully developed regions [18]:

$$\Delta p = \frac{2C_{f,app}\rho u_{avg}^2 x}{D_h} \tag{10.15}$$

This equation evaluates the pressure drop as a function of position x from the entrance of the channel. The difference between apparent and fully developed friction factor over a distance x can be expressed in terms of a pressure defect or incremental pressure-drop number, given by

$$K(x) = \left(C_{f,app} - C_f\right)\frac{4x}{D_h} \tag{10.16}$$

$K(x)$ accounts for the additional pressure drop due to momentum change and accumulated increment in wall shear stress in the developing region (Figure 10.2). Therefore, the total pressure drop can be expressed as

$$\Delta p = \frac{2Po\mu u_{avg}\, x}{D_h^2} + K(x)\frac{\rho u_{avg}^2}{2} \tag{10.17}$$

For $x \geq L_h$, the incremental pressure defect takes on a constant value of $K(\infty)$. Shah and London [17] postulated the following equation for $K(x^+)$ in circular tubes:

$$K(x^+) = 13.76(x^+)^{1/2} + \frac{1.25 + 64x^+ - 13.76(x^+)^{1/2}}{1 + 0.00021(x^+)^{-2}} - 64x^+ \tag{10.18}$$

where x has been replaced by the dimensionless axial distance x^+ given by

$$x^+ = \frac{x}{D_h\,\text{Re}} \tag{10.19}$$

Chen [19] proposed the following correlation for $K(\infty)$ in circular tubes

$$K(\infty) = 1.20 + \frac{38}{\text{Re}} \tag{10.20}$$

For further pressure-defect correlations applicable to different channel geometries, the reader is again referred to Shah and London [17].

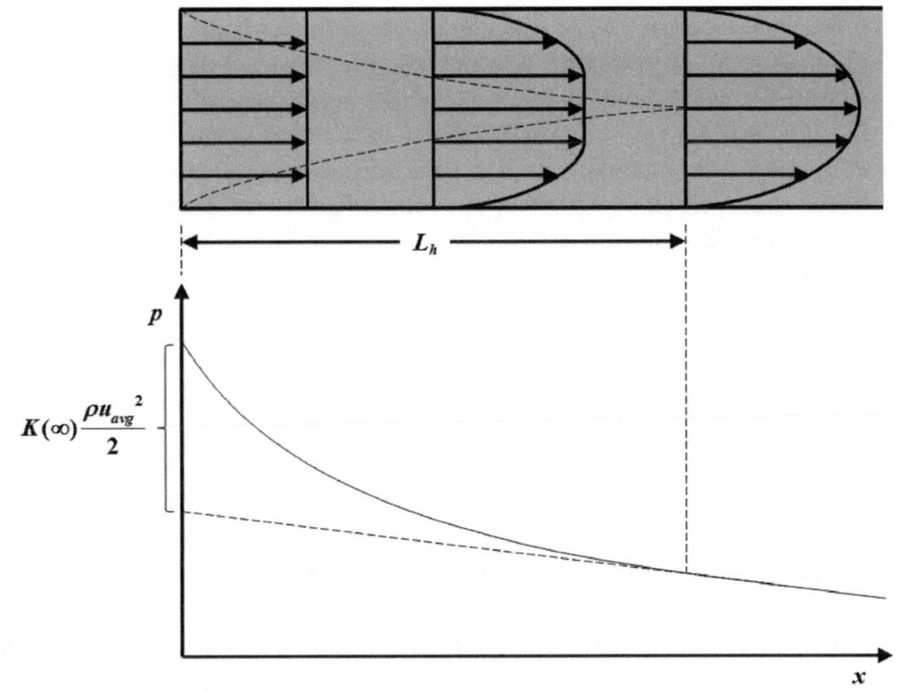

Figure 10.2 Developing velocity profile and pressure drop in the entry region of a circular cross section channel. For $x > L_h$, the flow is fully developed, and the pressure drop is linear. In the developing region, the pressure drop is nonlinear due to momentum change and accumulated increment in wall shear stress. Both of these effects are captured in terms of a pressure defect, $K(x)$, defined as the difference between the actual pressure drop and the fully developed equivalent pressure drop for a given length.

In addition to entrance-length effects, it is also important to consider other effects related to inlet and outlet ancillaries. These are called "minor losses" and can include the sudden contraction usually encountered at the microchannel inlet, the sudden expansion usually encountered at the microchannel outlet, and bends encountered in the system. All these can be accounted for through the use of loss coefficients K_c, K_e, and K_{90}, respectively, such that the overall pressure drop encountered in a microchannel heat sink system can be expressed as [18]:

$$\Delta p = \frac{\rho u_m^2}{2}\left[\left(A_c / A_p\right)^2 (2K_{90}) + (K_c + K_e) + \frac{4C_{f,app}L}{D_h}\right] \tag{10.21}$$

or equivalently

$$\Delta p = \frac{\rho u_m^2}{2}\left[\left(A_c / A_p\right)^2 (2K_{90}) + (K_c + K_e) + \frac{4C_f L}{D_h} + K(x)\right] \tag{10.22}$$

10.2.3 Turbulent Flow

Turbulence yields chaotic and stochastic flow behavior. Turbulent flow is characterized by rapid variations (locally and globally) in velocity and pressure in both space and time. It arises under conditions in which flow inertial forces are much larger than viscous forces and therefore dominate. The relative prevalence of these two forces is quantified through the Reynolds number (Re). For internal flows, transition from laminar to turbulent behavior occurs at Re ~ 2,300.

Due to the chaotic nature of the velocity field in turbulent flows, analytical or even numerical calculation of a generalized expression for the friction factor is nearly impossible. It is widely agreed, however, that the correlation derived by Prandtl in 1935 provides a good approximation of the Darcy friction factor for fully developed turbulent flow in a smooth circular channel [20]:

$$\frac{1}{\sqrt{f}} = 2.0\log\left(\text{Re}\sqrt{f}\right) - 0.8 \tag{10.23}$$

where the Darcy friction factor is given by

$$f = 4C_f \tag{10.24}$$

Other expressions exist that account for both the developing and fully developed regions and noncircular geometries [18, 20]. In addition, pressure drop in turbulent flow in ducts is greatly influenced by the roughness of the channel walls. Several correlations exist that take surface roughness effects into account through the use of a relative roughness term, ε/D_h (where ε is the absolute surface roughness), in the friction-factor formulation. Among them is the Colebrook equation, which provides a good representation of the Darcy friction factor for turbulent flow in circular tubes as depicted in the widely used Moody chart [20]:

$$\frac{1}{\sqrt{f}} = -2.0\log\left(\frac{\varepsilon / D}{3.7} + \frac{2.51}{\text{Re}\sqrt{f}}\right) \tag{10.25}$$

10.2.4 Steady-State Convective Heat-Transfer Equations: Constant Heat Flux and Constant-Temperature Boundary Conditions

We start the analysis of internal single-phase heat-transfer flow by introducing two key concepts. The first one is the thermal entry length. Steady-state, fully developed thermal conditions refer to those conditions in which the flow thermal (nondimensional temperature) profile is not a function of axial distance along the tube. Internal flows require a finite length of duct before reaching fully developed thermal conditions. As the flow enters the channel, the temperature profile is in constant change as the thermal effects of the wall are propagated into the bulk of the flow. Under the assumptions of uniform temperature with fully developed velocity-profile conditions at the entrance, the thermal entrance length is given by

$$\frac{L_t}{D_h} = 0.05\,\text{Re}\,\text{Pr} \tag{10.26}$$

where Pr compares viscous (momentum) diffusivity to thermal diffusivity and is therefore a ratio of how quickly the momentum (velocity) boundary layer develops in relationship to the thermal (temperature) boundary layer.

The second key concept is that of the mixed mean bulk-flow temperature. Instead of representing the mean or average value for the spatial temperature profile, the mixed mean bulk-flow temperature is defined in terms of the thermal energy transported by the fluid as it moves past a cross section of tube. As such, it is not only dependent on the temperature profile but also on the velocity profile, as the energy transport is a function of temperature (energy measure) and mass flow rate (flow advection measure). The mixed mean bulk-flow temperature T_m is defined as

$$T_m = \frac{\int_{A_c} \rho u c_v T dA_c}{\dot{m} c_v} = \frac{\int_{A_c} \rho u T dA_c}{\dot{m}} \text{ for constant } c_v \tag{10.27}$$

As can be seen from (10.27), the mixed mean temperature is the temperature that provides the same amount of energy transport by advection under a uniform temperature field as that transported by the actual flow with its velocity and temperature profiles.

The mixed mean temperature is used as the reference temperature for internal flows in Newton's law of cooling, which relates heat transfer to a convective heat-transfer coefficient and two reference temperatures. Thus,

$$q_s'' = h(T_s - T_m) \tag{10.28}$$

Due to their nature, internal flows are well suited to a fixed control volume, open system thermodynamic energy balance analysis. The difference in advection energy transport between inlet and outlet must be equal to the heat transfer and shaft work done on the fluid. Since there is no shaft work in simple pipe flow, a basic energy balance yields

$$q_{conv} = \dot{m} c_p (T_{m,o} - T_{m,i}) \tag{10.29}$$

Equation (10.29) is a general and powerful expression that applies to all internal heat-transfer flows irrespective of thermal or fluid flow conditions. An exception arises for incompressible flows when the pressure gradient is extremely large, in which case (10.29) is modified as

$$q_{conv} = \dot{m}\left[c_v (T_{m,o} - T_{m,i}) + \frac{p_o - p_i}{\rho}\right] \tag{10.30}$$

However, for the purposes of this chapter and in most microchannel cooling applications, (10.29) is valid. By casting and combining (10.28) and (10.29) in differential form, the following differential equation for the mean mixed temperature behavior as a function of axial location is obtained:

$$dq_{conv} = \dot{m} c_p dT_m = q_s'' P dx = h(T_s - T_m) P dx \tag{10.31}$$

$$\frac{dT_m}{dx} = \frac{h(T_s - T_m) P}{\dot{m} c_p} \tag{10.32}$$

Equation (10.32) provides the framework for the two fundamental types of internal flow convective heat transfer: constant surface heat flux and constant surface temperature.

Under constant surface heat flux, we can rewrite (10.32) as

$$\frac{dT_m}{dx} = \frac{q_s''P}{\dot{m}c_p} \neq f(x) \text{ (i.e., not a function of } x\text{)} \tag{10.33}$$

and therefore

$$T_m(x) = T_{m,i} + \frac{q_s''P}{\dot{m}c_p} x \tag{10.34}$$

In other words, under constant surface heat flux, the bulk mixed mean temperature increases linearly as a function of axial location. This applies irrespective of whether we have fully developed conditions or not.

For constant surface-temperature conditions, derivation of the bulk mixed mean temperature dependence on axial location is slightly more involved but still straightforward, leading to the following result [21]:

$$\frac{T_s - T_m(x)}{T_s - T_{m,i}} = \exp\left(-\frac{Px}{\dot{m}c_p}\bar{h}\right) \tag{10.35}$$

where

$$\bar{h} = \frac{1}{L}\int_0^L h(x)dx \tag{10.36}$$

is the average heat-transfer coefficient over the tube length.

Equation (10.35) depicts an exponential behavior of T_m as it tends toward a limiting value of T_s. Just as in the case of (10.34), (10.35) is a general equation that applies to all internal flows under constant surface temperature, irrespective of other flow conditions. However, there is still specificity related to each particular flow in terms of the heat-transfer coefficient, which influences the exponential behavior of the fluid mixed mean temperature. The same can be said for the case of constant surface heat flux. Although (10.34), and thus the flow mixed mean temperature, is independent of the convective heat-transfer coefficient, the surface temperature on the other hand is directly dependent on the value and behavior of the convective heat-transfer coefficient. The dependence of the convective heat-transfer coefficient on flow conditions is clearly illustrated by looking at its definition along with Fourier's law of heat conduction:

$$q_s'' = h(T_s - T_{char}) = -k_f \left.\frac{\partial T}{\partial n}\right|_{n=0} \tag{10.37}$$

$$h = \frac{-k_f \left.\frac{\partial T}{\partial n}\right|_{n=0}}{(T_s - T_{char})} \tag{10.38}$$

where T_{char} is the characteristic temperature in Newton's law of cooling (free stream temperature in external flows and mixed mean temperature in internal flows), and n is the direction normal to the surface. Equation (10.38) is a general equation that relates the convective heat-transfer coefficient to the heat conduction in the flow at the surface. Since the heat conduction in the flow at the surface is dependent on the flow temperature profile, the convective heat-transfer coefficient is therefore dependent on the flow thermal conditions.

For circular tubes with laminar flow under fully developed thermal and velocity-profile conditions, it can be shown that for the case of constant surface heat flux, the convective heat-transfer coefficient is equal to [21]

$$h = \frac{48}{11}\left(\frac{k_f}{D}\right) \tag{10.39}$$

It is usually more customary to express convective heat transfer in terms of the nondimensional Nusselt (Nu) number, which compares heat transfer by fluid convection to heat transfer by fluid thermal diffusion (equivalently, it can be thought of as a dimensionless temperature profile within the flow). In this case,

$$Nu = \frac{hD}{k_f} = 4.36 \tag{10.40}$$

A more elaborate analysis produces a similar result for the case of constant wall temperature, for which [21]

$$Nu = \frac{hD}{k_f} = 3.66 \tag{10.41}$$

Calculating convective heat-transfer coefficients under entry-length conditions is a more complicated problem, and two cases need to be considered. In the first case, we have fully developed velocity-profile conditions with a developing temperature profile. This is akin to starting the heat transfer after an unheated section beyond the tube inlet once the velocity profile has reached fully developed conditions. It is also representative of the flow of fluids with large Pr, where the velocity-profile entry length is much smaller than the thermal entry length. The second case considers concurrent momentum and thermal entry-length conditions, with both the velocity and temperature profiles simultaneously changing. Here, we present average Nu correlations for the two cases described above under constant surface-temperature conditions, given that the average convective heat-transfer coefficient is a required parameter in (10.35):

$$\overline{Nu} = 3.66 + \frac{0.0668(D/L)\mathrm{Re}\,\mathrm{Pr}}{1 + 0.04\left[(D/L)\mathrm{Re}\,\mathrm{Pr}\right]^{2/3}} \tag{10.42}$$

$$\overline{Nu} = 1.86\left(\frac{\text{Re}\,\text{Pr}}{L/D}\right)^{1/3}\left(\frac{\mu}{\mu_s}\right)^{0.14} \tag{10.43}$$

where μ_s is the viscosity of the fluid at the surface temperature. All other properties in (10.42) and (10.43) are evaluated at the average value of the mean temperature, $T_m = (T_{m,i} + T_{m,o})/2$. Turbulence increases convective heat transfer due to the higher momentum transport associated with this type of flow regime. Several heat-transfer correlations exist for turbulent flow. The reader is referred to the treatise by Shah and London [17] and the textbook by Incropera et al. [21]. Here, we present two of the most widely used:

$$Nu = 0.023\text{Re}^{4/5}\,\text{Pr}^{n} \tag{10.44}$$

where $n = 0.4$ for heating ($T_s > T_m$) and 0.3 for cooling ($T_s < T_m$).

$$Nu = 0.027\text{Re}^{4/5}\,\text{Pr}^{1/3}\left(\frac{\mu}{\mu_s}\right) \tag{10.45}$$

It is important to realize that most turbulent-based heat-transfer correlations are intrinsically empirical in nature and therefore are only applicable for very specific sets of conditions. For example, different correlations must be used under the same flow conditions, depending on whether there is heating or cooling of the flow, what the temperature difference between the wall and fluid is, and most important of all, what the level of turbulence is as characterized by the Re.

10.3 Two-Phase Convection in Microchannels

Two-phase flow cooling provides significant advantages over single-phase cooling in terms of heat-transfer rates and lower cooling temperatures. By taking advantage of the latent heat of phase change, boiling in microchannels can significantly increase the amount of heat removed from electronic components while sustaining a lower temperature, namely the saturation temperature. Two-phase flow boiling provides an enticing approach to achieve high heat-flux transfer rates on the order of 1 kW/cm^2.

Despite the potential heat-transfer benefits associated with two-phase flow cooling, several practical difficulties have prevented its use in actual microscale heat exchangers. Most of these difficulties are associated with the unstable nature of microchannel flow boiling at these scales.

10.3.1 Boiling Instabilities

Locally, some of the most stable boiling regimes present in macroscale flow are nonexistent in microscale flows. Namely, the incipient bubble boiling regime very common in macroscale duct flow is not possible in microchannel boiling. The small microchannel dimensions preclude the existence of multiple individual bubbles within the actual bulk flow. The same can be said for other macroscale flow boiling regimes, such as churn flow. Instead, the number of microscale flow boiling regimes is very limited with the formation of bubbles leading almost instantaneously to an elongated vapor plug and thin, annular film regime. Formation of independent bubbles within the bulk flow is possible but only under very low heat loads and high-volume flow regimes, where the potential advantages of flow phase change are not fully exploited. The inherent nature of these flow regimes leads to very unstable behavior with the microchannel flow transition changing stochastically from single-phase liquid flow to metastable thin, annular liquid flow and, ultimately, to burnout and dryout conditions in which the flow has essentially transitioned to a single-phase vapor flow with low convective heat-transfer capabilities.

This local flow-instability behavior can propagate into a global instability behavior, with large pressure fluctuations and almost instantaneous microchannel flow transitions in multichannel heat-exchanger systems. Qu and Mudawar [22] reported on boiling flow instabilities arising in parallel mini- and microchannels arrangements and previously reported by Kandlikar et al. [23] and Hetstroni et al. [24], classifying them into severe pressure-drop oscillations and mild parallel-channel instabilities. They concluded that severe pressure-drop oscillation was the result of interaction between the vapor generation in the microchannels and the compressible volume in the flow loop upstream of the heat sink. The parallel-channel instability only produced mild pressure fluctuations and was the result of density wave oscillation within each channel and feedback interaction between channels.

Peles et al. [25] developed and used a simplified, one-dimensional model of flow with a flat evaporation front dividing the liquid and vapor phases, along with experiments conducted on 16-mm-long, parallel triangular microchannels ranging in size from 50 to 200 μm to study the behavior of boiling two-phase flow in microchannel heat sinks. They concluded that the evaporating mechanism in two-phase flow in microchannels was considerably different from that observed in their macroscale counterparts. As stated previously, they observed that the most prevalent and characteristic flow regime in microchannels consisted of two distinct phase domains, one for the liquid and another for the vapor. A very short (on the order of the hydraulic diameter) section of two-phase mixture existed between the two. As such, they argued, the outlet vapor mass quality for a steady-state flow could only take on the values of zero (single-phase liquid flow) or unity (saturated or supersaturated vapor). Since the energy required for an outlet of quality zero is much larger than that for an outlet of quality one, an energy gap exists between those two levels, for which steady, evaporating two-phase flow is precluded in these microscale systems.

Peles et al.'s [25] approach looks at the instability problem from an energy barrier perspective. The heat-flux input is the driving force that can take the system from one state to the other and over the energy hump. Heat fluxes that bestow the system with energy levels in the gap region lead to instability. This is somewhat in contrast to the approach by Qu and Mudawar [22], who focus on the resistive-capacitive behavior of the upstream loop section as the key parameter for instability. The two approaches are complementary since it is the interaction between the vapor generation and the fluidic system characteristics that leads to the oscillations. The heat flux drives the evaporation fronts in both directions, therefore increasing the backward pressure of the system. The backward pressure leads to an increase in the pressure in the upstream loop section, which then feeds back into the microchannel section and leads to a forward pressure push that can lead to expulsion of the evaporating front. Since the heat flux remains fixed at the gap energy level, the expulsion of the evaporating front leads to an unstable condition.

Xu et al. [26] studied parallel multichannel instability in a heat sink consisting of 26 rectangular microchannels 300 μm in width and 800 μm in depth. They found that the onset of flow instability (OFI) occurred under an outlet temperature ranging between 93°C and 96°C, several degrees below the saturation temperature of 100°C corresponding to the exit pressure conditions. They also identified three types of oscillations: large-amplitude/long-period oscillations (LALPOs), small-amplitude/short-period oscillations (SASPOs), and thermal oscillations (TOs). Chang and Pan [27] also conducted work on two-phase flow instability in a microchannel heat sink consisting of 15 parallel microchannels of rectangular cross section 100 μm in width and 70 μm in depth. They identified two different two-phase flow patterns under stable or unstable conditions. For the stable two-phase flow oscillations, bubble nucleation, slug flow, and annular flow appeared sequentially in the flow direction (Figure 10.3). For the unstable case, forward or reversed slug or annular flows appeared alternatively in every channel. Intermittent reversed flow of the two-phase mixture to the inlet chamber was observed (Figure 10.4). They also found that the pressure-drop oscillations could be used as an index of the appearance of reversed flow. Pressure fluctuations above 6 kPa would lead to flow instability with reversed flow to the inlet chamber.

Despite their similarities, single-microchannel and parallel-multimicrochannel instabilities are inherently different in nature. Single-microchannel instabilities arise primarily due to interactions between pressure fluctuations in the upstream flow delivery systems and the rapid and explosive nature of the phase change during boiling in the microchannel. This resistive-capacitive oscillation is also known as a Ledinegg instability. One way to suppress the onset and lessen the fluctuations of Ledinegg instabilities is to increase the flow resistance upstream of the heat sink, thereby reducing the upstream propagation of the backpressure effects from the sudden vapor generation in the microchannels. On the other hand, parallel-multimicrochannel instability is primarily characterized by a rapid and random (in appearance) redistribution of the flow among the different microchannels. This flow redistribution is the result of the uniform pressure condition across the microchannels, along with the increase in the flow resistance in those microchannels undergoing boiling: flow increases in the low-resistance microchannels, while at the same time it decreases in the high-resistance ones (boiling microchannels) in order to maintain pressure equalization among them.

10.3.2 Pressure Drop and Heat-Transfer Coefficient

The two key parameters in the design of conventional convective-based cooling systems are the pressure drop and the convective heat-transfer coefficient. The same holds true for microchannel heat sinks. The pressure drop is one of the inputs (the other being volume flow rate) needed to assess the pumping-power requirements of the cooling system. Likewise, the convective heat-transfer coefficient provides a measurement of the cooling effectiveness of the system.

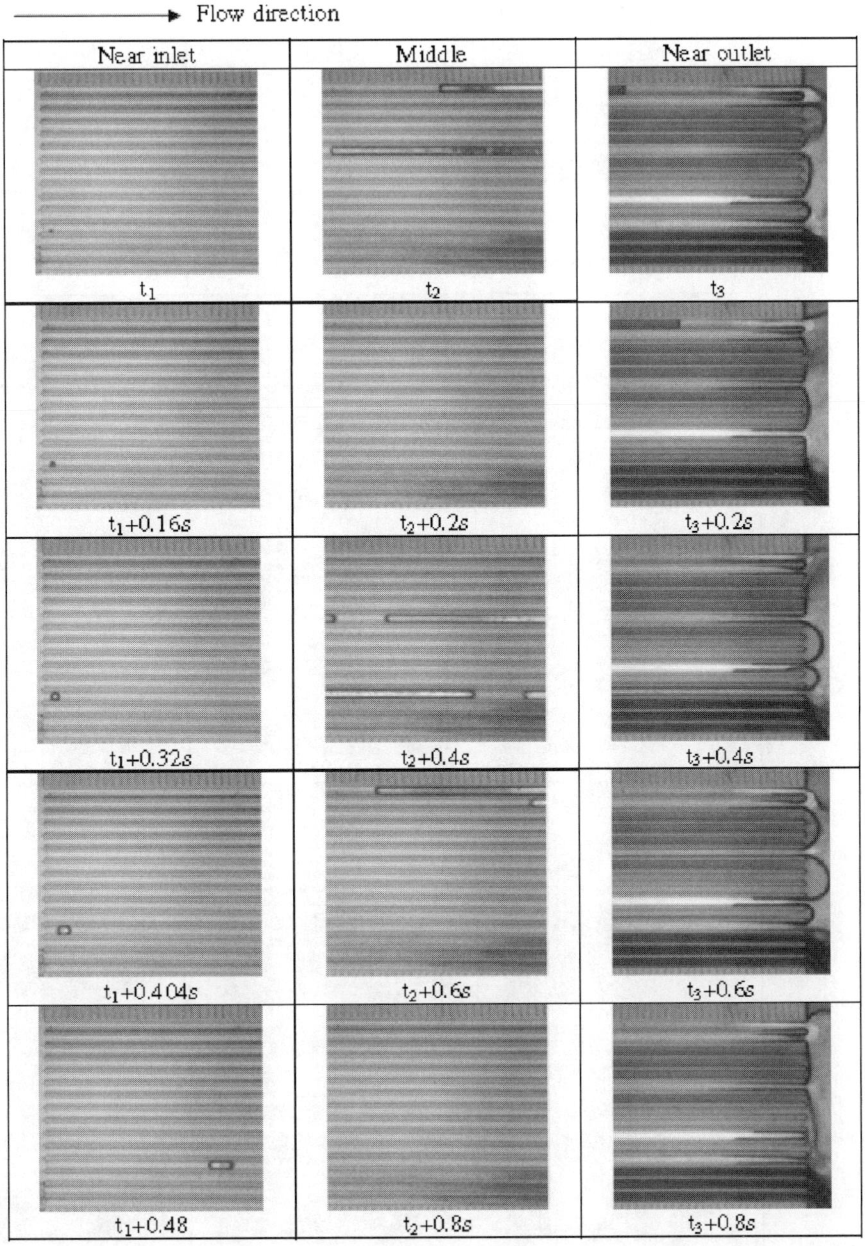

Figure 10.3 Evolution of two-phase flow patterns in the entrance, middle, and exit regions of a heat sink consisting of 15 parallel microchannels of rectangular cross section 100 μm in width and 70 μm in depth. In this stable two-phase flow, oscillations, bubble nucleation, slug flow, and annular flow appear sequentially in the flow direction [27]. (© 2007 Elsevier. Reprinted with permission.)

The characteristics of pressure drop in microchannel two-phase flows are very peculiar. In addition to the frictional component, two-phase-flow pressure drop is characterized by the presence of the acceleration component. As the liquid phase changes into vapor, there is a sudden decrease in fluid density accompanied by an increase in fluid volume. In order to maintain the prescribed mass flow rate, the lighter fluid must be accelerated, leading to an increase in pressure drop: additional

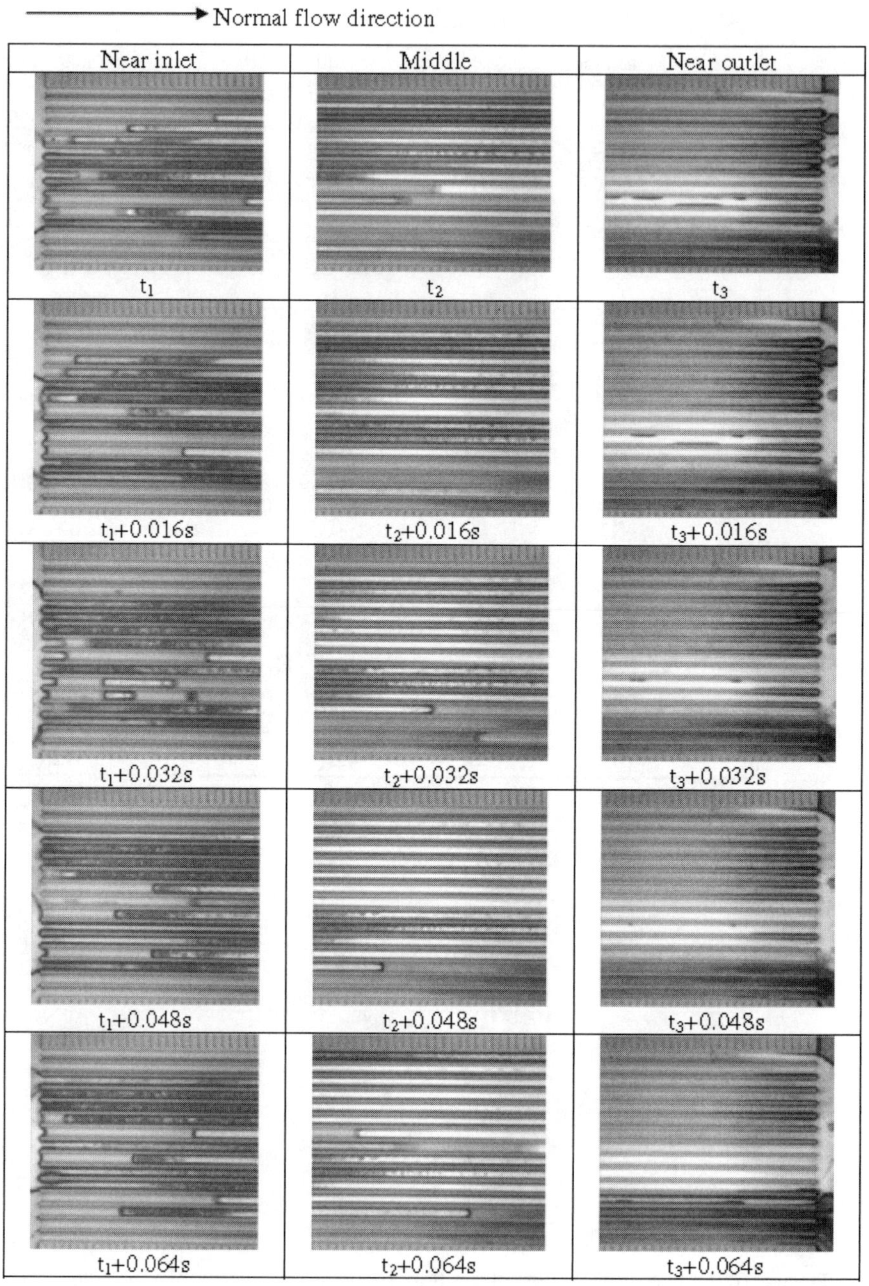

Figure 10.4 Evolution of two-phase flow patterns in the entrance, middle, and exit regions of a heat sink consisting of 15 parallel microchannels of rectangular cross section 100 μm in width and 70 μm in depth. In this unstable case, forward or reversed slug or annular flows appear alternatively in every channel. Intermittent reversed flow of the two-phase mixture to the inlet chamber is also observed [27]. (© 2007 Elsevier. Reprinted with permission.)

work must be done in order to accelerate the vapor to the required velocity. It is interesting to note that due to the higher kinematic viscosity of vapor as compared to liquid water, for a particular mass flow rate, the required pressure drop for vapor is higher than that for liquid water. Although this seems to fly in the face of or to contradict the stated lower-mass flow rates required for microchannel boiling cooling systems, it must be recalled that the benefits of boiling two-phase flow are associated with the latent heat of phase change and not with the use of single-phase vapor as the coolant.

Several studies exist on the topic of pressure drop in two-phase flows in microchannels. Most of them are experimental in nature and focus on the establishment of relationships between flow regime and pressure drop. They generally revolve around the use of an experimental sample retrofitted for optical access. Pressure-drop measurements between inlet and outlet are supplemented by white-light visualization studies of the different flow structures.

Koo et al. [28] highlighted the importance of pressure drop in the performance of microchannel heat sinks. It is demonstrated in this study how the wall-temperature distribution is governed in part by the coupling between the pressure drop and the saturation temperature and how it influences the overall performance of the microchannel heat sink. Employing a homogeneous two-phase flow model developed in an earlier work [10, 29], they investigated the effect that a one-dimensionally varying heat flux had on the temperature field. They found that the most advantageous configuration was to apply most of the heat to the latter part of the two-phase microchannel heat sink. Under this spatial arrangement, the temperature increases in the liquid phase region due to sensible heating are minimized by limiting the heat input in this upstream section. Lower temperature management is also achieved by placing the high-heat-input section downstream, where flow boiling occurs. In this two-phase region, fluid temperature is limited to the saturation value by the latent heat. As the flow pressure decreases in this section as it approaches the exit value, so does the fluid saturation temperature and, hence, the wall temperature. Interestingly enough, under this arrangement, the highest wall temperature is not located at the higher heat-flux region but rather near the inlet, which is the lower heat-flux region. Based on these results, Koo et al. [28] concluded that the pressure drop is the most critical factor in the design of microchannel heat sinks and that careful optimization should be performed in order to minimize pressure drop along the microchannels (higher pressures translate into higher saturation temperatures).

The number of heat-transfer studies is less prevalent, probably due in part to the difficulties associated with accurately measuring the heat-transfer coefficient. Properly measuring the convective heat-transfer coefficient requires knowledge of the heat transferred to the fluid in addition to the surface and bulk fluid temperatures. Measurement of the actual amount of heat transferred to the fluid is a difficult task. In most cases, a resistive heater is used as the heat source, and the total heat generation can be calculated from the joule heating equation. However, not all of the heat is convected by the microchannel flow, and environmental loses need to be accounted for. These include primarily heat losses by natural convection to the environment. Measuring the local wall and bulk-flow temperature is also an extremely difficult task. Local wall-temperature measurements can be achieved through the use of temperature sensors, such as microthermocouples or integrated resistor temperature detectors (RTDs), which can be incorporated into the fabrication of the test samples, such as those developed and employed by Zhang et al. [10]. RTDs rely on the dependence of resistivity on temperature. Zhang et al. [10] used microfabricated beam suspended silicon microchannels to investigate two-phase flow boiling behavior under constant heat-flux boundary conditions. RTDs and heaters were incorporated into the back side of the silicon beam by ion implantation, allowing for distributed temperature measurements and heating (Figure 10.5). By using deep reactive ion etching (DRIE), rectangular microchannels with hydraulic diameters in the range of 25 to 60 μm were fabricated on a suspended silicon bridge reducing the channel wall thickness and effectively preventing conduction heat losses. Both homogeneous and separated (annular) flow models were developed and validated against pressure and temperature measurements carried out in the microchannels sample. Measuring the local bulk-flow temperature is extremely difficult. Optical techniques such as fluorescence can be used for these purposes, but even in this case, measurements are normally qualitative at best. Under very specific instances and for very specific flow regimes, the convective heat-transfer coefficient can be inferred from the morphology of the flow structure. This is the case for the stratified and annular flow regimes, for which the convective heat-transfer coefficient can be determined from the thickness of the liquid film surrounding the heated wall. For this particular flow regime, the convective heat-transfer coefficient can be calculated from the following simple expression [30]:

$$h = \frac{k_f}{\delta} \tag{10.46}$$

The convective heat-transfer coefficient in two-phase flow is a very dynamic, if not outright unstable, quantity due to the unsteady nature of this type of flow. As can be seen from (10.46), the thinner the film thickness is, the higher the heat-transfer coefficient. However, thin liquid-film thickness can quickly lead to burnout or dryout, where the liquid film has completely disappeared, leaving only a vapor core filling the microchannel. Under these conditions, the flow has become single phase again, but with a much lower conductivity, namely, that of the vapor.

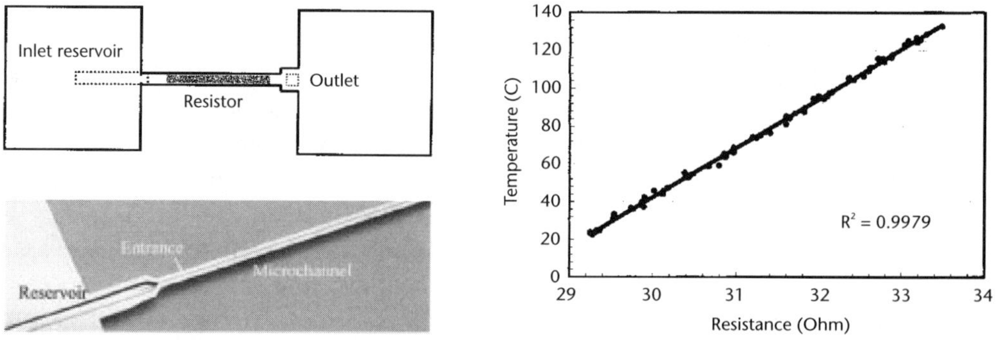

(a) (b)

Figure 10.5 (a) Microfabricated beam suspended silicon microchannels are used to investigate two-phase flow boiling behavior under constant heat-flux boundary conditions. Resistance temperature detectors (RTDs) and heaters were incorporated in the back side of the silicon beam by ion implantation, allowing for distributed temperature measurements and heating. (b) An RTD calibration curve shows the linear dependence between resistance and temperature [10]. (© 2002 IEEE. Reprinted with permission.)

10.4 Modeling

Unlike single-phase flow cooling, where macroscale correlations and models are still applicable, microchannel two-phase flow cooling requires the development of very specialized and specific models. The wealth of two-phase and boiling flow models available for macroscale systems have very limited applicability in microchannel systems. Surface tension has drastically different effects in microscale two-phase flow relative to its macroscale counterpart. Although surface tension is relevant in macroscale two-phase flow due to its influence in flow-transition criteria and bubble nucleation dimensions, its impact in microscale two-phase flow is significantly different and arguably more important. In microchannel two-phase flow, the characteristic bubble nucleation dimensions are comparable to the hydraulic diameter, leading to a cross section expansion confinement that abruptly redirects bubble growth to the axial direction. Capturing these confinement interactions is fundamental to the development of microscale boiling and phase-change flow models. Likewise, the gravitational effects included in macroscale models become almost irrelevant in microscale models. Determining whether these effects should be included in a particular model is important because it can greatly reduce the amount of computational complexity, allowing for better usage of the computational budget toward more relevant phenomena.

The two basic approaches to two-phase modeling are homogeneous flow modeling and separate flow modeling. As their names imply, homogeneous flow modeling assumes that the two phases intermix in a homogeneous configuration with identical properties. In a separate flow model, each phase is treated individually, while accounting for interactions between the two. Intermediate approaches can be considered a hybrid configuration of the two, but a clear distinction can be made in terms of the number of fundamental conservation equations used. Single mass, momentum, and energy equations are usually the norm in homogeneous flow models. Any implementation of more than one conservation equation for any of the three fundamental quantities of interest should be considered a separated flow model.

10.4.1 Homogeneous Flow Modeling

Homogeneous flow models rely on the assumption that flow behavior can be modeled by assuming that the bulk flow consists of a homogeneous intermix of the phases involved with equal property values between the two phases. Fundamentally, the velocity and temperature between the phases is considered the same. The homogeneous flow properties are normally characterized by the volume or mass average of the "coflowing" phases.

Homogeneous flow models have advantages in term of simplicity and are quite accurate in flows that are inherently "well" mixed or exhibit dispersed characteristics of one phase into the other, such as macroscale bubbly flows. However, it has been shown that they can also be applied to microscale two-phase flows, despite their inherent nonhomogeneous nature.

Homogeneous flow models can be successfully implemented in flows that are inherently nonhomogeneous by introducing constitutive equations that take into account the interactions between the two phases [31]. Despite the use of single-phase conservation equations, these flows are sometimes considered a type of separated flow model.

Koo et al. [29, 32] used a homogeneous flow model that employed the correlation of Kandlikar [33] for heat-transfer-coefficient calculations to simulate and compare the performance between a conventional heat sink and a parallel array of 18 microchannels for a 3D IC consisting of two functional silicon layers. The model was based on a two-phase flow regime consisting of a homogeneous core composed of a mist of liquid droplets and vapor moving at the same velocity, surrounded by a very thin, slow-moving liquid film (Figure 10.6). Central to the homogeneous model treatment is the assumption of identical velocities for the liquid and vapor phases in the core, along with neglect of the slower velocity in the thin surrounding film, treating it simply as a solid boundary with lower conductivity than the microchannel wall. Several nonuniform power distribution conditions were simulated by using different layouts of the logic and memory regions. It was found that a two-phase microchannel network outperforms current conventional heat sink cooling technology in terms of both junction temperature uniformity within each layer and temperature difference between the two layers.

Despite the success of two-phase homogeneous flow models in predicting flow behavior in microchannels, their applicability is somewhat limited. As we have seen from the previous examples, homogeneous flow models can be successfully applied to microchannel flows in which one of the phases is dominant or the two-phases flow at the same velocity. However, under highly nonuniform flow conditions, such as is the case with the bubble/slug flow regime, only separated flow models can capture the relevant physics of the flow.

10.4.2 Separated Flow Modeling

Separated flow models on the other hand, although more accurate, tend to be too computationally expensive, making them hard to implement on large systems. They are characterized by the use of independent conservation equations for each of the

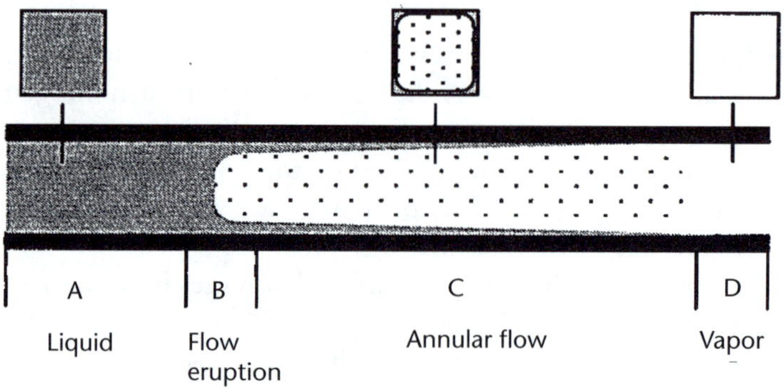

Figure 10.6 Two-phase flow regime used in the development of the homogeneous flow model by Koo et al. [29]. The flow structure consists of a homogeneous core composed of a mist of liquid droplets and vapor moving at the same velocity, surrounded by a very thin, slow-moving liquid film. (© 2001 IEEE. Reprinted with permission.)

phases. At the far end of the separated flow modeling spectrum, the phases are treated as completely distinct entities, allowing for exchanges between them and using a computational approach that keeps track of interface locations as well as properties for each of the phases. This sort of approach is extremely computationally intensive and almost impossible to implement in large-scale systems. Numerical simulations of this type are primarily used to study flow behavior locally.

By relaxing the interface tracking requirements and limiting the amount of interaction between the phases, separated flow models can take on simpler forms that are much easier to handle computationally, while retaining the key benefits associated with treating each phase independently.

The simplest form of separated flow models treats the two phases completely separately, restricting any type of interface exchange. In this type of model, the interaction between the phases is included through constitutive equations that are dependent on the relative velocity difference between the phases. Some specific examples include the Lockhart-Martinelli model for pressure-drop predictions in two-phase flows [34]. The key velocity-differential constitutive correlations in the Lockhart-Martinelli model are empirical in nature and rely on flow-regime maps for their proper implementation. Another example is the particle trajectory model, which is used in dispersed flows (i.e., one phase in the form of "particles" and the other phase as continuous) [35]. Despite the Lagrangian nature of this model, it is a simplified separate flow model because there is no tracking of the actual interfaces; rather, the particles are treated as "points" with relevant characteristics and forces acting upon them. One final example of this type of model is the drift-flux model, which has a very similar conception as the Lockhart-Martinelli model but with stronger coupling between the relative motions (velocity difference) of the phases. The drift-flux model can appropriately handle countercurrent flow and is thus particularly useful in dealing with the limitations of other models in this respect.

The next level of complexity in separated flow models is referred to as two-fluid models. These models treat each phase separately and do not keep track of interface locations, but they do allow for interface exchanges between the different conservation equations. The heart of these models is in the constitutive equations and relations used to model interface exchanges.

Finally, there are separated flow models that completely account for each phase and their interactions through interface tracking. These models are confined to the realm of computational fluid dynamics (CFD) codes and algorithms and, due to their computational complexity, are restricted to the study of local two-phase flow phenomena. Interface tracking is achieved through different methodologies, such as the volume-of-fluid (VOF) approach, where the void fraction of the computational cells is used as the parameter dictating interface location: value of one for an all-gas phase cell, value of zero for an all-liquid phase cell, and any other value in between for an interface cell (subcell interface location is computed through piecewise linearization with adjacent cells and void-fraction value). Implementing a model of this type on a global system of large scale is extremely impractical, if not outright impossible.

Garimella et al. [36] developed an experimentally validated, separated two-phase flow model for pressure drop during intermittent flow of condensing refrigerant R134a in horizontal microchannels. The model was based on the observed slug/bubble flow regime and the assumptions about the shape of the bubble, film, and slug regions proposed by Suo and Griffith [37] and Fukano et al. [38]. The slug/bubble flow regime is considered periodic, and a unit cell consists of a cylindrical bubble surrounded by a uniform annular film of liquid. The bubble/annular film section is bounded by liquid slugs on either side, with the bubble moving faster than the slugs (Figure 10.7). The liquid velocity in the annular film is much slower than both the bubble and slug velocities. The total pressure drop is calculated as the sum of the purely frictional pressure drops from the slug and bubble/film regions and the losses associated with the flow between the film and the slug. They found that the pressure drop for the same ratio of tube length to hydraulic diameter, L_{tube}/D_h, increases almost linearly with increasing quality and more sharply with decreasing tube diameter and increasing mass flux.

10.5 Pumping Considerations

Overcoming the large pressure drops associated with microchannel cooling in an efficient manner is one the major roadblocks to successful implementation of this technology. In order to realize the potential cooling performance of this promising technology, pumps capable of sustaining large pressure drops and flow rates are required. This necessitates the use of large pumps with substantial power requirements. This diminishes the size benefits of using microchannels for cooling, while the imposed taxing on the power greatly reduces the overall efficiency and heat-enhancement benefits of the technology. It also has negative implications in terms of noise issues. Thus, there is a great interest in finding a micropump technology capable of delivering the required performance needed to achieve the full potential of this technology. Although several potential candidates exist, there are no clear favorites, and, overall, none of the existent micropumps is capable of delivering the pressure and flow rates needed.

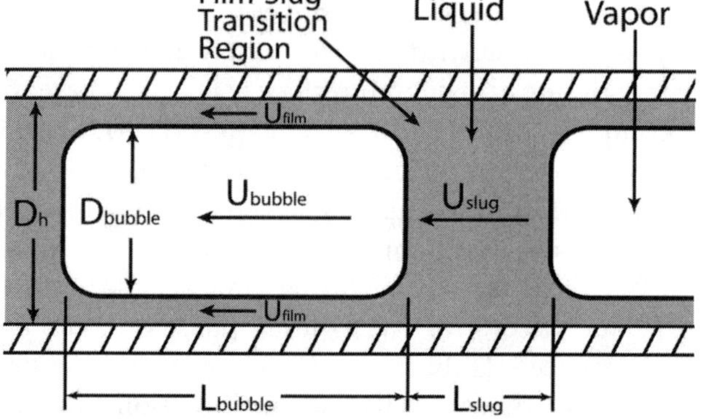

Figure 10.7 Two-phase flow regime used in the development of the separated flow model by Garimella et al. [36]. The slug/bubble flow regime is considered periodic, and a unit cell consists of a cylindrical bubble surrounded by a uniform annular film of liquid. The bubble/annular film section is bounded by liquid slugs on either side, with the bubble moving faster than the slugs. (© 2002 ASME. Reprinted with permission.)

Before reviewing the existing technologies, it is important to develop a reference framework on which the necessary pressure-drop and flow-rate requirements are based. Here, we review and present the first-order analysis performed by Singhal et al. [39, 40], which provides a good starting point. The analysis is based on the minimum pressure head and flow rate required under specific thermal constraints, namely, the maximum temperature at any point in the chip and the maximum-temperature gradient on the chip that can be tolerated.

Assuming fully developed velocity and temperature flow conditions under a constant heat-flux boundary condition, the temperature profiles of both the mixed mean bulk-flow temperature and the surface temperature are both linear. Under these conditions, it can be shown that the required liquid flow rate needed to sustain a given temperature gradient is given by

$$Q = \frac{q}{\rho c_p L_d (dT / dx)} \tag{10.47}$$

such that, for a given maximum allowable temperature gradient,

$$Q \geq \frac{q}{\rho c_p L_d (dT / dx)_{\max}} \tag{10.48}$$

Likewise, the required flow rate for a given maximum allowable temperature can be calculated using (10.34), (10.48), and Newton's law of cooling, (10.28), as well as from the notion that the maximum fluid and chip temperatures occur at the exit. Thus, we have

$$Q \geq \left(\frac{q}{\rho c_p}\right) \frac{1}{\left[T_{\max} - T_{f,i} - \dfrac{q(w_c + w_w)}{NukL_d W_d} \dfrac{\alpha}{(1+\alpha)^2}\right]} \tag{10.49}$$

Given that the flow rate and pressure drop are related through (10.3) and (10.13), the pressure-drop requirements can also be specified:

$$\Delta p \geq \left(\frac{q\mu}{8\rho c_p W_d (dT / dx)_{\max}}\right) \frac{(1+\alpha)^2}{\alpha^3} f\,\mathrm{Re} \frac{w_c + w_w}{w_c^4} \tag{10.50}$$

and

$$\Delta p \geq \left(\frac{q\mu L_d}{8\rho c_p W_d}\right) \frac{\dfrac{(1+\alpha)^2}{\alpha^3} f\,\mathrm{Re} \dfrac{w_c + w_w}{w_c^4}}{\left[T_{\max} - T_{f,i} - \dfrac{q(w_c + w_w)}{NukL_d W_d} \dfrac{\alpha}{(1+\alpha)^2}\right]} \tag{10.51}$$

depending on whether the maximum-temperature gradient or the maximum-temperature requirements are considered, respectively.

If the thickness of the microchannel walls (w_w) and aspect ratio (α) are assumed to be fixed, a valid constraint given microfabrication limitations, a pressure-drop versus flow-rate operating map can be constructed by plotting (10.48) against (10.50) and (10.49) against (10.51) for a given heat load (q) as functions of microchannel width (w_c). We can start with the maximum-temperature constraint. Plotting (10.49) against (10.51) for a given heat load and as a function of microchannel width, we get the dot-dashed line in Figure 10.8. All the points to the right of and above this line are within the "operating region," where the maximum-temperature constraint is satisfied. We will now consider the maximum-temperature-gradient constraint. Looking at (10.48), it is apparent that the required flow rate for a given temperature-gradient constraint is independent of microchannel width and only a function of the temperature gradient. However, the pressure drop, as given by (10.50), does depend on the microchannel width, and it decreases as the microchannel width increases. Thus, the "operating region" boundary line is defined by a vertical line crossing the flow-rate axis at the required value needed to satisfy the maximum-temperature-gradient constraint. All the points to the right of this line and above the minimum required pressure drop compose the "operating region" for a given temperature-gradient ceiling. As depicted in Figure 10.8, the intersection of these two regions comprises the global "operating region" of the microchannel heat sink for a set of given thermal constraints.

The suitability of a given pump for a particular microchannel heat sink design is assessed by superimposing the pump curve and corresponding pressure-head versus flow-rate load characteristics of the microchannel heat sink on top of the "operating region" map. The pump curve refers to the flow-rate versus pressure-head operating characteristics of the pump and is usually obtained experimentally. In general, the pressure head than can be sustained by the pump decreases as the flow rate increases. The pressure-head versus flow-rate load characteristics of the heat sink are obtained by using (10.3) or (10.13). The intersection of the pump curve and the heat sink load curve determine the "operating point" of the overall cooling system. Whether this point lies within the "operating region" determines the suitability of

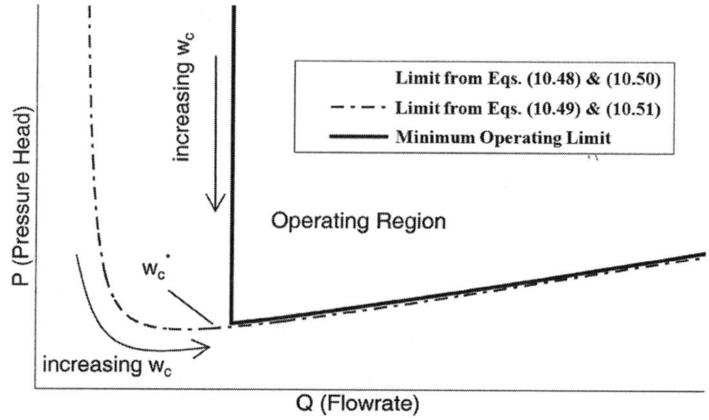

Figure 10.8 "Operating region" of a microchannel heat sink for a set of given thermal constraints. This map is constructed by plotting and combining the pressure-flow requirements, given maximum-temperature-gradient and maximum-temperature constraints for the microchannel heat sink [39, 40]. (© 2004 Taylor & Francis Ltd., http://www.informaworld.com.)

the system as a whole for achieving the desired thermal conditions. This is depicted in Figure 10.9, in which the open dots represent "operating points" of pump–heat sink combinations capable of dissipating the required heat to maintain suitable thermal operating conditions on the chip. On the other hand, the solid dots represent "operating points" for pump–heat sink combinations that would not meet the required thermal constraints.

The previous analysis is also useful as a tool for heat sink optimization. The apex of the "operating region" demarcation line represents the minimum pumping requirements in terms of both pressure head and volume flow rate capable of achieving the desired thermal conditions. The microchannel width corresponding to this point represents an optimal value for this parameter, given the fixed constraints on the microchannel wall width and aspect ratio. This optimal microchannel width can be obtained by equating the flow rates from (10.48) and (10.49):

$$w_c^* = \frac{NukL_dW_d}{q}\left(T_{\max} - T_{f,i} - L_d(dT/dx)_{\max}\right)\frac{(1+\alpha)^2}{\alpha} - w_w \tag{10.52}$$

Singhal et al. [39, 40] presented an illustrative example to assess the suitability of current miniature conventional pumps as well as several vanguard technology micropumps. The example is based on the use of water as the cooling liquid; the other inputs used in the example are presented in Table 10.1. The results of the exercise are summarized in Figure 10.10, which overlays the thermal-constraints "operating region" map, along with the load curves for two specific microchannel heat sinks (100 and 400 μm microchannel width) in the range considered ($50 \leq w_c \leq 800$), against the pump curves for conventional centrifugal, gear, and flexible-impeller miniature pumps, as well as the curves for several micropumps,

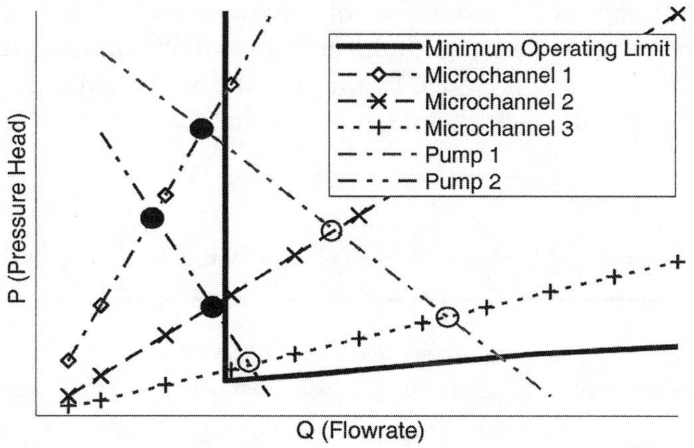

Figure 10.9 Overlay of thermal-constraints "operating region" for a microchannel heat sink of fixed aspect ratio, hypothetical pump curves, and corresponding pressure-head versus flow-rate load characteristics for the heat sink for different microchannel widths. The intersection of the pump curve and the heat sink load curve determine the "operating point" of the overall cooling system. Whether this point lies within the operating region determines the suitability of the system as a whole for achieving the desired thermal conditions. The open dots represent suitable pump–heat sink combinations, while the opposite holds for the solid dots [39, 40]. (© 2004 Taylor & Francis Ltd., http://www.informaworld.com.)

Table 10.1 Inputs Used in the Pump Suitability Example by Singhal et al. [39, 40]

Parameter	*Value*
Coolant	
Density, ρ	984.25 kg/m^3
Specific heat, c_p	4184 J/kgK
Viscosity, μ	4.89×10^{-4} N·s/m^2
Chip	
Length, L	1 cm
Width, W	1 cm
Microchannels	
Aspect ratio, μ	6
Channel width, w_c	50 to 800 μm
Wall thickness, w_w	100 μm
Thermal Parameters	
Heat load, q	100 W
Maximum chip temperature, T_{max}	80°C
Maximum chip-temperature gradient, $(dT/dx)_{max}$	5°C/cm

namely, a valveless (nozzle-diffuser) micropump using piezoelectric actuation, an injection-type electrohydrodynamic (EHD) micropump, an electroosmotic micropump, a rotary micropump, and a piezoelectric micropump. None of the micropumps presented in the literature can meet the thermal and load requirements of the microchannel heat sinks as stand-alone units, especially in terms of volume flow rate. Therefore, the pump curves depicted in Figure 10.10 represent parallel arrangements of several micropumps capable of achieving the required volume flow rates. The conventional pumps and parallel micropump arrangements are also compared in Table 10.2 in terms of maximum volume flow rate, maximum pressure head, and size. For the micropumps, the size of the individual micropumps is presented along with the number of micropumps in parallel needed to achieve the required volume flow rate (number in parentheses). Although the volume flow rate and pressure head provided by the conventional pumps are much larger than those provided by the micropump combinations, the opposite is true in terms of the size metric.

Table 10.2 Comparison of the Capabilities and Sizes of the Different Pumps Considered by Singhal et al. [39, 40]

Pump	*Maximum Flow Rate (L/min)*	*Maximum Pressure (kPa)*	*Size (mm^3)*
Miniature centrifugal magnetic drive pump	11.36	344.74	114.3 × 95.2 × 82.5
Miniature gear pump 1	1.5	144.79	101.6 × 44.5 × 66.7
Miniature gear pump 2	2.3	68.95	87.4 × 81.0 × 92.1
Flexible-impeller pump	14.38	68.95	152.4 × 114.3 × 107.9
Valveless micropump	0.345	74.53	15 × 17 × 1.4 (150x)
Injection-type EHD micropump	0.35	2.48	3 × 3 × 0.76 (25x)
Electroosmotic micropump	0.32	202.65	10 × 10 × 15 (400x)
Rotary micropump	0.35	1.4	3.17 × 3.17 ×0.6 (1,000x)
Piezoelectric micropump	0.375	1.7	5.3 × 5.3 × 1.5 (250x)

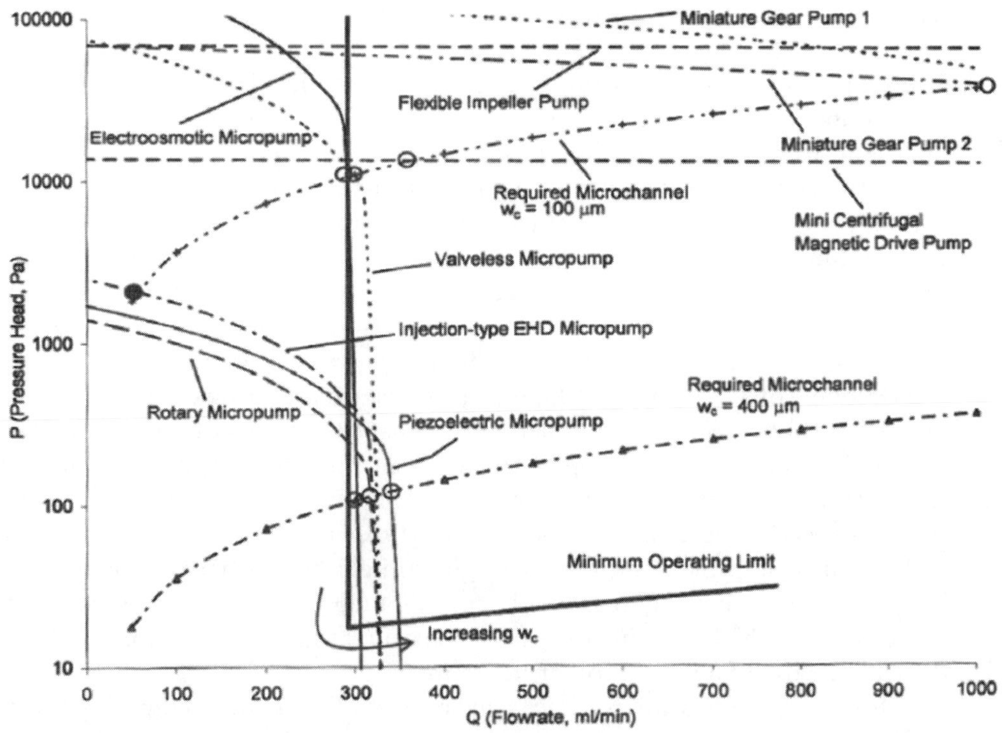

Figure 10.10 Overlay of the thermal-constraints "operating region" map, along with the load curves for two specific microchannel heat sinks (100 and 400 μm microchannel width) in the range considered (50 = wc = 800) by Singhal et al. [39, 40], against the pump curves for conventional centrifugal, gear, and flexible-impeller miniature pumps, as well as the curves for several micropumps, namely, a valveless (nozzle-diffuser) micropump using piezoelectric actuation, an injection-type electrohydrodynamic (EHD) micropump, an electroosmotic micropump, a rotary micropump, and a piezoelectric micropump. (© 2003 ASME. Reprinted with permission.)

From Figure 10.10, it is apparent that, for the microchannels of width 100 μm only, the conventional pumps and the electroosmotic and the valveless micropump combinations satisfy the pumping requirements of the microchannel heat sink design. For the heat sink with microchannel width of 400 μm, all the conventional pumps, as well as micropump combinations, provide the desired pumping requirements to satisfy the thermal constraints. Thus, from the previous analysis of Singhal et al. [39, 40], it is apparent that electroosmotic and valveless micropumps are the only feasible microscale technologies that would provide comparable pumping capabilities to their conventional counterparts, while maintaining their edge in terms of overall size advantage. However, the feasibility of these two micropump technologies is marginal at best, even with the relatively low heat load of 100 W/cm^2 used in this example. Projected future heat loads, and even existing ones in the microelectronics industry, exceed this value, making the current state of the art of these technologies unsuitable for the task, particularly in 3D IC architectures. Further research and development in micropump technologies is needed to achieve the pumping characteristics that would make them a realistic feasibility for current and future IC designs.

10.6 Optimal Architectures and 3D IC Considerations

The proper design of successful microchannel heat sinks is a typical example of an engineering trade-off. The small dimensions involved allow for higher heat-transfer rates due to the inherent shorter diffusion lengths and higher fluid velocities, which translate into large convective heat-transfer coefficients and lower thermal resistances in general. On the other hand, the use of microchannels introduces a challenge in terms of the pressure drops required to maintain the desired flow rates. Therefore, it is important to optimize the microchannel geometry and heat sink architecture in general so that the improvements in thermal performance overcompensate for the increased burden on pumping-power requirements.

Despite their proven cooling benefits, conventional parallel straight microchannel heat sinks have inherent drawbacks in terms of temperature uniformity and required pressure drops. Serpentine microchannel arrays can lead to better temperature uniformity, but they exacerbate the negative trend toward larger pressure drops due to increased channel lengths. These issues are compounded in 3D IC architectures, in which the use of single straight microchannels would make it impossible to achieve any reasonable degree of temperature uniformity and would incur extremely high pressure drops in any attempt to compensate for this drawback through the use of a large coolant flow rate.

Different optimization schemes have been studied in an attempt to improve the thermal performance and lessen the pumping requirements of microchannel-based heat sinks for cooling planar architectures. Wei and Joshi [41, 42] studied the use of stacked 2D microchannel heat sinks for cooling a single heat-generating device layer (Figure 10.11). They employed a simple 1D resistance network to evaluate the overall thermal performance of a stacked microchannel heat sink. They found that under fixed pressure-drop or pumping-power constraints, multilayered microchannel heat

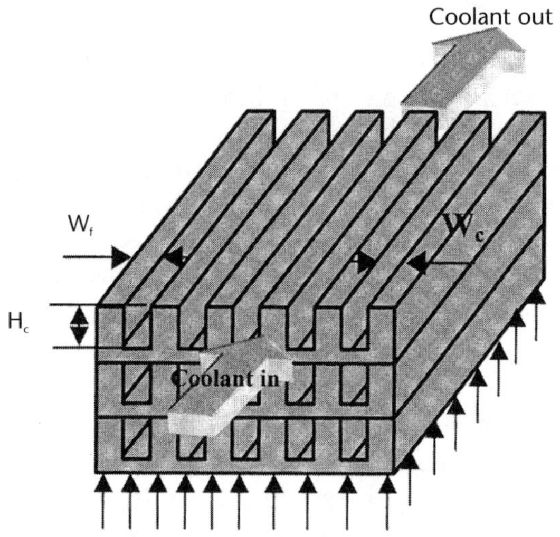

Figure 10.11 Schematic of a three-dimensional microchannel stack used in the optimization study of Wei and Joshi [41, 42]. (© 2005 ASME. Reprinted with permission.)

sink performance is superior to that of a single-layered one. However, under fixed flow-rate constraints, optimal thermal performance is achieved for a two-layered microchannel heat sink system. The multilayered architecture acts as a heat spreader, increasing the overall surface area over which heat is transferred to the fluid, thereby reducing the overall thermal resistance of the system under fixed pressure-drop constraints. Although the overall thermal resistance is reduced when two layers are used instead of one for the case of fixed-volume flow-rate conditions, adding further layers tends to increase the overall thermal resistance. The addition of more channels results in lower flow rate and velocity in each channel, resulting in a decrease in heat-transfer coefficient, even though there is an increase in heat-transfer surface area. Under fixed pumping-power constraints, the increase in heat-transfer surface-area effect overcompensates for the reduction in channel flow velocity, leading again to an overall reduction in thermal resistance.

Koo et al. [32] furthered the study of stacked microchannel heat sinks by adding several heat-generating device layers sandwiched in between the microchannel layers. They looked at a 3D IC with integrated microchannel cooling system consisting of three device layers and two microchannel layers arranged in an alternating pattern (Figure 10.12). In addition, they developed and implemented a two-phase flow model to account for boiling in the microchannels. Finally, they analyzed the effects of nonuniform power generation on the cooling of 3D ICs by looking at a simplified architecture consisting of a single-microchannel cooling layer stacked between two device layers. They split the device into logic circuitry, accounting for 90% of the power generation, and memory, accounting for 10% of the total 3D IC power generation. Four different stack schemes were analyzed (Figure 10.13). For case (a), the logic circuit occupied the whole of device layer 1, while the memory was on device layer 2. In the other cases, each layer was equally divided into memory and logic circuitry. For case (b), a high-heat-generation area was located near the inlet of the channels, while it was near the exit of channels for case (c). Case (d) had a combined thermal condition in which layer 1 had high heat flux, and layer 2 had low heat dissipation near the inlet. The total circuit area was 4 cm^2, while the total power generation was 150W. From their results, they concluded that the optimal configuration was to manage the higher power dissipation toward the microchannel heat sink outlet region since this would minimize the pressure

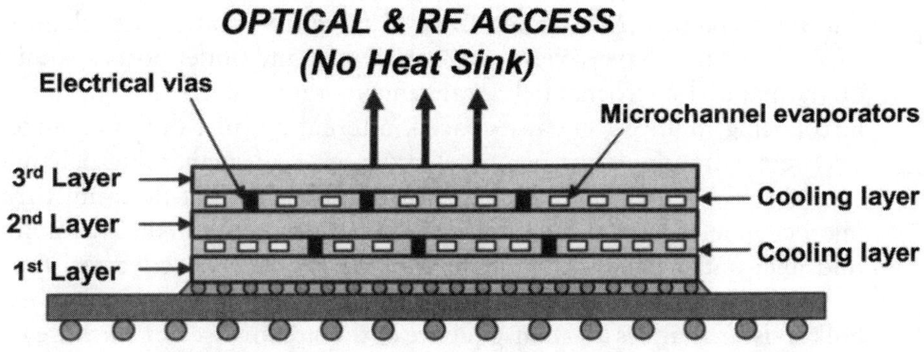

Figure 10.12 3D IC with integrated microchannel cooling system consisting of three device layers and two microchannel layers arranged in an alternating pattern used in the study by Koo et al. [32]. (© 2005 ASME. Reprinted with permission.)

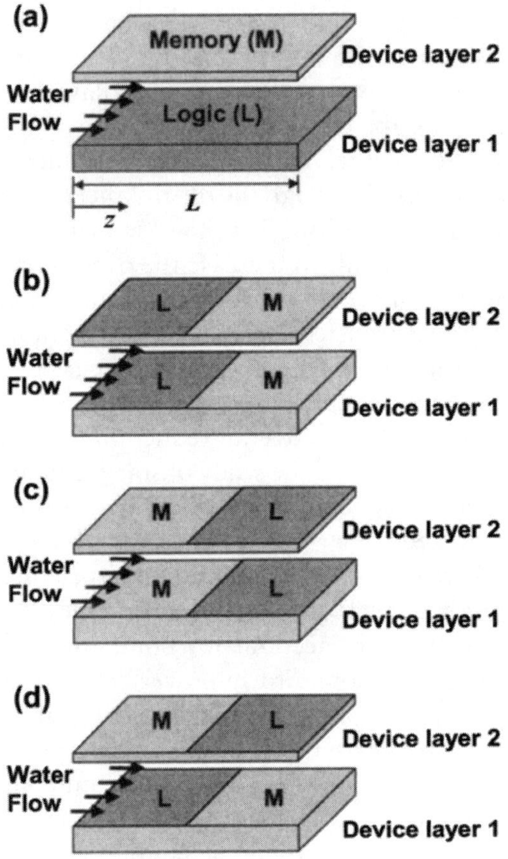

Figure 10.13 (a–d) Two-layer 3D circuit layouts for evaluating the performance of microchannel cooling. The areas occupied by memory and logic are the same, and the logic dissipates 90% of the total power consumption [32]. (© 2005 ASME. Reprinted with permission.)

drop of the two phase near the highest heat-flux regions, thereby decreasing the local wall temperature. They reasoned that if more heat is applied to the upstream region, boiling occurs earlier, resulting in increased pressure drop in the channel. Also, the average junction temperature was lower and the temperature field was more uniform employing this power-generation configuration.

Enhanced thermal and fluidic performance can also be attained by using manifold microchannel (MMC) heat sinks. Unlike traditional microchannel heat sinks, MMC heat sinks have several alternating inlet and outlet ports spanning the length of the parallel-microchannel arrangement, rather than a single inlet and outlet. Introducing multiple, equally spaced, alternating inlet-outlet arrangements effectively converts a long microchannel into a series of smaller microchannels. This has several effects. First, by reducing the overall length that fluid must traverse in the microchannel, pressure is reduced by a factor roughly equal to the number of manifold inlet-outlet pairs. Second, the bulk flow and convective resistances are both reduced, also consequences of the reduction in effective microchannel length. The bulk resistance arises as a consequence of the streamwise temperature increase in the bulk flow. The shorter distance traversed by the flow translates into lower bulk-temperature rise and therefore lower bulk resistance. The convective resistance is inversely proportional to the convective heat-transfer coefficient. Breaking up the single-microchannel flow into multiple smaller microchannel flows translates into a larger number of developing entry regions, where the Nusselt number and, consequently, the convective heat-transfer coefficient are higher.

The MMC concept was first introduced by Harpole and Eninger in 1991 [43]. They developed a complete two-dimensional single-phase flow/thermal model of the concept and optimized its design parameters for the case of a 1 kW/cm^2 heat flux with a top surface temperature of 25°C. They analyzed MMCs having between 10 and 30 inlet-outlet manifold pairs per centimeter, found that the optimal value was 30, meaning that performance was always improved by adding extra manifold pairs, and concluded that their number should only be limited by manufacturing constraints. Through their design-parameter optimization, they were able to achieve effective heat-transfer coefficients on the order of 100 W/cm^2K with a total pressure drop of only 2 bar. Copeland [44] and Copeland et al. [45] have further analyzed MMC heat sink performance through analytical, experimental, and numerical studies. Of particular interest are the thermal results in [45], in which the authors employed a 3D finite element model under isothermal wall conditions. They found that channel length or equivalent spacing between inlet and outlet for a manifold pair had almost no significant effect on thermal resistance, though the pressure drop was reduced considerably as this length was reduced.

Further improvements in terms of reduced pumping power and enhanced thermal performance of microchannel heat sinks can be achieved through the introduction of fractal and nonfractal tree-branching microchannel networks. Gosselin and Bejan [46] have demonstrated that tree-branching architectures can be used to optimize fluidic networks in terms of pumping-power requirement. Their findings can be summarized as follows: (1) pumping power is the appropriate cost function in the optimization of fluidic networks, not flow resistance or pressure drop, and minimization of each function leads to different ideal network architectures (in some special instances, pumping-power and pressure-drop optimization lead to the same solution); (2) minimum pumping-power networks do not exhibit loops; and, most importantly; and (3) under pumping-power constraints, spanning networks (point centered outwards) containing branching points (Gilbert-Steiner points) provide the best architecture. The second and third points are the basis for the ideal tree-branching geometry. Bejan [47] also extends the concept of using fluidic tree networks for the optimization of volumetric cooling problems. Invoking again the minimization of pumping power as the appropriate parameter for system optimization, he derives a three-quarters power law relating heat dissipation to volume ($q = V^{3/4}$) as the optimal relationship between the two, given the pumping-power constraints. Pence [48] employs the bioinspired analogy of the circulatory system as an efficient transport system to argue for the use of multiscaled branching flow networks in cooling applications. She points out that it would be inefficient for the heart to pump blood entirely through capillaries originating from the heart (source) and ending in the different body extremities (terminal points). Instead, the blood is first pumped through large-diameter arterial structures, which progressively branch out into smaller-diameter structures, finally ending in a fine web of tiny capillaries. Through the use of optimized, fractal-like, branching microchannel networks in heat sinks (Figure 10.14), she is able to achieve a 60% reduction in pressure drop

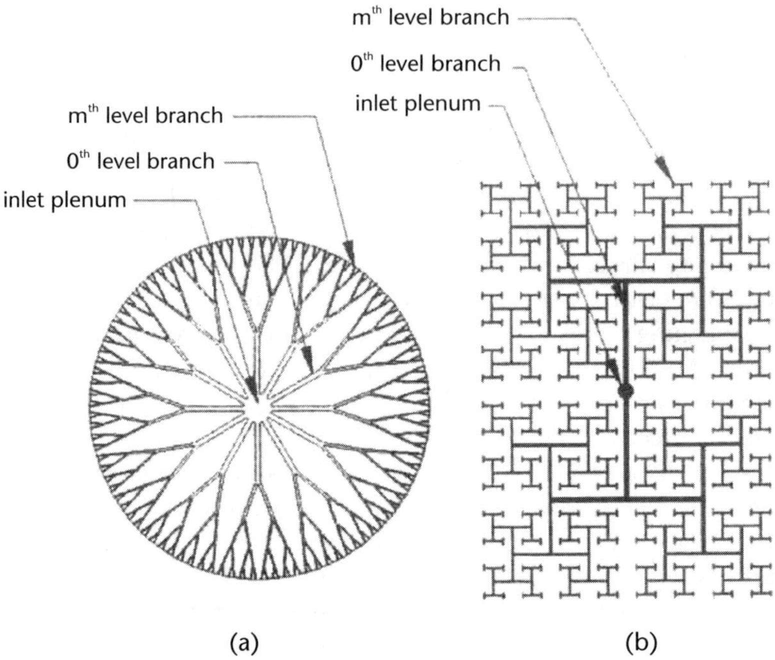

Figure 10.14 Fractal-like branching microchannel networks in heat sinks used in the study of Pence [48]. (© 2002 Taylor & Francis Ltd., http://www.informaworld.com.)

for the same flow rate and 30°C lower wall temperature under identical pumping-power conditions in comparison to the performance of an equivalent conventional parallel-microchannel heat sink arrangement. Enhanced thermal performance in a fractal-like microchannel branching network arises from the increase in convective heat-transfer coefficient associated with the smaller diameters and in the total number of developing entry regions resulting from the branching. This translates into lower average temperature and better temperature field uniformity. Wang et al. [49, 50] also advocate the use of tree-shaped and fractal-like microchannel nets for improved heat transfer and cooling for microelectronic chips. In addition to the thermal performance improvements, they emphasize the robustness of fractal-like networks as pertains to possible blockage of the fluid flow in the microchannels by particulate [50]. This is of particular importance in microelectronics cooling, which needs high reliability. They showed that tree-shaped microchannel networks were inherently more resilient to particle fouling, where a blocked channel results in the breakup of the system due to the increased temperature of the stagnant fluid, in comparison to straight microchannel networks. However, under very specific situations, the opposite might be true, especially if one of the main branches in the network were blocked, leading to major global failure.

It is evident that as the semiconductor industry moves toward 3D IC, optimization of microchannel architectures becomes ever more important. Bejan's [47] three-quarters power law and minimum pumping-power optimization approach becomes even more relevant for these heat-generating volumetric systems and should be the block around which 3D microchannel cooling architectures are developed. The work of Koo et al. [32] suggests that synergy between the chip and heat sink designers is crucial in order to achieve proper cooling of 3D semiconductor chips. In general, the optimal cooling architecture should incorporate cues from all the previous optimization schemes, with cascading manifold microchannel networks that more closely resemble some of the 3D examples found in nature, such as the respiratory and circulatory systems, river basins and deltas, and of course tree-branching morphology.

10.7 Future Outlook

Given the realm of cooling possibilities for microelectronic components, it is hard to predict what the future holds in terms of which technology will lead the way toward the cooling of 3D ICs. In this chapter, we have concentrated on active microfluidic cooling solutions, specifically those based on convective flow through microchannels. If not a clear leading contender, this is one of the only promising technologies that has seen direct market application [51]. The reality is that with the current increases in microelectronics power densities, the envelope of this technology must be pushed in order to achieve the required heat-flux removal. Using the current benchmark for future cooling technologies of 1 kW/cm^2, we can expect that as the industry moves into 3D architectures and the transistor size continues to decrease, volumetric cooling solutions will need to address dissipation of power densities on the order of at least 20 kW/cm^3. This will clearly require the development and implementation of very clever and novel cooling technologies. Microchannel convective cooling will most definitely have to rely on strategies that involve phase change and boiling, while addressing the issues associated with maintaining acceptable pressure-drop and pumping-power requirements, both of which become that much more challenging with the introduction of the increasing complexity associated with 3D microchannel networks.

One of the major limitations associated with boiling microchannel systems is the proper management and disposal of the vapor phase. The sudden and explosive phase change that occurs in the microchannel has detrimental effects in terms of increased pressure drop, the onset of instabilities, and, most importantly, the occurrence of burnout and dryout conditions. To overcome these limitations, vapor-phase management solutions must be developed that enable taking advantage of the latent heat associated with phase change without incurring the negative effects. An extremely attractive solution is the use of local vapor-management devices that can quickly and efficiently remove the vapor phase at the phase-change location. One specific example of a very promising technology is the use of a vapor escape membrane developed by David et al. [52]. This device consists of a hydrophobic membrane located on top of the microchannel heat sink (Figure 10.15). The hydrophobic nature of the membrane permits venting of the vapor phase to an escape chamber, while preventing liquid passage to this vapor reservoir. This is effectively akin to a vapor-phase stripper, which maintains a fully liquid phase moving through the microchannels. As such, the issues associated with phase-change-acceleration pressure drop and burnout/dryout conditions are effectively eliminated.

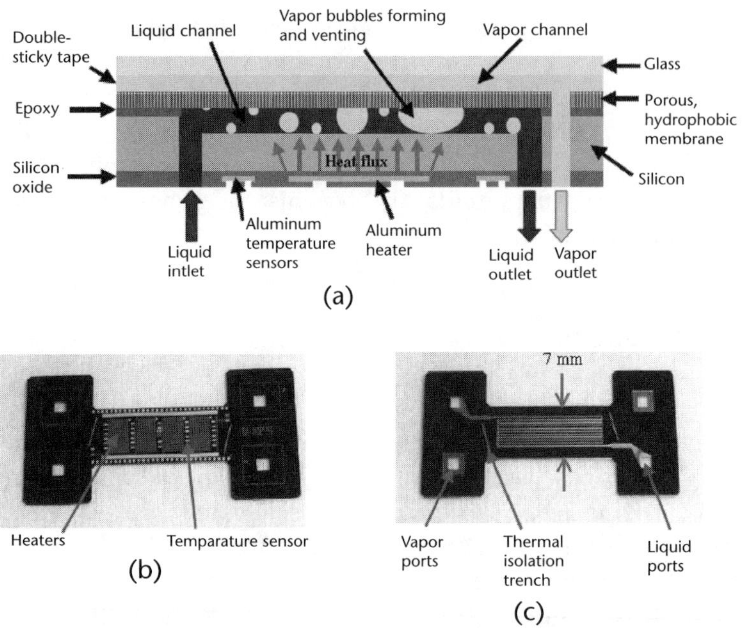

Figure 10.15 (a) Schematic of the vapor escape membrane concept being developed by David et al. [52]. The device consists of a hydrophobic membrane located on top of the microchannel heat sink. The hydrophobic nature of the membrane permits venting of the vapor phase to an escape chamber, while preventing liquid passage to this vapor reservoir. (b) Back side of the actual device, showing the integrated heaters and temperature sensors. (c) Front side of the actual device, showing the serpentine microchannel geometry. (© 2007 ASME. Reprinted with pemission.)

10.8 Nomenclature

A_c	Microchannel cross-sectional area
A_p	Heat sink plenum cross-sectional area
C_f	Fanning friction-factor coefficient
$C_{f,app}$	Apparent Fanning friction-factor coefficient
c_p	Specific heat capacity at constant pressure
c_v	Specific heat capacity at constant volume
D	Circular microchannel internal diameter
D_h	Noncircular microchannel hydraulic diameter
f	Darcy friction-factor coefficient
h	Convective heat-transfer coefficient
K	Hagenbach's factor (incremental pressure-defect coefficient)
K_{90}	Hagenbach's factor for a 90° bend
K_c	Hagenbach's factor for a sudden contraction
K_e	Hagenbach's factor for a sudden expansion
k_f	Fluid thermal conductivity
L	Microchannel length
L_d	Chip die length
L_h	Hydrodynamic entry length
L_t	Thermal entry length
$\dot{m}$	Liquid mass flow rate
n	Direction normal to the microchannel wall surface
Nu	Nusselt number
p	Pressure

P	Perimeter
p_i	Inlet pressure
Po	Poiseuille number
p_o	Outlet pressure
Pr	Prandtl number
P_w	Wetted perimeter
Q	Liquid volume flow rate
q_s''	Surface heat flux
q_{conv}	Microchannel convective heat transfer
R	Circular microchannel internal radius
r	Circular microchannel radial coordinate
Re	Reynolds number
Re_D	Reynolds number based on circular microchannel internal diameter
Re_{Dh}	Reynolds number based on noncircular microchannel internal hydraulic diameter
T	Temperature
T_{char}	Characteristic temperature
$T_{f,i}$	Inlet fluid temperature
T_m	Mixed mean bulk-flow temperature
$T_{m,i}$	Inlet mixed mean bulk-flow temperature
$T_{m,o}$	Outlet mixed mean bulk-flow temperature
T_{max}	Maximum allowable chip die temperature
T_s	Microchannel wall surface temperature
u	Microchannel axial fluid velocity
u_{avg}	Average microchannel axial fluid velocity
w_c	Microchannel width
W_d	Chip die width
w_w	Microchannel wall width
x	Microchannel axial coordinate
x^+	Nondimensional microchannel axial coordinate
α	Rectangular microchannel aspect ratio
δ	Liquid-film thickness
ε	Surface roughness
μ	Fluid viscosity
μ_s	Fluid viscosity evaluated at the microchannel wall temperature
ρ	Fluid density
τ_w	Microchannel wall shear stress

References

[1] Kandlikar, S. G., and A. V. Bapat, "Evaluation of Jet Impingement, Spray and Microchannel Chip Cooling Options for High Heat Flux Removal," *Heat Transfer Engineering,* Vol. 28, No. 11, 2007, pp. 911–923.

[2] Tuckerman, D. B., and R. F. W. Pease, "High-Performance Heat Sinking for VLSI," *IEEE Electron Device Letters,* Vol. 2, No. 5, 1981, pp. 126–129.

[3] Phillips, R. J., "Forced-Convection, Liquid-Cooled, Microchannel Heat Sinks," MSME Thesis, Cambridge, 1987.

[4] Samalam, V., "Convective Heat Transfer in Microchannels," *J. Electronic Materials,* Vol. 18, No. 5, 1989, pp. 611–617.

[5] Peng, X. F., and B. X. Wang, "Forced Convection and Flow Boiling Heat Transfer for Liquid Flowing through Microchannels," *Int. J. Heat and Mass Transfer,* Vol. 36, No. 14, 1993, pp. 3421–3427.

[6] Bowers, M. B., and I. Mudawar, "High Flux Boiling in Low Flow Rate, Low Pressure Drop Mini-Channel and Micro-Channel Heat Sinks," *Int. J. Heat and Mass Transfer,* Vol. 37, No. 2, 1994, pp. 321–332.

[7] Peles, Y. P., L. P. Yarin, and G. Hetsroni, "Heat Transfer of Two-Phase Flow in a Heated Capillary," *Proc. 11th International Heat Transfer Conference*, Kuongju, Korea, August 23–28, 1998, pp. 193–198.

[8] Lin, S., P. A. Kew, and K. Cornwell, "Two-Phase Heat Transfer to a Refrigerant in a 1 mm Diameter Tube," *Int. J. Refrigeration,* Vol. 24, No. 1, 2001, pp. 51–56.

[9] Qu, W., and I. Mudawar, "Experimental and Numerical Study of Pressure Drop and Heat Transfer in a Single-Phase Micro-Channel Heat Sink," *Int. J. Heat and Mass Transfer,* Vol. 45, No. 12, 2002, pp. 2549–2565.

[10] Zhang, L., et al., "Measurements and Modeling of Two-Phase Flow in Microchannels with Nearly Constant Heat Flux Boundary Conditions," *J. Microelectromechanical Systems,* Vol. 11, No. 1, 2002, pp. 12–19.

[11] Agostini, B., et al., "State of the Art of High Heat Flux Cooling Technologies," *Heat Transfer Engineering,* Vol. 28, No. 4, 2007, pp. 258–281.

[12] Goodling, J. S., "Microchannel Heat Exchangers: A Review," Proceedings of the SPIE: High Heat Flux Engineering II, July 12–13, San Diego, California, 1993, pp. 66–82.

[13] Hassan, I., P. Phutthavong, and M. Abdelgawad, "Microchannel Heat Sinks: An Overview of the State-of-the-Art," *Microscale Thermophysical Engineering,* Vol. 8, No. 3, 2004, pp. 183–205.

[14] Hidrovo, C. H., et al., "Two-Phase Microfluidics for Semiconductor Circuits and Fuel Cells," *Heat Transfer Engineering,* Vol. 27, No. 4, 2006, pp. 53–63.

[15] Gad-el-Hak, M., "Fluid Mechanics of Microdevices—The Freeman Scholar Lecture," *J. Fluids Engineering, Transactions of the ASME,* Vol. 121, No. 1, 1999, pp. 5–33.

[16] Colin, S., "Single-Phase Gas Flow in Microchannels," *Heat Transfer and Fluid Flow in Minichannels and Microchannels*, Elsevier, Kidlington, Oxford, 2006, pp. 9–86.

[17] Shah, R. K., and A. L. London, *Advances in Heat Transfer. Laminar Flow Forced Convection in Ducts. A Source Book for Compact Heat Exchanger Analytical Data*, New York: Academic Press, 1978.

[18] Kandlikar, S. G., "Single-Phase Liquid Flow in Minichannels and Microchannels," *Heat Transfer and Fluid Flow in Minichannels and Microchannels*, Elsevier, Kidlington, Oxford, 2006, pp. 87–136.

[19] Chen, R. Y., "Flow in the Entrance Region at Low Reynolds Numbers," *J. Applied Mechanics, Transactions ASME*, Vol. 95, No. 1, 1973, pp. 153–158.

[20] White, F. M., *Fluid Mechanics*, 6th ed., New York: McGraw-Hill, 2008.

[21] Incropera, F. P., et al., *Fundamentals of Heat and Mass Transfer*, 6th ed., New York: John Wiley & Sons, 2007.

[22] Qu, W., and I. Mudawar, "Measurement and Prediction of Pressure Drop in Two-Phase Micro-Channel Heat Sinks," *Int. J. Heat and Mass Transfer,* Vol. 46, No. 15, 2003, pp. 2737–2753.

[23] Kandlikar, S. G., et al., "High-Speed Photographic Observation of Flow Boiling of Water in Parallel Mini-Channels," Proceedings of the 35th National Heat Transfer Conference., June 10–12, Anaheim, California, 2001, pp. 675–684.

[24] Hetsroni, G., et al., "A Uniform Temperature Heat Sink for Cooling of Electronic Devices," *Int. J. Heat and Mass Transfer,* Vol. 45, No. 16, 2002, pp. 3275–3286.

[25] Peles, Y. P., L. P. Yarin, and G. Hetsroni, "Steady and Unsteady Flow in a Heated Capillary," *Int. J. Multiphase Flow,* Vol. 27, 2001, pp. 577–598.

[26] Xu, J., J. Zhou, and Y. Gan, "Static and Dynamic Flow Instability of a Parallel Microchannel Heat Sink at High Heat Fluxes," *Energy Conversion and Management,* Vol. 46, No. 2, 2005, pp. 313–334.

[27] Chang, K. H., and C. Pan, "Two-Phase Flow Instability for Boiling in a Microchannel Heat Sink," *Int. J. Heat and Mass Transfer,* Vol. 50, No. 11–12, 2007, pp. 2078–2088.

[28] Koo, J.-M., et al., "Convective Boiling in Microchannel Heat Sinks with Spatially-Varying Heat Generation," Proceedings of ITHERM 2002: The 8th Intersociety Conference on Thermal and Thermomechanical Phenomena in Electronic Systems, May 30–June 1, San Diego, California, 2002, pp. 341–346.

[29] Koo, J.-M., et al., "Modeling of Two-Phase Microchannel Heat Sinks for VLSI Chips," Proceedings of MEMS 2001: The 14th IEEE International Conference on Micro Electro Mechanical Systems, January 21–25, Interlaken, Switzerland, 2001, pp. 422–426.

[30] Qu, W., and I. Mudawar, "Flow Boiling Heat Transfer in Two-Phase Micro-Channel Heat sinks—II. Annular Two-Phase Flow Model," *Int. J. Heat and Mass Transfer,* Vol. 46, No. 15, 2003, pp. 2773–2784.

[31] Kleinstreuer, C., *Two-Phase Flow: Theory and Applications*, New York: Taylor & Francis, 2003.

[32] Koo, J.-M., et al., "Integrated Microchannel Cooling for Three-Dimensional Electronic Circuit Architectures," *J. Heat Transfer,* Vol. 127, No. 1, 2005, pp. 49–58.

[33] Kandlikar, S. G., "General Correlation for Saturated Two-Phase Flow Boiling Heat Transfer inside Horizontal and Vertical Tubes," *J. Heat Transfer, Transactions ASME,* Vol. 112, No. 1, 1990, pp. 219–228.

[34] Levy, S., *Two-Phase Flow in Complex Systems*, New York: John Wiley & Sons, 1999.

[35] Brennen, C. E., *Fundamentals of Multiphase Flow*, Cambridge: Cambridge University Press, 2005.

[36] Garimella, S., J. D. Killion, and J. W. Coleman, "Experimentally Validated Model for Two-Phase Pressure Drop in the Intermittent Flow Regime for Circular Microchannels," *J. Fluids Engineering, Transactions of the ASME,* Vol. 124, No. 1, 2002, pp. 205–214.

[37] Suo, M., and P. Griffith, "Two-Phase Flow in Capillary Tubes," *American Society of Mechanical Engineers—Transactions—J. Basic Engineering,* Vol. 86, No. 3, 1964, pp. 576–582.

[38] Fukano, T., A. Kariyasaki, and M. Kagawa, "Flow Patterns and Pressure Drop in Isothermal Gas-Liquid Concurrent Flow in a Horizontal Capillary Tube," *Nippon Kikai Gakkai Ronbunshu, B Hen/Transactions of the Japan Society of Mechanical Engineers, Part B,* Vol. 56, No. 528, 1990, pp. 2318–2325.

[39] Garimella, S. V., and V. Singhal, "Single-Phase Flow and Heat Transport and Pumping Considerations in Microchannel Heat Sinks," *Heat Transfer Engineering,* Vol. 25, No. 1, 2004, pp. 15–25.

[40] Singhal, V., D. Liu, and S. V. Garimella, "Analysis of Pumping Requirements for Microchannel Cooling Systems," Advances in Electronic Packaging 2003, Volume 2, Proceedings of the 2003 International Electronic Packaging Technical Conference and Exhibition, Maui, Hawaii, July 6–11, 2003, pp. 473–479.

[41] Wei, X., and Y. Joshi, "Optimization Study of Stacked Micro-Channel Heat Sinks for Micro-Electronic Cooling," *IEEE Transactions on Components and Packaging Technologies,* Vol. 26, No. 1, 2003, pp. 55–61.

[42] Wei, X., and Y. Joshi, "Stacked Microchannel Heat Sinks for Liquid Cooling of Microelectronic Components," *J. Electronic Packaging, Transactions of the ASME,* Vol. 126, No. 1, 2004, pp. 60–66.

[43] Harpole, G. M., and J. E. Eninger, "Micro-Channel Heat Exchanger Optimization," *SEMI-THERM VII: Proceedings of the 7th Annual IEEE Semiconductor Thermal Measurement and Management Symposium,* Phoenix, Arizona, February 12–14, 1991, pp. 59–63.

[44] Copeland, D., "Manifold Microchannel Heat Sinks: Analysis and Optimization," *Thermal Science Engineering,* Vol. 3, No. 1, 1995, pp. 7–12.

[45] Copeland, D., M. Behnia, and W. Nakayama, "Manifold Microchannel Heat Sinks: Isothermal Analysis," *IEEE Transactions on Components, Packaging, and Manufacturing Technology Part A,* Vol. 20, No. 2, 1997, pp. 96–102.

[46] Gosselin, L., and A. Bejan, "Tree Networks for Minimal Pumping Power," *Int. J. Thermal Sciences,* Vol. 44, No. 1, 2005, pp. 53–63.

[47] Bejan, A., "The Tree of Convective Heat Streams: Its Thermal Insulation Function and the Predicted 3/4-Power Relation Between Body Heat Loss and Body Size," *Int. J. Heat and Mass Transfer,* Vol. 44, No. 4, 2001, pp. 699–704.

[48] Pence, D. V., "Reduced Pumping Power and Wall Temperature in Microchannel Heat Sinks with Fractal-Like Branching Channel Networks," *Microscale Thermophysical Engineering,* Vol. 6, No. 4, 2002, pp. 319–330.

[49] Wang, X.-Q., A. S. Mujumdar, and C. Yap, "Thermal Characteristics of Tree-Shaped Microchannel Nets for Cooling of a Rectangular Heat Sink," *Int. J. Thermal Sciences,* Vol. 45, No. 11, 2006, pp. 1103–1112.

[50] Wang, X.-Q., A. S. Mujumdar, and C. Yap, "Numerical Analysis of Blockage and Optimization of Heat Transfer Performance of Fractal-Like Microchannel Nets," *J. Electronic Packaging, Transactions of the ASME,* Vol. 128, No. 1, 2006, pp. 38–45.

[51] Upadhya, G., et al., "Micro-Scale Liquid Cooling System for High Heat Flux Processor Cooling Applications," *SEMI-THERM 2006: Proceedings of the 22nd Annual IEEE Semiconductor Thermal Measurement and Management Symposium,* Dallas, Texas, March 14–16, 2006, pp. 116–119.

[52] David, M. P., et al., "Vapor-Venting, Micromachined Heat Exchanger for Electronics Cooling," *Proc. IMECE2007: 2007 ASME International Mechanical Engineering Congress and Exposition*, Seattle, Washington, November 11–15, 2007.

CHAPTER 11

Single and 3D Chip Cooling Using Microchannels and Microfluidic Chip I/O Interconnects

Bing Dang, Muhannad S. Bakir, Deepak Sekar, Calvin R. King Jr., and James D. Meindl

11.1 Introduction

The reliability, performance, and power dissipation of interconnects and transistors are a function of the operating temperature. As such, chip-level cooling will become more important in the future as the power consumption and power density of microprocessors increase. The International Technology Roadmap for Semiconductors (ITRS) projects that the power density of a single-chip package will increase to greater than 100 W/cm^2 for high-performance applications in 2018 [1] from the current power density of 60 to 80 W/cm^2. Historically, in order to maintain constant junction temperature with increasing power dissipation, the size of an air-cooled heat sink used to cool a microprocessor has been steadily increasing. Figure 11.1 illustrates the power dissipation and heat sink size (volume) of various Intel microprocessors [2]. It is clear that the size of the heat sink has been increasing with each new microprocessor, thus imposing limits on system size, chip packing efficiency, and interconnect length between chips. The typical junction-to-ambient thermal resistance of conventional air-cooled heat sinks, which constitutes the thermal interconnects of a system, has been larger than 0.5°C/W [3]. The scaling of conventional air-cooled thermal interconnects cannot meet ITRS-projected power dissipation at the end of the roadmap as well as meeting reasonable form-factor requirements, acoustic noise constraints on the fan, and chip junction-to-ambient thermal resistance. As illustrated in Figure 11.2 and discussed in Chapters 9 and 10, both single-phase and two-phase liquid cooling yield heat-transfer coefficients that are at least two orders of magnitude higher than what is achievable with forced air-cooling [4]. The low thermal resistance (<0.2 °C/W) projected by the ITRS for the cost-performance and high-performance single-chip packages in 2018 may only be achieved if single-phase or two-phase liquid cooling is used. Challenges in cooling become exacerbated with 3-D chip integration as discussed in Chapters 1 and 10.

An on-chip (monolithic) integrated microchannel heat sink eliminates the need for thermal interface materials (TIMs) and represents the highest level of process integration possible for liquid cooling. In 1981, Tuckerman and Pease demon-

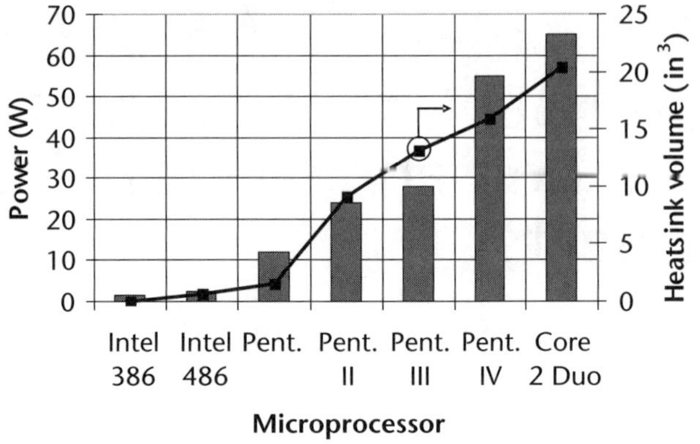

Figure 11.1 Power dissipation and heat sink size of various Intel microporcessors [2].

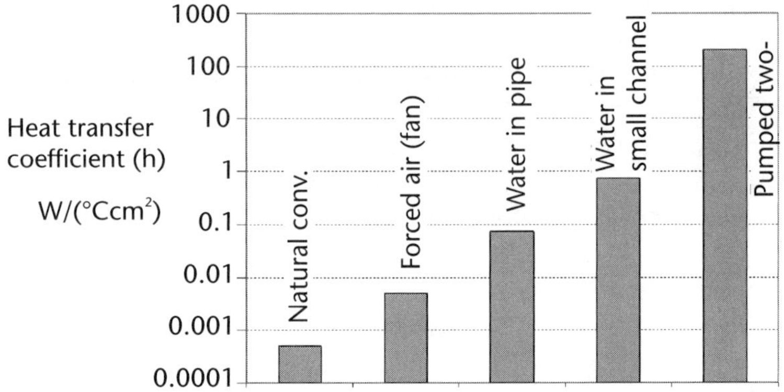

Figure 11.2 Plot of the heat transfer coeeficient for various cooling mechanisms [3].

strated that chip junction-to-ambient thermal resistance can be as low as 0.09°C-cm^2/W and that a power density as high as 790 W/cm^2 can be rejected when a microchannel heat sink is directly integrated into a Si chip [5]. Liquid cooling also promises to lower chip operating temperature, which can further improve the performance and reliability of microelectronic systems [6]. Moreover, an integrated microchannel heat sink can largely reduce the IC package footprint and volume to leverage form-factor and performance requirements of multichip and 3-D chip stacks. On-chip microfluidic channels may also enable direct hot spot cooling since the thermal interfaces are eliminated and provides very large heat transfer cofficient. Due to these benefits, microfluidic cooling has recently received increasing interest [7–10]. To date, most of the published research in this field has focused on characterization of microchannel dimensions or thermofluidic properties. Significant challenges remain in the integration, interconnection, and packaging of the microchannel heat sink.

In this chapter, we describe the complete set of technologies required to fabricate, interconnect, package, and flip-chip-assemble silicon chips with microchannel heat sinks. The chapter is organized as follows: Section 11.2 summarizes the history and status of the microchannel cooling technologies in IC applications. Section 11.3 describes the fabrication of CMOS-compatible microchannel heat sinks. Section 11.4 describes the fabrication and integration of fluidic chip I/Os (first-level interconnect) that provide inlet/outlet interconnections to the microchannel heat sink. Assembly of the "thermofluidic chips" is presented in Section 11.5. Section 11.6 reports thermal measurements and testing of the assembled chips. Hydraulic requirements of the microchannel heat sink and fluidic I/Os are described in Section 11.7. Section 11.8 describes the extending of the microfluidic interconnection to 3D chip stacks. Finally, Section 11.9 is the conclusion.

11.2 Summary of Microchannel Cooling Technologies for ICs

Many liquid-cooling approaches have already been employed to cool chips. For example, the electronic chips may be directly immersed in a pool of inert dielectric liquid [11]. Other examples include thermosyphons and heat pipes, where a liquid evaporates with applied heat and condenses elsewhere, in a closed system, to dissipate that heat [12, 13]. These techniques are usually bulky and are inherently unsuitable for today's compact and highly integrated systems, although such two-phase cooling has a very high heat-transfer coefficient, as discussed in Chapter 10.

It is well known that microfluidic channels can provide exceptionally high heat-transfer efficiencies and low thermal resistances [3, 5]. By definition, any channel with dimensions less than 1 mm and greater than 1 μm can be called a microchannel. The most important feature of microchannels is their large surface-to-volume ratio, which leads to a high rate of heat transfer (or heat-transfer coefficient). Currently, most microchannels fall into the range of 30 to 300 μm [14]. In 1982, Tuckerman and Pease [5] first studied a microchannel heat sink directly integrated into a Si chip, demonstrating that a heat flux as high as 790 W/cm^2 can be removed, and a thermal resistance as low as 0.09°C/W can be achieved with single-phase liquid cooling. Since then, microfluidic heat sinks have been extensively studied [7, 8, 15, 16]. As illustrated in Figure 11.3(a), with such an on-chip microfluidic configuration, direct hot spot cooling becomes possible since all thermal interfaces are eliminated. An integrated microfluidic heat sink can also significantly reduce the IC package area and space, which is desirable for the rapidly growing applications of multichip modules and 3D integration [10, 17]. Because of these benefits, microfluidic cooling has recently received increasing interest from major semiconductor manufacturers, including Intel and IBM [18, 19]. However, many technical challenges still remain. For example, in conventional fabrication methods for microchannels, the deep trenches are enclosed by direct wafer bonding (glass or Si), which requires ultrasmooth/clean bonding interfaces, high pressure, and high temperature (350°C ~ 1,000°C) and/or high voltage (500V ~ 1,000V) [20, 21]. These process conditions may not be compatible with structures fabricated during front- and back-end-of-the-line (FEOL and BEOL) semiconductor processes. As a result, most microchannel heat sinks that have been reported in the literature [7, 8] are fabricated separately and attached onto the back side of a chip, as shown

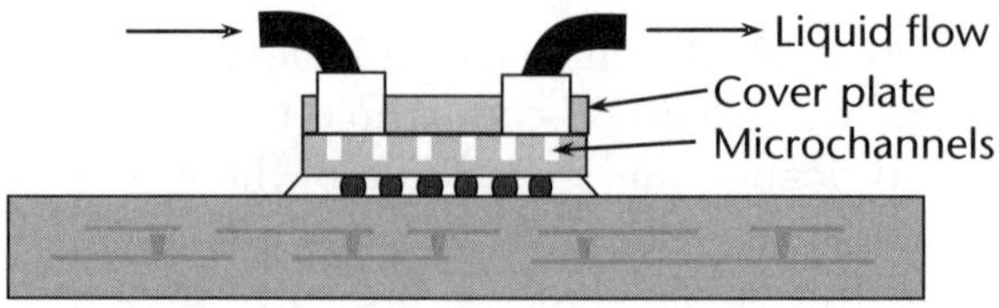

(a)

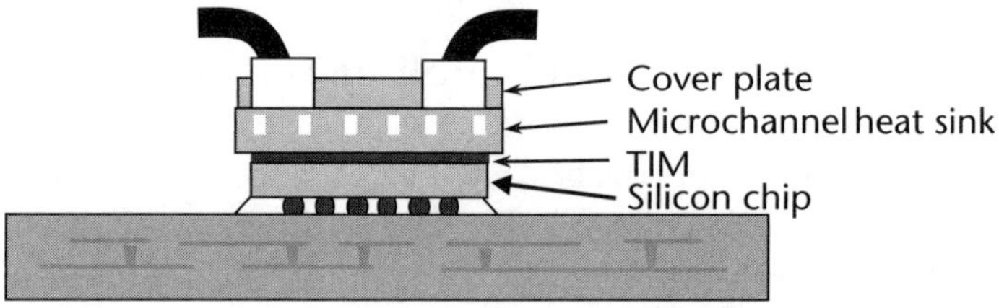

(b)

Figure 11.3 Schematic of typical microchannel heat sinks in literature: (a) the microchannel heat sink is integrated as part of a Si chip [5, 18], and (b) the microchannel heat sink is fabricated separately and attached onto the back side of a chip with a TIM [7, 8].

in Figure 11.3(b). Using this approach, high-performance TIMs are still a limiting factor.

The TIM may account for as much as 50% of the total junction-to-ambient thermal resistance [7]. This implies that even with significant advances to reduce the thermal resistance of a heat sink, unless the TIM is eliminated or reduced, the TIM may become the bottleneck in achieving low-thermal-resistance interconnects. Another challenge for microchannel heat sinks is fluid delivery and extraction (i.e., how to deliver the cold liquid to the microchannel heat sink and extract the hot liquid from the microchannel heat sink). This problem, of course, becomes exacerbated for a multichip module. The fluidic interconnection of multichip packages can be a significant challenge, as illustrated in Figure 11.4. Therefore, technologies to address the above list of challenges are needed to enable seamless integration, interconnection, and packaging for gigascale chips in microchannel heat sinks.

In addition to the liquid cooling at the back side of a chip, board-level liquid cooling has also been studied [22, 23]. Bauer et al. [22] described a concept of integrating a channel system directly into the ceramic multilayer wiring board based on low-temperature, cofired ceramic technology (LTCC). Ilgen [23] further reported a fabrication method for thermo-fluidic channels based on the low-cost FR4 organic laminate technology. Considering these technologies, there is an opportunity to implement an integrated thermofluidic packaging scheme that utilizes both chip-level and board-level cooling systems, as illustrated in Figure 11.5. Development of several key technologies is needed to enable compact packaging of GSI chips with thermo-fluidic heat-dissipation capabilities. Low-temperature CMOS-compatible processes are needed for on-chip microchannels. Novel thermofluidic

Figure 11.4 The conceived schematic for a multichip system with the chip-level liquid cooled heat sink and individual tubing at the back side of each chip.

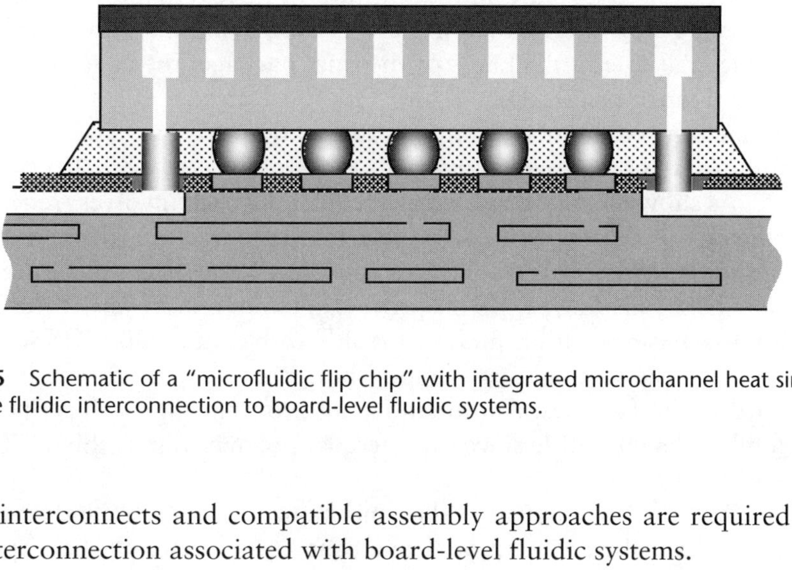

Figure 11.5 Schematic of a "microfluidic flip chip" with integrated microchannel heat sink and the massive fluidic interconnection to board-level fluidic systems.

chip I/O interconnects and compatible assembly approaches are required for the fluidic interconnection associated with board-level fluidic systems.

11.3 Fabrication of On-Chip Microfluidic Heat Sink

Microchannels can be fabricated using micromachining, molding, or embossing in a variety of materials, such as Si, glass, metals, and polymers [24–27]. In this chapter, two approaches are investigated to fabricate an on-chip microchannel heat sink. The first process relies on the polymer bonding of a cover plate at low temperature (<200°C). The second process utilizes a sacrificial polymer that can be thermally decomposed at low temperatures (<200°C). In both processes, the first step is to fabricate trenches ($h > 200\ \mu m$) into the back side of a Si wafer using deep reactive ion etching (DRIE).

As illustrated in Figure 11.6, for the first approach, the polymer glue is applied onto a cover plate by spin coating. The cover plate is then flipped and placed onto the back side of the Si wafer containing the preetched trenches. Next, a compressive force is applied on the cover to maintain intimate contact between the two sub-

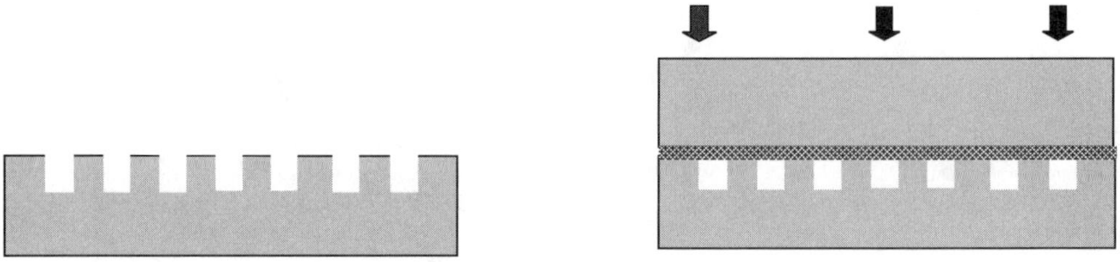

(a) (b)

Figure 11.6 (a, b) Illustration of formation of microchannels by glue bonding of a glass cover plate.

strates while the sample is placed in an oven to cure the polymer bonding film. The process step is followed by polymer cure. Following cure, a strong bond is formed between the cover plate and the Si wafer. The cover plate may be either a Si wafer or a Pyrex glass wafer, which has a CTE of 2.8 to match that of the Si substrate. To reduce the overall thickness of the final package, the cover plate may be mechanically thinned by grinding. Figure 11.7 illustrates cross-sectional SEM images of the formed microchannels using this fabrication process. The micrographs reveal the uniformity of the bonded interface.

As shown in Figure 11.8, the second approach involves trench filling with a sacrificial polymer by spin coating, surface planarization through mechanical polishing, overcoating with a layered structure, and decomposition of the sacrificial polymer by heating [27]. This process does not require any wafer bonding; thus, this approach is monolithic and compatible with common CMOS wafer processing. For instance, the spin coating and polishing process steps are readily available in standard BEOL facilities. Another advantage of this approach is the resulting thin chip (and package) profile since the overcoat polymer film is thin (<30 μm), as shown in Figure 11.9.

To further increase the mechanical strength and hermeticity of the microchannels, an additional overcoat layer can easily be applied. As shown in Figure 11.10, after a second layer of overcoat is applied and cured, the top surface of the polymer-based microchannels is still flat with an increased thickness (~50 μm in

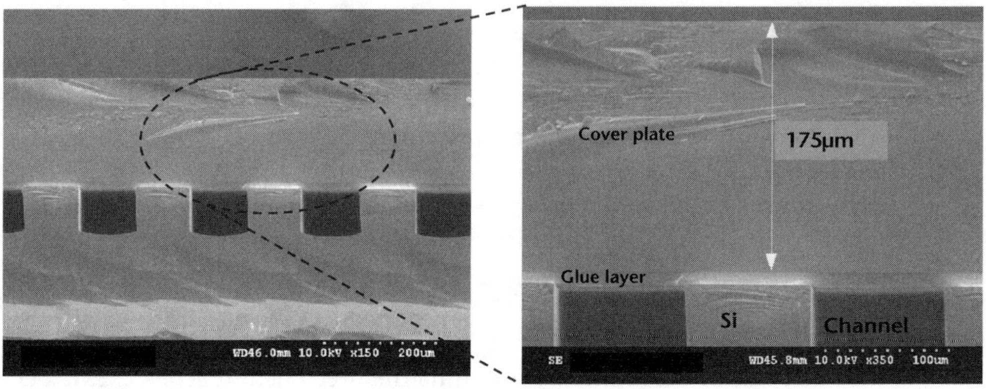

Figure 11.7 Cross-sectional SEM photographs of microchannels enclosed by glue bonding of a glass cover plate.

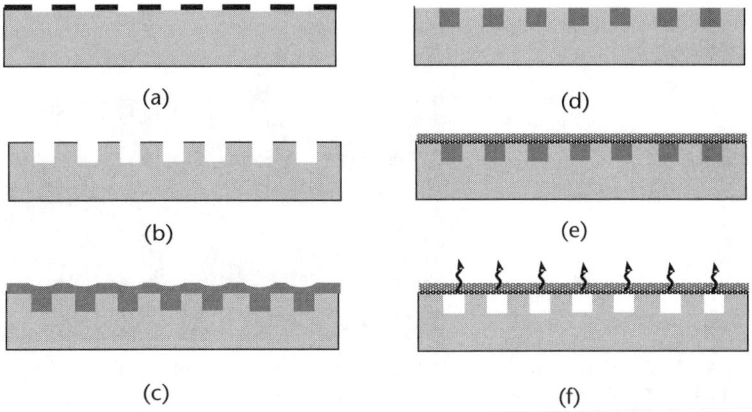

Figure 11.8 Schematic illustration of low-temperature monolithic process of micrchannel fabrication in Si wafer through a sacrificial polymer and an overcoat. (a) Trench patterning, (b) trench etching and surface cleaning, (c) trench refilling by spin-coat, (d) polishing and descum, (e) over-coating with a layered structure (a porous SiO_2 layer and a polymer layer), and (f) sacrificial polymer decomposition and overcoat curing.

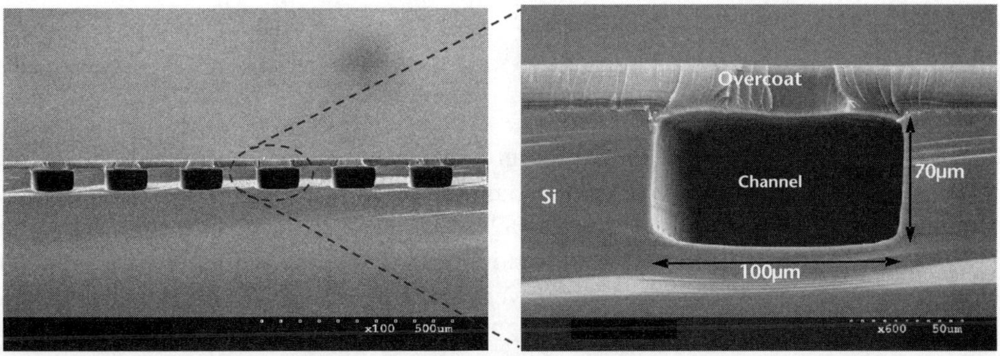

Figure 11.9 Cross-sectional SEM micrographs of microchannels enclosed by monolithic sacrificial polymer method.

total thickness). In addition to the dielectric polymer, many other materials may also be used as hermetic coating, such as metal layers deposited by dc sputtering and electro-/electroless plating, spin-on glass, and so forth.

The important requirements for the polymer-capped microchannels include strong adhesion, good hermeticity, and high mechanical strength under high pressure, which are critical for the reliability of the microchannels. Table 11.1 lists the basic physical properties of several dielectric polymers that are commonly used for wafer-level packaging. In this work, Avatrel 2195P (Promerus) was used as the overcoat polymer because it has been known to serve as a low-stress dielectric polymer in various packaging applications for several years [28–30]. Compared with other candidate polymers, Avatrel 2195P has the following advantages: it can be cured at low temperature and has a relatively high T_g; it is compatible with a variety of commonly used passivation films, such as SiO_2 and Si_3N_4, with good adhesion. More importantly, its moisture absorption rate is relatively low (<0.1% at 100% R.H.). Another reliability concern with polymer-capped microchannels is the inter-

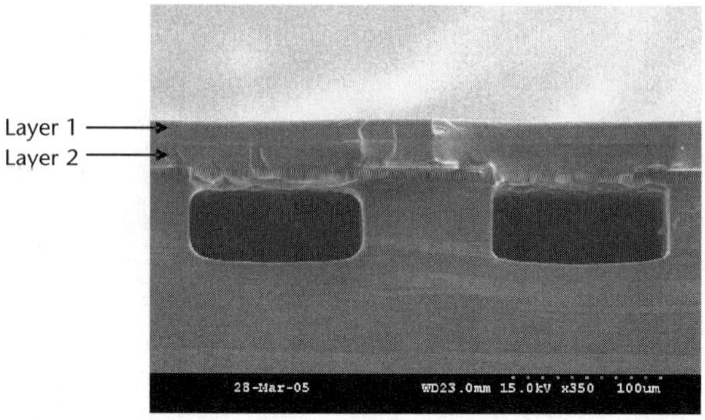

Figure 11.10 Cross-sectional SEM photograph of completed microfluidic channels enclosed by two layers of polymer overcoat.

Table 11.1 Physical Properties of Various Dielectric Polymers Used in Wafer-Level Packaging Applications [30, 31]

Polymer	*CTE (ppm)*	*Tensile Modulus (GPa)*	*Tg (°C)*	*Moisture Absorption rate*	*Curing Temperature (°C)*
Avatrel	~50	0.5	250~350	<0.1%	~160
Polyimide	~20	50	~430	2–3%	~300
SU-8	~42	4.7	>200	0.3%	~160
BCB	~52	2.9	~350	~0.23%	~250

facial integrity of the layered overcoat during thermal stressing. The overcoat polymer, Avatrel 2195P, reportedly has good adhesion with various substrates and a low residual stress [30]. A thermal-cycling test was performed for the microchannels that were tested with high-pressure flow described above. No cracking or delamination was observed at the polymer layer after 1,000 thermal cycles based on the JEDEC Standard (Condition J of JESD 22-A104-B, 0°C to 100°C, three cycles per hour).

11.4 Integration of Microfluidic and Electrical Chip I/O Interconnections

In addition to the microchannels, microfluidic interconnections to and from the microchannel heat sink are needed to enable on-chip microfluidic cooling. In this section, the concept of chip thermofluidic I/Os is introduced to enable chip-to-board fluidic interconnection. Figure 11.11 illustrates a process used to fabricate and integrate the microchannel heat sink, the thermofluidic chip I/O interconnections, and C4 solder bumps on a silicon die using wafer-scale batch processing. The process begins following the completion of semiconductor BEOL processing [Figure 11.11(a)]. A two-step Si etch process is performed to fabricate the through-vias and microchannels on the back side of the Si dice. In the first masking step [Figure 11.11(b)], vias are defined through the photoresist and etched partially into the Si wafer. Trenches are then patterned and etched using the second masking step [Figure 11.11(c)]. The vias obtained in the first etch step are aligned with the trenches such that the vias are etched further (and simultaneously) while the trenches are etched. Silicon etching is stopped by the SiO_2 passivation film on the front side of the wafer. In this manner, through-chip vias are produced to achieve

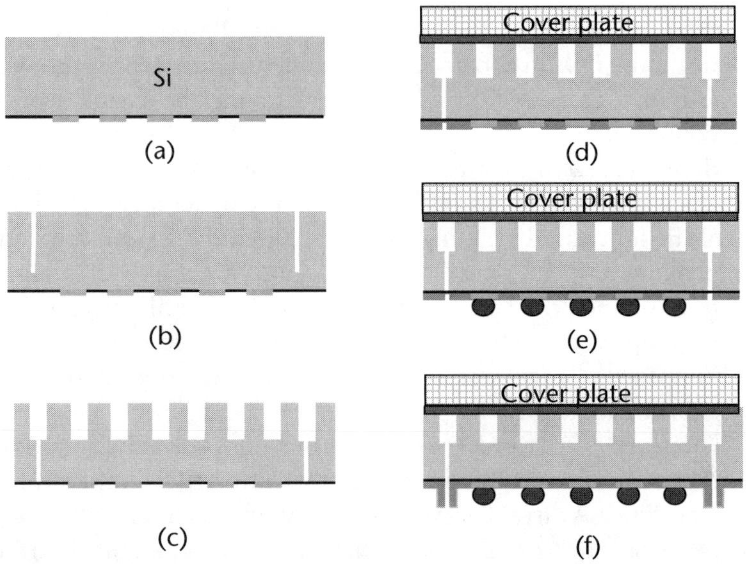

Figure 11.11 Schematic of the fabrication process used to integrate the microchannel heat sink, the thermofluidic chip I/O interconnections, and C4 solder bumps on a silicon die using wafer-scale batch processing: (a) the process begins after the completion of semiconductor BEOL processing, (b) through-wafer vias are patterned and partially etched, (c) trenches are patterned and simultaneously etched with the partially etched through-wafer vias, (d) microchannels are covered with glass or Si plate and a polymer passivation is deposited and pattern at the front-side, (e) solder bumps are deposited on the front-side using electrodeposition, and (f) polymer pipes are fabricated by a thick photodefinable polymer.

fluidic interconnection from the front side to the back side of the chip. Next, a glass plate with precoated polymer film (Avatrel 2195P) is bonded to the side of the wafer containing the trenches and cured under pressure in a N_2-purged oven [Figure 11.11(d)]. After curing, the bonded Si/glass wafer is flipped to begin processing of the front side of the wafer. A polymer passivation layer is first coated and patterned to form vias for both the fluidic through-wafer vias and electrical pads. At this stage in the process, the wafer is ready for conventional solder bumping processes to fabricate the chip-to-package electrical interconnections. After solder reflow with the aid of a liquid flux, spherical solder bumps are formed [Figure 11.11(e)]. In this work, Sn/Pb (60/40) solder is used due to availability and its excellent wettability, which usually requires a peak reflow temperature of 220°C. If necessary, other solders, including Pb-free solders like SnCu, SnAg, or SnAgCu, may also be used in response to the global transition to Pb-free electronics manufacturing. Next, using a thick layer of Avatrel 2195P, an area-array distribution of polymer micropipes is fabricated to function as fluidic I/O interconnections between the chip and the package-level fluidic networks [Figure11.11(f)]. The polymer micropipes are fabricated by photodefinition of the Avatrel polymer film [31]. The micropipes can also be fabricated using other photodefinable dielectric polymers such as SU-8. The micropipes are aligned with the through-chip fluidic vias in order to form an interconnection with the microchannel heat sink on the back side of the chip. After fabrication, the polymer micropipes are cured at 160°C for 1 hour. The passivation layer inside the pipes is next etched using wet etching to allow fluidic circulation. Following this process step, the wafer is diced. Based on this approach, wafer-level thermofluidic I/Os are formed and can be used to deliver the working fluid directly to the monolithically fabricated microchannel heat sink from board-level fluidic interconnects.

Figure 11.12 illustrates SEM images of a row of micropipes fabricated along with an area array of C4 solder bumps at the wafer level. The inner and outer diameters of the fabricated polymer pipe are 100 and 250 μm, respectively, and its height is approximately 100 μm. The solder bumps are approximately 65 μm in height and 80 μm in diameter. The micropipes are intentionally designed to be taller than the solder bumps because they will be partially inserted into orifices (vias) when the solder bumps make contact with the Cu pads on the top interconnect layer on the substrate during assembly.

The slope of the polymer micropipe sidewall may be modified by heating the polymer micropipes following their fabrication to a temperature above the T_g of the polymer. Since Avatrel 2195P has a T_g of approximately 250°C, the polymer structure deforms when heated above this temperature. Figure 11.13 illustrates the slight deformation of the polymer micropipe after heating to 260°C and 280°C. In practice, such tapered micropipes may be desirable to assist and enhance the alignment/insertion of the micropipes with orifices on the board.

Furthermore, the polymer micropipes can be metalized to improve mechanical strength. Solder may be capped on the top of the polymer (or metallic) micropipe I/Os, as shown in Figure 11.14, to seal the fluidic I/Os. The plating and reflow processes are compatible with the solder bumps fabricated for the electrical I/Os. Thus, this could potentially enable the use of conventional solder to enable the interconnection of the electrical and the fluidic chip I/Os.

11.5 Flip-Chip Assembly of Die with Electrical and Thermofluidic I/Os

Two distinct substrate-level coolant-delivery schemes are proposed and are illustrated in Figure 11.15. In Figure 11.15(a), the microfluidic channels are embedded on the front side of the substrate, and orifices are formed through the overcoat. One

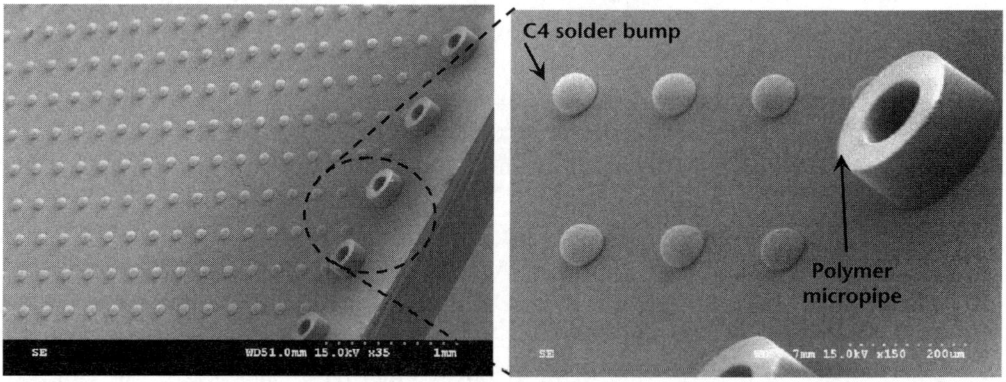

Figure 11.12 SEM micrographs of a row of polymeric micropipes fabricated adjacent to an area-array distribution of solder bumps.

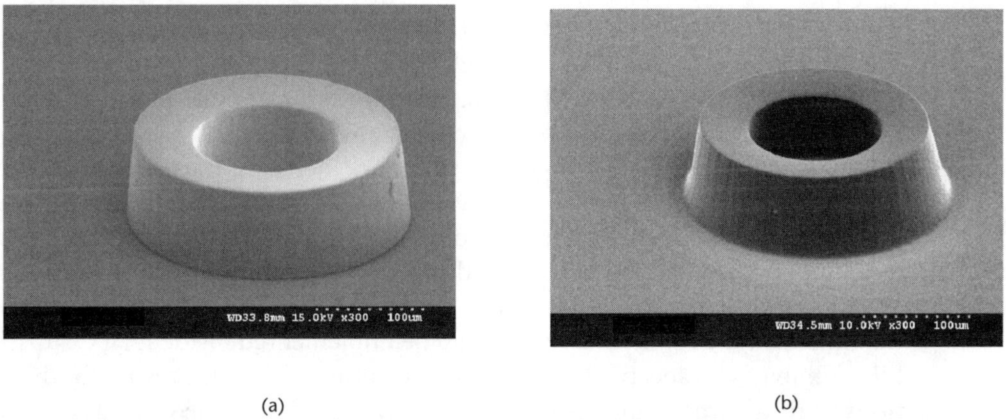

Figure 11.13 SEM micrographs of tapered polymer micropipes formed as a result of heating to a temperature above the T_g of the polymer: (a) 1 hr at 260°C, (b) 1 hr at 280°C.

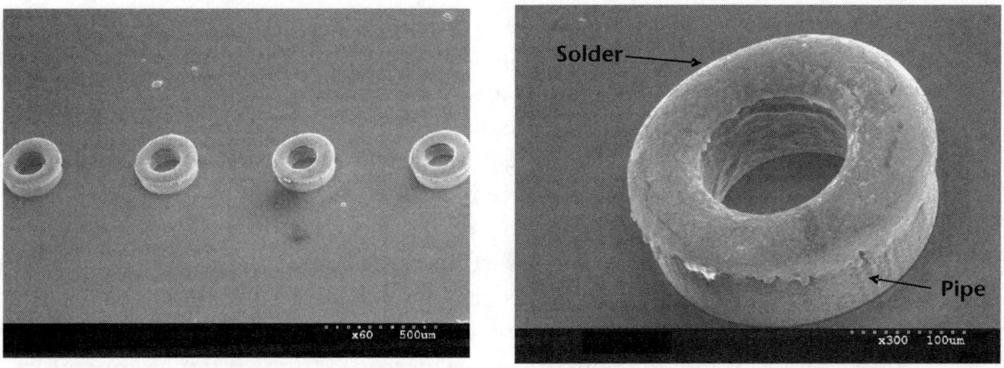

Figure 11.14 SEM micrograph of solder-capped polymer micropipe.

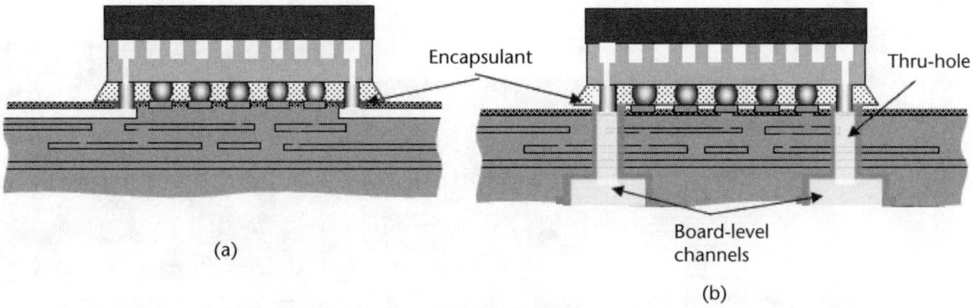

Figure 11.15 Schematic of two distinct chip coolant delivery schemes to enable the on-chip microchannel heat sinks: (a) manifold channels are located at the front-side of the board, and (b) manifold channels are located at the back-side of the board and large through-holes are used to interconnect with the micropipes, and thus, the on-chip microchannel heat sink.

challenge with the implementation of this scheme is that the available area on the chip-mounting side of the substrate is usually limited and occupied by signal routing. To address this potential incompatibility, a second configuration was developed and is illustrated in Figure 11.15(b). In this configuration, the manifold channels are embedded inside the multilayer board or fabricated at the opposite side of the board, and through-holes are utilized for fluidic interconnection. Since through-holes are commonly available in printed circuit boards (PCBs) and the channels can be fabricated through a variety of methods, the proposed processes can potentially be adopted into the PWB technologies—although preventing water absorption by the board materials is obviously important. For instance, the channels can simply be embedded or enclosed through the multilayer lamination process in the standard PCB fabrication. Bauer et al. [22] and Ilgen [23] have demonstrated that microfluidic channel networks can be implemented inside LTCC substrates and FR4 laminates, respectively. In this work, although substrates for both configurations were fabricated and tested, emphasis was placed on the second configuration.

Figure 11.16 illustrates the micrographs of the front side of a silicon test substrate where a peripheral distribution of fluidic vias is fabricated and integrated with the electrical traces. The copper traces are ~8 μm in thickness and were first patterned and etched. A polymer passivation film was applied to serve as the solder mask for the Cu pads. Since the overcoat polymer (Avatrel 2195P) is photofinable, vias are easily formed for both the fluidic and electrical interconnects to accommodate the micropipe inlets/outlets and solder bumps, respectively, on the test chips.

The chip-assembly method is developed based on common flip-chip assembly and encapsulation processes. The basic procedure of the thermofluidic I/O assembly is illustrated in Figure 11.17. Since the polymer micropipes are designed to be taller than the solder bumps, a portion of the polymer micropipes is inserted into the orifices when the solder bumps come into contact with the metal pads on the substrate. During the bonding process, both the chip and the substrate are heated to 150°C after alignment. A compression force (200g) is next applied to bring them into contact. While the temperature is increased to the reflow temperature (220°C), the compression force is maintained. Once reflowed, metallurgical joining forms between the solder bumps and copper pads on the substrate, thus forming mechani-

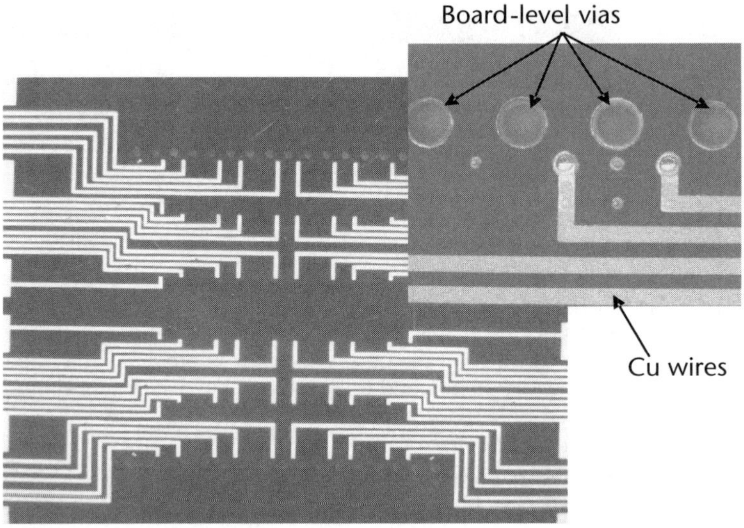

Figure 11.16 Optical micrographs of test board substrates containing copper interconnects on the front-side and through-vias for the thermofluidic I/O configuration in Figure 11.8(b).

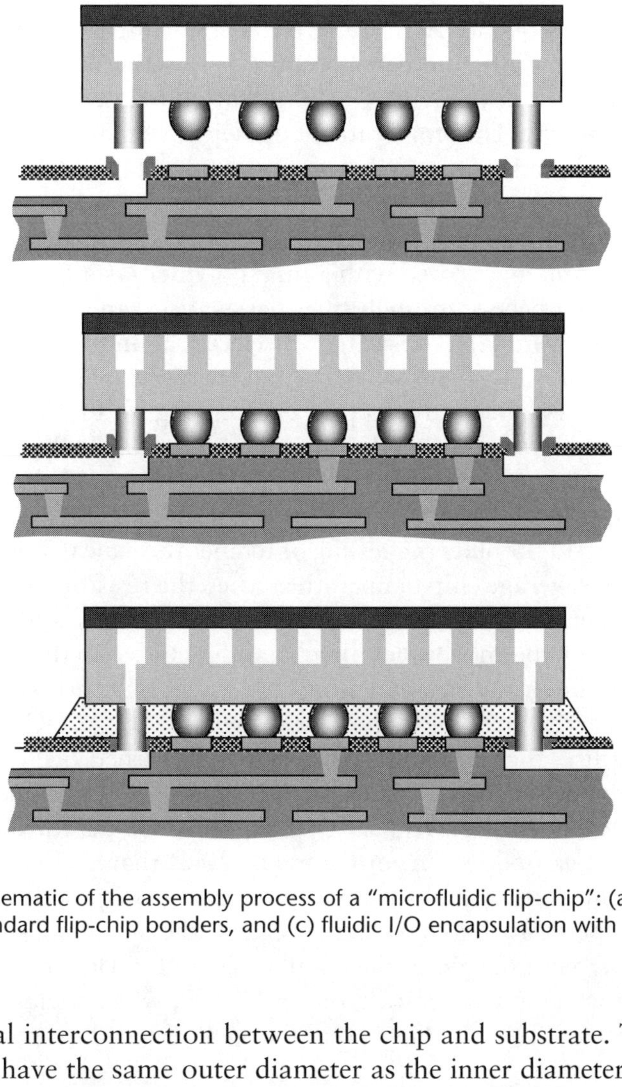

Figure 11.17 Schematic of the assembly process of a "microfluidic flip-chip": (a) alignment, (b) bonding using standard flip-chip bonders, and (c) fluidic I/O encapsulation with epoxy (underfill).

cal and electrical interconnection between the chip and substrate. The micropipes are designed to have the same outer diameter as the inner diameter of the orifices. However, since Avatrel 2195P is a negative tone photopolymer, the diameter of the micropipes is usually slightly larger than that of the orifices. Therefore, the micropipes are actually squeezed into the orifices during bonding, which is advantageous as it helps to create a tight seal. To ensure hermetic sealing of the micropipes and improve reliability of the solder bumps, a liquid epoxy is applied at the chip edge after assembly, which is reminiscent of conventional underfill processing. After curing of the underfill (for approximately 1 hour at 100°C), the polymer micropipes are joined with the orifices, and the hermetic fluidic path between the chip and the board is formed.

11.6 Thermal Measurements

This section will describe the details of the test chips, measurement methodologies, and demonstration of the cooling operation. Multifunction thin-film heaters/thermometers were designed and fabricated on the test chips. The heaters are made of thin-film Ti/Pt (~300Å/1,500Å) using a lift-off process. Pt was selected for the thin-film heater/thermometer because it is a thermally stable noble metal and its electrical resistivity is linearly dependent on temperature for the range of temperatures of interest. The temperature coefficient of the Ti/Pt film is 0.0033 ~ 0.0039 K^{-1} and can be used at temperatures up to 800°C [32]. Therefore, the heaters can also serve as thermometers for the purpose of temperature sensing. After the heaters/thermometers were fabricated, the wafer surface was passivated with a SiO_2 layer (0.2 μm) and coated with a final polymer layer (~10 μm thick Avatrel 2195P). Vias were opened through the passivation and solder bumps, and polymer micropipes were then fabricated for electrical and fluidic interconnections, respectively. Figure 11.18 shows SEM micrographs of the test die that contains the microchannels enclosed by a pyrex wafer [Figure 11.18(a)] and the front surface of the fabricated test die with the integrated thin-film Ti/Pt resistors, solder bumps, and polymer micropipes [Figure 11.18(b)]. The heights of the solder bumps and the micropipes are 75 and 90 μm, respectively.

Figure 11.19 plots the results of temperature measurements at the inlet and outlet and the average chip temperature when the resistors were heated with a power of 45W under flow rates of 65 and 104 mL. The inlet and outlet temperatures were measured by thermocouples directly and recorded in the computer software. Before each measurement, the circulating chiller was first turned on to maintain a low temperature at the metal block heat exchanger (20°C ~ 22°C). After attaining the target temperature, the DI water pump was then turned on for circulation through the microchannel heat sink. After the DI water flow rate was stabilized, the dc power supply was turned on to heat the chip, and temperature measurements were performed. It can be seen from Figure 11.20(a) that the on-chip temperature and the outlet temperature increase rapidly once power is applied to the heaters, while the inlet temperature does not change because the metal block heat exchanger dissipates the power with the aid of the circulating chiller. Under a flow rate of ~65 mL/min, the average chip temperature rise is 18.1°C, which yields a corresponding thermal resistance of 0.4°C/W. Given that the total tested chip area is 0.6 cm^2, the unit ther-

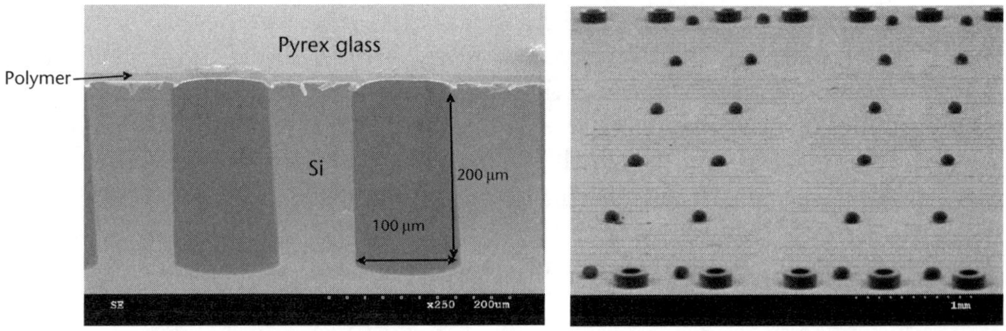

(a) (b)

Figure 11.18 SEM images of the test die used for thermal test measurements. (a) The Pyrex enclosed microfluidic heat sink and (b) front side of the test die are shown. The front side of the die contains the Pt heaters/thermometers, solder bumps, and fluidic micropipe I/Os. The microchannel heat sink contains 51 microchannels distributed across a chip area of 1 cm^2.

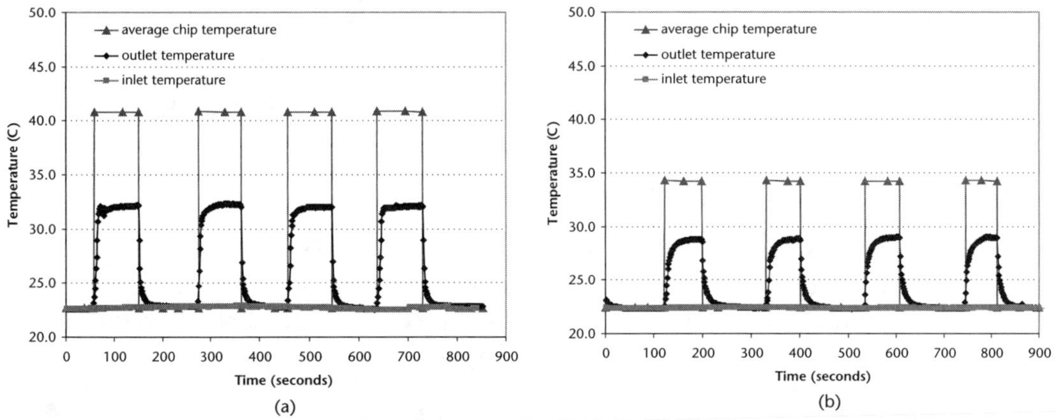

Figure 11.19 Measured temperatures at the inlet, outlet, and average chip temperature at a flow rate of 65 ml/min (a) and 104 ml/min (b). Under a flow rate of 65 ml/min, the measured unit thermal resistance is approximately 0.24°C-cm²/W, and is 0.17°C-cm²/W under a flow rate of 104 ml/min. Microchannel heat sink dimensions are shown in Figure 11.11.

mal resistance is approximately 0.24°C-cm²/W. Under a larger flow rate (~104 mL/min), the average temperature rise is 12.7°C, and the corresponding thermal resistance for the chip is 0.28°C/W, which yields a unit thermal resistance of 0.17°C-cm²/W.

As described previously, an important benefit of using an on-chip microchannel heat sink is to enable the cooling of highly localized power density (or hot spots). The heater design in this work allows the emulation of hot spots. Thus, experiments were performed by supplying a large localized power in a small area (~0.08 cm²). Figure 11.20(b) illustrates the measured temperature rise (at the heated area) as a function of localized power density. As expected, the measured temperature rise on the heaters increases with increasing power density. The temperature rise is reduced

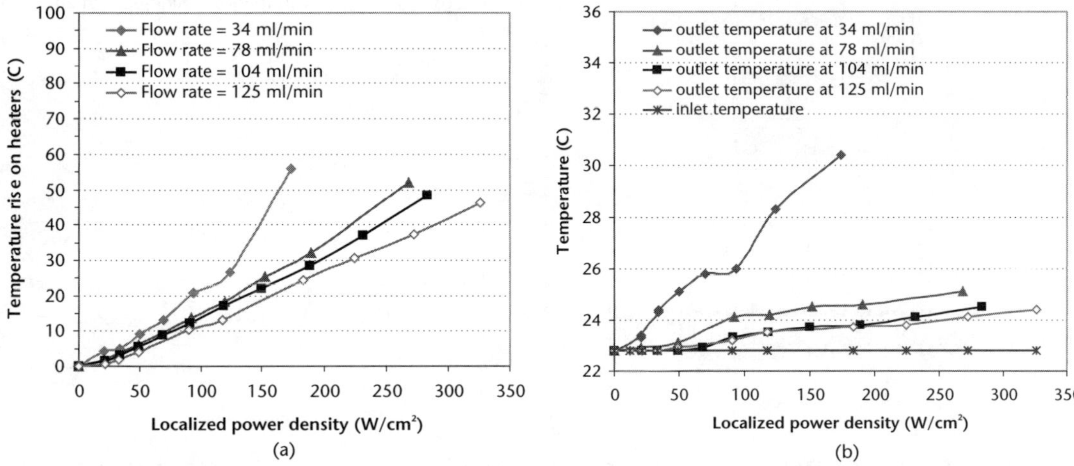

Figure 11.20 (a) On-chip temperature rise as a function of power density under various DI water flow rates (heating area is ~ 0.08 cm²). (b) Measured inlet temperature and outlet temperature as a function of power density for various DI water flow rates. Microchannel heat sink dimensions are shown in Figure 11.11.

as the DI water flow rate increases. Under a larger flow rate, the ΔT – power curve becomes nearly linear, and a localized power density greater than 325 W/cm^2 was tested with a temperature rise of 46°C at a flow rate of 125 mL/min.

11.7 Hydraulic Requirement Analysis

The principles of on-chip microchannel heat sinks were first studied by Tuckerman and Pease in 1981 [5]. More advanced closed-form models were also developed and can be found in [9]. It is generally accepted that the thermal resistance of a microchannel heat sink includes three physical contributions:

$$R_{th} = R_{cond} + R_{conv} + R_{heat} \tag{11.1}$$

where R_{cond} (K/W) is determined by the thermal conductivity (W/mK); R_{conv}, the convective thermal resistance, depends on the solid-liquid heat-transfer coefficient, h (W/Kcm^2), and overall channel surface area, A (cm^2); and R_{heat}, the thermal resistance due to the caloric heating of the working fluid, is inversely proportional to the liquid flow rate, $\dot{V}(m^3/s)$, for a given fluid. For any given microchannel dimensions, a larger flow rate yields lower thermal resistance. Unfortunately, large flow rates require high pressure to drive liquid flow. Assuming DI water is used as the working fluid and that a fully developed laminar flow is present in the microchannels, the required pressure drop can be estimated as a function of the chip power (heat flux) based on hydrodynamic equations in the literature [9, 33, 34]. As illustrated in Figure 11.21, the fluidic paths between an inlet and outlet for a straight channel pattern include the through-vias, the micropipes, and the microchannels. Since the micropipes and the vias are circular channels, the pressure drop (Pa) can be calculated using [33]

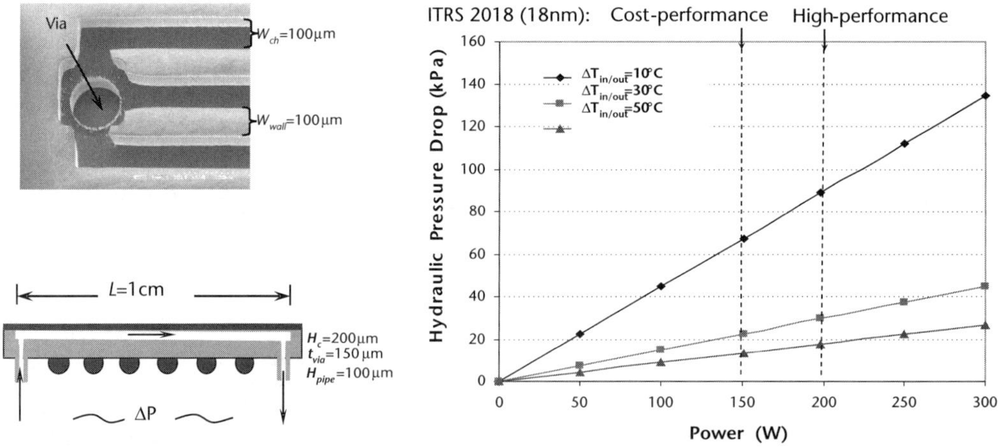

Figure 11.21 Illustration of the on-chip thermofluidic path and the calculated pressure drop as a function of power for various temperature rises between the inlets and the outlets with DI water as the coolant assuming fully developed flow (W_c =100 μm, H_c = 200 μm, $n_{channels}$ = 51, #_of_micropipes = 34, D_{via} =100 μm).

$$\Delta P = 2f\,\mathrm{Re}\frac{L_c}{D_h^2}\upsilon\mu \tag{11.2}$$

where the product of the friction (*f*) and the Reynolds number (Re) is a constant:

$$f\,\mathrm{Re} = 64 \tag{11.3}$$

and μ (Pa.s) is the viscosity of the coolant. Since the hydraulic diameter, D_h (μm), of a via equals its diameter, the velocity, ν (m/s), can be calculated as

$$\upsilon = \frac{flow\ rate}{overall\ cross-sectional\ area} = \frac{\dot{V}}{n_{via}\pi\left(\frac{D_{via}^2}{4}\right)} \tag{11.4}$$

Therefore, the pressure drop in the vias and micropipes can be expressed as a function of the number of vias (n_{via}), via height (H_{via}), micropipe height (H_{pipe}), and via diameter (D_{via}):

$$\Delta P_{via+microscope} = 512\cdot\frac{\mu\cdot\dot{V}\cdot\left(2H_{via}+2H_{pipe}\right)}{n_{via}\cdot\pi\cdot D_{via}^4} \tag{11.5}$$

The pressure drop, ΔP_{ch}, in the microchannels can be estimated using [34]

$$\Delta P_{ch} = \frac{1}{2}f\cdot\mathrm{Re}\cdot L_c\cdot\frac{\mu\cdot\dot{V}(1+W_c/H_c)^2}{n\cdot H_c\cdot W_c^3} \tag{11.6}$$

As a result, the total pressure drop between an inlet and outlet can be calculated as

$$\Delta P_{inlet/outlet} = \Delta P_{ch} + \Delta P_{via+micropipe} \tag{11.7}$$

Based on the above equations, the minimum hydraulic pressure drop can be calculated for given microchannel dimensions and power projection requirements. Figure 11.21 plots the calculated pressure drop as a function of the power dissipation and temperature between the inlets and outlets (caloric heating of the coolant).

Using the dimensions of the microchannels fabricated in this work (n_{ch} = 51; W_{ch} = 100 μm; H_{ch} = 200 μm) and the thermofluidic I/O interconnects in the present design, the model suggests that it is possible to cool >100W/cm^2 power dissipation with a pressure drop of less than 1 atm. It must be pointed out that the pressure drop in the above analysis is the minimum required pressure because other effects, such as flow splitting, combining, or turning, have not been included. These effects may contribute to the overall pressure drop as discussed in Chapter 10.

In practice, the additional pressure drop along the fluidic paths at the board level may also need to be considered, depending on their dimensions, their total length, and the nature of flow (turbulent or laminar). In this work, pressure loss in the board-level fluidic paths is negligible because of their short length and large hydraulic diameters. As indicated in (11.7), the pressure loss in fluidic I/Os (vias and micropipes) may play an important role in the overall pressure drop if their number is small. However, increasing the number of the fluidic I/Os requires more chip area for the vias and the micropipes. Therefore, chip-area overhead also should be considered for the through-vias and micropipes.

Figure 11.22 illustrates the calculated pressure drop on the thermofluidic I/Os in reference to the chip area they occupy. Obviously, the theoretical pressure drop on the fluidic I/Os can be largely reduced if the number of fluidic I/Os increases, while the chip area occupied by the fluidic I/Os increases linearly. Given that the microfluidic I/Os are batch-fabricated and are assembled by flip-chip bonding, increasing the number of fluidic I/Os will essentially have no impact on cost (for fabrication and assembly) and will not require any additional processes or steps to assemble.

Using the current design, the 17 pairs of through-silicon fluidic vias (D_{via} = 100 μm) occupy an area of 0.00267 cm^2, while the 17 pairs of micropipes (D_{pipe} = 250 μm) occupy an area of 0.0167 cm^2. For a chip that is 1.4 cm^2 (cost-performance applications projected by ITRS), these thermofluidic I/Os occupy less than 1.19% of the chip area. Therefore, the incurred chip-area penalty is a small fraction of the total chip area. The corresponding pressure loss in the fluidic I/Os is approximately 3.2 KPa. For a chip that is 3.1 cm^2 (high-performance applications projected by ITRS), more fluidic I/Os may potentially be accommodated with even smaller pressure loss in them. Of course, the number of thermofluidic I/Os and the size of the vias and micropipes can be further optimized to provide a sound trade-off between the pressure drop and the chip-area penalty in practice.

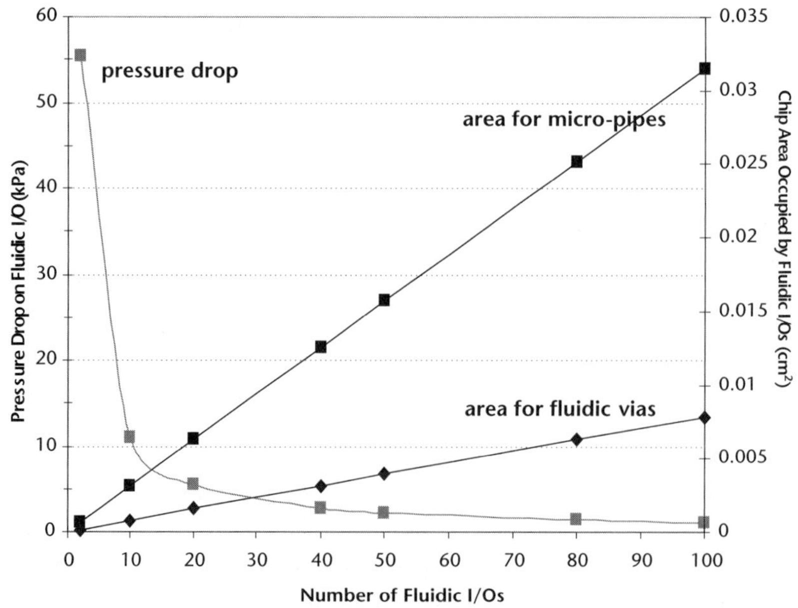

Figure 11.22 Illustration of the trade-off between the chip area penalty and pressure drop on the thermofluidic I/Os as a function of the number of thermofluidic I/Os. A power of 100W is assumed with a fluid temperature rise of 30°C. The dimensions are same as those in Figure 11.21.

11.8 Microfluidic Network to 3D Microsytems

The concept of 3D component stacking was proposed as early as the 1960s [35]. However, commercial products with chip stacking are mainly limited to low-power applications because effective cooling has been known to be one of the major challenges for 3D microsystems [10, 36, 37]. The on-chip microchannels, thermofluidic I/O interconnects, and board-assisted fluidic delivery developed in this work can enable an integrated liquid-cooling platform for a 3D stack of chips to enable cooling of >100 W/cm^2 for each high-power-density chip. Figure 11.23 shows a proposed microfluidic-network cooling scheme that has the potential to be used for cooling three-dimensional ICs. Each silicon die of the 3D stack contains the following features: (1) a monolithically integrated microchannel heat sink; (2) through-silicon electrical (copper) vias (TSEV) and through-silicon fluidic (hollow) vias (TSFV), the latter being used for fluidic routing in the 3D stack; and (3) solder bumps (electrical I/Os) and microscale polymer pipes (fluidic I/Os) on the side of the chip opposite to the microchannel heat sink [31]. Microscale fluidic interconnection between strata is enabled by through-wafer fluidic vias and polymer pipe I/O interconnects. The chips are designed such that when they are stacked, each chip makes electrical and fluidic interconnection to the dice above and below. Consequently, power delivery and signaling are supported by the electrical interconnects (solder bumps and copper TSVs), and heat removal for each stratum is supported by the fluidic I/Os and microchannel heat sinks. The process used for assembly of the 3D prototype is similar to that described in Section 11.5, thus compatible with conventional flip-chip bonding.

The process used to fabricate the dice is shown in Figure 11.24. The process begins by fabricating electrical TSVs [Figure 11.24(a)] followed by the fabrication of trenches and microfluidic TSVs into the silicon wafer, as shown in Figure 11.24(b). Next, the trenches are encapsulated to form the microchannels, as shown in Figure 11.24(c) and described earlier. Vias are next formed in the overcoat polymer to simultaneously expose the electrical TSVs and form fluidic vias that ultimately allow fluid flow to the upper and lower dice. Following this process step, copper pads are patterned above the electrical TSVs to facilitate solder bonding dur-

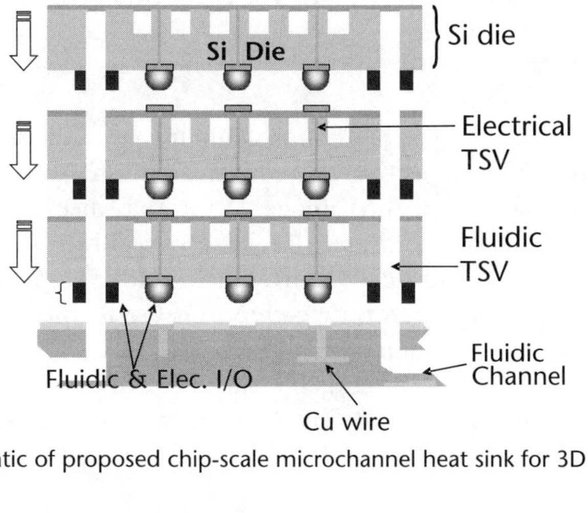

Figure 11.23 Schematic of proposed chip-scale microchannel heat sink for 3D integrated circuits.

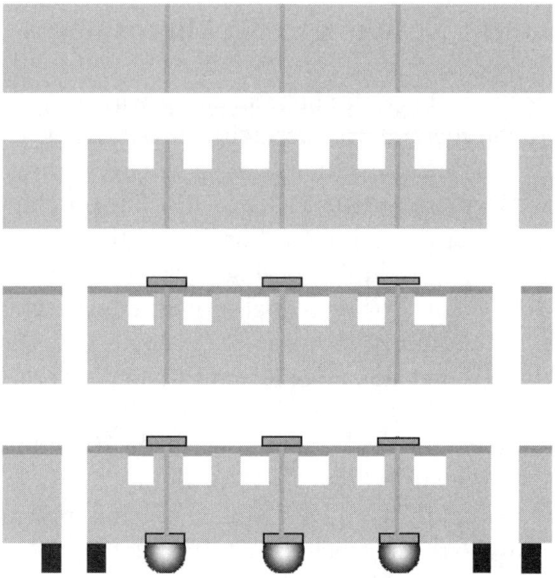

Figure 11.24 (a–d) Schematic illustration of the process used to fabricate silicon dice, at the wafer level, that each contain electrical and microfluidic TSVs and I/Os.

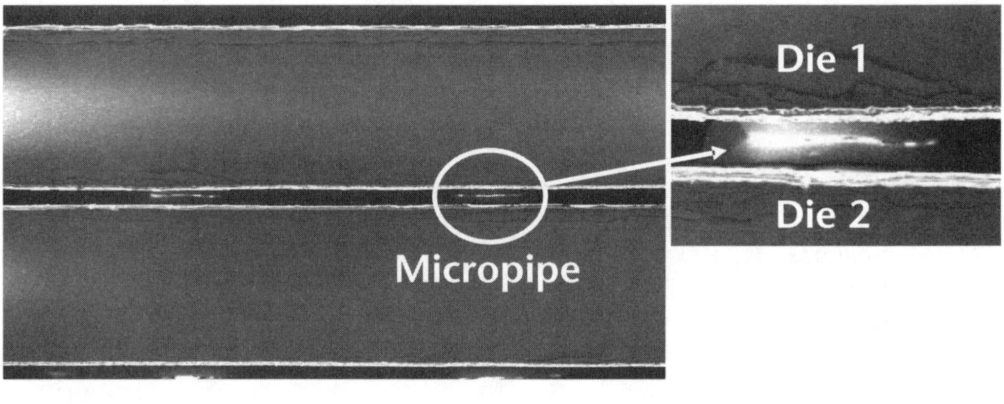

Figure 11.25 Cross-sectional SEM image of 3D microfludic chip-to-chip bonding.

ing assembly. Finally, solder bumps and microfluidic polymer micropipes (electrical and fluidic I/Os, respectively) are fabricated as described earlier in the chapter.

A cross-sectional SEM image of the two-dice stack highlighting the polymer micropipes is shown in Figure 11.25. Figure 11.26 shows a 3D stack of two chips assembled to a substrate and a 3D stack of four chips assembled to a substrate [38, 39]. Electrical and microfluidic chip I/Os are formed at each level in the 3D stack. Cross-sectional optical images of fabricated electrical TSVs in a silicon wafer without and with a microchannel heat sink are shown in Figure 11.27. In Figure 11.27(b), the microchannel is 200 μm tall and 100 μm wide, while in Figure 11.27(c), the microchannel is 300 μm tall and 100 μm wide. These fabrication

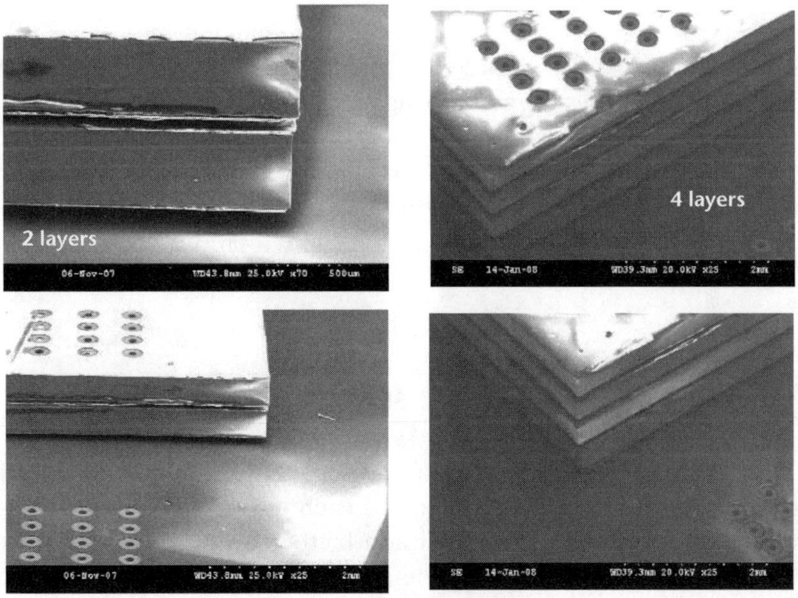

Figure 11.26 SEM images of 2-chip and 4-chip 3D stacks.

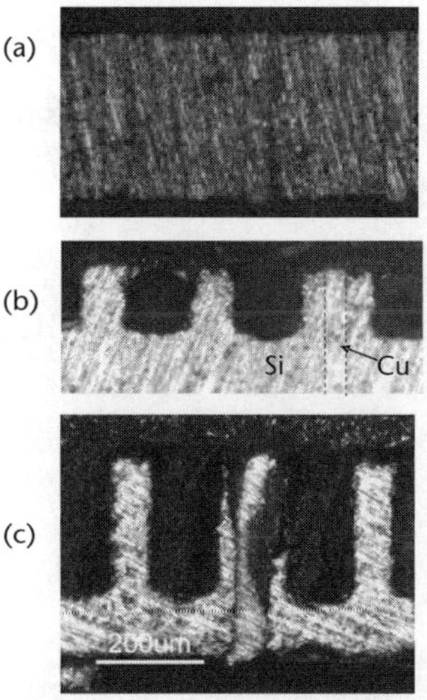

Figure 11.27 Optical images of a silicon wafer with through-silicon electrical vias (a) and the subsequent fabrication of two different aspect ratio microchannel heat sinks around the TSVs (b) and (c).

results demonstrate that the electrical TSVs, which are critical to 3D system integration, can be integrated with a wide range of microchannel geometry to meet a wide range of thermal-resistance and pressure-drop requirements. In both cases, the overall chip area is 1×1.2 cm^2, and the copper TSVs are 50 μm in diameter.

While underfill can be used as a sealing method for the fluidic interconnection, another option is solder ring–based hermetic sealing, as suggested in [40]. As shown in Figure 11.28, solder ring–based fluidic sealing has been successfully demonstrated to form compact 3D packages by stacking multilayer FR4 substrates. The same concept can be extended to the microscale with wafer-level batch fabrication. Figure 11.29 shows the solder rings that can be fabricated and reflowed with solder bump interconnects simultaneously to simplify the processing steps.

Compact models for microchannel heat sinks have been obtained in [5, 9, 31, 34]. The ratio of pressure drop in fluidic vias and pipes ($\Delta P_{via\text{-}pipe\text{-}total}$) to the pressure drop in microchannels ($\Delta P_{channels}$) for the two-chip stack can be expressed as

$$\frac{\Delta P_{channels}}{\Delta P_{via-pipe-total}} = \frac{f \operatorname{Re} L_c \left(1 + \frac{W_c}{H_c}\right)^2 n_{via/pipe} \pi D^4_{via/pipe}}{1024 n_c H_{via-pipe-total} H_c W_c^3} \tag{11.8}$$

Note that this equation assumes separate and parallel fluid pathways from the external fluidic tube from the top to each die in the two-chip 3D stack. Here, Re is the Reynolds number, f is a friction factor, W_c and H_c are the width and height of each microchannel, $n_{via/pipe}$ is the number of fluidic TSVs and pipes, $D_{via/pipe}$ is the inner diameter of fluidic TSVs and pipes, n_c is the number of microchannels on each chip, and $H_{via\text{-}pipe\text{-}total}$ is the total length of fluidic TSVs/pipes for the die having the longest fluidic pathway. Substituting L_c = 10 mm, W_c = 100 μm, H_c = 200 μm, $D_{via/pipe}$ = 250 μm, $H_{via\text{-}pipe\text{-}total}$ = 0.9 mm (for 400 μm silicon chips and two 50 μm micropipes), $n_c = 2n_{via/pipe} = 50$, we attain

$$\frac{\Delta P_{channel}}{\Delta P_{via-pipe-total}} \approx 12 \qquad \frac{Silicon\ area\ for\ fluidic\ TSVs}{Chip\ area} \approx 2.5\% \tag{11.9}$$

The above numbers reveal that fluidic TSVs and pipes consume minimal surface area for a two-chip, microchannel-cooled 3D IC; at the same time, they have negligible pressure drop through them. This is largely because the total length of fluidic vias and pipes is only 0.9 mm, while the length of microchannels is as high as 10 mm.

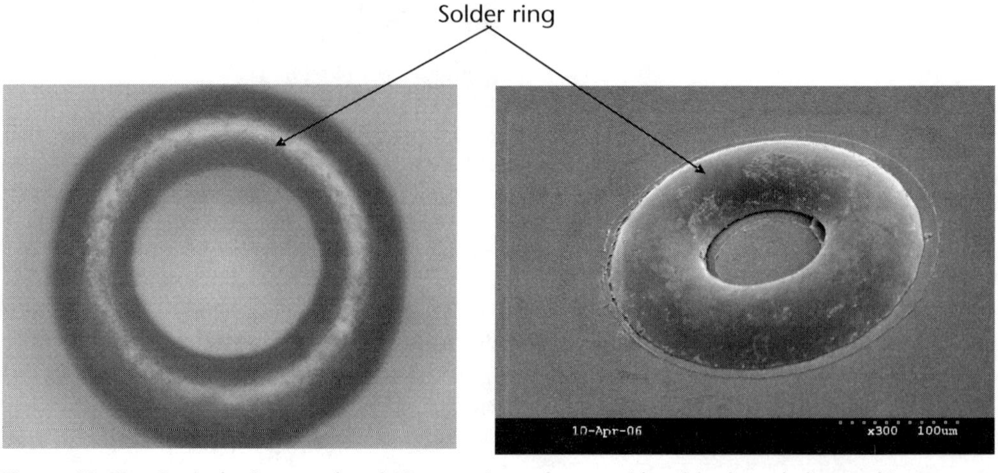

Figure 11.28 Optical micrographs of 3D organic packages with solder based electrical interconnection and thermal-fluidic sealing [39].

This means the fluidic network that provides liquid coolant to the microchannel heat sink does not impose any significant overhead for a two-chip stack. Essentially, integrated circuits in a multichip system will have the same thermal resistance irrespective of whether they are placed side by side or form part of two-chip 3D stacks.

Microchannel-cooled 3D ICs can provide significant benefits to high-performance servers. Figure 11.30 shows that a substantial improvement in chip-to-chip interconnect lengths can be obtained for a typical server when two microprocessor chips are stacked. Another benefit of microchannel-cooled servers is the reduced thermal resistance compared to today's air-cooled servers, which translates to improvements in chip performance and/or power. This can be studied with the simple model [41]:

$$P = \frac{T - T_{amb}}{R_{th}} \tag{11.10}$$

$$P = aC_{total}\left[V_{dd}(T)\right]^2 f + N_{gates} V_{dd}(T) I_{leak0}\, e^{\frac{V_t(T)+\Delta V_t}{nkT/q}} \tag{11.11}$$

Here, P is the chip power, T is the chip temperature, T_{amb} is the ambient temperature, R_{th} is the thermal resistance, a is the activity factor, C_{total} is the total capacitance, f is the clock frequency, I_{leak0} is the leakage current coefficient, n is the subthreshold slope factor, and ΔV_t is the threshold voltage variation factor. Nose and Sakurai derived models for the optimal supply voltage [$V_{dd}(T)$] and threshold voltage [$V_t(T)$] of an IC that minimize power for a certain performance constraint [42]. These V_{dd} and V_t values are functions of temperature and other device/system parameters. When these formulas for optimal V_{dd} and V_t are substituted into (11.11), (11.10) and (11.11) represent two equations with two unknowns, power and temperature. If they are solved and simplified, a cubic equation can be obtained that yields closed-form expressions for chip power and temperature for a given ther-

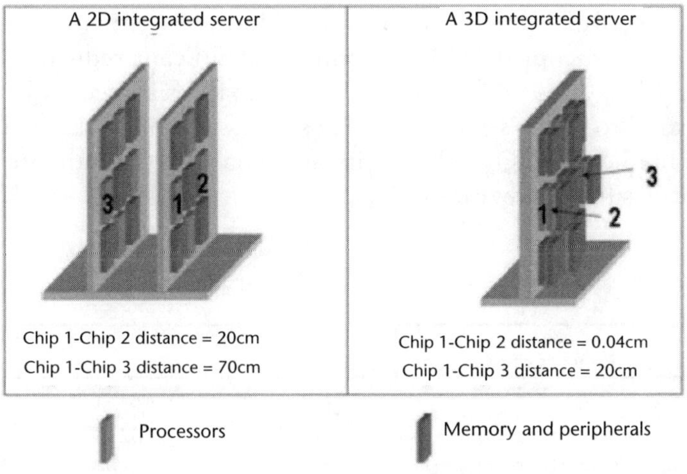

Figure 11.29 Chip-to-chip interconnect length reduction with microchannel cooled 3D-ICs.

mal resistance, device technology, and frequency. This compact model can be utilized to obtain the data in Table 11.2, which shows that significant advances in performance, power, and/or operation temperature can be obtained for each chip of a 65 nm–technology, microchannel-cooled 3D IC. Compared to the air-cooled heat sink case, the frequency of each chip in the 3D stack can be increased by approximately 12% if power is kept constant. The reduced thermal resistance can also lower the operating temperature of each chip in the 3D stack, in this case, from 88°C to 52°C, thus improving the reliability of the components. If the chip frequency is held constant, the power dissipation of each chip in the 3D stack can be reduced by 20%. This would also reduce the operating temperature from 88°C to 47°C, thereby improving the reliability of the components. Finally, if the power dissipation of each chip is allowed to increase, chip frequency can increase by more than 50%. Due to the liquid-cooled heat sink, the higher power associated with this increased frequency can be cooled and chip junction temperature maintained at 88°C.

11.9 Conclusion

A novel chip-scale microfluidic cooling scheme is presented using *wafer-scale batch fabrication* and promises to address the cooling and form-factor needs of future gigascale nanoelectronic systems. The key features of the proposed microfluidic cooling technology include: (1) the low-temperature, CMOS-compatible microchannel heat sink, (2) through-wafer electrical and microfluidic vias, (3) electrical and microfluidic chip I/O interconnects, and (4) flip-chip assembly processes and technology to hermetically seal and collectively bond the electrical and microfluidic chip I/O interconnects within a 3D stack and to the board. The assembly and testing of such chips was successfully demonstrated. It was shown that under a flow rate of ~104 mL/min, the unit thermal resistance of a fabricated microchannel heat sink was 0.17°C-cm^2/W. Further reduction in the thermal resistance can be attained by increasing the height of the channels, for example. Moreover, the challenging thermal interconnect problem of 3D ICs can be addressed using the microchannel-cooled 3D IC technology described in the chapter. Servers with microchannel-cooled 3D ICs promise a significant reduction in chip-to-chip interconnect length as well as improved thermal resistance and electrical performance compared to today's air-cooled 2D servers. In short, the wafer-level, batch- fabricated fluidic chip I/Os will be critical to enabling the ultimate performance gains of gigascale silicon integration.

Table 11.2 Benefits of Microchannel Cooled 3D-ICs

	Frequency	*Power*	*Temperature*
Air cooled processor R_{th} = 0.6° C/W	3 GHz	102 W	88°
Microchannel cooled processor	3 GHz	83 W	47°
R_{th} = 0.24° C/W	3.3 GHz	102 W	52°
	4.5 GHz	254 W	88°

Acknowledgments

The authors acknowledge the support of the Interconnect Focus Center. This work is also in part based upon work supported by the National Science Foundation under Grant No. 0701560. The authors are thankful to Dr. Paul Kohl, Dr. Paul Joseph, the School of Chemical Engineering at Georgia Tech, and Promerus, for providing the polymer materials. Many thanks are extended to Dr. Xiaojin Wei and Dr. Yogendra Joshi for their discussion and help with testing facilities.

References

[1] International Technology Roadmap for Semiconductors (ITRS), 2006 update.

[2] Various microprocessor datasheets; see Intel at www.intel.com.

[3] Gurrum, S. P., et al., "Thermal Issues in Next-Generation Integrated Circuits," *IEEE Transactions on Device and Materials Reliability*, Vol. 4, 2004, pp. 709–714.

[4] Zhou, Z. J., L. R. Hoover, and A. L. Philips, "An Integrated Thermal Architecture for Thermal Management of High Power Electronics," *Proc. International Conference on THERMES*, 2002, pp. 317–336.

[5] Tuckerman, and R. F. W. Pease, "High Performance Heat Sinking for VLSI," *IEEE Electron Device Letters*, Vol. EDL-2, 1981, pp. 126–129.

[6] Naeemi, A., and J. D. Meindl, "Impact of Deep Sub-Ambient Cooling on GSI Interconnect Performance," *Proc. IEEE International Interconnect Technology Conference*, 2005, pp. 156–158.

[7] Zhang, H. Y., et al., "Development of Liquid Cooling Techniques for Flip Chip Ball Grid Array Packages with High Heat Flux Dissipations," *IEEE Transactions on Components and Packaging Technologies*, Vol. 28, 2005, pp. 127–135.

[8] Jiang, L., et al., "Closed-Loop Electroosmotic Microchannel Cooling System for VLSI Circuits," *IEEE Transaction on Components and Packaging Technologies*, Vol. 25, No. 3, 2002, pp. 347–355.

[9] Liu, D., and S. V. Garimella, "Analysis and Optimization of the Thermal Performance of Microchannel Heat Sinks," *Int. J. Numerical Methods for Heat and Fluid Flow*, Vol. 15, 2005, pp. 7–26.

[10] Koo, J.-M., et al., "Integrated Microchannel Cooling for Three-Dimensional Electronic Circuit Architectures," *ASME J. Heat Transfer*, Vol. 127, No. 1, 2005, pp. 49–58.

[11] Nelson, R. D., S. Sommerfeldt, and A. Bar-Cohen, "Thermal Performance of an Integral Immersion Cooled Multichip Module Package," *IEEE Transactions on Components, Packaging, and Manufacturing Technology, Part A,* Vol. 17, 1994, pp. 405–412.

[12] Chu, R. C., et al., "Review of Cooling Technologies for Computer Products," *IEEE Transactions on Device and Materials Reliability*, Vol. 4, No. 4, 2004, pp. 568–585.

[13] Pal, A., et al., "Design and Performance Evaluation of a Compact Thermosyphon," *IEEE Transaction on Components and Packaging Technologies*, Vol. 25, No. 4, 2002, pp. 601–607.

[14] Gad-el-Hak, M., *MEMS Handbook*, 2nd ed., Boca Raton, FL: CRC/Taylor & Francis, 2006.

[15] Pijnenburg, R. H. W., et al., "Integrated Micro-Channel Cooling in Silicon," *Proc. 34th European Solid-State Device Research Conference*, 2004, pp. 129–132.

[16] Kandlikar, S. G., and H. R. Upadhye, "Extending the Heat Flux Limit with Enhanced Microchannels in Direct Single-Phase Cooling of Computer Chips," *Proc. 21st Annual IEEE Semiconductor Thermal Measurement and Management Symposium*, 2005, pp. 8–15.

[17] Hahn, R., et al., "High Power Multichip Modules Employing the Planar Embedding Technique and Microchannel Water Heat Sinks," *IEEE Transactions on Components, Packaging, and Manufacturing Technology, Part A*, Vol. 20, No. 4, 1997, pp. 432–441.

[18] Je-Young Chang, et al., "Convective Performance of Package Based Single Phase Microchannel Heat Exchanger," *Proc. ASME InterPACK, IPACK 2005-73126*, 2005.

[19] Colgan, E. G., et al., "A Practical Implementation of Silicon Microchannel Coolers for High Power Chips," *Proc. 21st Annual IEEE Semiconductor Thermal Measurement and Management Symposium*, 2005, pp. 1–7.

[20] Wallis, G., and D. I. Pomerantz, "Field-Assisted Glass-Metal Sealing," *J. Applied Physics*, Vol. 40, No. 10, 1969, pp. 3946–3949.

[21] Schmidt, M. A., "Wafer-to-Wafer Bonding for Microstructure Formation," *Proc. IEEE*, Vol. 86, 1998, pp. 1575–1585.

[22] Bauer, R., et al., "Investigation on an Integrated Liquid Cooling System in LTCC-Multilayer," *Proc. 24th International Spring Seminar on Electronics Technology: Concurrent Engineering in Electronic Packaging*, 2001, pp. 178–182.

[23] Ilgen, H., "Water-Cooled Multilayer Printed Circuit Board," *Elektronik*, Vol. 48, No.19, 1999, pp. 46–52.

[24] de Boer, M. J., et al., "Micromachining of Buried Micro Channels in Silicon," *J. Microelectromechanical Systems,* Vol. 9, No. 1, 2000, pp. 94–103.

[25] Papautsky, I., et al., "A Low-Temperature IC-Compatible Process for Fabricating Surface-Micromachined Metallic Microchannels," *J. Microelectromechanical Systems*, Vol. 7, No. 2, 1998, pp. 267–273.

[26] Jayachandran, J. P., et al., "Air-Channel Fabrication for Microelectromechanical Systems via Sacrificial Photosensitive Polycarbonates," *J. Microelectromechanical Systems,* Vol. 12, No. 2, 2003, pp. 147–159.

[27] Bhusari, D., et al., "Fabrication of Air-Channel Structures for Microfluidic, Microelectromechanical, and Microelectronic Applications," *J. Microelectromechanical Systems,* Vol. 10, No. 3, 2001, pp. 400–408.

[28] Promerus, www.promerus.com.

[29] Wong, C. P., "Polymers for Encapsulation: Materials Processes and Reliability," *ChipScale Review*, March 1998.

[30] Bakir, M. S., et al., "Sea of Leads (SoL) Ultrahigh Density Wafer Level Chip Input/Output Interconnections," *IEEE Trans. Electron Devices*, Vol. 50, No. 10, 2003, pp. 2039–2048.

[31] Dang, B., M. Bakir, and J. Meindl, "Integrated Thermofluidic I/O Interconnects for an On-Chip Microchannel Heat Sink," *IEEE Electron Device Letters*, Vol. 27, No. 2, 2006, pp. 117–119.

[32] Qu, W., R. Green, and M. Austin, "Development of Multi-Functional Sensors in Thick-Film and Thin-Film Technology," *Meas. Sci. Technol.*, Vol. 11, 2000, pp. 1111–1118 (printed in the United Kingdom).

[33] Shah, R. K., and A. L. London, *Laminar Flow Forced Convection in Ducts*, New York: Academic Press, 1978.

[34] Wei, X., "Stacked Microchannel Heat Sinks for Liquid Cooling of Microelectronics Devices," PhD thesis, Georgia Institute of Technology, 2004.

[35] Early, J., "Speed, Power and Component Density in Multi-Element High-Speed Logic Systems," *ISSCC Dig. Tech. Papers*, February 1960, pp. 78–79.

[36] Davis, J. A., and J. D. Meindl, *Interconnect Technology and Design for Gigascale Integration*, Boston: Kluwer Academic Publishers, 2003.

[37] Pozder, S., et al., "Back-End Compatibility of Bonding and Thinning Processes for a Wafer-Level 3D Interconnect Technology Platform," *Proc. IEEE International Interconnect Technology Conference*, 2004, pp. 102–104.

[38] King, C. K., et al., "3D Stacking of Chips with Electrical and Microfluidic I/O Interconnects," *Proc. Electronics Components and Technol. Conf.*, 2008.

[39] Bakir, M. S., et al., "3D Heterogeneous Integrated Systems: Liquid Cooling, Power Delivery, and Implementation," *Proc. IEEE Custom Integrated Circuits Conf.*, 2008, pp. 421–482.

[40] Schindler-Saefkow, F., et al., "A 3D-Package Technology for Fluidic Applications Based on Match-X," *Proc. 9th International Conference on New Actuators, ACTUATOR 2004*, 2004, pp. 577–580.

[41] Sekar, D., et al., "A 3D-IC Technology with Integrated Microchannel Cooling," *Proc. Int. Interconnect Technol. Conf.*, 2008, pp. 13–15.

[42] Nose, K., and T. Sakurai, "Optimization of VDD and VTH for Low-Power and High-Speed Applications", *Proc. Asia and South Pacific Design Automation Conference*, 2000, pp. 469–474.

CHAPTER 12

Carbon Nanotube Electrical and Thermal Properties and Applications for Interconnects

Lingbo Zhu, Dennis W. Hess, and C. P. Wong

12.1 Introduction

As originally proposed, Moore's law states that the number of transistors in semiconductor devices or integrated circuits (ICs) doubles approximately every 2 years [1]. Over the past 25 years, this prediction has been realized, largely due to device scaling. However, one of the historical consequences of increasing the number of devices on a chip, thus microprocessor performance, is an associated increase in power dissipation. Thus, more advanced semiconductor devices generally require enhanced heat dissipation; that is, thermal power-dissipation needs have increased steadily with increasing microprocessor performance, requiring increased focus on thermal management [2]. Furthermore, due to the nonuniform distribution of power on the die, thermal nonuniformity, usually referred to as hot spots where power density could be greater than 300 W/cm^2, must be taken into account in circuit design and operation. Although microprocessors, ICs, and other sophisticated electronic components are designed to operate at temperatures above ambient temperature (generally in the range of 60°C to 100°C), the hot spots in a device should be avoided to maintain system reliability and performance. Thus, excessive heat generated by electronic devices must be dissipated into the surrounding environment. However, the ever-increasing chip power creates thermal management challenges both within the chip and in packaging systems and requires the development of new cooling technologies and high-thermal-conductivity materials.

As noted above, thermal management has become an increasingly important element in the design of electronic products, as has been discussed in previous chapters. It has been reported that a reduction in the device operation temperature can correspond to an exponential increase in the reliability and life expectancy of the device [3]. Therefore, controlling the device temperature within the operational limits is critical. Currently, a number of cooling technologies are available for electronic devices and circuits; some of these can remove only a few watts of heat, while others can dissipate heat in the hundreds-of-watts range. As chip power increases above 125W to 150W, novel cooling techniques will be needed to maintain chip temperatures within functional temperature limits [3]. The development of liquid-cooling technologies, such as liquid microchannel cooling, represents a major evolution in the field [4]. However, the economics, simplicity, and unmatched reliability of convection methods underscore the desire to cool systems and components with natural air convection. For this purpose, heat sinks are usually implemented on the back side of a chip through the use of thermal interface materials (TIMs), as illustrated in Chapter 9. Heat must first flow from the chip to the heat spreader via TIM1, where a portion of the allowable temperature drop occurs. Heat then flows through TIM2 to the heat sink, which absorbs the heat and is cooled by air convection. The heat spreader disperses heat, reduces peak and average heat fluxes, and decreases associated temperature drops. A heat sink is employed to increase the surface area that is in direct contact with air, thereby decreasing the thermal resistance. The driving forces to enhance IC speed result in higher rates of heat generation and power density. Consequently, the temperature drop caused by thermal conduction between the chip and heat spreader is predicted to become comparable to the maximum allowed temperature drop (temperature difference at the chip and heat sink). As a result, the chip and heat spreader thermal resistance will be comparable to the total allowed resistance by 2010 [5], eliminating the possibility of a temperature drop at the heat sink. Heat spreaders are usually high-thermal-conductivity materials, such as Cu, diamond, AlN, and SiC. Thus, the spreaders themselves cause only a very small fraction of the total thermal resistance. Effective reduction of resistance between the chip and the spreader (the largest possible fraction of system resistance in the near future) is therefore limited to the development of high-conductivity TIMs and the suppression of interfacial thermal resistance between TIMs and substrates. Development of new TIMs is thus crucial to meet packaging thermal-performance requirements for future generations of high-performance IC chips.

Heat-dissipation challenges create opportunities for fundamental research in materials and thermal management strategies. Specifically, it has been suggested that future cooling approaches may be based on micro- and nanotechnologies [6]. For thermal management applications, the distinctive properties of one-dimensional structures and materials have gained much attention. Among such materials, carbon nanotubes (CNTs), due to their unique thermal properties, give rise to new opportunities in thermal management of microelectronic devices and ICs. Also, the extraordinary electrical and mechanical properties of CNTs make them a promising candidate for electrical interconnects [7, 8]. This chapter presents first the CNT synthesis methods with a focus on chemical vapor deposition (CVD) process. Next, the electrical and thermal properties of CNTs related to microelectronics applications are described. Then, the integration of CNTs into circuitry and packaging is discussed. Finally, the chapter concludes with a discussion of future research needs for CNT applications in microelectronics.

12.2 Carbon Nanotube Growth and Growth Mechanisms

12.2.1 Chirality of Carbon Nanotubes

The structure of a single-walled CNT (SWNT) is conveniently illustrated by rolling a graphene sheet along the vector C_h [9, 10]. The circumference of any carbon nanotube can be described in terms of the chiral vector:

$$C_h = m\alpha_1 + n\alpha_2 \tag{12.1}$$

where α_1 and α_2 are unit vectors, and m and n are integers. The chiral angle, θ, is determined relative to the direction defined by α_1. For instance, in Figure 12.1 the vector C_h has been constructed for $(m, n) = (3, 2)$. Different types of carbon nanotubes have different values of m and n. Zigzag nanotubes correspond to $(m, 0)$ or $(0, n)$ and have a chiral angle of 0°, armchair nanotubes have (n, n) and a chiral angle of 30°, while chiral nanotubes have general (n, m) values and a chiral angle between 0° and 30°.

The two-dimensional (2D) energy dispersion relations for π bands of graphite, E_{2D}, are expressed using the wave vectors (k_x, k_y) [11]:

$$E_{2D}\left(k_x, k_y\right) = \pm\gamma_0\left\{1 + 4\cos\left(\frac{\sqrt{3}k_x a}{2}\right)\cos\left(\frac{k_y a}{2}\right) + 4\cos^2\left(\frac{k_y a}{2}\right)\right\}^{1/2} \tag{12.2}$$

where γ_0 is the nearest-neighbor overlap integral, and $a = 0.246$ nm is the in-plane lattice constant. When graphene is rolled over to form a nanotube, the periodic boundary condition, $C_h \cdot k = 2\pi q$, is imposed to eliminate the wave vectors $k = \left(k_x, k_y\right)$, where q is an integer. In this way, 1D energy bands can be obtained by slicing the 2D energy dispersion relations using the periodic boundary condition. This leads to the following results [9, 10]: SWNTs are metallic if

$$(n - m) = 3q \tag{12.3}$$

where q is an integer, while for those with

$$(n - m) = 3q \pm 1 \tag{12.4}$$

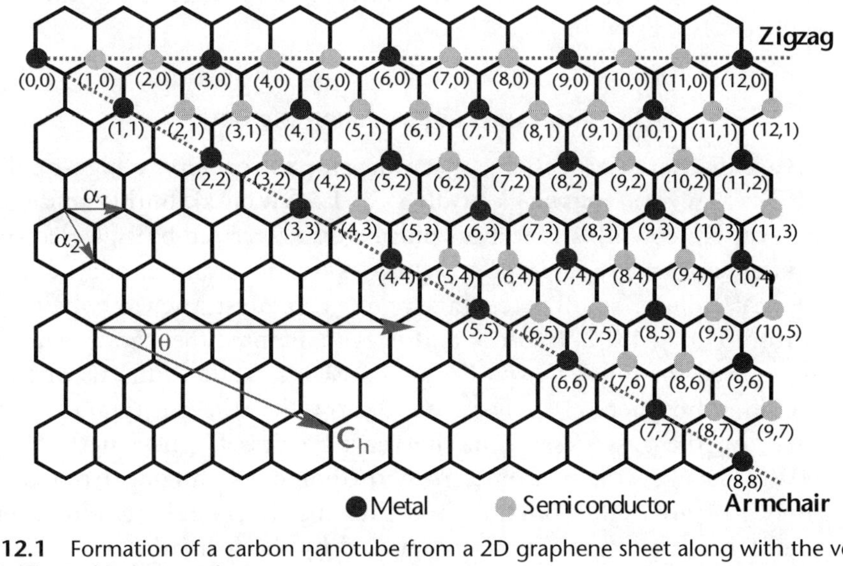

Figure 12.1 Formation of a carbon nanotube from a 2D graphene sheet along with the vector that specifies a chiral nanotube.

the SWNTs are semiconducting. This suggests that one-third of the nanotubes grown are metallic, while two-thirds are semiconducting. The band gap for a semiconducting nanotube is given by [12]

$$E_g = 2d_{cc}\gamma / D \tag{12.5}$$

where d_{cc} is the C-C distance, γ is the nearest-neighbor overlap integral, and D is the nanotube diameter. Therefore, the band gap of a 1 nm semiconducting nanotube is between 0.7 and 0.9 eV, which is in good agreement with Scanning Tunneling Microscopy (STM) measurements for SWNTs [12, 13].

The geometry of the graphene lattice and the chiral vector of the tube determine the structural parameters, such as diameter, unit cell, and number of carbon atoms. The diameter of the nanotube is given by [9]

$$d = \frac{|C_b|}{\pi} = \frac{\sqrt{3}a_{c-c}\sqrt{m^2 + mn + n^2}}{\pi} \tag{12.6}$$

where a_{c-c} is the C-C bond length (1.42Å).

12.2.2 Nanotube Growth Methods

The proposed applications of CNTs will not be realized until nanotube growth is optimized and well controlled. For composite applications, high-quality CNTs are required at the kilogram or ton level, which requires growth methods that are simple, efficient, and cost-effective. For device applications, the layout of CNTs will rely on self-assembly or controlled-growth strategies on substrates combined with micromachining technologies. In general, CNTs can be grown by arc-discharge, laser ablation, and chemical vapor deposition (CVD) methods [11]. However, for device applications, growth of CNTs by CVD methods is particularly attractive, due to features such as selective spatial growth, large-area deposition capabilities, and aligned CNT growth.

12.2.2.1 Arc-Discharge and Laser-Ablation Methods

An arc-discharge process has been developed to prepare high-quality multiwalled CNTs (MWNTs) and single-walled CNTs (SWNTs). In this process, carbon atoms are evaporated with an inert gas plasma characterized by high electric currents passing between opposing carbon electrodes (cathode and anode). Usually, the carbon anode contains a small percentage of metal catalyst, such as cobalt (Co), nickel (Ni), or iron (Fe). In 1992, Ebbesen and Ajayan adapted the standard arc-discharge technique used for fullerene synthesis to the large-scale synthesis of MWNTs under a helium atmosphere [14]. The results show that the purity and yield depend sensitively on the gas pressure in the reaction vessel. The length of the synthesized MWNTs is several micrometers with diameters ranging from 2 to 20 nm. The nanotubes are highly oriented and thus highly crystalline. However, such synthesized CNTs are inevitably accompanied by the formation of carbon particles that are attached to the nanotube walls. A subsequent purification process is necessary to achieve high-purity nanotubes. In 1993, Bethune et al. reported the production of SWNTs by an arc-discharge method using a carbon electrode that contained ~2 atomic % cobalt [15]. At high temperature, the carbon and metal catalyst are covaporized into the arc, leading to the formation of carbon nanotubes with very uniform diameter (~1.2 nm). However, fullerenes (by-products of the arc-discharge process) also form readily in this process. In order to obtain high-purity SWNTs, a purification process is therefore necessary.

In 1996, the Smalley group reported the synthesis of high-quality SWNTs with yields greater than 70% using the laser-ablation method [16]. This method utilized double-pulse lasers to evaporate graphite rods doped with 1.2 atomic % of a 50:50 mixture of Co and Ni powder, which was placed in a tube furnace heated to 1,200°C in flowing argon at 500 Torr; this process was followed by heat treatment in vacuum at 1,000°C to sublime C_{60} and other small fullerenes. The resulting SWNTs were quite uniform, had a diameter of ~1.38 nm, and formed ropes consisting of tens of individual SWNTs closely packed into hexagonal crystal structures that were stabilized through van der Waals forces.

The success in producing large quantities of high-quality CNTs by arc-discharge and laser ablation offers a wide availability of CNTs for fundamental studies and exploration of potential applications. However, there are several concerns associated with these two growth methods. First, both methods rely on evaporating carbon atoms from a solid carbon source at temperatures greater than 3,000°C, which limits the scale-up of CNT production. Second, the CNTs synthesized by these methods are entangled, making purification, manipulation, and assembly difficult. Furthermore, the CNTs produced by these two methods are accompanied by by-products, including fullerenes, graphitic polyhedrons, and amorphous carbon in the form of particles or overcoats on the sidewalls of the nanotubes [17].

12.2.2.2 Chemical Vapor Deposition

As noted above, chemical vapor deposition (CVD) methods are particularly attractive for CNT growth for electronic device applications. The CVD growth process involves heating the catalyst to a high temperature and introducing hydrocarbon gas or carbon monoxide (CO) into the reactor. The mechanism for CNT growth has been generally assumed to be a dissociation-diffusion-precipitation process in which elemental carbon is formed on the surface of a metal particle, followed by diffusion and precipitation in the form of cylindrical graphite [18, 19]. The critical parameters in CVD growth of CNTs are the carbon precursor, catalyst, reactor chamber pressure, and growth temperature. The effect of these process parameters on CNT growth has been investigated extensively, and the trends observed provide additional insight into nanotube growth. However, the detailed nanotube growth mechanisms are still not well understood.

Influence of temperature. Temperature has a significant effect on CNT formation and growth. Depending on the nanotube growth method, the CNT growth temperature can range from 400°C [20, 21] to 3,600°C [22]. MWNTs are generally favored at temperatures between 500°C and 1,000°C, while SWNTs tend to grow at higher temperatures, although recently growth of aligned SWNTs at 750°C has been reported [23, 24]. Lee at al. examined the temperature dependence of CNT growth from 800°C to 1,100°C and observed an increase in nanotube diameter and growth rate with increasing temperature, which was ascribed to the increased bulk-diffusion rate of carbon in the metal catalyst particles.

Influence of carbon precursors. It has been reported that CNTs can be grown from numerous carbon sources, including hydrocarbon, polymer, and organometallic compounds [e.g., iron(II) phthalocyanine] [25]. Hydrocarbons include methane, ethylene, acetylene, and some aromatic compounds, such as benzene and xylene. Favorable conditions for CNT growth for individual carbon precursors depend strongly on the temperature, pressure, carrier gas, and choice of metal catalyst. The optimal experimental conditions for deposition of CNTs using a particular precursor may be ineffective with an alternate precursor.

Influence of metal catalyst. The nanotube density and minimum growth temperature may be improved if a better understanding of the catalytic process were available. For microelectronics applications, the thin catalyst layer would likely be deposited by electron-beam evaporation or sputtering. Then, the catalyst layer is transformed into nanoparticles by heat treatment at high temperature (>500°C). This process is controlled by film stress or dewetting and depends on the nature of the underlying support. The catalyst does not need to melt to form nanoparticles; rather, particle formation can occur as a result of strain at a rather low temperature [26]. Indeed, the temperature should not be so high that catalyst atoms are mobile on the support surface since this allows the average catalyst particle size to increase by the process of Ostwald ripening in which larger droplets grow at the expense of smaller droplets. Iron, nickel, and cobalt are the most commonly used catalysts for CNT growth. One of the reasons for choosing these metals as catalysts lies in the metal-carbon phase diagrams. Figure 12.2 illustrates an Fe-C phase diagram, which shows that α-Fe can, at most, dissolve 0.028 wt% C at 738°C and that γ-Fe can dissolve 2.08 wt% C at 1,154°C. Clearly, at high temperature, carbon has finite solubility in these metals, which leads to the formation of metal-carbon solutions and precipitation of carbon atoms when supersaturated.

Influence of support materials. Catalyst-support interactions not only determine the nanotube growth mode but affect CNT growth rate and quality. A support material must display thermal and chemical stability under synthesis conditions. Hernadi et al. prepared supported Fe catalysts by different methods (impregnation and ion-adsorption precipitation) on various supports [27]. Iron on graphite showed low activity with low CNT yield. Zeolite-supported catalysts prepared by ion exchange were inactive in the formation of nanotubes, while the catalysts prepared by impregnation showed higher activity. Fe/zeolite Y catalysts gave better results than Fe/ZSM-5. However, the Fe/silica catalysts led to even better results. Su et al. reported improved nanotube production using a novel aerogel supported Fe/Mo catalyst [28]. Nanotube production on Al_2O_3-supported catalysts is much higher than that on the SiO_2-supported catalysts with other growth conditions held constant. Colomer et al. also found that the synthesis of SWNTs is more efficient for

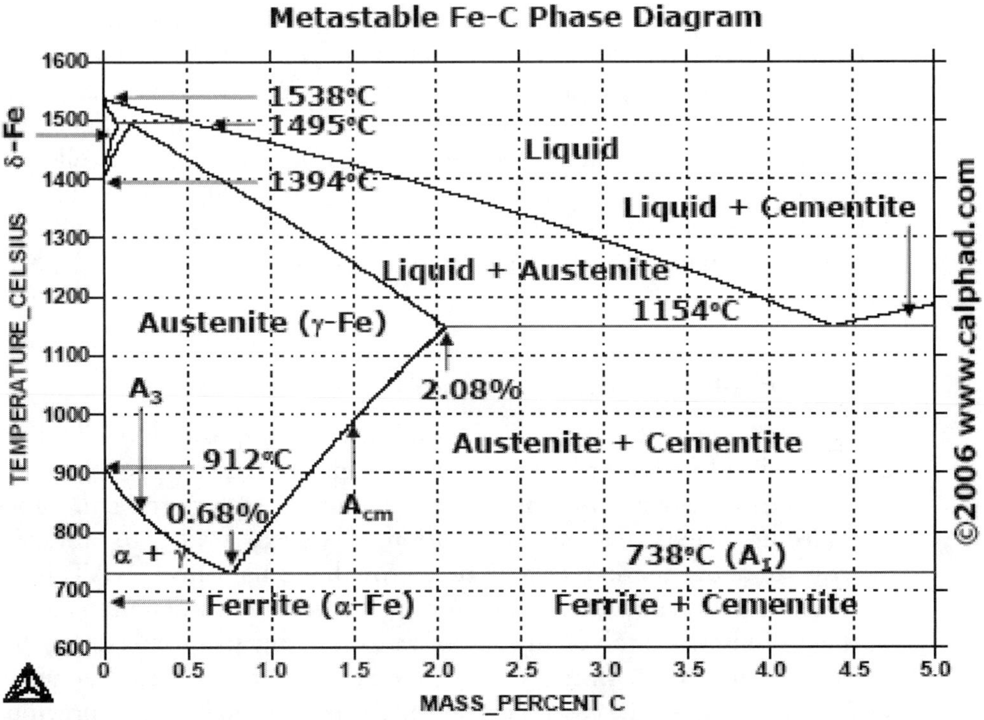

Figure 12.2 Phase diagram of Fe-C. (*Source:* www.calphad.com © 2006.)

metal catalysts supported on alumina than on fumed silica [29]. Other reports have indicated that the support should be carefully chosen specifically for different metals [30–33] since the metal-support interactions are critical in nanotube synthesis by CVD processes.

A key factor in the fabrication of interconnects is that they are a "back-end-of-line" (BEOL) process, for which the upper temperature is limited to 400°C to 500°C. PECVD can lower the CNT growth temperature, compared to a thermal CVD process. That is, the plasma dissociates hydrocarbon molecules and promotes surface diffusion of carbon on the catalyst droplet, thereby allowing nanotube growth to occur at lower temperatures than are possible with CVD [34, 35]. PECVD has some particular advantages for the growth of CNT vias because the electrical field at the plasma sheath promotes aligned growth in the vertical direction [36]. Use of a remote plasma can be advantageous in order to minimize ion bombardment, which can damage nanotube walls [37].

A MWNT can be considered an array of concentrically nested SWNTs. One of their common features is that the CNTs are close-ended, and the catalyst particle is encapsulated by one nanotube cap, as shown in Figure 12.3(a). Thus, when the MWNT is connected to an electrode for electrical studies, only the outermost shell of the MWNT serves as an electrical transport channel. If the caps of MWNTs can be removed to permit good contact with electrodes, thus with the internal walls of these MWNTs, the walls form parallel quantum conductors that participate in electrical transport, thereby enabling large current-carrying capacity. In this way, CNT electrical conductance could be dramatically improved, which would allow CNTs to serve as conductive nanowires for the replacement of copper and aluminum films used in state-of-the-art circuits. Such nanowires will be less susceptible to electromigration under high current density than are Cu and Al. Recently, Zhu el al. developed a simple and efficient method for in situ growth of aligned open-ended MWNTs by water-assisted selective etching [38]. When a controlled amount of water (~775 ppm) was introduced into the CVD tube during nanotube growth, the water appeared to etch away the nanotube caps while keeping the tube walls intact, as shown in Figure 12.3(b). The layers of as-grown CNTs have high purity and open ends.

12.2.3 Nanotube Growth Mechanisms

The catalyzed growth of CNTs by chemical vapor deposition (CVD) has been discussed in detail because this process offers a promising route for the fabrication of microelectronic devices and the bulk production of high-purity nanotubes that can be scaled up to achieve commercialization. However, in order to optimize the CVD process, an understanding of nanotube growth mechanisms by CVD is necessary.

The formation and growth of nanotubes has been postulated to be an extension of the well-known catalytic growth of carbon filaments on metals, such as cobalt, nickel, and iron [39], since catalyst particles were observed on either end of the CNTs, analogous to the case of carbon-filament studies. Based on previous research conducted on the formation of graphitic carbon over various metals, the form of the graphite produced is closely related to the physical dimensions of metal catalyst particles [40–42]. The appropriate metals for this filamentary growth have been consistently shown to be cobalt, nickel, and iron. The particular ability of these metals to

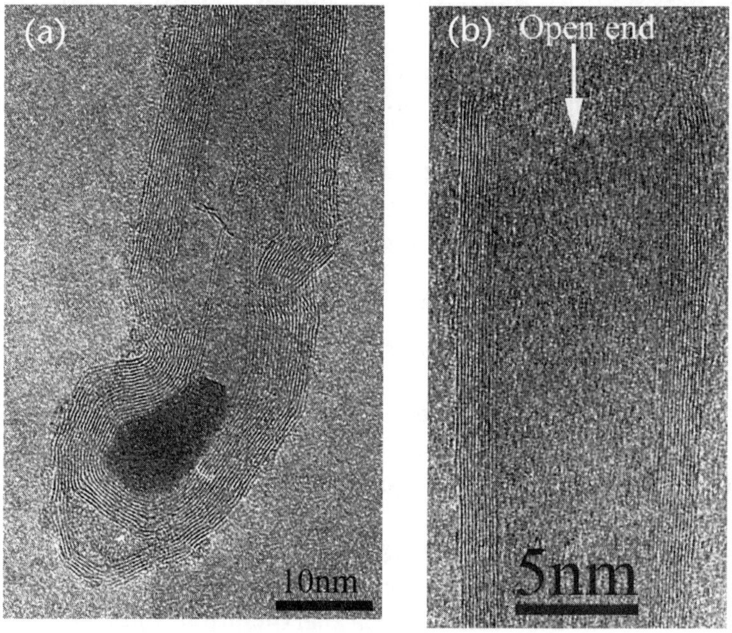

Figure 12.3 (a) Catalyst particle capped in one end of MWNT, and (b) open-ended MWNT etched by water vapor during nanotube growth. (*From:* [38]. © 2005 ACS. Reprinted with permission.)

form ordered carbon structures from decomposed hydrocarbon appears to be due to the following factors:

- Specific catalytic activity for the decomposition of hydrocarbons;
- Finite solubility of carbon in these metals;
- Ability to form metastable carbides;
- Ability of carbon to diffuse rapidly on and through these metals.

These properties allow ordered carbon structures to form by a diffusion-precipitation mechanism. When the pyrolysis of hydrocarbons occurs over certain transition metals, the metal can serve as a solvent in which carbon dissolves to form a solid solution; graphite is then formed by crystallization. When the metal is present as nanoparticles, the carbon forms filaments with a diameter similar to that of the catalyst particles. The metal particles can be supported on substrates or introduced in the vapor (e.g., catalyst precursors) within the CVD chamber. A generally accepted mechanism describing the growth of carbon filaments or nanotubes from the catalyst surface is dissolution of carbon in and diffusion through the metal particle [43]. However, the nature of the driving force for carbon diffusion through the catalyst particle is a subject of debate. The driving force could be temperature gradient [41, 44, 45] or concentration gradient [46] within the particle. For CNT growth, these two mechanisms may operate in parallel to describe CNT growth in various CVD environments.

Interaction between the metal catalyst particle and the support material is a key factor influencing the CNT growth mode. Figure 12.4 is a schematic representation of the two typical CNT growth modes: base growth and tip growth. If the particle adheres to the surface of support materials strongly, carbon precipitates from the front of the particle, and CNT growth continues with the particle attached to the substrate (base growth). In contrast, when the interaction between the metal catalyst and the support material is weak, carbon precipitates at the opposite surface of the particles, and the growing CNT lifts the particles as it grows (tip growth). In both cases, the critical steps in the CNT growth process are carbon solubility and carbon diffusion through the catalyst particles.

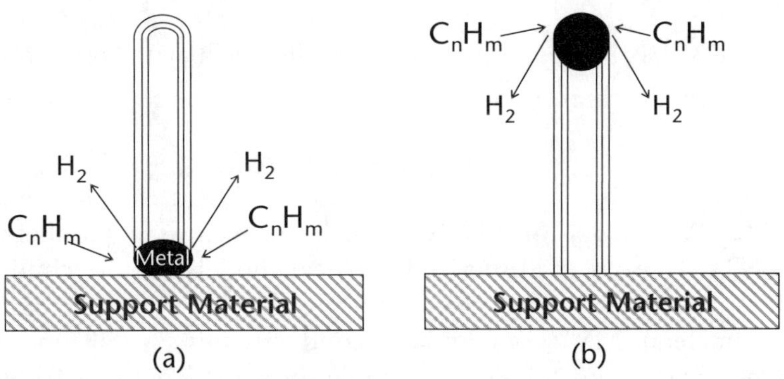

Figure 12.4 Typical CNT growth modes in CVD: (a) base-growth mode, and (b) tip-growth mode.

12.3 Carbon Nanotubes for Interconnect Applications

12.3.1 Electrical Properties of Carbon Nanotubes

Previous studies have demonstrated that a carbon nanotube behaves like a quantum wire due to geometrical confinement of the tube circumference [47]. The conductance of an MWNT or SWNT is determined by two factors: the conducting channels per shell and the number of shells. An SWNT consists of one shell. For an MWNT, the number of shells is diameter dependent [48]; that is,

$$N_{shell} = 1 + \frac{D_{outer} - D_{inner}}{2\delta} \tag{12.7}$$

where $\delta = 0.34$ nm is the nanotube intershell distance, and D_{outer} and D_{inner} are external and internal nanotube diameter, respectively. Assuming that the metallic-to-semiconducting tube ratio is one-third, the appropriate number of conducting channels per shell can be written as [49]

$$\tilde{N}_{chan/shell} = \begin{cases} (ad + b)/3, & d > 6\text{nm} \\ 2r, & d < 6\text{nm} \end{cases} \tag{12.8}$$

where $a = 0.1836$ nm^{-1}, $b = 1.275$, r is the metallic nanotube ratio, and d is the diameter of the nanotube shell. One conducting channel provides either quantized conductance ($G_0 = (2e^2/h) = (12.9\ \text{k}\Omega)^{-1}$) or ohmic conductance ($G_i$), depending upon the tube length i. For the low-bias situation $\left(V_b <\approx 0.1V\right)$ that is suitable for interconnect applications, the diameter-dependent channel conductance for one shell is [48]

$$G_{shell}(d,l) = \begin{cases} G_0 \tilde{N}_{chan/shell}, & l \le \lambda \\ G_i \tilde{N}_{chan/shell}, & l > \lambda \end{cases} \tag{12.9}$$

where λ is the mean free path. Thus, the conductance of a metallic SWNT $\left(\tilde{N}_{chan/shell} = 2\right)$ is

$$G_{shell}(d,l) = \begin{cases} 2G_0 = 1/6.45K\Omega, & l \le \lambda \\ 2G_i, & l > \lambda \end{cases} \tag{12.10}$$

An MWNT consists of several shells, each with its own d, λ, and $\tilde{N}_{chan/shell}$. Therefore, the total conductance is the sum of conductances for each shell:

$$G_{MWNT} = \sum_{N_{shell}} G_{shell}(d,l) \tag{12.11}$$

A SWNT rope or MWNT can be viewed as a parallel assembly of single SWNTs. Naeemi et al. derived physical models for the conductivity of MWNT interconnects [49]. The results indicate that for long interconnects (hundreds of micrometers), MWNTs may have conductivities several times larger than that of copper or even SWNT bundles, while for short lengths (<10 μm), SWNT bundles offer twofold higher conductivity than MWNTs.

Due to the structural imperfection of grown CNTs, the conductance for a SWNT, a SWNT rope, or MWNT can written as

$$G = G_0 M = \left(2e^2 / h\right) MT \tag{12.12}$$

where M is an apparent number of conducting channels, and T is the transmission probability for an electron through the contacts and the tube. Ideally, T is unity, and $M = 2$ for a perfect ballistic SWNT less than 1 μm long. In actual operation, T may be significantly lower than one due to electron-electron coupling, intertube coupling effects, scattering from defects and impurities, structural distortions, and coupling with substrates or contact pads. Therefore, the experimentally measured conductance is much lower than the quantized value. As a result, the high electrical resistance of a single nanotube necessitates the use of nanotube bundles aligned in parallel.

12.3.2 Carbon Nanotubes as Interconnects

Two types of interconnects are employed in microelectronic devices: horizontal and vertical. Horizontal interconnects link transistors in different locations on an integrated circuit; many layers of these horizontal interconnects (up to 12) can exist on a state-of-the-art circuit [50]. Each layer is then separated by an interlevel dielectric, generally porous SiO_2 or SiO_2 doped with C or F, to lower the dielectric constant [51]. These materials are rather weak mechanically and are thermally unstable above ~450°C. As dimensions decrease for on-chip interconnects, the current density carried by each interconnect increases. The International Technology Roadmap for Semiconductors (ITRS) predicts that in 2010, the current density will reach 5×10^6 A/cm^2 [51], a value that can only be supported by CNTs since they are capable of a current density of ~10^9 A/cm^2 [52]. Naeemi et al. suggested that mono- or bilayer metallic SWNTs may be promising candidates for short local interconnects [53]. Over short lengths, driver resistance is dominant, and latency is determined by interconnect capacitance. The average capacitance per unit length of these nanotube interconnects can be 50% smaller than that of copper interconnects, which leads to significant savings in power dissipation.

Vertical interconnects pass through holes (vias) in the dielectrics to connect horizontal interconnects to the source, drain, or gate metallization of transistors. In existing microelectronic technology, the vias are fabricated from copper. Via regions are the most common source of failures in interconnect structures due to the high current densities and heterogeneous current distributions that cause electron-induced material transport (electromigration) [8, 54]. Carbon nanotubes are expected to offer a substantially higher resistance to electromigration than copper lines. Thus, CNT connections between metallization layers may solve the problems of electromigration and heat removal. Researchers from Fujitsu and Infineon have investigated this area extensively [8, 55–57]. In one approach, a hole is etched in the interlayer dielectric, and catalyst is deposited into the bottom of the hole; excess catalyst is removed from the top of the hole. Alternatively, a catalyst layer is deposited under the interlayer dielectric and exposed by etching a hole in the dielectric. In both approaches, CNTs are then grown within the hole by CVD or by plasma-enhanced CVD (PECVD).

For microelectronics applications, it is important to grow high-quality CNTs at specific locations at low temperature. Awano et al. have reported a size-classified catalyst-nanoparticle CVD process for controlled growth of CNTs with uniform diameter and high density at low temperatures [57, 58]. This process differs from a conventional CVD process in the catalyst preparation step. Normally, catalyst particles for CNT growth are formed from a deposited thin catalyst film by a high-temperature treatment that causes segregation of the film into small particles or islands. Under these conditions, the diameter and density of CNTs is not well controlled because the size and density of the catalyst particles are not uniform. However, in a size-classified catalyst-nanoparticle CVD process, the catalyst-particle generation and deposition process is distinct from CNT growth, as shown in Figure 12.5. Laser ablation is used to evaporate catalyst particles from a catalyst metal target. The resulting nanoparticles are size separated by a low pressure differential mobility analyzer in terms of their movement within an electric field. The size-classified nanoparticles are then deposited onto substrates in a deposition chamber. Finally, the substrates with deposited nanoparticles are placed in a hot-filament CVD chamber for nanotube growth. The nanoparticles present will not coalesce on the substrate during CNT growth due to the low CNT growth temperature.

A schematic of a CNT via fabrication process is shown is shown in Figure 12.6 [59]. The substrate was composed of a 100 nm Cu film (wiring), a 5 nm Ta barrier layer, a 5 nm TiN contact layer, and a 350 nm SiO_2 dielectric layer. Via holes were formed using conventional photolithography and subsequent wet etching with a buffered HF solution. The TiN layer was deposited on the bottom of the holes by sputtering, followed by a lift-off process. Size-controlled Co particles with an average size of ~4 nm were then deposited onto the substrate using the method described above. The substrate was then placed in a hot-filament CVD chamber; CNT growth

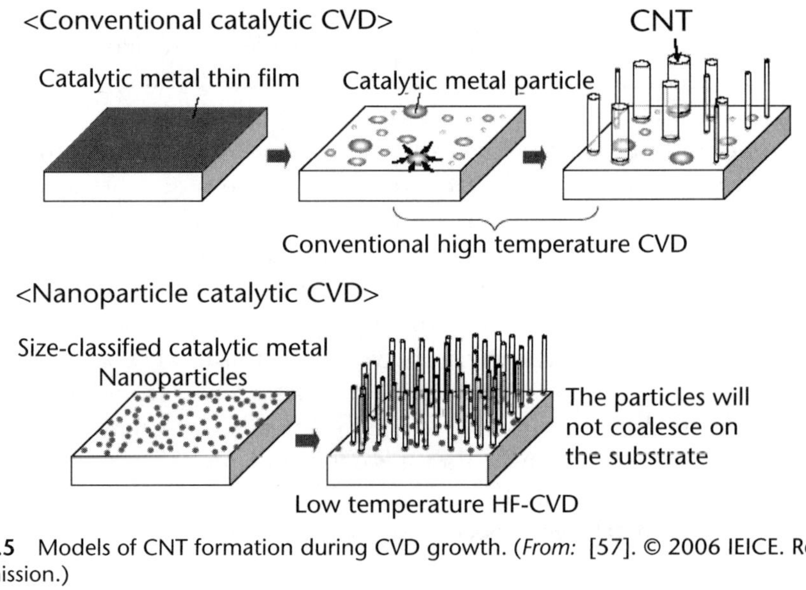

Figure 12.5 Models of CNT formation during CVD growth. (*From:* [57]. © 2006 IEICE. Reprinted with permission.)

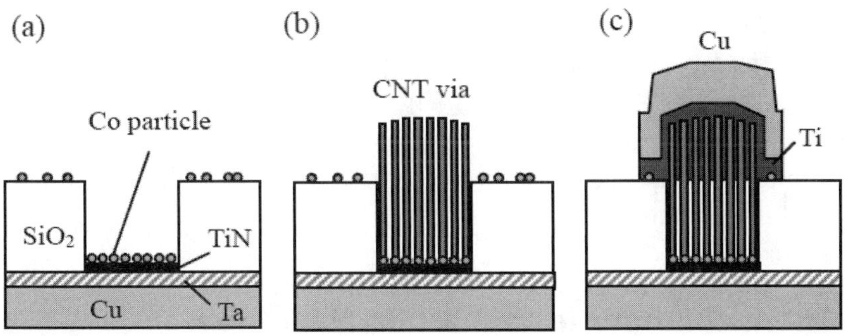

Figure 12.6 CNT via fabrication: (a) fabrication of via hole structure and deposition of cobalt particles, (b) CNT growth in vias, and (c) deposition and patterning of top layers. (*From:* [59]. © 2006 IEEE. Reprinted with permission.)

was performed using a 1:9 mixture of C_2H_2 and Ar at 510°C for ~1 minute. Finally, top contact layers of 50 nm Ti and 300 nm Cu were deposited to form connections to the CNTs. These metal layers were then patterned, and the Co particles on the SiO_2 layer were simultaneously removed during the etching step. Figure 12.7 shows an SEM image of CNT vias with a diameter of 2 μm grown by a size-classified catalyst-nanoparticle CVD process.

Via resistance was measured for 2-μm diameter CNT structures with a four-point probe using kelvin patterns and a TiN contact layer. The measured resistance of CNT vias is ~0.59Ω at room temperature [59]. This value is of the same order as the theoretical value of W plugs. The resistance is still one order of magnitude higher than the theoretical value of Cu vias, but it is expected to decrease further with additional improvement of CNT growth from the catalyst particles (e.g., catalyst-particle density, CNT growth conditions). In contrast to this process, the resistance of via structure that utilizes a 5 nm Ti contact layer shows a higher value (~1 MΩ), perhaps due to Ti oxidation prior to nanoparticle deposition.

When interconnects and vias are further reduced in size to meet requirements for future ICs, CNT vias will offer still more advantages. Vias consisting of only one MWNT are conceivable since multiwalled nanotubes can be produced with diameters from 5 to 100 nm; indeed, Infineon has demonstrated such a process [54].

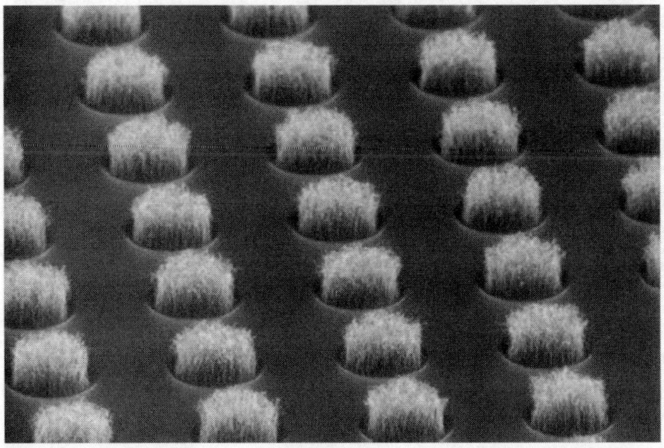

Figure 12.7 MWNT vias fabricated by a size-classified catalyst-nanoparticle CVD process. (*From:* [57]. © 2006 IEICE. Reprinted with permission.)

Figure 12.8(a) shows an individual nanotube grown in a hole with a diameter of 20 nm. The holes were fabricated with conventional lithography using a spacer-reduction method. Figure 12.8(b) shows the corresponding I-V characteristics of this structure. The investigators indicated that this characteristic is not exactly ohmic, possibly due to tunneling at the contacts and the availability of additional conduction states at higher biases.

12.4 Thermal Properties of Carbon Nanotubes

12.4.1 Thermal Properties of Individual Carbon Nanotubes

Rolling a graphene sheet into a nanotube has two major effects on the phonon dispersion and thus heat conduction [60]. First, the 2D phonon band structure of a graphene sheet is folded into the 1D band structure of the tube. Second, the cylindrical shape of the tube renders it stiffer than the graphene sheet, thereby rearranging the low-energy acoustic modes. For a nanotube, there are four acoustic modes: a longitudinal-acoustic (LA) mode, corresponding to motion of atoms along the tube axis; two degenerate transverse-acoustic (TA) modes, corresponding to atomic displacements perpendicular to the nanotube axis; and a twist mode, corresponding to a torsion around the tube axis. For graphite and related systems, the heat capacity can be written as

$$C = C_{ph} + C_e \tag{12.13}$$

where C_{ph} is the contribution from phonons, and C_e is the contribution from electrons. In the low-temperature regime, only the acoustic bands will be populated. If $T << hv / \kappa_B R$, then C_{ph} can be written [61]

$$C_{ph} = \frac{3L\kappa_B^2 T}{\pi\eta v} \times 3292 \tag{12.14}$$

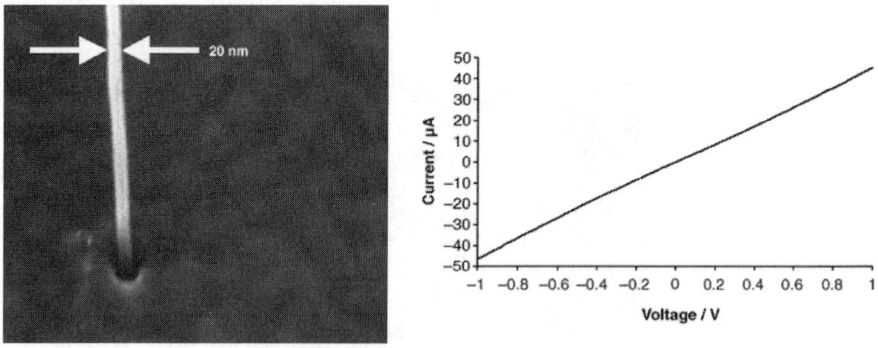

(a) (b)

Figure 12.8 (a) Directed growth of a MWNT in a lithographically defined nanohole in SiO_2, and (b) I-V curve is not quite linear, probably as a result of tunneling at the contacts or an increased availability of conducting states at higher biases. (*From:* [54]. © 2005 Wiley. Reprinted with permission.)

where κ_B is Boltzmann's constant, v is the velocity of sound (~10^4 m/s), and L is the tube length. The electronic specific heat for a metallic SWNT is also linear with temperature for $T << \hbar v_f / \kappa_B R$:

$$Ce = \frac{4\pi L \kappa_B^2 T}{3\eta v_F} \tag{12.15}$$

where v_F is the Fermi velocity (~10^6 m/s).

The ratio between phonon and electron contributions to the specific heat is

$$\frac{C_{ph}}{C_e} = \frac{v_F}{v} = 100 \tag{12.16}$$

Even for metallic tubes, phonon conduction dominates down to $T = 0$.

Figure 12.9 shows the calculated low-temperature specific heat of an isolated nanotube and nanotube ropes. Because only acoustic modes are populated, the specific heat displays linear temperature dependence at low temperature. Above ~6K, the slope of $C(T)$ increases as the optical subbands become populated. This linear behavior, with an increase in slope near 6K, is the expected signature for a 1D quantized phonon spectrum in single-walled nanotubes [60]. For MWNTs, phonon dispersion should occur along the tube radial direction; phonon dispersion of MWNTs has not been addressed theoretically [13]. Strong phonon coupling between the MWNT shells should cause behavior similar to that of graphite. A

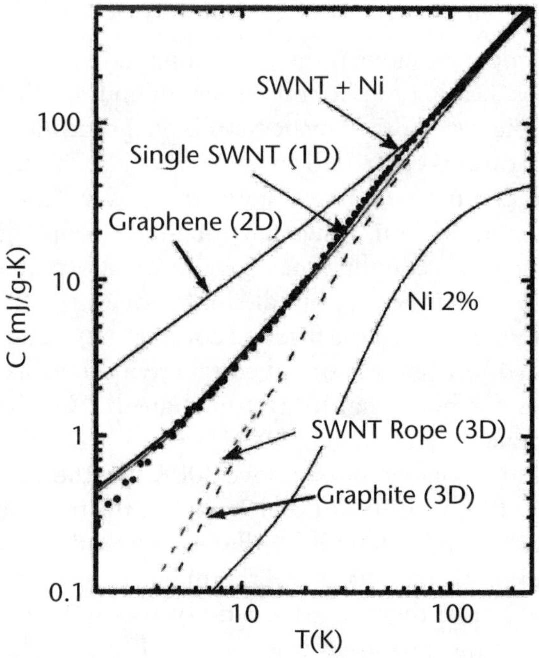

Figure 12.9 Measured specific heat of SWNTs (solid dots) compared with calculations for 2D graphene, 3D graphite, isolated tube, and strongly coupled ropes. Agreement is improved at high temperature by including the contribution of 2 atomic % nickel impurities. (*From:* [63]. © 2000 AAAS. Reprinted with permission.)

thermal relaxation technique has been used to measure the specific heat of MWNTs from 0.6K to 210K [62]; results indicate that MWNTs have a similar specific heat to that of SWNT ropes and graphite.

Several investigations have indicated that CNTs have unusually high thermal conductivity in the axial direction. For example, molecular dynamics simulations of a SWNT by Berber et al. indicated that the thermal conductivity of a SWNT can be as high as 6,600 W/m.K at room temperature [64]. Dai et al. presented a method for extracting the thermal conductivity of an individual SWNT from high-bias electrical measurements in the temperature range from 300K to 800K by reverse-fitting the data to an existing electrothermal transport model [65]. The thermal conductivity measured was nearly 3,500 W/m.K at room temperature for a SWNT of length 2.6 ìm and diameter 1.7 nm. Kim et al. developed a microfabricated suspended device hybridized with MWNTs (~1 μm) to allow the study of thermal transport where no substrate contact was involved [66]. The thermal conductivity and thermoelectric power of a single carbon nanotube were measured, and the observed thermal conductivity is greater than 3,000 W/mK at room temperature. Choi et al. obtained the thermal conductivity of individual MWNTs (outer diameter of ~45 nm) by employing the 3ω method [67]. In this approach, the third-harmonic amplitude determined from the applied alternating current at the fundamental frequency (ω) is expressed in terms of thermal conductivity. A microfabricated device composed of a pair of metal electrodes 1 ìm apart was used to position a single nanotube across the designated metal electrodes by utilizing the principle of dielectrophoresis. The thermal conductivity was reported to be 650 to 830 W/m.K at room temperature.

12.4.2 Thermal Properties of Carbon Nanotube Bundles

In a SWNT rope, phonons propagate along individual tube axes as well as between parallel tubes, leading to phonon dispersion in both the longitudinal and transverse directions. The net effect of dispersion is a significant reduction of the specific heat at low temperatures compared to an isolated nanotube [63]. However, as shown in Figure 12.9, the measured specific heat of SWNTs agrees with the isolated tube model down to 5K, indicating that tube-tube coupling is weaker than theoretical estimates; the weak coupling may suggest no substantial reduction in thermal conductivity when the tubes are bundled into ropes.

Hone et al. measured the thermal conductivity of aligned and unaligned SWNTs from 10K to 400K [68]. Thermal conductivity increased smoothly with increasing temperatures for both aligned and unaligned SWNTs. At room temperature, the thermal conductivity of aligned SWNTs was greater than 200 W/m.K, compared to ~30 W/m.K of unaligned ones; above 300K, the thermal conductivity increased and then leveled off near 400K. In such samples, the thermal conductivity is likely to be limited by tube-tube junctions so that a single tube should have high thermal conductivity. Yi et al. measured the thermal conductivity of millimeter-long aligned MWNTs [69]. The thermal conductivity was low; at room temperature, thermal conductivity of these samples was only ~ 25 W/m.K due to a large number of CNT defects. However, thermal conductivity could reach ~2,000 W/m.K if the aligned MWNTs were annealed at 3,000°C to remove the defects. Yang et al. investigated the thermal conductivity of MWNT films prepared by microwave CVD using a pulsed photothermal reflectance technique [70]. The average thermal conductivity of carbon nanotube films, with a film thickness from 10 to 50 μm, was ~15 W/m.K at room temperature, independent of tube length. However, by taking into account a small volume filling fraction of CNTs, the effective nanotube thermal conductivity can reach 200 W/m.K.

12.5 Carbon Nanotubes as Thermal Interface Materials

With the continual increase in cooling demand for microprocessors, an increased focus on developing thermal solutions has developed within the microelectronics industry. As noted above, thermal interface materials (TIMs) play a key role in thermally interconnecting various components. Developing new TIMs is thus a key activity to ensure that packaging thermal-performance requirements are met for future generations of high-performance chips. Common TIMs are polymer-based composites with high-thermal-conductivity fillers. The effective thermal conductivities of particle-filled polymer are only ~7 W/m.K. Moreover, the resistance in commercial products can be substantially larger than anticipated values due to resistances at TIM boundaries and to the existence of small voids. Solders, such as AuSn, InPb, and In, are now being actively pursued as TIMs due to their high thermal conductivity (>40 W/m.K) [71]. However, in many applications, solders as TIMs are undesirable, due to the effect of thermal fatigue, solder voiding, cost, and process complexity. As a result, much interest exists in developing new TIMs.

12.5.1 Carbon-Nanotube-Based Thermal Interface Materials

Because CNTs have very high thermal conductivity and excellent mechanical properties, CNT-based composites or structures have been proposed for TIM applications. Biercuk et al. mixed SWNTs into epoxy to enhance the thermal transport properties of the composites [72]. Epoxy loaded with 1 wt% unpurified SWNT material showed a 70% increase in thermal conductivity at 40K, rising to 125% at room temperature. By comparison, the enhancement due to 1 wt% loading of vapor-grown carbon fibers was three times smaller than that of SWNT samples. Both thermal and mechanical properties of SWNT-epoxy composites are improved, without the need to chemically functionalize the nanotubes. Yu et al. prepared SWNT/epoxy composites using purified functionalized SWNTs [73]. The purified SWNTs provided approximately fivefold greater enhancement of thermal conductivity than did the impure SWNT samples with the same loading, demonstrating the significance of SWNT quality for thermal management.

In addition to the use of CNT-based composites as TIMs, aligned CNTs have been grown directly on silicon surfaces for thermal management. Xu et al. grew aligned CNTs on silicon wafers using plasma-enhanced CVD [74, 75]. The thermal testing performed was based on a one-dimensional reference bar method in high vacuum with radiation shielding; temperature measurements were carried out with an infrared camera. Dry CNT arrays have a minimum thermal interface resistance of 19.8 mm^{2}K/W, while CNT arrays with a phase-change material (PCM) produced a minimum resistance of 5.2 mm^{2}K/W, as shown in Figure 12.10. Xu et al. used a

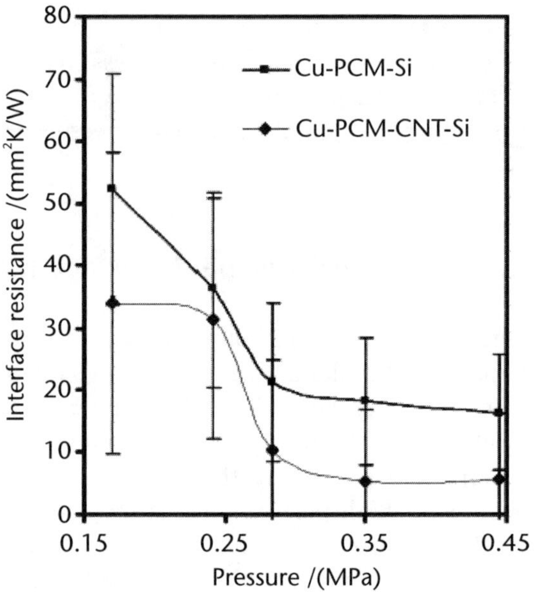

Figure 12.10 Thermal resistance of copper-silicon interfaces with a CNT array (Cu-PCM-CNT-Si) compared with PCM (Cu-PCM-Si) as a function of pressure. (*From:* [74]. © 2005 Elsevier. Reprinted with permission.)

photothermal metrology to evaluate the thermal conductivity of aligned CNT arrays grown on silicon substrates by plasma-enhanced CVD [76]. The effective thermal resistance was 12~16 mm^2K/W, which is comparable to the resistance of commercially available thermal grease.

12.5.2 Thermal Interfacial Resistance of CNT-Based Materials

Preliminary results of CNT-based materials for TIMs are encouraging and indicate the possibility of achieving a percolation threshold at very small volume fractions. However, CNT-based composites show thermal conductivities far below expected values, despite the excellent thermal conductivity of CNTs. This discrepancy can be attributed to several issues. First, the intrinsic thermal conductivity of nanotubes used in the experiment may be much lower than expected due to the defects that act as scattering centers for phonons. Another possibility is that the interactions between nanotubes and the surrounding matrix result in significant phonon scattering.

Nan et al. formulated a model demonstrating that the thermal-conductivity enhancement in the nanotube composites is mainly limited by interface thermal resistance [77]. A large thermal resistance across the nanotube-matrix interface will cause a significant degradation in the thermal-conductivity enhancement. For instance, with the same loading, the composites prepared by SWNTs with higher intrinsic thermal conductivity than MWNTs will have a lower thermal conductivity than those by MWNTs. Shenogin et al. used classical molecular dynamics simulations to study the interfacial resistance to heat flow between a carbon nanotube and liquid octane [78]. A large portion of the interfacial resistance was associated with weak coupling between the rigid tube structure and the soft organic liquid. The thermal resistance of the interface was calculated to be 3.3×10^{-8} m^{2}K/W. These results imply that the thermal conductivity of carbon-nanotube polymer composites and organic suspensions will be limited by the interface thermal resistance; such conclusions are consistent with recent experiments. Huxtable et al. used picosecond transient absorption to measure the interface thermal conductance of carbon nanotubes suspended in surfactant micelles in water [79]. Results indicated that heat transport in a nanotube composite material will be limited by the exceptionally large interface thermal resistance (~8.3×10^{-8} m^{2}K/W) and that the thermal conductivity of the composite will be much lower than the value estimated from the intrinsic thermal conductivity of nanotubes and their volume fraction.

12.5.3 Thermal Constriction Resistance Between Nanotube and Substrate

When the heat transfers at the interfaces between a nanotube and a planar substrate, the constriction resistance is developed due to the nanometer-scale contact area and enhanced phonon-boundary scattering at the nanocontacts [80]. Bahadur et al. modeled the interfacial thermal resistance of a nanowire/plane interface using continuum principles combined with van der Waals theory. The model estimates numerical values of constriction and gap resistances for several nanowire-substrate combinations with water and air as the surrounding media. For a low-conductivity medium, the interface resistance is dominated by the constriction resistance, which itself depends significantly on nanowire and substrate conductivities. Prasher developed a more accurate model for calculating the constriction resistance of nanometer scale by taking into account the ballistic nature of thermal transport in the vicinity of the constrictions since ballistic transport of phonons is very important at the constriction if the constriction size is comparable to the mean free path of carriers such as phonons or electrons [81]. The model accounts for the effects of appropriate length scales and shows that for constriction formed by the same materials, the Knudsen number is the appropriate dimensionless parameter, whereas for constriction formed by dissimilar materials, the microscopic Biot number is the appropriate parameter.

12.6 Integration of Carbon Nanotubes into Microsystems for Thermal Management

12.6.1 Integration Approaches for Carbon Nanotubes

For electronic device applications, CVD methods are particularly attractive. However, the CVD technique suffers from several drawbacks. One of the main challenges for applying CNTs to circuitry is the high growth temperature (>600°C). Such temperatures are incompatible with microelectronic processes, which are typically performed below 400°C to 500°C in BEOL sequences. Another issue is the poor adhesion between CNTs and the substrates, which will result in long-term reliability issues and high contact resistance. At the device level, CNTs must be integrated and interconnected with metal electrodes to allow signal input and output. Typical approaches for CNT growth on such substrates involve the deposition of catalysts, such as Fe or Ni, on metal layers, such Ti or Ti/Au. Unfortunately, results indicate that electrical contact is not necessarily improved, suggesting that attachment of CNTs onto the electrodes produces poor mechanical and electrical properties, yielding high contact resistance. On the other hand, to meet manufacturing requirements and throughput for IC applications, a large number of CNTs must be positioned simultaneously rather than aligned one by one.

For horizontal interconnects, CNTs must grow in the horizontal direction. However, there has been limited success in achieving controlled CNT growth only in the horizontal direction. Dai et al. demonstrated electric-field-directed growth of single-walled carbon nanotubes by CVD [82]. The field-alignment effect arises from the high polarizability of single-walled nanotubes. Large induced dipole moments result in large aligning torques and forces on the nanotube and thus prevent randomization of nanotube orientation due to thermal fluctuations and gas flows. Another approach is to use gas flow to control horizontal CNT growth [83]; here, the majority of the SWNTs are well oriented, and the length of the nanotubes can be greater than 2 mm for a 10 minute growth. The general problem with horizontal growth is not whether it can be performed but whether adequate yield can be achieved. The yield must be very close to 100% for a circuit, but to date, horizontal growth appears to be a low-yield process [84].

12.6.2 CNT Transfer Process

To overcome the above disadvantages, Zhu et al. proposed a methodology termed "CNT transfer technology," which is enabled by open-ended CNT structures [85]. This technique is similar to flip-chip technology as illustrated schematically in Figure 12.11.

The substrates for this process can be FR4 boards coated with copper foil or other materials with different moduli, such as heat sinks. To improve the adhesion and wetting of solder on the substrates, under bump metallization (UBM) layers are sputtered onto the substrate metallization. The eutectic tin-lead paste (100 μm) is then stencil-printed on the UBM. After reflow, the tin-lead solder is polished to 30 μm. The silicon substrates with CNTs are then flipped and aligned to the corre-

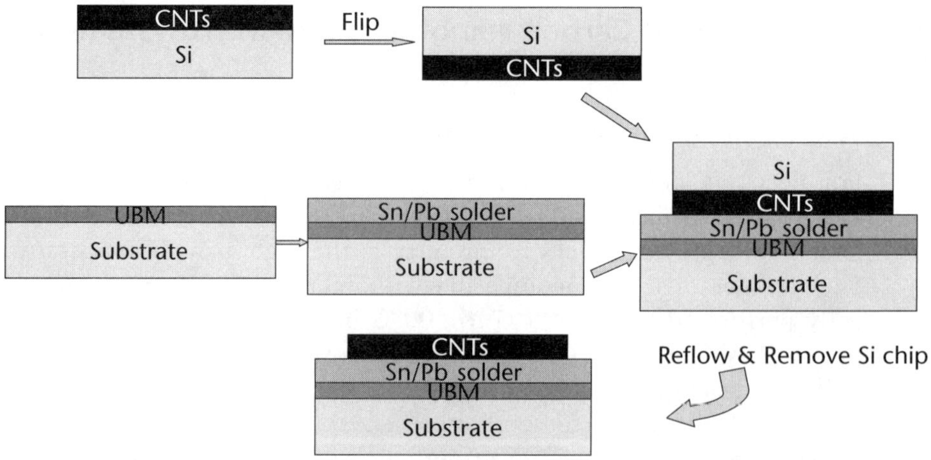

Figure 12.11 Schematic diagram of "CNT transfer technology." (*From:* [85]. © 2006 ACS. Reprinted with permission.)

sponding copper substrates and reflowed in a reflow oven at higher temperatures (peak temperature 270°C) than those typically used (220°C) to simultaneously form electrical and mechanical connections. This process is straightforward to implement and offers a strategy for both assembling CNT devices and scaling up a variety of devices fabricated using nanotubes (e.g., flat panel displays). This process offers an approach to overcome the serious obstacles of integration of CNTs into integrated circuits and microelectronic device packages by offering low process temperatures and improved adhesion of CNTs to substrates. Figure 12.12(a) indicates that the entire CNT film (1.5 × 1.5 cm) has been transferred to the substrate (2.54 × 2.54 cm), since CNTs are not evident on the silicon chip. To qualitatively demonstrate the bonding strength of the CNTs on the copper substrate that results from the solder reflow process, a section of the assembled CNTs was pulled from the surface by tweezers. Figure 12.12(b) shows the demarcation between the broken CNTs and the intact and connected ones. When pulled from the substrate, the CNTs break along the axis rather than at the CNT-solder interface. The excellent mechanical bonding strength of CNTs on the substrate anchors the CNTs and thereby improves the CNT/substrate interfacial properties.

The CNT transfer process can be used to assemble fine-pitch CNT bundles. Figure 12.13(b) shows the result of transferring the CNT bundles shown in Figure 12.13(a) onto a copper substrate. Clearly, the CNT bundle structures remain intact after transfer with respect to the bundle size, aspect ratio, and pitch. The CNT bundles are fine-pitched structures with diameter, aspect-ratio, and pitch of 25μm, 4, and 80μm, respectively.

Kordás et al. demonstrated a simple and scalable nanotube-on-chip assembly process using a transfer technique to integrate nanotube structures onto the chip, where the CNT fins are exploited to remove heat from silicon chip components, as shown in Figure 12.14 [86]. The nanotube fin arrays have a size of ~1.2 × 1.0 × 1.0 mm^3. For a 1 mm^2 test chip to reach the same temperature, the applied power can be ~1W larger when a CNT cooler is applied compared to that of the situation with a bare chip. Testing indicated that the nanotube fin structures allowed heat dissipa-

Figure 12.12 (a) Photograph of an open-ended CNT film transferred onto a copper substrate coated with eutectic tin-lead solder. (b) SEM of the copper substrates on which the CNTs were assembled after some CNTs were pulled from the surface by tweezers. This figure demonstrates the excellent mechanical bond strength of CNTs transferred to the copper substrate by the solder reflow process. (*From:* [85]. © 2006 ACS. Repeinted with permission.)

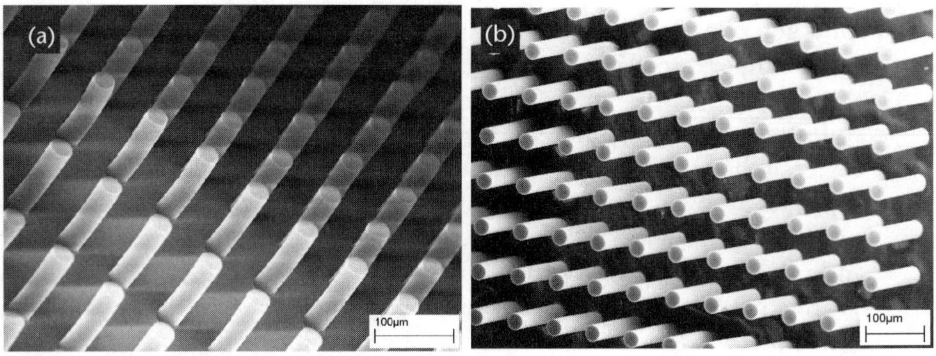

Figure 12.13 (a) SEM image of aligned CNT bundles grown on a silicon substrate with size, aspect ratio, and pitch of 25 μm, 4, and 80 μm, respectively, and (b) SEM image of the CNT bundles in (a) transferred onto a copper substrate.

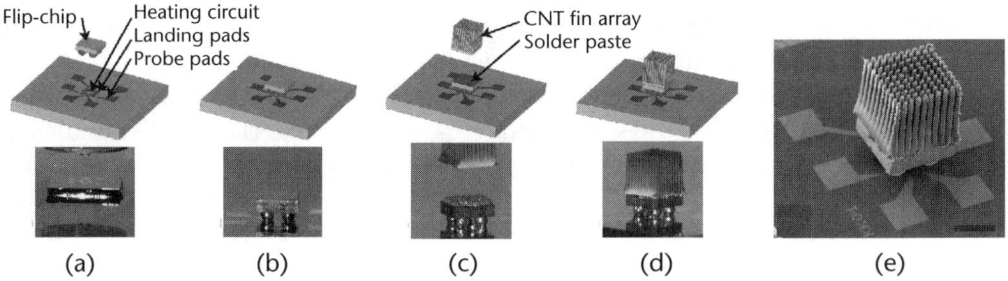

Figure 12.14 Steps of substrate/flip-chip/CNT-cooler assembly: (a) positioning and (b) soldering the flip-chip on the Cu landing pads of the substrate (this structure also served as a reference), (c) solder paste dispensing and CNT array positioning, (d) soldering on the Cu-coated back side of the chip, and (e) field-emission scanning electron microscopy image of an assembled structure (scale bar: 500 μm). (*From:* [86]. © 2006 AIP. Reprinted with permission.)

tion of ~30 and ~100 Wcm^{-2} more power at 100°C from a hot chip for the cases of natural and forced convections, respectively. The cooling performance of the nanotube fin structures combined with their low weight, mechanical robustness, and ease of fabrication make them possible candidates for on-chip thermal management applications. Mo et al. demonstrated the integration of carbon nanotubes in the liquid microchannel coolers for enhancement in cooling capability, as shown in Figure 12.15 [87]. By using lithography techniques, chemical vapor deposition, and adhesive bonding, a microcooler with two-dimensional nanotube fins was manufactured. Though the array of fins decreases the flow rate by 12%, the cooler with nanotube fins has much higher cooling capability. With 23% higher input power (8.9 W versus 7.2 W), the nanotube cooler could keep the transistor temperature 6°C lower than the reference cooler. This superiority became more significant as the input power increased.

12.6.3 Direct Growth of Carbon Nanotubes on Metal Substrates

Usually, aligned CNTs are grown either by using thin catalyst layers predeposited on substrates or through concurrent vapor-phase catalyst (such as ferrocene) deliv-

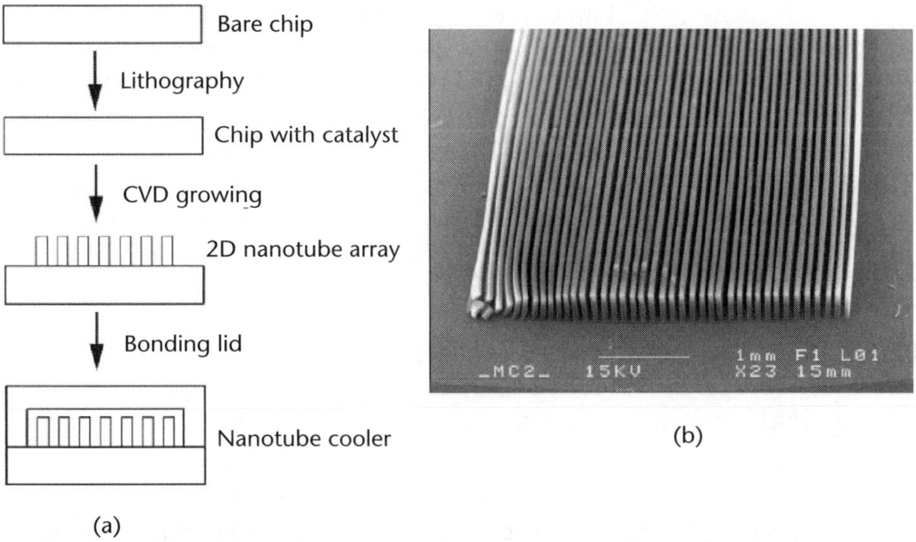

Figure 12.15 (a) Fabrication sequences for making carbon nanotube cooler, and (b) carbon nanotube fins on silicon chips. (*From:* [87]. © 2005 IEEE. Reprinted with permission.)

ery. For microelectronics applications such as field emitter and gigascale interconnects, appropriate CNT structures with direct metal contacts are required. Thus, the ability to grow CNTs directly onto metal surfaces with robust CNT-metal contacts is desirable. However, CNTs grow predominantly on nonmetallic substrates; due to the easy formation of alloys between the metal catalyst nanoparticles and the metal substrates, it is challenging to grow CNTs directly on metal substrates. Thus, it is important either to develop new processes for nanotube growth on metals or to identify suitable metals on which nanotubes can be grown easily with available technologies.

Some progress has been reported on direct CNT growth on metal substrates. For example, aligned CNTs are controllably synthesized by pyrolysis of iron(II) phthalocyanine (FePc), which contains both the metal catalyst and the carbon source required for nanotube growth, on Ni, Ti, and Ta sheets [88]. Xu et al. used voltage bias in methane inverse diffusion flames to grow CNTs on metal alloys [89]. Karwa et al. used the iron nanostructures present in stainless steel as the catalyst to grow aligned CNTs by CVD using ethylene as the carbon source [90]. Ajayan et al. reported the direct growth of aligned nanotubes on Inconel (major constituents: 72% Ni, 16% Cr, and 8% Fe) by a vapor-phase catalyst-delivery CVD method [91]. Arrays of well-aligned MWNTs were grown on as-received Inconel substrates; the CNTs are well anchored to the substrate. The CNT-Inconel interfaces exhibit good electrical contact and strong mechanical adhesion. Nanotube growth on Inconel by a vapor-phase catalyst-delivery CVD method has at least two major advantages [91]. First, the growth is not spatially restricted by the presence of a catalyst, thus offering a way to fabricate three-dimensional aligned nanotube arrays on metals in a single step. Second, the ability to grow nanotubes on any shape or size of substrate provides tremendous flexibility for developing applications in which the morphology of the conductive substrates is critical.

Recently, Ng et al. developed a process to grow aligned MWNTs on both sides of a metallic or metal-coated substrate by water-vapor-assisted CVD. Aligned CNT films can be grown with lengths greater than 100 μm [92]. This technique was used to fabricate a CNT-based thermal interface material comprising a thin copper foil covered on both sides with Cr-Au-MWNTs. The thermal resistance of a 12.5 μm copper foil coated with 50 μm MWNTs on each side was 12 mm^2 K/W using a steady-state measurement system that was designed in accordance with standard ASTM D5470 under an applied contact pressure of 0.3 MPa.

12.7 Summary and Future Needs

Scaling of microelectronic devices has led to an interest in utilizing carbon nanotubes for electrical interconnects and thermal management approaches. This chapter briefly reviewed carbon nanotube growth methods, electrical and thermal properties of nanotubes, and integration issues and applications for CNTs in microelectronic devices as interconnects and thermal management layers. Carbon nanotubes are also promising for vertical interconnects (for on-chip or packaging levels) and heat removal for microelectronics packaging.

CNTs may be able to meet some of the ITRS projections for device interconnects and thermal requirements. However, a number of materials and CNT process-integration issues need to be addressed before a CNT technology platform can be developed. Future work on CNTs should include

- *Growth of structurally perfect carbon nanotubes:* The projected excellent CNT properties are based on perfect CNT structures. The existence of defects along CNT walls greatly degrades CNT performance. A recent study has shown that the CNT resistance increases exponentially with length, although structural imperfections were uniformly created along the CNT length [93]. Simulation studies demonstrate that di-vacancies primarily contribute to the resistance due to localization effects and that the presence of 0.03% di-vacancies can produce an increase of three orders of magnitude in the resistance of a metallic SWNT of 400 nm length.
- *Chirality control of carbon nanotubes:* Interconnects clearly require metallic nanotubes; however, no proven method to control nanotube chirality (resulting in semiconducting or metallic CNTs) currently exists. The coexistence of metallic or semiconducting nanotubes within commercially available SWNTs has been a bottleneck for fundamental research and fabrication of high-performance devices. It is therefore desirable to develop selective synthesis or purification methods that allow the growth of exclusively metallic or semiconducting SWNT samples.
- *Positioning of carbon nanotubes in predefined locations simultaneously:* Integration of CNTs into scalable integrated circuits requires approaches to the difficult problem of simultaneously synthesizing and accurately positioning large numbers of individual, electrically homogeneous tubes with linear geometries and nearly 100% yield and reproducibility.

The barriers to CNT implementation in the packaging of microelectronic devices and ICs offer numerous opportunities for new developments and approaches. Clearly, more effort is required in order to take CNT technologies from the research laboratory to high-volume production.

References

[1] Moore, G. E., "Progress in Digital Integrated Electronics," *International Electron Devices Meetings,* Washington, D.C., December 2–4, 1975, pp. 11–13.

[2] Mahajan, R., C. P. Chiu, and G. Chrysler, "Cooling a Microprocessor Chip," *Proc. IEEE*, Vol. 94, No. 8, 2006, pp. 1476–1486.

[3] Schmidt, R., "Challenges in Electronic Cooling—Opportunities for Enhanced Thermal Management Techniques—Microprocessor Liquid Cooled Minichannel Heat Sink," *Heat Transfer Engineering*, Vol. 25, No. 3, 2004, pp. 3–12.

[4] Azar, K., "Power Consumption and Generation in the Electronics Industry—A Perspective," *20th IEEE SEMI-Therm Symposium*, San Jose, California, March 21–23, 2004, pp. 201–212.

[5] Schelling, P. K., L. Shi, and K. E. Goodson, "Managing Heat for Electronics," *Materials Today*, Vol. 8, No. 6, 2005, pp. 30–35.

[6] Prasher, R. S., et al., "Nano and Micro Technology Based Next Generation Package-Level Cooling Solutions," *Intel Technology Journal*, Vol. 9, No. 4, 2005, pp. 285–292.

[7] Kreupl, F., et al., "Carbon Nanotubes in Interconnect Applications," *Microelectronic Engineering*, Vol. 64, No. 1–4, 2002, pp. 399–408.

[8] Graham, A. P., et al., "How Do Carbon Nanotubes Fit into the Semiconductor Roadmap?" *Applied Physics A—Materials Science and Processing*, Vol. 80, No. 6, 2005, pp. 1141–1151.

[9] Dresselhaus, M. S., G. Dresselhaus, and R. Saito, "Carbon-Fibers Based on C-60 and Their Symmetry," *Physical Review B*, Vol. 45, No. 11, 2002, pp. 6234–6242.

[10] Hamada, N., S. Sawada, and A. Oshiyama, "New One-Dimensional Conductors—Graphitic Microtubules," *Physical Review Letters*, Vol. 68, No. 10, 1992, pp. 1579–1581.

[11] Dresselhaus, M. S., G. Dresselhaus, and P. Avouris, *Carbon Nanotubes—Synthesis, Structure, Properties and Applications*, Berlin: Springer-Verlag, 2001.

[12] Odom, T. W., et al., "Atomic Structure and Electronic Properties of Single-Walled Carbon Nanotubes," *Nature*, Vol. 391, No. 6662, 1998, pp. 62–64.

[13] Wildoer, J. W. G., et al., "Electronic Structure of Atomically Resolved Carbon Nanotubes," *Nature*, Vol. 391, No. 6662, 1998, pp. 59–62.

[14] Ebbesen, T. W., and P. M. Ajayan, "Large-Scale Synthesis of Carbon Nanotubes," *Nature*, Vol. 358, No. 6383, 1992, pp. 220–222.

[15] Bethune, D. S., et al., "Cobalt-Catalyzed Growth of Carbon Nanotubes with Single-Atomic-Layerwalls," *Nature*, Vol. 363, No. 6430, 1993, pp. 605–607.

[16] Thess, A., et al., "Crystalline Ropes of Metallic Carbon Nanotubes," *Science*, Vol. 273, No. 5274, 1996, pp. 483–487.

[17] Liu, J., et al., "Fullerene Pipes," *Science*, Vol. 280, No. 5367, 1998, pp. 1253–1256.

[18] Amelinckx, S., et al., "A Structure Model and Growth-Mechanism for Multishell Carbon Nanotubes," *Science*, Vol. 267, No. 5202, 1995, pp. 1334–1338.

[19] Amelinckx, S., et al., "A Formation Mechanism for Catalytically Grown Helix-Shaped Graphite Nanotubes," *Science*, Vol. 265, No. 5172, 1994, pp. 635–639.

[20] Shyu, Y. M., and F. C. N. Hong, "Low-Temperature Growth and Field Emission of Aligned Carbon Nanotubes by Chemical Vapor Deposition," *Materials Chemistry and Physics*, Vol. 72, No. 2, 2001, pp. 223–227.

[21] Shyu, Y. M., and F. C. N. Hong, "The Effects of Pre-Treatment and Catalyst Composition on Growth of Carbon Nanofibers at Low Temperature," *Diamond and Related Materials*, Vol. 10, No. 3–7, 2001, pp. 1241–1245.

[22] Laplaze, D., et al., "Carbon Nanotubes: Dynamics of Synthesis Processes," *Carbon*, Vol. 40, No. 10, 2002, pp. 1621–1634.

[23] Hata, K., et al., "Water-Assisted Highly Efficient Synthesis of Impurity-Free Single-Walled Carbon Nanotubes," *Science*, Vol. 306, No. 5700, 2004, pp. 1362–1364.

[24] Kim, M. J., et al., "Efficient Transfer of a VA-SWNT Film by a Flip-Over Technique," *J. American Chemical Society*, Vol. 128, No. 29, 2006, pp. 9312–9313.

[25] Sen, R., A. Govindaraj, and C. N. R. Rao, "Carbon Nanotubes by the Metallocene Route," *Chemical Physics Letters*, Vol. 267, No. 3–4, 1997, pp. 276–280.

[26] Hofmann, S., et al., "Effects of Catalyst Film Thickness on Plasma-Enhanced Carbon Nanotube Growth," *J. Applied Physics*, Vol. 98, No. 3, 2005, pp. 034308-1–034308-8.

[27] Hernadi, K., et al., "Fe-Catalyzed Carbon Nanotube Formation," *Carbon*, Vol. 34, No. 10, 1996, pp. 1249–1257.

[28] Su, M., B. Zheng, and J. Liu, "A Scalable CVD Method for the Synthesis of Single-Walled Carbon Nanotubes with High Catalyst Productivity," *Chemical Physics Letters*, Vol. 322, No. 5, 2000, pp. 321–326.

[29] Colomer, J. F., et al., "Synthesis of Single-Wall Carbon Nanotubes by Catalytic Decomposition of Hydrocarbons," *Chemical Communications*, Vol. 71, No. 14, 1999, pp. 1343–1344.

[30] Vander Wal, R. L., T. W. Ticich, and V. E. Curtis, "Substrate-Support Interactions in Metal-Catalyzed Carbon Nanofiber Growth," *Carbon*, Vol. 39, No. 15, 2001, pp. 2277–2289.

[31] Willems, I., et al., "Control of the Outer Diameter of Thin Carbon Nanotubes Synthesized by Catalytic Decomposition of Hydrocarbons," *Chemical Physics Letters*, Vol. 317, No. 1–2, 2000, pp. 71–76.

[32] Su, M., et al., "Lattice-Oriented Growth of Single-Walled Carbon Nanotubes," *J. Physical Chemistry B*, Vol. 104, No. 28, 2000, pp. 6505–6508.

[33] Kukovecz, A., et al., "Catalytic Synthesis of Carbon Nanotubes over Co, Fe and Ni Containing Conventional and Sol-gel Silica-Aluminas," *Physical Chemistry Chemical Physics*, Vol. 2, No. 13, 2000, pp. 3071–3076.

[34] Choi, Y. C., et al., "Growth of Carbon Nanotubes by Microwave Plasma-Enhanced Chemical Vapor Deposition at Low Temperature," *J. Vacuum Science and Technology A—Vacuum Surfaces and Films*, Vol. 18, No. 4, 2000, pp. 1864–1868.

[35] Hofmann, S., et al., "Low-Temperature Growth of Carbon Nanotubes by Plasma-Enhanced Chemical Vapor Deposition," *Applied Physics Letters*, Vol. 83, No. 1, 2003, pp. 135–137.

[36] Bower, C., et al., "Plasma-Induced Alignment of Carbon Nanotubes," *Applied Physics Letters*, Vol. 77, No. 6, 2000, pp. 830–832.

[37] Hiramatsu, M., et al., "High-Rate Growth of Films of Dense, Aligned Double-Walled Carbon Nanotubes Using Microwave Plasma-Enhanced Chemical Vapor Deposition," *Japanese J. Applied Physics Part 2—Letters and Express Letters*, Vol. 44, No. 22, 2005, pp. L693–L695.

[38] Zhu, L. B., et al., "Aligned Carbon Nanotube Stacks by Water-Assisted Selective Etching," *Nano Letters*, Vol. 5, No. 12, 2005, pp. 2641–2645.

[39] Sinnott, S. B., et al., "Model of Carbon Nanotube Growth Through Chemical Vapor Deposition," *Chemical Physics Letters*, Vol. 315, No. 1, 1999, pp. 25–30.

[40] Baker, R. T. K., "Catalytic Growth of Carbon Filaments," *Carbon*, Vol. 27, No. 3, 1989, pp. 315–323.

[41] Baker, R. T. K., et al., "Nucleation and Growth of Carbon Deposits from Nickel Catalyzed Decomposition of Acetylene," *J. Catalysis*, Vol. 26, No. 1, 1972, pp. 51–62.

[43] Keep, C. W., R. T. K. Baker, and J. A. France, "Origin of Filamentous Carbon Formation from Reaction of Propane over Nickel," *J. Catalysis*, Vol. 47, No. 2, 1977, pp. 232–238.

[45] Yang, R. T., and S. J. Doong, "Gas Separation by Pressure Swing Adsorption—A Pore-Diffusion Model for Bulk Separation," *AIChE Journal*, Vol. 31, No. 11, 1985, pp. 1829–1842.

[44] Yang, R. T., and K. L. Yang, "Evidence for Temperature-Driven Carbon Diffusion Mechanism of Coke Deposition on Catalysts," *J. Catalysis*, Vol. 93, No. 1, 1985, pp. 182–185.

[45] Baker, R. T. K., et al., "Formation of Filamentous Carbon from Iron, Cobalt and Chromium Catalyzed Decomposition of Acetylene," *J. Catalysis*, Vol. 30, No. 1, 1973, pp. 86–95.

[46] Nielsen, J. R., and D. L. Trimm, "Mechanisms of Carbon Formation on Nickel-Containing Catalysts," *J. Catalysis*, Vol. 48, No. 1–3, 1977, pp. 155–165.

[47] Frank, S., et al., "Carbon Nanotube Quantum Resistors," *Science*, Vol. 280, No. 5370, 1998, pp. 1744–1746.

[48] Haruehanroengra, S., and W. Wang, "Analyzing Conductance of Mixed Carbon-Nanotube Bundles for Interconnect Applications," *IEEE Electron Device Letters*, Vol. 28, No. 8, 2007, pp. 756–759.

[49] Naeemi A., and J. D. Meindl, "Compact Physical Models for Multiwall Carbon-Nanotube Interconnects," *IEEE Electron Device Letters*, Vol. 27, No. 5, 2006, pp. 338–340.

[50] Wu, T. W., and E. C. Chen, "Crystallization Behavior of Poly(Epsilon-Caprolactone)/Multiwalled Carbon Nanotube Composites," *J. Polymer Science Part B—Polymer Physics*, Vol. 44, No. 3, 2006, pp. 598–606.

[51] Mizuno, S., et al., "Dielectric Constant and Stability of Fluorine Doped PECVD Silicon Oxide Thin Films," *Thin Solid Films*, Vol. 283, No. 1, 2006, pp. 30–36.

[52] Wei, B. Q., R. Vajtai, and P. M. Ajayan, "Reliability and Current Carrying Capacity of Carbon Nanotubes," *Applied Physics Letters*, Vol. 79, No. 8, 2001, pp. 1172–1174.

[53] Naeemi, A., and J. D. Meindl, "Monolayer Metallic Nanotube Interconnects: Promising Candidates for Short Local Interconnects," *IEEE Electron Device Letters*, Vol. 26, No. 8, 2005, pp. 544–546.

[54] Graham, A. P., et al., "Carbon Nanotubes for Microelectronics?" *Small*, Vol. 1, No. 4, 2005, pp. 382–390.

[55] Nihei, M., "Electrical Properties of Carbon Nanotube Bundles for Future Via Interconnects," *Japanese J. Applied Physics Part 1—Regular Papers Short Notes and Review Papers*, Vol. 44, No. 4A, 2005, pp. 1626–1628.

[56] Hoenlein, W., et al., "Carbon Nanotubes for Microelectronics: Status and Future Prospects," *Materials Science and Engineering C—Biomimetic and Supramolecular Systems*, Vol. 23, No. 6, 2003, pp. 663–669.

[57] Awano, Y., "Carbon Nanotube Technologies for LSI Via Interconnects," *IEICE Transactions on Electronics*, Vol. E89-C, No. 11, 2006, pp. 1499–1503.

[58] Sato, S., et al., "Growth of Diameter-Controlled Carbon Nanotubes Using Monodisperse Nickel Nanoparticles Obtained with a Differential Mobility Analyzer," *Chemical Physics Letters*, Vol. 382, No. 3, 2003, pp. 361–366.

[59] Sato, S., et al., "Novel Approach to Fabricating Carbon Nanotube Via Interconnects Using Size-Controlled Catalyst Nanoparticles," *Proc. 2006 International Interconnect Technology Conference*, Burlingame, California, June 5–7, 2006, pp. 230–232.

[60] Hone, J., et al., "Thermal Properties of Carbon Nanotubes and Nanotube-Based Materials," *Applied Physics A—Materials Science and Processing*, Vol. 74, No. 3, 2002, pp. 339–343.

[61] Benedict, L. X., S. G. Louie, and M. L. Cohen, "Heat Capacity of Carbon Nanotubes," *Solid State Communications*, Vol. 100, No. 3, 1996, pp. 177–180.

[62] Mizel, A., et al., "Analysis of the Low-Temperature Specific Heat of Multiwalled Carbon Nanotubes and Carbon Nanotube Ropes," *Physical Review B*, Vol. 60, No. 5, 1999, pp. 3264–3270.

[63] Hone, J., et al., "Quantized Phonon Spectrum of Single-Wall Carbon Nanotubes," *Science*, Vol. 289, No. 5485, 2000, pp. 1730–1733.

[64] Berber, S., Y. K. Kwon, and D. Tomanek, "Unusually High Thermal Conductivity of Carbon Nanotubes," *Physical Review Letters*, Vol. 84, No. 20, 2000, pp. 4613–4616.

[65] Pop, E., et al., "Thermal Conductance of an Individual Single-Wall Carbon Nanotube above Room Temperature," *Nano Letters*, Vol. 6, No. 1, 2006, pp. 96–100.

[66] Kim, P., et al., "Thermal Transport Measurements of Individual Multiwalled Nanotubes," *Physical Review Letters*, Vol. 87, No. 21, 2001, p. 215502.

[67] Choi, T. Y., et al., "Measurement of Thermal Conductivity of Individual Multiwalled Carbon Nanotubes by the 3-Omega Method," *Applied Physics Letters*, Vol. 87, No. 1, 2005, p. 013108.

[68] Hone, J., et al., "Electrical and Thermal Transport Properties of Magnetically Aligned Single Wall Carbon Nanotube Films," *Applied Physics Letters*, Vol. 77, No. 5, 2000, pp. 666–668.

[69] Yi, W., et al., "Linear Specific Heat of Carbon Nanotubes," *Physical Review B*, Vol. 59, No. 14, 1999, pp. R9015–R9018.

[70] Yang, D. J., et al., "Thermal Conductivity of Multiwalled Carbon Nanotubes," *Physical Review B*, Vol. 66, No. 16, 2002, p. 165440.

[71] Chiu, C. P., J. G. Maveety, and Q. A. Tran, "Characterization of Solder Interfaces Using Laser Flash Metrology," *Microelectronics Reliability*, Vol. 42, No. 1, 2002, pp. 93–100.

[72] Biercuk, M. J., et al., "Carbon Nanotube Composites for Thermal Management," *Applied Physics Letters*, Vol. 80, No. 15, 2002, pp. 2767–2769.

[73] Yu, A. P., et al., "Effect of Single-Walled Carbon Nanotube Purity on the Thermal Conductivity of Carbon Nanotube-Based Composites," *Applied Physics Letters*, Vol. 89, No. 13, 2006, p. 133102.

[74] Xu, J., and T. S. Fisher, "Enhancement of Thermal Interface Materials with Carbon Nanotube Arrays," *Int. J. Heat and Mass Transfer*, Vol. 49, 2006, pp. 1658–1666.

[75] Xu, J., and T. S. Fisher, "Enhanced Thermal Contact Conductance Using Carbon Nanotube Array Interfaces," *IEEE Transactions on Components and Packaging Technologies*, Vol. 29, No. 9–10, 2006, pp. 261–267.

[76] Xu, Y., et al., "Thermal Properties of Carbon Nanotube Array Used for Integrated Circuit Cooling," *J. Applied Physics*, Vol. 100, No. 7, 2006, p. 074302.

[77] Nan, C. W., et al., "Interface Effect on Thermal Conductivity of Carbon Nanotube Composites," *Applied Physics Letters*, Vol. 85, No. 16, 2004, pp. 3549–3551.

[78] Shenogin, S., et al., "Role of Thermal Boundary Resistance on the Heat Flow in Carbon-Nanotube Composites," *J. Applied Physics*, Vol. 95, No. 12, 2004, pp. 8136–8144.

[79] Huxtable, S. T., et al., "Interfacial Heat Flow in Carbon Nanotube Suspensions," *Nature Materials*, Vol. 2, No. 11, 2003, pp. 731–734.

[80] Yu, C. H., et al., "Thermal Contact Resistance and Thermal Conductivity of a Carbon Nanofiber," *J. Heat Transfer—Transactions of the ASME*, Vol. 128, No. 3, 2006, pp. 234–239.

[81] Prasher, V., "Predicting the Thermal Resistance of Nanosized Constrictions," *Nano Letters*, Vol. 5, No. 11, 2005, pp. 2155–2159.

[82] Zhang, Y. G., et al., "Electric-Field-Directed Growth of Aligned Single-Walled Carbon Nanotubes," *Applied Physics Letters*, Vol. 79, No. 19, 2001, pp. 3155–3157.

[83] Huang, S. M., X. Y. Cai, and J. Liu, "Growth of Millimeter-Long and Horizontally Aligned Single-Walled Carbon Nanotubes on Flat Substrates," *J. American Chemical Society*, Vol. 125, No. 19, 2003, pp. 5636–5637.

[84] Robertson, J., "Growth of Nanotubes for Electronics," *Materials Today*, Vol. 10, No. 1–2, 2006, pp. 36–43.

[85] Zhu, L. B., et al., "Well-Aligned Open-Ended Carbon Nanotube Architectures: An Approach for Device Assembly," *Nano Letters*, Vol. 6, No. 2, 2006, pp. 243–247.

[86] Kordas, K., et al., "Chip Cooling with Integrated Carbon Nanotube Microfin Architectures," *Applied Physics Letters*, Vol. 90, No. 12, 2007, p. 123105.

[87] Mo, Z. M., "Integrated Nanotube Microcooler for Microelectronics Applications," *Proc. 55th Electronic Components and Technology Conference*, Orlando, Florida, May 31–June 3, 2005, pp. 51–54.

[88] Wang, B. A., et al., "Controllable Preparation of Patterns of Aligned Carbon Nanotubes on Metals and Metal-Coated Silicon Substrates," *J. Materials Chemistry*, Vol. 13, No. 5, 2003, pp. 1124–1126.

[89] Xu, F. S., X. F. Liu, and S. D. Tse, "Synthesis of Carbon Nanotubes on Metal Alloy Substrates with Voltage Bias in Methane Inverse Diffusion Flames," *Carbon*, Vol. 44, No. 3, 2006, pp. 570–577.

[90] Karwa, M., Z. Iqbal, and S. Mitra, "Scale-Up Self-Assembly of Carbon Nanotubes inside Long Steel Tubing," *Carbon*, Vol. 44, No. 7, 2006, pp. 1235–1242.

[91] Talapatra, S., et al., "Direct Growth of Aligned Carbon Nanotubes on Bulk Metals," *Nature Nanotechnology*, Vol. 1, No. 2, 2006, pp. 112–116.

[92] Wang, H., et al., "Synthesis of Aligned Carbon Nanotubes on Double-Sided Metallic Substrate by Chemical Vapor Deposition," *J. Physical Chemistry C*, Vol. 111, No. 34, 2007, pp. 12617–12624.

[93] Gomez-Navarro, C., et al, "Tuning the Conductance of Single-Walled Carbon Nanotubes by Ion Irradiation in the Anderson Localization Regime," *Nature Materials*, Vol. 4, No. 7, 2005, pp. 534–539.

CHAPTER 13

3D Integration and Packaging for Memory

Soon-Moon Jung and Kinam Kim

13.1 Introduction

As silicon technology moves to deep nanoscale dimensions (sub 30 nm), there are growing concerns as to whether the semiconductor industry can maintain the continuous growth trends and rate of growth of those achieved over the past several decades (Figure 13.1). The future of the Si industry will be uncertain due to the ultimate physical limits of planar CMOS transistor scaling and a decreasing margin in manufacturing resulting from technical complexities and ever-increasing investment in fabrication foundries [1, 2]. The dimension of the limits is predicted to be near 20 nm where a transistor contains less than 10 electrons, as shown in Figure 13.2, which is believed to be the minimum number of stored charges needed to

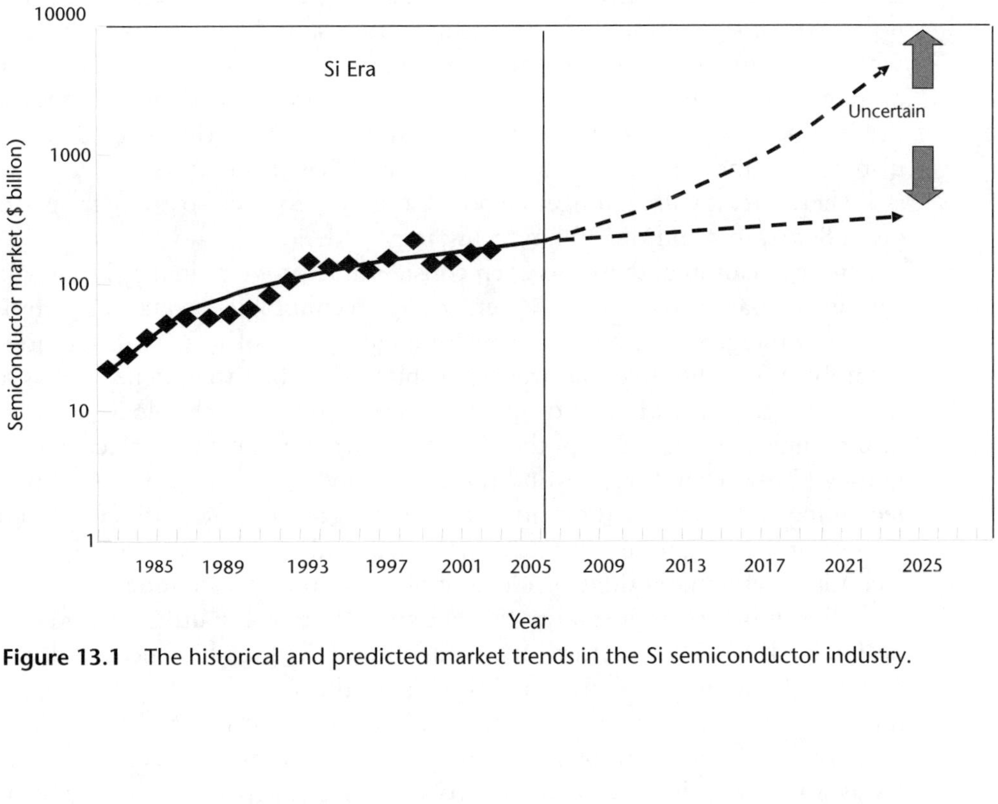

Figure 13.1 The historical and predicted market trends in the Si semiconductor industry.

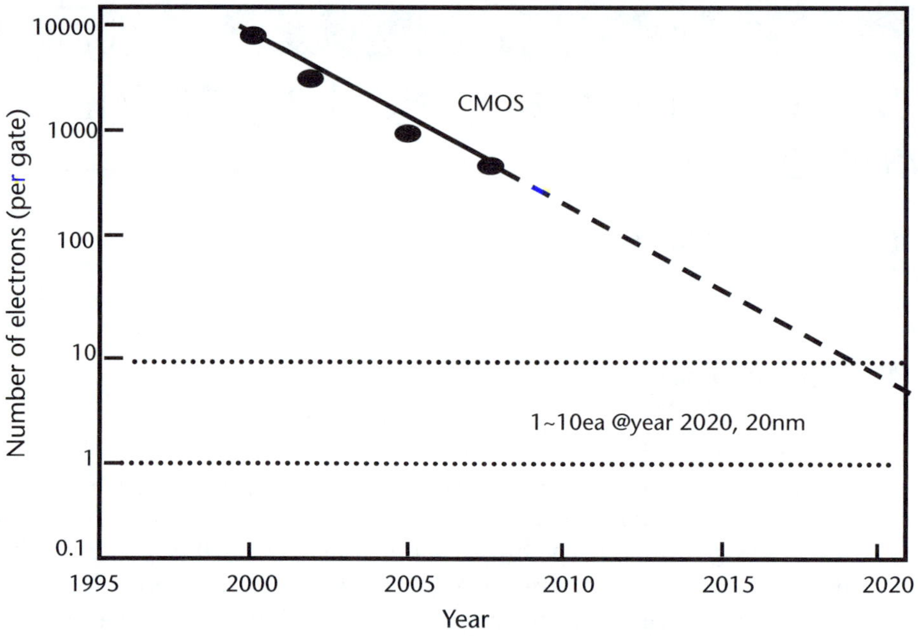

Figure 13.2 The historical and predicted trends in the number of electrons in a transistor.

avoid random and unpredicted errors resulting from so-called random telegraph noises, small signal noise, statistical fluctuations, and so forth. On the other hand, when the number of transistors per chip is so massive that its integration is extremely dense, unwanted errors in data processing may occur due to the influence originating from too-close spacing between the devices or the charged nodes, such as disturbance and interference between the switches (transistors).

Therefore, it will no longer be possible simply to pack so many transistors into a given Si area beyond the 20 nm technology node.

In the meantime, there has been considerable progress in developing alternative solutions that may use new, emerging technologies, such as nanosciences, nanotechnologies, biomolecular technologies, and others, without utilizing the incumbent Si technology. However, in spite of the fact that some studies have suggested rough ideas about how to integrate nano-/biotechnological embodiments into a single convenient chip, they still seem far from able to replace planar Si technology [3, 4]. Therefore, beyond the 20 nm node, there appears to be no concrete technology to overcome these limits and create great opportunities for various applications and emerging markets as silicon has had in the past. Probably, the silicon technology will maintain its leadership down to the 20 nm node.

Thus, it is very important to find the solution that can utilize the present Si technology and the infrastructure already invested in, which has successfully been used and well developed for more than 30 years for the information technology (IT) revolution. In practice, only three-dimensional (3D) integration of silicon technology can be a possible solution to solving the limits of the incumbent Si semiconductor industry as well as keeping the historical growth rate of the packing density below 10 nm dimensions.

13.2 Evolution of Memory Technology

Since Intel's first demonstration of a 1 Kbit dynamic random access memory (DRAM) chip in the early 1970s, the transistor packing density of DRAM chips has doubled every 18 months, as predicted by Moore's law. Today, DRAM technology has evolved to around 60 nm in minimum dimensions and 2 Gbit in packing density in 2007, as shown in Figure 13.3 [5]. Moreover, DRAM technology for the 50 nm node is being actively developed in order to continue increasing data-retention time and data rates. Increasing data retention in DRAM chips presents many significant challenges because it requires shrinking the dimensions while keeping both a sufficient amount of capacitance in a memory cell and an extremely low level of leakage current from a storage junction.

First, in order to meet the requirement of the cell capacitance, the cell capacitor has evolved to 3D structures, such as stacked capacitors or trench capacitors, from conventional 2D structures (e.g., planar capacitors), as shown in Figure 13.3.

Second, the DRAM cell transistor is designed to have a high threshold voltage in order to reduce subthreshold leakage current. This has resulted in high doping concentrations in the channel of the DRAM cell transistor, which not only increases leakage current in the storage junction but also produces less current drivability of the cell transistor. Therefore, the DRAM cell is designed to have an optimized channel and junction doping concentrations to suppress the subthreshold leakage current through the channel of the cell transistor and the storage-junction leakage current, simultaneously. Since the 100 nm node, the cell transistor has faced the limits of optimizing the channel and the junction by controlling the dopant concentration in a planar cell transistor. Therefore, a 3D cell transistor, called a recess

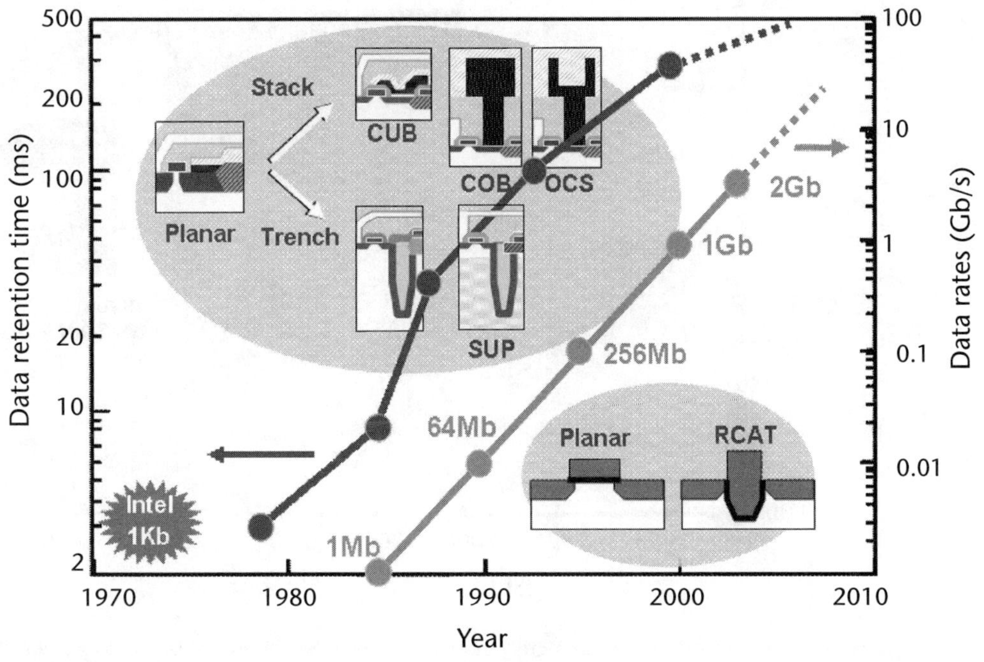

Figure 13.3 History and trends of the DRAM technology evolution in density, performance, retention time, capacitor type, and cell transistor structure [1].

channel array transistor (RCAT), has been adopted to increase the length of the channel to suppress the subthreshold current without increasing the channel dopant concentration, which causes the junction leakage [6].

NAND flash memory has the smallest cell size among silicon-based memory devices commercially available due to its simple one-transistor structure and serial connection of multiple cells in a string. Because of these merits, NAND flash memory can produce the largest bit density among solid-state memory devices. This can be achieved using the same patterning (lithography) technology. NAND flash has been used as a mass-data-storage memory for portable electronic applications such as MP3 players, digital still cameras, and cellular telephones since its first appearance in the middle of the 1980s. Furthermore, it is expected to find additional large-market applications, such as replacing magnetic hard disks in PCs. The need for NAND flash memory has skyrocketed with the recent advent of the mobile era [7, 8]. Therefore, ever-higher density, along with improved programming throughput, for NAND flash memory have been pursued. As a consequence, NAND flash memory has evolved toward ever-smaller cell size in two ways: by increasing string size and by developing two bits per cell, while at the same time increasing page depth, as shown in Figure 13.4. Current state-of-the-art NAND flash memory reaches 40 nm dimensions in a half pitch of the unit cell in technology and 32 Gbit in packing density [9]. In addition, NAND technology beyond 40 nm is now under development at research-and-development centers across the world. Along with the recent development of two bits per cell, the introduction of multibit cells, such as three bits and four bits, will further expedite this trend.

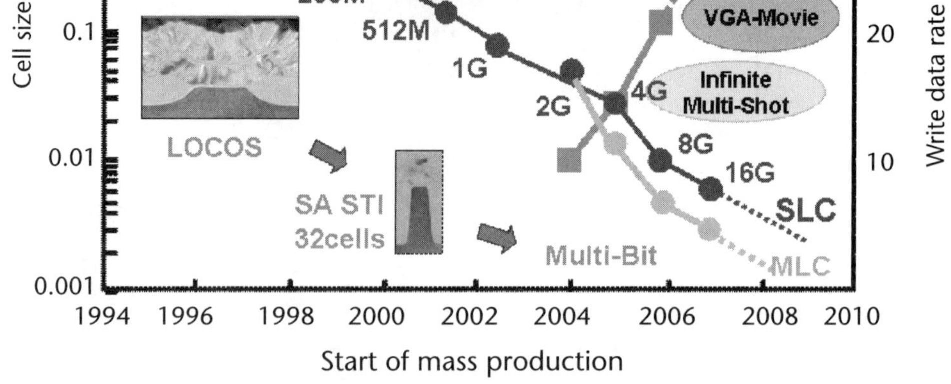

Figure 13.4 Historical and projected trends in NAND flash memory technology: density, performance, cell structure, cell size, and applications [1].

13.2.1 Challenges for Linear Shrinkage

As the dimensions of planar Si technology are reduced to below 30 nm, there are growing concerns regarding whether the future semiconductor industry can keep pace with the historical growth rate of the industry. The incumbent planar silicon technology has been so successful in increasing the packing density of memory devices and CPUs mainly due to the so-called linear shrink technology, by which the basic transistors of CPUs and/or memory cells are scaled down successfully without dramatic changes in each structure.

The linear shrink technology has thrived on two momentums, which are the timely evolution of optical lithography and extended transistor-scaling engineering without sacrificing the fabrication cost and performance of the CMOS transistor. It is widely predicted that the semiconductor industry can continue to use optical lithography down to the 32 nm node, as shown in Figure 13.5. However, beyond 32 nm, it is uncertain whether the industry can still utilize optical lithography. It is also not clear whether we have to introduce entirely different infrastructures, such as 13 nm–wavelength extreme ultraviolet (EUV) lithography, nanoimprint, and so on, because of a shallow process margin in manufacturing, ever-increasing fabrication costs, and technical complexities due to overhaul of all related materials and the methodologies of patterning. Nevertheless, the semiconductor industry will have to prepare a proper lithography beyond 32 nm.

Transistor-scaling theory basically tells us that when a transistor is scaled down by $1/K$ ($K > 1$), its power-delay product is improved to $1/K^3$, as shown in Figure 13.6. As a result, transistor scaling gives us both high performance and high density. The expected advantages of the linear scaling of the transistors, however, cannot be completely produced any more as the device dimension shrinks down to the nanoscale. This is because of nonscalable physical parameters, such as mobility,

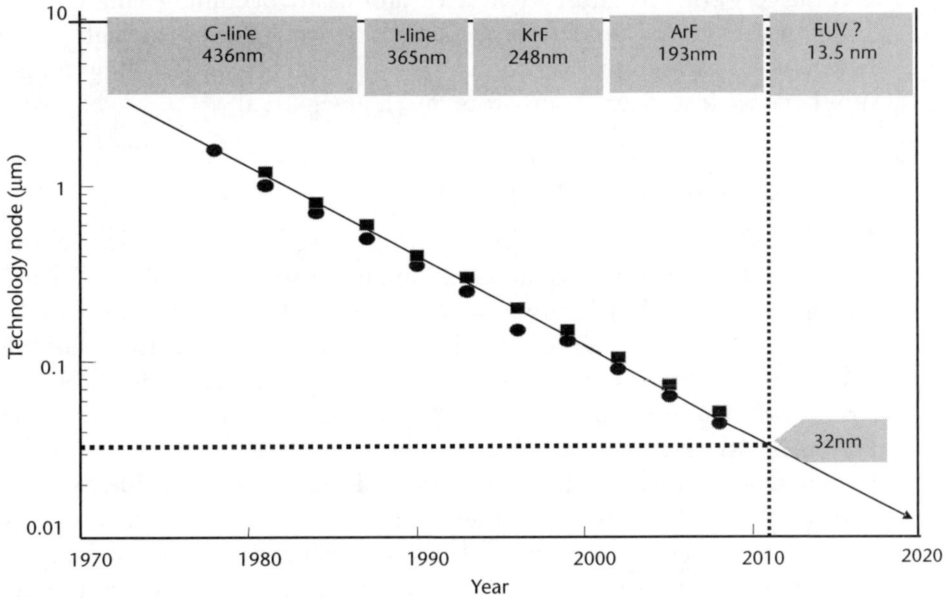

Figure 13.5 Historical and predicted trends in optical lithography tools and resolution.

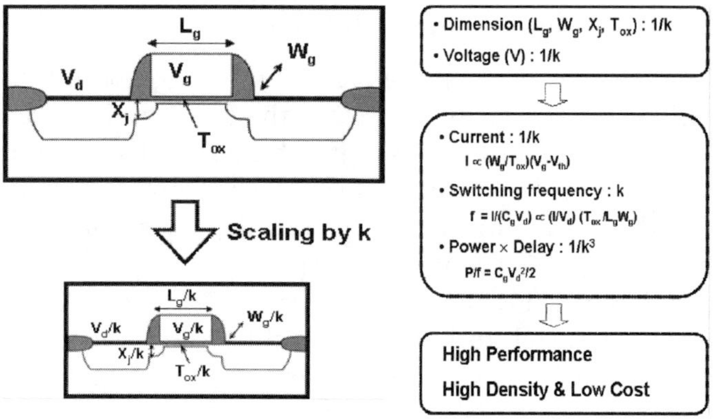

Figure 13.6 Rules in scaling CMOS transistors with improvement in performance and cost.

subthreshold swing, and other effects. For example, further increase of the operational frequency of a CPU by shrinking the gate length of the transistor cannot reduce the power of the chip. As a result, the speed of the chip cannot be improved by simply scaling down the gate length of the transistor as was done in the past due to the heating of the chip. That is why the multicore architecture has been adopted in CPU design instead of simply increasing the operation frequency.

Consequently, one might ask, what technology node will be the limit for silicon technology to be achievable. We may not have a proper answer yet. As mentioned in the previous section, when the dimensions of the transistor are scaled down to near 20 nm, the transistor contains as few as around 10 electrons. When the number of electrons is reduced to such level, random telegraph noise, the random fluctuation of the transistor channel impurity, and the loss probability of the data storage charge become very critical. Thus, all these variations are becoming a matter of probability, and unwanted errors in data processing may occur due to the high packing density. In practice, technology projection at the near–20 nm node will no longer permit us simply to pack so many transistors into a given area.

13.2.2 Scaling Limits in Flash Memory

NAND cells, which form a NAND string, have been so aggressively scaled down that the cell dimension has reached 40 nm for current technology. When NAND cell dimension approaches 30 nm or less, NAND flash memories based on floating-gate (FG) structure will face serious scaling issues in the physical aspects of cell structures and electrical performance. The typical vertical structure of the NAND flash cell string is shown in Figure 13.7. The physical constraints to build an extremely tiny FG-NAND flash cell are divided into three points.

First, both tunnel oxide and interpoly dielectric oxide/nitride/oxide (ONO) have reached their lower thickness limit. In order to preserve electrons for keeping the data, there is no room for further thickness reduction for either the tunnel oxide or the ONO. This means that the program and erase voltage cannot be reduced, and the electric field between the nodes will become so intense that it will cause adverse effects on reliability and performance below the 30 nm dimension.

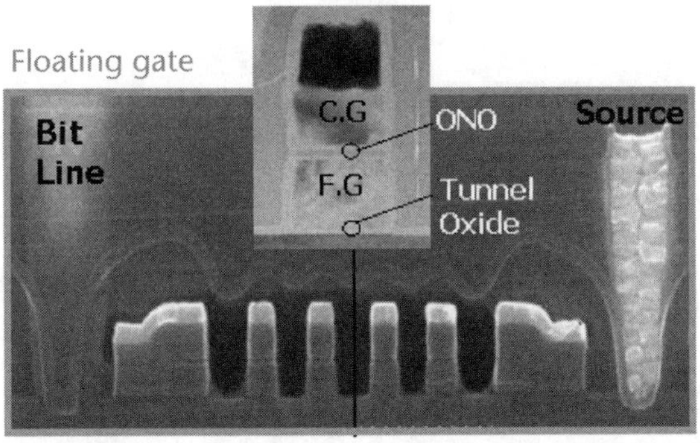

Figure 13.7 Vertical SEM images of NAND flash memory in a floating-gate cell.

Second, when dimensions scale down, the distance between the floating gates becomes so close that cell-to-cell interference becomes large. When the space between the data nodes gets close, the capacitive interference from the unrelated nodes increases. Figure 13.8 shows the cell-to-cell coupling interference of the FG cell for the various technology nodes. As shown in the figure, the floating gate can be interfered with by unrelated nodes. When the minimum dimension decreases, the coupling effect is intensified and causes additional V_{th} variation in the cells. Therefore, it is inevitable to reduce the FG height to reduce the coupling effect, as shown in Figure 13.9. In the multibit cell (two bits), the problem is exacerbated. The decreased coupling ratio means higher program or erase voltage for writing and reading. Such high voltage degrades the reliability of the NAND flash cell, reducing, for example, endurance and data retention. Therefore, in order to overcome this limit, the Si/SiO_2/Si_3N_4/SiO_2/Si (SONOS)–structured NAND cell is being explored. The floating gate can be replaced by a dielectric-charge trap layer, such as Si_3N_4 and

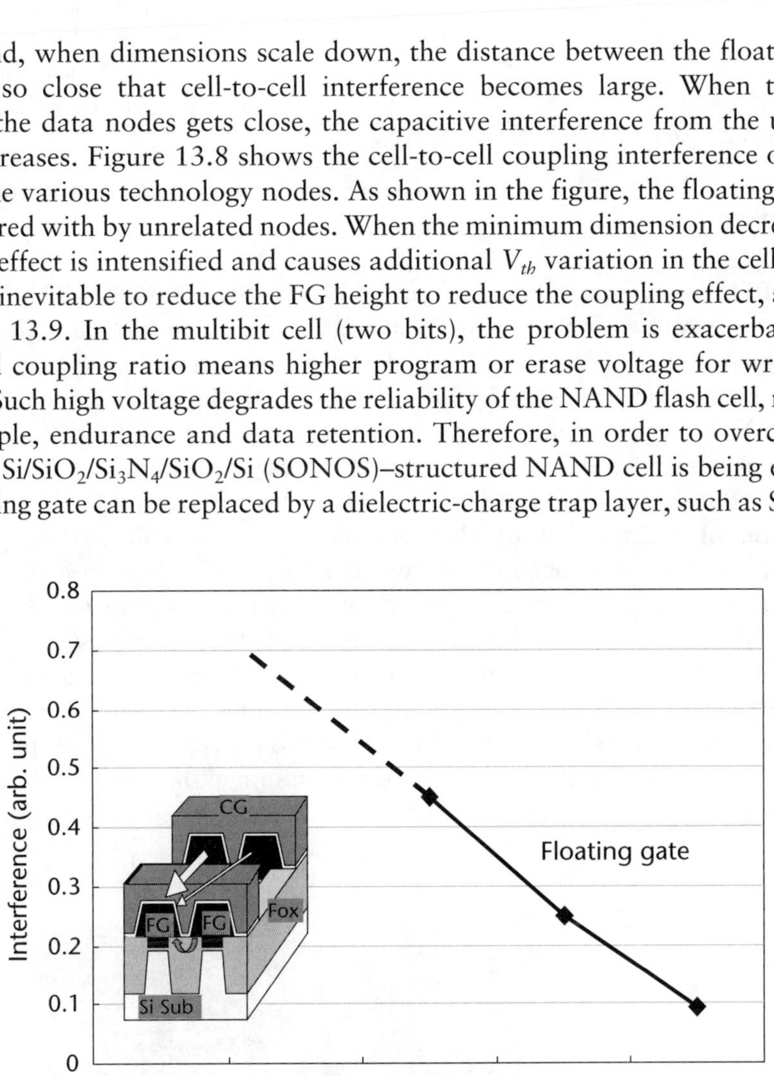

Figure 13.8 Trends in coupling interference of the NAND flash memory in a floating-gate cell.

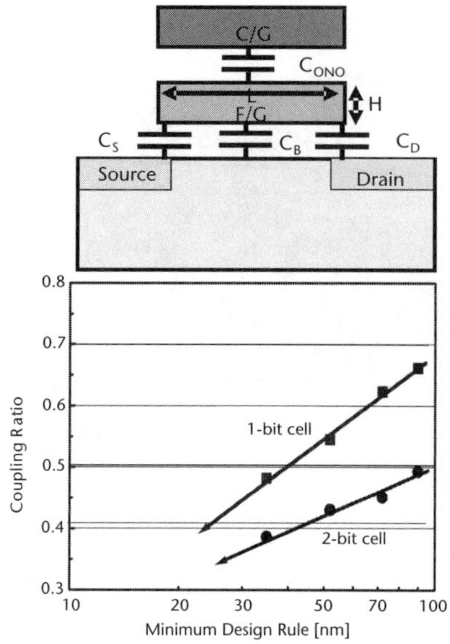

$$Vfg = Vcg \times \alpha cg$$

$$\alpha cg = \frac{C_{ONO}}{(C_D + C_S + C_B + C_{ONO})}$$

αcg: Coupling ratio

C_{ONO} = P x C_{ONO} / unit area

H : Height of floating gate

L : Length of floating gate

P= 2 x H + L : Peripheral area of floating gate faced with control gate

Vfg: Voltage in floating gate

Vcg: Voltage in control gate

Figure 13.9 Calculation of and trends in coupling ratio in a floating-gate cell of the NAND flash memory.

HfO. For example, the so-called $TaN/Al_2O_3/Si_3N_4/SiO_2/Si$ (TANOS)–structured cell consists of tunnel SiO_2, a Si_3N_4 layer as the trapping layer, a high-k aluminum oxide layer as a blocking oxide instead of the conventional ONO layer, and TaN as the gate electrode (Figure 13.10). A nitride (Si_3N_4) layer can store many electrons. This Si_3N_4-charge trapping layer is free from the cell-to-cell interference. However, the charge-trapped device still has many issues yet to solve.

Lastly, the most fundamental issue with FG-NAND flash memory is the reduction of the number of electrons as the design rule shrinks. It is expected in FG-NAND flash memory that we will see a serious electron-storage problem from the 30 nm node because of the lack of electrons (less than 100) being stored in a unit memory bit. The number of electrons needed to distinguish the data levels and the tolerable loss or variation of electrons for keeping the data are estimated for the FG cell in Figure 13.11. In the graph, the left *y*-axis represents the number of electrons, and the right-axis is the number of tolerable electrons for keeping the data level in one- or two-bit cells. It is optimistically assumed that the coupling ratio will be kept

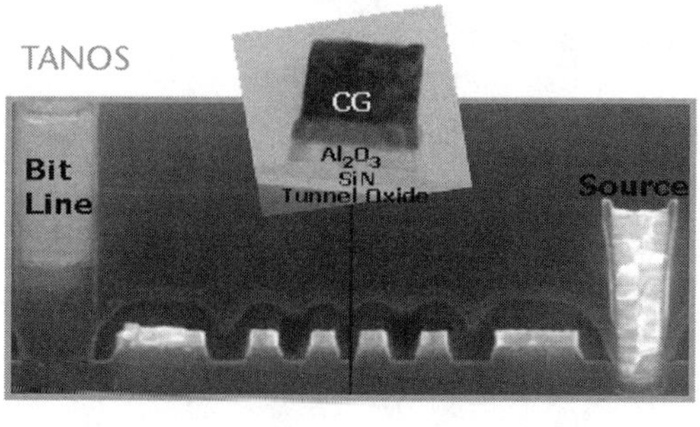

Figure 13.10 Vertical SEM images of NAND flash memory in a SONOS-like TANOS gate cell.

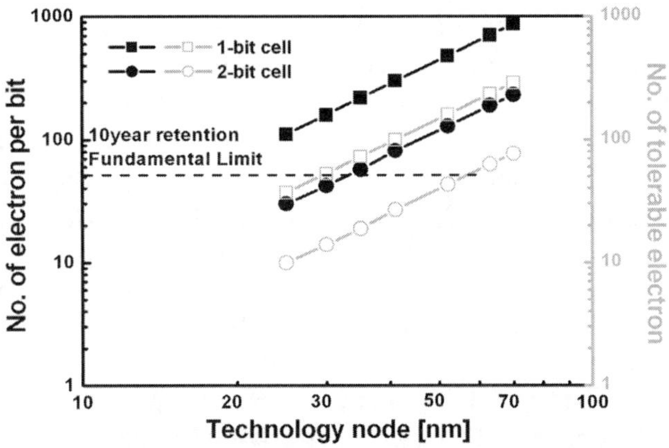

Figure 13.11 Number of electrons for distinguishing the data level and tolerable loss of or variation in electrons for keeping the data in NAND flash memory estimated from the floating cell.

the same at 0.6, and the minimum voltage shift for losing data will be 0.6V. Also, the retention time limit is calculated with the probability of losing electrons from the data nodes, which varies according to the total number of stored electrons. For example, at the 25 nm node, the predicted number of stored electrons per bit for a two-bit cell is approximately 30. The tolerated loss of electrons is approximately 10. At such low values, 10 year data retention is not guaranteed. Therefore, below the 25 nm node, data retention as nonvolatile memory will be uncertain. This is a very serious fundamental limit that we will face pretty soon if the technology nodes advance at the present speed. As the dimension further shrinks, 3D technology will be needed to continuously increase density.

13.2.3 Scaling Limits in DRAM and SRAM

DRAM has often struggled to increase data-retention times, which are a key parameter in power consumption as well as performance. It is essential to minimize leakage current, which predominantly comes from the storage junction, in order to meet the requirement for ever-increasing retention time. Thus, the most critical factor in DRAM scaling stems from how to design a cell array transistor for which transistor dimension strongly influences data-retention times. In general, data retention in DRAM is inversely proportional to electric-field strength induced across the junction of a cell transistor due to high junction-leakage current. High electric field is caused by high doping concentration across the junction. Unfortunately, dimension scaling of planar-based cell transistors in DRAM is inevitably accompanied by a severe increase in doping concentration underneath the channel region, to a certain degree, to reduce the short channel effect (SCE). Typically, degradation of the retention time becomes significant below 100 nm due to a rapid increase in the junction electric-field. This issue can be overcome by introducing 3D cell transistors, whose junction electric field can be greatly suppressed due to a lightly doped channel region.

As an attempt to improve the retention period in DRAM, the gate in the RCAT detours around some part of the Si substrate so that an elongated channel can be formed to provide great immunity against SCE. When we look at historical trends in the DRAM cell, conventional DRAM technology has been extended down to the 50 nm node through the adoption, with minor modifications, of the RCAT structure, as shown in Figure 13.12. Beyond 50 nm, another breakthrough for array transistors may be needed to suppress the ever-increasing leakage current in DRAM. One approach is to form a 3D vertical cell array transistor.

There have been great demands for higher-density SRAM in all areas of SRAM application, such as network and cache stand-alone memory and embedded memory in logic devices. However, the 6T full-CMOS SRAM has a basic limitation because it needs six transistors on a Si substrate compared to one transistor in a DRAM cell. The typical cell area of the 6T SRAM is 80~100F^2 (F means a minimum pattern size) compared to 6~9F^2 for a DRAM cell. The 6T full-CMOS SRAM has two types of wells (N-well and P-well) in a cell area and thus requires a good well-to-well isolation that can be scaled as shown in Figure 13.13. Further shrinkage of planar 6T full-CMOS SRAM encounters a barrier below the 45 nm dimension based on planar Si technology. Therefore, various alternative embedded-memory solutions, such as capacitorless 1T DRAM, thyristor-type RAM, and magnetic RAM, have been proposed to replace planar 6T full-CMOS SRAM. Their feasibility for real mass production is still very uncertain due to the need to adopt new materials and new operational device physics.

13.3 3D Chip-Stacking Package for Memory

Packaging technology provides one of the simplest ways to stack chips vertically (3-D), leading to increased memory density and functionality. This packaging process consists of three key technologies: (1) thinning of the wafer, (2) bonding of the

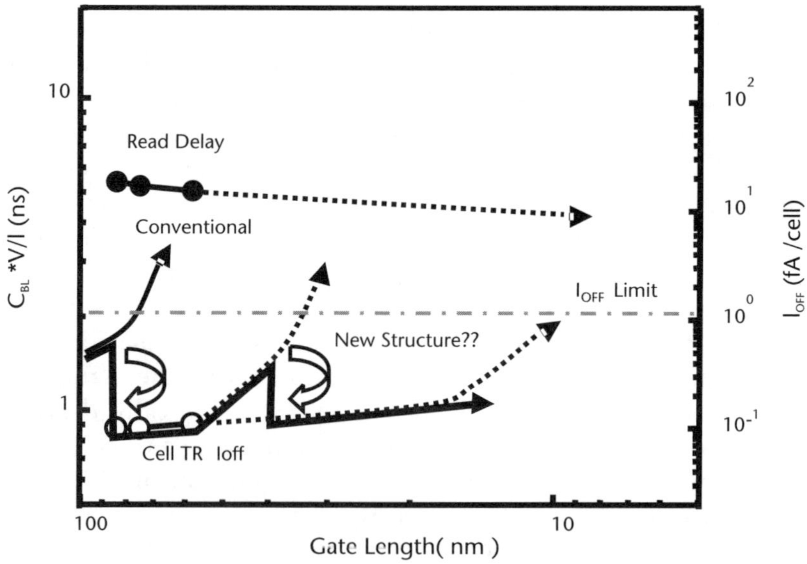

Figure 13.12 Trends in DRAM cell transistor leakage current and read delay time.

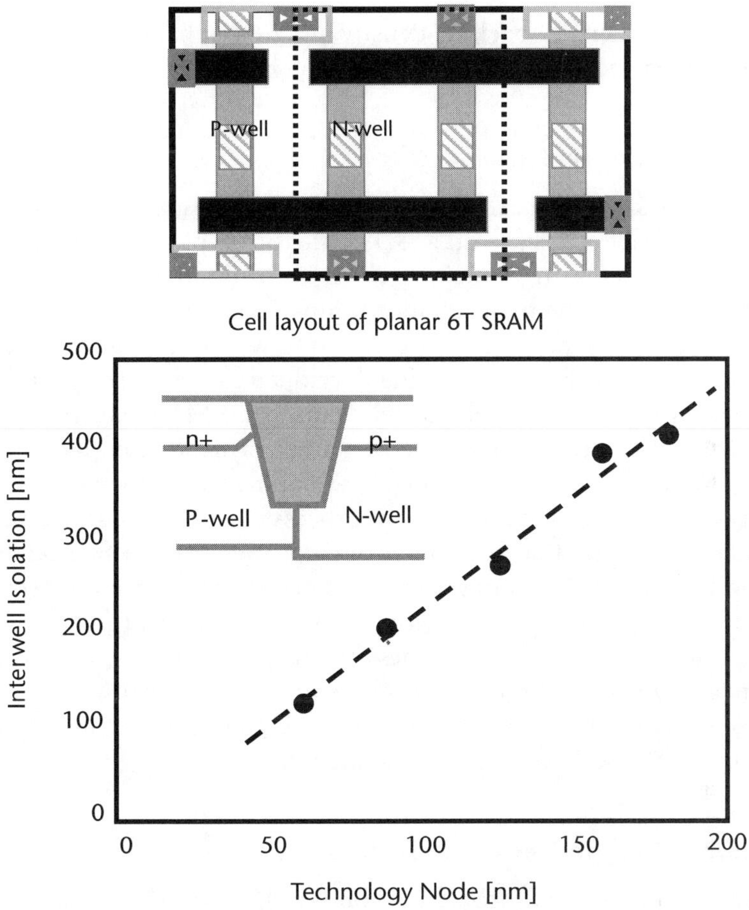

Figure 13.13 Trends in interwell isolation of the planar 6T full-CMOS SRAM cell and cell layout of 6T SRAM.

chips to form the 3D stack, and (3) forming the interconnections between separated stacked chips and the package substrate or pins. A typical solution for the latter (the interconnection) is wire bonding. It is a very simple and low-cost process but yields interconnections with large parasitic inductance, capacitance, and resistance, which produce delay and noise in the signal. Therefore, in order to take full advantage of 3D chip stacking, shorter-length interconnection is very critical. Flip-chip technology, which utilizes an area-array distribution of solder bumps, is one such solution [10]. The bump-shaped metal ball is used to form electrical and mechanical interconnections between the pads of the stacked chips or the chip and the printed circuit board (PCB) substrate. The bumps reduce the parasitic RC components, leading to an improvement in the internal signal integrity of the packaged chips. However, it is not enough to improve the performance of the stacked chips to the level of an integrated system-on-a-chip (SoC). Therefore, through-silicon via (TSV) technology is a newly developed technology to achieve SoC-like system performance in a package with multistack chips instead of the planar CMOS SoC, which is facing the limits of scaling. However, 3D package technology has fundamental issues, such as yield loss resulting from the stacking process and no cost-reduction effect (due to simply adding chips to increase packing density and functionality) compared to the linear scaling of integrated CMOS devices on a Si chip.

13.3.1 Multichip Package

The simplest and lowest-cost 3D chip package technology is the multichip package (MCP), which consists of a 3D stack of chips in a single package to achieve multifunctionality (by stacking chips with different functions) or higher memory density (by stacking many memory chips). Each functional chip is stacked and interconnected to the package substrate using wire bonding, as shown in Figure 13.14. Basically, this technology is being developed to reduce the package area for portable applications, such as a cellular telephones and MP3 players. The advantages of the MCP are its small footprint and better performance compared to a single-chip solution. For multichip stacking, a wafer-thinning technology with stress relief and without warpage is the most important process to enable the stacking of more chips for a given height. For a large number of chips in a stack, a fine-pitch wire bonding and long loop wiring are important. It is necessary to interconnect the topmost chips without overhanging of the wire bonds. However the fundamental limitation of the MCP will be its cost-effectiveness because of yield loss due to nonexistent redundancy for repair and no per-bit cost reduction. In this respect, 3D-device-integration technology definitely overcomes the MCP limitation because it is easy to implement redundancy for repair and cost reduction.

13.3.2 Through-Silicon Via Technology

As stated previously, through-silicon via (TSV) technology can provide high performance to the 3D chip-stack package system. With respect to performance, it will be possible to replace planar CMOS SoC with system-in-package (SIP). The contact vias are formed through the substrate of the stacked chips, as shown in Figure 13.15.

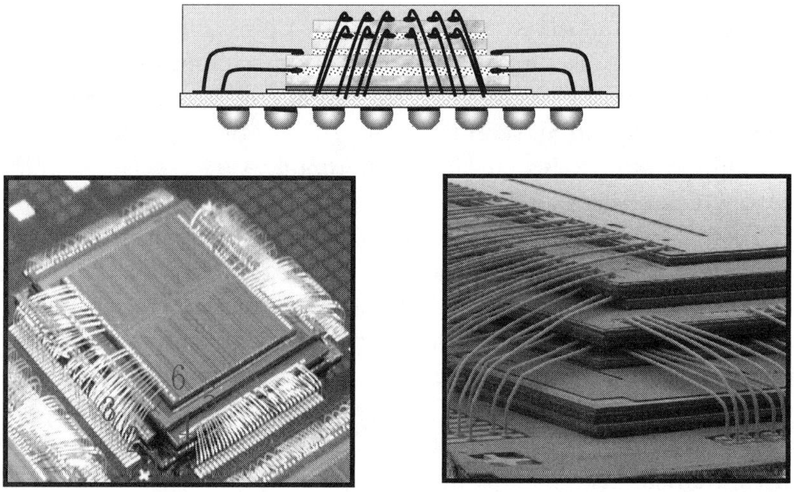

Figure 13.14 Photograph of an MCP package and SEM side-view image of the stacked chips in an MCP [1].

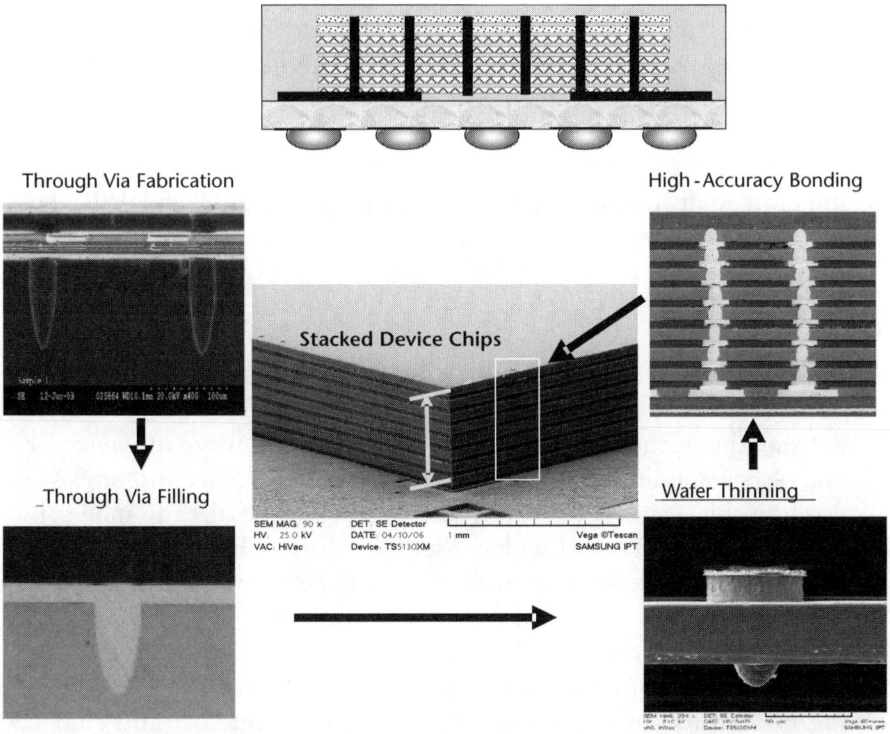

Figure 13.15 Schematic illustration and SEM image of an MCP implemented using TSV technology and photographic illustration of the fabrication sequence in TSV technology [1].

It can provide many advantages in connecting the stacked chips for system integration. For example, the total distance of the wiring interconnection will be greatly reduced by the point-to-point interconnection that TSV offers between chips at the module or block-circuitry level of the chips as compared to wire bonding. Such interconnect reduction leads to a dramatic decrease in the propagation delay and operational current without the signal noise due to the reduced electrical parasitic. Moreover, the operating power at the same frequency can be dramatically decreased due to reduced I/O and busing capacitances.

There are many ways to form TSV and stack chips. For example, there are three methods to stack chips: chip-on-chip (CoC), chip-on-wafer (CoW), and wafer-on-wafer (WoW). Each method has pros and cons in terms of packaging process cost and yield loss. With respect to packaging cost, WoW is the cheapest technique. Stacking wafers is much easier compared to stacking individual chips, especially since a thinned chip is very difficult to handle in the stacking process. However, in WoW stacking, the yield loss is significant because it cannot use known good die (KGD). As expected, CoC stacking has the smallest loss in yield after stacking, even though the cost is high. The chip size is a key factor in determining the total cost-effectiveness among the three methods.

The sequences of steps for TSV formation in the WoW method are shown in Figure 13.15. First, deep vias are formed on the already-fabricated wafers by dry etching of Si. The typical depth of the vias is more than 50 μm. These vias are then filled with copper to form an electrically good contacting layer, which becomes important later in the process. Next, each wafer is thinned to a thickness of 50 μm and bonded with the other wafers. Good alignment prior to wafer bonding is critical for accurate stacking of the TSVs. Finally, the stacked and bonded wafers are diced for packaging.

As a package solution, TSV technology will be one of the most effective ways to integrate multifunctional chips, such as memory, CPU, and ASIC devices, for system integration. It will provide all advantages, such as high speed, low power, small form factor, and design flexibility, except cost-effectiveness.

13.4 3D Device-Stacking Technology for Memory

3D stacking technology might be one of the best ways to overcome the patterning and physical limitations based on well-established Si technology. As discussed previously in this chapter, however, simple stacking of already-made chips or packages has been developed and widely used to increase the packing bit density or to combine different functional chips in one package to save package area using, for example, MCP or package on package (PoP) technology [11].

However, in terms of the cost per bit, such simple chip stacking cannot reduce the cost because it does not reduce the fabrication-process cost for increasing packing bit density. In order to reduce the fabrication cost and chip size, device-level stacking technology is necessary instead of merely package-type chip stacking.

Device stacking or cell-array stacking can save additional processes for interconnection layers and peripheral logic devices of stacked chips when compared to simple package-based chip stacking as illustrated in Figure 13.16. For example, in

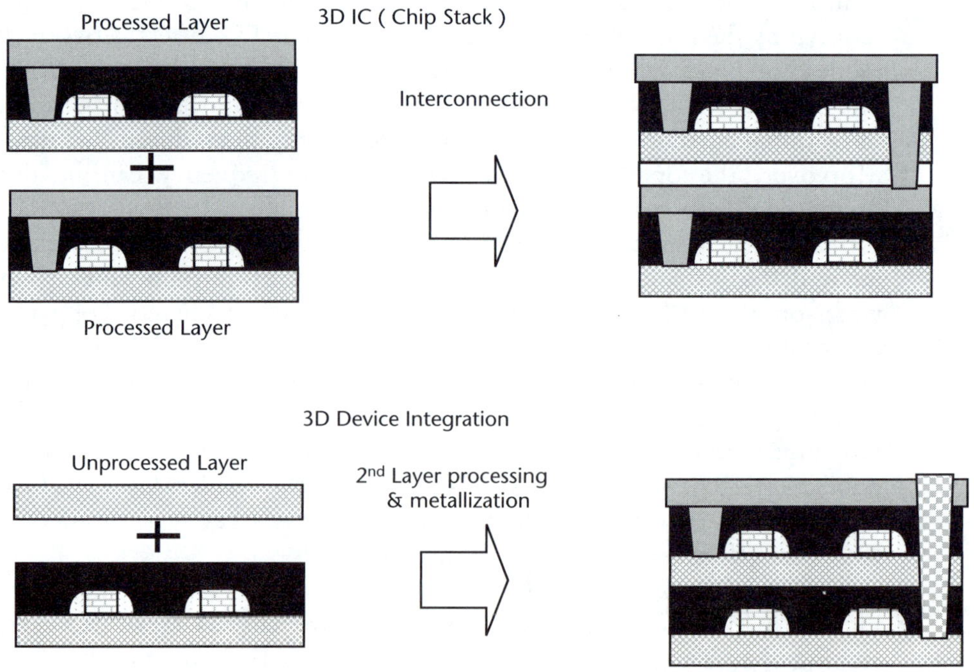

Figure 13.16 Schematic illustrations of 3D chip-stacking and 3D device-integration processes.

the case of chip stacking, already fully processed wafers or chips are bonded and interconnected. However, in device-stacking technology, an unprocessed Si layer or active layer is added on the bottom device layer before it is processed. Whole stacked device layers are interconnected simultaneously using the same metal layer at the end of the fabrication process. Such processing can save many lithographic and interconnection layers compared to 3D chip stacking or 3D package stacking. In addition to the fabrication cost, 3D device-stacking technology offers additional benefits, such as lower power, higher speed, greater design flexibility, and higher yield.

Three-dimensional device-stacking integration in memory has begun recently with SRAM to reduce its large cell size. The stacking of transistors, combined with no need for well-to-well isolation, reduces the SRAM cell size of $84F^2$ to an extremely small cell of $25F^2$ [12]. Encouraged by this successful approach in SRAM, researchers have also pursued stacked flash memory because incumbent planar flash memory will soon reach the limitations of increasing density, as mentioned earlier in this chapter. For example, the SRAM cell consists of six transistors. The conventional planar six transistors of SRAM are dissolved, and each type of SRAM cell transistor is separated into three different Si layers to reduce the cell area to less than one-third. This was made possible by developing the technology to stack 3D perfect single-crystal Si layers on amorphous interlayer dielectric (ILD) layers.

The 3D device-integration technology has numerous advantages over current planar technology. These are essentially: (1) elimination of uncertainty in deep nanoscale transistors; (2) extendable use of silicon infrastructures, especially optical lithography tools; and (3) formation of a baseline for multifunctional electronics in the future with a facilitating hierarchical architecture, where each layer is dedicated to its specific functional purpose (e.g., the first layer for data processing, the second layer for data storage, the third layer for data sensing, and so on).

13.4.1 3D Stack of DRAM and SRAM

As mentioned, 3D device-stacking technology in memory has begun recently with SRAM in order to reduce its large cell size and overcome the limits in its shrinkage. The area penalty of embedded SRAM cache in a CPU chip can reach up to 80%. So, the need for a relatively cheap, external cache memory is growing. But the planar 6T full-CMOS SRAM cell size of ~$84F^2$ is too large to achieve the appropriate high-density SRAM. Moreover, the simple linear shrinkage is not so easy in SRAM due to well-to-well isolation and too many contact holes and local interconnections. By implementing stacked single-crystal Si (S^3) double-stack technology, the load PMOS and the access NMOS are stacked up on the second and third device layers, respectively, over the first bulk pull-down NMOS, and the $84F^2$ cell size can be implemented in an area of $25F^2$ with the additional benefit of eliminating the well isolation limit, as shown in Figure 13.17. By mixing device-stacking technology and chip-stacking technology, it is possible to make a high-performance CPU chip in the near future by coupling the CPU core chip with a 3D device-stacked SRAM cache.

In S^3 SRAM cell technology, the most important process step is the formation of the single-crystal Si thin-film layers on the amorphous ILD to yield a stacked single-crystal thin-film transistor (SSTFT) cell. The easiest way to form the Si layers on

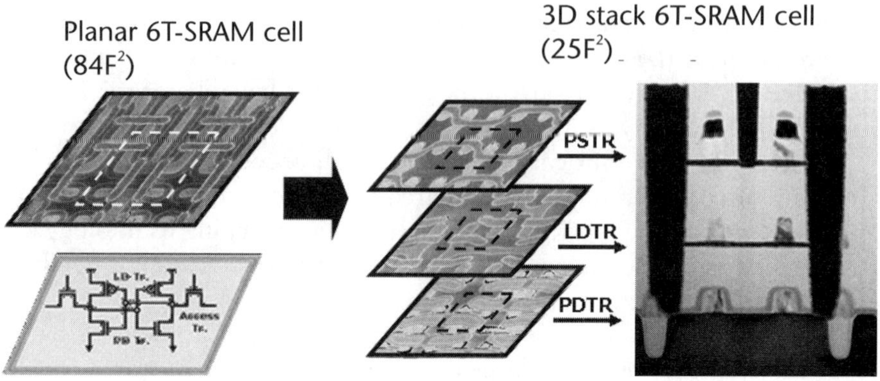

Figure 13.17 3-D device-stacking technology for the 6T SRAM cell. The cell size can be reduced dramatically to $25F^2$ from $84F^2$ of planar 6T SRAM [12].

the amorphous dielectrics is to deposit polycrystalline Si films or amorphous Si films as used in a thin-film transistor (TFT) in an active matrix liquid crystal display (AM LCD). However, TFTs are not applicable to the fabrication of ultra-high-density memory because the polycrystalline films or the amorphous films have too many crystal defects that degrade carrier mobility and induce leakage current, as well as causing other undesirable effects. Even if they can operate well as small-scale devices, such poor electrical characteristics cannot be tolerated for high-density and highly reliable memory products. Therefore, the stacked Si layers must have perfect, single-crystal quality [similar to silicon-on-insulator (SOI) defect density]. For many years, the formation of perfect, single-crystal Si on amorphous layers has been dreamed of and researched by many people [12–15]. One of the crystallization techniques is selective epitaxial growth (SEG) from the Si wafer via small seed contacts. When the SEG technique is used for stacking Si layers on the ILD, defect-free seeding and perfect epitaxial growth control are essential.

First, seed contact holes are made through the ILD oxide layer by lithographic patterning with a certain periodic distance, as shown in Figure 13.18. Next, epitaxial Si crystal is grown vertically and selectively from the seeding contact holes, and when it reaches the top of the contact holes, it can grow laterally and epitaxially along the oxide surface in all directions from the holes. It has facets due to the growth-rate difference between the growing planes. The laterally grown layers, from each seed contact hole, meet with each other as the process progresses. Finally, the whole surface of the wafer becomes covered with the epitaxially grown Si crystals, which exhibits a rough topology (hills and valleys) resulting from the growth-rate differences between the growing directions of the Si crystals. The surface can be flattened by a planarization process with the chemical mechanical polishing (CMP) technique. The crystal quality of the stacked Si films was analyzed with TEM. The top right-hand image of Figure 13.18 shows the bright field TEM image of the film demonstrating an almost perfect single crystal without grain boundaries or defects. The electron-diffraction pattern of the film is also shown in Figure 13.18 (bottom right-hand image). It is the pattern of a perfect single crystal.

Another example of the crystallization technique is the laser crystallization method, which is shown Figure 13.19. This technique uses epitaxial crystallization

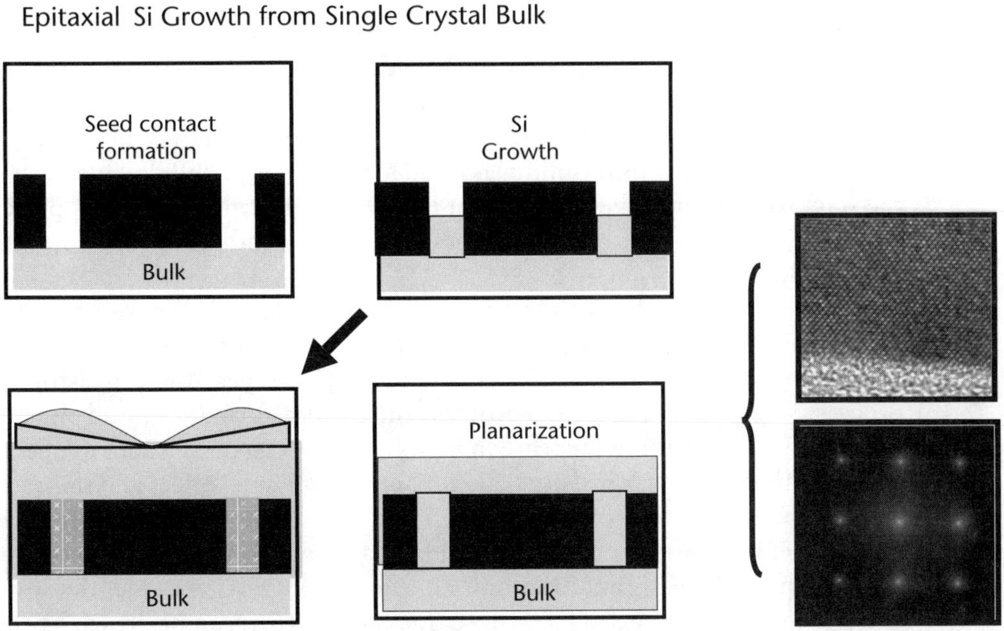

Figure 13.18 Schematic illustrations and TEM images of the selective epitaxial growth (SEG) technique.

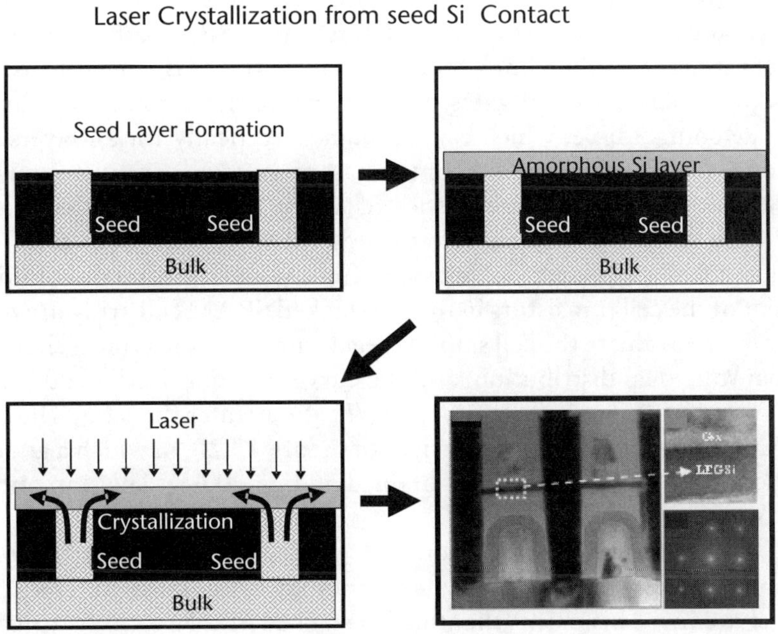

Figure 13.19 Schematic illustrations and TEM images of the laser crystallization technique.

of amorphous Si films by single-crystal seeding and melting with laser energy. It also needs selective epitaxial growth from the Si wafer through the seed process described above. Since the formed silicon layer has facets and topology, it is polished completely with the CMP process. Amorphous Si films are deposited on the seed Si of the contacts and the ILD layers. When a laser illuminates these amorphous Si layers with enough energy to melt them down, heat is conducted through the seed contact holes, and the crystallization process spreads from the seed Si. Therefore, single-crystal Si is grown epitaxially from the single-crystal seed Si through the crystallization process. This technique is good for achieving small thickness variations in the stacked Si layers. The TEM bright image and electron-diffraction pattern of the stacked Si layers made by laser crystallization are shown on the bottom right of Figure 13.19.

In addition to the S^3 stack technology, low-thermal and low-resistance processes for the high-performance transistor are necessary. In order to stack the SSTFT cell transistors for the S^3 SRAM cell, it is necessary to repeat the transistor formation process, such as the oxidation process, the Si film formation, and the activation, three times. The resulting increase in high-heat processing degrades the performance of the peripheral bulk transistors due to the short channel effect and the deactivation of the dopants. Therefore, it is crucial to minimize the total heat budget. A low thermal budget requires low-temperature plasma gate oxidation, low thermal thin-film deposition, and spike rapid thermal anneal (RTA). For example, the low thermal gate oxide layers of the SSTFT, whose thickness is 16Å, can be grown by the plasma oxidation method at 400°C. Also, other process temperatures are maintained to be below 650°C after forming the bulk transistor.

In addition to the SSTFT, the other key factor in process integration of the 3D stacked SRAM cell is forming the node contacts, which are vertically and laterally contacted. In this cell, for the latch function of SRAM, the local interconnection layers for the cross-coupling of the nodes and gates of the cell transistors are not needed because all nodes and gates of the transistors are connected through just a single node contact hole, which can be aligned vertically for all layers (from the bottom active node to the top node of the pass transistor), as shown in Figure 13.17.

The electrical characteristics of the SSTFT pass NMOS, and the SSTFT load PMOS should be comparable to those of the planar bulk transistor because their channel Si is a perfect single-crystal film. When the total distributions of the on current of the cell transistors in the 3D stacked SRAM cell array are plotted on the same graph to evaluate the cell stability, each curve has the typical characteristics of a normal statistical distribution, and the curves are not overlapped. This is illustrated in Figure 13.20. The cell ratio ($I_{pull\ down}/I_{pass}$) is greater than 2.5. The static noise margin (SNM) curve is shown on the right of Figure 13.20, in which a good noise margin for the SRAM cell operation is obtained at V_{dd} = 0.6V. The static noise margin is 282 mV at V_{cc} = 1.2V.

When we look at the nature of logic devices, where transistors and interconnections are key elements, the logic technology is similar to 3D stacked SRAM memory technology. This means that this 3D stacked SRAM technology can be easily implemented into the logic technologies because of their similarity. For example, the 3-D device-stacking technology will move to merge a memory device and a logic device into a single chip by hierarchical stacking technology. Especially since most of silicon area in a future SoC will be allocated to memory even in terms of cost-effectiveness alone, the 3D stacked SRAM technology will be very important. Furthermore, it could provide a new solution to overcome the physical and lithographic limits of linear shrinkage of the planar Si CMOS logic technology.

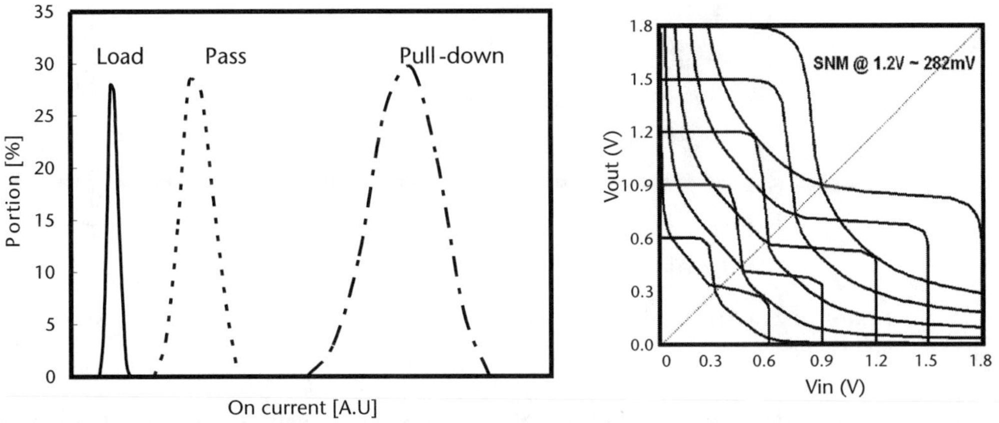

Figure 13.20 Current distributions of each cell transistor and static noise margin in the S^3 SRAM cell array.

Basically, the DRAM cell with a capacitor is not adequate for stacking because the capacitor is too tall in the stack capacitor type or too deep in the trench capacitor type. In order to stack DRAM cells three-dimensionally, a capacitorless DRAM cell or a 1T DRAM cell is necessary. The capacitorless DRAM cell is being studied intensively by many researchers as an enabler for embedded memory in the logic devices due to the limits of the planar 6T SRAM as cache memory [16].

In the capacitorless DRAM, electrical charges are stored in the electrically floated body of the cell transistor. The stored charges can change the potential of the body of the cell transistor and cause a shift in the threshold voltage of the cell transistor. Therefore, we can store data in the body of the cell transistor instead of the capacitor of the conventional DRAM cell. This 1T DRAM cell needs an SOI wafer for the thin floated body. This structure is very desirable for Si device-stacking technology because of its simplicity. Stacking cell transistors, then interconnecting the sources and drains of the stacked transistors with the bit lines and the source lines comprises the whole process for 3D stacked memory cells, as shown in Figure 13.21. The 3D stacked SRAM technology can be fully utilized to fabricate the 3D stacked 1T DRAM cell.

When the conventional DRAM cell faces the linear shrink limit in the future, this technology can provide the solution for higher packing density as embedded memory or stand-alone memory. However, it will have to sacrifice the data-retention time of the DRAM cell due to the limited number of stored charges in the floated body of the 1T DRAM cell compared to the conventional DRAM cell with a large capacitor.

13.4.2 3D Stacked NAND Flash Memory

Recently, the great demand for higher density and lower bit cost in NAND flash memory is growing because it is the key device for the mass-data-storage applications in various portable electronic products. The price per bit has been decreased by 70% per year, and the bit density has doubled annually. In order to maintain such trends, the linear shrink of the patterns has been aggressively driven by devel-

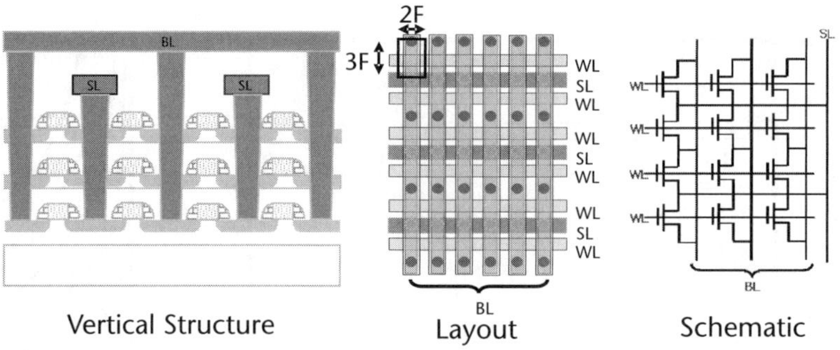

Figure 13.21 Schematics of the 3D stacked, capacitorless 1T DRAM: vertical structure, layout, and circuit.

oping multilevel cell (MLC) technology and early adoption of advanced lithographic tools [7]. However, the linear scaling of NAND flash memory is approaching physical, electrical, and reliability limits, especially as the technology advances to near–30 nm dimensions. First, continued linear scaling will have to use extreme ultraviolet (EUV) lithography, which is expected to be available after 2010, according to the ITRS roadmap. Even if the tool is prepared, the cost of the tool will be much higher and its throughput will not be comparable to that of ArF lithography. From the standpoint of the bit cost, this means that, even if the dimensions are shrunk and the density is increased, the bit-cost-reduction trends will not match the historical trends any more, as shown in Figure 13.22. This projection implies that the economic motivation driving linear device shrinkage for the higher density will diminish, and the bit growth rate in data-storage applications will slow down.

Second, as mentioned previously, from the electrical and reliability perspective, shrinkage below the 30 nm dimension will cause serious problems, such as electrical

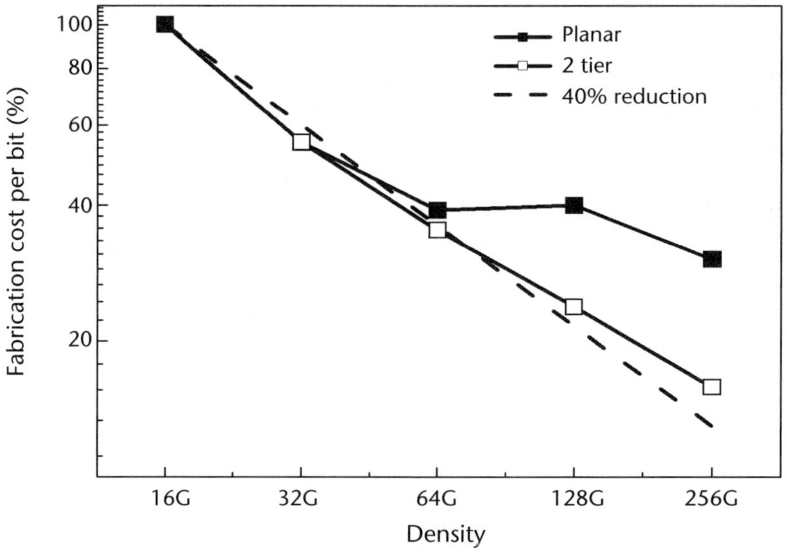

Figure 13.22 Predictions for fabrication cost of NAND flash memory as the bit density increases for the planar cell and the stacked cell.

isolation between word lines (WLs) and cell nodes, the short channel effect, cell current reduction, and tolerable charge losses of the stored charges for data retention, which are less than 10~100 in round numbers. Furthermore, these problems result from the fundamental physical limits, which are impossible to overcome with the conventional modifications generally used in the past nodes.

Therefore, one of the best ways to circumvent these barriers caused by simple conventional linear shrink technology is to stack the cell arrays using the fewest additional processes possible. The simplest solution to increasing memory density is the stacking of chips or packages. However, this simple stacking process cannot reduce the bit cost or the fabrication cost because chips that are already completely integrated are stacked. However, in 3D device-stacking technology, the Si active layers are stacked with minimum processes and are interconnected simultaneously with the bottom cell arrays and the peripheral circuits [17].

Schematic examples of doubling the density of the NAND flash cell arrays are shown in Figure 13.23. As shown in these circuit schematics, the NAND cell strings are made on the first bulk Si substrate and stacked on the second layer. The cell strings of the upper layers are stacked over the bottom layers, which are already formed on the bulk Si substrate. The upper cell array is made on an SOI-like Si layer on the ILD.

In order to achieve the same electrical characteristics of the cell strings in both layers, perfect SOI-like single-crystal Si layers are formed on the ILD layers with various crystallization technologies (discussed above). The gate stack of the cell strings is a kind of SONOS (Si/Oxide/Nitride/Oxide/Si) structure for reducing the

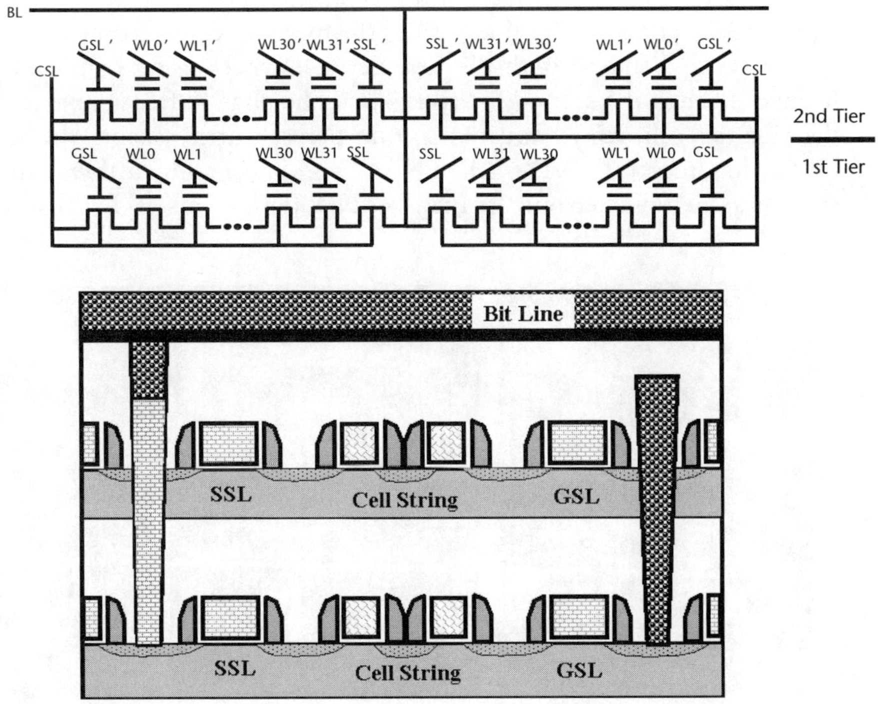

Figure 13.23 Schematic circuit diagram and vertical structures of the doubly stacked NAND flash cell arrays.

total height of the stacked layers. Every string has 32 NAND cell transistors, one string selection transistor, and one ground selection transistor. The cell strings of both layers are connected by the same bit line and a common ground line with a single contact hole to save the additional area, respectively. It means they are structured by sharing bit-line (BL) and source-line schemes. These schemes can minimize not only the layout area but also the bit-line loading, such as the resistance and capacitance. The bit-line contacts and the common source lines are patterned simultaneously on both layers of the cell string by etching vertically through the upper level Si layers to the bottom active layers, as shown in Figure 13.23. The bit-line contact holes are filled sequentially with the N-doped poly-Si and W. Therefore, both of the cell strings are connected through a single contact hole to the same bit line. Also, the common source line (CSL) is formed through the second active layer and is contacted on the bottom active layer. The CSL lines are electrically tied to the p-well of the Si layers. Therefore, the body of the strings is electrically and physically tied with the common source line, and the well bias is simultaneously applied by CSL (source-body tied). Cross-sectional SEM images of the 3D stacked NAND flash memory are shown in Figure 13.24.

As shown in the cross-sectional SEM images, the upper and lower cell strings have exactly the same gate stack patterns. It means that the active and gate layout patterns of both layers overlap perfectly. Therefore, the perfectly overlapped word lines (WLs) of both cell strings should be separated by different WL decoders and connected to the different WL decoders at the other end of the cell array. The WL decoders of the upper and lower cell arrays are laid out separately at the ends of the cell arrays.

An example architecture of the 3D stacked NAND memory chip is shown in Figure 13.25 [18]. The bit line (BL) of the memory cell is formed on the second-tier cell arrays and shared with the first-tier cell arrays using contact vias, which are formed through the second Si layer. Thus, the page buffer is able to access both of the first-tier cell arrays and the second-tier cell arrays using the shared BL. The bit-line loading of the 3-D stacked NAND is almost comparable with the loading of conventional first-tier-only planar NAND flash. This is because the additional

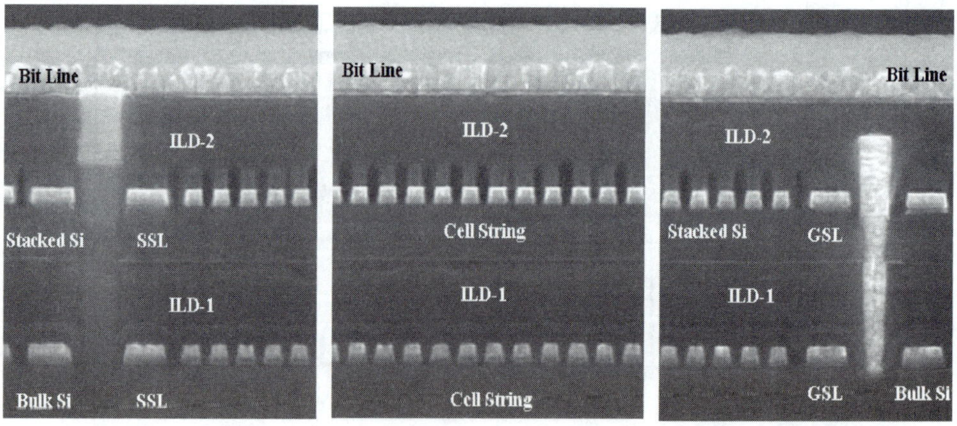

Figure 13.24 Cross-sectional SEM images of 3D stacked NAND flash memory cell strings, which were fabricated with 63 nm node technology. It has 32 cells per string and TANOS gate stack structures.

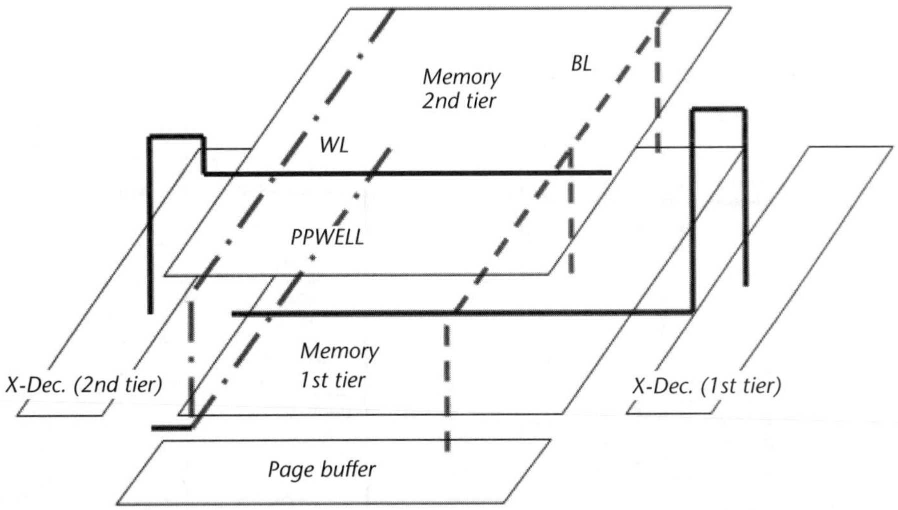

Figure 13.25 Chip architecture of 3-D stacked NAND flash memory: bit lines are shared with both first tier and second tier, but *x* decoders are separated for individual tier cell arrays in the double-stacked NAND flash memory cell array [18].

loading effect by stacking of second tier memory cell arrays is as small as below 3% of total bit-line loading. Therefore, there is no performance penalty from the stacking cell arrays in the 3D stacked NAND flash memory.

Word lines (WLs) of memory arrays of both first-tier and second-tier cell arrays are driven independently by their own individual tier-dedicated WL decoders. Stacking also provides exactly the same WL loading as the conventional first-tier-only planar NAND flash memory. The common source line (CSL) of a NAND cell string and wells of memory arrays are electrically connected by the same structure of contact via of the shared BL and are driven by CSL and well-driver circuits, respectively.

Figure 13.26 shows the operational-bias conditions of the stacked NAND flash memory cell. In read and program operations, required bias voltages are applied only to the selected block of the MATs, while the string select lines (SSLs) of unselected blocks, including all blocks of unselected MATs, are biased to ground. Particularly during erase operation, since wells of memory arrays of both MAT1 and MAT2 are electrically connected together, a high voltage, such as 18V, is applied to both the selected and unselected MATs. Thus, in order to avoid Fowler-Nordheim (F-N) tunneling erasing in the unselected MATs, all WLs of the unselected MATs are floated just like WLs of the unselected blocks in the selected MAT. This can be realized by the tier-dedicated WL decoder, which can control the voltages of each tier of MATs independently.

In fabrication of the 3D stacked NAND structure, forming a high-aspect, deep contact hole is inevitable and inherent because the height of the total stacked layers is increased (doubling of the number of the layers) compared to with planar technology. As discussed earlier, the upper and lower cell strings overlap to reduce the cell layout area. Both cell strings are simultaneously connected to the bit line or the common source line by one contact hole. Therefore, one must etch the stacked layer

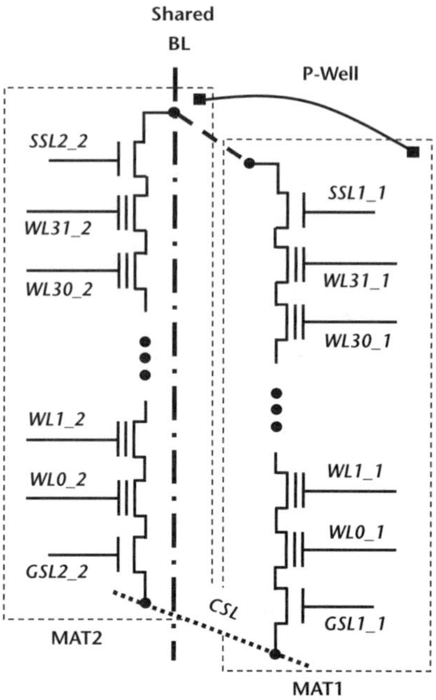

	Read	PGM	Erase
Shared BL	Vpc	0V/Vcc	Floating
SSL_2	Vread	Vcc	Floating
SSL_1	0V	0V	Floating
Selected WL_in 2nd tier	Vread	Vpgm	0V
Unselected WL_in 2nd tier	Vread	Vpass	0V
Unselected WL_in 1st tier	Floating	Floating	Floating
GSL_2	Vread	0V	Floating
GSL_1	0V	0V	Floating
CSL	0V	Vcc	Floating
P-Well	0V	0V	Verase

Figure 13.26 Schematic circuit diagram of the shared bit-line scheme in the 3D stacked NAND flash memory and the operational-bias-conditions table for the bit-line shared-scheme cell strings [18].

sequentially until the via reaches the active bulk Si. The aspect ratio of the deep contact hole is larger than 20. It is very important to reduce the thickness of the stacked layer as much as possible. As one of the possible solutions, the thickness of the top Si layers is minimized by introducing a novel well-bias scheme, the so-called source-body-tied (SBT) scheme.

In stacking NAND cell arrays, thinner stacked Si layers are desirable for simplicity in the integration of the processes. However, when a thin body is used, the body of the cell strings is disconnected from the other cell strings by trench isolations and is floated electrically. The cell strings with the floated body have a disadvantage in the erase operation, as shown in Figure 13.27. In the conventional body-bias scheme, a block of all cells of the string can be erased simultaneously because the body is biased negatively and all word lines are grounded. However, in the floating-thin-body scheme, only one cell of the string can be erased at a time because the word lines of other cells are biased at V_{pass} to connect the channel of the selected cell. The erase time of the product will be increased by 32 times. Therefore, in order to solve the erase problem in SOI-type thin-body structures, a novel operational scheme is suggested. The common source of the cell string is tied with the body of the string electrically. This scheme can erase the cell strings in exactly the same way as the conventional body-bias scheme does.

In the stacking technology, during the formation of the upper cells, the already-made bottom cells and peripheral transistors have to endure additional thermal cycles. Therefore, the additional thermal effect is a very important factor to consider in 3D device integration. Diffusion of dopants in the bottom devices should be suppressed for given thermal budgets. The thermal endurance of the bottom cell transistor is expressed in terms of the effective channel length of the transistor in

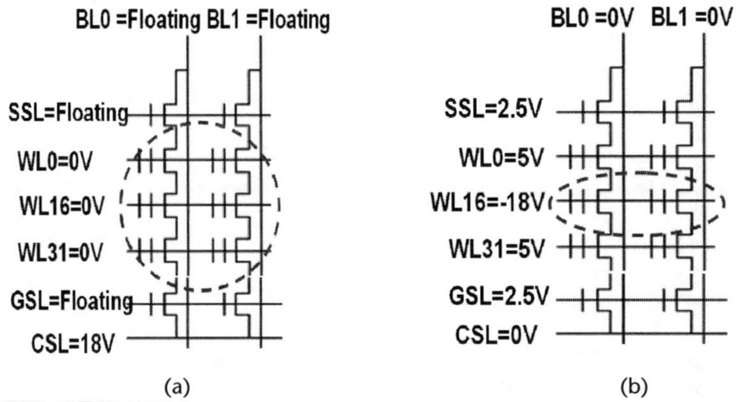

Figure 13.27 Comparisons of the erase operation: (a) erase by page without well bias, and (b) erase by block with well bias.

Figure 13.28. If the temperature exceeds a certain level, the effective channel length is decreased dramatically. Therefore, we use only low thermal processes and tools to deposit thin films to oxidize the Si layers and to activate the dopants after forming the bottom transistors. In the 3D device-stacking technology, the total thermal budget should be tightly controlled. A list of the key processes of 3D stacked NAND flash memory is shown in Table 13.1.

Basically, the upper and lower cell array MATs are supposed to have the same electrical characteristics. However, since the 3D stacked NAND flash memory cells are formed on different Si layers, the electrical characteristics of the memory cells, such as program, erase, and natural cell V_{th} (threshold voltage) distributions, can be different or shifted between the first and second tiers.

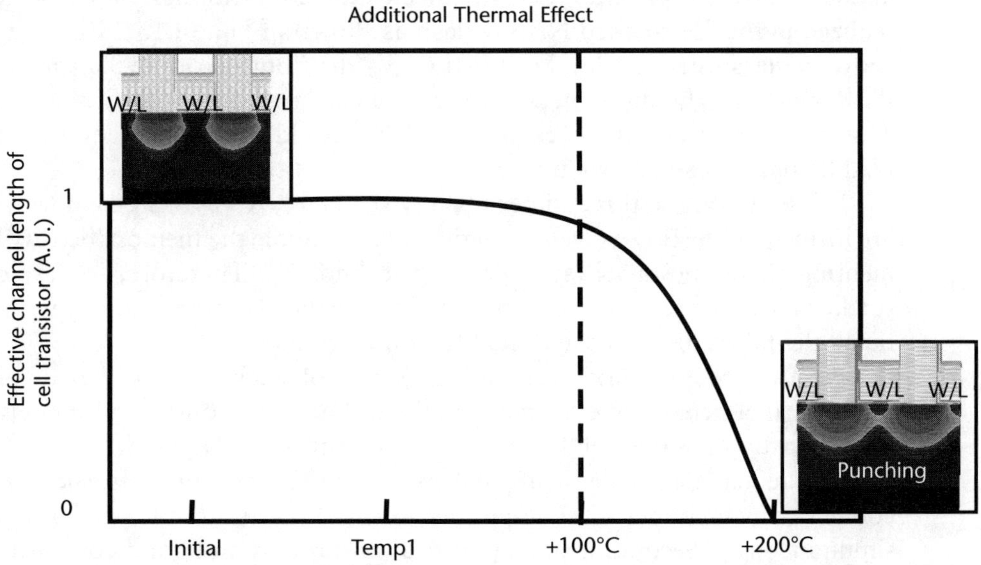

Figure 13.28 The simulated additional thermal effect after forming the cell transistor: effective channel length as a function of temperature.

Table 13.1 Summary of Key Process Flow Sequence for the 3D Stacked NAND Flash Memory

Well & Vth Adjust Implant
Active (Dual Trench Isolation)
1st Gate Stack Structure ; (Tunnel Ox/Trap SiN/Blocking Ox/TaN/WN+W)
Gate-1 Poly Patterning
Halo/LDD Implant, 1st Spacer, S/D Implant and RTA
1st ILD/ILD CMP
Formation of Single Crystal Active Si for 2nd Cell String
2nd Gate Stack Structure ; (Tunnel Ox/Trap SiN/Blocking Ox/TaN/WN+W)
Gate-2 Poly Patterning
Halo/LDD Implant, 1st Spacer, S/D Implant and RTA
2nd ILD/ILD CMP
Cell Penetration Contact (for Source-Body Tied) Formation
Other Contacts and Metal (Bit Line)

As seen in the figure, two different V_{th} distributions are measured from both tiers of MATs. This results in wider dispersion of the total V_{th} distribution curve and eventually causes more degradation of programming performance with the conventional program method. This is because the start voltage of the incremental step program pulse (ISPP) is typically determined by the V_{th} of the fastest cell, which is located at the right edge of the V_{th} distribution curve. Thus, it causes an increase in the required number of ISPPs, which is linearly proportional to the V_{th} distribution of the device. In order to minimize the degradation due to an increase in the ISSP, a layer-by-layer compensated program scheme is needed. For example, based on the V_{th} variations of each tier of MATs, program parameters such as start voltage, stepping voltage of ISPPs, and maximum number of ISPPs, should be set optimally in the 3D stacked NAND flash memory, as shown in Figure 13.30. After adjusting the parameters of the ISPP with the implemented layer-compensated control circuit, almost equivalent program performance to the conventional planar device can be realized in the 3D stacked NAND flash as shown in Figure 13.29(b). Figure 13.31 shows a measured multilevel cell (MLC) V_{th} distribution of the 3D stacked NAND flash memory. The differences between the cell strings of both tiers are negligible. This proves that the 3D stacked NAND cell could satisfy the requirements of the NAND flash memory product.

In summary, the three-dimensionally stacked NAND flash memory cell arrays are formed on the ILD as well as on the bulk to double the memory density by implementing single-crystal Si-layer stacking technology. Therefore, in developing 3D stacked flash memory, it is important to replace the planar technology without sacrificing the fabrication cost and quality of the product.

First, in order to maximize the area benefit of stacking the cell array or to minimize the area penalty due to stacking, the technology should be ultimately pursued using exactly the same cell layout as the conventional 2D planar cell. For example, this can be made possible by implementing new process concepts, such as through-contact, which connects the upper layer cell string and the lower layer cell string simultaneously. Second, a simple fabrication process and good compatibility with the present Si technology are needed. For that, additional patterning layers should be minimized by adopting a simple gate stack structure, such as a SONOS-like

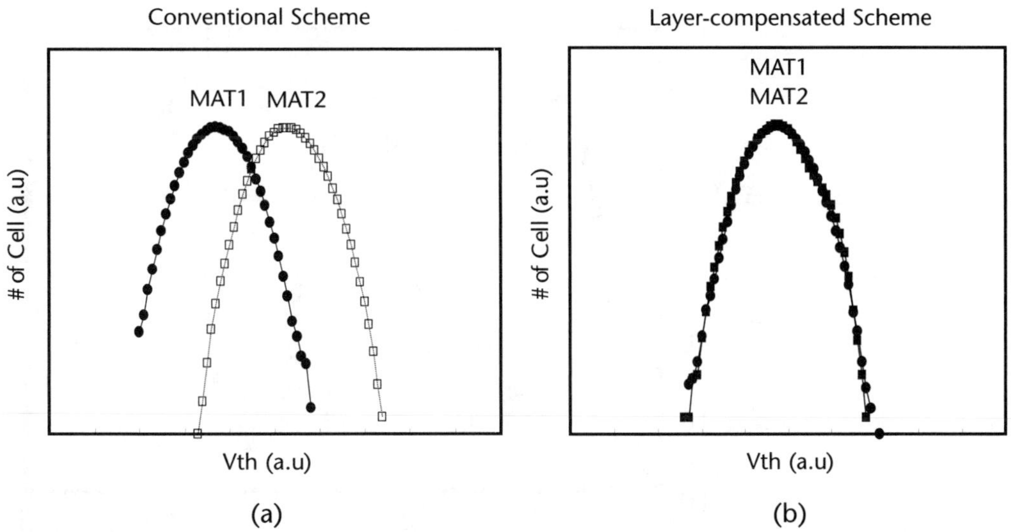

Figure 13.29 (a, b) V_{th} (threshold voltage) distributions of the 3D stacked NAND flash memory with double-stacked cell MATs after applying a program pulse in conventional and layer-compensated schemes [18].

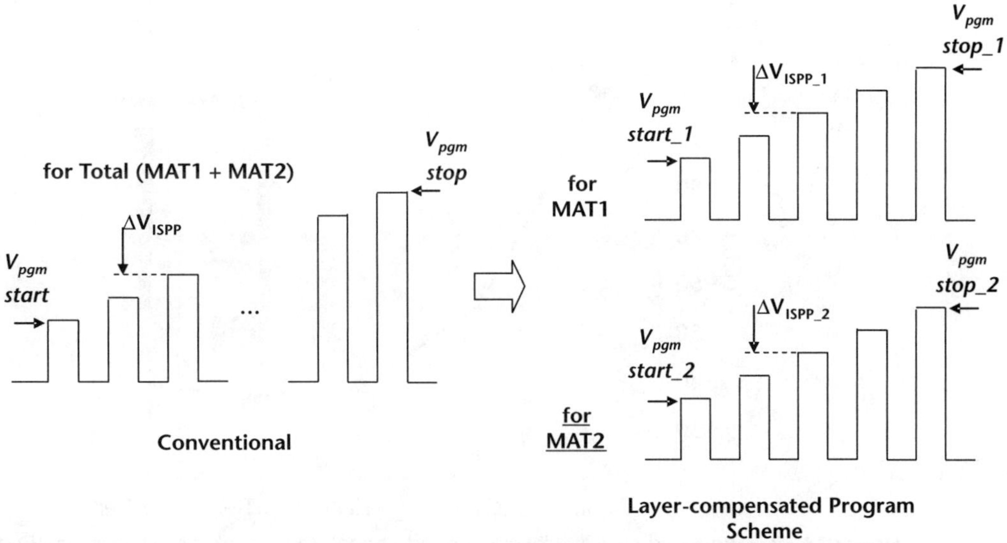

Figure 13.30 Conceptual examples of programming in the conventional and layer-compensated schemes [18].

structure. Third, the same electrical quality and reliability as the present NAND product, which is made by planar technology, should be achieved to replace the incumbent planar technology.

When the cost-effectiveness of the 3D device-stacking technology is considered, there is an optimal number of device layers, as shown in Figure 13.32. If we use 30-nm node technology for fabricating 256 Gbit density NAND flash memory, four

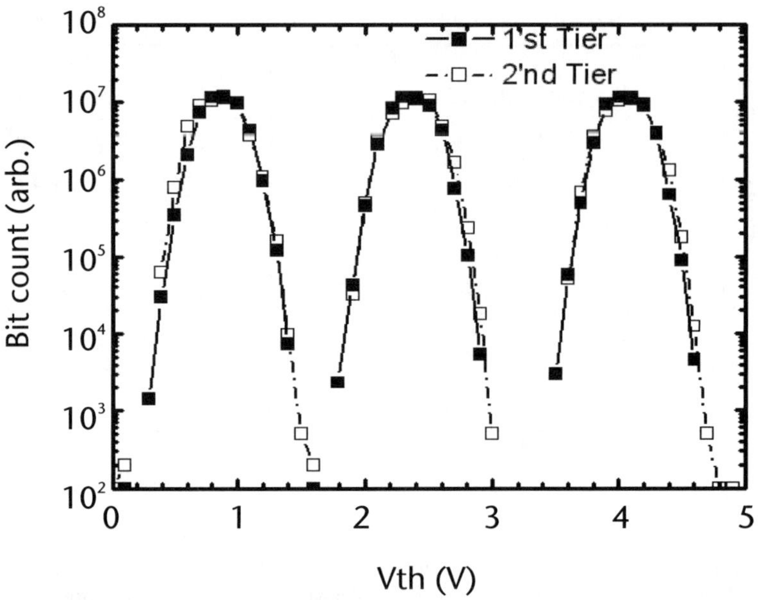

Figure 13.31 Measured V_{th} (threshold voltage) distributions from two-bit NAND flash memory cells fabricated using 3D device-stacking technology.

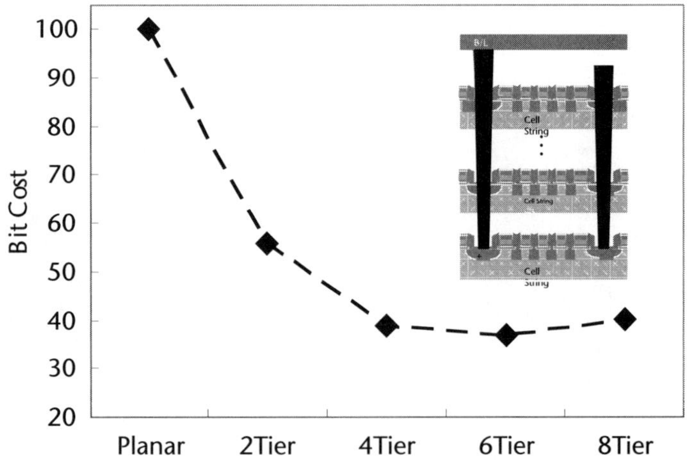

Figure 13.32 Simulated cost-effectiveness of additional stacking of the cell layer in 3D stacked NAND flash memory.

stacking layers are the most cost-effective according to the bit-cost simulation based on fabrication cost, number of process steps, and chip sizes.

There is another idea for 3D device integration for NAND flash memory. Instead of stacking cell transistors, it uses vertical or pillar-shaped cell transistors [19]. Theoretically, as the number of stacking devices increases, this kind of transistor can be fabricated with fewer photolithography layers and process steps compared to the device-layer stacking technology. However, the vertical transistor has limits to overcome in many perspectives, such as growing good uniform dielectric

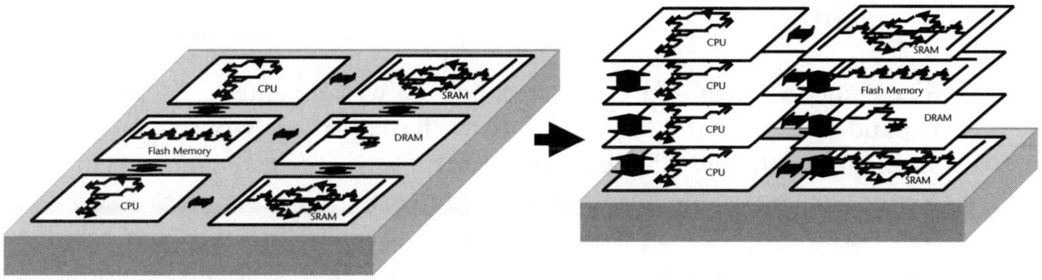

Figure 13.33 Schematic illustrations of 3-D integration of silicon-on-chip (SoC).

layers at the sidewall of the vertical channel, developing the perfect vertical profile of the Si pillar or the Si hole from the top to the bottom, and adjusting the doping concentration of the channel and the source/drain regions with ion implantation technology or other doping techniques. All of these issues are fundamentally difficult to achieve at the levels of uniformity and reproducibility needed for mass production.

13.5 Other Technologies

In addition to 3D integration in the memory, it is predicted that logic technology will move to 3D device integration because of its many advantages, such as small footprint, reduction of metallization length, and ease in combining multifunctionality, to name a few. It should be noted that for logic devices, both transistors and interconnections are key elements. Therefore, 3D logic technology is different from stacking memory cells. It may be even more advantageous compared to memory technology because of its simplicity. For example, when 3D integration technology is used to implement logic, a vertical way of interconnection will be more efficient compared to the lateral way of the planar SoC in terms of speed and power consumption due to reduction in parasitic RC components.

In addition, 3D device-integration technology will make it easy to combine a memory device and a logic device by hierarchical stacking. Because most of the silicon area in an SoC is occupied by memory, this kind of 3D integration will be a major trend. Therefore, even in terms of cost-effectiveness alone, 3D device-integration technology seems to be essential and unavoidable.

Furthermore, after the 3D stacking integration of logic and memory devices on one Si chip, the next step will be to stack multifunctional electronics such as radio frequency (RF) modules, CMOS image sensors (CISs), biosensors (e.g., lab-on-a-chip), and so forth, over the logic and memory device layers. The advantages of stacking multifunctional electronics are numerous: power savings due to the elimination of external wiring, higher packing density due to a tiny footprint, better performance due to diminishing wiring distance, and, most importantly, cost reduction in fabrication. The gains of 3D device-stacking technology will be especially intensified when it meets new materials and new concepts because this will create more values and enrich various multifunctional electronics, which will strongly boost the silicon industry.

13.6 Conclusion

Since silicon integrated circuit technology was invented, the silicon industry has expanded exponentially according to the so-called Moore's law through linear shrinkage of the planar silicon transistor. However, as the incumbent planar silicon technology enters the deep nanoscale dimension, it faces many issues that are very difficult to solve based on the conventional planar silicon technology. Even though many new concepts, new materials, and new technologies are explored to substitute for planar silicon technology, they seem to be too immature to take over the incumbent silicon technology in the near future. Therefore, 3D silicon-integration technology might be the only solution to overcome the physical and lithographical limits of planar silicon technology. Fortunately, 3D integration technology can fully utilize the knowledge and experiences gleaned from 2D planar technology over the past 30 years, which will help to quickly bring the technology to high-volume manufacturing. Furthermore, when 3D silicon technology interacts with new materials and concepts, it will be the center technology in merging NT, BT, IT and others.

References

[1] Hwang, C.-G., "New Paradigms in the Si Industry," *IEDM Tech. Dig.*, December 11–13, 2006, San Francisco, CA, pp. 19–26.

[2] Kim, K., et al., "Memory Technology for Sub-40 nm Node," *IEDM Tech. Dig.*, December 10–12, 2007, Washington, DC, pp. 27–30.

[3] Whang, D., et al., "Large-Scale Hierarchical Organization of Nanowire Arrays for Integrated Nanosystems," *Nano Letters*, Vol. 3, No. 9, 2003, pp. 1255–1259.

[4] Wada, Y., "Prospects for Single-Molecule Information-Processing Devices for the Next Paradigm," *Ann. New York Acad. Sci.*, Vol. 960, 2002, pp. 39–61.

[5] Kim, K., et al., "Memory Technologies in Nano-Era: Challenges and Opportunities," *Digest of Technical Papers of 2005 ISSCC*, Vol. 48, 2005, pp. 576–577.

[6] Kim, J. Y., et al., "The Breakthrough in Data Retention Time of DRAM Using Recess-Channel-Array Transistor (RCAT) for 88 nm Feature Size and Beyond," *VLSI Technical Digest*, June 10–12, 2003, Kyoto, Japan, pp. 11–12.

[7] Kim, K., et al., "Future Outlook of NAND Flash Technology for 40 nm Node and Beyond," *Technical Digest of 21st NVSMW*, 2006, Monterey, CA, pp. 9–11.

[8] Kim, K., et al., "The Future Prospect of Nonvolatile Memory," *Proc. Technical Papers of 2005 IEEE VLSI-TSA*, April 25–27, 2005, Hsinchu, Taiwan, pp. 88–94.

[9] Park, Youngwoo, et al., "Highly Manufacturable 32Gb Multi-Level NAND Flash Memory with 0.0098 mm^2 Cell Size Using TANOS(Si-Oxide-Nitride-Al_2O_3-TaN) Cell Technology," *IEDM Tech. Dig.*, December 11–13, 2006, San Francisco, CA, pp. 29–32.

[10] Ahn, E. C., et al., "Reliability of Flip Chip BGA Package on Organic Substrate," *Proc. 50th Electronic Components and Technology Conference*, May 21–24, 2000, Las Vegas, NV, pp. 1215–1220.

[11] Shin, D. K., et al., "Development of Multi Stack Package with High Drop Reliability by Experimental and Numerical Methods," *Proc. 56th Electronic Components and Technology Conference*, May 30–June 2, 2006, San Diego, CA, pp. 377–382.

[12] Jung, Soon-Moon, et al., "The Revolutionary and Truly 3-Dimensional $25F^2$ SRAM Cell Technology with the Smallest S^3 (Stacked Single-Crystal Si) Cell, 0.16um^2, and SSTFT (Stacked Single-Crystal Thin Film Transistor) for Ultra High Density SRAM," *Technical*

Digest of 2004 VLSI Technology Symposium, June 15–17, 2004, Honolulu, HI, pp. 228–229.

[13] Akasaka, Y., et al., "Concept and Basic Technologies for 3D IC Structure," *IEDM Tech. Dig.*, 1986, Vol. 32, p. 488.

[14] Neudeck, G. W., et al., "Novel Silicon Epitaxy for Advance MOSFET Devices," *IEDM Tech. Dig.*, December 10–13, 2000, San Francisco, CA, p. 169.

[15] Kim, S. K., et al., "Low Temperature Silicon Layering for Three-Dimensional Integration," *IEEE International SOI Conference Proceeding*, October 4–7, 2004, Charleston, SC, pp. 136–138.

[16] Shino, T., et al., "Floating Body RAM Technology and Its Scalability to 32 nm Node and Beyond," *IEDM Tech. Dig.*December 11–13, 2006, San Francisco, CA, pp. 569–572.

[17] Jung, Soon-Moon, et al., "Three Dimensionally Stacked NAND Flash Memory Technology Using Stacking Single Crystal Silicon Layers on ILD and TANOS Structure for beyond 30 nm Node," *IEDM Tech. Dig.*, December 11–13, 2006, San Francisco, CA, pp. 37–40.

[18] Park, K. T., et al., "A 45 nm 4-Gigabit Three Dimensional Double Stacked Multi-level NAND Flash Memory with Shared Bit-Line Structure," *2008 ISSCC Dig.*, 2008, Vol. 52, pp. 9–11.

[19] Tanaka, H., et al., "Bit Cost Scalable Technology with Punch and Plug Process for Ultra High Density Flash Memory," *Technical Digest of 2007 VLSI Technology Symposium*, June 12–14, 2007, Kyoto Japan, pp. 14–15.

CHAPTER 14

3D Stacked Die and Silicon Packaging with Through-Silicon Vias, Thinned Silicon, and Silicon-Silicon Interconnection Technology

J. Knickerbocker, P. Andry, Bing Dang, R. Horton, G. McVicker, C. Patel, R. Polastre, K. Sakuma, E. Sprogis, S. Sri-Jayantha, C. Tsang, B. Webb, and S. Wright

14.1 Introduction

Three-dimensional interconnections for improving transistor circuit density in system applications have evolved for over 50 years, leveraging technology advances in semiconductor wafers, packaging products, and printed wiring boards [1]. Circuit interconnection advance in semiconductor scaling, including lithography scaling, increased die size, and increased wiring layers, has far surpassed packing and printed wiring board interconnection density, creating a gap in off-chip interconnection. New form factors of products began to take advantage of thinned-silicon wafers and die stacking during the last 15 years. However, most applications had a limitation of two layers for area-array flip-chip interconnection or limitations in off-chip bandwidth due to the use of wire bonding or package-on-package (PoP) peripheral input/output (I/O) interconnections between dice and the package(s), which limits performance and application. There have been applications where face-to-face die-attach or use of advanced packaging has helped die-to-die interconnectivity performance; however, these technologies have generally been limited to off-chip package interconnection and assembly, which is orders of magnitude lower in density compared to the emerging 3D fine-pitch die-level integration. Figure 14.1 shows a comparison for interconnection density for traditional packaging and printed wiring boards (PWB) compared to the high interconnection density possible with TSV and SSI to connect thinned die. Two examples of high interconnection density structures include (i) use TSV and SSI for short vertical interconnections between stacked die and (ii) use of a silicon package with TSV and fine pitch SSI to provide high interconnection horizontal wiring between die or die stacks. This can be in the form of silicon packages, chip stacking, or IC wafer-level 3D fabrication. Schematic cross sections illustrating these emerging structures are illustrated in Figure 14.2, including silicon packages with TSV and fine-pitch

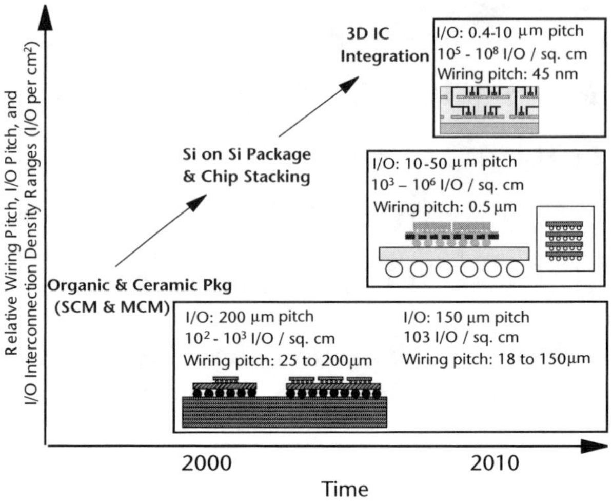

Figure 14.1 Shows relative comparison wiring pitch, I/O pitch, and I/O densities.

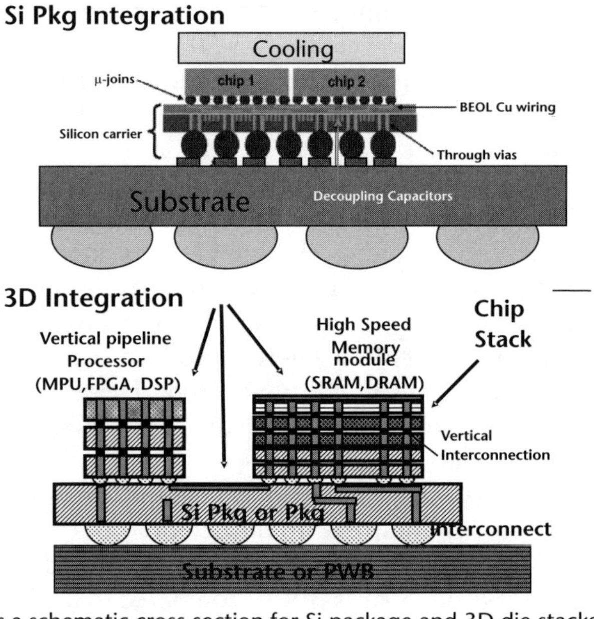

Figure 14.2 Shows a schematic cross section for Si package and 3D die stacks or 3D integrated circuits. (© 2008 IEEE.)

interconnection as well as die stacks or integrated 3D circuits, which can be considered to be dependent on fabrication approach, die size, and integration density.

Universities, consortia, and industry have driven research and early demonstrations toward this new emerging technology with TSV, thinned silicon, and high-density SSI. Unlike prior off-die integration technologies, these new 3D structures offer the opportunity for superior electrical characteristics and high-density vertical interconnection between circuits on silicon dice or strata levels by reducing interconnection distance and electrical parasitics. This new technology offers many potential advantages compared to traditional system-on-a-chip (SoC) or system-in-a-package (SIP) technologies. Moreover, the short distance between circuits can permit silicon dice or strata to be specialized and thereby simplify wafer processing and reduce wafer costs. For example, fabrication of individual microprocessor wafers, memory wafers, I/O communication wafers, digital wafers, analog wafers, optical communication wafers, and high performance silicon package wafers could each be fabricated with fewer manufacturing process steps compared to integrated System on Chip (SOC) wafers. The heterogeneous die could then be integrated into 3D structures to support a wide variety of product applications. TSV, thinned silicon, and SSI scaling from under 10 to over 10^8 I/Os per square centimeter, depending on structure (see Figure 14.1), compared to traditional off-chip integration scaling up to about 10^3 cm^2 I/O fills the gap between "on-chip" integration density and traditional "off-chip" I/O interconnection. The wide range of TSV and SSI densities applied to heterogeneous chip integration comes at a time when Moore's law for semiconductor chip scaling is slowing down or reaching an end as the technology scales to atomic dimensions [2, 3]. Therefore, these newly created, high-density, 3D-technology-integration options offer potential for new applications from lower-cost, simple products to highly integrated 3D products. The design, architecture, and form factors for this technology can be prioritized toward a number of product benefits, such as performance enhancement, power efficiency, low cost, time to market, smaller size, and other attributes that bring value to the application.

Research on 3D integration with TSV, thinned silicon, and fine-pitch silicon-silicon integration has been evolving for more than a decade. 3D test-vehicle designs have been followed by build, assembly, and characterization studies to provide an understanding of structure and process-integration capabilities and limitations. Results from these technology studies provide guidance in terms of 3D design rules, structures, processes, test, and reliability, which can support a growing variety of product requirements and provide "data" toward "preferred technology decisions." Practical technology fabrication and integration approaches need to be considered for targeted TSV and SSI interconnection density, silicon thickness, and power densities. Options such as TSV conductor material and SSI integration material and approach between die-on-die, die-on-wafer, and wafer-to-wafer processes should be under consideration. At the same time, one must consider not only the specific new 3D technology features of TSV, thinned silicon, and silicon-silicon-interconnection but a whole range of technology elements when developing 3D technology such as shown in Table 14.1. In this approach, considerations of application design objectives and high-yielding processes, such as including feature redundancy for interconnections if needed, die size, manufacturing throughput, cost, and test methodology, are also important for specific applications. In this chapter, we report on examples of test-vehicle designs, fabrication, and characterization from research.

14.2 Industry Advances in Chip Integration

Over the last decade, publications have described research including approaches for 3D integrated circuits and chip stacking where vertical vias and interconnections

Table 14.1 3D Technology Intergration Technology Elements

System Technology Element	*3D Consideration, Compatibility, or Approach*
1. Design: library	Performance; power efficiency; low cost
2. Architecture	System assessments
3. Design tools	EDA, design kits
4. Chip technology	CMOS, low-k; SOI, power SOC compatibility
5. Package	Si package, Si stack organic, ceramic
6. Assembly	Solder, metal-metal, oxide-oxide
7. Test	Wafer level test, self test, KGD, cost
8. Module	Power, cooling
9. Reliability	Exceeds application requirements

permit silicon-on-silicon stacking and high-bandwidth interconnection. Since the late 1990s and early 2000s, many researchers studying 3D silicon integration have generated technical publications reporting results and research progress from organizations like ASET Consortia of Japan, Fraunhofer-Institute of Germany, and the Massachusetts Institute of Technology (MIT) in the United States. Research investigations have explored a wide variety of structures, processes, and bonding approaches. Researchers recognize the importance of developing fine-pitch vertical interconnections using through-silicon vias, thinning technology for silicon wafers and interconnection technology for joining thinned-silicon dice into die stacks, die to-silicon packages, and wafer-to-wafer bonding technologies. In addition to power delivery and signal interconnections, investigators have also studied approaches for thermal cooling and modeled heat removal from thinned silicon die stacks and structures containing fine pitch interconnections. Each 3D application will, of course, have its own integration challenges. Common elements in the technology will generally apply to many 3D applications. Understanding of efficient power delivery and heat removal is one example. Another is the interconnection density between silicon layers, which can support short-distance interconnection and high bandwidth. Another is the density and distribution of TSV versus active circuit locations, such as TSVs through chip macros or in dense regions between chip macros, or some combination thereof. For 3D die stacks and modules, the approach for assembly and test must be considered. Die-on-die assembly compared to die-on-wafer or wafer-on-wafer assembly may be a better solution, depending on die size, die yield within a wafer, assembly yield, and the reliability resulting from the associated structures and processes,

From through-silicon via investigations, technical reports included studies of submicron TSV diameters for compatibility with front-end-of-line (FEOL) and back-end-of-line (BEOL) wafer fabrication or alternatively for silicon-based package solutions. TSV diameter and pitch studies have ranged from large sizes, such as about 10 μm to over 100 μm via diameter and silicon thickness of about 50 to 300 μm, down to via diameters of less than 1 μm to 10 μm with silicon thickness of less than 50 μm, down to silicon thickness such as SOI (<1 to 10 μm). TSV conductors reported include tungsten, copper, composite, paste, doped polysilicon, and other electrical conductors. One study by K. Takahaski et al. evaluated 10 μm copper conductors utilizing through-silicon vias for electrical interconnection at 20

μm pitch (Figure 14.3) [4]. Silicon-to-silicon fine-pitch interconnection with TSV has also been reported (Figure 14.4) [5]. Numerous bonding and electrical interconnection structures, materials, and process approaches have been evaluated between silicon dice or strata layers in thinned-silicon die stacks or from silicon-to-silicon packages as shown in Figure 14.5 [6, 7]. In one of these examples, anisotropic conductive polymers were used to bond 25 μm thinned dice using 50 μm–pitch AuSn bumps. Other technical publications reported fine-pitch solder connections to copper as a means to stack thinned-silicon dice to either other silicon dice or to join dice to silicon packages [5–8]. An application leveraging TSV and fine-pitch interconnection for memory die stacks has also been presented as illustrated in Figure 14.6 [9]. The silicon-silicon interconnection structure, material, and process options need to be able to meet product specifications and manufacturing yields and to be compatible with wafer and module fabrication and assembly objectives. As 3D technology and product shipments grow with time, more significant statistical data on reliability and yield will be reported. Relative costs for TSV, wafer thinning, and

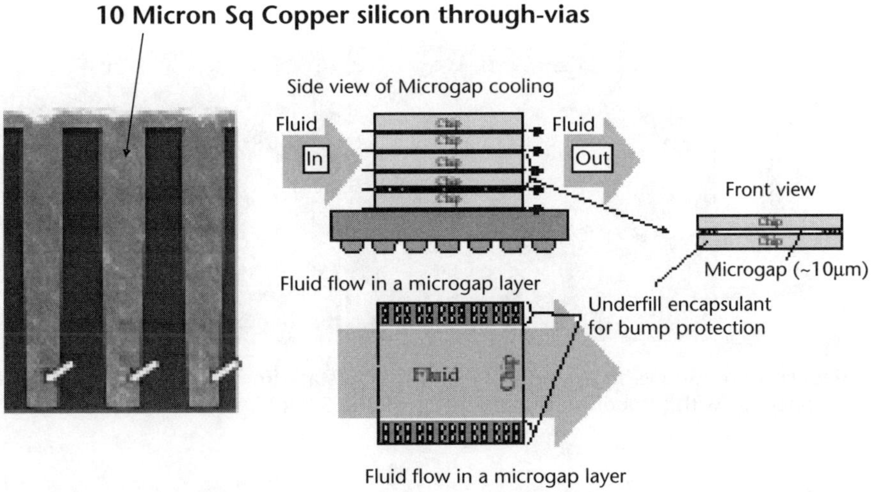

Figure 14.3 Shows an example of copper used for through silicon vias (TSV) from ASET 2004 as reported at ECTC 2004 by K. Takahaski et al. [4]. (© 2004 IEEE.)

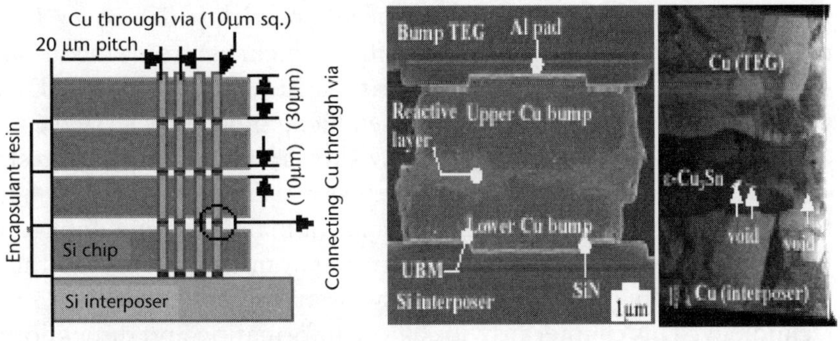

Figure 14.4 Shows an example of fine pitch interconnection of TSVs by M. Umemoto et al. from ECTC 2004 [5]. (© 2004 IEEE.)

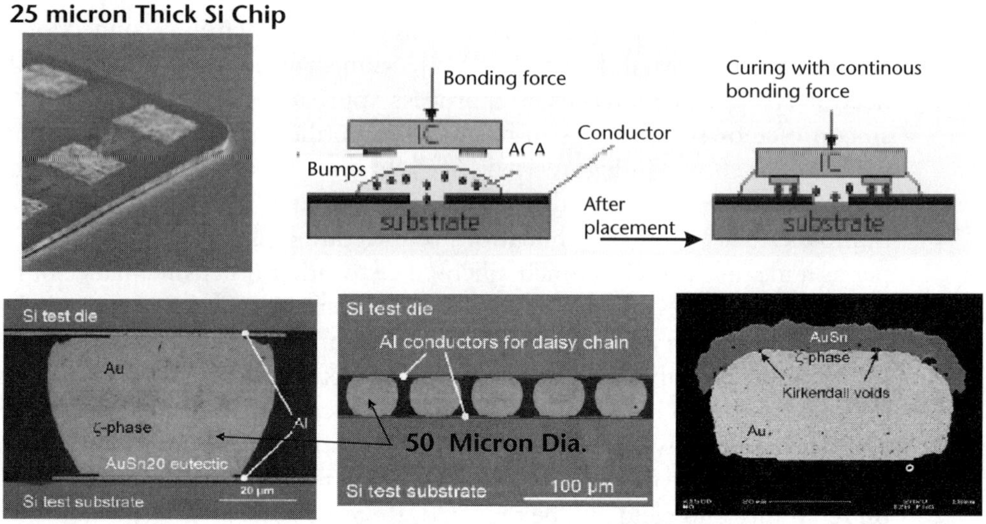

Figure 14.5 Shows an example of thinned silicon and fine pitch interconnections from Fraunhofer-institute as shown by [7] and [8], respectively. (© 2004 IEEE.)

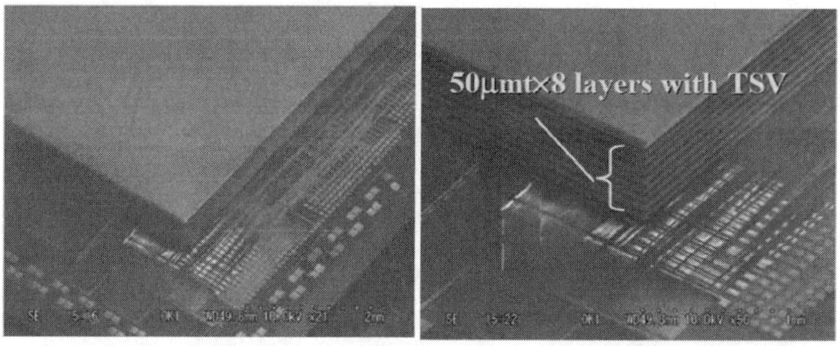

Figure 14.6 Shows an example of memory chip stack from H. Ikeda of Elpida with eight 50 μm thinned die with through-silicon-vias [9]. (© 2005 SEMI.)

silicon-silicon interconnection will be reduced with volume production, improved tool utilization, and tool depreciation.

Technical progress in silicon packaging, die stacking, and integrated 3D structures has been reported [10–14]. More recently, TSV, silicon thinning, and silicon-to-silicon interconnection have been reported, which cover TSV sizes from less than 0.2 μm to 70 to 90 μm size and silicon thickness from silicon-on-insulator (SOI, <1 to 10 μm) to silicon thickness of less than 150 μm [1, 15–18]. Demonstration test vehicles are designed, fabricated, assembled, and characterized as a means to help develop a new technology. Figure 14.7 shows a few examples of test vehicles used to study TSV, electrical parametrics (such as resistance, inductance, capacitance, cross talk, and signal integrity using frequency domain and time domain measurements), I/O, interconnection, assembly, thermal parametrics (such as thermal resistance, thermal spreading and thermal transmission measurements), and reliability for 3D technology. This chapter provides more information and discussion associated with 3D technology, including design and application considerations, through-silicon

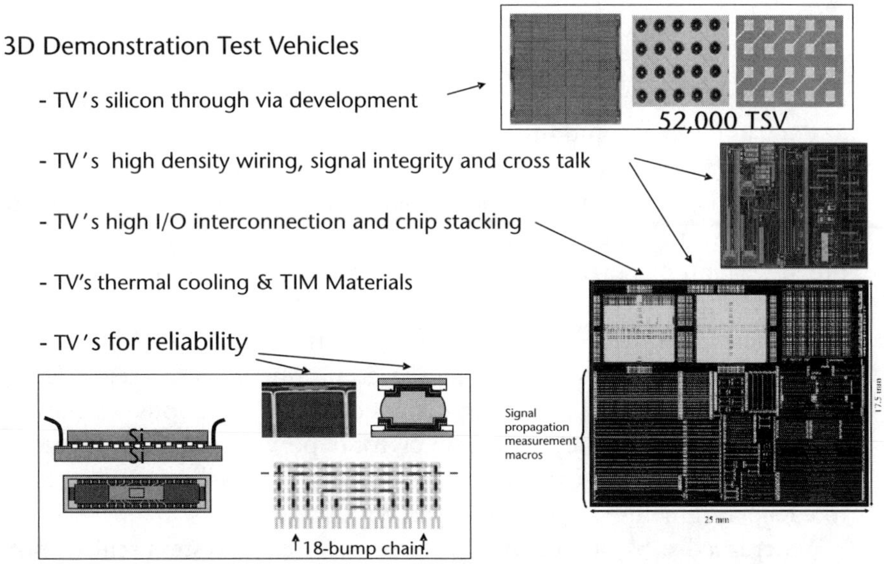

Figure 14.7 Shows variety of 3D and silicon package technology demonstration test vehicles.

vias (TSVs), wafer thinning, silicon-silicon interconnection (SSI), assembly and test methodology, and cooling investigations.

14.3 2D and 3D Design and Application Considerations

Traditional 2D semiconductor manufacturing meets significant technical challenges at each step forward in lithographic node. For example, device variability, leakage current, and new lithographic challenges need to be addressed at each new technology node. Manufacturing device variability is becoming more challenging due both to control of dopant levels at smaller dimensions and to low dopant concentrations. Leakage currents are increasing as gate dielectrics reduce in size and have led to use of high-k dielectric gates. Semiconductor lithographic requirements are also challenging, leading to use of immersion lithography for the smaller semiconductor nodes. As semiconductor manufacturing moves from 65 and 45 nm and leading edge products toward 32, 22, and 15 nm dimensions, we might expect that Moore's law [2] and Dennard's semiconductor scaling rules [19] will slow down and reach an end. As we approach semiconductor limits, quantum mechanical properties take over and thus traditional semiconductor CMOS scaling based on silicon node advances will reach its limits. Potential post-CMOS technologies such as carbon nanotubes, graphene, and quantum computing may provide continued scaling for advanced computing [3].

System-performance scaling challenges are also growing. Historical system trends have benefited from more than 60% system performance advantage due in part to processor chip performance improvement per generation to 65 nm technology. System scaling has also been limited in part by less than 15% improvements in memory access time, less than 10% growth in off-chip interconnection density, and limitations in chip cooling [20, 21]. It is expected that direct system performance scaling from semiconductor processor node advances will likely be less than 20% for each new semiconductor generation. Even with lithography scaling giving circuit density growth for each new technology node, challenges such as leakage power, latency in memory, and interconnection bandwidth to memory could impact future systems performance [20, 21]. In order to achieve continued system performance scaling at better than 60% per generation, the attributes of a system stack will require advances. For example, increasing the numbers of software threads, combined with an increasing number of cores per processor die, is expected to help system performance gains. Today, gate leakage limits power efficiency at higher frequencies, but the introduction of high-k metal gates will reduce leakage current. For smaller lithographic nodes, increased numbers of chip repeaters are needed due to smaller wire sizes and long-distance interconnections at these smaller lithographic nodes, but they can impact total power and power efficiency. As these multicore processors scale in systems usage, so too will they will need to scale access to cache and memory with increased bandwidth.

Increased subsystem communications can help system scaling both through the use of 3D interconnection for shorter paths between circuits and through the use of optical communications. In current planar logic and memory dice, the interconnection length for domains can range in length up to a few or even tens of millimeters. Use of 3D integration can offer domain interconnections at lengths of a few tens of microns, thereby providing two to three orders of magnitude of reduction in wire length for reduced latency. 3D also offers opportunities for significant interconnection density increases vertically between silicon layers in a stack or horizontally with the use of an added 3D layer silicon packaging with TSV, each approach can directly improve interconnection bandwidth by two to four orders of magnitude or more compared with traditional off-chip interconnection density or bandwidth [11, 14, 21]. Thus, system challenges and semiconductor scaling limits need not limit system scaling, power reduction, and miniaturization. For example, to compliment semiconductor advances, subsystem scaling using three-dimensional, high- performance die stacks and packages can benefit system advances [3]. The subsystem scaling can include 3D integration with high-bandwidth electrical and optical interconnection between large numbers of multicore processors and cache/memory [1, 3, 20, 21]. Appropriate architectures and software, which can leverage 3D multicore, multithreaded structures, will also be critical to optimize the benefits for system scaling during the next decade. It is expected that a combination of semiconductor scaling, 3D subsystem scaling (use of 3D electrical and optical interconnection), the use of multicore, multithread processors, system accelerators, and system software enhancements will be necessary to support system performance scaling at greater than 60% per generation. These are needed to make the potential future post-CMOS technologies and quantum computing technology solutions of the next decade practical and cost-effective product solutions.

14.4 Through-Silicon Vias

Applications may need only a few thousand or million vertical interconnections, which are very product dependent and link to architecture, desired product specifications, silicon thickness, materials, structures/sizes, and processes. Examples of TSV structures fabricated and characterized include TSVs that have a wide range of sizes, heights, aspect ratios, materials, densities, and fabrication processes [13–15, 22].

For example, the range in TSV size includes diameters, or x-y sizes, from less than 1 to 90 μm. The silicon thickness includes silicon-on-insulator (SOI) where thickness of silicon for transistors and back-end-of-line wiring can be less than 10 mm and up to full wafer thickness of 730 μm for the thickest silicon TSVs (although most studies have been at 300 μm or less). TSV aspect ratios include evaluations from a low aspect ratio of about 2:1 to values of over 20:1 for copper fill and over 40:1 for tungsten conductor fill. The ratio represents the thickness of silicon to diameter (or width) of the TSV. These structures, sizes, and aspect ratios impact the physical process as well as the resulting electrical, mechanical, thermal, and reliability characteristics of the TSV.

14.4.1 TSV Process Sequence

A process sequence comparison for via formation in thinned-silicon structures is shown in Figure 14.8. In this example, the TSV process sequence is compared for via-first and via-last processes. In a TSV process, typically an isotropically etched hole is formed in silicon, often followed by deposition of a dielectric layer for electrical insulation of the via from the silicon, followed by metallization, which includes a liner deposition, seed, and metal fill operations. When performed during in the first stages of wafer fabrication, the process is often referred to as vias first or vias middle, following FEOL circuit fabrication to be compatible with CMOS wafer- fabrication steps. When formed after complete wafer fabrication and thin-

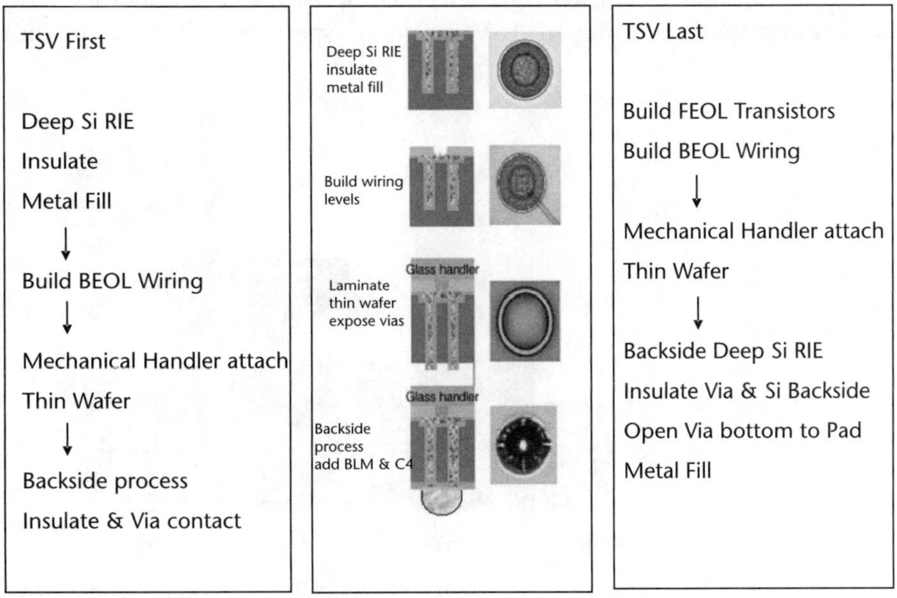

Figure 14.8 Shows a process flow comparison for TSV first along with via schematics / photos and vias last processes [15]. (© 2006 IEEE; ©2008 IBM J. Res. Dev. [15, 22].)

ning, the process is often called TSV-last processing. Next steps for vias-first or -middle processing would include BEOL processes, attaching a mechanical wafer such as glass or silicon to provide support while the first wafer with TSV fabrication is thinned from standard wafer thickness of about 730 μm (200 mm) or 785 μm (300 mm) to the desired product thickness. The mechanical wafer can be attached by means of a polymer adhesive, which can later be removed by means of chemical etch, heat, laser, or an alternative method. Adhesive properties can limit subsequent process steps, depending on the adhesive material and mechanical handle wafer used. The wafer can then be thinned by mechanical grinding in one or more steps, polishing or chemical-mechanical polishing, chemical etching (wet or dry), or a combination of mechanical and etch methods, followed by back-side metallization and redistribution wiring for back-side electrical insulation around the TSV and electrical interconnection to the TSV.

TSV-first or -middle processing may include TSV prior to FEOL or post-FEOL but prior to BEOL. TSV-last processing may be done from the top side of the wafer prior to thinning, following FEOL and BEOL processing. Alternatively, TSV-last processing may be driven with via formation following wafer thinning from the back side of the wafer and would require a landing site and layer to connect to wiring or circuits on the top of a wafer with transistor circuits. There are many combinations of processes among TSV first, middle, and last that can also lead to robust structures. Simple silicon packages can have TSV formed first or last without constraints for compatibility with active circuits. Additional perspectives on silicon-through vias have been reported [15]. Figure 14.9 shows cross sections of through-silicon vias for 40 μm–, 50 μm–, and 150 μm–thick silicon for copper and tungsten metallurgies, respectively, and composite via at 300 μm silicon thickness. A dielectric insulator and adhesion layer/liner are also used in preparation of the TSV conductor fill. A variety of TSV shapes can be considered in order to be compatible with polysilicon, tungsten, copper, or composite TSV conductors. Examples include annular TSV, bar-shaped TSV, or high-aspect-ratio TSV as previously reported

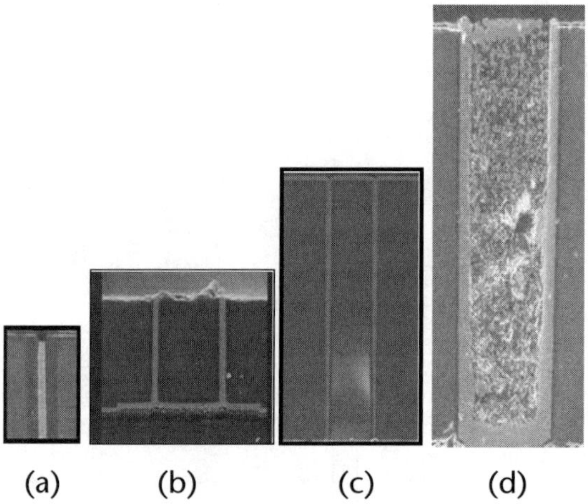

Figure 14.9 (a–d) Examples of TSV cross sections with: 14.9A 2 μm diameter via and 40 μm silicon thickness, 14.9B 70 μm annular via and 50 μm silicon thickness, 14.9C 70 μm annular via and 150 μm silicon thickness, and 14.9D 70 μm composite via and 300 μm silicon thickness. (© 2005 IBM J. Res. Dev. [11].)

[15, 22, 23]. The shape and size of the TSV can be determined by a combination of considerations, such as cross-sectional area required to meet electrical resistance specifications, current-carrying specifications, and taking into consideration the manufacturing yield and thermomechanical stress of TSV based on shape, conductor, and dielectric. Looking at one parameter, such as TSV cross-sectional area, which would directly impact the electrical resistance per micron of TSV thickness, a comparison for an annular-shaped conductor compared to a multiple bar-shaped conductor design, each with 1μ metal conductor width, could each give about 30 μm^2 of metal area for a 10μ TSV, assuming one annular ring versus three parallel bars each at 10 μm length. Conductor area comparison for a 50 μm–size TSV shows that approximately 153 and 300 μm^2 conductor metal area could be obtained, respectively, for one annular ring versus six parallel bars of conductor, each at 50 μm length. For thick silicon such as 300 or 150 μm, many many BEOL and wafer-handling or dicing process steps, a freestanding silicon wafer can be processed. In contrast, a handle wafer is used to support wafer processing and is later laser-released or debonded for wafers requiring many process steps or when thinner silicon structures (<100 μm thickness) are desired. The thinned silicon can later be bonded to a module component (silicon, ceramic, or organic), which can provide mechanical integrity in addition to providing other electrical, optical, or thermal benefits to the product followed by mechanical handle wafer removal.

14.4.1.1 TSV Etch Process

TSV structures can be fabricated by a number of methods of silicon removal, such as wet processing using a chemical etch, or by dry etching, such as laser drilling, reactive ion etch (RIE), or deep reactive ion etching (DRIE). Combinations of these etch processes are also possible. Etching often begins with photolithography and the use of a photoresist or a "hard mask," such as a patterned oxide or nitride layer to define the features to be etched.

Wet etching of silicon can use a mixture of nitric acid (HNO_3) and hydrofluoric acid (HF) diluted with water or acetic acid and be performed at room temperature for isotropic etching. Anisotropic wet etching can be achieved using diluted potassium hydroxide (KOH) or other etch chemistries, such as tetramethylammonium hydroxide (TMAH), where preferential etching in the (110) and (100) silicon crystal planes compared to the (111) silicon crystal orientationoccurs. These different etch rates can lead to many different etch morphologies, depending on wafer bulk crystalline orientation and mask and wafer orientation. However, for many 3D applications, isotropic vertical etched shapes are desired; thus, wet etching may provide shape adjustment in combination with a dry etch.

Dry etching using a laser can be used to create high-aspect-ratio holes. Alternatively, silicon can be etched using reactive ion etch (RIE), which uses a plasma etch consisting of source gases broken down in a plasma to create highly reactive electrons, photons, neutral species, and positive ions that impinge with surface atoms, both removing them and forming stable compounds. The chemical and physical reactions etch a surface, and remaining by-products can be removed, causing the plasma to continue to react. RIE has been used for semiconductor processing and can create high-aspect-ratio structures or TSV, but it can be slow. An enhanced process called time-multiplexed deep etching, or just deep reactive ion etching (DRIE), was patented by Bosch (U.S. 5501892), which combines cycles of deep silicon plasma etching with passivation sequentially to etch very high aspect ratios at high etch rates on the order of up to 10 μm/min. The sequential operation with plasma radicals, such as CF_x from octobluorocyclobutane (C_4F_8), followed by passivation can also lead to a scalloped surface in the TSV.

14.4.1.2 TSV Dielectric

TSVs may be fabricated with or without dielectric insulation from the silicon. For simple structures with a desired short to the base silicon, no dielectric is required. For application requiring electrical isolation from the silicon, a polymer dielectric or oxide or nitride dielectric may be used.

Polymer dielectrics can be processed by means of spray, spin coat, or alternate deposition method, where temperature limitations may be required, such as less than 200°C or 300°C. Robust fill and uniform isolation can be dependent on TSV diameter and aspect ratio and processing conditions. Following deposition, polymers are cured. For via-last, post-FEOL and post-BEOL wafer finishing for TSV, polymer dielectrics may be appropriate. Similarly for larger-size TSV, a polymer dielectric may be a low-cost method to isolate the conductor from the silicon.

Oxide and/or nitride depositions can be completed, such as thermal oxide processing at temperatures between about 900°C and 1,100°C, whereas plasma-assisted deposition of oxide or nitrides may be completed at temperatures down to about 200°C to 250°C. Other deposition techniques may also be employed, and, again, TSV diameter, size, and aspect ratio can impact the uniformity of dielectric isolation.

14.4.1.3 TSV Conductor

Deposition of a conductor begins with deposition of a thin conductive liner and seed layer. Similar to BEOL processing of copper lines, TSV can use a physical vapor deposition (PVD) of tantalun nitride tantalum (TaNTa) and copper (Cu), followed by a full metal deposition. The liner-seed deposition should be uniform; thus, deposition quality can be dependent on TSV shape and aspect ratio. One should avoid a top opening that is smaller than the corresponding vertically etched hole as this can lead to the top of the via's later being filled prior to complete fill of the entire TSV, leaving a central void or defect in the TSV. Defects in the liner-seed layer can also lead to defects like voids along the sidewall when filling a TSV; thus, surface defects or disruptive morphologies should be avoided. Other liner-seed deposition materials and methods may be used that provide the same function of a continuous electrically conducting layer inside the TSV surfaces with uniformity and proper morphology.

Examples of TSV conductor fill materials include polysilicon, copper, tungsten, and composite materials. Doped polysilicon used in TSV can have the advantage of coefficient of thermal expansion (CTE) match to silicon, compatibility with CMOS processing, and simplified process sequences; however, it has the disadvantage of orders-of-magnitude-higher electrical resistance and limitations on carrying significant current per unit area. Depending on product specifications—such as multiple high chip stacks and thick silicon strata levels, for instance, greater than 30 μm—doped polysilicon may not be the conductor of choice. For lower electrical resistance and higher current-carrying applications, a metal conductor may be a better choice. Conductors such as copper, tungsten, or metal-composite-filled conductor may be preferred. Copper has desirable low electrical resistance and high current per unit cross section. The disadvantages of copper include a high CTE of 18 ppm versus silicon's 3 ppm, which can give rise to high thermomechanical stresses during processing or in product use, depending on the size and shape of the TSV. In addition, use of copper conductors may lead to added processing steps or the choice of when to apply a conductor to be compatible with CMOS processing. Tungsten can also be considered as a TSV conductor and offers the advantages of a CTE of about 4.5 ppm, which is closer to silicon and can provide lower thermal-mechanical stress compared to copper. Tungsten electrical resistance is three times higher than copper's, but it is far better than doped polysilicon, and high current per area cross section is possible. Another benefit of tungsten is that it can be used to fill high-aspect-ratio vias even in excess of 40:1. Metal composites may also be used for TSV conductors, and their properties are dependent on the choice of conductor, matrix fill material, and the composite microstructure. Silver-filled pastes, copper annular structures with central filled composites, and other TSV structures can also serve application requirements depending on size and desired properties. Electrical inductance values of about 0.15 pH per each micron of silicon thickness have been measured. Similarly, electrical resistance values of 0.2 mΩ per each micron of silicon thickness have been measured. For the TSV conductors such as doped polysilicon, copper, tungsten, and metal conductors in a composite TSV, application considerations and specifications need to be weighed, depending on the 3D structure being fabricated.

Other TSV conductor considerations include cost, compatibility with the desired semiconductor or package, and processing of the TSV prior to FEOL, post-FEOL, or post-FEOL and post-BEOL from the top or bottom surface. Another question may include other limitations, such as temperature limitations based on application, mechanical handler, or compatibility with semiconductor processes for one wafer or subsequent stacking or process operations. For example, for thin-silicon and high-density TSV applications, doped polysilicon, tungsten, or copper may be considered. For low-temperature limitations, a copper or composite via may be desired due to higher-temperature deposition processes for polysilicon or tungsten, which may be in excess of 300°C. For thicker silicon layers on the order of 100 to 300 μm, copper, tungsten, or a composite TSV may serve the product, depending on required TSV specifications. Ultimately, the TSV material choice is dependent on the resulting microstructure of the TSV, including size and manufacturing process, as well as considerations of yield and specifications, such as electrical specifications, mechanical specifications, and reliability specifications.

For robust TSVs demonstrated by research, process build demonstrations were used to ensure low cost, CMOS-compatible process with high yield and reliability. Demonstration test structures were characterized for the best structures for various size TSV and were measured for electrical parametrics (such as resistance, inductance, capacitance, cross talk, and signal integrity using frequency domain and time domain measurements) and reliability. Mechanical modeling and simulations were used to compliment experimental measurements. 3D stress models were able to identify high-stress regions in a TSV structure and led to modified TSV structures for process compatibility and high reliability. Results have shown up to 100% yields in test samples, elimination of defects, demonstration of stable structures during processing, and demonstration of reliability.

14.5 BEOL, Signal Integrity, and Electrical Characterization

14.5.1 BEOL

For 3D structures and wiring, standard back-end-of-line (BEOL) processes include dielectric insulation, photolithography, metal deposition, chemical-mechanical polish, and dielectric deposition. Dielectric insulation layers such as plasma-enhanced chemical vapor deposition (PECVD) or a polymer-based dielectric can be deposited on the wafer surface by a spin-on process and bake. Next, during photolithographic processing, spin-on photo resists are deposited at a given thickness, after which the film is baked to remove solvents and then exposed to ultraviolet light. Postexposure, the film is developed and baked to create a stable film ready for subsequent processing, such as etch of an dielectric film.

Similar to the TSV deposition described previously, BEOL metal wiring and vias are formed using PVD liner-seed deposition and copper electroplating. Electroplating of copper has been used for more than a decade in semiconductor processing, which involves the use of an electrolyte against the wafer as well as of a voltage potential applied between the seed layer (cathode) and an anode. The voltage potential drives a current in the anode or copper ions contained in a chemical bath to be reduced to copper metal on the surface of the cathode or seed layer. The plating bath can have many components, including copper electrolyte, such as copper sulfate and sulfuric acid as well as chloride ions. Other additives in the plating bath can be accelerators, suppressors, and levelers, which can aid in the deposition rate and quality of the copper film being deposited. Copper-from-copper anodes can be dissolved into the chemical bath, replenishing the ions driven to the cathode seed layer. The copper surface deposition fills the etched photoresist, which later becomes the vias or wires in the BEOL, as well as the surface of the photoresist. The quality of deposition also depends on the current flow, uniformity, tool, and waveform, such as a pulsed waveform during deposition, which can accelerate deposition rates.

Chemical-mechanical planarization (CMP) can be used to remove the over-plated copper. The copper damascene wiring levels use this CMP process, which combines mechanical abrasion of the surface copper with a mechanical polishing pad and abrasive colloidal slurry, which cause a combined mechanical material removal and a chemical reaction to remove excess copper. The polishing action of the slurry and polishing pad aids in material removal and creating a more uniform flat surface. During this operation, the wafer is held with a backing film, and pressure is applied to remove the excess copper uniformly to create a more planar surface.

14.5.2 Signal Integrity and Electrical Characterization

For 3D structures, electrical characteristics of wiring and via structures need to be measured and correlated with electrical models and simulations. For example, the *x-y* wiring and *z*-direction TSV, silicon-silicon interconnections, and vertical buses through multiple layers of thin stacked silicon should be understood. Figure 14.10 shows examples of demonstration structures that have allowed characterization of trench decoupling capacitors [Figure 14.10(a)]; signal-integrity characterization at different line lengths, line widths, and spaces [Figure 14.10(b)]; vertical TSV, SSI structures, and vertical buses [Figure 14.10(c)]; and reliability characterization [Figure 14.10(d)]. For example, signal integrity has been characterized using frequency- and time-domain measurements [10, 24–26]. Results were measured for signal transmission from 2.5 to 75 mm line length and for frequency between 1 and 10 GHz. Examples of a few test vehicles used in electrical characterization are shown in Figure 14.10(d). For example, integrated decoupling capacitance using deep trench capacitors and TSVs was shown to have 12 to 14 $\mu F/cm^2$ capacitance. TSV electrical resistance was measured at 0.2 mΩ per micron of silicon thickness, and inductance was measured at 0.15 pH per micron of silicon thickness. Examples of short-distance measurements for signal integrity at 2.5 and 7.5 mm line lengths across a variety of line widths and spaces from 0.8 to 3 μm and different signal-to-ground structures showed that open-eye diagrams could be obtained for frequencies from 3 to 8.5 GHz, where signal delay was measured from less than 25 ps up to 95 ps and far-end crosstalk noise was measured to be from 1.2% to 35%. Microbump interconnection characterization showed direct current electrical resistance under 18 mΩ for 25μ diameter solder between surface pads on each silicon strata layer. Signal integrity and chip-to-chip I/O link-characterization measurements are also important in order to increase bandwidth between dice in a die stack or between dice on a silicon package. I/O driver circuits can be reduced in size

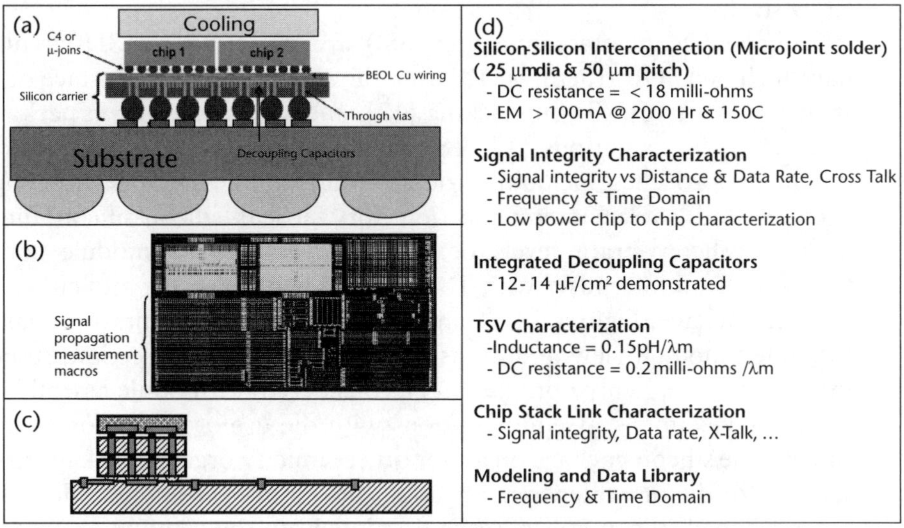

Figure 14.10 (a) A multichip test vehicle with integrated decoupling capacitors; (b) a signal integrity test vehicle frequency and time domain characterization; (c) a chip stack characterization test vehicle; and (d) examples of electrical characterization results which flow into a data library/database.

for short-reach communications and have been demonstrated to support over 10-fold reduced power compared to traditional off-chip I/O driver circuits.

14.6 Silicon-Silicon Interconnections, Microbumps, and Assembly

14.6.1 Interconnection Material, Structure, and Processes

Traditional off-chip interconnection has used wire bonding or area-array flip-chip solder bonding to provide interconnection to a package. Aluminum or gold wire bonds have often provided one to four perimeter rows of interconnection with a pitch of bond pads down to about 30 μm and a total number of connections per die up to about 800 I/Os. Area-array off-chip interconnections such as those using lead tin solder have provided I/O pads at a pitch of 200 μm and, more recently, of about 150 μm, where total I/O per die has been to about 8,000 I/O. The peripheral wire bond or area-array solder interconnections have usually provided not only power and signal interconnections but have accommodated many product form factors and processes during assembly for high reliability in products. Heterogeneous die stacking, 3D integration with organic and ceramic packages for high yield, cost effective test methodology and reliability are important in this emerging technology. Thermally enhanced overmolds and underfills may also be needed to meet product application.

For newly developing 3D die stacking and silicon packaging, higher-density interconnection densities are desired in many applications. In these cases, fine-pitch interconnection may be necessary for high bandwidth and low latency in a die stack. These interconnections need to consider power delivery and distribution, electrical signal transmission, thermal transmission, mechanical integrity, and reliability, while also supporting these new 3D product form factors as well as achieving high assembly yield and wafer test requirements for "known good die" (KGD) and KGD stacks.

Silicon-silicon interconnection (SSI) investigations for 3D circuit integration have included SOI structures [13], thin die stacks using face-to-face or face-to-back structures [14, 16], silicon stacking [12], and silicon-on-silicon package and interconnection investigations [17, 18]. These investigations have included different bonding approaches, including oxide-to-oxide bonding, copper-to-copper interconnection, and solder interconnection for vertical silicon-silicon interconnection between silicon strata levels or silicon layers. For a module composed of a thinned-silicon interposer with TSVs placed between a die and either a ceramic or organic package, Figures 14.11 and 14.12 show a representative unassembled and expanded module view and an assembled module cross section, respectively. For these structures, many options can be considered for module assembly.

In one example, assembly options with the use of solder for module assembly can include silicon package on a support ceramic or organic package first, silicon die stacking or die-on-wafer first, or module assemblies with one reflow. In one example, a thinned-silicon interposer is first joined to the ceramic or organic substrate, followed by placement and reflow for attachment of one or more additional silicon dice. This option of joining a thinned-silicon structure to a ceramic or organic substrate first, followed by addition of multiple dice or die stacks onto the silicon struc-

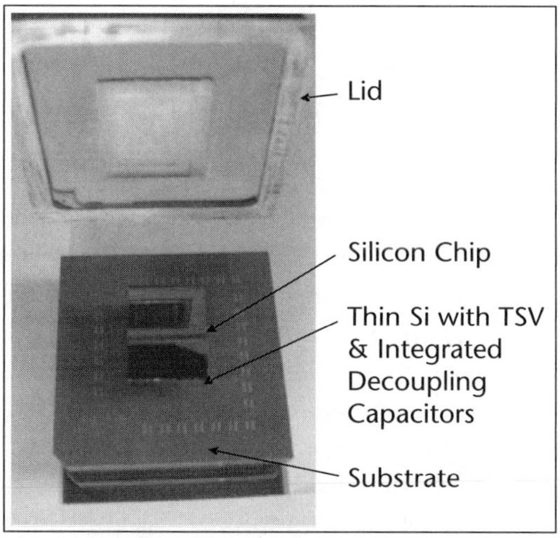

Figure 14.11 Shows an unassembled, expanded view of a package, thinned silicon decoupling capacitor, die, and lid prior to assembly. (© 2008 IBM J Res Dev [1].)

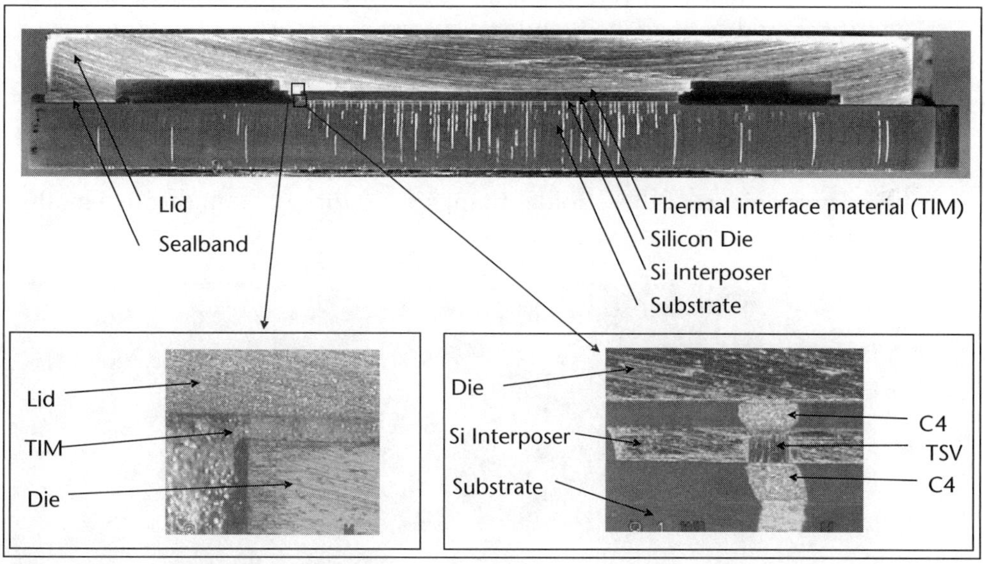

Figure 14.12 Shows a cross section of assembled die stack with integrated Si decoupling capacitor. (© 2008 IBM J Res Dev [1].)

ture may be important for creating a multichip module using a thinned-silicon package structure. In this example, the silicon package can provide voltage segmentation and support heterogeneous dice or die stacks with high bandwidth between them, as well as assembly of different die sizes. In addition, for very thin silicon packages (<100 μm), the added mechanical integrity available from joining the silicon layer to the ceramic or organic substrate along with an underfill or adhesive layer may be important to control the planarity of the thin silicon and meet application-reliability specifications. For a silicon package layer that has a much greater thickness (e.g., >200 μm), the package in some applications may be a stand-alone package and could, for example, be joined to a board using surface-mount technology, such as a ball grid array (BGA) connection.

In a second example, thinned-silicon dice can be stacked using die-on-die or die-on-wafer process flows with one reflow to join all dice in one die stack or multiple die stacks onto a wafer. In this second assembly approach, stacking may use similar *x-y* physical sizes for thinned-silicon dice. The dice may have been pretested at wafer level to have known good die (KGD) for assembly. In this example, capillary underfill materials or injection molding materials, seal bands integrated into the structure during assembly, or no-flow underfill materials can be employed to improve mechanical integrity, thermal conductivity, and reliability. Generally, for thinned-silicon dice, care must be taken in handling the dice to avoid edge damage and breakage. Also for fine-pitch silicon interconnections, a seal band or underfill may be employed to reduce the potential for joint degradation during use due to corrosion or fatigue. Another example of assembly process flow for die-on-die or die-on-wafer technology has been summarized by Sakuma et al. [18] using a die stack mold for low-cost and high-yield precision assembly of die stacks. In this example, depending on silicon thickness for each die in a stacked assembly, appropriate care must be given to handling the silicon layer; control of silicon layer planarity should be managed, and structures, processes, and circuits need to consider electrostatic discharge to be successful.

For fine-pitch solder microbumps used for silicon-silicon package interconnection or SSI for die-to-die or die-to-wafer bonding, Figure 14.13 shows examples for 25 μm–diameter solder bumps, including an individual solder bump, and area-array views of these solder bumps at 50 μm pitch up through a 300 μm wafer.

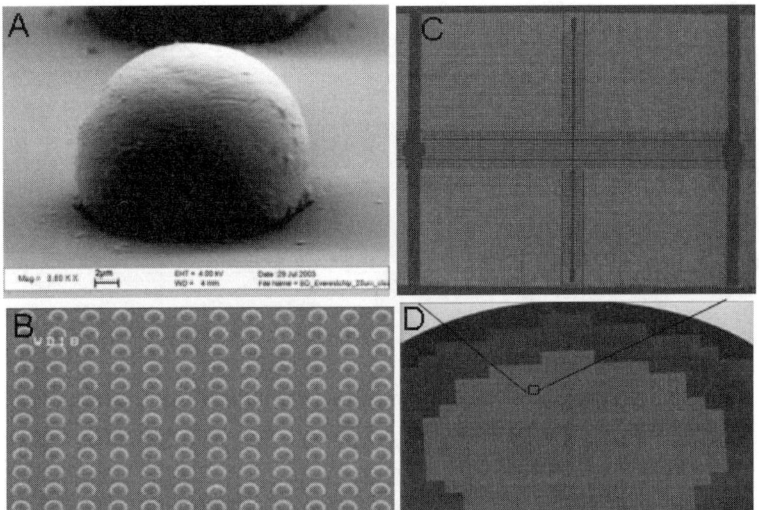

Figure 14.13 (a) A single ~25 μm diameter micro-bump; (b) array of ~192 micro-bumps; (c) shows an array of ~40,000 micro-bumps /cm^2; and (d) a 300mm wafer with micro-bumps at 50 μm pitch formed by C4NP with potential of ~28 million micro-bumps. (© 2008 IBM J. Res. Dev; © 2007 IEEE [1,12].)

Some test-vehicle demonstrations included dice with I/O ranging from 2,160 to over 10,000 interconnections. For test dice with 5,200 microbumps, either 37 or 63 PbSn eutectic solder or lead-free solder of SnCu 99.3 to 0.7 was used for solder reflow as previously reported [25, 26]. Assemblies of test vehicles or dice with approximately 2,160, 5,000, or over 11,000 connections were demonstrated and, in each case, were able to show 100% assembled yields. Solder microbumps have been fabricated using plating processes on top of ball-limiting metal pads or by use of solder injection [17].

Assembly and stacking demonstrations for 200 μm controlled-collapse chip connection (C-4) are shown in Figure 14.14(a, b), where a thinned-silicon interposer with TSV is assembled to a package and silicon die. Figure 14.14(c) and Figure 14.14(d) [18] show examples of stacked die assemblies utilizing "thin solder interconnection such as less than 8 μm height" and integrated TSV structures which can provide high density vertical integration. Test-vehicle demonstrations such as these permit assembly process learning to compare process options, study interconnection metallurgies, define process specifications, and support assembly characterization, including yield statistics, alignment and physical characterization, electrical parametric characterization, thermal characterization, and reliability characterization.

For high-bandwidth silicon-silicon interconnection, fine-pitch interconnection such as solder at 50 μm pitch can be used, or much finer pitch, such as 20, 10, or 4 μm, or even finer pitch assemblies can be employed. For these interconnections, the assembly and test methodology need to be considered as do form factor or approach for fabrication of the wafer. Assembly-technology approaches can be based on chip-to-chip, chip-to-wafer, or wafer-to-wafer assembly, as shown in Figure 14.15. The die-on-die stacking approach offers benefits like the ability to

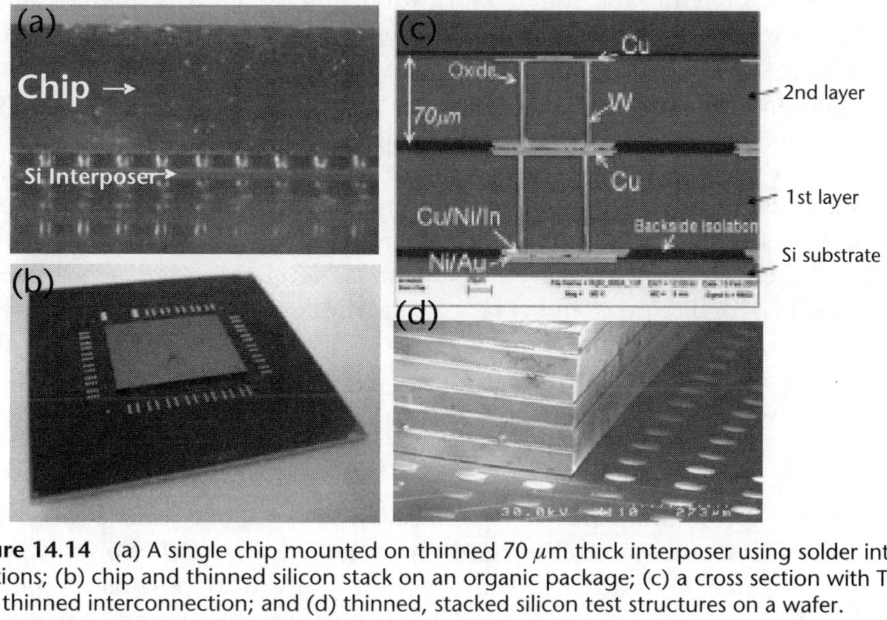

Figure 14.14 (a) A single chip mounted on thinned 70 μm thick interposer using solder interconnections; (b) chip and thinned silicon stack on an organic package; (c) a cross section with TSV and thinned interconnection; and (d) thinned, stacked silicon test structures on a wafer. (© 2008 IBM J. Res. Dev; © 2007 IEEE [1,12].)

	Chip to Chip	Chip to Wafer	Wafer to wafer
Pro	Flexible, use of KGD	Flexible, use of KGD	Low cost
Con	Handling and bonding	Handling and bonding	Overall yield, chip size
Die or Wafer thickness	< 4μm to > 150μm	< 4μm to > 150μm	< 4μm to > 150μm
Bonding technology	Solder Metal to Metal Adhesive	Solder Metal to Metal Adhesive	Solder or Metal Oxide bonding Adhesive

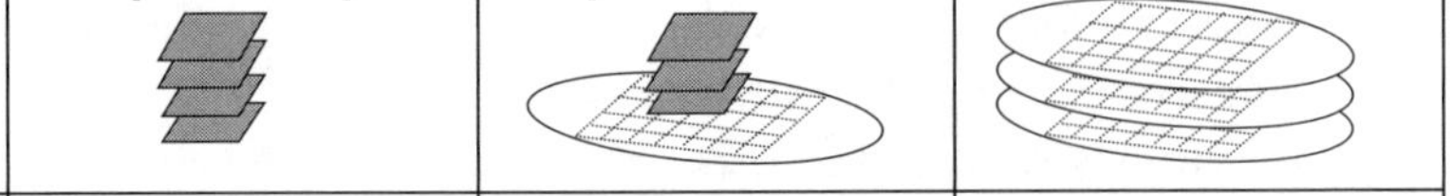

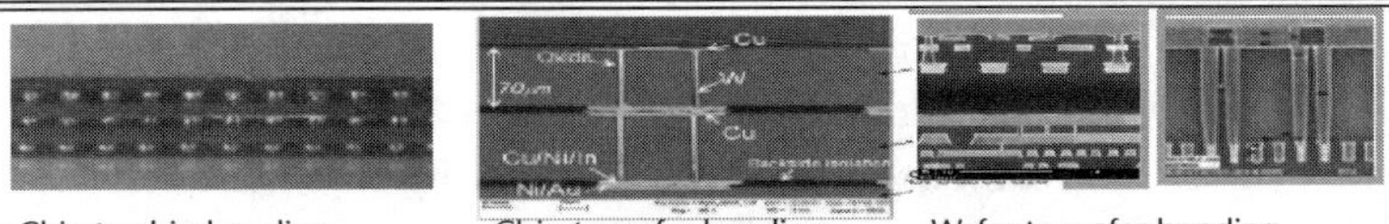

IBM Test Demonstrations: Chip to chip bonding — C4 Solder or micro-bumps; Chip to wafer bonding — Thin Solder /Intermetallic; Wafer to wafer bonding — Cu to Cu versus Oxide Bonding

Figure 14.15 Silicon-silicon interconnection comparison for chip-to-chip, chip-to-wafer and wafer-to-wafers with solder, thin metal, copper–copper and oxide to oxide bonding. (© 2008 IBM J. Res. Dev [1].)

stack known good die, to achieve fine-pitch alignment, to use dice of the same or different sizes, and with a good assembly yield, to achieve good die stack yield. Similarly, die-to-wafer processing can use known good die to fabricate known good die stacks that can be tested after assembly. Die-on-wafer-level processing can provide a mechanical platform for stacking thinned dice and provide a common industry platform for assembly. In either of these cases, the assembly and testing approach needs to be factored into the design, fabrication, and assembly to enable robust manufacturing toward integrated product modules. It is important to consider the compatibility of design across the silicon strata levels, factoring in TSV structure, process, and sequence (such as TSV-first or -last process sequence), silicon-silicon interconnection, die size, test, thermal requirements, and yield.

For wafer-to-wafer assembly processing, the challenges to achieve high yield in stacked structures can be significant. For example, the yield at wafer level needs to be high for each die in order to not lose product during wafer-to-wafer assembly due to defective dice at any given silicon strata level. Alternatively, depending on the yield loss for dice or assembly processing, the design may consider redundancy in the stack structure or spare strata levels to achieve higher yield. These and alternate means may be employed to aide in wafer-stacking yield.

14.6.2 Future Fine-Pitch Interconnection

Silicon-silicon interconnection, as discussed above, may utilize solder interconnections as is common for chip assembly in the industry. Other options for die-to-die, die-to-wafer, and wafer-to-wafer assembly are also possible. Copper-to-copper bonding and oxide-to-oxide bonding are two examples that may provide a path for high interconnection density and low cost if these approaches can provide high-yield assembly. For wafer-scale integration of circuits, including wafer thinning, alignment and bonding technical results for the interconnections between silicon levels have been reported with dimensions as small as approximately 0.14 μm diameter, 1.6 μm height, 0.4 μm pitch, and density of interconnections of 10^8/cm^2 (Figure 14.15) [14, 16]. Application requirements, along with process-integration maturity, will over time be expected to support interconnection densities from traditional packaging levels from less than 10^3/cm^2 to as much as 10^8/cm^2. For 200 or 300 mm wafers, as with wafer processing, wafer-stacking processes will indeed need to be robust to support thousands, millions, or perhaps even billions of interconnections between each silicon strata level.

14.7 Known Good Die and Reliability Testing

Known good die can be obtained from pretesting dice at wafer level or from statistical testing. Alternatively, die stacks can be created with use of redundant interconnections to aid in wafer stacking or die stack yields. To demonstrate a path forward for "known good die" with fine-pitch interconnections, test probes were fabricated at a 50 μm pitch, and corresponding microbumps were successfully contacted as previously reported [1, 11]. Known good die or die stacks can be obtained from wafers or wafer stacks using built in self-test (BIST), with use of pretesting of die or die stacks (such as with use of socket test, wafer-level probe testing, noncontact testing, or temporary chip-attach testing), and through other test assessment options.

Reliability testing for fine-pitch interconnections has also continued to be studied through use of demonstration test vehicles. For example, electrical continuity tests of microbump chains showed the 20 to 25 μm–diameter microbumps to have approximately 5 to 26 mΩ resistance, depending on the test structure used [27]. Reliability studies for 50 μm pitch solder microbumps have results showing electromigration results of over 2,000 hours for 100 mA current at 125°C and 150°C; deep thermal cycle results of over 25,000 cycles of –55°C to +125°C; temperature-humidity bias of over 1,000 hours for 85°C, 85% relative humidity, and 1.5V; and over 2,000 hours of high-temperature storage at 150°C [11, 27]. Results indicate that fine-pitch interconnections can be fabricated and meet typical product-reliability stress requirements. A reliability data summary is shown in Figure 14.16(a).

Figure 14.16(b, c) summarizes results from microbump electrical and mechanical shear testing as a function of pad size [28]. Further studies of multichip and die stack test structures with increased interconnection densities between 10^3 to 10^8 cm^2 for TSV and SSI are at various stages of design, build, and characterization and will permit ongoing experiments and data to be investigated, including design rules, process, bonding, test structures/methodology, and characterization. These ongoing investigations in wafer-to-wafer processing, as well as chip-to-wafer and chip-to-chip interconnection, will continue to provide data that will permit interconnection density, materials, structures, and processes to be optimized for manufacturing consideration against applications. Data collected can provide guidance to meet application-reliability objectives for TSV and SSI and in a variety of integrated-module form factors that permit system miniaturization.

A Si on Si Microbump & C-4 reliability stressing

PbSn solder & SnCu w/ 25 micron diameter @
*** shows accelerated stress conditions**
**** shows data for reduced pad area**

Test condition	Sample / Condition	Results
Electro-migration*	100 mA @ 150C 100 mA @ 125C	> 2000 Hr > 2000 Hr
Deep Thermal Cycle*	-55 to +125C	> 25,000 Cycles
Temp-Humidity-Bias*	85C, 85%RH, 1.5V	> 1000 Hr
High Temperature Storage*	150C	> 2000 Hrs
Contact Resistance	Function of pad size	See 16B
Mechanical Shear	Function of pad size	See 16C

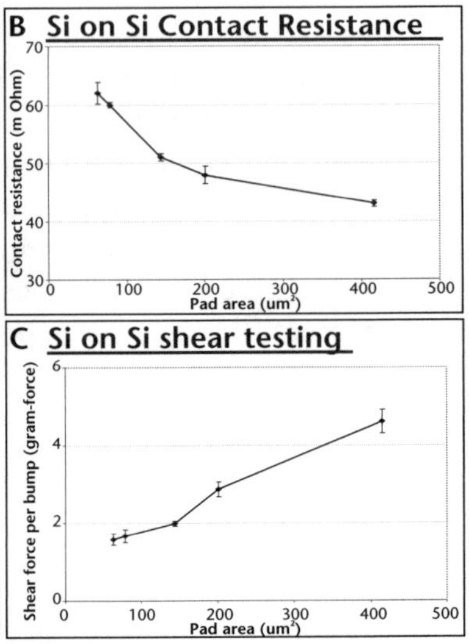

Figure 14.16 (a) 25 μm micro-bump characterization and reliability stress data; (b) electrical contact resistance when jointed to smaller pad sizes; and (c) mechanical shear results when joined to smaller pad sizes. (© 2005 IBM J Res Dev; © 2007 IEEE [12,28].)

14.8 3D Modeling

Many models and simulations exist in 2D, and some also exist in 3D tools. For example, within a chip design, 3D modeling tools exist that permit electrical design, electrical transmission modeling, and simulations. However for 3D structures with multiple levels of silicon strata or layers and module form factors, full 3D designs, models, and simulations are not as easily obtained. Tools exist for mechanical and thermal modeling in 3D structures, but tools for performance simulations, full design, and comparison are not broadly available. In time, design-modeling tools will become available to support 2D versus 3D comparisons, 3D electrical transmission models and simulation tools for die stacks and module structures, 3D performance modeling, power delivery, and distribution. In addition, greater experience with 3D processing, yield understanding, and cost models will help to optimize 3D structures for applications.

Design, architecture, and performance modeling provides a great opportunity to improve system solutions using 3D structures. Examples of architecture considerations and performance benefits for 3D have been reported by Emma et al. [21] and Joseph et al. [23], respectively. Examples of electrical transmission measurements, modeling, and simulation have been reported by Patel et al. [24]. Mechanical models have been reported for TSV, SSI using small solder bumps, and thermomechanical evaluations of 3D structures [11, 29]. Stress and deformation have been evaluated at each stage of the TSV manufacturing process at the temperature for each operation. Elastic properties were characterized by the elastic modulus and Poisson ratio [15]. For some materials, such as copper, the yield strength of the material is likely to be exceeded, and the nonlinear properties need to be included. A stress- strain curve can be incorporated, but simple yield stress is usually sufficient. The range of process temperatures drove inclusion of the coefficient of thermal expansion as discussed above. In addition, shear stresses needed to be evaluated at material interfaces and compared with the adhesion strengths between materials. The highest-stress conditions are generally seen at the via to adjacent wiring and dielectric layers. Understanding mechanical aspects of via structure and process flow can be leveraged to minimize the maximum vertical stress for use in silicon-based technology. From the understanding of structure and stresses, such as for die stack or die-on-silicon package, the electrical and mechanical design specifications for the product application can be satisfied utilizing through-silicon vias.

For SSI modeling, initial development of a model to understand stress and strain levels in a solder μ-C4 began with the use of a macro-micro model [11]. In the finite element model, the macro characteristics of the structure could be considered while still providing microlevel detailed understanding for the high volume of small features needed to understand mechanical characteristics. For example, the model would address the large quantity of microjoints used in the structure, while being able to begin to understand actual stress and strain on an individual solder μ-C4 level. Further, x and y displacements could show the relative pressure loads in the macro model and distributed the stress to the solder interconnections for relative comparison. The macro and micro mechanical modeling of stress in solder μ-C4 could then be evaluated across the various ball-limiting- metallurgy (BLM) and solder interconnections or compared to alternative fine-pitch interconnections, such as with copper-to-copper bonding or when using oxide-oxide bonding for fine-pitch interconnections.

3D power delivery, distribution, and cooling models and demonstration vehicles are under further investigation and should lead to improved understanding and application to products in time. Similarly, 3D knowledge of wafer build, assembly, yield, and cost models is also leading to improved understanding with time.

14.9 Trade-Offs in Application Design and Integration

Leverage of TSV, thinned-silicon, and silicon-silicon interconnections for system integration permits a wide range of products covering varied interconnection densities. For example, simple wireless dice with fewer than 10 TSVs to applications for high-performance computing that may require more than 10^6 TSVs and SSIs between silicon layers may each leverage the emerging 3D technology.

Another consideration regarding 3D integration is the form factor in which the product is designed. Figure 14.17 shows a schematic for two approaches that could be considered as part of 3D system integration. One leverages high-bandwidth silicon interconnection by means of a vertical stack of silicon dice, and the other leverages high-bandwidth silicon interconnection by means of a silicon package combined with die stacks. In the case of Figure 14.17(a) (die stack only), advantages can include shortest wire length between dice and the opportunity to reduce power for signal communications due to reduced capacitance and resistance in the wire lengths and sizes. Wire lengths for die-to-die in a die stack may be tens of micro-meters. However, the design also has challenges, including delivery of power to each level within the stack, the circuit density on any given layer's being lost to TSV for power delivery, and signals. In addition, for a

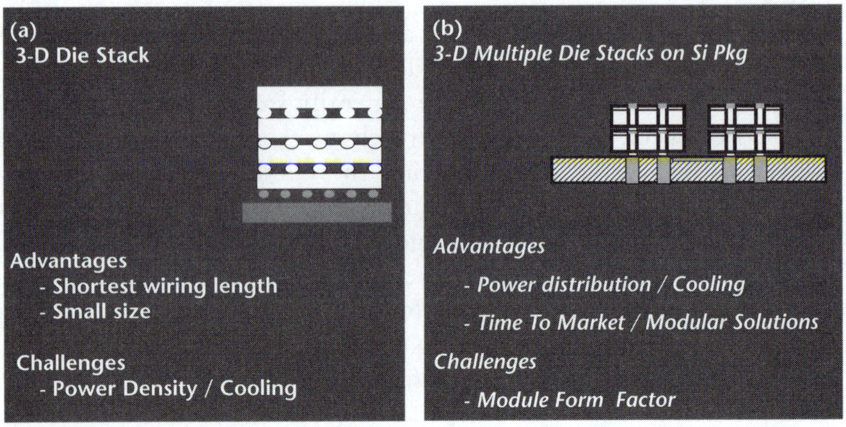

Figure 14.17 (a) Schematic cross section comparisons for high bandwidth vertical interconnection; and (b) for high bandwidth between die or die stacks on a silicon package.

vertical stack, removing heat from the stack can also lead to power density or operational performance limits, depending on the type of dice being stacked. A design with combined die stacks on a silicon package can spread out power delivery for multiple die stacks while maintaining high bandwidth between die stacks and spreading cooling requirements across multiple die stacks. However, increased wire length and latency, which are associated with this approach [Figure 14.17(b)], may limit which applications can consider this technology. Wire lengths for die to die interconnections across a silicon package may range from less than 50 micrometers length to interconnection lengths of several thousand of micro-meters. Figure 14.18 shows examples of thermal modeling and thermal-mechanical modeling similar to the two structure approaches discussed in Figure 14.17. From the modeling results, details of power levels, heat transfer, and stress levels could identify technical challenges and limitations for each form factor. For example, a focus like hotspot power density, impact of die position or location in a stack or structure, heat transport through interconnections, and localized stresses can ultimately provide

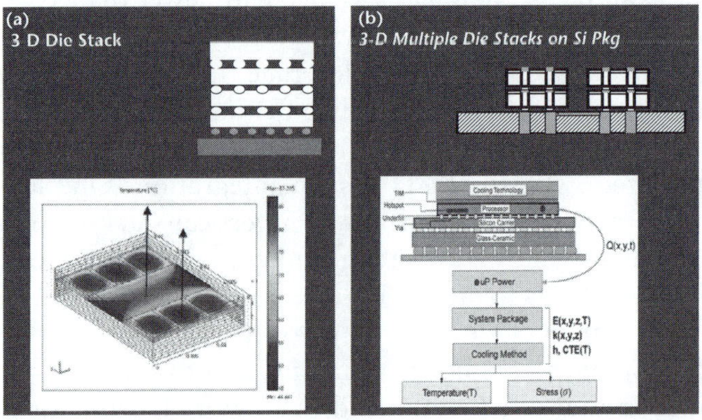

Figure 14.18 (a) Schematic cross section comparisons and thermal modeling for high bandwidth vertical interconnection; and (b) for high bandwidth between die or die stacks on a silicon package [11, 29]. (© 2005 IBM .J Res. Dev; © 2008 IBM J. Res. Dev [12,29].)

the necessary data to select the best design approach to meet the desired application requirements. In 3D structures using die stacking and silicon packaging, stress reduction for silicon-to-silicon interconnections can be realized due to the coefficient of thermal expansion match of materials, which can reduce stresses by 10% to 30% and thereby help lower the modulus of low-k dielectrics and assist in module integration with package materials with higher coefficients of thermal expansion.

Examples of die stack only and dice or die stacks on silicon or high-bandwidth package structures have been reported in technical conferences [30, 31]. Figure 14.19 shows an application example of 3D memory chip stacks that take advantage of TSVs and fine-pitch interconnection to integrate multiple chips for high memory density [30]. Another example of this structure could be processor-to-memory die stacks for high bandwidth and performance [21]. Figure 14.20 shows an example of integrating large scale integration (LSI) dice by means of high-density interconnections between dice [31], and similar high-density interconnection using a silicon package or stacked silicon packages has been reported [8].

Press announcements and technical publications have begun to show that the first TSV and 3D products are entering production in 2008, including a die manufactured for wireless applications [23] and image sensors from Toshiba [32, 33]. Wider industry adoption and acceleration of product applications is likely with 300 mm tools, such as those that have become available for deep silicon reactive ion etch, thin wafer handling, alignment, and bond. To gain the greatest leverage for

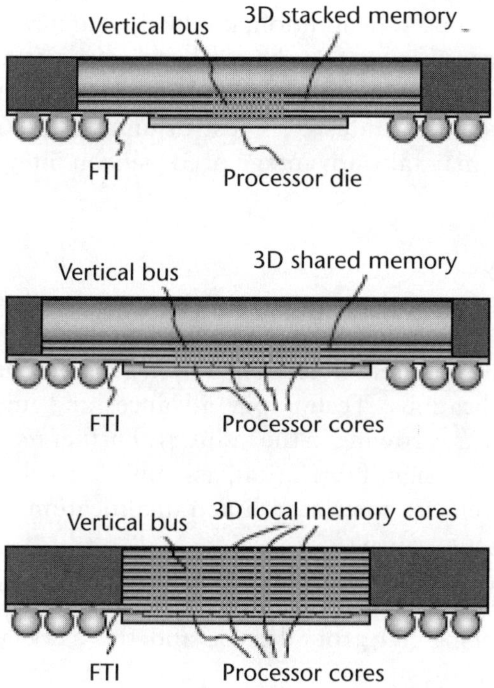

Figure 14.19 Shows an example of 3D stacked memory integrated on a logic device presented by Y. Kurita et al. representing collective efforts of NEC, Oki and Elpida corporations at ECTC 2007 [30]. (© 2007 IEEE.)

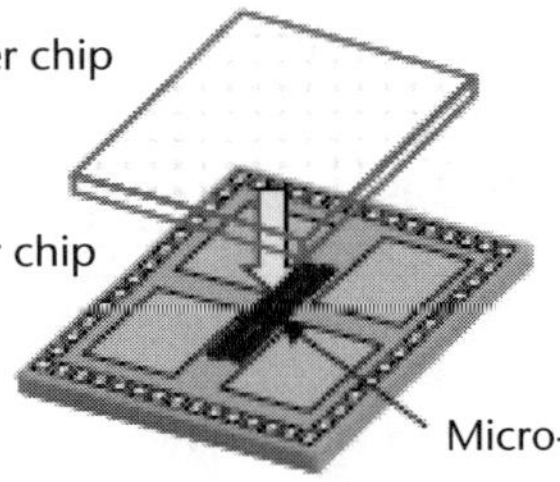

Figure 14.20 Shows an example of chip on chip technology as shown by S. Wakiyama et al. of Sony Corporation at the ECTC 2007 [31]. (© 2007 IEEE.)

more complex products, product architects will need to understand how to leverage the full potential for 3D silicon integration for specific applications. Meanwhile, process engineers will need to develop processes and corresponding design rules permitting high yield and 3D integration approaches that can support the targeted range of product applications at competitive costs.

Applications for 3D can be expected to be far reaching with time. Examples might include portable electronics, such as cell phones, portable medical products, and portable sensors. With reduced power consumption, portable products may benefit from enhanced battery life, not to mention significantly more compact products with scaling functional capabilities. Additional applications could include military, information technology, communications, automotive, and space applications. For computing applications, memory chip stacks for high-bandwidth integration with microprocessors could provide reduced power, system performance scaling, and smaller products. In addition, it is likely that new applications and products will emerge between advances in these microelectronics and nanoelectronics technologies and emerging biotechnology, as well as other emerging nanotechnologies. It seems clear that the industry is just beginning to consider new applications and products that may take advantage of 3D silicon integration.

14.10 Summary

Emerging 3D silicon integration using through-silicon vias (TSVs), thinned silicon, and silicon-silicon interconnection (SSI) has the potential to become used in a broad range of applications. Technology advances and implementations using 200 and 300 mm tools are growing in the industry. Further technology advances include new 3D, finer-pitch design, fabrication, assembly, and characterization of these demonstration test vehicles for research and qualification in collaboration with development and manufacturing.

Future product applications will depend on:

1. Advancing 3D ground rules and the associated tools and processes that support them;
2. New architectures to achieve higher performance, improve power efficiency, lower costs compared to alternative product solutions, and miniaturize product form factors;

3. The ability to create business value.

Acknowledgments

This work has been partially supported by DARPA under the Chip-to-Chip Optical Interconnects (C2OI) Program, agreement MDA972-03-3-0004. This work has also been partially supported by DARPA under the PERCS program, agreement NBCH30390004, and the Maryland Procurement Office (MPO), contracts H98230-04-C-0920 and H98230-07-C-0409. The authors wish to acknowledge support from IBM Research Materials Research Laboratory, Central Services Organization, and collaboration with System and Technology Group. In addition, the authors wish to thank management for support, including T. Chainer and T. C. Chen.

References

[1] Knickerbocker, J. U., et al., "3D Silicon Integration," submitted to *IBM J. Res. Dev.*, Vol. 52, 2008.

[2] Moore, G. E., "Cramming More Components onto Integrated Circuits," *Electronics*, Vol. 38, No. 8, April 19, 1965.

[3] Chen, T. C., "Where Si-CMOS Is Going: Trendy Hype vs. Real Technology," Keynote ISSCC 2006.

[4] Takahaski, K., et al., "Process Integration of 3D Chip Stack with Vertical Interconnection," *Electronic Components and Technology Conference* 2004, pp. 601–609.

[5] Umemoto, M., et al., "High Performance Vertical Interconnection for High-Density 3D Chip Stacking Package," *Electronic Components and Technology Conference* 2004, pp. 616–623.

[6] Anisotropic Conductive Adhesive (ACA): Feil, M., et al., "The Challenge of Ultra Thin Chip Assembly," ECTC 2004.

[7] Hunter, M., et al., "Assembly and Reliability of Flip Chip Solder Joints Using Miniaturized Au/Sn Bumps," ECTC 2004.

[8] Kripesh, V., et al., "Three Dimensional System-in-Package Using Stacked Si Platform Technology," *IEEE Transactions on Advanced Packaging*, Vol. 28, No. 3, August 2005.

[9] Ikeda, H., M. Kawano, and T. Mitsuhashi, "Stacked Memory Chip Technology Development," *SEMI Technology Symposium (STS) 2005 Proceedings*, Session 9 pp. 37–42.

[10] Patel, C. S., et al., "Silicon Carrier with Deep Through-Vias, Fine Pitch Wiring, and Through Cavity for Parallel Optical Transceiver," *55th Electronic Components and Technology Conference*, 2005.

[11] Knickerbocker, J. U., et al., "Development of Next-Generation System-on-Package (SOP) Technology Based on Silicon Carriers with Fine Pitch Chip Interconnection," *IBM J. Res. Dev.*, Vol. 49, No. 4/5, 2005.

[12] Sakuma, K., et al., "3D Chip Stacking Technology with Low-Volume Lead-Free Interconnections," *Electronic Components and Technology Conference*, 2007, pp. 627–632.

[13] Guarini, K. W., et al., "Electrical Integrity of State-of-the-Art 0.13 um SOI CMOS Devices and Circuits Transferred for Three-Dimensional (3D) Integrated Circuit (IC) Fabrication," *IEDM Tech. Digest*, 2002, p. 943.

[14] Topol, A. W., et al., "Three-dimensional Integrated Circuits," *IBM J. Res. Dev.* Vol. 50, No. 4/5, 2006.

[15] Andry, P. S., et al., "Design and Fabrication of Robust Through-Silicon Vias," submitted to *IBM J. Res. Dev.*, Vol. 52, 2008.

[16] Koester, S., et al., "Wafer Level—Three Dimension Integration Technology," Submitted to *IBM J. Res. Dev.*, Vol. 52, 2008.

[17] Dang, B., et al., "3D Chip Stacking with C4 Technology," submitted to *IBM J. Res. Dev.*, Vol. 52, 2008.

[18] Sakuma, K., et al., "3D Chip-Stacking Technology with Through Silicon Vias and Low-Volume Lead-Free Interconnections," submitted to *IBM J. Res. Dev.*, Vol. 52, 2008.

[19] Dennard, R. H., et al., "Design of Ion-Implanted MOSFETs with Very Small Physical Dimensions,", *IEEE J. Solid State Circuits*, 1974.

[20] Agerwala, T., and M. Gupta, "Systems Research Challenges: A Scale-Out Perspective," *IBM J. Res. Dev.*, Vol. 50, No. 2/3, 2006.

[21] Emma, Philip, and Eren Kursun, "Is 3D Silicon the Next Growth Engine after Moore's Law, or Is It Different?" *IBM J. Res. Dev.* 2008.

[22] Andry, P., et al, "A CMOS-Compatible Process for Fabricating Electrical Through-Vias in Silicon," ECTC 2006.

[23] Joseph, A., et al., "Novel Through-Silicon Vias Enable Next Generation Silicon Germanium Power Amplifiers for Wireless Communications," submitted to *IBM J. Res. Dev.*, Vol. 52, 2008.

[24] Patel, C. S., "Silicon Carrier for Computer Systems," *Proc. Design Automation Conference*, July 24–28, 2006.

[25] Knickerbocker, J. U., et al., "3D Silicon Integration and Silicon Packaging Technology Using Silicon Through-Vias," JSSC, 2006.

[26] Knickerbocker, J. U., et al., "System-on-Package (SOP) Technology, Characterization and Applications," ECTC 2006, pp. 414–421.

[27] Wright, S. L., et al., "Characterization of Micro-Bump C4 Interconnections for Si-Carrier SOP Applications," *Electronic Components and Technology Conference*, 2006, pp. 633–640.

[28] Dang, B., et al., "Assembly, Characterization, and Reworkability of Pb-Free Ultra-Fine Pitch C4s for System-on-Package," *Electronic Components and Technology Conference*, 2007, pp. 42–48.

[29] Sri-Jayantha, S. M., et al., "Thermomechanical Modeling of 3D Electronic Packages," submitted to *IBM J. Res. Dev.*, Vol. 52, 2008.

[30] Kurita, Y., et al., "A 3D Stacked Memory Integrated on a Logic Device Using SMAFTI Technology," *Electronic Components and Technology Conference,* 2007, pp. 821–829.

[31] Wakiyama, S., et al., "Novel Low-Temperature CoC Interconnection Technology for Multichip LSI (MCL)," *Electronic Components and Technology Conference*, 2007

[32] Vardaman, J., "3D Through-Silicon Vias Become a Reality," *TechSearch International,* Austin, Texas, June 1, 2007.

[33] Takahashi, K., and M. Sekiguchi, "Through Silicon Via and 3D Wafer/Chip Stacking Technology," *VLSI Circuits, 2006. Digest of Technical Papers.* 2006 Symposium on Vol., No., 2006, pp. 89–92.

CHAPTER 15

Capacitive and Inductive-Coupling I/Os for 3D Chips

Noriyuki Miura and Tadahiro Kuroda

15.1 Introduction

Three-dimensional (3D) system integration is one of the key enabling technologies to realize "More than Moore" [1] system integration. As discussed in previous chapters, 3D integration enables chips to be stacked vertically in a package and thus communicate through vertical I/O interconnections. This is a sharp contrast to "horizontal" (planar) placement of chips, which is the most common system configuration today. Since the communication distance between the stacked chips is very short (less than 5 μm in some stacks), high-speed I/Os can be developed with minimum power and area overhead. In addition, the vertical I/Os can be distributed across the entire chip area to enhance parallelism of the I/Os, while conventional I/Os with bonding wires can be placed only in a chip periphery. Moreover, chip thinning, together with device scaling, will further improve the density and performance of the vertical I/Os. As a result, because of the larger I/O count possible in 3D integration and the short length of the interconnections, it is expected that the vertical I/Os between stacked chips will be able to provide the high data bandwidth required by Moore's law with the benefit of low-power signaling. Recall in Chapter 6 that such interconnections are not available in conventional horizontally distributed chips. These performance advantages of 3D system integration strongly motivate research into how to form the vertical interconnections and I/O circuit technologies for the stacked chips.

As discussed in Chapters 13 and 14, through-silicon via (TSV) technology is a mechanical wired solution whereby the stacked chips are connected by metal via holes through the Si substrate. Of course, TSVs require additional wafer-level fabrication processes that also typically require mechanical polishing, resulting in additional cost at the semiconductor foundry. Moreover, protection circuits for electrostatic discharge (ESD) are needed for the wired I/O because it is physically connected. The ESD protection circuits limit the operation speed and increase the power dissipation and layout area.

Capacitive and inductive-coupling I/Os are emerging *noncontact* (wireless) parallel links for stacked chips. Capacitive coupling utilizes a pair of electrodes that are formed using conventional IC fabrication (each electrode is essentially a metal pad). The inductive-coupling I/O is formed by placing two planar coils (planar inductors) above each other and is also made using conventional IC fabrication. The inductive-coupling I/O is essentially a transformer at the microscale. No additional wafer or mechanical processes are required to fabricate either; hence, they are low cost. In addition, since there is no pad exposed for possible contact, ESD protection circuitry can be removed; hence, they yield low power, high speed, and a small-area I/O cell. Furthermore, chips operating under different supply voltages can be simply interconnected without using a level shifter since the I/Os under consideration are ac coupled. These are the advantages over the TSV technology. However, optimization in both electromagnetics and circuits is required. This chapter introduces capacitive- and inductive-coupling I/Os and describes electromagnetic and circuit codesign for high performance and reliable operation. The modeling and design of channel layout and transceiver circuits are presented and examined by test-chip measurements. Future challenges and opportunities are also discussed.

15.2 Capacitive-Coupling I/O

15.2.1 Configuration

Figure 15.1 illustrates an overview of the capacitive-coupling I/O, which can be fabricated in a standard digital CMOS process without any additional wafer and mechanical processes. The electrodes are formed using IC interconnections in each chip. By stacking the chips in a "face-to-face" configuration, the pair of electrodes is capacitively coupled, providing a wireless channel between the stacked chips. The capacitive-coupling channel is voltage driven. A transmitter applies a voltage on transmitter electrode V_T, according to transmit digital data *Txdata*. The V_T signal generates an electric field between the electrodes E, which induces a voltage in receiver electrode V_R. A receiver detects the V_R changes and recovers digital data *Rxdata*. Due to the face-to-face chip stack, the communication distance X is shorter than 5 μm, providing strong capacitive coupling between the electrodes. It guarantees high signal-to-noise ratio (SNR), even if wideband radio frequency is used. Therefore, pulse-based communication is employed instead of carrier-based communication. Complicated analog circuits, such as a voltage-controlled oscillator, low-noise amplifier, mixer, or filter, can be removed, and only simple digital circuits are used in the transceiver. Figure 15.2 depicts the first capacitive-coupling transceiver that was proposed by S. Kuhn et al. in [2]. Note that the transmitter is just a CMOS inverter buffer. Without any modulations, it directly drives the transmitter electrode by *Txdata*. The receiver electrode follows changes in *Txdata* (V_T), and positive or negative pulse-shaped voltage V_R is generated. A positive pulse is generated when *Txdata* transits from low to high, and a negative pulse is generated when *Txdata* transits from high to low. The receiver consists of a gain stage and a latch. A self-biased inverter in the gain stage amplifies the V_R signal, and it drives the succeeding latch to switch and recover *Rxdata*. No additional control circuits or signals are required for the data recovery.

15.2.2 Channel Modeling

As described above, the capacitive-coupling I/O can be realized in very simple digital circuits. However, a large-swing received voltage V_R is required for a reliable opera-

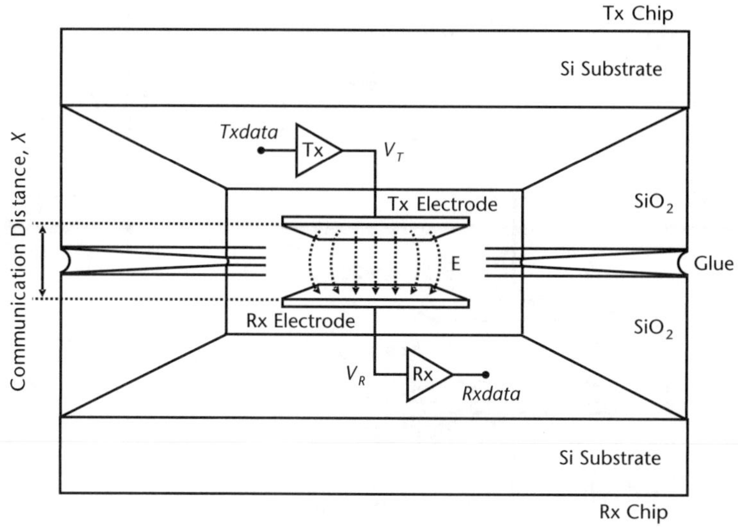

Figure 15.1 Capacitive-coupling I/O.

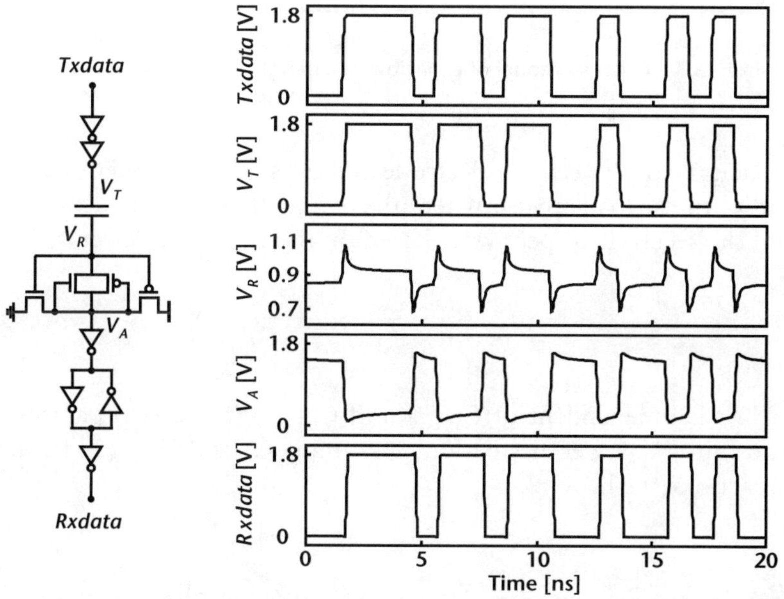

Figure 15.2 First capacitive-coupling transceiver (Kuhn transceiver) and its simulated waveforms.

tion. For example, practically in the transceiver circuit shown in Figure 15.2, the receiver's sensitivity is reduced to improve noise immunity. Also, considering further sensitivity reduction due to transistor mismatch and process variation, the pulse amplitude of V_R should be at least higher than 200 mV or 10% of V_{DD}. In order to secure such large-swing V_R, we need to model carefully the capacitive-coupling channel and design dimensions and distances of the electrodes.

Figure 15.3 depicts an equivalent circuit of the capacitive-coupling channel: C_C is the coupling capacitance between the electrodes, $C_{SUB,\,T}$ and $C_{SUB,\,R}$ are parasitic

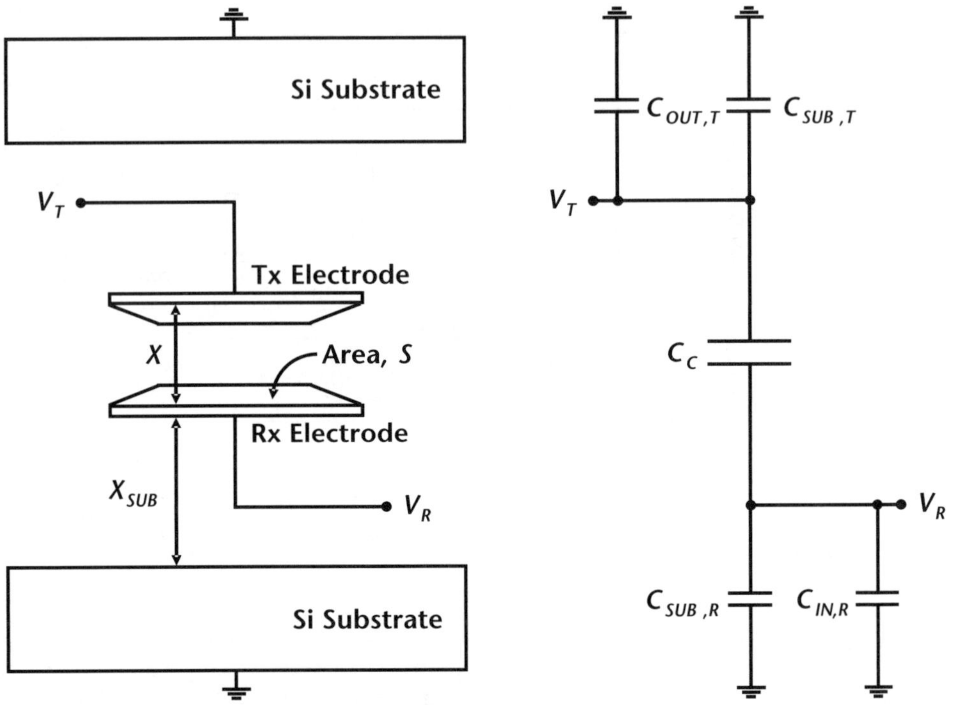

Figure 15.3 Channel model of capacitive coupling.

capacitances between the electrode and the substrate, and $C_{IN,R}$ and $C_{OUT,T}$ are input and output capacitances, where the T and R subscripts denote for the transmitter and the receiver, respectively. Based on the equivalent circuit, V_R is given by

$$V_R = \frac{C_C}{C_C + C_{SUB,R} + C_{IN,R}} V_T \tag{15.1}$$

V_R is independent of $C_{SUB,T}$ and $C_{OUT,T}$ (these two parameters determine the transmitter's power dissipation). Assuming C_C and $C_{SUB,R}$ to be simple parallel-plate capacitors, we have

$$V_R = \frac{\frac{\varepsilon}{X} S}{\left(\frac{\varepsilon}{X} + \frac{\varepsilon_{SUB}}{X_{SUB}}\right) S + C_{IN,R}} V_T \tag{15.2}$$

where S is an electrode size, X and X_{SUB} are distances, and ε and ε_{SUB} are dielectric constants between the electrodes and between the electrode and the substrate, respectively. In order to achieve higher V_R, X should be small, and X_{SUB} should be large. X can be reduced to 1 μm in face-to-face chip stacks [3, 4]. X_{SUB} can be increased to 5~10 μm when the top metal layer is used for the electrodes. The dielectric constants also affect V_R. In a standard CMOS technology, ε_{SUB} is equal to ε_{SiO2} (dielectric constant of SiO_2). Increasing ε provides large coupling capacitance between the electrodes and, hence, V_R. It is effective to fill a gap between two stacked chips with a high-ε adhesive, such as in [3–5]. The electrode size S is a layout parameter. The minimum electrode size is restricted by the input capacitance of the receiver circuit $C_{IN,R}$. Technology (device) scaling causes $C_{IN,R}$ to decrease, thereby allowing the electrode size to be scaled (the scaling scenario of the capacitive-coupling I/O will be discussed later in Section 15.7.1).

15.2.3 Crosstalk

Array-area distribution of the capacitive- and inductive-coupling I/Os increases data bandwidth. However, since these two technologies employ wireless communications, crosstalk between neighboring channels may degrade the performance. This section and Section 15.3.3 will discuss crosstalk in capacitive- and inductive-coupling I/Os, respectively.

In the capacitive-coupling I/O, V_R is induced by the electric field between the transmitter and the receiver electrodes. In a single channel [Figure 15.4(a)], all the electric field lines from the transmitter electrode terminate on the receiver electrode. On the other hand, in a channel array [Figure 15.4(b)], some fringing field lines are terminated onto adjacent receiver electrodes (Rx1, Rx2), which causes the crosstalk (by causing voltage on the unwanted electrode). A guard-ring structure [5] effectively reduces the capacitive-coupling crosstalk by shielding neighboring channels from the fringing field lines [Figure 15.4(c)]. Moreover, the crosstalk between the electrode and circuits can be reduced by a guard ring. In the capacitive-coupling I/O, crosstalk is not a serious issue.

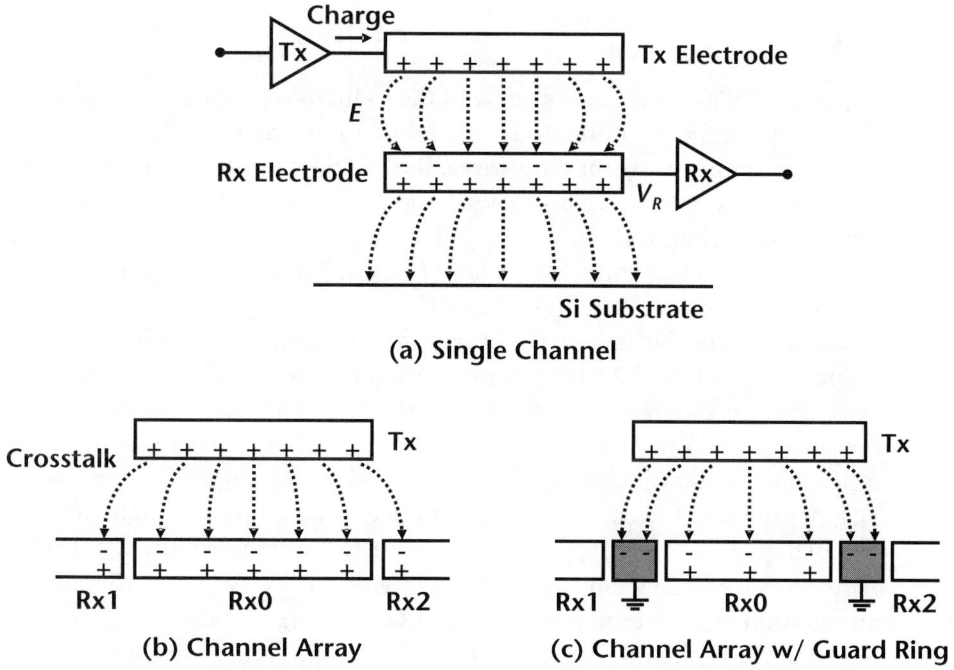

Figure 15.4 Electrical field of capacitive coupling in (a) single channel, (b) channel array, and (c) channel array with guard ring.

15.3 Inductive-Coupling I/O

The inductive-coupling I/O is another wireless vertical interconnection. It was firstly introduced by D. Mizoguchi et al. [6] in order to overcome two limitations of capacitive-coupling I/Os. One is short communication distances: capacitive-coupling I/Os can be used only at distances shorter than 10 μm [7]. Equations (15.1) and (15.2) indicate that V_R is reduced for long-distance communication since C_C decreases with increasing X, and V_T is limited under the supply voltage V_{DD}. Even if the electrode size is enlarged, V_R hardly increases because both C_C and C_{SUB} increase in a similar way. Supposing the electrode size S is large enough in (15.2), we have

$$V_R < \frac{\frac{\varepsilon}{X}}{\frac{\varepsilon}{X} + \frac{\varepsilon_{SUB}}{X_{SUB}}} V_{DD} \tag{15.3}$$

Equation (15.3) implies that, even if the electrode size is enlarged, V_R remains constant, and V_R simply reduces with increasing X. The other limitation of the capacitive-coupling I/O is weak coupling strength through the Si substrate: The capacitive-coupling channel cannot communicate through the Si substrate. Capacitive coupling utilizes a vertical electric field for signal transmission, which would be strongly attenuated in the Si substrate. Due to this drawback, the capacitive-coupling I/O can be applied only to face-to-face chip stacks and cannot be applied either to face-up or to back-to-back chip stacks. The inductive-coupling I/O introduced in this section solves these problems.

15.3.1 Configuration

Figure 15.5 illustrates an overview of the inductive-coupling I/O, which is also fabricated using standard semiconductor fabrication processes. The coils are formed on each chip simply using on-chip wires. By stacking the chips, the pair of coils is inductively coupled, providing a wireless channel between the stacked chips. The inductive-coupling channel behaves just like a transformer. It is driven by transmit current I_T, according to transmit digital data *Txdata*. The I_T signal generates magnetic field H, and it induces voltage in the receiver coil V_R, which is proportional to dI_T/dt. A receiver detects the induced voltage and converts it to digital data *Rxdata*. The inductive-coupling I/O can communicate through the substrate since it utilizes magnetic field for signal transmission, and the magnetic field is minimally attenuated in the substrate. As mentioned above, the electrical field of capacitive coupling on the other hand is significantly reduced in the substrate. Figure 15.6 presents calculated S21 parameters for capacitive and inductive coupling through the substrate. The S21 of capacitive coupling is significantly attenuated when the substrate resistivity is reduced to 1~0.1 Ωcm (typical resistivity of p+ Si). On the other hand, the attenuation in the inductive-coupling I/O is negligible; thus, the inductive-coupling I/Os can be applied to "face-to-face," "face-up," and "back-to-back" chip stacks. Figure 15.5 depicts the inductive-coupling I/O in the face-up chip stack. The communication distance between the coils X is determined by the thicknesses of the

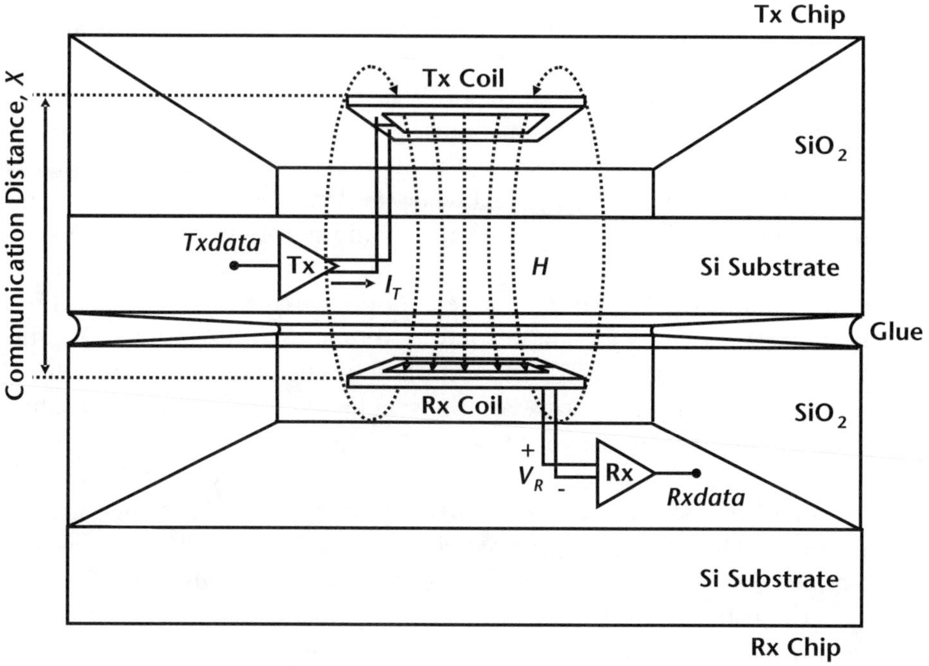

Figure 15.5 Inductive-coupling I/O.

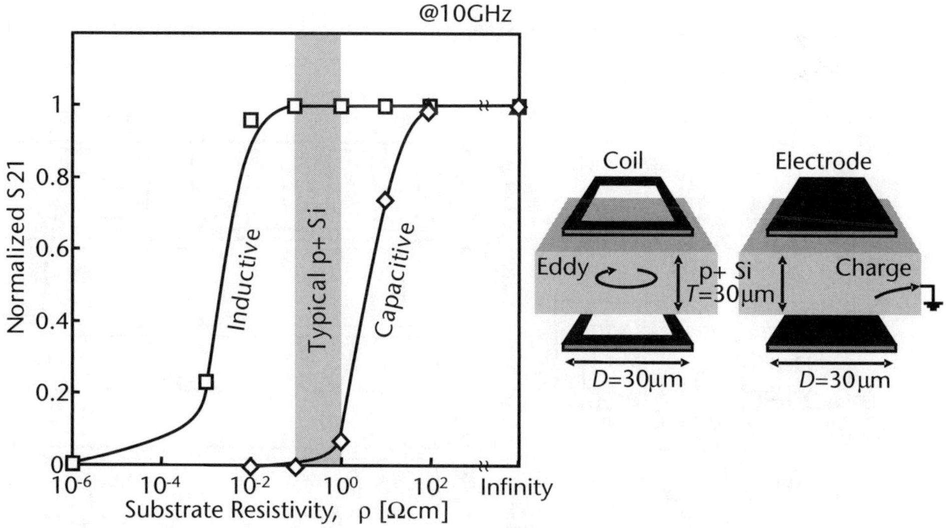

Figure 15.6 Simulated S21 dependence on substrate resistivity.

stacked chip T_{chip} and the adhesive $T_{adhesive}$. X can reach up to several hundred micrometers, which is much greater than that of the capacitive-coupling I/O. However, the inductive-coupling I/O can communicate at longer distances since the reduced voltage can be increased by increasing the coil size (details will be explained later), while the capacitive-coupling I/O cannot extend the communication distance even if the electrode size is increased.

The transceiver circuit for the inductive-coupling I/O is as simple as that for capacitive coupling. Figure 15.7 depicts the inductive-coupling transceiver that was presented by N. Miura et al. in [8, 9]. At the rising edge of the transmitter clock *Txclk*, an H-bridge driver with a pulse generator produces positive or negative pulse current I_T, according to *Txdata*. A positive pulse is generated when *Txdata* is high, and a negative pulse is generated when *Txdata* is low. The I_T signal induces a positive or negative pulse-shaped voltage V_R in the receiver coil. The receiver is a latch comparator that directly samples V_R by the receiver clock *Rxclk* and recovers digital data *Rxdata*. Unlike the capacitive-coupling transceiver in Figure 15.2, a synchronous scheme is employed in this inductive-coupling transceiver. Additional circuits, such as a clock transceiver and a sampling timing controller, are required, increasing hardware complexity. However, the synchronous transceiver consumes less power than the asynchronous transceiver since static power dissipation is eliminated. In addition, the power overhead due to the additional circuits can be eliminated by sharing these circuits among parallel data transceivers, and the total power dissipation is finally reduced in the synchronous scheme [8, 9]. Further details will be explained in Sections 15.4 and 15.6. An asynchronous inductive-coupling transceiver is also possible [10], which is employed in high-speed communications. It will be introduced in Section 15.5.

15.3.2 Channel Modeling

Channel modeling is a critical issue in the inductive-coupling I/O as it is in the capacitive-coupling I/O. Moreover, the layout structure of the coil in the induc-

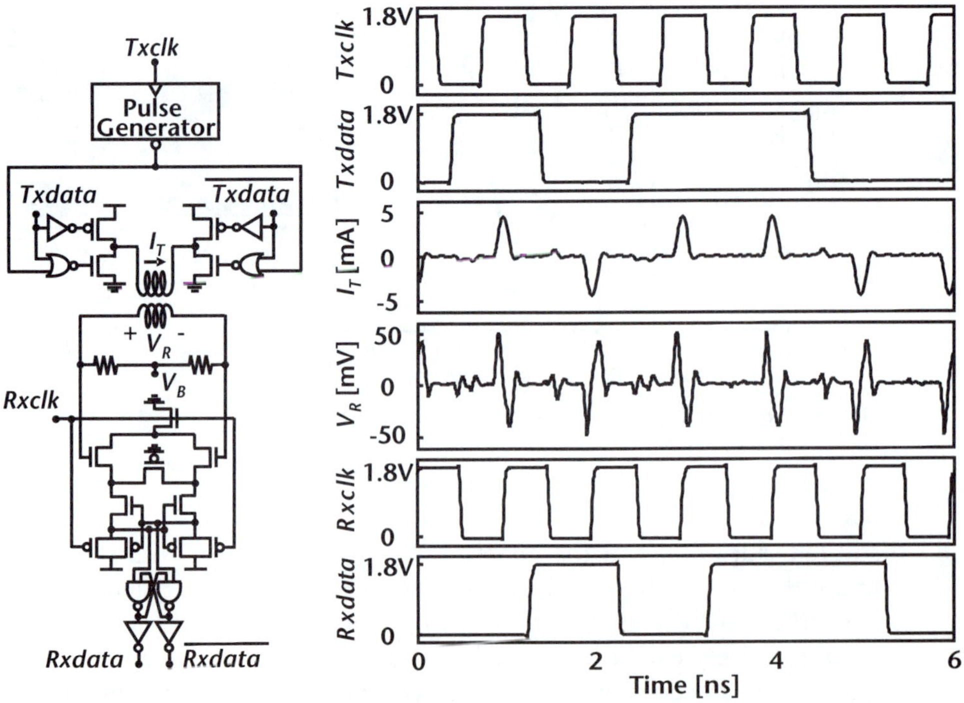

Figure 15.7 Prototype inductive-coupling transceiver and simulated waveforms. (© 2007 IEEE.)

tive-coupling I/O is more complicated than that of the electrode in the capacitive-coupling I/O. Not only the diameter but the number of coil turns has to be optimized. More dedicated channel modeling is required in the inductive-coupling I/O design.

Figure 15.8 depicts an equivalent circuit of the inductive-coupling channel. The transmitter and the receiver coils can be modeled as an *LCR* parallel resonator where *L* is self-inductance, *R* is the parasitic resistance of the coil, and *C* is the parasitic capacitance of the coil and I/O capacitance of the transceiver. Magnetic coupling between the coils is modeled by a mutual inductance *M*. Based on the equivalent circuit, V_R is given by

$$V_R = \frac{1}{\left(1-\omega^2 L_R C_R\right) + j\omega R_R C_R} \cdot j\omega M \cdot \frac{1}{\left(1-\omega^2 L_T C_T\right) + j\omega R_T C_T} \cdot I_T \qquad (15.4)$$

The second term denotes the magnetic coupling. It generates the received voltage as a time derivative form of the transmit current (dI_T/dt). The mutual inductance *M* determines the gain of the magnetic coupling and is expressed as

$$M = k\sqrt{L_T L_R} \qquad (15.5)$$

where *k* is a coupling coefficient defined by the ratio between the amount of transmitted and received magnetic flux. *M* is only determined by *k* because *L* is mostly constant for the same operating frequency of the channel (further detail will be provided later). The coupling coefficient *k* is approximately calculated by the communication distance *X* and the coil diameter *D* as

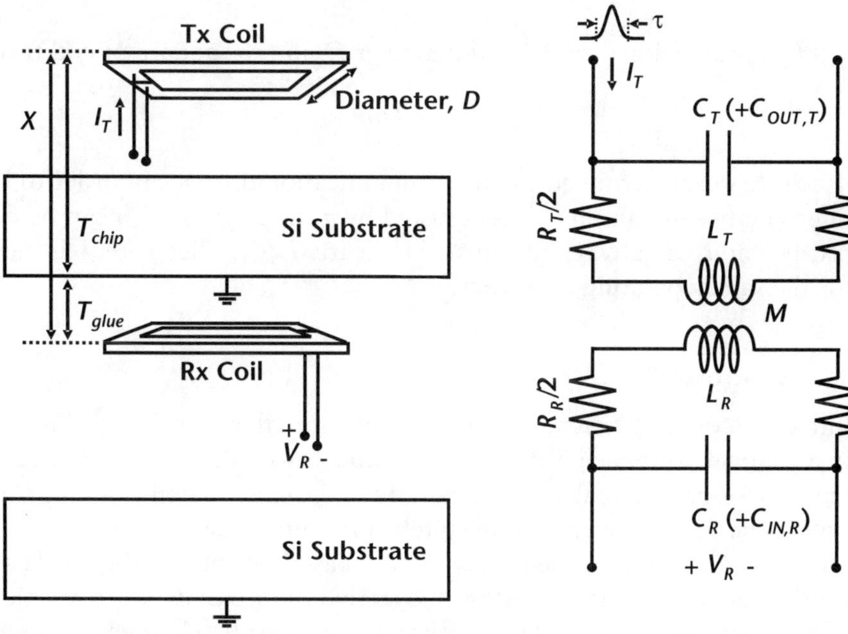

Figure 15.8 Channel model of inductive coupling.

$$k = \left\{ \frac{0.25}{(X/D)^2 + 0.25} \right\}^{1.5} \tag{15.6}$$

When the communication distance is equal to the coil diameter ($X/D = 1$), k is approximately 0.1. The comparator-only receiver in Figure 15.7 correctly operates within the range of $X/D < 1$. For the range of $X/D > 1$, a gain stage is inserted to amplify the small received signal [11]. From (15.6), we can see that the communication distance can be linearly extended by increasing the coil diameter. This is a contrast to the capacitive-coupling I/O. Recall from (15.3) that the capacitive-coupling I/O cannot extend the communication distance even by increasing the electrode size.

The first and the third fractions in (15.4) represent the parasitic effect of the coils, which act as second-order low-pass filters whose cutoff frequency is given by a self-resonant frequency of the coil:

$$f_{SR} = 1/2\pi\sqrt{LC} \tag{15.7}$$

f_{SR} limits the operating frequency of the channel:

$$f_{CH} = 1/\tau \tag{15.8}$$

where τ is a pulse width of I_T. In order to suppress ringing in received pulses and hence intersymbol interference (ISI), f_{SR} should be higher than f_{CH}. The self-inductance L is maximized while keeping $f_{SR} > f_{CH}$. The maximum allowable inductance L_{MAX} can be derived from (15.7) and (15.8):

$$L_{MAX} = \frac{\pi^2}{4Cf_{CH}^2} = \frac{\pi^2\tau^2}{4C} \tag{15.9}$$

L is proportional to the coil diameter D and the square of coil turns n^2:

$$L \propto Dn^2 \tag{15.10}$$

Since D is determined by the communication distance in order to keep the coupling coefficient constant, L is adjusted by n. In most cases, n can be designed arbitrarily, and L can always be adjusted to nearly L_{MAX}. Therefore, L is mostly constant for the same operating frequency.

15.3.3 Crosstalk

Inductive-coupling crosstalk is induced by mutual magnetic coupling between coils. In a channel array, neighboring transmitter coils induce crosstalk in a receiver coil. Figure 15.9 presents calculated crosstalk in the receiver coil from the transmitter coil array. The coil diameter, channel pitch, and communication distance are 30, 40, and 20 μm, respectively. A theoretical model based on the Biot-Savart law [12] is used for the calculation. Figure 15.9(a) shows that the crosstalk is attenuated by the cubic of horizontal distance Y^3. The number of the crosstalk channels only increases Y^2 by increasing the array size. Therefore, aggregated crosstalk in the channel array is rap-

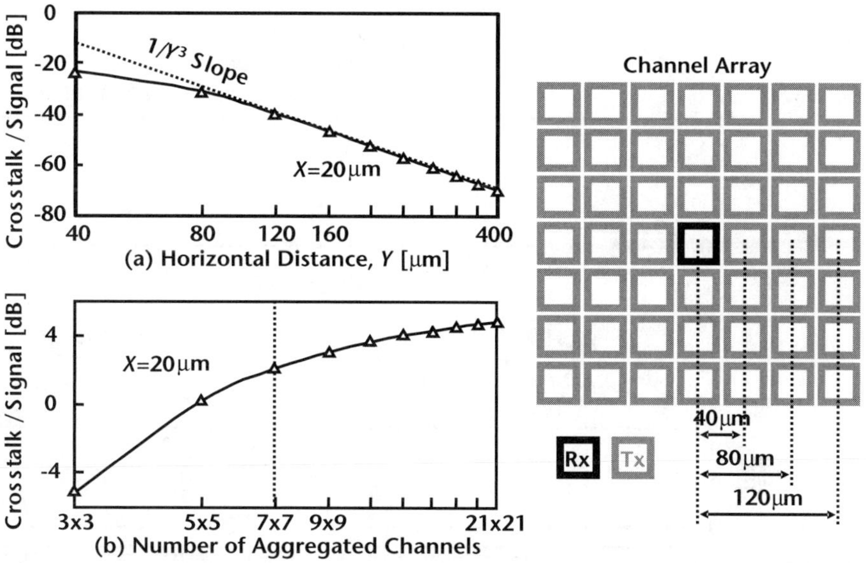

Figure 15.9 Calculated crosstalk between inductive-coupling I/Os dependence on (a) horizontal distance and (b) number of aggregated channels of channel array. (©2007 IEEE.)

idly saturated, as shown in Figure 15.9(b). However, the crosstalk from two surrounding channels must be reduced. Unfortunately, the guard ring cannot reduce the inductive-coupling crosstalk effectively. A crosstalk-reduction technique is required for high-density channel arrangement. A circuit solution based on time-division multiplexing (TDM) is presented in [8, 9, 12]. Circuit details will be introduced in Section 15.6.

Crosstalk between the coil and circuits is another issue. Figure 15.10 shows the calculated mutual inductance between a 30 μm–diameter coil and a 1 mm signal line. The mutual inductance increases to 0.25 nH when the signal line runs below the coil wire. Even in this worst case, the crosstalk voltage induced by the signal line in the receiver coil is negligible (~1 mV) because current in the signal line I_{LINE} is very small (even if it is 1 mm long, the load capacitance is as high as 50 fF, so large current does not flow). On the other hand, the crosstalk voltage that is induced by the transmitter coil in the signal line is relatively large (~10 mV) because the large transmit current I_T flows in the transmitter coil for interchip communications. This crosstalk voltage, although negligible for digital signals, may not be negligible for low-swing analog signals. In this case, the signal line must be placed away from the coil. When the signal line is placed at a distance equal to twice the coil diameter, the mutual inductance between them is reduced to 1/10 of the worst case, and the crosstalk voltage can be suppressed to less than 1 mV. Measurement results reported in [13] show that the crosstalk from the coils is neglected even to SRAM. Crosstalk between the coil and circuits is not a problem.

15.3.4 Advantages and Disadvantages

Basics of the inductive-coupling I/Os are briefly overviewed so far. This section will review and summarize advantages and disadvantages of the inductive-coupling I/O

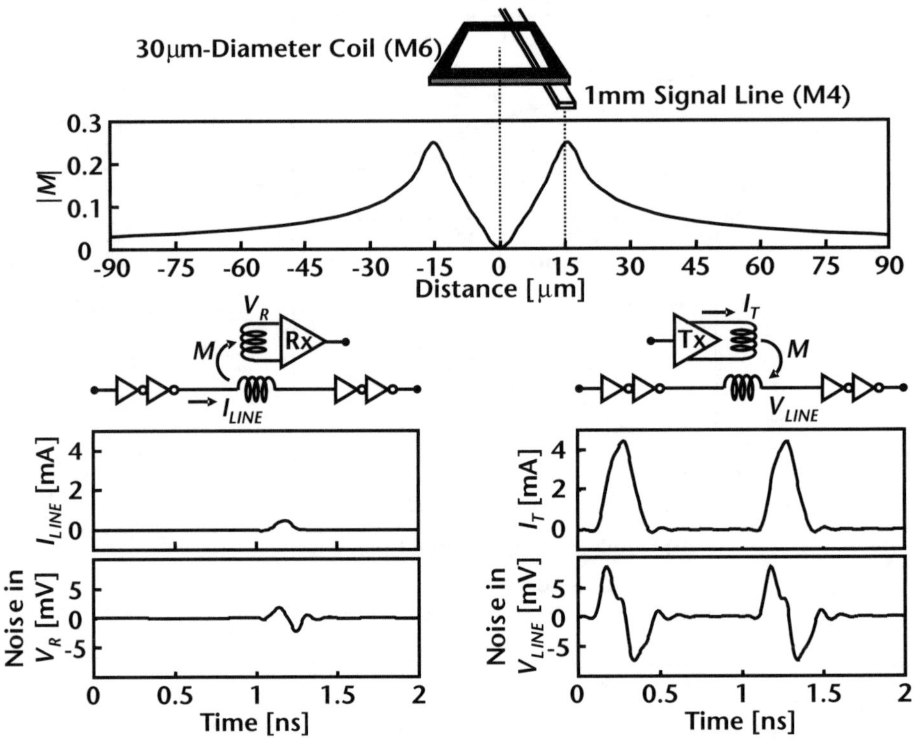

Figure 15.10 Calculated crosstalk between inductive coupling and signal line.

over the capacitive-coupling I/O. As mentioned in the introduction of this section, the inductive-coupling I/O has the following two advantages over the capacitive-coupling I/O:

1. Inductive coupling can communicate at longer distances than capacitive coupling. As discussed in the introduction of this section, the received voltage of capacitive coupling hardly increases even if the electrode size is enlarged. The received voltage is simply reduced when the communication distance is extended. On the other hand, in inductive coupling, even if the communication distance is extended, the received voltage can be kept constant by increasing the coil diameter proportionally.
2. Inductive coupling can communicate through the substrate. As shown in Figure 15.6, the magnetic field of inductive coupling is hardly attenuated in the Si substrate (loss due to the eddy current is negligible), while the electric field of capacitive coupling is shielded by the Si substrate.

Exploiting the above two advantages, the inductive-coupling I/O can be applied to "face-to-face," "face-up," and "back-to-back" chip stacks. It can also be used for communication in more than three stacked chips. On the other hand, the capacitive-coupling I/O can only communicate between two face-to-face stacked chips. In the face-to-face chip stack, a new packaging technology is required for power delivery. Figure 15.11 illustrates the power delivery to the face-to-face, face-up, and back-to-back chip stacks. For the face-up and back-to-back chip stacks, conven-

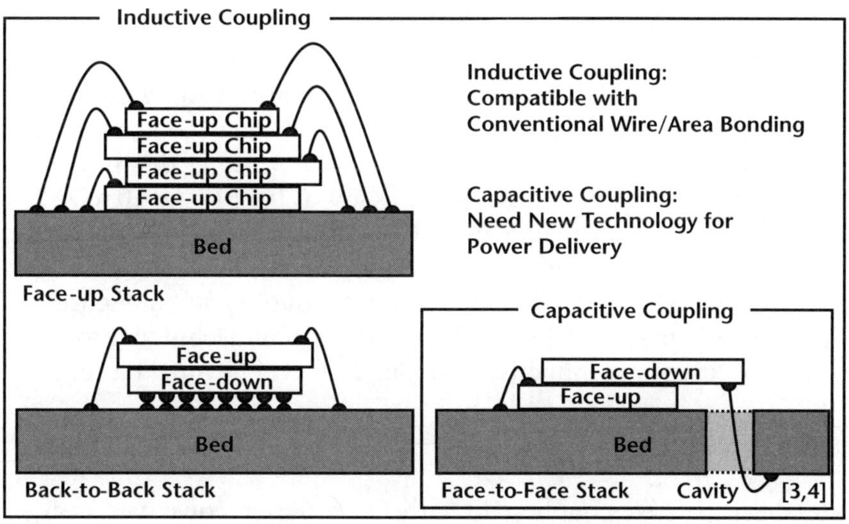

Figure 15.11 Power delivery for capacitive and inductive-coupling I/Os.

tional wire or area bonding can be used. However, the face-to-face chip stacks requires new packaging technologies, such as back-side bonding through a board cavity [3, 4], buried microbump [14], or TSV technology, resulting in higher cost.

The inductive-coupling I/O has the following two disadvantages over the capacitive-coupling I/O:

1. The crosstalk in inductive coupling is stronger than that in capacitive coupling. In capacitive coupling, as explained in Section 15.2.3, crosstalk from the adjacent channels is small and can be reduced simply by the guard ring. On the other hand, in inductive coupling, the crosstalk from the two surrounding channels should be considered and cannot be reduced by the guard ring. Crosstalk-reduction techniques are required to solve this problem.
2. The scalability of the inductive-coupling I/O is relatively worse. Compared with the electrode in capacitive coupling, the coil layout is more complicated because multiple turns of metal wires are required to provide high self-inductance. When the communication distance is reduced, the coil diameter can be reduced to keep the coupling coefficient constant, while the number of turns should be increased to keep the self-inductance constant. Such small coils with a large number of turns cannot be fabricated due to process limitations. The performance of the inductive-coupling I/O may be limited in face-to-face chip stacks.

15.4 Low-Power Design

Applications for 3D chips include both high-performance and low-power systems in battery-powered devices, such as HDTV camcorders, mobile game players, and cellular phones. The capacitive- and inductive-coupling I/Os can be employed between processors and memory. In such battery-powered devices, the interface should provide high memory bandwidth with low power dissipation. For example, in HDTV systems, H.264 video decoding requires memory bandwidth of up to 20 Gbps for 1080 HDTV resolutions [15], while the decoder chip consumes only 100 mW power dissipation [16]. In order to keep the total I/O power dissipation down to 10 mW, the I/O energy dissipation should be as low as 0.5 pJ/b (= 10 mW/20 Gbps). The previously introduced capacitive- and inductive-coupling I/Os exceed this power budget: the capacitive-coupling I/O in Figure 15.2 has an energy dissipation of 4.6 pJ/b in 350 nm CMOS [5], while the inductive-coupling I/O in Figure 15.7 consumes 2.8 pJ/b in 180 nm CMOS [8]. In this section, circuit techniques for energy reduction in the capacitive- and inductive-coupling I/Os will be introduced.

15.4.1 Circuit Design

In the capacitive-coupling I/O, the transmitter consumes charge and discharge energy of CV_{DD}^2. It can be reduced effectively by device scaling. On the other hand, the receiver consumes dc current $I_{DC,\ R}$ because the inverters in the gain stage are self-biased at the logic threshold to provide high gain [Figure 15.12(a)]. Due to the static current consumption, it is difficult to reduce the energy dissipation only by device scaling. A. Fazzi et al. presented a double-feedback topology to cut this dc current. Figure 15.12(b) depicts a low-power double-feedback receiver [17]. Inverter X1 amplifies the received signal V_R and causes M5 or M6 to switch on. It provides positive feedback to the receiver electrode, and V_R is charged to the voltage level for the data recovery. After a certain delay, *Rxdata* is decided, and the second feedback turns off M3 or M4 to cut the dc current in the front-end circuit.

In the inductive-coupling I/O, the power dissipation in the transmitter is more dominant. The inductive-coupling I/O in 180 nm CMOS [8] consumes an energy of 2.2 pJ/b in the transmitter and 0.6 pJ/b in the receiver. In addition, the energy dissi-

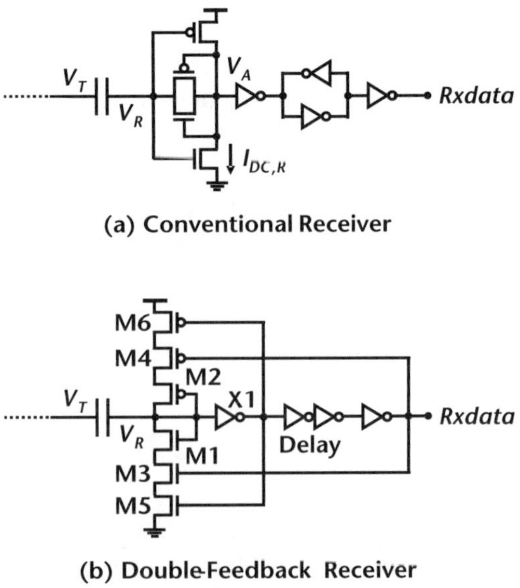

Figure 15.12 Schematic diagram of (a) conventional capacitive-coupling receiver, and (b) low-power double-feedback receiver.

pation in the receiver (latch comparator in Figure 15.7) is only charge and discharge energy CV_{DD}^2, which is effectively reduced by device scaling. The transmitter mostly consumes energy in the H-bridge driver to produce the transmit pulse current I_T. In order to reduce the energy dissipation, the pulse shape of I_T should be optimized under variations in the process, voltage, and temperature (PVT). The pulse shape of I_T is modeled as a triangular pulse with pulse width τ and pulse amplitude I_P (Figure 15.13). The energy dissipation E_{TX} is proportional to the total electric charge carried from V_{DD} to ground. $I_P\tau/2$ denotes the area of the pulse and also the total electric charge. Thus, E_{TX} is approximately given by

$$E_{TX} = V_{DD} I_P \tau / 2 \tag{15.11}$$

The received voltage V_R is induced through inductive coupling as a derivative of I_T (Figure 15.13). The pulse amplitude V_P is approximately given by $2MI_P/\tau$. When coil size and communication distance are given (i.e., M is given), pulse slew rate $S_P = 2I_P/\tau$ determines V_P and hence the bit-error rate (BER). By using S_P, (15.11) is expressed as

$$E_{TX} = V_{DD} S_P \tau^2 / 4 \tag{15.12}$$

Equation (15.12) indicates that by reducing τ while keeping the slew rate S_P, E_{TX} is reduced by τ^2 with constant V_P and BER. However, when reducing τ, the received pulse becomes narrower, and the receiver's timing margin is reduced. In order to maintain BER even with the narrower pulse, a robust timing design is necessary against PVT variations. N. Miura et al. presented a digitally controlled pulse-shaping circuit and a timing-control circuit in [18, 19] (Figure 15.14). The pulse-shaping circuit consists of pulse-width, pulse-slew-rate, and pulse-amplitude controls. In the pulse-width control, a four-phase clock generator provides 0°, 45°,

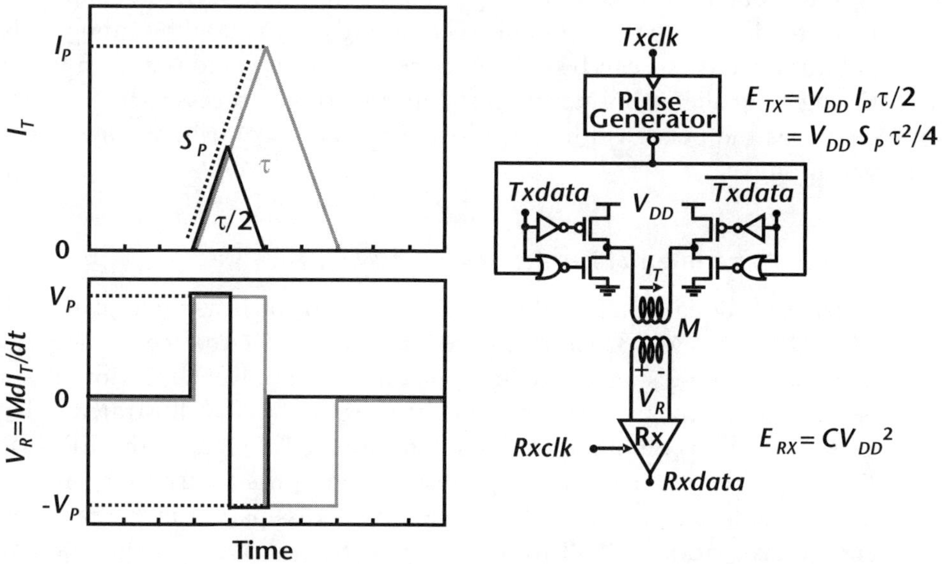

Figure 15.13 Transmit current and received voltage in inductive-coupling I/O. (© 2007 IEEE.)

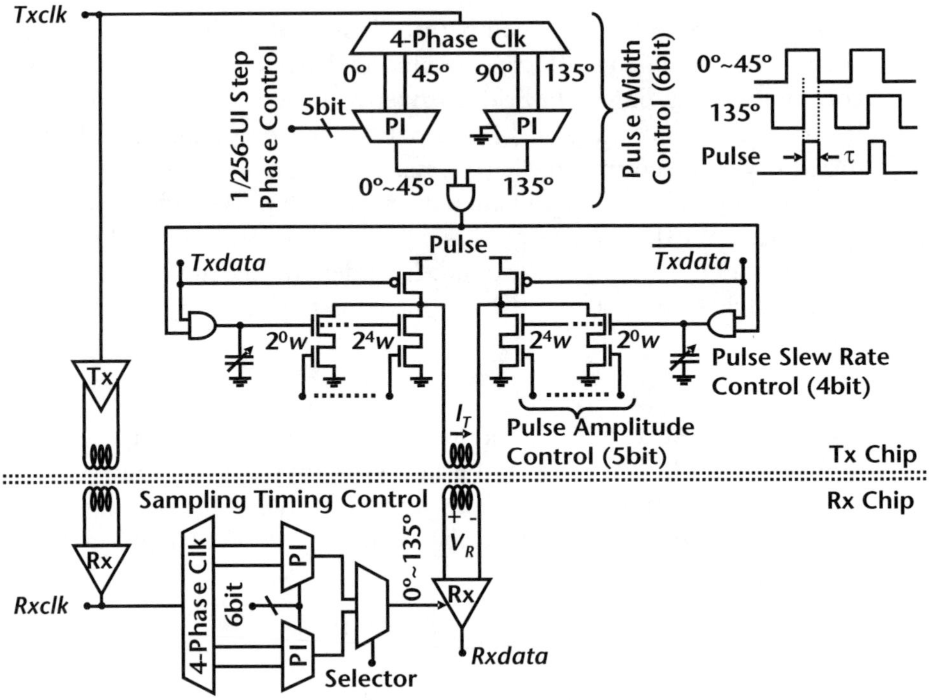

Figure 15.14 Pulse-shaping circuit. (© 2007 IEEE.)

90°, and 135° clocks to two phase interpolators (PIs). One of the PIs interpolates a clock phase between 0° and 45° by 4 ps step. Another PI is a dummy circuit that always outputs a 135° clock. A succeeding AND gate generates a pulse clock that determines the pulse width τ. The pulse slew rate is digitally controlled by variable capacitors. The pulse amplitude is digitally controlled by changing the channel width of the NMOS in the H-bridge driver. In the timing design, a source-synchronous transmission is employed where the inductive-coupling clock link is located adjacent to the data link. The timing jitter caused by supply noise and temperature variations can be effectively rejected as common-mode noise. A sampling timing controller calibrates timing shift due to the process variations. It is the same circuit as the pulse-width controller, which also contributes to reduce the timing variation.

15.4.2 Experimental Results

Fazzi et al. designed and fabricated a test chip for low-power capacitive-coupling I/O in 130 nm CMOS. The chips are stacked in a face-to-face configuration, and the communication distance is reduced to 1 μm. The power supply of the stacked chips is delivered by the back-side bonding through the cavity illustrated in Figure 15.11. The transceiver communicates at a data rate of 1.7 Gbps with BER less than 10^{-12}. Device scaling reduces the energy dissipation in the transmitter, and the proposed double-feedback receiver reduces the energy dissipation in the receiver. The total energy dissipation is 0.08 pJ/b. Further details about the experimental setup and measurement results are reported in [17].

N. Miura et al. designed and fabricated test chips of the low-power inductive-coupling I/O in 180 nm CMOS. Figure 15.15 is the microphotograph of the stacked test chips. The transmitter chip is placed on top of the receiver chip (face-up configuration). The transmitter chip is thinned down to 10 μm thickness. The communication distance, including the 5 μm–thick adhesive, is 15 μm. The coil diameters are 30 and 200 μm for the data and the clock link, respectively. The operating frequency is 1 GHz. The communication distance, channel size, and operating frequency are identical with the measurement setup for the 2.8 pJ/b inductive-coupling I/O [8]. Figure 15.16 presents the measurement results for the data transceiver with the pulse-shaping circuit. Figure 15.16(a) is measured bathtub curves with the pulse-amplitude control. The pulse width is set to 120 ps. It shows that the minimum pulse amplitude required for BER < 10^{-12} is 60 mV because the timing margin and BER are significantly degraded when the pulse amplitude is lower than 60 mV. Figure 15.16(b) is the measured bathtub curves with the pulse-width control. The pulse amplitude is set to a minimum amplitude of 60 mV. It is confirmed that E_{TX} is reduced by τ^2. When τ is set to the minimum pulse width of 60 ps, E_{TX} is reduced to 0.13 pJ/b, which is 17 times lower than the previous design, and the timing margin for BER < 10^{-12} is 25 ps. Supply noise immunity has been measured to evaluate the robustness of the timing design. An individual load is connected to the local supply of each transmitter and receiver chip (Figure 15.17). The load is randomly changed at various frequencies to generate supply noise intentionally. The data transceiver communicates at 1 Gbps with the minimum E_{TX} of 0.13 pJ/b. Figure 15.17 presents measured BER dependence on supply noise voltage. Due to the clock link located adjacent to the data link, timing jitter caused by the supply noise is suppressed within the timing margin of 25 ps. As a result, the data transceiver exhibits sufficiently high immunity against supply noise of 350 mV. In addition, a test chip is fabricated in 90 nm CMOS and measured in the same experimental setup. By the

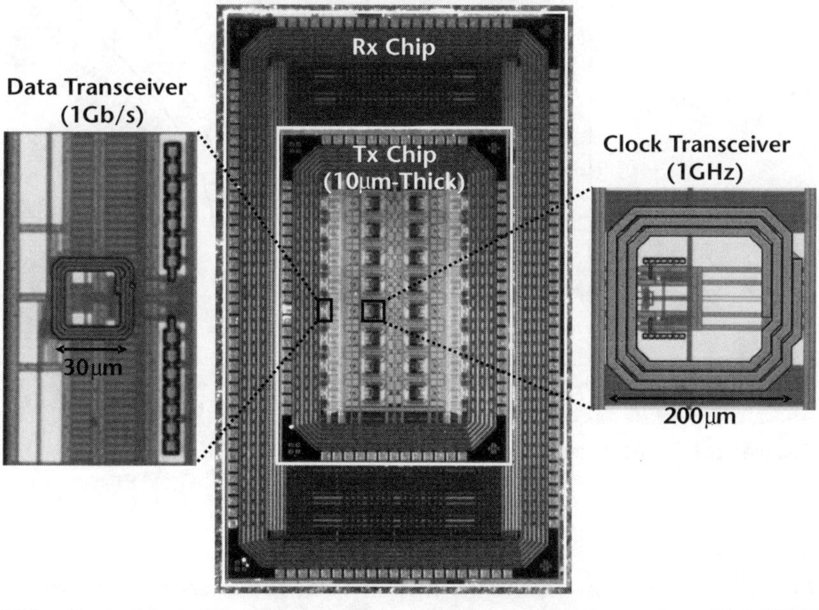

Figure 15.15 Stacked test chips of low-power inductive-coupling transceiver. (© 2007 IEEE.)

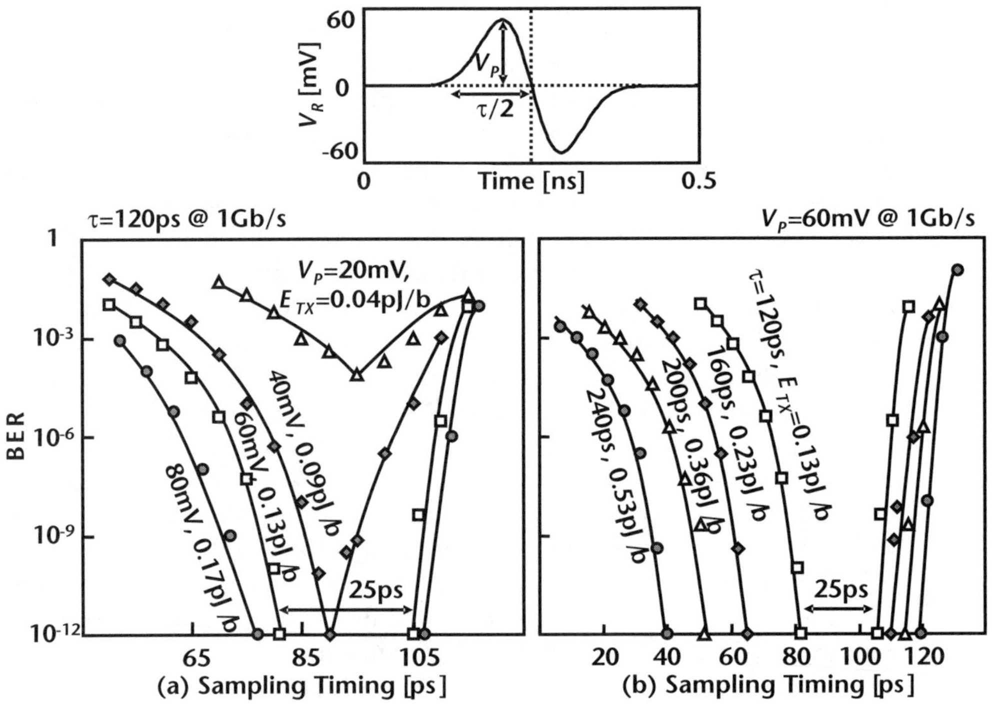

Figure 15.16 Measured timing bathtub curves dependence on (a) pulse amplitude and (b) pulse width. (© 2007 IEEE.)

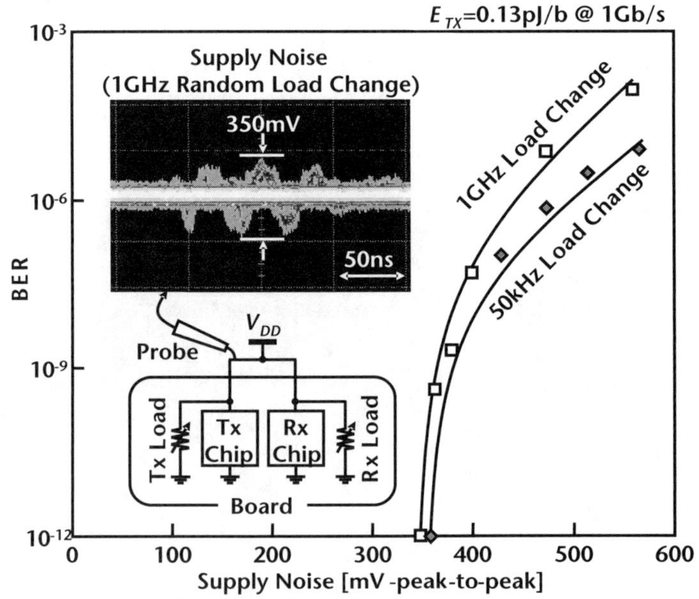

Figure 15.17 Measured supply noise immunity. (© 2007 IEEE.)

device scaling from 180 to 90 nm CMOS, the energy dissipation in the receiver is reduced to 0.03 pJ/b without degrading the data rate, BER, or timing margin. The energy dissipation in the transmitter is also reduced to 0.11 pJ/b. The total energy dissipation is reduced to 0.14 pJ/b, which is 1/20 of the result in [8].

15.5 High-Speed Design

Compared with wired I/Os, the capacitive- and inductive-coupling I/Os have an advantage in high-speed design because they can eliminate the highly capacitive ESD protection circuits, thereby improving the bandwidth of the channel. A 30 μm–diameter microbump is equivalent to 50 fF load capacitance, including the ESD protection circuits [20], while the load capacitance of the capacitive- and inductive-coupling channel can be reduced to less than 10 fF due to the absence of the ESD protection circuits [17]. Furthermore, as the capacitive- and inductive-coupling channels are formed by IC interconnections, the bandwidth can be further improved by device scaling. The capacitive- and inductive-coupling channels do not limit the data rate of the I/Os. By optimizing the transceiver circuit topology, the data rate can be maximized up to the performance limitations of the transistors. Q. Gu et al. presented a prototype transceiver for an 11 Gbps high-speed capacitive-coupling I/O [21]. N. Miura et al. modified the transceiver and developed an 11 Gbps high-speed inductive-coupling I/O for the same layout area and BER [10]. This section will introduce the high-speed inductive-coupling I/O and burst transmission utilizing the I/Os for reducing the number of data links and layout area.

15.5.1 Circuit Design

Figure 15.18 depicts the proposed high-speed, low-latency, asynchronous inductive-coupling transceiver. An H-bridge driver in the transmitter generates I_T from *Txdata* and drives the transmitter coil. Positive or negative small pulse-shaped voltage V_R is induced in the receiver coil. A hysteresis comparator detects the small pulse V_R and converts it to digital data *Rxdata*. An asynchronous scheme is employed for the data link. No clock is needed for the data recovery. Since complicated timing control in the synchronous scheme by using multiphase clocks and a high-precision phase interpolator [18, 19] is not needed, operation speed is improved. Instead, coil size should be increased to improve SNR in order to compensate for the weak noise immunity of the asynchronous receiver. This area overhead can be eliminated by burst transmission, which will be introduced later. The modulation scheme is also modified such that *Txdata* drives the H-bridge directly to generate I_T. A pulse generator in the conventional transmitter is removed. The number of circuit stages in the transmitter is reduced, resulting in small link latency. Instead, the transmitter consumes dc current, but it is negligibly small in high-speed operation. The simulated latency in 180 nm CMOS is 36 ps, which is equivalent to 0.5 FO4 delay. This short latency enables high-speed burst transmission, which will be discussed later. The maximum data rate of the inductive-coupling transceiver is determined by the transition frequency of the transistors f_T. A self-resonant frequency of the coil can be designed to be higher than 100 GHz, even in 180 nm CMOS, and it does not limit the data rate. Circuit simulation shows that the induc-

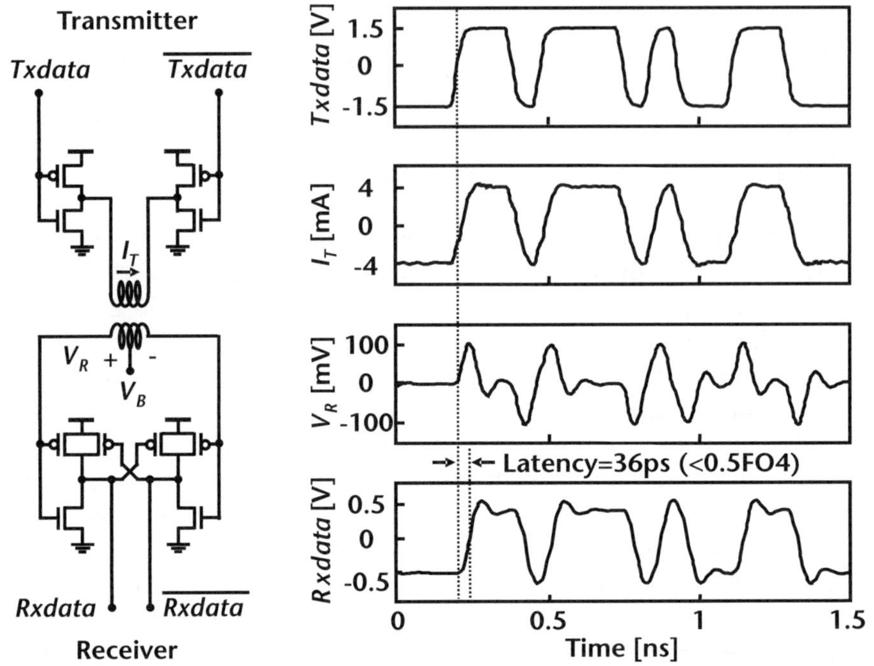

Figure 15.18 High-speed low-latency asynchronous inductive-coupling link and its simulated waveforms. (© 2008 IEEE.)

tive-coupling transceiver operates at a data rate of 11 Gbps in 180 nm CMOS (f_T = 60 GHz).

Burst transmission utilizing the high-speed inductive-coupling transceiver reduces the number of data links and layout area. Figure 15.19 depicts a block diagram of a transceiver that supports burst transmission. Multibit data *Ltxdata* are multiplexed into burst data *Txdata* and transmitted by the high-speed inductive-coupling transceiver. The high-frequency clock *Txclk* provides timing to the multiplexer (MUX) and is generated by a local ring oscillator (OSC) and a counter. The counter generates the same number of clock waves as the number of data bits after the rising of a system clock. *Txclk* is transmitted by another inductive-coupling transceiver along with the data transceiver and used for demultiplexing the received burst data *Rxdata*. Large jitter in the ring OSC can be cancelled out by the source-synchronous transmission. In addition, since both clock and data are transmitted by the same inductive-coupling transceivers whose latency is as small as 36 ps, variation in sampling timing, t_{sample}, caused by PVT changes can be largely suppressed. A delay is inserted in the clock path to the demultiplexer in order to latch the data in the middle of the data cycle. No other timing control is needed.

Miura et al. designed the burst transmission for 400 MHz system clock with the goal of meeting the needs of a processor for mobile phones such as in [22]. MUX/DEMUX are designed for 6.4 Gbps operation, since upper limit of the operating frequency of the MUX/DEMUX is around 10 Gbps in 180 nm CMOS [23]. 16-bit at 400 Mbps *Ltxdata* are multiplexed by 3.2 GHz *Txclk* into 6.4 Gbps burst *Txdata*. Both *Txdata* and *Txclk* are transmitted by the inductive-coupling transceivers and *Rxdata* is demultiplexed by *Rxclk*. The sampling timing, t_{sample}, is set in the

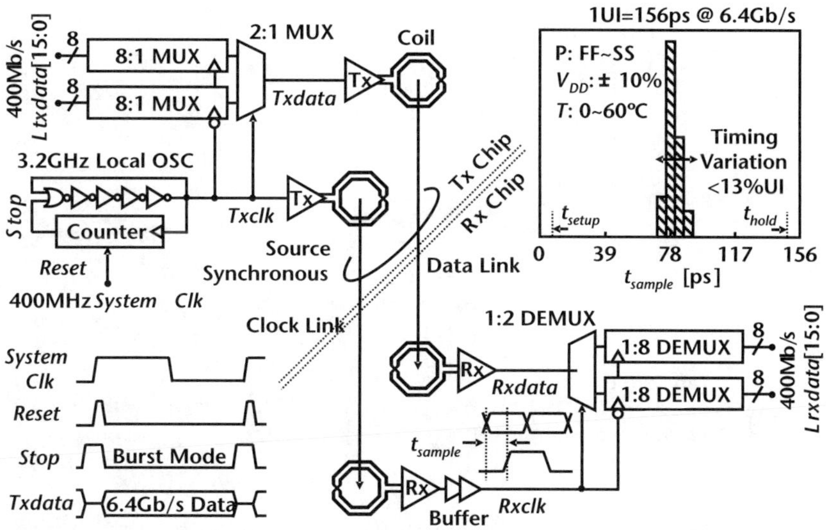

Figure 15.19 Burst transmission for reducing number of data link. (© 2008 IEEE.)

middle of the cycle. Simulated PVT variation in t_{sample} is less than 20 ps (<13% UI). Wide design margin is obtained by the source-synchronous transmission with the low-latency inductive-coupling transceiver.

15.5.2 Experimental Results

Miura et al. designed and fabricated transmitter and receiver chips using 180 nm CMOS. The transmitter chips were thinned down to 40, 25, and 10 μm and stacked over the receiver chip, all in face-up configurations, by using a 5 μm–thick adhesive (Figure 15.20). As a result, the communication distance tested is 45, 30, and 15 μm. The diameter of the coils in the fabricated chips is 120 μm. The total layout area, including MUX/DEMUX and OSC, is 0.1 mm^2. Circuits for evaluation are also embedded in the test chips. Measured BER dependence on the data rate is presented in Figure 15.21. For the communication distance of 15 μm, the maximum data rate is 11 Gbps with BER < 10^{-14}. For a distance of 45 μm, the maximum data rate is 8.5 Gbps with BER < 10^{-14}.

Measured tolerance against supply voltage change in the burst transmission is shown in Figure 15.22. In 6.4 Gbps burst transmission, BER < 10^{-14} is achieved for ±10% variations of the supply voltage. It is confirmed that the source-synchronous transmission by the low-latency inductive-coupling transceiver provides strong immunity against supply voltage change.

Figure 15.23 summarizes measured performance in 180 and 90 nm CMOS. In 180 nm CMOS, the maximum data rate of a data link is 11 Gbps, and it is 6.4 Gbps in burst transmission. Conventional synchronous parallel links would have required 16 data links [8, 9] for an aggregated data rate of 6.4 Gbps. Even though the coil diameter can be reduced to 90 μm in the synchronous scheme, a total layout area of 0.3 mm^2 is needed. The burst transmission requires only two links (data and clock). Layout area, including MUX/DEMUX and OSC, is reduced to 0.1 mm^2, which is one-third of that in the synchronous scheme. In 90 nm CMOS (f_T = 150

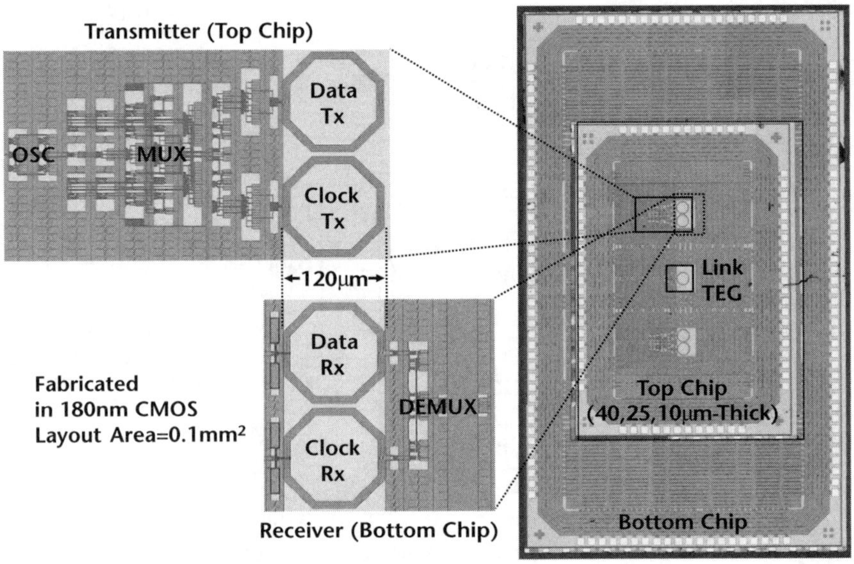

Figure 15.20 Stacked test chips of burst transceiver with high-speed inductive-coupling link. (© 2008 IEEE.)

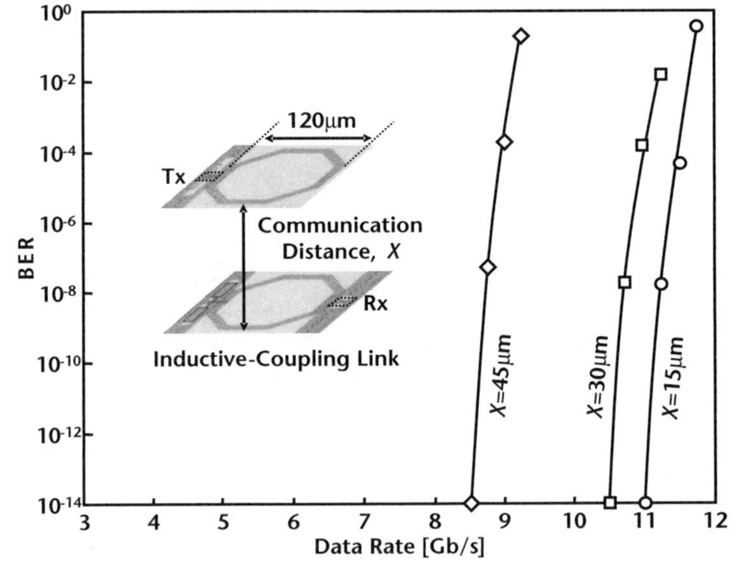

Figure 15.21 Measured BER dependence on data rate. (© 2008 IEEE.)

GHz), the maximum data rate is increased to 15.2 Gbps in burst transmission. It can multiplex 38 data links so that the total area can be reduced by 1/9.

15.6 High-Density Design

The burst transmission reduces the number of data links and the layout area. However, since the energy efficiency is very low due to large static current consumption

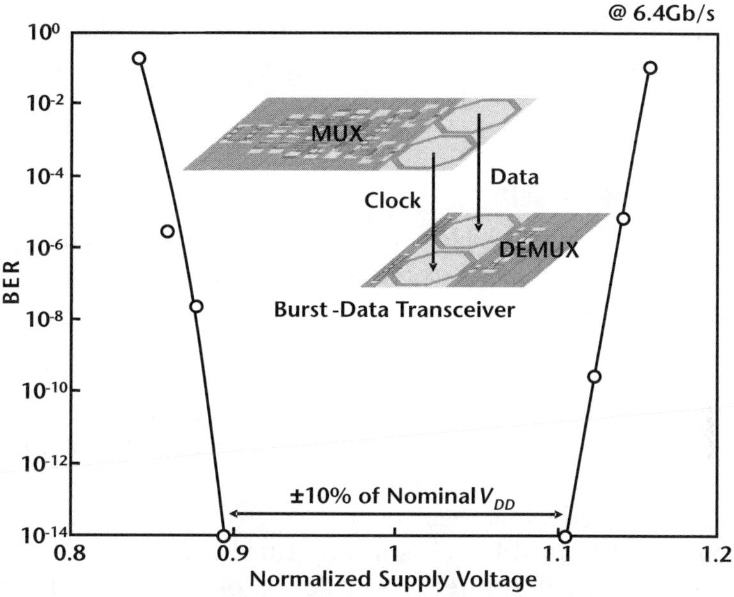

Figure 15.22 Measured BER dependence on supply voltage in burst transmission. (© 2008 IEEE.)

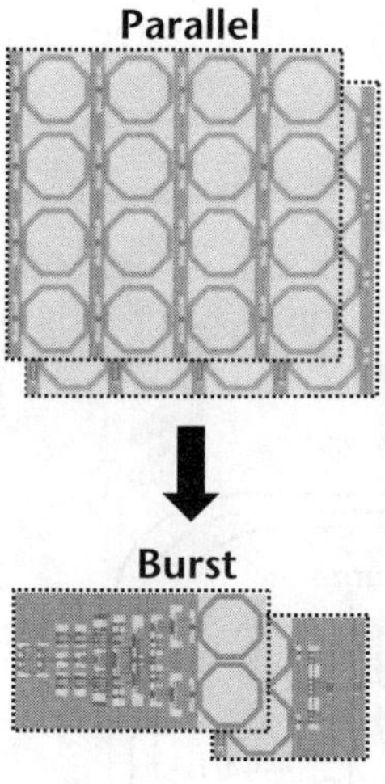

		180nm CMOS	90nm CMOS
Aggregated Data Rate		6.4Gb/s	15.2Gb/s
Number of Links	Parallel*	16 Links	38 Links
	Burst	2 Links	2 Links
Area	Parallel*	0.3mm^2	0.72mm^2
	Burst	0.1mm^2	0.08mm^2
Area Reduction by Burst Transmission		1/3	1/9

*Plural Use of 400Mb/s Data Links with 400MHz System Clock

Figure 15.23 Performance summary in 180 nm and 90 nm CMOS. (© 2008 IEEE.)

in the MUX/DEMUX, it cannot be applied to over–terabit per second high-bandwidth communications. The energy efficiency can be improved in the parallel communication, while the area overhead should be minimized by arranging the data links in high density. As mentioned earlier, the crosstalk between the inductive-coupling channels is more serious, and a crosstalk-reduction technique is required for high-density channel arrangement. A circuit solution based on time-division multiplexing (TDM) was proposed by Miura et al. [8, 9]. In this section, a high-density and high-bandwidth parallel inductive-coupling transceiver utilizing TDM will be introduced.

15.6.1 Circuit Design

Figure 15.24 depicts the block diagram of the proposed parallel inductive-coupling transceiver. The transceiver comprises 16 slices of a 64 channel block, yielding 1,024 channel data transceivers in total. Each of 64 channel blocks consists of 64 data transceivers, one clock transceiver, and a phase interpolator (PI). Clock for the transmitter, *Txclk,* is transmitted through inductive coupling, and clock for the receiver, *Rxclk,* is recovered by the clock transceiver. The clock frequency is 1 GHz. The PI generates four time slots in one clock cycle by creating four-phase clocks from both *Txclk* and *Rxclk* for TDM. Data transceivers are divided into the time slots to reduce crosstalk. Each data transceiver communicates at 1 Gbps/channel. A 1 Tbps data bandwidth is obtained by 1,024 parallel data links.

Figure 15.25 illustrates circuit details of TDM. The PI generates four-phase clocks that are assigned like a checkerboard pattern in the data transceiver array. Simulated waveforms of the received signal and crosstalk are presented on the left. When the channel pitch is reduced down to 30 μm, crosstalk increases to the same level of the signal. Although two-phase TDM reduces crosstalk to half of the signal level, this value is still not low enough for communication with BER lower than 10^{-13}. The four-phase TDM reduces crosstalk to 10 mV peak voltage and enables BER lower than 10^{-13}.

15.6.2 Experimental Results

N. Miura et al. designed and fabricated test chips in 180 nm CMOS. Figure 15.26 shows microphotographs of the test chips. The transmitter chip is placed on top of

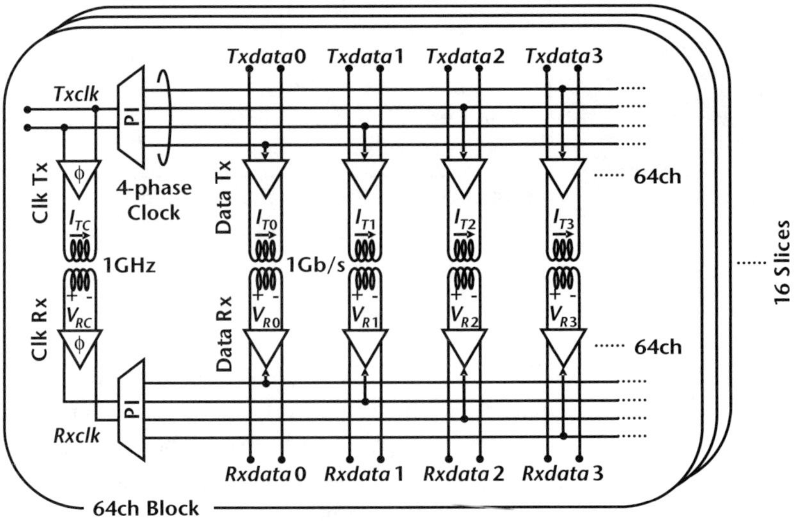

Figure 15.24 Block diagram of synchronous parallel inductive-coupling transceiver. (© 2007 IEEE.)

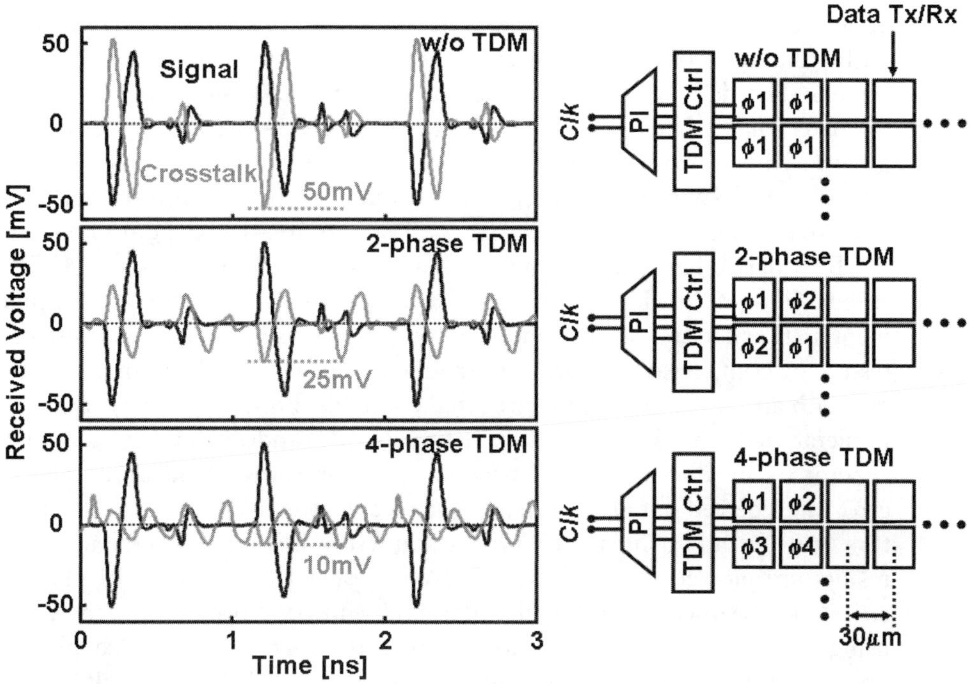

Figure 15.25 Time division multiplexing. © 2007 IEEE.

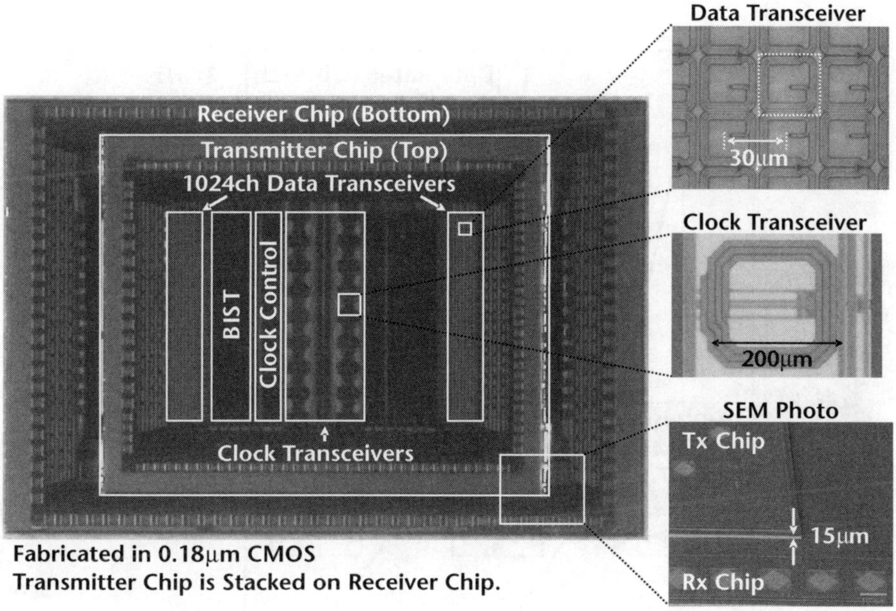

Figure 15.26 Stacked test chips of 1 Tb/s parallel inductive-coupling transceiver. © 2007 IEEE.

the receiver chip, with both chips in face-up configuration. Both chips are polished down to 10 μm in thickness. The resulting communication distance, including an adhesive layer, is 15 μm. The clock transceiver transmits 1 GHz clock by the coil with a diameter of 200 μm. The clock transceiver is set up for every 64 data transceivers. The data transceiver communicates at 1 Gbps/channel by the coil with a diameter of 29.5 μm, and 1,024 data transceivers are arranged with a pitch of 30 μm. The transmitter and receiver circuits are placed under the coils to save layout area. Experiments indicate that influence from the transceiver circuits to the inductive channel is negligible. The total layout area for the data link is 1 mm^2.

BER dependence on channel pitch and the number of phases in TDM was measured. An on-chip TDM controller changes the number of phases and phase assignment so that the transceiver with four- or two-phase TDM or without TDM can be tested for comparison. A pitch controller selects activated channels to change channel pitch and the number of aggregated channels. Built-in-self-test (BIST) circuits are implemented for BER measurement. Pseudorandom binary sequence (PRBS) generators produce $2^{23} - 1$ word patterns for transmitted data. The number of errors in received data is counted in the receiver. Scan chain initializes PRBS generators and outputs a measured errors count for BER measurement. The measured results are plotted in Figure 15.27. By increasing the number of phases in TDM, crosstalk is reduced, and the channel pitch can be shortened for the same BER. By using the four-phase TDM, 1,024 transceivers arranged with a pitch of 30 μm operate at 1 Gbps/channel for BER lower than 10^{-13}. As a result, aggregate data bandwidth of 1 Tbps is achieved with a layout area of 1 mm^2 and an area efficiency pf 1 mm^2/Tbps. The transceiver chip consumes 3W at 1.8V, and the energy efficiency is 3 pJ/b.

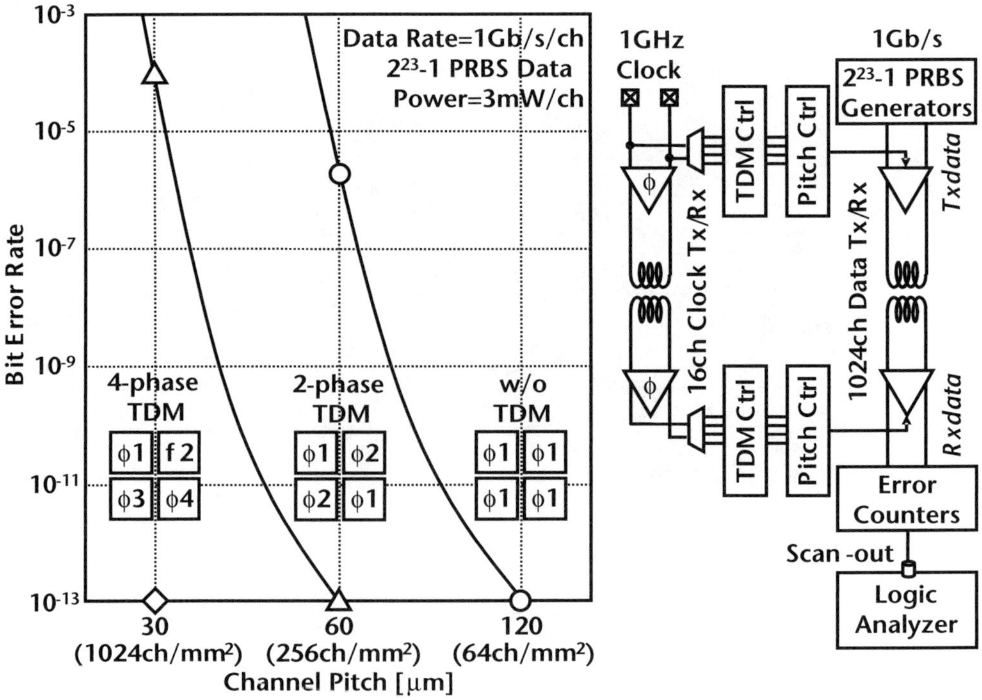

Figure 15.27 Measured BER dependence on channel pitch. (© 2007 IEEF.)

15.7 Challenges and Opportunities

15.7.1 Scaling Scenario

Figure 15.28 summarizes the scaling scenario of the capacitive-coupling I/O. The $1/\alpha$ device scaling reduces the receiver's input capacitance to $1/\alpha$. Although the communication distance is already at a minimum (<1 μm), the electrode diameter can be scaled to $1/\alpha^{0.5}$ while keeping the electric coupling coefficient constant. In this scaling scenario, both the data rate and the channel density increase by α, and the data rate per unit area increases by α^2.

Figure 15.29 summarizes the scaling scenario of the inductive-coupling I/O. The inductive-coupling I/O is used in face-up or back-to-back chip stacks. The communication distance is proportional to the chip thickness. Together with the $1/\alpha$ device scaling, the chip thickness is scaled to $1/\alpha$ in order to reduce the communication distance to $1/\alpha$. As a result, the coil diameter can be scaled to $1/\alpha$ while keeping the magnetic coupling coefficient constant. The coil turns are constant so that both the self-inductance and the parasitic capacitance are reduced by $1/\alpha$. Therefore, the self-resonant frequency as well as the data rate can be improved by α. In this scaling scenario, the channel density increases by α^2, and the data rate per unit area increases by α^3.

15.7.2 Wireless Power Delivery

Wireless power delivery is one of the future research challenges. If both data and power can be transmitted wirelessly, all the mechanical contact can be removed from the stacked chips, and the chips can be detachable, providing new 3D system

Transistor Size (Scaling Factor)		$1/\alpha$
Power Supply Voltage	$[V]$	$1/\alpha$
Current	$[I]$	$1/\alpha$
Circuit Delay Time	$[t]\sim[CV/I]$	$1/\alpha$
Rx Input Capacitance	$[C_{IN,R}]$	$1/\alpha$
Communication Distance	$[X]$	1
Electrode Diameter	$[D]$	$1/\alpha^{0.5}$
Coupling Capacitance	$[C_C]$	$1/\alpha$
Parasitic Capacitance	$[C_{SUB}]$	$1/\alpha$
Electric Coupling Coefficient	$[k]$	1
Received Signal	$[V_R]\sim[kV_T]$	$1/\alpha$
Crosstalk	$[V_{RS}/V_{RN}]$	1
Data Rate / Channel	$[1/t]$	α
Channel Number / Area	$[1/D^2]$	α
Aggregated Data Rate / Area	$[1/tD^2]$	α^2
Energy / Bit	$[CV^2]$	$1/\alpha^3$

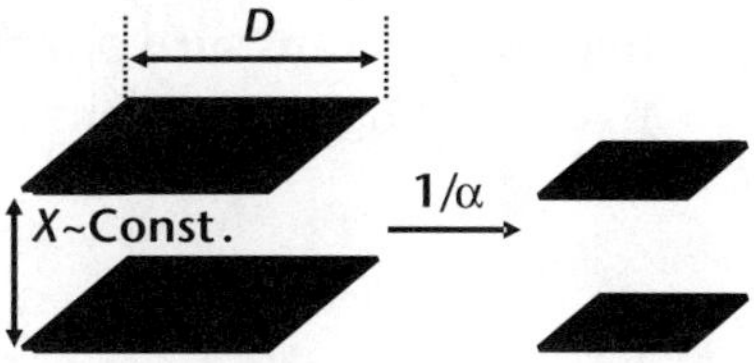

Coupling Capacitance

$$C_C \propto D^2/X$$

Received Signal Voltage

$$V_R = kV_T = \frac{C_C}{C_C + C_{SUB} + C_{IN,R}} V_T$$

Figure 15.28 Scaling scenario of capacitive coupling I/O.

Transistor Size (Scaling Factor)		$1/\alpha$
Power Supply Voltage	$[V]$	$1/\alpha$
Current	$[I]$	$1/\alpha$
Circuit Delay Time	$[t]\sim[CV/I]$	$1/\alpha$
Chip Thickness	$[T_{chip}]\sim[X]$	$1/\alpha$
Coil Diameter	$[D]\sim[1/X]$	$1/\alpha$
Coil Turn Number	$[n]$	1
Self Inductance	$[L]\sim[Dn^2]$	$1/\alpha$
Self Resonant Freq.	$[f_{SR}]\sim[1/(LC)^{0.5}]$	α
Magnetic Coupling Coefficient	$[k]$	1
Received Signal	$[V_R]\sim[kDn^2(I/t)]$	$1/\alpha$
Crosstalk	$[V_{RS}/V_{RN}]$	1
Data Rate / Channel	$[1/t]$	α
Channel Number / Area	$[1/D^2]$	α^2
Aggregated Data Rate / Area	$[1/tD^2]$	α^3
Energy / Bit	$[ItV]$	$1/\alpha^3$

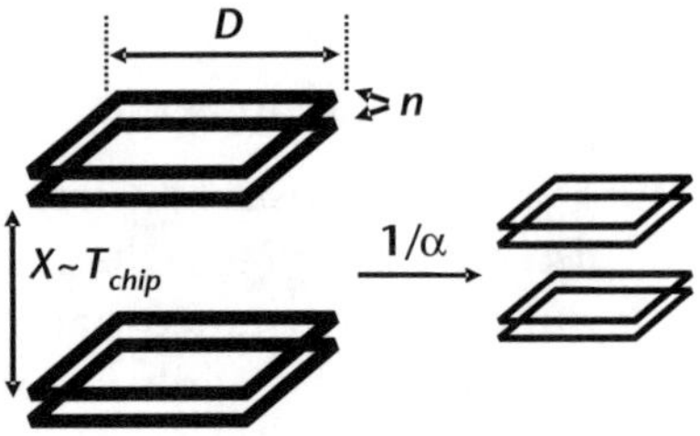

Self Inductance

$$L \propto Dn^2$$

Received Signal Voltage

$$V_R \propto M\left|\frac{dI_T}{dt}\right|_{max} = k\sqrt{L_T L_R}\left|\frac{dI_T}{dt}\right|_{max}$$

$$= kL\frac{I_D}{t_{pd}} = kDn^2\frac{I_D}{t_{pd}}$$

Figure 15.29 Scaling scenario of inductive coupling I/O.

applications. Figure 15.30 depicts the simplified schematics of the wireless power delivery. Capacitive and inductive coupling deliver ac power to the receiver chip. A rectifier in the receiver chip converts the ac power to the dc power supply of the chip core. Although the circuit topology is the same for both capacitive and inductive coupling, the available supply voltages are limited in capacitive coupling. As (15.1) indicates, the received voltage is always below the transmit voltage. On the other hand, the received voltage in inductive coupling is given by $j\omega MI_T$, and it is not limited by the transmit voltage. Y. Yuxiang et al. presented a prototype of the inductive-coupling power delivery system in 180 nm CMOS [24]. It delivers 36 mW with

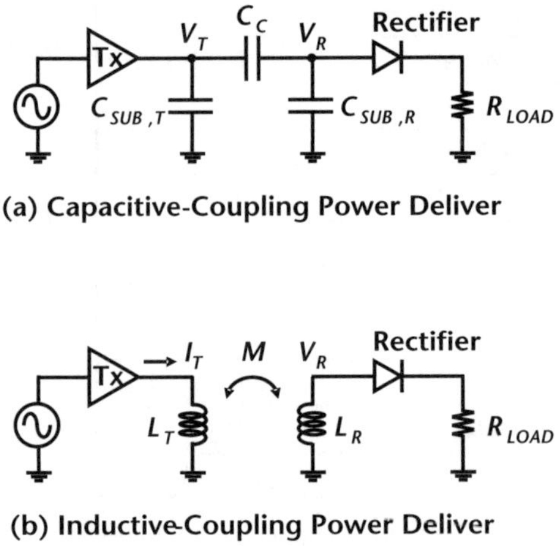

Figure 15.30 Wireless power delivery through (a) capacitive coupling and (b) inductive coupling.

a power transfer efficiency of 10%. Future work remains to be done in improving power efficiency and capacity. Further optimizations of the rectifier circuit and the coil layout are required for high efficiency. Utilizing high-μ magnetic materials, such as [25], or multichannel power delivery may improve capacity.

15.8 Conclusion

This chapter describes the capacitive- and inductive-coupling I/Os for 3-D chips. They are both simple wireless circuit solutions made from standard IC fabrication processes. Unlike with TSV technology, no additional wafer and mechanical processes are required, enabling a considerable cost reduction. Modeling and design of the channel layout and transceiver circuit are also presented. Electromagnetic and circuit codesign realizes high-performance and reliable operation that is competitive with conventional wired I/Os. Test-chip measurements demonstrate low-power (<1 pJ/b), high-speed (>10 Gbps/channel), and high-density (>1 Tbps/mm^2) interchip communications with product-level reliability (BER < 10^{-14}). The capacitive- and inductive-coupling I/Os are ready for practical use. The capacitive-coupling I/O exhibits better performance in face-to-face chip stacks because of its simple layout structure and thus good scalability. It can be applied in compact and low-power consumer products, such as mobile phones, digital cameras, and mobile game players. The inductive-coupling I/O provides a wider communication range and stacking variety. It can be applied in highly integrated and high-performance systems, such as cubic processors and memories.

References

[1] International Technology Roadmap for Semiconductors (ITRS), "ITRS 2007 Edition Executive Summary," December 2007, www.itrs.net.

[2] Kuhn, S., et al., "Vertical Signal Transmission in Three-Dimensional Integrated Circuits by Capacitive Coupling," *Proc. ISCAS*, April 1995, pp. 37–40.

[3] Fazzi, A., et al., "A 0.14mW/Gbps High-Density Capacitive Interface for 3D System Integration," *Proc. CICC*, September 2005, pp. 101–104.

[4] Fazzi, A., et al., "3D Capacitive Interconnections for Wafer-Level and Die-Level Assembly," *IEEE JSSC*, Vol. 42, No. 10, October 2007, pp. 2270–2282.

[5] Drost, R., et al., "Proximity Communication," *IEEE JSSC*, Vol. 39, No. 9, September 2004, pp. 1529–1535.

[6] Mizoguchi, D., et al., "A 1.2 Gb/s/pin Wireless Superconnect Based on Inductive Inter-Chip Signaling (IIS)," *ISSCC Dig. Tech. Papers*, February 2004, pp. 142–143.

[7] Hopkins, D., et al., "Circuit Techniques to Enable 430 Gb/s/mm^2 Proximity Communication," *ISSCC Dig. Tech. Papers*, February 2007, pp. 368–369.

[8] Miura, N., et al., "A 1Tb/s 3W Inductive-Coupling Transceiver for Inter-Chip Clock and Data Link," *ISSCC Dig. Tech. Papers*, February 2006, pp. 424–425.

[9] Miura, N., et al., "A 1Tb/s 3W Inductive-Coupling Transceiver for 3D-Stacked Inter-Chip Clock and Data Link," *IEEE JSSC*, Vol. 42, No. 1, pp. 111–122, January 2007, pp. 111–122.

[10] Miura, N., et al., "An 11 Gb/s Inductive-Coupling Link with Burst Transmission," *ISSCC Dig. Tech. Papers*, February 2008, pp. 298–299.

[11] Ishikuro, H., et al., "An Attachable Wireless Chip-Access Interface for Arbitrary Data Rate Using Pulse-Based Inductive-Coupling through LSI Package," *ISSCC Dig. Tech. Papers*, February 2007, pp. 360–361.

[12] Miura, N., et al., "Crosstalk Countermeasures for High-Density Inductive-Coupling Channel Array," *IEEE JSSC*, Vol. 42, No. 2, February 2007, pp. 410–421.

[13] Niitsu, K., et al., "Interference from Power/Signal Lines and to SRAM Circuits in 65 nm CMOS Inductive-Coupling Link," *ASSCC Dig. Tech. Papers*, November 2007, pp. 131–134.

[14] Wilson, J., et al., "Fully Integrated AC Coupled Interconnect Using Buried Bumps," *IEEE TADVP*, Vol. 30, No. 2, May 2007, pp. 191–199.

[15] Kumagai, K., et al., "System-in-Silicon Architecture and Its Application to H.264/AVC Motion Estimation for 1080HDTV," *ISSCC Dig. Tech. Papers*, February 2006, pp. 430–431.

[16] Lin, C. C., et al., "A 160K Gates/4.5 KB SRAM H.264 Video Decoder for HDTV Applications," *IEEE JSSC*, Vol. 42, No. 1, January 2007, pp.170–182.

[17] Fazzi, A., et al., "3D Capacitive Interconnections with Mono- and Bi-Directional Capabilities," *ISSCC Dig. Tech. Papers*, February 2007, pp. 356–357.

[18] Miura, N., et al., "A 0.14pJ/b Inductive-Coupling Inter-Chip Data Transceiver with Digitally-Controlled Precise Pulse Shaping," *ISSCC Dig. Tech. Papers*, February 2007, pp. 358–359.

[19] Miura, N., et al., "A 0.14pJ/b Inductive-Coupling Transceiver with Digitally-Controlled Precise Pulse Shaping," *IEEE JSSC*, Vol. 43, No. 1, January 2008, pp. 285–291.

[20] Ezaki, T., et al., "A 160 Gb/s Interface Design Configuration for Multichip LSI," *ISSCC Dig. Tech. Papers*, February 2004, pp. 140–141.

[21] Gu, Q., et al., "Two 10 Gb/s/pin Low-Power Interconnect Methods for 3D ICs," *ISSCC Dig. Tech. Papers*, February 2007, pp. 448–449.

[22] Ito, M., et al., "A 390 MHz Single-Chip Application and Dual-Mode Baseband Processor in 90 nm Tripple-Vt CMOS," *ISSCC Dig. Tech. Papers*, February 2007, pp. 274–275.

[23] Tanabe, A., et al., "A 10 Gb/s Demultiplexer IC in 0.18 μm CMOS Using Current Mode Logic with Tolerance to the Threshold Voltage Fluctuation," *ISSCC Dig. Tech. Papers*, February 2000, pp. 62–63.

[24] Yuxiang, Y., et al., "Non-Contact 10% Efficient 36mW Power Delivery Using On-Chip Inductor in 0.18 μm CMOS," *ASSCC Dig. Tech. Papers*, November 2007, pp. 115–118.

[25] Crawford, A., et al., "High-Frequency Microinductors with Amorphous Magnetic Ground Planes," *IEEE Trans. on Magnetics*, Vol. 38, No. 5, September 2002, pp. 3168–3170.

CHAPTER 16

Wafer-Level Testing of Gigascale Integrated Circuits

Hiren D. Thacker and Wilmer R. Bottoms

16.1 Introduction

The sustained exponential improvement in the performance of semiconductor-based electronic systems over the past five decades has been nothing short of incredible. This advancement has been driven by a combination of scaling down the size of individual MOS transistors and increasing their packing density. Unfortunately, a tougher path lays ahead for the traditional scaling paradigm as devices begin to approach physical limits. Innovation in system integration and packaging, such as 3D chip stacking, is a promising solution to continue along the path of historical performance improvement.

Emerging system architectures are driven by the consumer dominance of the electronics industry (Figure 16.1). Consumer electronics today demand not only more computing per unit volume than ever before but also tight integration and operation of computing logic and memory with other technologies, such as optoelectronics, microelectromechanical systems (MEMS), and radio-frequency (RF) communications. For these purposes, the system-in-package (SiP) approach—whereby logic and memory ICs may be assembled (in two or three dimensions) with other technologies, such as RF, MEMS, or optics on a common substrate—holds a lot of promise. However, at this nascent stage, a number of technological challenges lay ahead. These include: (1) a cost-effective thermal management system for an SiP or multichip structure, which may not be coplanar; (2) implementation of a stable power delivery network; (3) mechanical robustness in a multidie stack of ultrathin ICs with low-k dielectrics; (4) limited repairability; (5) high-frequency operation; (6) interconnect crosstalk; and (7) testability.

The proliferation of SiP components implies a strong demand for *known good die* (KGD)—either bare, unpackaged die or chip-scale packages.

This chapter aims to introduce the reader to the current status quo in wafer-level testing of gigascale integrated circuits (ICs). A special emphasis is placed on the challenges and opportunities for advanced probe cards built for massively parallel testing. In addition, the final section deals with prospects for wafer-level testing of gigascale chips with electrical and optical input/output (I/O) interconnects.

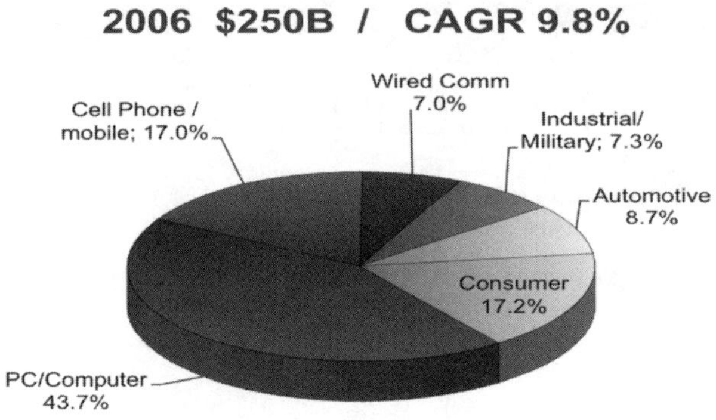

Product Class	2006 Shipment
Mobile phones	+ 20 %
Personal computers	+ 10~12 %
Digital cameras	+ 16 %
MP3 players	+ 52 %
Digital televisions	+ 52 %

Figure 16.1 The pie chart depicts the estimated size of the semiconductor industry as of 2006. Of the various market segments, consumer products such as digital televisions, MP3 players, and mobile phones saw the largest growth. (*Source:* SIA International Technology Roadmap for Semiconductors.)

16.2 Wafer-Level Testing of Gigascale Integrated Circuits

The elementary purpose of testing in IC manufacturing is to ensure that only KGDs are shipped to a customer. Unfortunately, the process of screening bad dice from good ones is time-consuming and increasingly difficult [1–3]. Shrinking device geometries, increasing frequencies of operation, and the sheer magnitude of transistors and I/Os on a chip are all factors contributing to the increasing complexity of IC testing.

Figure 16.2 illustrates a typical sequence of tail-end-of-line (TEOL) processes. The *rule of ten* is a common adage stating that the approximate cost of repairing a defect increases tenfold at each level [4]. As a result, ICs are tested repeatedly following every major process in the TEOL. A probe card is an enabling consumable component that provides the temporary physical and electrical link between any device under test (DUT) and automatic test equipment (ATE) during testing. While the thought of probe cards may conjure up images of a few needles or leads connected to a measurement instrument, in this era of parallel testing of multiple, multigigahertz, gigascale integration (GSI) chips having low-k interlayer dielectrics and sub–100 μm pitch die pads, probe cards are complex electromechanical assemblies.

A typical IC manufacturing process is illustrated in Figure 16.3. A bare silicon wafer, at the conclusion of back-end-of-line (BEOL) processing, is transformed into an area array of GSI chips. This is the first opportunity for testing the completely fabricated integrated circuit; up until this point, the wafer may have only undergone

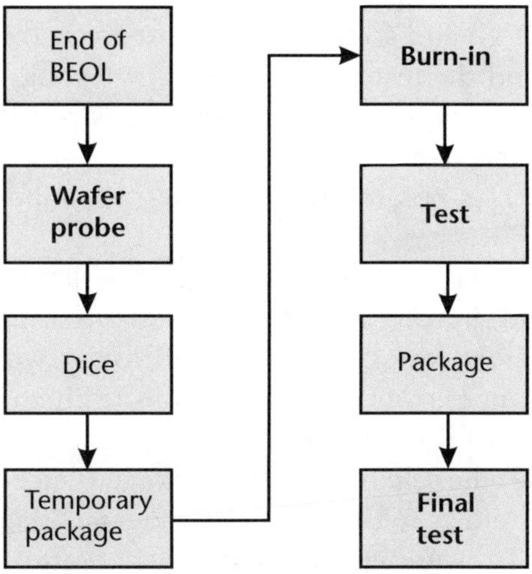

Figure 16.2 Flowchart of typical tail-end-of-line (TEOL) processes.

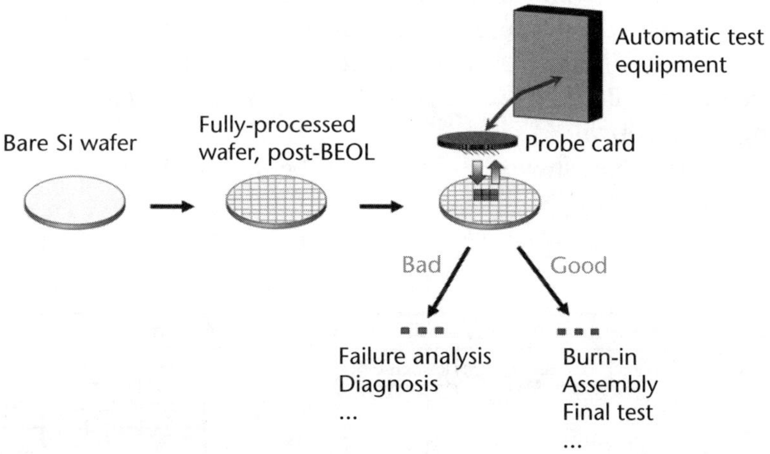

Figure 16.3 Schematic illustration of the typical manufacturing process for building a GSI chip. Wafer-sort testing is performed immediately following the end of BEOL processing.

some in-line parametric testing useful for monitoring the front-end and back-end processing. At this time, a probe card, under the control of ATE, sequentially contacts each die on the wafer in a step-and-repeat manner, and a series of short, decisive tests is conducted at each touchdown. This is commonly referred to as *wafer-sort testing* or *wafer probing*. Ideally, all bad dice would be identified during this step and discarded; all good dice would be sent onward for packaging, assembly, and final testing.

Manufacturers use wafer-sort not only to screen nonfunctional dice, but also to *bin* them into different performance categories. Testing at extreme temperatures may be required for military, space, and automotive applications, as well as to accelerate and detect defects in marginal dice. As such, wafer-sort oftentimes includes multiple iterations of wafer probing carried out at different temperature

(–40°C to +150°C) and stress (e.g., elevated voltage) conditions, depending on the class of DUT and the test plan. Wafer-sort testing has a direct impact on product yield and therefore needs to be as accurate as possible. Poor contact between even one DUT pad and its corresponding probe can result in a chip being marked as nonfunctional. This possibility underlines the importance of having reliable probe hardware.

From a test-plan perspective, a good balance must exist between the cost-effectiveness and thoroughness of wafer-sort testing. Reduced testing at wafer-sort would assure short wafer-test times and lower wafer-probe costs, but this could allow a significant percentage of bad chips (as high as 25%) to be sent onward for packaging [5]. This is especially unacceptable for semiconductor companies whose business includes the sale of bare (unpackaged) dice—an increasingly common situation with the rapid growth of 3D SiP products. Longer tests are required to improve fault coverage. To overcome the cost implications of this strategy, multidie site testing [6] has become quite common. Short, inadequate tests on individual dice are being replaced by marginally longer, but more efficient, tests on multiple dice using the same test equipment (Figure 16.4). The desire for parallelism in wafer-sort has, in turn, created a demand for multidie site probe modules. The level of parallelism varies across different product categories with DRAM and flash memory manufacturers demanding probe cards capable of contacting *all* dice on a 300 mm wafer in a single touchdown [6].

Figure 16.5 shows a typical probe test cell. It consists of four primary components:

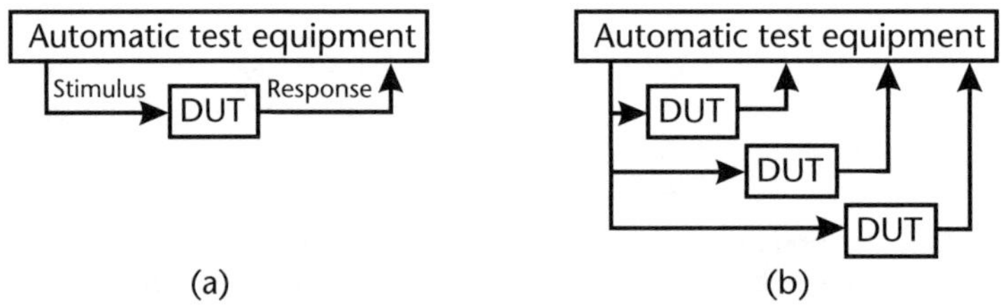

Figure 16.4 Implementation of (a) single DUT and (b) multi-DUT testing using the same ATE resources [7].

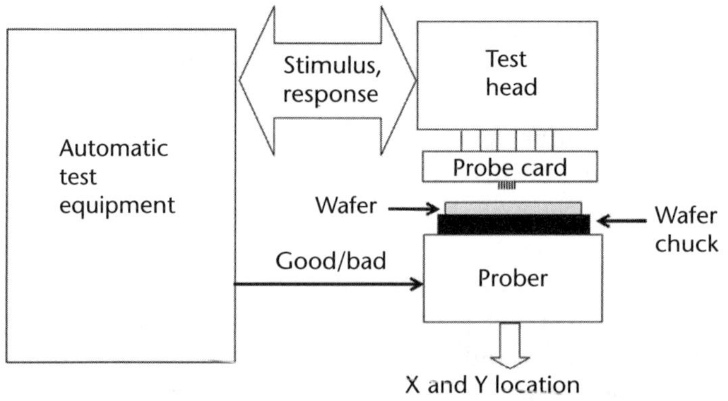

Figure 16.5 Schematic of a typical wafer-probe test cell.

1. *Automatic test equipment or ATE* is the hardware used to perform testing of integrated circuits. It comprises the instrumentation to generate test stimuli and to sense and store the response from the device(s)-under-test. Typically, ATEs are built on a per-pin architecture with each pin, or test channel, having its own signal generation and sensing circuitry. This architecture allows the testers to be scalable, but cost also scales linearly at the rate of a few thousand dollars per additional channel. The testers operate under the control of test programs authored by manufacturing test engineers. Figures of merit for an ATE include the number of test channels, the signal data-rate per channel, and the allowable parallelism.
2. A *prober* is the piece of equipment where wafers are physically loaded during wafer testing (Figure 16.5). The probe card is fastened to the probe ring and stays at this fixed position; the wafer under test is placed on a vacuum chuck within the prober and positioned underneath the probe card by precision x, y, and z motors. On command, the prober raises this chuck to bring the wafer into contact with the probes, and the test program is executed. After one set of dice has been tested, the prober lowers the chuck to disengage the wafer from the probe card and steps over such that a set of previously untested dice is now positioned below the probe card contacts. Then the chuck is driven up, and the testing resumes. This step-and-repeat procedure continues until all dice on the wafer have been tested. Most present-day probers use a vision-based alignment system to align DUT pads to probe pins. This alignment system includes at least one downward-looking (wafer) camera and one upward-looking (probe card) camera. The downward-looking camera is also referred to as the *bridge* camera as it is physically located in between the prober's load-lock area and the main probing enclosure. Probers are generally also equipped with a couple of different illumination systems (normally incident and/or ring lighting) and pattern-recognition software modules. The motors that drive the wafer from one location to another are both extremely precise (better than $\pm 2\ \mu m$ accuracy) and very fast (to minimize nontest time). Another feature in probers is that the wafer-chucks are configured to be heated or cooled as the test may require. External temperature controllers are employed for this purpose and can accurately maintain the chuck temperature within a few tenths of a degree. For operation at temperatures below 0°C, probers are designed and constructed to maintain a moisture-free environment to prevent water vapor condensation on the wafers, wafer chuck, probes, and other components.
3. *Test control software* is the brain behind wafer probing. The software consists of programmed interfaces that control the prober and the ATE hardware. Using the manufacturing test program as an input, the test software coordinates the prober, tester, and data-collection operations. In the early days, a bad die on a wafer would be marked with a drop of ink as a visual indicator for the test floor personnel. Today, test software generates wafer maps to track and store test results (e.g., bin1, bin2, fail).
4. *Probe cards* are essentially high-quality interposers that link the wafer under test to the ATE interface during testing. These are custom-designed

components driven by the electrical, geometrical, and mechanical I/O pad configuration of the IC product to be tested. I/O pad configurations vary not only for different classes of ICs but also from manufacturer to manufacturer and product to product. As such, new probe cards are typically more expensive than repeat orders to offset nonrecurring engineering (NRE) costs. The actual design and construction of a probe card assembly entails careful consideration of electrical, mechanical, and thermomechanical issues. The inset in Figure 16.6 illustrates the major components of a typical advanced probe card used in IC testing: a printed circuit board (PCB) or motherboard, redistribution substrates, and a probe substrate. The probe substrate contains probes for contacting the pads of the DUT; the redistribution substrate(s) and motherboard route test, power, and ground signals between the probes and the ATE interface. The tester side of the motherboard is built with a standard physical interface to fit the particular tester to be used. A pogo block on the tester head contacts the probe card and serves as a vertical interposer to connect the probe card to the tester interface board (TIB), which in turn provides access to the tester's pin electronics. The probe card stack-up may consist of multiple redistribution substrates for various reasons, such as to meet the thickness requirement of the card or to provide the ability to planarize the card. Figures 16.7 and 16.8 show images of an epoxy cantilever probe card and an advanced MEMS probe card, respectively. A detailed discussion of probe cards follows in the next section.

16.3 Probe Cards for Wafer-Level Testing

The physical interface between a DUT and ATE has consistently been identified as a challenge for testing in future generations of ICs [6]. High pin counts, tight pitch, high bandwidth, and low cost of fabrication and operation are just some of the key

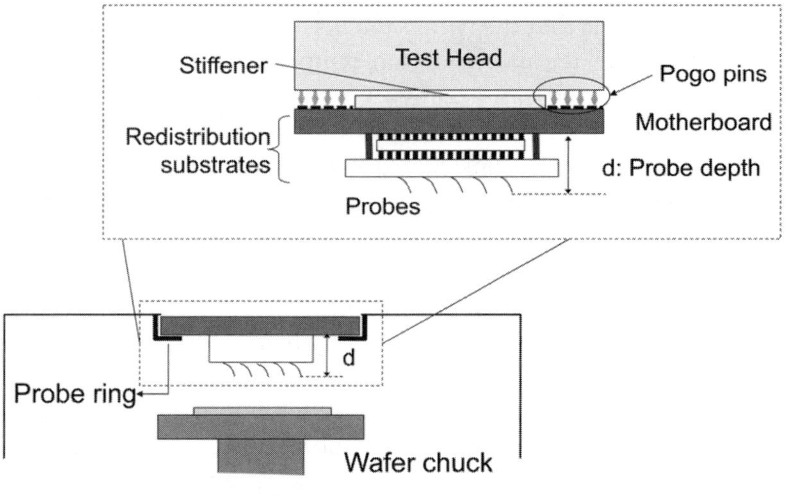

Figure 16.6 Schematic showing the interface between a probe card and an ATE. As shown in the inset drawing, a probe card generally comprises a probe substrate, redistribution substrates, and a motherboard.

Figure 16.7 Photograph of an epoxy cantilever probe card (probe side up).

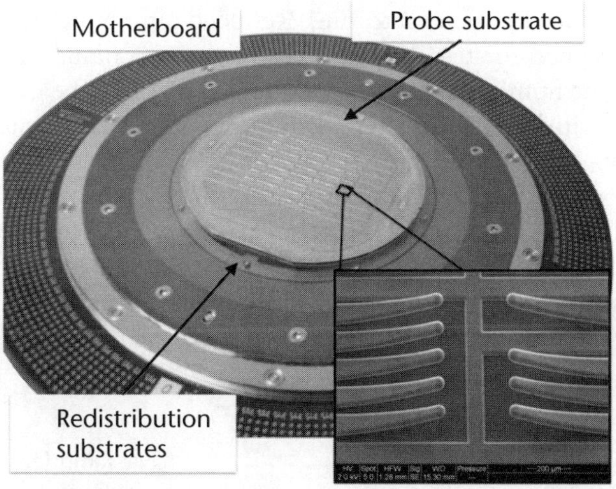

Figure 16.8 Photograph of a multidie–site, thin-film MEMS probe card. The inset shows a scanning electron microscope image of lithographically patterned, stressed metal MEMS probes (*Source*: NanoNexus, Inc.)

requirements for probe technologies. In addition, to reduce the overall cost of wafer testing, there has been an industrywide shift to implement massive parallelism in the process. The time to test a wafer in terms of the number of dice being probed and tested in parallel can be expressed as

$$T_{wafer\_test} = \frac{N_{die}}{N_{probe}}\left(T_{align} + T_{step}\right) + \frac{N_{die}}{N_{test}} T_{test} \quad (16.1)$$

where T_{wafer_test} is the time to test a wafer, N_{die} is the number of ICs on a wafer, N_{probe} is the number of ICs that are contacted in parallel, N_{test} is the number of ICs being tested in parallel, T_{align} is the time for probe to DUT pad alignment, T_{step} is the average time for the wafer chuck to position a new set of DUTs underneath the probe card, and T_{test} is the time to test a single chip. Additional contributors to T_{wafer_test} would include initial test setup time and time for periodic in situ probe mark inspection and probe-to-pad alignment adjustments. As these factors are dependent on the tester type and the test program, they are not included in (16.1). This simple model also assumes that at every touchdown, the probe card makes good contact with the target DUTs. In reality, if one or more probe sites has poor contact during a touchdown, then that die will be retested once testing of all other dice on the wafer is complete, resulting in a slightly longer test time compared to the time estimated by (16.1). The worst-case scenario for T_{wafer_test} occurs when a wafer is probed and tested one die at a time; the best-case scenario is when all the dice on the wafer are contacted and tested in parallel (wafer scale). Figure 16.9 plots wafer test time as a function of the number of DUTs being tested in parallel (with $N_{probe} = N_{test}$).

As the number of channels available on ATE systems is limited by the capital investment required, massively parallel testing is most popular in low pin-count products, such as commodity DRAM and flash memory. If the DUT is a high pin-count device, such as a microprocessor (MPU) or an application-specific IC (ASIC), a high level of parallelism will only be possible if these devices employ some form of reduced pin-count testing. Reduced pin-count testing may be achieved if the chip is designed-for-testability (DFT) using scan chains and other methods [4]. Additionally, the number of I/Os needed for testing may be greatly reduced if the chip is designed with built-in-self-test (BIST) modules. As the name suggests, when using BIST, some functionality of the tester hardware—that is, pattern generation and comparison of output to expected results—is added directly onto the silicon alongside the logic to be tested. If ICs are designed to be completely self-testable, then only a serial communications bus is needed to allow the external test equipment to initiate the self-test and to capture the final result (pass/fail response or more detailed diagnostic data). One such bus is defined by the IEEE 1149.1 standard (also known

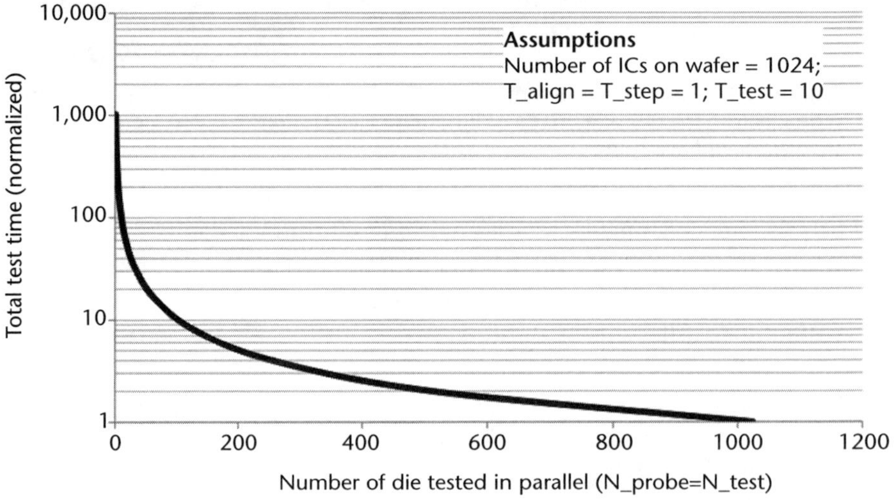

Figure 16.9 Plot of normalized test time per wafer as a function of the number of dice being tested in parallel.

as boundary scan). The details of DFT and BIST are adequately described in books related to digital testing, such as [4], and will not be covered here. It suffices to know that high pin-count ICs may also be tested in parallel if a DFT architecture is adopted up front.

16.3.1 Requirements

From an IC manufacturer's point of view, the basic function of a probe card is to interface with I/Os on the DUT. It should not load the DUT or cause any signal degradation. In addition, it should be able to do this repeatedly (hundreds of thousands of touchdowns) [6] without damaging the I/Os. Of course, this delicate combination of high-quality electrical design and mechanical robustness needs to be achieved at the minimum possible cost. The technology requirements of such multi-DUT probe cards is described next. These are broadly categorized under electrical, mechanical, and reliability requirements.

16.3.1.1 Electrical Requirements

1. *Probe card design architecture:*

   a. *Multidie/wafer-scale contact:* Multidie site testing is already being used extensively for testing memory device wafers. Keeping with this trend, modern-day probe cards absolutely must be scalable for multidie site testing. The projected number of probe points per touchdown for different classes of integrated circuits is summarized in Table 16.1.

   b. *High-density signal redistribution:* As probe cards are the mechanical and electrical interface between the DUT and the tester, it is essential that the probes (on the wafer side) have the same footprint (pitch and location) as the I/O pattern of the DUT; similarly, the pads on the probe card motherboard (on the tester side) must follow the footprint of the pogo pins and/or tester interface board pads. The pad pitch of I/Os on an IC today may be as low as 25 μm. Conversely, the pad pitch on tester interface boards, typically built using printed circuit board (PCB) fabrication methods, is measured in hundreds of micrometers. This pad pitch mismatch must be overcome using multilayered interconnect routing in the probe card assembly.

Table 16.1 Projected Trends for Multi-DUT Testing in Different Classes of Digital ICs [6]

	2009	*2012*	*2015*	*2018*	*2020*
Memory					
Signal probe points/touchdown	17,000	20,000	20,000	20,000	20,000
DUTs in parallel	1,024	1,024	2,048	2,048	2,048
Microprocessor					
Signal probe points/touchdown	1,024	2,000	2,000	6,000	6,000
DUTs in parallel	16	32	64	256	512
Application-specific IC					
Signal probe points/touchdown	1,500	3,000	3,000	3,000	3,000
DUTs in parallel	16	32	64	256	512

Signal fan-out and routing in a single-DUT probe card is quite simple; fan-out complexity increases manifold for multi-DUT probe cards. When a probe card is designed to contact an array of DUTs immediately adjacent to each other, it is said to have a *no-skip* configuration [see Figure 16.10(a)]. In most cases, a no-skip configuration minimizes the number of touchdowns required to completely test all dice on a wafer and is desirable from a product test engineer's perspective. However, in cases where the DUTs have a pin density (probe pads per unit area) greater than the native pin density of the probe card technology, then a skip configuration [skip row(s) and/or column(s)] may be needed to accommodate the probes and the trace routing [Figure 16.10(b)].

State-of-the-art probe cards today are designed to contact 256 DUTs (and more) in a single touchdown. Assuming each DUT has about 40 I/Os at a 75 μm pitch, this amounts to over 10,000 probes on a single probe card. It is easy to see that as the demand for such massively parallel probe cards increases, high-density interconnection technologies are needed to redistribute these signals from the probe points.

c. *Pin-count reduction:* Owing to the high incremental cost of adding signal channels to a tester, it is not expected that ATE pin counts will keep up with increasing DUT I/O counts or massively parallel testing schemes. This mismatch calls for some pin-count-reduction techniques to be implemented on the probe card. The use of common power and ground planes, combination of other common control signals, and driver-pin sharing are some ways of achieving this.

d. *Contact resistance:* At the microscopic level, every material, no matter how clean or smooth, has some surface roughness. As a result, when two surfaces contact each other, it is the asperities of the two surfaces, called a-spots, that are actually in contact [7]. The true contact area is thus smaller

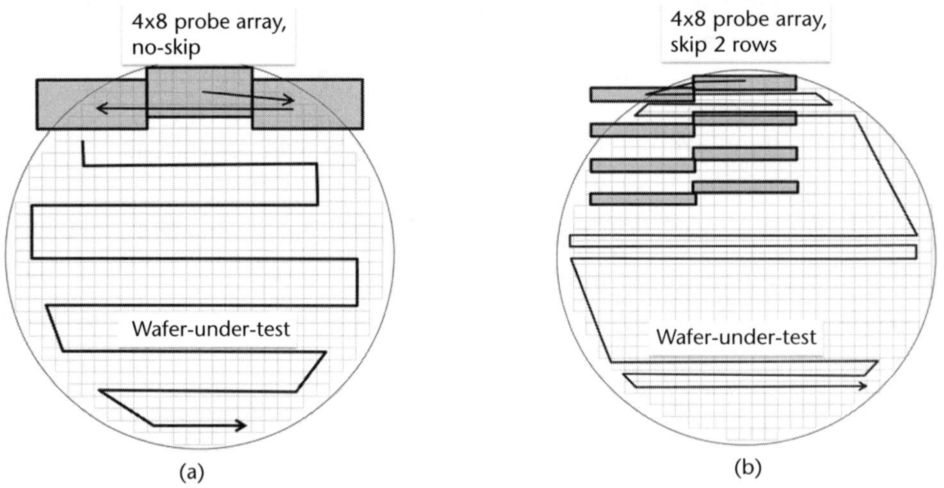

Figure 16.10 Schematic illustrating the step-and-repeat touchdown pattern during wafer testing when using (a) a no-skip probe card, and (b) a "skip 2 rows" probe card. While adding skips to a multisite probe array is likely to increase the number of touchdowns needed to test all ICs on a wafer, it may be necessary for electrical routing within the probe card.

than the geometric area, which appears to be in contact. Demountable probes contacting DUT pads are also governed by contact physics. Any electrical current passing through this interface passes through these tiny a-spots, which cause current crowding and are responsible for constriction resistance. In addition, any thin layers of an insulating material present at the interface also contribute to electrical resistance. The sum of constriction resistance, as well as any resistance owing to the presence of an insulator, equals the contact resistance at that interface. A large contact resistance may cause a sizable voltage (IR) drop at the point of contact and therefore degrade the transmitted signal.

When a mechanical load is applied between the two surfaces, the applied force spreads over the asperities in contact and can cause them to undergo plastic deformation. Force and contact area are related by a simple expression shown in (16.2) [7]:

$$F = A_c H \tag{16.2}$$

where F is the force applied to the interface in *Newtons*, A_c is the contact area in millimeters squared, and H is the hardness of the softer material in Megapascal. Assuming plastic deformation, increasing the applied force or adding overdrive to the probe enlarges the area of the a-spots, reduces any current crowding, and therefore reduces the constriction resistance. The applied force also assists with penetrating through any thin insulating layers and reducing the insulator-related resistance. This is the primary reason why demountable contacts are designed to scrub against the target surface. In the International Technology Roadmap for Semiconductors (ITRS), the expectation for contact resistance is less than 1ohm.

e. *Signal integrity*: Testing at high speeds requires that the probe card assembly not load the electrical signals as they are transferred between the DUTs and the tester electronics. The trace resistance of the wiring in the probe card assembly is a contributor to the interconnect RC-delay and should be minimized through design and construction. While some testers are able to adjust for differences in trace resistance between channels, when this option is not available, customers (that is, test engineers) may require electrical trace-length matching in the probe card, especially in those intended for high-speed testing. For the same reason, probe card interconnects should be built to minimize trace capacitance.

f. *Power integrity:* Each DUT being contacted by a probe card is assigned to its own power supplies on the tester. These power supplies are subsequently routed to power and ground planes on the probe card motherboard. In a worst-case situation, if all drivers on the DUT switch at once, the simultaneous switching noise on the power supplies can cause a glitch to affect the output of the test. As such, probe cards need to have the same electrical design integrity and checks as high-speed electronic packages. This would include designing the interconnects to have low inductance and placing decoupling capacitors close to the probes to filter out the impact of the inductance when a current spike does occur.

Also, to account for any such current spikes during testing, the probes and interconnects in the probe card assembly must have a current-carrying capacity that exceeds the maximum expected current. Not doing so could likely lead to catastrophic failure by way of burnt probe tips, which are the regions experiencing the highest current density owing to their small cross section area.

16.3.1.2 Mechanical Requirements

1. *Mechanical compatibility:* Probe cards are used in conjunction with production probers and testers. As such, their mechanical form factor must be compatible with the equipment on which they will be used. Probe card customers provide this information up front. This determines the size of the motherboard, the probe depth, and the layout and type of connectors on the tester side of the probe card. Probe depth is defined as the height of the probe card stack-up, including the probes themselves, measured from the base of the motherboard (see Figure 16.6). If the probe depth is too small, the wafer chuck may not have sufficient motion range in the $+Z$ (upward) direction to establish contact between the probe card and the wafer under test. Conversely, if the probe depth is too large, the probe card may crash into the wafer chuck. The acceptable variation of probe depth from the customer-specified target is around ±0.3 mm.

   While these specifications vary from customer to customer, there are only a few major manufacturers of equipment for wafer probing [Electroglas, Accretech, Tokyo Electron Limited (TEL)] and testing (Advantest, Verigy, Yokogawa, Teradyne). As such, probe card makers can quickly develop standardized solutions to cover most prober-tester combinations.
2. *Planarity:* Probes for electrical I/Os need to be in physical contact with the pads in order to complete an electrical path.[1] While setting up a probe card to test a wafer, the user will keep raising the wafer chuck (hence, the wafer under test) until electrical contact is achieved between the probes and the wafer under test. This point of first contact is determined electrically. When the wafer and probe card are not in contact, the tester's pin electronics measures an open at the tip of the individual probes. When the probe contacts the DUT pad, the tester senses a voltage drop across the electrostatic protection diode (ESD) connected to the chip's I/O, thereby determining electrical contact. The chuck is overdriven upwards until all probes are in contact (to the last-contact or all-touch position). In a perfect situation, all probes would make contact at the same height. However, in the real world, nonplanarity exists on the probe card, the wafer, and the wafer chuck. All of these add to nonplanarity of the setup and, in turn, to the amount of chuck overdrive needed from first touch to reach all-touch. In addition, owing to the nonplanarity of the wafer chuck and the wafer itself, the chuck overdrive

1. While this is most common, noncontact electrical testing of ICs using RF coupling between the probe card and chip I/Os was recently reported [29].

at which all-touch is recorded on one portion of the wafer may produce opens when the probe card is contacting a different area on the same wafer. To prevent such missed or intermittent contacts, test engineers add a small amount of additional overdrive to the all-touch z-position.

Electrical probes are basically mechanically compliant springs that provide the necessary out-of-plane compliance to absorb this overdrive. Figure 16.11 shows schematics of three different probe spring technologies: cantilever, buckling beam, and membrane. All three are vertically compliant, demountable contactors. The force exerted by these springs on a DUT pad is proportional to the amount they are physically overdriven. This contact force, specified in grams per mil of vertical displacement (overdrive), should be just enough to make good electrical contact but not enough to damage the I/Os or the interconnect stack and devices underneath. Thus, the relationship between planarity, overdrive, and contact force necessitates tight tolerance on planarity specifications for probe cards and prober chucks. For advanced probe cards, probe tip planarity should be within 25 μm and the desired spring compliance between 20 and 100 μm [6].

The metallurgy of the I/O pads on the DUT also drives the requirement for probe force. ICs may have I/O pads made of aluminum (and its variants such as aluminum-silicon or aluminum-copper), gold, copper, or solder (lead-based or lead-free). On exposure to air, aluminum forms an insulating oxide; probes must therefore exert 3 to 5 gram-force (gf) to break through this oxide and establish a stable contact. On the other hand, probing a gold pad requires very little force (~0.1 to 1 gf). While overdrive is one control to change force, reducing it may not be an option (for nonplanarity). In this case, the solution is to modify the spring constant of the probe spring itself. This may be achieved by tuning the geometrical design of the probe.

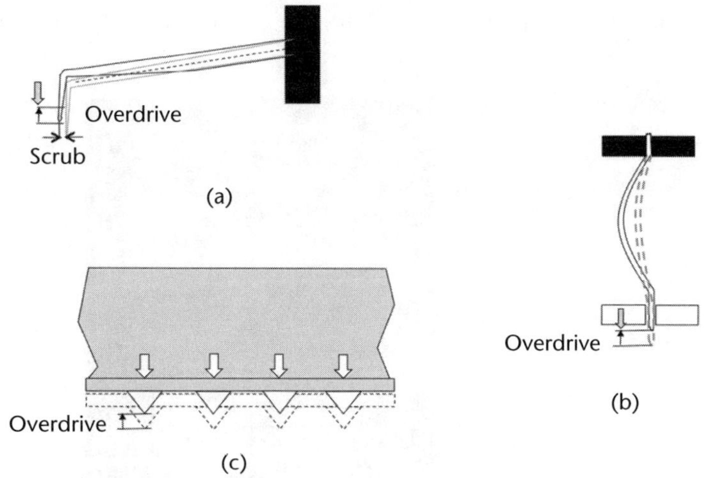

Figure 16.11 Illustration of the working principle of (a) cantilever probes, (b) cobra probes, and (c) membrane probes. Each of these probe technologies is designed to provide out-of-plane compliance.

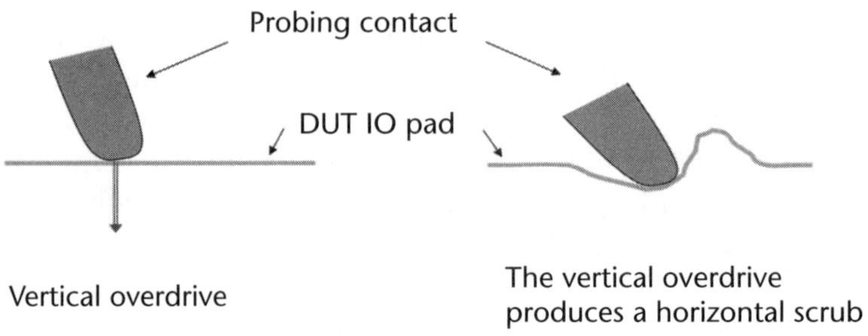

Figure 16.12 Overdriving a probe into the pad beyond first contact creates a horizontal scrub as the contact slides across the pad material. The vertical component of the force also causes the probe to penetrate some distance into the pads. The scrub length, width, and depth are all characteristic of a probe contact's shape and size, spring constant, contact material, and the material of the DUT pad.

3. *Scrub parameters:* Figure 16.12 illustrates a probe touching down on a DUT pad. The vertical overdrive of the probe produces not only penetration into the pad material but also a scrub in the horizontal plane as the spring slides across the pad surface. The shape and size of scrub marks (e.g., Figure 16.13) are characteristic of the spring technology. Scrub parameters such as scrub width, scrub length, and scrub depth are often used as figures of merit for contacts.

   The footprint of a scrub mark must be smaller than the scrub area specified by the customer (typically expressed as a percentage of the pad area). If the scrub is too large, then the resulting rough and uneven pad surface is likely to cause problems during packaging chip-attach (wire-bond or solder-bump attach).

   A wafer-probe test program often consists of multiple wafer-sort iterations as the same wafer is tested and stressed under different conditions in an effort to increase the confidence that only good dice are sent onwards for further processing. The die pads are therefore subjected to multiple

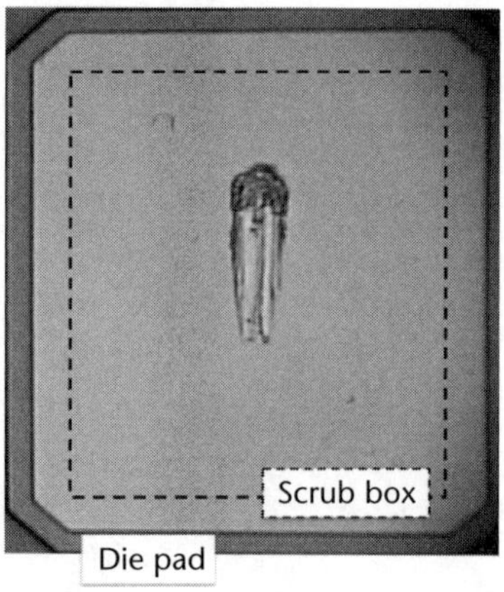

Figure 16.13 Microscope image of a scrub mark created on an Al pad by a NanoSpring contactor from NanoNexus. During probing, it is essential that the scrub mark lands inside the scrub box. (*Source:* NanoNexus, Inc.)

touchdowns from the same or different probe cards. If the scrub depth is uncontrollably large, then over multiple touchdowns, the probe card may probe through the pad and damage the interconnect layers underneath. It is therefore not unusual for IC manufacturers to specify a limit for the scrub depth after not one but *multiple* probe touchdowns. For these reasons, probe card manufacturers must carefully characterize the impact of their spring design and overdrive on scrub parameters. The scrub parameters of a NanoSpring probe (from NanoNexus) are plotted as a function of overdrive in Figure 16.14.

4. *In-plane (x-y) positional accuracy:* As chip complexity increases, so does the number of I/Os. At the same time, the I/O pad pitch and pad size are continuing to decrease [6]. The ITRS projects that in 2009 (50 nm technology node) I/O pad pitch will be in the range of 30 to 80 μm; beyond 2012 (36 nm technology node), the I/O pad pitch will reduce to 30 to 60 μm. These numbers could be even lower if a design employs a staggered pad configuration. From a probing perspective, this means that the target on which the probes have to land is becoming smaller. Smaller pads impose tighter tolerance on the in-plane (x-y) positional accuracy of the probe tips. The problem is further magnified for large-area, massively parallel probe cards. In Figure 16.13, the dashed line encloses a scrub box calculated from the IC manufacturer's information about the device. To prevent damage to the pad or the surrounding passivation material, the probe tip must scrub within this boundary.
5. *Contact properties:* As demountable contacts, the probe tips on wafer-sort probe cards are expected to mechanically withstand hundreds of thousands

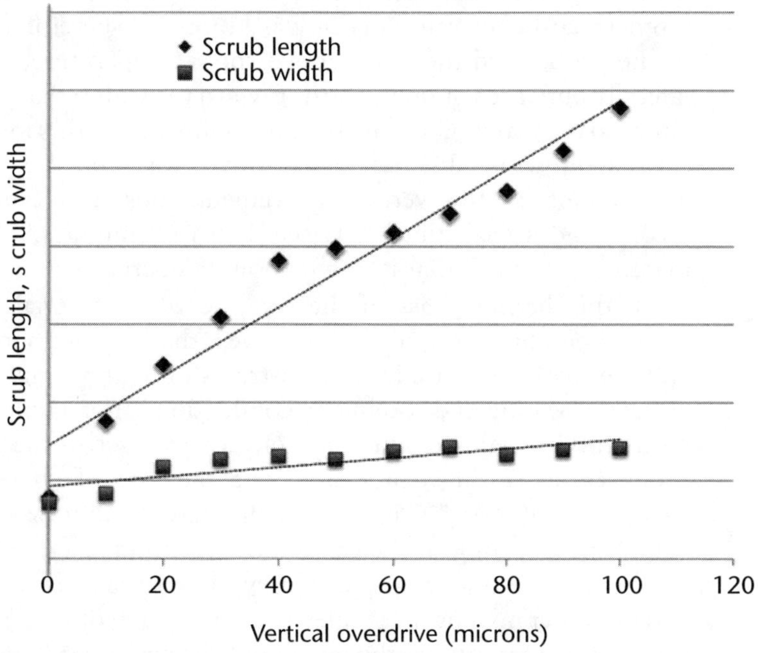

Figure 16.14 Data plots showing trends of scrub length and scrub width versus vertical overdrive for a representative advanced probe card contact. (*Source:* NanoNexus, Inc.)

of touchdowns to DUT pads. The repeated frictional contact with the metal pads eventually erodes away the probe tips. To overcome tip wear, most probe card manufacturers today coat probe tips with harder, wear-resistant materials. Common probe tip materials include tungsten rhenium (WRe), palladium cobalt (PdCo), and Paliney (an alloy comprising primarily palladium with small percentages of silver, gold, zinc, and platinum).

Another common cause for degradation of probe contacts over time is the accumulation of tiny amounts of materials from repeated contact with the DUT I/Os. For example, Al shavings from Al pads may stick to the probe tips. If these Al stringers are long enough, they may cause shorts, or if they oxidize, they may increase the contact resistance of the probe. The performance of tips may be restored by a cleaning process, either by way of mechanical scrubbing or chemical etching [8, 9].

6. *Thermal stability:* As discussed earlier in this chapter, a modern production probe card consists of multiple substrates assembled together (Figure 16.6). These substrates and all the electrical wiring within are constructed using a variety of materials, all likely to have different coefficients of thermal expansion (CTE). This may not have been of any concern if the probe card were always maintained at a constant temperature during use—but this is not the case. Different test programs during wafer probing call for testing to occur at temperatures ranging from –40°C to +150°C; that is, the prober chuck (and hence the wafer that is in intimate contact with the metal chuck) is maintained at the test temperature. During testing, only the probes are in contact with the hot (or cold) wafer. This contact provides a thermal conduction path through the probe card. The CTE mismatch between the substrates causes them to expand (or contract) at different rates. Once the probe card has been placed in the test cell, the large test head contacting the motherboard constrains any upward motion. Therefore, if the temperature of the probe card increases above room temperature, it can only expand laterally outwards or downwards toward the wafer; that is, the probe tips get closer to the wafer. Conversely, if the temperature is lowered, then the probe card assembly shrinks, and the tips move away from the surface of the wafer. To account for this vertical movement, most probe card manufacturers recommend a soak time, between 15 to 60 minutes, when a card is first placed in a test cell that is not at room temperature.

   If the thermal mass of the probe card is minimal, then it will reach thermal equilibrium quickly. However, the large amounts of metal wiring and the thick dielectric in the substrates present a significant thermal mass such that heating and cooling by conduction is not a quick process. As such, a vertical thermal gradient will typically exist across the cross section of the probe card. Testing can begin once the rate of change of temperature becomes small (<0.1°C/min). The thermal-stability problem does not stop here. During testing, as the wafer moves around in a step-and-repeat pattern directed by the test program, there will be situations where the probe card and the wafer no longer completely overlap each other (Figure 16.10). In this case, the temperature of the portion of the probe card not directly above the wafer chuck will begin to decrease, leading to some vertical motion of the probe card. In extreme cases, such motion may lead to intermittent contacts if the card moves upwards or to damaged pads if the motion is toward the wafer under test. Carefully selecting the overdrive can ensure that the probe card can continue to operate in a safe window even under these dynamic conditions.

The ideal solution to the thermal-stability problem would be to build a CTE-matched probe card with low thermal mass. Until this is realized, probe card manufacturers and test floor engineers must pay careful attention to how the cards react to temperature change.

16.3.1.3 Reliability Requirements

1. *Repeatability and contact life:* Probe cards are mission-critical components for IC manufacturers as they stand in the path of manufactured products ready for shipment to customers, pending the results of the testing. As such, manufacturers demand a high level of reliability. The selected probe technology must be able to repeatedly make reliable contact with hundreds of thousands of DUTs before replacement becomes necessary. Among other things, this means that the assembly as a whole must maintain its planarity and integrity; the probe tips should not show wear (if they do, the quality of the electrical contact will be reduced), and the probe fingers must continue to operate in the elastic regime (if they do not, any plastic deformation will be apparent in the inability of the probe array to be planarized).
2. *Cleaning mechanism:* Oftentimes, the cause for degradation of probe contacts is the accumulation of tiny amounts of materials that attach to them during repeated contact with the DUT I/Os. Probe lifetimes can be extended by a brief cleaning process, which can be either mechanical or chemical. In a mechanical cleaning process, the probes are lightly scrubbed against an abrasive pad, such as a lapping film, a tungsten carbide wafer, or a piece of ceramic [8]. Mechanical cleaning imparts high stress to the probe tips and can change their shape, eventually wearing them out over time. Mechanical cleaning methods are therefore not suitable for probe cards with tips manufactured using thin-film MEMS-based processes. Two kinds of chemical cleaning methods are widely used in industry. Using the first method, the probe card is made to touch down on a wafer coated with the cleaning chemical (Probe Polish™, Probe Clean™, Probe Scrub™ [10]). The polymeric material on the wafer is designed to dissolve away any debris that may have accumulated on the probe tips [8]. In the second method, the probe cards are cleaned by rinsing with a proprietary deoxidizing solution aptly named ProbeWash™, which removes any aluminum and aluminum oxide contaminants [11]. Therefore, as a value-adding feature, a probe technology should also be easy to clean. Oftentimes, probers have a cleaning block or pad installed adjacent to the wafer chuck, and probe card cleaning may be carried out periodically during the course of a test (online); cleaning can also be off-line and performed as part of a probe card maintenance program.

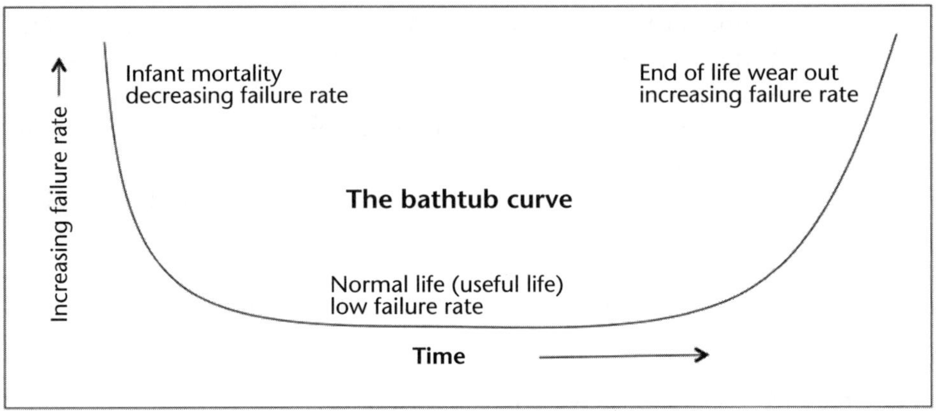

Figure 16.15 The reliability of a population of manufactured components generally follows the bathtub curve.

3. *Repairability:* The failure rate of a population of probe cards will generally follow the bathtub curve (Figure 16.15), commonly used to depict the reliability of manufactured components. Even when a probe technology has been consistently demonstrated to have extremely low failure rates, there will be some cards with infant-mortality failures, whereas others will eventually succumb to end-of-life failures. Infant-mortality failures may be due to a nonapparent manufacturing defect. In addition, a good card may fail or be damaged due to an operation error on the test floor. It is important to keep in mind that even if only a single probe among thousands fails, it amounts to a failing probe card. The small pad size and high-density I/Os on the DUTs makes probe redundancy (i.e., multiple probes for each DUT pad) a tall order. The impact of a failed probe card on the throughput of testing wafers on the IC test floor can be reduced if the probe card is designed to be repairable. Probe cards built by assembling individual probe contacts one at a time (such as cantilever cards) may be repaired by replacing only the failed probes. In advanced cards built using monolithic fabrication techniques, single pin replacement is not possible—the subcomponent as a whole must be replaced.

16.4 What Lies Ahead

The previous section summarized some of the challenging requirements probe card manufacturers aim to satisfy. Each technology provider continues to attack the problems in different ways. For example, to probe 30 μm pitch pads, cantilever card makers have adopted novel design architectures in which they vertically stack interwoven rows of cantilever contactors. Another probe card manufacturer may aim to achieve the same tight pitch by using photolithography-based microfabrication techniques to build tinier probes.

Looking ahead, mainstream IC manufacturers are likely to push for more parallelism and higher speed during testing. This is aligned with their interest to reduce cost of testing and to target the market for KGDs. Wafer-level packaging (WLP) is the fastest growing packaging technology gaining adoption in the industry today. Not only does a wafer-level chip-scale package provide the smallest form factor for a packaged device, but it also uses lower-cost batch-processing techniques for packaging all dice on the wafer at once. On completion of the WLP processes, the wafer is ready for dicing and assembly onto a SiP or other board. As such, WLP typically entails fabrication of compliant I/O interconnects on the die pads. If sufficient compliance is built into the I/Os themselves, then it is conceivable that these very I/Os could also serve as demountable contactors during testing—the probe side of a probe card would then simply be a large land grid array. This could drastically reduce the complexity and cost associated with probe cards. Solutions such as Sea-of-Leads I/O interconnects [12] are promising, and similar compliant I/Os have also been demonstrated in a manufacturing environment.

A major contributor to the cost of testing is the high-precision testing equipment itself. This is also a physical and economic limiter to massively parallel testing. The high cost of test equipment could be reduced if some of the test equipment functionality (e.g., pattern generation, timing, output comparison) were embedded directly on the DUT (BIST). Alternatives and supplements to BIST include mounting test-support processors on the probe card [13]; building active probe cards (where active test circuitry is built into the probe cards; this approach implicitly points to use of silicon as a probe substrate); or in a SiP scenario, including an ATE chip within the SiP.

On the other hand, the integration of silicon nanoelectronics with a potpourri of other technologies, such as MEMS, RF, and optoelectronics, is an indicator of the future requirements for wafer-level test methods, test equipment, and probe cards. The next section describes one such novel approach for testing ICs with electrical and optical components.

16.5 Prospects for Wafer-Level Testing of Gigascale Chips with Electrical and Optical I/O Interconnects

The superior performance of optical interconnects over their electrical counterparts, especially in terms of aggregate bandwidth [14], makes it a very attractive interconnection technology for future generations of high-performance gigascale chip-based systems [15]. The DUT, in the context of this section, is a GSI chip with electrical *and* optical I/O interconnects. As can be seen from the schematic shown in Figure 16.16, the logic remains at the core of the chip and is surrounded by electrical and optical I/Os. An optoelectronic (OE) device and transmitter/receiver circuitry lie between the optical I/O and the logic. The optical I/O interconnect is essentially a very short optical link, ranging from 100 μm to 1m in length, depending on whether it is a chip-to-chip or a board-to-board link. It comprises active optoelectronic devices (emitters, detectors, and/or modulators) as well as passive light-guiding components (e.g., waveguides, mirrors, diffractive elements, optical I/O). Based on the design, emitters and detectors are placed either on the board or on the chip. Passive optical elements direct light from the source to the detectors. Placement of the active devices is dependent on the application as well as the available technology.

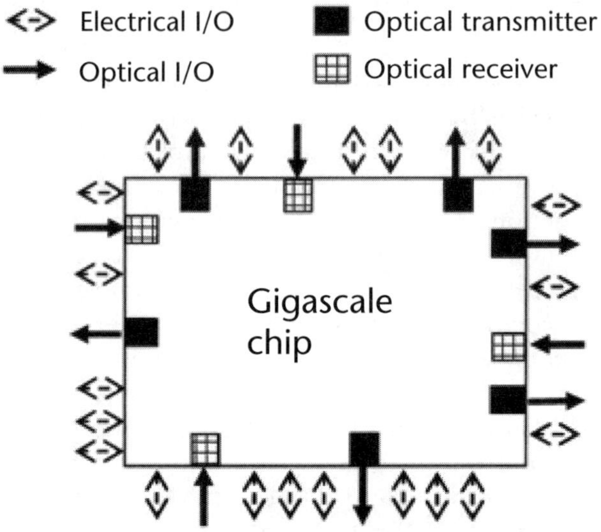

Figure 16.16 Schematic of a GSI chip with electrical and optical I/O interconnects. Architecturally, OE devices and the appropriate receiver/transmitter circuitry lie between the CMOS logic and the optical I/Os.

Owing to monolithic integration, optical components will exist on the chip and will thus need to be tested during wafer probing.

The number of optical I/Os at the time of the introduction of OE-GSI chips may be limited to the single digits. However, the enabling integration technology is expected to be scalable to allow hundreds of optical I/Os in a chip-size area for terabits-per-second aggregate bandwidth [16, 17].

16.5.1 Testing an Optoelectronic-GSI Chip

The manufacturing process flow for an optoelectronic-GSI (OE-GSI) chip would be similar to that shown in Figure 16.3 with the difference that the end result is a wafer of OE-GSI chips; it is expected that these will be tested by a probe card contacting them in a step-and-repeat manner.

Having optical I/Os on a gigascale chip does not change the singular purpose of wafer testing: to sort the good dice from the bad ones. However, the tests performed during wafer probing must now include testing of the optical I/Os. Assuming that DFT is employed, only a subset of the signal I/Os is needed, in addition to power and ground pins, for testing the logic [4, 18]. Then, one method for testing OE-GSI chips is to bypass the optical elements and to use only electrical I/Os to test the logic [Figure 16.17(a)]. However, this is not sufficient and will need to be followed or preceded by tests designed to verify the operation of the non-DFT electrical I/Os as well as the on-chip optical elements (I/Os, OE devices) [Figure 16.17(b)]. This disjoint approach is similar to the test strategy for testing optical channels recommended in [19–21].

Conversely, the test could also be performed simultaneously using both electrical and optical I/O interconnects. By this method, correct operation of the logic

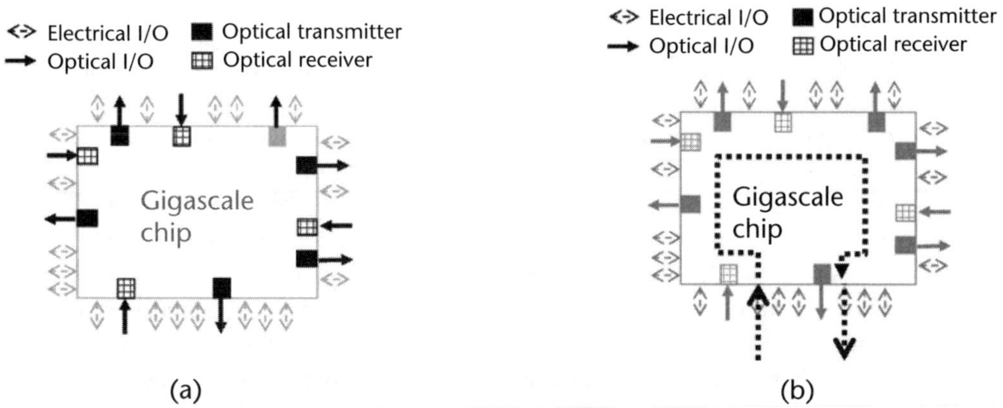

Figure 16.17 Two-step process for testing an OE-GSI chip. (a) First, the optical elements are bypassed, and only the CMOS logic is tested using DFT methods. (b) Following this, electrical testing is disengaged, and only the optical I/Os (hence, the associated OE devices and circuitry) are tested.

implies that the I/Os (electrical and optical) used during testing are performing to specification. Additional testing to verify the functionality of the non-DFT I/Os (electrical and optical) would be needed.

16.5.2 OE-GSI Testing: A New Domain for Manufacturing Testing

16.5.2.1 OE Testing: Background

A primary application of OE devices, so far, has been for use in high-speed fiber-optic communications networks. These networks comprise discretely packaged OE devices interconnected with optical fibers. Akin to Si ICs, semiconductor OE devices, such as photodetectors (PDs) and vertical cavity surface emitting lasers (VCSELs), are also built and tested at the wafer scale [22, 23]. However, these are individual devices, and testing them requires no more than a handful of electrical probes and a single optical probe. The purpose of such testing is to characterize the OE device. Tests may include measurement of responsivity, response time, and leakage currents for photodetectors and light-current-voltage (L-I-V) characterization, optical spectrum analysis, and optical-power measurement for photoemitters. Once the OE devices are packaged for use in large-scale fiber-optic systems, a different suite of tests is performed on them and on the optical link as a whole. Device testing may include modulation analysis, jitter testing, estimation of extinction ratio, and response time measurements. The quality of the link is assessed by measuring bit error rate (BER) and jitter [24].

It is clear from the above discussion that a vast amount of know-how exists for testing OE device-based fiber-optic communication systems. While the fundamentals behind testing such discrete OE devices and fiber-optic systems would also hold for testing OE-GSI chips, simply scaling these optical test strategies down to the chip level will not solve the problem. OE-GSI testing is a different playing field with a different set of constraints.

16.5.2.2 Why Is OE-GSI Testing Different?

When testing OE-GSI chips, measurement of OE device parameters would be confined to characterization testing during the initial phases of product development or to in-line testing during high-volume manufacturing. Even a simple measurement of the responsivity of an optical input, consisting of an optical I/O, a photodetector, and a receiver circuit, would require at least one optical probe and two electrical probes (for current/voltage sensing). This model may not scale: it may not be feasible to insert two additional electrical I/Os for every optical I/O on the chip solely to make it testable. An optical DFT method, which would perhaps reduce the total number of I/Os needed for testing the on-chip optical elements, is desirable.

In fiber-optic links, BER is an all-inclusive metric for link performance. BER is a statistical parameter, and stable measurements are achieved by continually monitoring this metric over a gating time, t, which is long enough to record 50 to 100 errors. For data communication at the chip level, only a BER better than 10^{-15} (that is, less than 1 error in 10^{15} bits transmitted) is acceptable [24]. Even if a pseudorandom bit stream is input at 1 Gbps, measuring 50 errors could take more than 13,000 hours (if the BER of the link is indeed 10^{-15}). Clearly, this is not feasible. Therefore, only highly reduced BER testing appears to be feasible at the chip-level. Owing to the short length of chip-to-chip optical links, BER may not even be an appropriate test. There is much room for research in this area.

Finally, the difference in physical scale between a fiber-optic link and an on-chip optical I/O is another reason that necessitates innovation in test methods. High-speed optical test equipment can be quite expensive. Therefore, manufacturers will welcome test methods that minimize this overhead.

16.5.2.3 Challenges for Wafer-Level Testing

Wafer-level testing of chips with electrical and optical I/O interconnects presents numerous challenges in addition to those for all-electrical ICs as described before.

1. *Tests for optical I/Os:* It was pointed out earlier in this section that numerous tests can be carried out to ascertain the performance of OE devices, optical couplers, and optical interconnects. Perhaps a majority of these may be used during the research-and-development phase of an OE-GSI product chip. However, during production, only rapid high-confidence tests are acceptable. In addition, these tests need to be integrated with the test plan for the entire chip. Some wafer-level test methods are discussed later.
2. *Optical I/O probing:* Probing an optical I/O is completely different from probing an electrical I/O and not as straightforward. The simplest way to interface an optical I/O is to place a photodetector or photoemitter directly above the I/O. In this case, the optical-to-electrical (O/E) or electrical-to-optical (E/O) conversion occurs immediately at the point of contact and eliminates the need for any optical signal redistribution. It also means that chips with electrical and optical I/Os could be tested with conventional electrical test equipment. However, integration of high-density OE devices on the probe substrate could be quite challenging, and there would be some concern as to whether these heterogeneous assemblies could be reworked if

they happened to fail during test operation. On the other hand, passive optical elements feeding into an optical waveguide network could be used to transfer unadulterated optical signals between the DUT and ATE. Specifically, these could include reflective [metallic or total internal reflection (TIR) mirror] or diffractive (volume- or surface-grating) elements.

3. *Optical signal distribution:* The function of an optical input on the chip is to convert the incoming optical signal to a logic-level voltage. Based on the efficiency of the photodetector and transfer characteristics of the receiver circuitry, one can estimate the minimum amount of optical power that must be coupled to the detector to achieve a logic-level voltage swing. The power-loss budget of the probe substrate (that is, the sum of all optical distribution elements) can then be estimated as the difference between the optical power output of the source (on the ATE or the probe substrate) and the minimum power required at the optical input to the chip. Failure to remain well within this loss budget would mean that the optical input on the chip will not receive enough power to generate the logic-level voltage swing. This scenario can be equated to a missed contact. Therefore, it is important to pick the optical redistribution elements with enough tolerance to prevent this from happening. Of course, a similar calculation for the reverse optical path can also be done.
4. *Repeatable acceptable optical alignment:* Repeatable and accurate probe-to-pad alignment is an important requirement during electrical testing, especially as target I/O dimensions become smaller. Alignment is even more critical when trying to probe optical I/Os. The ITRS [6] projects that wafer probers ought to have no more than 3.5 μm probe-to-pad misalignment beyond 2009. While this level of accuracy would certainly be acceptable for aligning multimode optical components, it may be too large for probing single-mode optical elements. Therefore, either probers will need to be built with tighter alignment tolerances, or misalignment compensation will need to be built into the probe module.
5. *Test equipment:* The kinds of tests on chips with optical I/Os are closely associated with the equipment that will be needed. Typically, high-precision optical and high-speed test equipment is very expensive. The challenge lies in testing the chip with as low equipment capital costs as possible.
6. *Thermal effects:* A temperature difference between the probe card and DUT can cause them to expand at different rates if there is a mismatch in their respective coefficients of thermal expansion (CTE). In the case of probing electrical I/Os, missed contacts due to CTE mismatch can be overcome by using compliant probes. However, there is no such simple solution to avoid a missed optical contact. In addition, thermal variations will also affect optical signal transmission and devices.

16.5.3 Probe Module for Testing Chips with Electrical and Optical I/O Interconnects

A probe card for OE-GSI devices has two primary purposes: (1) to contact the I/Os of the DUT, and (2) to route test signals and DUT responses between the DUT I/Os and ATE pins. Of course, the probe card also provides power and ground connections between the ATE and the DUT. The exact design of optical I/Os that would be implemented in an OE-GSI chip is still a subject of research. For the probe module described here, the DUT is a chip having polymer pillar–based I/O interconnects [16]. Polymer pillars are compliant, high-density (>10^5 I/Os/cm^2), electrical and optical chip-to-module I/O pins [16]. They are batch-fabricated at the wafer scale immediately following BEOL processing. Individual pillars can serve as air-clad microoptical fibers for guided-wave optical transmission, or they can be metalized for electrical-only connection (Figure 16.18). Figure 16.19 shows a schematic of a probe card that was designed for interfacing chips with polymer pillar–based pins. Each probe on this card resembles a microsocket and consists of a through-substrate interconnect with inward-facing, thin-film cantilevers fabricated at one end. During testing, the pillars are inserted into these sockets. The cantilevers bend to give way and in turn exert a lateral reaction force on the pillars. This force, along with the vertical motion of the pillar, creates a vertical sliding contact. Contact between metalized cantilevers and a metal-lined polymer pillar completes an electrical path. For optical I/Os, the cantilevers serve as placeholders that latch onto the sides of the pillar. Optical signals can be collected from or supplied to the DUT in a number of ways (e.g., photodetectors, lasers, mirror-terminated waveguides).

The aforementioned through-substrate interconnects serve to transfer both electrical and optical signals between the DUT and the redistribution substrate. Traditional multilevel electrical interconnects on the redistribution substrate are used to route electrical signals to the ATE. There are a number of ways to redistribute optical test and response signals. As shown in Figure 16.19, optical redistribution can be done on either side of the probe substrate, on the redistribution substrate, or on a separate optical redistribution interposer. Various implementations of the probe card and microsocket design are discussed in [25]. The details of the fabrication of a *prototype* probe substrate are available in [26].

Figure 16.20 shows scanning electron microscope (SEM) images of area arrays of batch-fabricated probe microsockets. In this prototype, the microsockets were fabricated at a 100 μm pitch. The MEMS cantilevers in each microsocket were

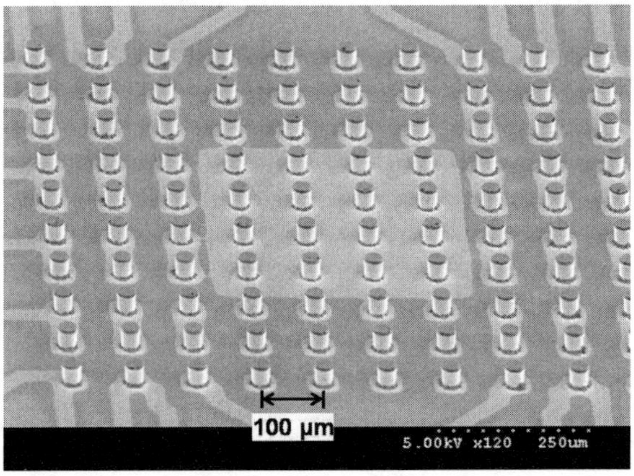

Figure 16.18 Scanning electron microscope (SEM) image of an array of polymer pillar–based I/O pins.

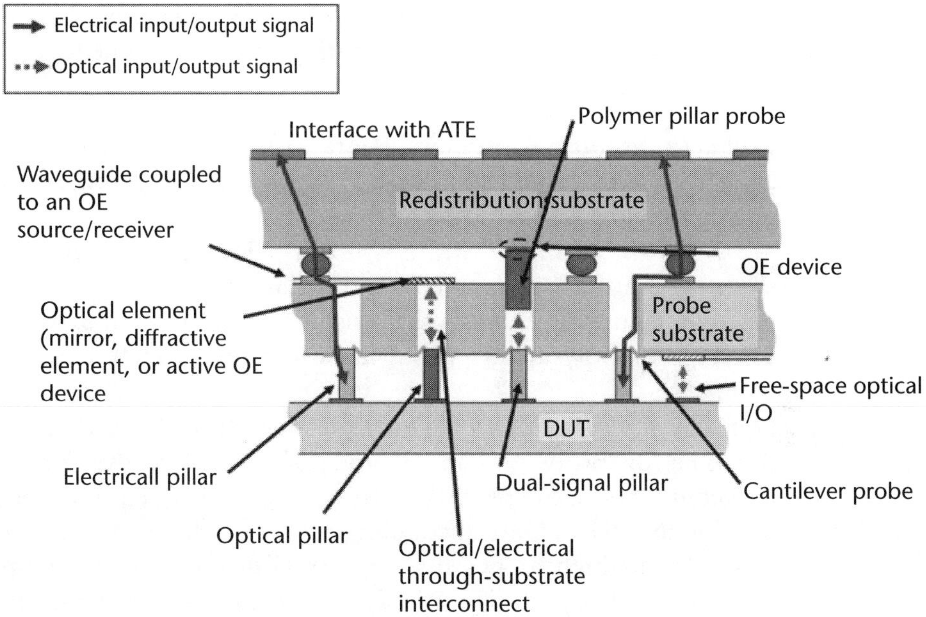

Figure 16.19 Probe module for interfacing chips with electrical and optical polymer pillar–based I/O interconnects. A few different ways of probing the I/Os and redistributing the signals are shown.

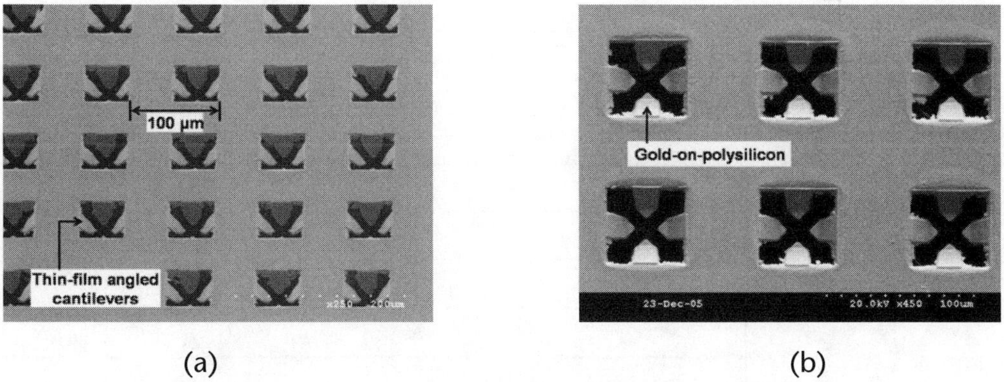

Figure 16.20 (a) SEM image of an area array of probe microsockets on the prototype probe substrate. The microsockets are at a 100 μm pitch. (b) SEM image showing the top view of an array of probes. Each microsocket probe consists of thin-film angled cantilevers fabricated directly above a through-substrate interconnect.

made of gold on doped polysilicon and extended out at a 54.7° angle below the horizontal.

16.5.3.1 Probe Card Demonstration

The prototype probe card described above, and in further detail in [27], was successfully used to contact chips with electrical and optical polymer pins. In addition, mechanical characterization of the probe microsocket cantilevers was performed to demonstrate viability of use in a manufacturing-type environment. The key experimental results are presented below.

1. *Contact resistance*: A probe substrate with 1 μm of gold on the surface was aligned to, and brought in contact with, a dummy DUT having electrical polymer pillar–based I/Os.

   When the probe substrate is brought into contact with this DUT, an electrical path is completed, and measurement of resistance becomes possible. Figure 16.21 plots the resistance as a function of probe overdrive. The gold-on-polysilicon probes have a typical contact resistance of 0.52 ohm.
2. *Optical coupling efficiency:* During testing, optical signals from sources on the ATE or probe module would pass through the through-substrate interconnects (on the probe substrate) and be coupled into optical polymer pins (on the DUT). Optical coupling at the detachable probe-to-I/O interface should be as high as possible to maximize the amount of optical power reaching the detector. Figure 16.22 plots the optical coupling efficiency for a 4 × 4 array of probes being used to contact a corresponding array of optical polymer pillar I/Os on the DUT. The optical coupling efficiency was calculated as the percentage of optical power transmitted into an individual through-substrate interconnect that gets coupled into an optical pillar during probing. The x-axis on this plot corresponds to each column of the measured

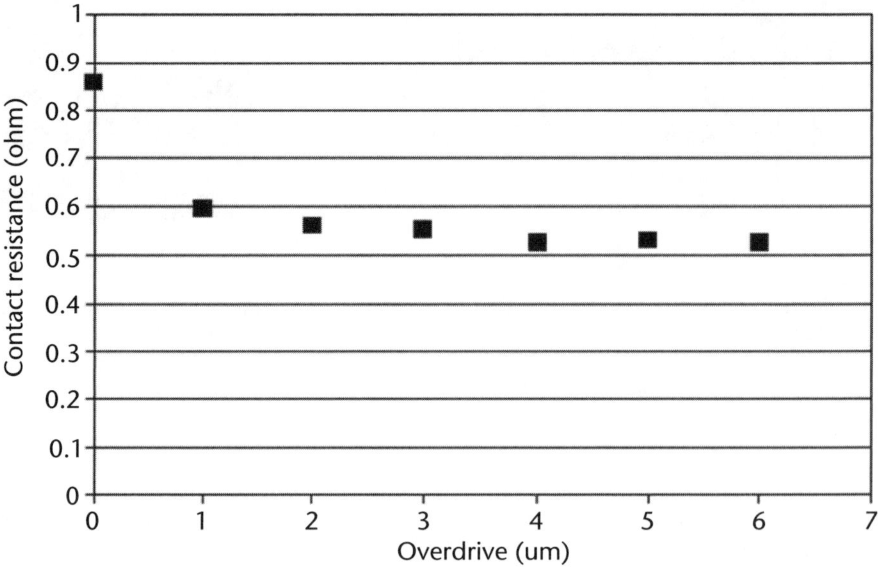

Figure 16.21 Plot of probe contact resistance as a function of vertical overdrive. The MEMS cantilevers were made of gold-on-polysilicon (35 μm long × 30 μm wide × 3 μm thick), and the pillars were approximately 30 μm in diameter by 39 μm tall. 0 μm on the x-axis represents the point of first contact. As expected, the contact resistance decreases and then stabilizes with increasing vertical overdrive.

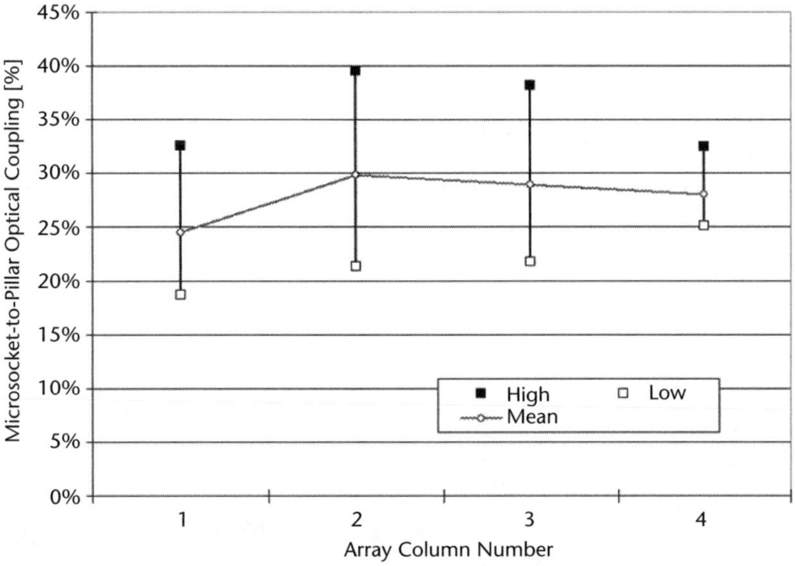

Figure 16.22 Plot showing the percentage of optical power coupled between through-substrate interconnects (52 μm wide) on the probe substrate and optical polymer pillar I/Os (28 μm diameter) on a dummy DUT. Data for a 4 × 4 array of coupled structures is shown. The *x*-axis corresponds to the columns in this array, and the range of values represents the rows in each column.

4 × 4 array. The range of values for each column corresponds to the coupling efficiency for the probe-to-pillar coupled structures in different rows of that column. This chart quantifies optical probing of a DUT with polymer pillar–based optical inputs.

In this test, an average of around 27% (and a maximum of 38%) optical power was coupled from a through-substrate interconnect to a pillar. In this measurement, the through-substrate interconnects were ~52 μm wide and the pillars were only 28 μm in diameter. The coupling efficiency could be improved by sizing the through-substrate interconnects to be comparable to the polymer pillar diameter.

3. *Mechanical performance*: The lifetime of a probe card is measured in terms of the number of test touchdowns the probes can withstand without failing. A nanoindentation tool fitted with a 30-μm diameter diamond flat-punch tip, designed to resemble a polymer pillar I/O, was used to test the mechanical performance of the angled cantilevers such as the ones shown in Figure 16.20. The tip was programmed to repeatedly load and unload a specified probe microsocket, and the change in stiffness was continually monitored. Figure 16.23 plots the stiffness as a function of number of tip touchdowns. After 80,000 indentation cycles, the probes show a negligible change in performance.

16.5.4 Radical Test Methods

This new domain of manufacturing testing of OE-GSI chips is a unique opportunity for test engineers to think outside the box and devise high-confidence yet inexpen-

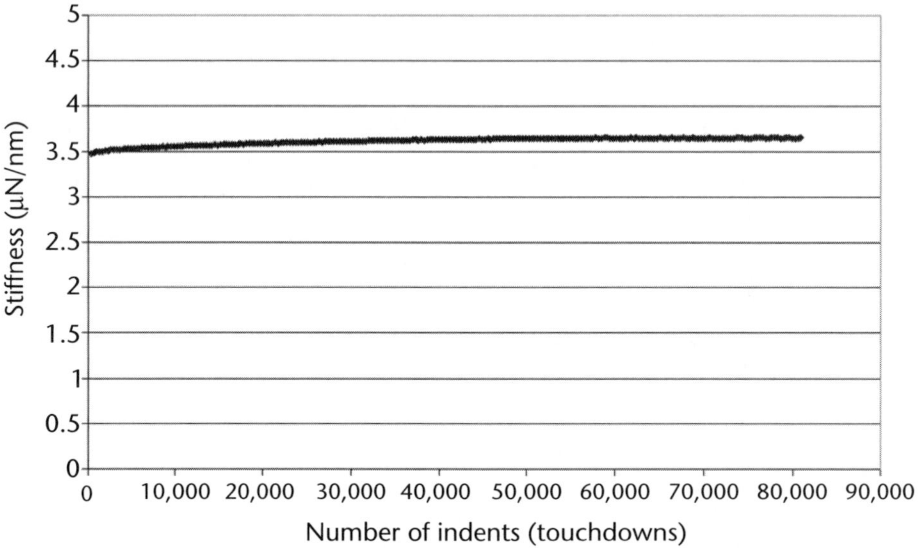

Figure 16.23 Plot of microsocket cantilever stiffness versus number of touchdowns. The data corresponds to a microsocket with four angled cantilevers made of gold on doped polysilicon (35 μm long × 15 μm wide × ~3 μm thick). The cantilevers exhibited negligible change in mechanical performance even after 80,000 indents.

sive test methods for OE components in a conventional semiconductor testing framework. For example, just two such ideas are proposed below. When a conventional probe card is brought in contact with an all-electrical DUT, among the first tests performed is a continuity check to verify that the probes are in contact with the I/Os of the DUT. Continuity tests make use of the electrostatic discharge (ESD) protection circuits on the chip [28]. The development of an analogous contact test for optical I/Os would be beneficial. This optical contact test could be used to verify not only that the optical components on the probe substrate are achieving good optical coupling with the I/Os of the DUT but also that the optical I/Os are functioning to specification.

1. Optical time-domain reflectometry (OTDR) is a technique commonly used for assessing optical links that have numerous optical interfaces and connectors. An optical signal is fed into a golden optical link, which is representative of an actual working link. Owing to discontinuities in the optical link (interfaces and connectors), some amount of the incident light will be back-scattered to the input. The reflected light and the round-trip delay in the calibration link serve as a reference to which other links can be compared. For example, if a large amount of inserted power is back-scattered, then it signals a bad optical interface or connector. Therefore, a measurement of this sort can be extrapolated to determine whether a link is performing within specification and if it is not, then the location of the failure can also be determined.
2. Another contact test for optical outputs on the DUT could employ an image-processing element. During the test, electrical signals are input to the DUT so that all the optical outputs begin emitting light. An image capture

device, such as a charge-coupled-device camera, could then be used to take a picture of the sample, and image-processing methods could be employed to determine not only whether all the outputs did indeed light up but also whether the output intensity was in the expected range.

16.6 Summary

The relentless innovation in processing technologies aided by ingenious package-level architectures is expected to prolong the juggernaut of performance improvement in gigascale chip-based systems. The integrated circuit has become omnipresent: its application space now runs the gamut from high-end computing and data storage to automotive control, optical communications, biological applications, and more. This implies that methods for wafer-level testing and the application space of probe cards must also continually evolve to allow testability of these products. This chapter has aimed to provide a glance at some of the prospects and opportunities for wafer-level testing and probe cards for the nanoelectronics era.

References

[1] Maly, W., "The Design and Test Cost Problem," *IEEE Design and Test of Computers*, November/December 2001, p. 6.

[2] Nigh, P., "In Search of the Best IC Test Method," *IBM MicroNews*, 1999.

[3] Zorian, Y., "Emerging Trends in VLSI Test and Diagnosis," *IEEE Comp. Soc. Workshop on VLSI*, 2000, pp. 21–27.

[4] Bushnell, M. D., and V. D. Agrawal, *Essentials of Electronic Testing for Digital, Memory and Mixed-Signal VLSI Circuits*. Boston: Kluwer Academic Publishers, 2000.

[5] Forster, J., "Single Chip Test and Burn-In," *Proc. IEEE Electron. Comp. and Technol. Conf.*, 2000, pp. 810–814.

[6] Semiconductor Industry Association, *International Technology Roadmap for Semiconductors*, 2006.

[7] Slade, P. G., *Electrical Contacts,* Boca Raton, FL, CRC Press, 1999.

[8] Broz, J., G. Humphrey, and W. Fitzgerald, "Probe Card Cleaning: A Short Tutorial," *IEEE Southwest Test Workshop,* 2007.

[9] Humphrey, G., "Probe Card On-line Cleaning," *IEEE Southwest Test Workshop*, 2000.

[10] See International Test Solutions at www.inttest.net/probe_cleaning.htm.

[11] See ProbeWash at www.rdchem.com/test-equipment-restoration/probewash-rdz-1730.html.

[12] Bakir, M., et al., "Sea of Leads (SoL) Ultrahigh-Density Wafer Level Chip Input/Output Interconnections," *IEEE Trans. Electr. Dev.*, Vol. 50, No. 10, October 2003, pp. 2039–2048.

[13] Keezer, D. C., and Q. Zhou, "Test Support Processors for Enhanced Testability of High Performance Circuits," *Proc. IEEE International Test Conf.*, 1999.

[14] Naeemi, A., et al., "Optical and Electrical Interconnect Partition Length Based on Chip-to-Chip Bandwidth Maximization," *Photon. Technol. Lett.*, Vol. 16, No. 4, 2004, pp. 1221–1223.

[15] Miller, D., "Rationale and Challenges for Optical Interconnects to Electronic Chips," *Proc. IEEE*, Vol. 88, No. 6, 2000, pp. 728–749.

[16] Bakir, M., et al., "Electrical and Optical Chip I/O Interconnections for Gigascale Systems," *IEEE Trans. on Electron Dev.*, Vol. 54, No. 9, 2007, pp. 2426–2437.

[17] Kash, J., et al., "Terabus: A Chip-to-Chip Parallel Optical Interconnect," *Ann. Mtg. of the Lasers and Electro-Optics Soc.*, 2005, pp. 363–364.

[18] Nagle, H. T., et al., "Design for Testability and Built-In Self Test: A Review," *IEEE Trans. Ind. Electron.*, Vol. 36, No. 2, 1989, pp. 129–140.

[19] Cook, C., et al., "A 36-Channel Parallel Optical Interconnect Module Based on Optoelectronics-on-VLSI Technology," *IEEE J. Select. Topics Quantum Electron.*, Vol. 9, No. 2, 2003, pp. 387–399.

[20] Venditti, M. B., et al., "Design and Test of an Optoelectronic-VLSI Chip with 540-Element Receiver-Transmitter Arrays Using Differential Optical Signaling," *IEEE J. Select. Topics Quantum Electron.*, Vol. 9, No. 2, 2003, pp. 361–379.

[21] Gagnon, M., and B. Kaminska, "Optical Communication Channel Test Using BIST Approaches," *Proc. Int. Test Conf.*, 1997, pp. 626–635.

[22] Rogers, D. L., et al., "Design, Fabrication, and Automated Testing of 32-Channel Integrated MSM/MESFET Optoelectronic Integrated Circuit (OEIC) Receiver Arrays," *Optoelectronic Interconnects*, Vol. 2400, 1995, pp. 296–299.

[23] Heffner, W. R., and A. Robertson, "Wafer Level Optoelectronic Testing for DFB Laser Diodes," *Int. Conf. InP and Related Materials*, 2000, pp. 435–438.

[24] Derickson, D., *Fiber Optic Test and Measurement*, Upper Saddle River, NJ: Prentice Hall, 2000.

[25] Thacker, H., et al., "Probe Module for Wafer-Level Testing of Gigascale Chips with Electrical and Optical I/O Interconnects," *Proc. of ASME Interpack*, Part B, 2005.

[26] Thacker, H., et al., "High-Density Probe Substrate for Testing Optical Interconnects," *Proc. IEEE Int. Interconnect Technol. Conf.*, 2005, pp. 159–161.

[27] Thacker, H., "Probe Modules for Wafer-Level Testing of Gigascale Chips with Electrical and Optical I/O Interconnects," PhD thesis, Georgia Institute of Technology, 2006.

[28] Burns, M., and G. W. Roberts, *An Introduction to Mixed-Signal IC Test and Measurement*, New York: Oxford University Press, 2001.

[29] Moore, B., et al., "High throughput Noncontact SIP Testing," Proc. IEEE International Test Conf., 2007.

About the Editors

Muhannad S. Bakir is a research faculty at the Microelectronics Research Center at the Georgia Institute of Technology. He has published more than 60 refereed and invited publications in conference proceedings and journals and has received five best and outstanding paper awards from international IEEE conferences. He holds nine U.S. patents in the area of silicon ancillary technologies. Dr. Bakir received a B.S. in electrical engineering from Auburn University and a Ph.D. in electrical and computer engineering from the Georgia Institute of Technology.

James D. Meindl is the director of the Microelectronics Research Center and is the founding director of the Nanotechnology Research Center at the Georgia Institute of Technology, where he is also the Joseph M. Pettit Chair Professor of Microelectronics. Prior to his tenure at Georgia Tech, he served as senior vice president for academic affairs and provost of Rensselaer Polytechnic Institute. Dr. Meindl received a Ph.D. in electrical engineering at Carnegie Institute of Technology (Carnegie-Mellon University) and is a Life Fellow of the IEEE and a member of the National Academy of Engineering. He was awarded the 2006 IEEE Medal of Honor.

List of Contributors

P. Andry
IBM—T.J. Watson Research Center
System on Package & 3D Integration
1101 Kitchawan Road
06-116
Yorktown Heights, NY 10598
United States

Muhannad S. Bakir
Microelectronics Research Center
Georgia Institute of Technology
791 Atlantic Drive NW
Atlanta, GA 30332-0269
United States

Alan F. Benner
IBM Corporation
Mail Stop P932, Bldg. 416 Ofc. 3-01
2455 South Road
Poughkeepsie, NY 12601
United States

W. R. Bottoms
NanoNexus
3101 Alexis Drive
Palo Alto, CA 94304
United States

Bing Dang
IBM—T.J. Watson Research Center
System on Package & 3D Integration
1101 Kitchawan Road
06-116
Yorktown Heights, NY 10598
United States

Carl L. Dohrman
Materials Science and Engineering
MIT
Room 13-5153
77 Massachusetts Avenue
Cambridge, MA 02139
United States

John C. Eble III
Rambus, Inc.
1512 East Franklin Street
Suite 200
Chapel Hill, NC 27514
United States

Andrei G. Fedorov
Georgia Institute of Technology
George W. Woodruff School of
Mechanical Engineering
771 Ferst Drive
Atlanta, GA 30332-0405
United States

Eugene A. Fitzgerald
Materials Science and Engineering
MIT
Room 13-5153
77 Massachusetts Avenue
Cambridge, MA 02139
United States

Kenneth E. Goodson
Mechanical Engineering Department
Stanford University
Building 530, Room 214
Stanford, CA 94305-3030
United States

Siva P. Gurrum
Semiconductor Packaging
Texas Instruments Inc.
13536 N. Central Expressway
MS 940
Dallas, TX 75243
United States

Peter Hazucha
Intel Corporation
2111 NE 25th Avenue
Mail Stop: JF2-04
Hillsboro, OR 97124
United States

Dennis W. Hess
Georgia Institute of Technology
School of Chemical & Biomolecular Engineering
311 Ferst Drive, NW
Atlanta, GA 30332-0100
United States

Carlos H. Hidrovo
Mechanical Engineering Department
The University of Texas at Austin
ETC 7.148A
1 University Station C2200
Austin, TX 78712-0292
United States

Paul S. Ho
Interconnect and Packaging Group
The University of Texas at Austin
PRC, Bldg. 160
10100 Burnet Road
Austin, TX 78758
United States

R. Horton
IBM—T.J. Watson Research Center
System on Package & 3D Integration
1101 Kitchawan Road
06-116
Yorktown Heights, NY 10598
United States

Gang Huang
Microelectronics Research Center
Georgia Institute of Technology
791 Atlantic Drive NW
Atlanta, GA 30332-0269
United States

Rui Huang
Aerospace Engineering and Engineering Mechanics
The University of Texas at Austin
1 University Station, C0600
Austin, TX 78712
United States

Se Hyuk Im
Aerospace Engineering and Engineering Mechanics
The University of Texas at Austin
1 University Station, C0600
Austin, TX 78712
United States

Yogendra Joshi
Georgia Institute of Technology
George W. Woodruff School of Mechanical Engineering
771 Ferst Drive
Atlanta, GA 30332-0405
United States

Soon-Moon Jung
System LSI Division, Semiconductor Business
Samsung Electronics Co., Ltd.
San#24 Nongseo-Dong, Giheung-Gu
Yongin-City, Gyeonggi-Do, 446-711
Korea

Karan Kacker
The George W. Woodruff School of
Mechanical Engineering
Georgia Institute of Technology
813 Ferst Drive
Atlanta, GA 30332-0405
United States

Tanay Karnik
Intel Corporation
2111 NE 25th Avenue
Mail Stop: JF2-04
Hillsboro, OR 97124
United States

Kinam Kim
Memory Division, Semiconductor Business
Samsung Electronics Co., Ltd.
San#24 Nongseo-Dong, Giheung-Gu
Yongin-City, Gyeonggi-Do, 446-711
Korea

Calvin R. King, Jr.
Microelectronics Research Center
Georgia Institute of Technology
791 Atlantic Drive NW
Atlanta, GA 30332-0269
United States

J. Knickerbocker
IBM—T.J. Watson Research Center
System on Package & 3D Integration
1101 Kitchawan Road
06-116
Yorktown Heights, NY 10598
United States

Tadahiro Kuroda
Keio University
Department of Electronics and Electrical Engineering
3-14-1 Hiyoshi, Kohoku-ku
Yokohama
Kanagawa 223-8522
Japan

G. McVicker
IBM—T.J. Watson Research Center
System on Package & 3D Integration
1101 Kitchawan Road
06-116
Yorktown Heights, NY 10598
United States

James D. Meindl
Microelectronics Research Center
Georgia Institute of Technology
791 Atlantic Drive NW
Atlanta, GA 30332-0269
United States

Noriyuki Miura
Keio University
Department of Electronics and Electrical Engineering
3-14-1 Hiyoshi, Kohoku-ku, Yokohama
Kanagawa 223-8522
Japan

Michael J. Mori
Materials Science and Engineering
MIT
Room 13-5153
77 Massachusetts Avenue
Cambridge, MA 02139
United States

Azad Naeemi
Microelectronics Research Center
Georgia Institute of Technology
791 Atlantic Drive NW
Atlanta, GA 30332-0269
United States

Oluwafemi O. Ogunsola
IBM Corporation
Mail Drop 42J, Bldg.300, Ofc. 3A6-3
2070 Route 52
Hopewell Junction, NY 12533-6683
United States

Fabrice Paillet
Intel Corporation
2111 NE 25th Avenue
Mail Stop: JF2-04
Hillsboro, OR 97124
United States

C. Patel
IBM—T.J. Watson Research Center
System on Package & 3D Integration
1101 Kitchawan Road
06-116
Yorktown Heights, NY 10598
United States

R. Polastre
IBM—T.J. Watson Research Center
System on Package & 3D Integration
1101 Kitchawan Road
06-116
Yorktown Heights, NY 10598
United States

Kaladhar Radhakrishnan
Intel Corporation
5000 W. Chandler Blvd.
Mail Stop: CH5-157
Chandler, AZ 85226
United States

K. Sakuma
IBM—T.J. Watson Research Center
System on Package & 3D Integration
1101 Kitchawan Road
06-116
Yorktown Heights, NY 10598
United States

Gerhard Schrom
Intel Corporation
2111 NE 25th Avenue
Mail Stop: JF2-04
Hillsboro, OR 97124
United States

Deepak Sekar
Microelectronics Research Center
Georgia Institute of Technology
791 Atlantic Drive NW
Atlanta, GA 30332-0269
United States

Kaveh Shakeri
Cypress Semiconductor
Technology R&D
195 Champion Court
San Jose, CA 95134
United States

Suresh K. Sitaraman
The George W. Woodruff School of
Mechanical Engineering
Georgia Institute of Technology
813 Ferst Drive
Atlanta, GA 30332-0405
United States

E. Sprogis
IBM—T.J. Watson Research Center
System on Package & 3D Integration
1101 Kitchawan Road
06-116
Yorktown Heights, NY 10598
United States

S. Sri-Jayantha
IBM—T.J. Watson Research Center
System on Package & 3D Integration
1101 Kitchawan Road
06-116
Yorktown Heights, NY 10598
United States

Hiren D. Thacker
NanoNexus
39380 Civic Center Drive #417
Fremont, CA 94538
United States

C. Tsang
IBM—T.J. Watson Research Center
System on Package & 3D Integration
1101 Kitchawan Road
06-116
Yorktown Heights, NY 10598
United States

B. Webb
IBM—T.J. Watson Research Center
System on Package & 3D Integration
1101 Kitchawan Road
06-116
Yorktown Heights, NY 10598
United States

Xiaojin Wei
Mail Stop 87P
IBM System and Technology
2070 Route 52
Hopewell Junction, NY 12533
United States

C. P. Wong
Georgia Institute of Technology
School of Materials Science & Engineering
771 Ferst Drive, NW
Atlanta, GA 30332-0245
United States

S. Wright
IBM—T.J. Watson Research Center
System on Package & 3D Integration
1101 Kitchawan Road
06-116
Yorktown Heights, NY 10598
United States

Xuefeng Zhang
Interconnect and Packaging Group
The University of Texas at Austin
PRC, Bldg. 160
10100 Burnet Road
Austin, TX 78758
United States

Lingbo Zhu
The Dow Chemical Company
1776 Bldg., F25C
Midland, MI 48674
United States

Index

J

K

L

P

T

Recent Titles in the Artech House Integrated Microsystems Series

Adaptive Cooling of Integrated Circuits Using Digital Microfluidics, Philip Y. Paik, Krishnendu Chakrabarty, and Vamsee K. Pamula

Fundamentals and Applications of Microfluidics, Second Edition, Nam-Trung Nguyen and Steven T. Wereley

Integrated Interconnect Technologies for 3D Nanoelectronic Systems, Muhannad S. Bakir and James D. Meindl, editors

Introduction to Microelectromechanical (MEM) Microwave Systems, Héctor J. De Los Santos

An Introduction to Microelectromechanical Systems Engineering, Nadim Maluf

MEMS Mechanical Sensors, Stephen Beeby et al.

Micro and Nano Manipulations for Biomedical Applications, Tachung C. Yihllie Talpasanu

Microfluidics for Biotechnology, Jean Berthier and Pascal Silberzan

Organic and Inorganic Nanostructures, Alexei Nabok

Post-Processing Techniques for Integrated MEMS, Sherif Sedky

Pressure-Driven Microfluidics, Václav Tesař

RF MEMS Circuit Design for Wireless Communications, Héctor J. De Los Santos

Wireless Sensor Network, Nirupama Bulusu and Sanjay Jha

For further information on these and other Artech House titles, including previously considered out-of-print books now available through our In-Print-Forever® (IPF®) program, contact:

Artech House
685 Canton Street
Norwood, MA 02062
Phone: 781-769-9750
Fax: 781-769-6334
e-mail: artech@artechhouse.com

Artech House
46 Gillingham Street
London SW1V 1AH UK
Phone: +44 (0)20 7596-8750
Fax: +44 (0)20 7630-0166
e-mail: artech-uk@artechhouse.com

Find us on the World Wide Web at: www.artechhouse.com